BEGINNING AND INTERMEDIATE ALGEBRA

SECOND EDITION

STUDENT'S SOLUTIONS MANUAL

JEFFERY A. COLE
Anoka-Ramsey Community College

Margaret L. Lial
American River College

John Hornsby
University of New Orleans

Addison
Wesley
Longman

An imprint of Addison Wesley Longman, Inc.

Reading, Massachusetts • Menlo Park, California • New York • Harlow, England
Don Mills, Ontario • Sydney • Mexico City • Madrid • Amsterdam

Table of Contents

Preface

This *Student's Solutions Manual* contains solutions to selected exercises in the text *Beginning and Intermediate Algebra, Second Edition* by Margaret L. Lial and E. John Hornsby, Jr. It contains solutions to the odd-numbered exercises in each section, as well as solutions to all the exercises in the review sections, the chapter tests, and the cumulative review sections.

This manual is a text supplement and should be read along *with* the text. You should read all exercise solutions in this manual because many concept explanations are given and then used in subsequent solutions. All concepts necessary to solve a particular problem are not reviewed for every exercise. If you are having difficulty with a previously covered concept, refer back to the section where it was covered for more complete help.

A significant number of today's students are involved in various outside activities, and find it difficult, if not impossible, to attend all class sessions; this manual should help meet the needs of these students. In addition, it is my hope that this manual's solutions will enhance the understanding of all readers of the material and provide insights to solving other exercises.

I appreciate feedback concerning errors, solution correctness or style, and manual style. Any comments may be sent directly to me at the address below or in care of the publisher: Addison Wesley Longman, One Jacob Way, Reading, MA 01867.

I would like to thank Ken Grace, of Anoka–Ramsey Community College, for typesetting the manuscript and providing invaluable help with many features of the manual; Marv Riedesel and Mary Johnson, of Inver Hills Community College, for their careful accuracy checking; my wife, Joan, for proofreading the manuscript; Ruth Berry, of Addison Wesley Longman, for facilitating the production process; and the authors and Jennifer Crum, of Addison Wesley Longman, for entrusting me with this project.

Jeffery A. Cole
Anoka-Ramsey Community College
11200 Mississippi Blvd. NW
Coon Rapids, MN 55433

CHAPTER 1   THE REAL NUMBER SYSTEM

Section 1.1

1. True; the number above the fraction bar is called the numerator and the number below the fraction bar is called the denominator.

3. False; the fraction $\frac{17}{51}$ can be

   reduced to $\frac{1}{3}$ since $\frac{17}{51} = \frac{17 \cdot 1}{17 \cdot 3} = \frac{1}{3}$.

5. False; *product* refers to multiplication, so the product of 8 and 2 is 16. The *sum* of 8 and 2 is 10.

7. Since 19 has only itself and 1 as factors, it is a prime number.

9. $64 = 2 \cdot 32$
   $= 2 \cdot 2 \cdot 16$
   $= 2 \cdot 2 \cdot 2 \cdot 8$
   $= 2 \cdot 2 \cdot 2 \cdot 2 \cdot 4$
   $= 2 \cdot 2 \cdot 2 \cdot 2 \cdot 2 \cdot 2$

   Since 64 has factors other than itself and 1, it is a composite number.

11. $3458 = 2 \cdot 1729$
    $= 2 \cdot 7 \cdot 247$
    $= 2 \cdot 7 \cdot 13 \cdot 19$

    Since 3458 has factors other than itself and 1, it is a composite number.

13. As stated in the text, the number 1 is neither prime nor composite, by agreement.

15. $30 = 2 \cdot 15$
    $= 2 \cdot 3 \cdot 5$

    Since 30 has factors other than itself and 1, it is a composite number.

17. $500 = 2 \cdot 250$
    $= 2 \cdot 2 \cdot 125$
    $= 2 \cdot 2 \cdot 5 \cdot 25$
    $= 2 \cdot 2 \cdot 5 \cdot 5 \cdot 5,$

    so 500 is a composite number.

19. $124 = 2 \cdot 62$
    $= 2 \cdot 2 \cdot 31,$

    so 124 is a composite number.

21. Since 29 has only itself and 1 as factors, it is a prime number.

23. $\frac{8}{16} = \frac{1 \cdot 8}{2 \cdot 8} = \frac{1}{2}$

25. $\frac{15}{18} = \frac{3 \cdot 5}{3 \cdot 6} = \frac{5}{6}$

27. $\frac{15}{45} = \frac{1 \cdot 15}{3 \cdot 15} = \frac{1}{3}$

29. $\frac{144}{120} = \frac{6 \cdot 24}{5 \cdot 24} = \frac{6}{5}$

31. $\frac{16}{24} = \frac{2 \cdot 8}{3 \cdot 8} = \frac{2}{3}$

   Therefore, (c) is correct.

33. $\frac{4}{5} \cdot \frac{6}{7} = \frac{4 \cdot 6}{5 \cdot 7} = \frac{24}{35}$

35. $\frac{1}{10} \cdot \frac{12}{5} = \frac{1 \cdot 12}{10 \cdot 5} = \frac{1 \cdot 2 \cdot 6}{2 \cdot 5 \cdot 5} = \frac{6}{25}$

37. $\frac{15}{4} \cdot \frac{8}{25} = \frac{15 \cdot 8}{4 \cdot 25}$

    $= \frac{3 \cdot 5 \cdot 4 \cdot 2}{4 \cdot 5 \cdot 5}$

    $= \frac{3 \cdot 2}{5}$

    $= \frac{6}{5}$ or $1\frac{1}{5}$

39. $2\frac{2}{3} \cdot 5\frac{4}{5}$

    Change both mixed numbers to improper fractions.

    $2\frac{2}{3} = 2 + \frac{2}{3} = \frac{6}{3} + \frac{2}{3} = \frac{8}{3}$

    $5\frac{4}{5} = 5 + \frac{4}{5} = \frac{25}{5} + \frac{4}{5} = \frac{29}{5}$

    $2\frac{2}{3} \cdot 5\frac{4}{5} = \frac{8}{3} \cdot \frac{29}{5}$

    $= \frac{8 \cdot 29}{3 \cdot 5}$

    $= \frac{232}{15}$ or $15\frac{7}{15}$

41. $\frac{5}{4} \div \frac{3}{8} = \frac{5}{4} \cdot \frac{8}{3}$ *Multiply by the reciprocal of the second fraction.*

    $= \frac{5 \cdot 8}{4 \cdot 3}$

    $= \frac{5 \cdot 4 \cdot 2}{4 \cdot 3}$

    $= \frac{5 \cdot 2}{3}$

    $= \frac{10}{3}$ or $3\frac{1}{3}$

43. $\dfrac{32}{5} \div \dfrac{8}{15} = \dfrac{32}{5} \cdot \dfrac{15}{8}$ *Multiply by the reciprocal of the second fraction.*

$$= \dfrac{32 \cdot 15}{5 \cdot 8}$$

$$= \dfrac{8 \cdot 4 \cdot 3 \cdot 5}{1 \cdot 5 \cdot 8}$$

$$= \dfrac{4 \cdot 3}{1} = 12$$

45. $\dfrac{3}{4} \div 12 = \dfrac{3}{4} \cdot \dfrac{1}{12}$ *Multiply by the reciprocal of 12.*

$$= \dfrac{3 \cdot 1}{4 \cdot 12}$$

$$= \dfrac{3 \cdot 1}{4 \cdot 3 \cdot 4}$$

$$= \dfrac{1}{4 \cdot 4} = \dfrac{1}{16}$$

47. $2\dfrac{5}{8} \div 1\dfrac{15}{32}$

Change both mixed numbers to improper fractions.

$$2\dfrac{5}{8} = 2 + \dfrac{5}{8} = \dfrac{16}{8} + \dfrac{5}{8} = \dfrac{21}{8}$$

$$1\dfrac{15}{32} = 1 + \dfrac{15}{32} = \dfrac{32}{32} + \dfrac{15}{32} = \dfrac{47}{32}$$

$$2\dfrac{5}{8} \div 1\dfrac{15}{32} = \dfrac{21}{8} \div \dfrac{47}{32}$$

$$= \dfrac{21}{8} \cdot \dfrac{32}{47}$$

$$= \dfrac{21 \cdot 32}{8 \cdot 47}$$

$$= \dfrac{21 \cdot 8 \cdot 4}{8 \cdot 47}$$

$$= \dfrac{21 \cdot 4}{47}$$

$$= \dfrac{84}{47} \text{ or } 1\dfrac{37}{47}$$

49. To multiply two fractions, multiply their numerators to get the numerator of the product and multiply their denominators to get the denominator of the product. For example,

$$\dfrac{2}{3} \cdot \dfrac{8}{5} = \dfrac{2 \cdot 8}{3 \cdot 5} = \dfrac{16}{15}.$$

To divide two fractions, replace the divisor with its reciprocal and then multiply. For example,

$$\dfrac{2}{5} \div \dfrac{7}{9} = \dfrac{2}{5} \cdot \dfrac{9}{7} = \dfrac{2 \cdot 9}{5 \cdot 7} = \dfrac{18}{35}.$$

51. $\dfrac{7}{12} + \dfrac{1}{12} = \dfrac{7+1}{12}$

$$= \dfrac{8}{12}$$

$$= \dfrac{2 \cdot 4}{3 \cdot 4} = \dfrac{2}{3}$$

53. $\dfrac{5}{9} + \dfrac{1}{3}$

Since $9 = 3 \cdot 3$, and 3 is prime, the LCD (least common denominator) is $3 \cdot 3 = 9$.

$$\dfrac{1}{3} = \dfrac{1}{3} \cdot \dfrac{3}{3} = \dfrac{3}{9}$$

Now add the two fractions with the same denominator.

$$\dfrac{5}{9} + \dfrac{1}{3} = \dfrac{5}{9} + \dfrac{3}{9} = \dfrac{8}{9}$$

55. $3\dfrac{1}{8} + \dfrac{1}{4}$

$$3\dfrac{1}{8} = 3 + \dfrac{1}{8} = \dfrac{24}{8} + \dfrac{1}{8} = \dfrac{25}{8}$$

$$3\dfrac{1}{8} + \dfrac{1}{4} = \dfrac{25}{8} + \dfrac{1}{4}$$

Since $8 = 2 \cdot 2 \cdot 2$ and $4 = 2 \cdot 2$, the LCD is $2 \cdot 2 \cdot 2$ or 8.

$$3\dfrac{1}{8} + \dfrac{1}{4} = \dfrac{25}{8} + \dfrac{1 \cdot 2}{4 \cdot 2}$$

$$= \dfrac{25}{8} + \dfrac{2}{8}$$

$$= \dfrac{27}{8} \text{ or } 3\dfrac{3}{8}$$

57. $\dfrac{7}{12} - \dfrac{1}{9}$

Since $12 = 2 \cdot 2 \cdot 3$ and $9 = 3 \cdot 3$, the LCD is $2 \cdot 2 \cdot 3 \cdot 3 = 36$.

$$\dfrac{7}{12} = \dfrac{7}{12} \cdot \dfrac{3}{3} = \dfrac{21}{36} \text{ and } \dfrac{1}{9} \cdot \dfrac{4}{4} = \dfrac{4}{36}$$

Now subtract fractions with the same denominator.

$$\dfrac{7}{12} - \dfrac{1}{9} = \dfrac{21}{36} - \dfrac{4}{36} = \dfrac{17}{36}$$

59. $6\dfrac{1}{4} - 5\dfrac{1}{3}$

$$6\dfrac{1}{4} = 6 + \dfrac{1}{4} = \dfrac{24}{4} + \dfrac{1}{4} = \dfrac{25}{4}$$

$$5\dfrac{1}{3} = 5 + \dfrac{1}{3} = \dfrac{15}{3} + \dfrac{1}{3} = \dfrac{16}{3}$$

Since $4 = 2 \cdot 2$, and 3 is prime, the LCD is $2 \cdot 2 \cdot 3 = 12$.

continued

$$6\frac{1}{4} - 5\frac{1}{3} = \frac{25}{4} - \frac{16}{3}$$

$$= \frac{25 \cdot 3}{4 \cdot 3} - \frac{16 \cdot 4}{3 \cdot 4}$$

$$= \frac{75}{12} - \frac{64}{12}$$

$$= \frac{11}{12}$$

61. $\frac{5}{3} + \frac{1}{6} - \frac{1}{2}$

Since 2 and 3 are prime and $6 = 2 \cdot 3$, the LCD is $2 \cdot 3 = 6$. Write $\frac{5}{3}$ and $\frac{1}{2}$ as equivalent fractions with 6 as denominator.

$$\frac{5}{3} = \frac{5}{3} \cdot \frac{2}{2} = \frac{10}{6} \text{ and } \frac{1}{2} = \frac{1}{2} \cdot \frac{3}{3} = \frac{3}{6}$$

Now add and subtract; then write the answer in lowest terms.

$$\frac{5}{3} + \frac{1}{6} - \frac{1}{2} = \frac{10}{6} + \frac{1}{6} - \frac{3}{6}$$

$$= \frac{10 + 1 - 3}{6}$$

$$= \frac{8}{6} = \frac{4 \cdot 2}{3 \cdot 2} = \frac{4}{3} \text{ or } 1\frac{1}{3}$$

63. Multiply the number of cups of water per serving by the number of servings.

$$\frac{3}{4} \cdot 8 = \frac{3}{4} \cdot \frac{8}{1}$$

$$= \frac{3 \cdot 8}{4 \cdot 1}$$

$$= \frac{3 \cdot 2 \cdot 4}{4 \cdot 1}$$

$$= \frac{3 \cdot 2}{1} = 6 \text{ cups}$$

For 8 microwave servings, 6 cups of water will be needed.

65. Since

closing price − gain = opening price

we have

$$38\frac{5}{8} - 4\frac{5}{8} = 34 \text{ dollars}$$

as the opening price.

67. The difference between the two measures is found by subtracting, using 16 as the LCD.

$$\frac{3}{4} - \frac{3}{16} = \frac{3 \cdot 4}{4 \cdot 4} - \frac{3}{16}$$

$$= \frac{12}{16} - \frac{3}{16}$$

$$= \frac{12 - 3}{16} = \frac{9}{16}$$

The difference is $\frac{9}{16}$ inch.

69. The perimeter is the sum of the measures of the 5 sides.

$$196 + 98\frac{3}{4} + 146\frac{1}{2} + 100\frac{7}{8} + 76\frac{5}{8}$$

$$= 196 + 98\frac{6}{8} + 146\frac{4}{8} + 100\frac{7}{8} + 76\frac{5}{8}$$

$$= 196 + 98 + 146 + 100 + 76 + \frac{6 + 4 + 7 + 5}{8}$$

$$= 616 + \frac{22}{8} \quad \left(\frac{22}{8} = 2\frac{6}{8} = 2\frac{3}{4} \right)$$

$$= 618\frac{3}{4} \text{ feet}$$

The perimeter is $618\frac{3}{4}$ feet.

71. Divide the total board length by 3.

$$15\frac{5}{8} \div 3 = \frac{125}{8} \div \frac{3}{1}$$

$$= \frac{125}{8} \cdot \frac{1}{3}$$

$$= \frac{125 \cdot 1}{8 \cdot 3}$$

$$= \frac{125}{24} = 5\frac{5}{24}$$

The length of each of the three pieces must be $5\frac{5}{24}$ inches.

73. Divide the total amount of tomato sauce by the number of servings.

$$2\frac{1}{3} \div 7 = \frac{7}{3} \div \frac{7}{1} = \frac{7}{3} \cdot \frac{1}{7} = \frac{7 \cdot 1}{3 \cdot 7} = \frac{1}{3}$$

For 1 serving of barbeque sauce, $\frac{1}{3}$ cup of tomato sauce is needed.

75. We can multiply the fraction in the chart $\left(\frac{13}{100} \right)$ by the number of people in the group (5000).

$$\frac{13}{100} \cdot 5000 = \frac{13 \cdot 50 \cdot 100}{100 \cdot 1} = \frac{13 \cdot 50}{1} = 650$$

About 650 people in the group are employed in the service occupations.

77. **(a)** 12 is $\frac{1}{3}$ of 36, so Crum got a hit in exactly $\frac{1}{3}$ of her at-bats.

(b) 5 is a little smaller than $\frac{1}{2}$ of 11, so Jordan got a hit in just less than $\frac{1}{2}$ of his at-bats.

(c) 1 is a little smaller than $\frac{1}{10}$ of 11, so Jordan got a home run in just less than $\frac{1}{10}$ of his at-bats.

(d) 9 is a little smaller than $\frac{1}{4}$ of 40, so Baldock got a hit in just less than $\frac{1}{4}$ of her at-bats.

(e) 8 is $\frac{1}{2}$ of 16, and 10 is $\frac{1}{2}$ of 20, so Tobin and Perry each got hits $\frac{1}{2}$ of the times they were at bat.

79. $\frac{14}{26} + \frac{98}{99} + \frac{100}{51} + \frac{90}{31} + \frac{13}{27}$

Estimate each fraction.

$\frac{14}{26}$ is about $\frac{1}{2}$.

$\frac{98}{99}$ is about 1.

$\frac{100}{51}$ is about 2.

$\frac{90}{31}$ is about 3.

$\frac{13}{27}$ is about $\frac{1}{2}$.

Therefore, the sum is approximately

$\frac{1}{2} + 1 + 2 + 3 + \frac{1}{2} = 7$.

The correct choice is (b).

Section 1.2

1. False; $4 + 3(8 - 2) = 4 + 3 \cdot 6 = 4 + 18 = 22$. The common error leading to 42 is adding 4 to 3 and then multiplying by 6. One must follow the rules for order of operations.

3. False; the correct interpretation is $4 = 16 - 12$.

5. $7^2 = 7 \cdot 7 = 49$

7. $12^2 = 12 \cdot 12 = 144$

9. $4^3 = 4 \cdot 4 \cdot 4 = 64$

11. $10^3 = 10 \cdot 10 \cdot 10 = 1000$

13. $3^4 = 3 \cdot 3 \cdot 3 \cdot 3 = 81$

15. $4^5 = 4 \cdot 4 \cdot 4 \cdot 4 \cdot 4 = 1024$

17. $\left(\frac{2}{3}\right)^4 = \frac{2}{3} \cdot \frac{2}{3} \cdot \frac{2}{3} \cdot \frac{2}{3} = \frac{16}{81}$

19. $(.04)^3 = (.04)(.04)(.04) = .000064$

21. To evaluate 6^3, multiply the base, 6, by itself 3 times. The exponent, 3, indicates the number of times to multiply the base by itself.

23. $\begin{aligned} 9 \cdot 5 - 13 &= 45 - 13 &\quad \textit{Multiply} \\ &= 32 &\quad \textit{Subtract} \end{aligned}$

25. $\begin{aligned} \frac{1}{4} \cdot \frac{2}{3} + \frac{2}{5} \cdot \frac{11}{3} &= \frac{1}{6} + \frac{22}{15} &\quad \textit{Multiply} \\ &= \frac{5}{30} + \frac{44}{30} &\quad \textit{LCD = 30} \\ &= \frac{49}{30} \text{ or } 1\frac{19}{30} &\quad \textit{Add} \end{aligned}$

27. $\begin{aligned} 9 \cdot 4 - 8 \cdot 3 &= 36 - 24 &\quad \textit{Multiply} \\ &= 12 &\quad \textit{Subtract} \end{aligned}$

29. $\begin{aligned} (4.3)(1.2) &+ (2.1)(8.5) \\ &= 5.16 + 17.85 &\quad \textit{Multiply} \\ &= 23.01 &\quad \textit{Add} \end{aligned}$

31. $\begin{aligned} 5&[3 + 4(2^2)] \\ &= 5[3 + 4(4)] &\quad \textit{Use the exponent} \\ &= 5(3 + 16) &\quad \textit{Multiply} \\ &= 5(19) &\quad \textit{Add} \\ &= 95 &\quad \textit{Multiply} \end{aligned}$

33. $\begin{aligned} 3^2&[(11 + 3) - 4] \\ &= 3^2[14 - 4] &\quad \textit{Add inside parentheses} \\ &= 3^2[10] &\quad \textit{Subtract} \\ &= 9[10] &\quad \textit{Use the exponent} \\ &= 90 &\quad \textit{Multiply} \end{aligned}$

35. Simplify the numerator and denominator separately; then divide.

$$\frac{6(3^2 - 1) + 8}{3 \cdot 2 - 2} = \frac{6(9 - 1) + 8}{6 - 2}$$
$$= \frac{6(8) + 8}{4}$$
$$= \frac{48 + 8}{4}$$
$$= \frac{56}{4} = 14$$

37. $\dfrac{4(6 + 2) + 8(8 - 3)}{6(4 - 2) - 2^2} = \dfrac{4(8) + 8(5)}{6(2) - 2^2}$

$$= \frac{4(8) + 8(5)}{6(2) - 4}$$
$$= \frac{32 + 40}{12 - 4}$$
$$= \frac{72}{8} = 9$$

39. Begin by squaring 2. Then subtract 1, to get a result of $4 - 1 = 3$ within the parentheses. Next, raise 3 to the third power to get $3^3 = 27$. Multiply this result by 3 to obtain 81. Finally, add this result to 4 to get the final answer, 85.

continued

$$4 + 3\left(2^2 - 1\right)^3 = 4 + 3(4 - 1)^3$$
$$= 4 + 3\left(3^3\right)$$
$$= 4 + 3(27)$$
$$= 4 + 81 = 85$$

41. The statement $5 < 6$ is true since 5 is less than 6.

43. "$8 \geq 17$" means "8 is greater than or equal to 17." The statement is false since 8 is less

than 17.

45. $17 \leq 18 - 1$
$17 \leq 17$

This statement is true since $17 = 17$ is true.

47. $6 \cdot 8 + 6 \cdot 6 \geq 0$
$48 + 36 \geq 0$
$84 \geq 0$

The statement is true since 84 is greater than zero.

49. $6[5 + 3(4 + 2)] \leq 70$
$6[5 + 3(6)] \leq 70$ *Add inside parentheses*
$6[5 + 18] \leq 70$ *Multiply*
$6[23] \leq 70$ *Add*
$138 \leq 70$ *Multiply*

Both $138 < 70$ and $138 = 70$ are false, so $138 \leq 70$ is false.

51. $\dfrac{9(7 - 1) - 8 \cdot 2}{4(6 - 1)} > 3$

$\dfrac{9 \cdot 6 - 8 \cdot 2}{4(5)} > 3$

$\dfrac{54 - 16}{20} > 3$

$\dfrac{38}{20} > 3$

$\dfrac{19}{10} > 3$

The statement is false since $\dfrac{19}{10}$ or $1\dfrac{9}{10}$ is

less than 3.

53. $8 \leq 4^2 - 2^2$
$8 \leq 16 - 4$ *Use the exponents*
$8 \leq 12$ *Subtract*

Since $8 < 12$, the statement is true.

55. "Fifteen is equal to five plus ten" is written

$$15 = 5 + 10.$$

57. "Nine is greater than five minus four" is written

$$9 > 5 - 4.$$

59. "Sixteen is not equal to nineteen" is written

$$16 \neq 19.$$

61. "Two is less than or equal to three" is written

$$2 \leq 3.$$

63. "$7 < 19$" means "seven is less than nineteen." The statement is true.

65. "$3 \neq 6$" means "three is not equal to six." The statement is true.

67. "$8 \geq 11$" means "eight is greater than or equal to eleven." The statement is false.

69. Answers will vary. One example is

$$5 + 3 \geq 2 \cdot 2.$$

The statement is true since $8 > 4$.

71. $5 < 30$ becomes $30 > 5$ when the inequality symbol is reversed.

73. $12 \geq 3$ becomes $3 \leq 12$ when the inequality symbol is reversed.

75. In comparing age, "is younger than" expresses the idea of "is less than."

77. $12 \geq 12$

The inequality symbol $\geq$ implies a true statement if 12 equals 12 *or* if 12 is greater than 12. Although $12 > 12$ is a false statement, 12 is equal to 12, so $12 \geq 12$ is a true statement.

79. Look for the bars that are lower than the preceding bars. The corresponding months are:

December 1996, January 1997, May 1997,

June 1997, November 1997.

81. **(a)** We need to find the difference in the index levels. The index decline is $53.1 - 52.4 = .7$.

(b) The percent of decline is

$$\dfrac{.7}{53.1} \cdot 100 \approx 1.3\%.$$

Section 1.3

1. If $x = 3$, then the value of $x + 7$ is $3 + 7$, or 10.

3. The sum of 12 and x is represented by the expression $12 + x$. If $x = 9$, then the value of $12 + x$ is $12 + 9$, or 21.

5. This question is equivalent to asking "is a number ever equal to four more than itself?" Since that never occurs, the answer is no.

7. $2x^3 = 2 \cdot x \cdot x \cdot x$, while $2x \cdot 2x \cdot 2x = (2x)^3$. The last expression is equal to $8x^3$.

9. The exponent 2 applies only to its base, which is x. (The expression $(4x)^2$ would require multiplying 4 by $x = 3$ first.)

11. (Answers will vary.) Two such pairs are $x = 0$, $y = 6$ and $x = 1$, $y = 4$. To determine them, choose a value for x, substitute it into the expression $2x + y$, and then subtract the value of $2x$ from 6.

In part (a) of Exercises 13–26, replace x with 4. In part (b), replace x with 6. Then use the order of operations.

13. **(a)** $x + 9 = 4 + 9$
$\qquad = 13$

 (b) $x + 9 = 6 + 9$
$\qquad = 15$

15. **(a)** $5x = 5(4) = 20$

 (b) $5x = 5(6) = 30$

17. **(a)** $4x^2 = 4 \cdot 4^2$
$\qquad = 4 \cdot 16$
$\qquad = 64$

 (b) $4x^2 = 4 \cdot 6^2$
$\qquad = 4 \cdot 36$
$\qquad = 144$

19. **(a)** $\dfrac{x+1}{3} = \dfrac{4+1}{3}$
$\qquad = \dfrac{5}{3}$

 (b) $\dfrac{x+1}{3} = \dfrac{6+1}{3}$
$\qquad = \dfrac{7}{3}$

21. **(a)** $\dfrac{3x-5}{2x} = \dfrac{3 \cdot 4 - 5}{2 \cdot 4}$
$\qquad = \dfrac{12-5}{8}$
$\qquad = \dfrac{7}{8}$

 (b) $\dfrac{3x-5}{2x} = \dfrac{3 \cdot 6 - 5}{2 \cdot 6}$
$\qquad = \dfrac{18-5}{12}$
$\qquad = \dfrac{13}{12}$

23. **(a)** $3x^2 + x = 3 \cdot 4^2 + 4$
$\qquad = 3 \cdot 16 + 4$
$\qquad = 48 + 4 = 52$

 (b) $3x^2 + x = 3 \cdot 6^2 + 6$
$\qquad = 3 \cdot 36 + 6$
$\qquad = 108 + 6 = 114$

25. **(a)** $6.459x = 6.459 \cdot 4$
$\qquad = 25.836$

 (b) $6.459x = 6.459 \cdot 6$
$\qquad = 38.754$

In part (a) of Exercises 27–42, replace x with 2 and y with 1. In part (b), replace x with 1 and y with 5.

27. **(a)** $8x + 3y + 5 = 8(2) + 3(1) + 5$
$\qquad = 16 + 3 + 5$
$\qquad = 19 + 5$
$\qquad = 24$

 (b) $8x + 3y + 5 = 8(1) + 3(5) + 5$
$\qquad = 8 + 15 + 5$
$\qquad = 23 + 5$
$\qquad = 28$

29. **(a)** $3(x + 2y) = 3(2 + 2 \cdot 1)$
$\qquad = 3(2 + 2)$
$\qquad = 3(4)$
$\qquad = 12$

 (b) $3(x + 2y) = 3(1 + 2 \cdot 5)$
$\qquad = 3(1 + 10)$
$\qquad = 3(11)$
$\qquad = 33$

31. **(a)** $x + \dfrac{4}{y} = 2 + \dfrac{4}{1}$
$\qquad = 2 + 4$
$\qquad = 6$

 (b) $x + \dfrac{4}{y} = 1 + \dfrac{4}{5}$
$\qquad = \dfrac{5}{5} + \dfrac{4}{5}$
$\qquad = \dfrac{9}{5}$

33. **(a)** $\dfrac{x}{2} + \dfrac{y}{3} = \dfrac{2}{2} + \dfrac{1}{3}$
$\qquad = \dfrac{6}{6} + \dfrac{2}{6}$
$\qquad = \dfrac{8}{6} = \dfrac{4}{3}$

 (b) $\dfrac{x}{2} + \dfrac{y}{3} = \dfrac{1}{2} + \dfrac{5}{3}$
$\qquad = \dfrac{3}{6} + \dfrac{10}{6}$
$\qquad = \dfrac{13}{6}$

35. **(a)** $\dfrac{2x + 4y - 6}{5y + 2} = \dfrac{2(2) + 4(1) - 6}{5(1) + 2}$
$\qquad = \dfrac{4 + 4 - 6}{5 + 2}$
$\qquad = \dfrac{8 - 6}{7}$
$\qquad = \dfrac{2}{7}$

(b) $\dfrac{2x + 4y - 6}{5y + 2} = \dfrac{2(1) + 4(5) - 6}{5(5) + 2}$

$= \dfrac{2 + 20 - 6}{25 + 2}$

$= \dfrac{22 - 6}{27}$

$= \dfrac{16}{27}$

37. (a) $2y^2 + 5x = 2 \cdot 1^2 + 5 \cdot 2$

$= 2 \cdot 1 + 5 \cdot 2$

$= 2 + 10$

$= 12$

(b) $2y^2 + 5x = 2 \cdot 5^2 + 5 \cdot 1$

$= 2 \cdot 25 + 5 \cdot 1$

$= 50 + 5$

$= 55$

39. (a) $\dfrac{3x + y^2}{2x + 3y} = \dfrac{3(2) + 1^2}{2(2) + 3(1)}$

$= \dfrac{3(2) + 1}{4 + 3}$

$= \dfrac{6 + 1}{7}$

$= \dfrac{7}{7}$

$= 1$

(b) $\dfrac{3x + y^2}{2x + 3y} = \dfrac{3(1) + 5^2}{2(1) + 3(5)}$

$= \dfrac{3(1) + 25}{2 + 15}$

$= \dfrac{3 + 25}{17}$

$= \dfrac{28}{17}$

41. (a) $.841x^2 + .32y^2$

$= .841 \cdot 2^2 + .32 \cdot 1^2$

$= .841 \cdot 4 + .32 \cdot 1$

$= 3.364 + .32$

$= 3.684$

(b) $.841x^2 + .32y^2$

$= .841 \cdot 1^2 + .32 \cdot 5^2$

$= .841 \cdot 1 + .32 \cdot 25$

$= .841 + 8$

$= 8.841$

43. "Twelve times a number" translates as $12 \cdot x$ or $12x$.

45. "Added to" indicates addition. "Seven added to a number" translates as $x + 7$.

47. "Two subtracted from a number" translates as $x - 2$.

49. "A number subtracted from seven" translates as $7 - x$.

51. "The difference between a number and 6" translates as $x - 6$.

53. "12 divided by a number" translates as $\dfrac{12}{x}$.

55. "The product of 6 and four less than a number" translates as $6(x - 4)$.

57. No, *and* is a connective word that joins the two factors: the number, and 6.

59. $5m + 2 = 7$; 1

$5(1) + 2 = 7$?   *Let m = 1*

$5 + 2 = 7$?

$7 = 7$ *True*

Because substituting 1 for m results in a true statement, 1 is a solution of the equation.

61. $2y + 3(y - 2) = 14$; 3

$2 \cdot 3 + 3(3 - 2) = 14$?   *Let y = 3*

$2 \cdot 3 + 3 \cdot 1 = 14$?

$6 + 3 = 14$?

$9 = 14$ *False*

Because substituting 3 for y results in a false statement, 3 is not a solution of the equation.

63. $6p + 4p + 9 = 11$; $\dfrac{1}{5}$

$6\left(\dfrac{1}{5}\right) + 4\left(\dfrac{1}{5}\right) + 9 = 11$?   *Let p = $\dfrac{1}{5}$*

$\dfrac{6}{5} + \dfrac{4}{5} + 9 = 11$?

$\dfrac{10}{5} + 9 = 11$?

$2 + 9 = 11$?

$11 = 11$ *True*

The true result shows that $\dfrac{1}{5}$ is a solution of the equation.

65. $3r^2 - 2 = 46$; 4

$3(4)^2 - 2 = 46$?   *Let r = 4*

$3 \cdot 16 - 2 = 46$?

$48 - 2 = 46$?

$46 = 46$ *True*

The true result shows that 4 is a solution of the equation.

67. $\dfrac{z+4}{2-z} = \dfrac{13}{5}; \dfrac{1}{3}$

$\dfrac{\dfrac{1}{3}+4}{2-\dfrac{1}{3}} = \dfrac{13}{5}$? *Let* $z = \dfrac{1}{3}$

$\dfrac{\dfrac{1}{3}+\dfrac{12}{3}}{\dfrac{6}{3}-\dfrac{1}{3}} = \dfrac{13}{5}$?

$\dfrac{\dfrac{13}{3}}{\dfrac{5}{3}} = \dfrac{13}{5}$?

$\dfrac{13}{3}\cdot\dfrac{3}{5} = \dfrac{13}{5}$?

$\dfrac{13}{5} = \dfrac{13}{5}$ *True*

The true result shows that $\dfrac{1}{3}$ is a solution of the equation.

69. "The sum of a number and 8 is 18" translates as

$$x + 8 = 18.$$

Try each number from the given set,

$\{2, 4, 6, 8, 10\}$, in turn.

$$
\begin{aligned}
x + 8 &= 18 \quad \textit{Given equation}\\
2 + 8 &= 18 \quad \textit{False}\\
4 + 8 &= 18 \quad \textit{False}\\
6 + 8 &= 18 \quad \textit{False}\\
8 + 8 &= 18 \quad \textit{False}\\
10 + 8 &= 18 \quad \textit{True}
\end{aligned}
$$

The only solution is 10.

71. "Sixteen minus three-fourths of a number is 13" translates as

$$16 - \dfrac{3}{4}x = 13.$$

Try each number from the given set,

$\{2, 4, 6, 8, 10\}$, in turn.

$$
\begin{aligned}
16 - \dfrac{3}{4}x &= 13 \quad \textit{Given equation}\\
16 - \dfrac{3}{4}(2) &= 13 \quad \textit{False}\\
16 - \dfrac{3}{4}(4) &= 13 \quad \textit{True}
\end{aligned}
$$

$$
\begin{aligned}
16 - \dfrac{3}{4}(6) &= 13 \quad \textit{False}\\
16 - \dfrac{3}{4}(8) &= 13 \quad \textit{False}\\
16 - \dfrac{3}{4}(10) &= 13 \quad \textit{False}
\end{aligned}
$$

The only solution is 4.

73. "One more than twice a number is 5" translates as $2x + 1 = 5$.

Try each number from the given set. The only resulting true equation is

$$2 \cdot 2 + 1 = 5,$$

So the only solution is 2.

75. "Three times a number is equal to 8 more than twice the number" translates as

$$3x = 2x + 8.$$

Try each number from the given set.

$$
\begin{aligned}
3x &= 2x + 8 \quad \textit{Given equation}\\
3(2) &= 2(2) + 8 \quad \textit{False}\\
3(4) &= 2(4) + 8 \quad \textit{False}\\
3(6) &= 2(6) + 8 \quad \textit{False}\\
3(8) &= 2(8) + 8 \quad \textit{True}\\
3(10) &= 2(10) + 8 \quad \textit{False}
\end{aligned}
$$

The only solution is 8.

77. There is no equals sign, so $3x + 2(x - 4)$ is an expression.

79. There is an equals sign, so $7t + 2(t + 1) = 4$ is an equation.

81. There is an equals sign, so $x + y = 3$ is an equation.

83. $\begin{aligned} y &= .319x - 624.31\\ &= .319(1990) - 624.31\\ &= 10.5 \end{aligned}$

The approximation \$10.50 is \$.33 less than the actual earnings of \$10.83.

85. $\begin{aligned} y &= .319x - 624.31\\ &= .319(1995) - 624.31\\ &= 12.095 \approx 12.10 \end{aligned}$

The approximation \$12.10 is \$.27 less than the actual earnings of \$12.37.

Section 1.4

1. The only integer between 3.5 and 4.5 is 4.

3. There is only one whole number that is not positive and less than 1: the number 0.

5. An irrational number that is between $\sqrt{11}$ and $\sqrt{13}$ is $\sqrt{12}$. There are others.

7. True; every natural number is positive.

9. True; every integer is a rational number. For example, 5 can be written as $\dfrac{5}{1}$.

11. $\left\{-9, -\sqrt{7}, -1\dfrac{1}{4}, -\dfrac{3}{5}, 0, \sqrt{5}, 3, 5.9, 7\right\}$

(a) The natural numbers in the given set are 3 and 7, since they are in the natural number set $\{1, 2, 3, \dots\}$.

(b) The set of whole numbers includes the natural numbers and 0. The whole numbers in the given set are 0, 3, and 7.

(c) The integers are the set of numbers $\{\dots, -3, -2, -1, 0, 1, 2, 3, \dots\}$. The integers in the given set are -9, 0, 3, and 7.

(d) Rational numbers are the numbers which can be expressed as the quotient of two integers, with denominators not equal to 0.

We can write numbers from the given set in this form as follows:

$$-9 = \frac{-9}{1}, -1\frac{1}{4} = \frac{-5}{4}, -\frac{3}{5} = \frac{-3}{5}, 0 = \frac{0}{1}$$

$$3 = \frac{3}{1}, 5.9 = \frac{59}{10}, \text{ and } 7 = \frac{7}{1}.$$ Thus, the rational numbers in the given set are -9, $-1\dfrac{1}{4}$, $-\dfrac{3}{5}$, 0, 3, 5.9, and 7.

(e) Irrational numbers are real numbers that are not rational. $-\sqrt{7}$ and $\sqrt{5}$ can be represented by points on the number line but cannot be written as a quotient of integers. Thus, the irrational numbers in the given set are $-\sqrt{7}$ and $\sqrt{5}$.

(f) Real numbers are all numbers that can be represented on the number line. All the numbers in the given set are real.

13. The *natural numbers* are the numbers with which we count. An example is 1. The *whole numbers* are the natural numbers with 0 also included. An example is 0. The *integers* are the whole numbers and their negatives. An example is -1. The *rational numbers* are the numbers that can be represented by a quotient of integers with denominator not 0, such as $\dfrac{1}{2}$. The *irrational numbers*, such as $\sqrt{2}$, cannot be represented as a quotient of integers. The *real numbers* include all positive numbers, negative numbers, and zero. All the numbers listed are reals.

15. An increase of 93,000 can be represented by a positive number, 93,000.

17. Thirty degrees below zero can be represented by a negative number, -30.

19. A decrease of 31,532 can be represented by a negative number, $-31,532$.

21. Eight feet below sea level can be represented by a negative number, -8.

23. Graph 0, 3, -5, and -6.

Place a dot on the number line at the point that corresponds to each number. The order of the numbers from smallest to largest is -6, -5, 0, 3.

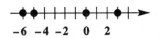

25. Graph -2, -6, -4, 3, and 4.

27. Graph $\dfrac{1}{4}$, $2\dfrac{1}{2}$, $-3\dfrac{4}{5}$, -4, and $-1\dfrac{5}{8}$.

29. **(a)** $|-7| = 7$ (A)

The distance between -7 and 0 on the number line is 7 units.

(b) $-(-7) = 7$ (A)

The opposite of -7 is 7.

(c) $-|-7| = -(7) = -7$ (B)

(d) $-|-(-7)| = -|7|$ *Work inside absolute value symbols first*

$$= -(7)$$
$$= -7 \quad \text{(B)}$$

31. **(a)** The opposite of -2 is found by changing the sign of -2. The opposite of -2 is 2.

(b) The absolute value of -2 is the distance between 0 and -2 on the number line.

$$|-2| = 2$$

The absolute value of -2 is 2.

33. **(a)** The opposite of 6 is -6.

(b) The distance between 0 and 6 on the number line is 6 units, so the absolute value of 6 is 6.

35. $7 - 4 = 3$

(a) The opposite of 3 is -3.

(b) The absolute value of 3 is 3.

37. $7 - 7 = 0$

(a) The opposite of 0 is -0, which is 0.

(b) The absolute value of 0 is 0.

39. If $a - b > 0$. then the absolute value of $a - b$ in terms of a and b is $a - b$.

41. $-12, -4$

Since -12 is to the left of -4 on the number line, -12 is smaller than -4.

43. $-8, -1$

Since -8 is located to the left of -1 on the number line, -8 is smaller.

45. $3, |-4|$

Since $|-4| = 4$, 3 is the smaller of the two numbers.

47. $|-3|, |-4|$

Since $|-3| = 3$ and $|-4| = 4$, $|-3|$ or 3 is smaller.

49. $-|-6|, -|-4|$

Since $-|-6| = -6$ and $-|-4| = -4$, $-|-6|$ is to the left of $-|-4|$ on the number line, so $-|-6|$ or -6 is the smaller of the two numbers.

51. $|5 - 3|, |6 - 2|$

Since $|5 - 3| = |2| = 2$ and $|6 - 2| = |4| = 4$, $|5 - 3|$ or 2 is the smaller of the two numbers.

53. $6 > -(-2)$

Since $-(-2) = 2$ and $6 > 2$, $6 > -(-2)$ is true.

55. $-4 \leq -(-5)$

Since $-(-5) = 5$ and $-4 < 5$, $-4 \leq -(-5)$ is true.

57. $|-6| < |-9|$

Since $|-6| = 6$ and $|-9| = 9$, $|-6| < |-9|$ is true.

59. $-|8| > |-9|$

Since $-|8| = -8$ and $|-9| = -(-9) = 9$, $-|8| < |-9|$, so $-|8| > |-9|$ is false.

61. $-|-5| \geq -|-9|$

Since $-|-5| = -5$, $-|-9| = -9$, and $-9 < -5$, $-|-5| \geq -|-9|$ is true.

63. $|6 - 5| \geq |6 - 2|$

Since $|6 - 5| = |1| = 1$ and $|6 - 2| = |4| = 4$, $|6 - 5| < |6 - 2|$, so $|6 - 5| \geq |6 - 2|$ is false.

65. The negative number with the largest absolute value in the table is -14.1, so the greatest drop is softwood plywood from 1995 to 1996.

67. The absolute values for softwood plywood are 11.3 and 14.1 for the first and second columns. The statement is true since $11.3 < 14.1$.

69. $|a + b| = |a - b|$

A pair of values that makes the statement true is $a = 5$, $b = 0$. The statement will be true when $a = 0$ or $b = 0$ or both $a = 0$ and $b = 0$. A pair of values that makes the statement false is $a = 5$, $b = 4$. Any values for a and b such that neither a nor b is zero will make the statement false.

71. $|a + b| = -|a + b|$

A pair of values that makes the statement true is $a = 5$, $b = -5$. The statement will be true when $a + b = 0$ ($|0| = -|0|$), or , equivalently, when $a = -b$, that is, a and b are opposites. A pair of values that makes the statement false is $a = 3$, $b = -5$. If $a \neq -b$, the statement is false.

73. Three examples of positive real numbers that are not integers are $\frac{1}{2}, \frac{5}{8}$, and $1\frac{3}{4}$. Other examples are $.7$, $4\frac{2}{3}$, and 5.1.

75. Three examples of real numbers that are not whole numbers are $-3\frac{1}{2}, -\frac{2}{3}$, and $\frac{3}{7}$. Other examples are $-4.3, -\sqrt{2}$, and $\sqrt{7}$.

77. Three examples of real numbers that are not rational numbers are $\sqrt{5}, \pi$, and $-\sqrt{3}$. All irrational numbers are real numbers that are not rational.

79. The statement "Absolute value is always positive." is not true. The absolute value of 0 is 0, and 0 is not positive. A more accurate way of describing absolute value is to say that *absolute value is never negative,* or *absolute value is always nonnegative.*

Section 1.5

1. The sum of two negative numbers will always be a *negative* number.

3. To simplify the expression $8 + [-2 + (-3 + 5)]$, I should begin by adding -3 and 5, according to the rules for order of operations.

5. To add two numbers with the same sign, add their absolute values and keep the same sign for the sum. For example, $3 + 4 = 7$ and $-3 + (-4) = -7$. To add two numbers with different signs, subtract the smaller absolute value from the larger absolute value, and use the sign of the number with the larger absolute value. For example, $6 + (-4) = 2$ and $(-6) + 4 = -2$.

7. $6 + (-4)$

To add $6 + (-4)$, find the difference between the absolute values of the numbers.

$$|6| = 6 \text{ and } |-4| = 4$$

continued

$$6 - 4 = 2$$

Since $|6| > |-4|$, the sum will be positive:

$$6 + (-4) = 2.$$

9. $7 + (-10)$

Since the numbers have different signs, find the difference between their absolute values:

$$10 - 7 = 3.$$

Because -10 has the larger absolute value, the sum is negative:

$$7 + (-10) = -3.$$

11. $-7 + (-3)$

The sum of two negative numbers is negative.

$$-7 + (-3) = -10$$

13. $-10 + (-3)$

Because the numbers have the same sign, add their absolute values:

$$10 + 3 = 13.$$

Because both numbers are negative, their sum is negative:

$$-10 + (-3) = -13.$$

15. $-12.4 + (-3.5)$

The sum of two negative numbers is negative.

$$-12.4 + (-3.5) = -15.9.$$

17. $-8 + 7 = -1$

19. $5 + [14 + (-6)]$

Perform the operation inside the brackets first, then add.

$$5 + [14 + (-6)] = 5 + 8 = 13$$

In Exercises 21–50, use the definition of subtraction to find the differences.

21. $4 - 7 = 4 + (-7) = -3$

23. $6 - 10 = 6 + (-10) = -4$

25. $-7 - 3 = -7 + (-3) = -10$

27. $-10 - 6 = -10 + (-6) = -16$

29. $7 - (-4) = 7 + (4) = 11$

31. $6 - (-13) = 6 + 13 = 19$

33. $-7 - (-3) = -7 + (3) = -4$

35. $3 - (4 - 6) = 3 - [4 + (-6)]$
$$= 3 - (-2)$$
$$= 3 + 2$$
$$= 5$$

37. $-3 - (6 - 9) = -3 - [6 + (-9)]$
$$= -3 - (-3)$$
$$= -3 + 3$$
$$= 0$$

39. $\dfrac{1}{2} - \left(-\dfrac{1}{4}\right) = \dfrac{1}{2} + \dfrac{1}{4}$
$$= \dfrac{2}{4} + \dfrac{1}{4} = \dfrac{3}{4}$$

41. $-8 + [3 + (-1) + (-2)]$
$$= -8 + [2 + (-2)]$$
$$= -8 + 0$$
$$= -8$$

43. $\dfrac{5}{8} - \left(-\dfrac{1}{2} - \dfrac{3}{4}\right)$
$$= \dfrac{5}{8} - \left[-\dfrac{1}{2} + \left(-\dfrac{3}{4}\right)\right]$$
$$= \dfrac{5}{8} - \left[-\dfrac{2}{4} + \left(-\dfrac{3}{4}\right)\right]$$
$$= \dfrac{5}{8} - \left(-\dfrac{5}{4}\right)$$
$$= \dfrac{5}{8} + \dfrac{5}{4}$$
$$= \dfrac{5}{8} + \dfrac{10}{8}$$
$$= \dfrac{15}{8}$$

45. $[(-3.1) - 4.5] - (.8 - 2.1)$
$$= [(-3.1) + (-4.5)] - [.8 + (-2.1)]$$
$$= -7.6 - (-1.3)$$
$$= -7.6 + 1.3$$
$$= -6.3$$

47. $[-5 + (-7)] + [-4 + (-9)]$
$$+ [13 + (-12)]$$
$$= -12 + (-13) + 1$$
$$= -25 + 1 = -24$$

49. $-4 + [(-6 - 9) - (-7 + 4)]$
$$= -4 + \{[-6 + (-9)] - (-3)\}$$
$$= -4 + (-15 + 3)$$
$$= -4 + (-12) = -16$$

51. No, it is not possible to get a positive number when a negative number is added to another negative number. When both numbers are negative, the sign of the sum is negative.

53. The difference between two negative numbers can be positive or negative. In the following examples, the difference is a negative number.

$$-8 - (-2) = -8 + (2) = -6$$
$$-6 - (-5) = -6 + (5) = -1$$

55. $|3 - 8| - |-2 + 8| = |-5| - |6|$
$$= -(-5) - 6$$
$$= 5 - 6$$
$$= -1$$

57. $\left|\dfrac{2}{3} - \dfrac{7}{3}\right| + 2\left|-\dfrac{4}{9} + \dfrac{5}{9}\right|$

$= \left|-\dfrac{5}{3}\right| + 2\left|\dfrac{1}{9}\right|$

$= -\left(-\dfrac{5}{3}\right) + 2\left(\dfrac{1}{9}\right)$

$= \dfrac{5}{3} + \dfrac{2}{9}$

$= \dfrac{5 \cdot 3}{3 \cdot 3} + \dfrac{2}{9}$

$= \dfrac{15}{9} + \dfrac{2}{9}$

$= \dfrac{17}{9}$ or $1\dfrac{8}{9}$

59. "The sum of -5 and 12 and 6" is written

$-5 + 12 + 6$.

$-5 + 12 + 6 = [-5 + 12] + 6$
$$= 7 + 6 = 13$$

61. "14 added to the sum of -19 and -4" is written $[-19 + (-4)] + 14$.

$[-19 + (-4)] + 14 = (-23) + 14$
$$= -9$$

63. "The sum of -4 and -10, increased by 12," is written $[-4 + (-10)] + 12$.

$[-4 + (-10)] + 12 = -14 + 12$
$$= -2$$

65. "4 more than the sum of 8 and -18" is written $[8 + (-18)] + 4$.

$[8 + (-18)] + 4 = (-10) + 4$
$$= -6$$

67. "The difference between 4 and -8" is written $4 - (-8)$.

$4 - (-8) = 4 + 8 = 12$

69. "8 less than -2" is written $-2 - 8$.

$-2 - 8 = -2 + (-8) = -10$

71. "The sum of 9 and -4, decreased by 7" is written $[9 + (-4)] - 7$.

$[9 + (-4)] - 7 = 5 + (-7) = -2$

73. "12 less than the difference between 8 and -5" is written $[8 - (-5)] - 12$.

$[8 - (-5)] - 12 = [8 + (5)] - 12$
$$= 13 - 12$$
$$= 13 + (-12)$$
$$= 1$$

75. The outlay for 1988 is \$290.4 billion and the outlay for 1989 is \$303.6 billion. Thus the *change in outlay* is

$303.6 - 290.4 = 303.6 + (-290.4)$
$$= 13.2$$

billion dollars (an increase).

77. The outlay for 1990 is \$299.3 billion and the outlay for 1991 is \$273.3 billion. Thus the *change in outlay* is

$273.3 - 299.3 = 273.3 + (-299.3)$
$$= -26.0$$

billion dollars (a decrease since it is negative).

79. $17{,}400 - (-32{,}995) = 17{,}400 + 32{,}995$
$$= 50{,}395$$

The difference between the height of Mt. Foraker and the depth of the Philippine Trench is 50,395 feet.

81. $-23{,}376 - (-24{,}721) = -23{,}376 + 24{,}721$
$$= 1345$$

The Cayman Trench is 1345 feet deeper than the Java Trench.

83. **(a)** From 1991 to 1992, the change was

$48 - 58 = 48 + (-58)$
$$= -10.$$

(b) From 1992 to 1993, the change was

$53 - 48 = 53 + (-48)$
$$= 5.$$

(c) From 1989 to 1990, the change was

$65 - 77 = 65 + (-77)$
$$= -12.$$

(d) From 1993 to 1994, the change was

$22 - 53 = 22 + (-53)$
$$= -31.$$

85. To find the new temperature two minutes later, add 49 to -4.

$-4 + 49 = 45$

The temperature rose to $45°$F.

87. $14°$F lower than $-27°$F can be represented as

$-27 - 14 = -27 + (-14)$
$$= -41.$$

The record low in Huron is $-41°$F.

89. $14,494 - (-282) = 14,494 + 282$
$$= 14,776$$

The difference between these two elevations is 14,776 feet.

91. First find the total number of pounds lost. Since
$$45 + 45 + 205 = 295,$$

the polar bear lost 295 pounds over the winter. Now find the bear's weight when she left her den. Consider the loss as equivalent to a gain of a negative number of pounds. Since
$$660 + (-295) = 365,$$

the bear weighed 365 pounds when she left her den in March.

93. Use negative numbers to represent amounts Kim owes and positive numbers to represent payments and credits.

$-870.00 + 35.90 + 150.00 + (-82.50)$
$+ (-10.00) + (-10.00) + 500.00$
$+ (-37.23) = -323.83$

Kim still owes $323.83.

Section 1.6

1. A positive number is greater than 0 .

3. The product or the quotient of two numbers with the same sign is greater than 0 , since the product or quotient of two positive numbers is positive and the product or quotient of two negative numbers is positive.

5. If three negative numbers are multiplied together, the product is less than 0 , since a negative number times a negative number is a positive number, and that positive number times a negative number is a negative number.

7. If a negative number is squared and the result is added to a positive number, the final answer is greater than 0 , since a negative number squared is a positive number, and a positive number added to another positive number is a positive number.

9. If three positive numbers, five negative numbers, and zero are multiplied, the product is equal to 0 . Since one of the numbers is zero, the product is zero (regardless of what the other numbers are).

11. $(-4)(-5) = 20$

Note that the product of two negative numbers is positive.

13. $(-7)(4) = -(7 \cdot 4) = -28$

Note that the product of a negative number and a positive number is negative.

15. $(-4)(-20) = 4 \cdot 20 = 80$

17. $(-8)(0) = 0$

Note that the product of any number and 0 is 0.

19. $\left(-\dfrac{3}{8}\right)\left(-\dfrac{20}{9}\right) = \left(\dfrac{3}{8}\right)\left(\dfrac{20}{9}\right)$

$$= \dfrac{3 \cdot 20}{8 \cdot 9}$$

$$= \dfrac{3 \cdot (4 \cdot 5)}{(4 \cdot 2) \cdot (3 \cdot 3)}$$

$$= \dfrac{3 \cdot 4 \cdot 5}{4 \cdot 2 \cdot 3 \cdot 3}$$

$$= \dfrac{5}{2 \cdot 3} = \dfrac{5}{6}$$

21. $(-6)\left(-\dfrac{1}{4}\right) = 6\left(\dfrac{1}{4}\right) = \dfrac{6}{4} = \dfrac{2 \cdot 3}{2 \cdot 2} = \dfrac{3}{2}$

23. Using only positive integer factors, 32 can be written as $1 \cdot 32$, $2 \cdot 16$, or $4 \cdot 8$. Including the negative integer factors, we see that the integer factors of 32 are $-32, -16, -8, -4, -2, -1, 1, 2, 4, 8, 16$, and 32.

25. The integer factors of 40 are $-40, -20, -10, -8, -5, -4, -2, -1, 1, 2, 4, 5, 8, 10, 20$, and 40.

27. The integer factors of 31 are $-31, -1, 1$, and 31.

29. $\dfrac{-15}{5} = -\dfrac{3 \cdot 5}{5} = -3$

Note that the quotient of two numbers having different signs is negative.

31. $\dfrac{20}{-10} = -\dfrac{2 \cdot 10}{10} = -2$

33. $\dfrac{-160}{-10} = \dfrac{10 \cdot 16}{10} = 16$

Note that the quotient of two numbers having the same sign is positive.

35. $\dfrac{0}{-3} = 0$, because 0 divided by any nonzero number is 0.

37. $\dfrac{-10.252}{-.4} = 25.63$

Note that dividing by a number with absolute value between 0 and 1 gives us a number *larger* than the original numerator.

39. Dividing by a fraction $\left(\text{in this case } -\dfrac{1}{2}\right)$ is the same as multiplying by the reciprocal of the fraction $\left(\text{in this case } -\dfrac{2}{1}\right)$.

$$\left(-\frac{3}{4}\right) \div \left(-\frac{1}{2}\right) = \left(-\frac{3}{4}\right) \cdot \left(-\frac{2}{1}\right)$$
$$= \frac{3 \cdot 2}{2 \cdot 2 \cdot 1}$$
$$= \frac{3}{2}$$

41. To multiply two signed numbers, multiply their absolute values. If the signs of the numbers are the same, the product is positive. If the signs of the numbers are different, the product is negative. For example, $3 \cdot 5 = 15$, $(-3)(-5) = 15$, and

$(-3)(5) = -15$.

In Exercises 43–56, use the order of operations.

43. $7 - 3 \cdot 6 = 7 - 18$
$ = -11$

45. $-10 - (-4)(2) = -10 - (-8)$
$ = -10 + 8$
$ = -2$

47. $-7(3 - 8) = -7[3 + (-8)]$
$ = -7(-5) = 35$

49. $(12 - 14)(1 - 4) = (-2)(-3)$
$ = 6$

51. $(7 - 10)(10 - 4) = (-3)(6)$
$ = -18$

53. $(-2 - 8)(-6) + 7 = (-10)(-6) + 7$
$ = 60 + 7$
$ = 67$

55. $3(-5) + |3 - 10| = -15 + |-7|$
$ = -15 + 7$
$ = -8$

57. $\dfrac{-5(-6)}{9 - (-1)} = \dfrac{30}{10}$
$\phantom{\dfrac{-5(-6)}{9 - (-1)}} = \dfrac{3 \cdot 10}{10} = 3$

59. $\dfrac{-21(3)}{-3 - 6} = \dfrac{-63}{-3 + (-6)}$
$\phantom{\dfrac{-21(3)}{-3 - 6}} = \dfrac{-63}{-9} = 7$

61. $\dfrac{-10(2) + 6(2)}{-3 - (-1)} = \dfrac{-20 + 12}{-3 + 1}$
$\phantom{\dfrac{-10(2) + 6(2)}{-3 - (-1)}} = \dfrac{-8}{-2} = 4$

63. $\dfrac{-27(-2) - (-12)(-2)}{-2(3) - 2(2)} = \dfrac{54 - 24}{-6 - 4}$
$\phantom{\dfrac{-27(-2) - (-12)(-2)}{-2(3) - 2(2)}} = \dfrac{30}{-10}$
$\phantom{\dfrac{-27(-2) - (-12)(-2)}{-2(3) - 2(2)}} = -3$

65. $3x + 2y$

To evaluate this expression with $x = -3$ and $y = 4$, replace x with -3 and y with 4. Next find the two products, which are $3(-3) = -9$ and $2(4) = 8$. Finally, find their sum, which is $-9 + 8 = -1$.

In Exercises 67–78, replace x with 6, y with -4, and a with 3. Then use the order of operations to evaluate the expression.

67. $5x - 2y + 3a = 5(6) - 2(-4) + 3(3)$
$ = 30 - (-8) + 9$
$ = 30 + 8 + 9$
$ = 38 + 9$
$ = 47$

69. $(2x + y)(3a) = [2(6) + (-4)][3(3)]$
$ = [12 + (-4)](9)$
$ = (8)(9)$
$ = 72$

71. $\left(\dfrac{1}{3}x - \dfrac{4}{5}y\right)\left(-\dfrac{1}{5}a\right)$
$= \left[\dfrac{1}{3}(6) - \dfrac{4}{5}(-4)\right]\left[-\dfrac{1}{5}(3)\right]$
$= \left[2 - \left(-\dfrac{16}{5}\right)\right]\left(-\dfrac{3}{5}\right)$
$= \left(2 + \dfrac{16}{5}\right)\left(-\dfrac{3}{5}\right)$
$= \left(\dfrac{10}{5} + \dfrac{16}{5}\right)\left(-\dfrac{3}{5}\right)$
$= \left(\dfrac{26}{5}\right)\left(-\dfrac{3}{5}\right)$
$= -\dfrac{78}{25}$

73. $(-5 + x)(-3 + y)(3 - a)$
$ = (-5 + 6)[-3 + (-4)][3 - 3]$
$ = (1)(-7)(0)$
$ = 0$

75. $-2y^2 + 3a = -2(-4)^2 + 3(3)$
$ = -2(16) + 9$
$ = -32 + 9$
$ = -23$

77.
$$\frac{2y^2 - x}{a + 10} = \frac{2(-4)^2 - (6)}{3 + 10}$$
$$= \frac{2(16) - 6}{13}$$
$$= \frac{32 - 6}{13}$$
$$= \frac{26}{13}$$
$$= 2$$

79. "The product of -9 and 2, added to 9" is written $9 + (-9)(2)$.
$$9 + (-9)(2) = 9 + (-18)$$
$$= -9$$

81. "Twice the product of -1 and 6, subtracted from -4" is written $-4 - 2[(-1)(6)]$.
$$-4 - 2[(-1)(6)] = -4 - 2(-6)$$
$$= -4 - (-12)$$
$$= -4 + 12 = 8$$

83. "Nine subtracted from the product of 1.5 and -3.2 is written $(1.5)(-3.2) - 9$.
$$(1.5)(-3.2) - 9 = -4.8 - 9$$
$$= -4.8 + (-9)$$
$$= -13.8$$

85. "The product of 12 and the difference between 9 and -8" is written $12[9 - (-8)]$.
$$12[9 - (-8)] = 12[9 + 8]$$
$$= 12(17) = 204$$

87. "The quotient of -12 and the sum of -5 and -1" is written
$$\frac{-12}{-5 + (-1)},$$
and
$$\frac{-12}{-5 + (-1)} = \frac{-12}{-6} = 2.$$

89. "The sum of 15 and -3, divided by the product of 4 and -3" is written
$$\frac{15 + (-3)}{4(-3)},$$
and
$$\frac{15 + (-3)}{4(-3)} = \frac{12}{-12} = -1.$$

91. "The product of $-\frac{1}{2}$ and $\frac{3}{4}$, divided by $-\frac{2}{3}$" is written
$$\frac{\left(-\frac{1}{2}\right)\left(\frac{3}{4}\right)}{-\frac{2}{3}},$$

and
$$\frac{\left(-\frac{1}{2}\right)\left(\frac{3}{4}\right)}{-\frac{2}{3}} = \frac{-\frac{3}{8}}{-\frac{2}{3}} = \frac{3}{8} \cdot \frac{3}{2} = \frac{9}{16}.$$

93. "Six times a number is -42" is written
$$6x = -42.$$
The solution is -7, since
$$6(-7) = -42.$$

95. "The quotient of a number and 3 is -3" is written
$$\frac{x}{3} = -3.$$
The solution is -9, since
$$\frac{-9}{3} = -3.$$

97. "6 less than a number is 4" is written
$$x - 6 = 4.$$
The solution is 10, since
$$10 - 6 = 4.$$

99. "When 5 is added to a number, the result is -5" is written
$$x + 5 = -5.$$
The solution is -10, since
$$-10 + 5 = -5.$$

101. Add the numbers and divide by 5.
$$\frac{(23 + 18 + 13) + [(-4) + (-8)]}{5}$$
$$= \frac{54 - 12}{5}$$
$$= \frac{42}{5} \text{ or } 8\frac{2}{5}$$

103. Add the integers from -10 to 14.
$$(-10) + (-9) + \cdots + 14 = 50$$
[the 3 dots indicate that the pattern continues]

There are 25 integers from -10 to 14 (10 negative, zero, and 14 positive). Thus, the average is $\frac{50}{25} = 2$.

105. Adding the hourly earnings gives us
$15.3 + 15.08 + 12.37 + 14.23 + 12.43 + 7.69 + 12.33 + 11.39 = 100.82$

To find the average of the eight groups, divide by 8.
$$\frac{100.82}{8} = 12.6025$$
So, rounded to the nearest cent, the average is $12.60.

107. The average of a group of numbers is the sum of all the numbers divided by the number of numbers. If the average is 0, then the sum of all the numbers must be 0 since the only way to make a quotient 0 is to have its numerator equal to 0.

109. (a) 3,473,986 is divisible by 2 because its last digit, 6, is divisible by 2.

(b) 4,336,879 is not divisible by 2 because its last digit, 9, is not divisible by 2.

111. (a) 6,221,464 is divisible by 4 because the number formed by its last two digits, 64, is divisible by 4.

(b) 2,876,335 is not divisible by 4 because the number formed by its last two digits, 35, is not divisible by 4.

113. (a) 1,524,822 is divisible by 2 because its last digit, 2, is divisible by 2. It is also divisible by 3 because the sum of its digits,
$$1 + 5 + 2 + 4 + 8 + 2 + 2 = 24,$$
is divisible by 3.

Because 1,524,822 is divisible by *both* 2 and 3, it is divisible by 6.

(b) 2,873,590 is divisible by 2 because it last digit, 0, is divisible by 2. However, it is not divisible by 3 because the sum of its digits,
$$2 + 8 + 7 + 3 + 5 + 9 + 0 = 34,$$
is not divisible by 3.

Because 2,873,590 is not divisible by *both* 2 and 3, it is not divisible by 6.

115. (a) 4,114,107 is divisible by 9 because the sum of its digits,
$$4 + 1 + 1 + 4 + 1 + 0 + 7 = 18,$$
is divisible by 9.

(b) 2,287,321 is not divisible by 9 because the sum of its digits,
$$2 + 2 + 8 + 7 + 3 + 2 + 1 = 25,$$
is not divisible by 9.

Section 1.7

1. B, since 0 is the identity element for addition.

3. C, since $-a$ is the additive inverse of a.

5. B, since 0 is the only number that is equal to its negative; that is, $0 = -0$.

7. B, since the multiplicative inverse of a number a is $\frac{1}{a}$ and the only number that we *cannot* divide by is 0.

9. G, since we can consider $(5 \cdot 4)$ to be one number $(5 \cdot 4) \cdot 3$ is the same as $3 \cdot (5 \cdot 4)$ by the commutative property.

11. $7 + 18 = 18 + 7$

The order of the two numbers has been changed, so this is an example of the commutative property of addition: $a + b = b + a$.

13. $5(13 \cdot 7) = (5 \cdot 13) \cdot 7$

The numbers are in the same order but grouped differently, so this is an example of the associative property of multiplication: $(ab)c = a(bc)$.

15. $-6 + (12 + 7) = (-6 + 12) + 7$

The numbers are in the same order but grouped differently, so this is an example of the associative property of addition:
$(a + b) + c = a + (b + c)$.

17. $-6 + 6 = 0$

The sum of the two numbers is 0, so they are additive inverses (or opposites) of each other. This is an example of the additive inverse property: $a + (-a) = 0$.

19. $\left(\frac{2}{3}\right)\left(\frac{3}{2}\right) = 1$

The product of the two numbers is 1, so they are multiplicative inverses (or reciprocals) of each other. This is an example of the multiplicative inverse property: $a \cdot \frac{1}{a} = 1 \, (a \neq 0)$.

21. $2.34 + 0 = 2.34$

The sum of a number and 0 is the original number. This is an example of the identity property of addition: $a + 0 = a$.

23. $(4 + 17) + 3 = 3 + (4 + 17)$

The order of the numbers has been changed, but not the grouping, so this is an example of the commutative property of addition: $a + b = b + a$

25. $6(x + y) = 6x + 6y$

The number 6 outside the parentheses is "distributed" over the x and y. This is an example of the distributive property.

27. $-\dfrac{5}{9} = -\dfrac{5}{9} \cdot \dfrac{3}{3} = -\dfrac{15}{27}$

$\dfrac{3}{3}$ is a form of the number 1. We use it to rewrite $-\dfrac{5}{9}$ as $-\dfrac{15}{27}$. This is an example of the identity property of multiplication.

29. $5(2x) + 5(3y) = 5(2x + 3y)$

This is an example of the distributive property. The number 5 is "distributed " over $2x$ and $3y$.

31. Jack recognized the identity property of addition.

33. ADDITION:

(i) The commutative property of addition states that if you add two numbers in the reverse order, you will get the same sum.

(ii) The associative property of addition states that when adding three numbers, it doesn't matter which two numbers are added first.

(iii) The identity property of addition states that adding zero to any number leaves it unchanged.

(iv) The inverse property of addition states that the sum of a number and its opposite is zero.

MULTIPLICATION:

(i) The commutative property of multiplication states that if you multiply the same two numbers in the reverse order, you will get the same product.

(ii) The associative property of multiplication states that when multiplying three numbers, it doesn't matter which two numbers are multiplied first.

(iii) The identity property if multiplication states that multiplying any number by 1 leaves the number unchanged.

(iv) The inverse property of multiplication states that the product of any nonzero number and its reciprocal is 1.

35. $r + 7$; commutative

$$r + 7 = 7 + r$$

37. $s + 0$; identity

$$s + 0 = s$$

39. $-6(x + 7)$; distributive

$$-6(x + 7) = -6(x) + (-6)(7)$$
$$= -6x + (-42)$$
$$= -6x - 42$$

41. $(w + 5) + (-3)$; associative

$$(w + 5) + (-3) = w + [5 + (-3)]$$
$$= w + 2$$

43. $6t + 8 - 6t + 3$

$$
\begin{aligned}
&= 6t + 8 + (-6t) + 3 && \textit{Definition of} \\
& && \textit{subtraction} \\
&= (6t + 8) + (-6t) + 3 && \textit{Order of} \\
& && \textit{operations} \\
&= (8 + 6t) + (-6t) + 3 && \textit{Commutative} \\
& && \textit{property} \\
&= 8 + [6t + (-6t)] + 3 && \textit{Associative} \\
& && \textit{property} \\
&= 8 + 0 + 3 && \textit{Inverse} \\
& && \textit{property} \\
&= (8 + 0) + 3 && \textit{Order of} \\
& && \textit{operations} \\
&= 8 + 3 && \textit{Identity} \\
& && \textit{property} \\
&= 11 && \textit{Add}
\end{aligned}
$$

45. $\dfrac{2}{3}x - 11 + 11 - \dfrac{2}{3}x$

$$= \dfrac{2}{3}x + (-11) + 11 + \left(-\dfrac{2}{3}x\right)$$

Definition of subtraction

$$= \left[\dfrac{2}{3}x + (-11)\right] + 11 + \left(-\dfrac{2}{3}x\right)$$

Order of operations

$$= \dfrac{2}{3}x + (-11 + 11) + \left(-\dfrac{2}{3}x\right)$$

Associative property

$$= \dfrac{2}{3}x + 0 + \left(-\dfrac{2}{3}x\right) \qquad \textit{Inverse}$$
$$\textit{property}$$

$$= \left(\dfrac{2}{3}x + 0\right) + \left(-\dfrac{2}{3}x\right)$$

Order of operations

$$= \dfrac{2}{3}x + \left(-\dfrac{2}{3}x\right) \qquad \textit{Identity}$$
$$\textit{property}$$

$$= 0 \qquad \textit{Inverse}$$
$$\textit{property}$$

47. $\left(\dfrac{9}{7}\right)(-.38)\left(\dfrac{7}{9}\right)$

$$= \left[\left(\dfrac{9}{7}\right)(-.38)\right]\left(\dfrac{7}{9}\right) \qquad \textit{Order of}$$
$$\textit{operations}$$

$$= \left[(-.38)\left(\dfrac{9}{7}\right)\right]\left(\dfrac{7}{9}\right) \qquad \textit{Commutative}$$
$$\textit{property}$$

continued

$= (-.38)\left[\left(\dfrac{9}{7}\right)\left(\dfrac{7}{9}\right)\right]$ *Associative property*

$= (-.38)(1)$ *Inverse property*

$= -.38$ *Identity property*

49. $t + (-t) + \dfrac{1}{2}(2)$

$= t + (-t) + 1$ *Inverse property*

$= [t + (-t)] + 1$ *Order of operations*

$= 0 + 1$ *Inverse property*

$= 1$ *Identity property*

51. $25 - (6 - 2) = 25 - (4)$
$= 21$
$(25 - 6) - 2 = 19 - 2$
$= 17$

Since $21 \neq 17$, this example shows that subtraction is not associative.

53. $-3(4 - 6)$

When distributing a negative number over a quantity, be careful not to "lose" a negative sign. The problem should be worked in the following way.

$-3(4 - 6) = -3(4) - 3(-6)$
$= -12 + 18$
$= 6$

55. $5x + x = 5x + 1x$
$= (5 + 1)x$
$= 6x$

57. $4(t + 3) = 4 \cdot t + 4 \cdot 3$
$= 4t + 12$

59. $-8(r + 3) = -8(r) + (-8)(3)$
$= -8r + (-24)$
$= -8r - 24$

61. $-5(y - 4) = -5(y) + (-5)(-4)$
$= -5y + 20$

63. $-\dfrac{4}{3}(12y + 15z)$

$= -\dfrac{4}{3}(12y) + \left(-\dfrac{4}{3}\right)(15z)$

$= \left[\left(-\dfrac{4}{3}\right) \cdot 12\right]y + \left[\left(-\dfrac{4}{3}\right) \cdot 15\right]z$

$= -16y + (-20)z$

$= -16y - 20z$

65. $8 \cdot z + 8 \cdot w = 8(z + w)$

67. $7(2v) + 7(5r) = 7(2v + 5r)$

69. $8(3r + 4s - 5y)$
$= 8(3r) + 8(4s) + 8(-5y)$
      Distributive property
$= (8 \cdot 3)r + (8 \cdot 4)s + [8(-5)]y$
      Associative property
$= 24r + 32s - 40y$ *Multiply*

71. $q + q + q = 1 \cdot q + 1 \cdot q + 1 \cdot q$
$= (1 + 1 + 1)q$
$= 3q$

73. $-5x + x = -5x + 1x$ *Identity property*

$= (-5 + 1)x$ *Distributive property*

$= -4x$ *Add*

75. $-(4t + 3m)$

$= -1(4t + 3m)$ *Identity property*

$= -1(4t) + (-1)(3m)$ *Distributive property*

$= (-1 \cdot 4)t + (-1 \cdot 3)m$ *Associative property*

$= -4t - 3m$ *Multiply*

77. $-(-5c - 4d)$

$= -1(-5c - 4d)$ *Identity property*

$= -1(-5c) + (-1)(-4d)$ *Distributive property*

$= (-1 \cdot -5)c + (-1 \cdot -4)d$ *Associative property*

$= 5c + 4d$ *Multiply*

79. $-(-3q + 5r - 8s)$
$= -1(-3q + 5r - 8s)$
$= -1(-3q) + (-1)(5r) + (-1)(-8s)$
$= (-1 \cdot -3)q + (-1 \cdot 5)r + (-1 \cdot -8)s$
$= 3q - 5r + 8s$

81. Answers will vary. For example, "putting on your socks" and "putting on your shoes" are everyday operations that are not commutative.

83. $-3[5 + (-5)] = -3(0) = 0$

84. $-3[5 + (-5)] = -3(5) + (-3)(-5)$

85. $-3 \times 5 = -15$

86. We must interpret $(-3)(-5)$ as 15, since it is the additive inverse of -15.

Section 1.8

1. $6t$ and $5t^2$ are unlike terms and cannot be added together, so $6t + 5t^2 = 11t^3$ is a *false* statement.

3. $8r^2$ and $-12r^2$ are like terms, as are $3r$ and $4r$. Thus,

$$8r^2 + 3r - 12r^2 + 4r$$
$$= (8r^2 - 12r^2) + (3r + 4r)$$
$$= -4r^2 + 7r,$$

and the statement is *true*.

5. Since the statement must be true for all real numbers x, we'll substitute 0 for x and see which statements are true.

(a) $6 + 2x = 8x$
$$6 + 2(0) = 8(0)$$
$$6 = 0 \quad \text{*False*}$$

(b) $6 - 2x = 4x$
$$6 - 2(0) = 4(0)$$
$$6 = 0 \quad \text{*False*}$$

(c) $6x - 2x = 4x$
$$6(0) - 2(0) = 4(0)$$
$$0 = 0 \quad \text{*True*}$$

(d) $3 + 8(4x - 6) = 11(4x - 6)$
$$3 + 8[4(0) - 6] = 11[4(0) - 6]$$
$$3 + 8(-6) = 11(-6)$$
$$3 - 48 = -66$$
$$-45 = -66 \quad \text{*False*}$$

So the only statement that could possibly be true is (c).

$$\begin{aligned}
6x - 2x &= 4x \quad \text{*Given*} \\
(6 - 2)x &= 4x \quad \text{*Distributive*} \\
&\qquad\qquad \text{*property*} \\
4x &= 4x \quad \text{*Subtract*}
\end{aligned}$$

Since the left side is equal to the right side, statement (c) is true for all real numbers x.

7. The numerical coefficient of $5x^3y^7$ is 5. Therefore, the correct response is (a). In (b), the numerical coefficient of x^5 is 1. In (c), the numerical coefficient of $\dfrac{x}{5} = \dfrac{1}{5}x$ is $\dfrac{1}{5}$. In (d), the numerical coefficient of 5^2xy^3 is $5^2 = 25$.

9. $4r + 19 - 8 = 4r + 11$

11. $5 + 2(x - 3y) = 5 + 2(x) + 2(-3y)$
$$= 5 + 2x - 6y$$

13. $-2 - (5 - 3p) = -2 - 1(5 - 3p)$
$$= -2 - 1(5) - 1(-3p)$$
$$= -2 - 5 + 3p$$
$$= -7 + 3p$$

15. The numerical coefficient of the term $-12k$ is -12.

17. The numerical coefficient of the term $5m^2$ is 5.

19. Because xw can be written as $1 \cdot xw$, the numerical coefficient of the term xw is 1.

21. Since $-x = -1x$, the numerical coefficient of the term $-x$ is -1.

23. The numerical coefficient of the term 74 is 74.

25. Answers will vary. One such pair is $-4x$ and $7x$ since $-4x$ has a negative numerical coefficient and $7x$ has a positive numerical coefficient. The sum, $-4x + 7x = 3x$, has a positive numerical coefficient.

27. $8r$ and $-13r$ are like terms since they have the same variable with the same exponent (which is understood to be 1).

29. $5z^4$ and $9z^3$ are unlike terms. Although both have the variable z, the exponents are not the same.

31. All numerical terms (constants) are considered like terms, so 4, 9, and -24 are like terms.

33. x and y are unlike terms because they do not have the same variable.

35. Apples and oranges are examples of unlike fruits, just like x and y are unlike terms. We cannot add x and y to get an expression any simpler than $x + y$; we cannot add, for example, 2 apples and 3 oranges to obtain 5 fruits that are all alike.

37. The commutative and associative properties can be used to rearrange terms so that like terms are grouped together. Then combine like terms.

$$4k + 3 - 2k + 8 + 7k - 16$$
$$= (4k - 2k + 7k) + (3 + 8 - 16)$$
$$= (4 - 2 + 7)k + (3 + 8 - 16)$$
$$\qquad\qquad \text{*Distributive property*}$$
$$= 9k - 5 \quad \text{*Add and subtract*}$$

39. $-\dfrac{4}{3} + 2t + \dfrac{1}{3}t - 8 - \dfrac{8}{3}t$

$$= \left(2t + \dfrac{1}{3}t - \dfrac{8}{3}t\right) + \left(-\dfrac{4}{3} - 8\right)$$
$$\qquad\qquad \text{*Group like terms*}$$
$$= \left(2 + \dfrac{1}{3} - \dfrac{8}{3}\right)t + \left(-\dfrac{4}{3} - 8\right)$$
$$\qquad\qquad \text{*Distributive property*}$$
$$= \left(\dfrac{6}{3} + \dfrac{1}{3} - \dfrac{8}{3}\right)t + \left(-\dfrac{4}{3} - \dfrac{24}{3}\right)$$
$$\qquad\qquad LCD = 3$$
$$= -\dfrac{1}{3}t + \left(-\dfrac{28}{3}\right)$$
$$= -\dfrac{1}{3}t - \dfrac{28}{3}$$

41. $-5.3r + 4.9 - 2r + .7 + 3.2r$

$= (-5.3r - 2r + 3.2r) + (4.9 + .7)$

  Group like terms

$= (-5.3 - 2 + 3.2)r + (4.9 + .7)$

  Distributive property

$= -4.1r + 5.6$ *Add and subtract*

43. $2y^2 - 7y^3 - 4y^2 + 10y^3$

$= (2y^2 - 4y^2) + (-7y^3 + 10y^3)$

  Group like terms

$= (2 - 4)y^2 + (-7 + 10)y^3$

  Distributive property

$= -2y^2 + 3y^3$

45. $13p + 4(4 - 8p)$

$= 13p + 4(4) + 4(-8p)$ *Distributive property*

$= 13p + 16 - 32p$ *Multiply*

$= -19p + 16$ *Combine like terms*

47. $-4(y - 7) - 6$

$= -4(y) - (-4)(7) - 6$

$= -4y - (-28) - 6$

$= -4y + 28 - 6$

$= -4y + 22$

49. $-5(5y - 9) + 3(3y + 6)$

$= -5(5y) - 5(-9) + 3(3y) + 3(6)$

  Distributive property

$= -25y + 45 + 9y + 18$ *Multiply*

$= -16y + 63$ *Combine like terms*

51. $-4(-3k + 3) - (6k - 4) - 2k + 1$

$= -4(-3k + 3) - 1(6k - 4) - 2k + 1$

$= 12k - 12 - 6k + 4 - 2k + 1$

  Distributive property

$= (12k - 6k - 2k) + (-12 + 4 + 1)$

  Group like terms

$= 4k - 7$ *Combine like terms*

53. $-7.5(2y + 4) - 2.9(3y - 6)$

$= -7.5(2y) - 7.5(4) - 2.9(3y) - 2.9(-6)$

  Distributive property

$= -15y - 30 - 8.7y + 17.4$ *Multiply*

$= -23.7y - 12.6$ *Combine like terms*

55. "Five times a number, added to the sum of the number and three" is written $(x + 3) + 5x$.

$(x + 3) + 5x = x + 3 + 5x$

$= (x + 5x) + 3$

$= 6x + 3$

57. "A number multiplied by -7, subtracted from the sum of 13 and six times the number" is written $(13 + 6x) - (-7x)$.

$(13 + 6x) - (-7x) = 13 + 6x + 7x$

$= 13 + 13x$

59. "Six times a number added to -4, subtracted from twice the sum of three times the number and 4" is written $2(3x + 4) - (-4 + 6x)$.

$2(3x + 4) - (-4 + 6x)$

$= 2(3x + 4) - 1(-4 + 6x)$

$= 6x + 8 + 4 - 6x$

$= 6x + (-6x) + 8 + 4$

$= 0 + 12 = 12$

61. $9x - (x + 2)$

Wording will vary. One example is "the difference between 9 times a number and the sum of the number and 2." Another example is "the sum of a number and 2 subtracted from 9 times a number."

63.

x	$x + 2$
0	2
1	3
2	4
3	5

64. For every increase of 1 unit for x, the value of $x + 2$ increases by _1_ unit.

65. **(a)**

x	$x + 1$
0	1
1	2
2	3
3	4

  (b)

x	$x + 3$
0	3
1	4
2	5
3	6

  (c)

x	$x + 4$
0	4
1	5
2	6
3	7

66. For any value of b, as x increases by 1 unit, the value of an expression of the form $x + b$ also increases by 1 unit.

67. **(a)**

x	$2x + 2$
0	2
1	4
2	6
3	8

(b)

x	$3x+2$
0	2
1	5
2	8
3	11

(c)

x	$4x+2$
0	2
1	6
2	10
3	14

68. For every increase of 1 unit for x, the value of $mx + 2$ increases by __m__ units.

69. (a)

x	$2x+7$
0	7
1	9
2	11
3	13

(b)

x	$3x+5$
0	5
1	8
2	11
3	14

(c)

x	$4x+1$
0	1
1	5
2	9
3	13

In comparison, we see that while the values themselves are different, the number of units of increase is the same as in the corresponding parts of Exercise 67.

70. For every increase of 1 unit in x, the value of $mx + b$ increases by __m__ units.

Chapter 1 Review Exercises

1. $$\frac{8}{5} \div \frac{32}{15} = \frac{8}{5} \cdot \frac{15}{32}$$
$$= \frac{8 \cdot (3 \cdot 5)}{5 \cdot (8 \cdot 4)}$$
$$= \frac{8 \cdot 3 \cdot 5}{5 \cdot 8 \cdot 4}$$
$$= \frac{3}{4}$$

2. $$\frac{3}{8} + 3\frac{1}{2} - \frac{3}{16} = \frac{3}{8} + \frac{7}{2} - \frac{3}{16}$$
$$= \frac{3 \cdot 2}{8 \cdot 2} + \frac{7 \cdot 8}{2 \cdot 8} - \frac{3}{16} \quad LCD = 16$$
$$= \frac{6}{16} + \frac{56}{16} - \frac{3}{16}$$
$$= \frac{62}{16} - \frac{3}{16}$$
$$= \frac{59}{16} \text{ or } 3\frac{11}{16}$$

3. $$\frac{3}{8} + \frac{2}{5} = \frac{3 \cdot 5}{8 \cdot 5} + \frac{2 \cdot 8}{5 \cdot 8} \quad LCD = 40$$
$$= \frac{15}{40} + \frac{16}{40}$$
$$= \frac{31}{40}$$

Since the entire pie chart represents $\frac{40}{40}$, this leaves $\frac{9}{40}$ unaccounted for. Thus, $\frac{9}{40}$ of the group did not have an opinion.

4. $\frac{3}{8}$ of the 800 people responded "yes."

$$\frac{3}{8} \cdot 800 = \frac{3}{8} \cdot \frac{800}{1}$$
$$= \frac{3 \cdot (8 \cdot 100)}{8 \cdot 1}$$
$$= \frac{3 \cdot 8 \cdot 100}{8 \cdot 1}$$
$$= 300$$

300 people responded "yes."

5. $5^4 = 5 \cdot 5 \cdot 5 \cdot 5 = 625$

6. $\left(\frac{3}{5}\right)^3 = \frac{3}{5} \cdot \frac{3}{5} \cdot \frac{3}{5} = \frac{27}{125}$

7. $(.02)^5 = (.02)(.02)(.02)(.02)(.02)$
$= .0000000032$

8. $(.001)^3 = (.001)(.001)(.001)$
$= .000000001$

9. $8 \cdot 5 - 13 = 40 - 13 = 27$

10. $7[3 + 6(3^2)] = 7[3 + 6(9)]$
$= 7(3 + 54)$
$= 7(57)$
$= 399$

11. $$\frac{9(4^2 - 3)}{4 \cdot 5 - 17} = \frac{9(16 - 3)}{20 - 17}$$
$$= \frac{9(13)}{3}$$
$$= \frac{3 \cdot 3 \cdot 13}{3} = 39$$

12. $$\frac{6(5 - 4) + 2(4 - 2)}{3^2 - (4 + 3)} = \frac{6(1) + 2(2)}{9 - (4 + 3)}$$
$$= \frac{6 + 4}{9 - 7}$$
$$= \frac{10}{2} = 5$$

13. $12 \cdot 3 - 6 \cdot 6 = 36 - 36 = 0$

Since $0 = 0$ is true, so is $0 \le 0$, and therefore, the statement "$12 \cdot 3 - 6 \cdot 6 \le 0$" is true.

14. $3[5(2) - 3] = 3(10 - 3) = 3(7) = 21$

Therefore, the statement "$3[5(2) - 3] > 20$" is true.

15. $4^2 - 8 = 16 - 8 = 8$

Since $9 \le 8$ is false, the statement "$9 \le 4^2 - 8$" is false.

16. "Thirteen is less than seventeen" is written $13 < 17$.

17. "Five plus two is not equal to 10" is written $5 + 2 \ne 10$.

18. **(a)** The years in which there were *fewer than* 13.4 billion catalogs mailed are 1983–1988 and 1993–1995.

(b) The years in which there were *at least* 12.8 billion catalogs mailed are 1987–1992 and

1994–1996.

(c) The five years having the largest numbers of mailings are 1989 (13.4 billion), 1990 (13.7), 1991 (13.4), 1992 (13.5), and 1996 (13.4). The *total* number of catalogs mailed in those 5 years is $13.4 + 13.7 + 13.4 + 13.5 + 13.4$

$= 67.4$ billion.

In Exercises 19–22, replace x with 6 and y with 3.

19. $\begin{aligned} 2x + 6y &= 2(6) + 6(3) \\ &= 12 + 18 = 30 \end{aligned}$

20. $\begin{aligned} 4(3x - y) &= 4[3(6) - 3] \\ &= 4(18 - 3) \\ &= 4(15) = 60 \end{aligned}$

21. $\begin{aligned} \frac{x}{3} + 4y &= \frac{6}{3} + 4(3) \\ &= 2 + 12 = 14 \end{aligned}$

22. $\begin{aligned} \frac{x^2 + 3}{3y - x} &= \frac{6^2 + 3}{3(3) - 6} \\ &= \frac{36 + 3}{9 - 6} \\ &= \frac{39}{3} = 13 \end{aligned}$

23. "Six added to a number" translates as $x + 6$.

24. "A number subtracted from eight" translates as $8 - x$.

25. "Nine subtracted from six times a number" translates as $6x - 9$.

26. "Three-fifths of a number added to 12" translates as $12 + \frac{3}{5}x$.

27. $5x + 3(x + 2) = 22$; 2
$$\begin{aligned} 5x + 3(x + 2) &= 5(2) + 3(2 + 2) \quad Let\ x = 2 \\ &= 5(2) + 3(4) \\ &= 10 + 12 = 22 \end{aligned}$$

Since the left side and the right side are equal, 2 is a solution of the given equation.

28. $\dfrac{t + 5}{3t} = 1$; 6
$$\begin{aligned} \frac{t + 5}{3t} &= \frac{6 + 5}{3(6)} \quad Let\ t = 6 \\ &= \frac{11}{18} \end{aligned}$$

Since the left side, $\dfrac{11}{18}$, is not equal to the right side, 1, 6 is not a solution of the equation.

29. "Six less than twice a number is 10" is written

$$2x - 6 = 10.$$

Letting x equal 0, 2, 4, 6, and 10 results in a false statement, so those values are not solutions.

Since $2(8) - 6 = 16 - 6 = 10$, the solution is 8.

30. "The product of a number and 4 is 8" is written

$$4x = 8.$$

Since $4(2) = 8$, the solution is 2.

31. $-4, -\dfrac{1}{2}, 0, 2.5, 5$

Graph these numbers on a number line. They are already arranged in order from smallest to largest.

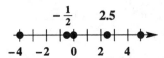

32. $-2, |-3|, -3, |-1|$

Recall that $|-3| = 3$ and $|-1| = 1$. From smallest to largest, the numbers are $-3, -2, |-1|, |-3|$.

33. Since $\dfrac{4}{3}$ is the quotient of two integers, it is a *rational number*. Since all rational numbers are also real numbers, $\dfrac{4}{3}$ is a *real number*.

34. Since the decimal representation of $\sqrt{6}$ does not terminate nor repeat, it is an *irrational number*. Since all irrational numbers are also real numbers, $\sqrt{6}$ is a *real number*.

35. $-10, 5$

Since any negative number is smaller than any positive number, -10 is the smaller number.

36. $-8, -9$

Since -9 is to the left of -8 on the number line, -9 is the smaller number.

37. $-\dfrac{2}{3}, -\dfrac{3}{4}$

To compare these fractions, us a common denominator.

$$-\frac{2}{3} = -\frac{8}{12}, \quad -\frac{3}{4} = -\frac{9}{12}$$

Since $-\dfrac{9}{12}$ is to the left of $-\dfrac{8}{12}$ on the number line, $-\dfrac{3}{4}$ is the smaller number.

38. $0, -|23|$

Since $-|23| = -23$ and $-23 < 0$, $-|23|$ is the smaller number.

39. $12 > -13$

This statement is true since 12 is to the right of -13 on the number line.

40. $0 > -5$

This statement is true since 0 is to the right of -5 on the number line.

41. $-9 < -7$

This statement is true since -9 is to the left of -7 on the number line.

42. $-13 \geq -13$

This is a true statement since $-13 = -13$.

43. **(a)** The opposite of the number -9 is its negative; that is, $-(-9) = 9$.

(b) Since $-9 < 0$, the absolute value of the number -9 is $|-9| = -(-9) = 9$.

44. 0

(a) $-0 = 0$

(b) $|0| = 0$

45. 6

(a) $-(6) = -6$

(b) $|6| = 6$

46. $-\dfrac{5}{7}$

(a) $-\left(-\dfrac{5}{7}\right) = \dfrac{5}{7}$

(b) $\left|-\dfrac{5}{7}\right| = -\left(-\dfrac{5}{7}\right) = \dfrac{5}{7}$

47. $|-12| = -(-12) = 12$

48. $-|3| = -3$

49. $-|-19| = -[-(-19)] = -19$

50. $-|9 - 2| = -|7| = -7$

51. $-10 + 4 = -6$

52. $14 + (-18) = -4$

53. $-8 + (-9) = -17$

54. $\dfrac{4}{9} + \left(-\dfrac{5}{4}\right) = \dfrac{4 \cdot 4}{9 \cdot 4} + \left(-\dfrac{5 \cdot 9}{4 \cdot 9}\right) \quad LCD = 36$

$$= \frac{16}{36} + \left(-\frac{45}{36}\right)$$

$$= -\frac{29}{36}$$

55. $-13.5 + (-8.3) = -21.8$

56. $(-10 + 7) + (-11) = (-3) + (-11)$
$$= -14$$

57. $[-6 + (-8) + 8] + [9 + (-13)]$
$$= \{[-6 + (-8)] + 8\} + (-4)$$
$$= [(-14) + 8] + (-4)$$
$$= (-6) + (-4) = -10$$

58. $(-4 + 7) + (-11 + 3) + (-15 + 1)$
$$= (3) + (-8) + (-14)$$
$$= [3 + (-8)] + (-14)$$
$$= (-5) + (-14) = -19$$

59. $-7 - 4 = -7 + (-4) = -11$

60. $-12 - (-11) = -12 + (11) = -1$

61. $5 - (-2) = 5 + (2) = 7$

62. $-\dfrac{3}{7} - \dfrac{4}{5} = -\dfrac{3 \cdot 5}{7 \cdot 5} - \dfrac{4 \cdot 7}{5 \cdot 7}$

$$= -\frac{15}{35} - \frac{28}{35} \quad LCD = 35$$

$$= -\frac{15}{35} + \left(-\frac{28}{35}\right)$$

$$= -\frac{43}{35}$$

63. $2.56 - (-7.75) = 2.56 + (7.75)$
$$= 10.31$$

64. $(-10 - 4) - (-2) = [-10 + (-4)] + 2$
$$= (-14) + (2)$$
$$= -12$$

65. $(-3 + 4) - (-1) = (-3 + 4) + 1$
$$= 1 + 1$$
$$= 2$$

66. $-(-5 + 6) - 2 = -(1) + (-2)$
$$= -1 + (-2)$$
$$= -3$$

67. "19 added to the sum of -31 and 12" is written

$$(-31 + 12) + 19 = (-19) + 19$$
$$= 0.$$

68. "13 more than the sum of -4 and -8" is written

$$[-4 + (-8)] + 13 = -12 + 13$$
$$= 1.$$

69. "The difference between -4 and -6" is written

$$-4 - (-6) = -4 + 6$$
$$= 2.$$

70. "Five less than the sum of 4 and -8" is written

$$[4 + (-8)] - 5 = (-4) + (-5)$$
$$= -9.$$

71. $x + (-2) = -4$

Because

$$(-2) + (-2) = -4,$$

the solution is -2.

72. $12 + x = 11$

Because

$$12 + (-1) = 11,$$

the solution is -1.

73. $-23.75 + 50.00 = 26.25$

Kareem now has a positive balance of $26.25.

74. $-26 + 16 = -10$

The high temperature was $-10°F$.

75. $-28 + 13 - 14 = (-28 + 13) - 14$
$$= (-28 + 13) + (-14)$$
$$= -15 + (-14)$$
$$= -29$$

His present financial status is $-$29$.

76. $-3 - 7 = -3 + (-7)$
$$= -10$$

The new temperature is $-10°$.

77. $3 - 12 + 13 = [3 + (-12)] + 13$
$$= -9 + 13$$
$$= 4$$

The team gained 4 yards.

78. $4,480,000 + 759,000 - 530,000$
$$= 5,239,000 - 530,000$$
$$= 4,709,000$$

There were 4,709,000 employees in the construction industry in 1994, an increase of 229,000 form 1985.

79. $(-12)(-3) = 36$

80. $15(-7) = -(15 \cdot 7)$
$$= -105$$

81. $\left(-\dfrac{4}{3}\right)\left(-\dfrac{3}{8}\right) = \dfrac{4}{3} \cdot \dfrac{3}{8}$

$$= \dfrac{4 \cdot 3}{3 \cdot 8}$$

$$= \dfrac{4}{8} = \dfrac{1}{2}$$

82. $(-4.8)(-2.1) = 10.08$

83. $5(8 - 12) = 5[8 + (-12)]$
$$= 5(-4) = -20$$

84. $(5 - 7)(8 - 3) = [5 + (-7)][8 + (-3)]$
$$= (-2)(5) = -10$$

85. $2(-6) - (-4)(-3) = -12 - (12)$
$$= -12 + (-12)$$
$$= -24$$

86. $3(-10) - 5 = -30 + (-5) = -35$

87. $\dfrac{-36}{-9} = \dfrac{4 \cdot 9}{9} = 4$

88. $\dfrac{220}{-11} = -\dfrac{20 \cdot 11}{11} = -20$

89. $-\dfrac{1}{2} \div \dfrac{2}{3} = -\dfrac{1}{2} \cdot \dfrac{3}{2} = -\dfrac{3}{4}$

90. $-33.9 \div (-3) = \dfrac{-33.9}{-3} = 11.3$

91. $\dfrac{-5(3) - 1}{8 - 4(-2)} = \dfrac{-15 + (-1)}{8 - (-8)}$

$$= \dfrac{-16}{8 + 8}$$

$$= \dfrac{-16}{16} = -1$$

92. $\dfrac{5(-2) - 3(4)}{-2[3 - (-2)] - 1} = \dfrac{-10 - 12}{-2(3 + 2) - 1}$

$$= \dfrac{-10 + (-12)}{-2(5) - 1}$$

$$= \dfrac{-22}{-10 + (-1)}$$

$$= \dfrac{-22}{-11} = 2$$

93. $\dfrac{10^2 - 5^2}{8^2 + 3^2 - (-2)} = \dfrac{100 - 25}{64 + 9 + 2}$

$$= \dfrac{75}{75} = 1$$

94. $\dfrac{(.6)^2 + (.8)^2}{(-1.2)^2 - (-.56)} = \dfrac{.36 + .64}{1.44 + .56}$

$$= \dfrac{1.00}{2.00} = .5$$

In Exercises 95–98, replace x with -5, y with 4, and z with -3.

95. $6x - 4z = 6(-5) - 4(-3)$
$$= -30 - (-12)$$
$$= -30 + 12 = -18$$

96. $5x + y - z = 5(-5) + (4) - (-3)$
$$= (-25 + 4) + 3$$
$$= -21 + 3 = -18$$

97. $5x^2 = 5(-5)^2$
$$= 5(25)$$
$$= 125$$

98. $z^2(3x - 8y) = (-3)^2[3(-5) - 8(4)]$
$$= 9(-15 - 32)$$
$$= 9[-15 + (-32)]$$
$$= 9(-47) = -423$$

99. "Nine less than the product of -4 and 5" is written
$$-4(5) - 9 = -20 + (-9)$$
$$= -29.$$

100. "Five-sixths of the sum of 12 and -6" is written
$$\frac{5}{6}[12 + (-6)] = \frac{5}{6}(6)$$
$$= 5.$$

101. "The quotient of 12 and the sum of 8 and -4" is written
$$\frac{12}{8 + (-4)} = \frac{12}{4} = 3.$$

102. "The product of -20 and 12, divided by the difference between 15 and -15" is written
$$\frac{-20(12)}{15 - (-15)} = \frac{-240}{15 + 15}$$
$$= \frac{-240}{30} = -8.$$

103. "8 times a number is -24" is written
$$8x = -24.$$
If $x = -3$,
$$8x = 8(-3) = -24.$$
The solution is -3.

104. "The quotient of a number and 3 is -2" is written
$$\frac{x}{3} = -2.$$
If $x = -6$,
$$\frac{x}{3} = \frac{-6}{3} = 2.$$
The solution is -6.

105. (a) To find the average of the 5 payrolls, add the 5 *payrolls* and divide by 5.

$$\begin{array}{r} \$68,988,134 \\ 63,460,567 \\ 59,583,500 \\ 59,536,000 \\ +\ 55,304,595 \\ \hline \$306,872,796 \end{array}$$

Dividing by 5 gives us \$61,374,559.20, or \$61,374,559 rounded to the nearest dollar.

(b) Work as in part (a) with the *average salaries*.

$$\begin{array}{r} \$2,555,116 \\ 2,440,791 \\ 2,127,982 \\ 2,126,286 \\ +\ 1,975,164 \\ \hline \$11,225,339 \end{array}$$

Dividing by 5 and rounding gives us \$2,245,068.

106. Find the average of the four numbers.
$$\frac{2854.6 + 2483.5 + 2808.1 + 1924.2}{4}$$
$$= \frac{10,070.4}{4} = 2517.6$$

The average of the sales was \$2517.6 million.

107. $6 + 0 = 6$

This is an example of an identity property.

108. $5 \cdot 1 = 5$

This is an example of an identity property.

109. $-\frac{2}{3}\left(-\frac{3}{2}\right) = 1$

This is an example of an inverse property.

110. $17 + (-17) = 0$

This is an example of an inverse property.

111. $5 + (-9 + 2) = [5 + (-9)] + 2$

This is an example of an associative property.

112. $w(xy) = (wx)y$

This is an example of an associative property.

113. $3x + 3y = 3(x + y)$

This is an example of the distributive property.

114. $(1 + 2) + 3 = 3 + (1 + 2)$

This is an example of a commutative property.

115. $7y + y = 7y + 1y = (7 + 1)y = 8y$

116. $-12(4 - t) = -12(4) - (-12)(t)$
$$= -48 + 12t$$

117. $3(2s) + 3(5y) = 3(2s + 5y)$

118. $-(-4r + 5s) = -1(-4r + 5s)$
$$= (-1)(-4r) + (-1)(5s)$$
$$= 4r - (1)(5s)$$
$$= 4r - 5s$$

119. $25 - (5 - 2) = 25 - 3 = 22$
$(25 - 5) - 2 = 20 - 2 = 18$

When there are three numbers involved in subtractions, you get different answers depending on which subtraction you perform first. For this reason, subtraction is not associative.

120. $180 \div (15 \div 5) = 180 \div 3 = 60$
$(180 \div 15) \div 5 = 12 \div 5 = 2.4$

When there are three numbers involved in divisions, you get different answers depending on which division you perform first. For this reason, division is not associative.

121. $2m + 9m = (2 + 9)m$ *Distributive property*
$$= 11m$$

122. $15p^2 - 7p^2 + 8p^2$
$$= (15 - 7 + 8)p^2 \quad \textit{Distributive property}$$
$$= 16p^2$$

123. $5p^2 - 4p + 6p + 11p^2$
$$= (5 + 11)p^2 + (-4 + 6)p$$
Distributive property
$$= 16p^2 + 2p$$

124. $-2(3k - 5) + 2(k + 1)$
$$= -6k + 10 + 2k + 2$$
Distributive property
$$= -4k + 12$$

125. $7(2m + 3) - 2(8m - 4)$
$$= 14m + 21 - 16m + 8$$
Distributive property
$$= (14 - 16)m + 29$$
$$= -2m + 29$$

126. $-(2k + 8) - (3k - 7)$
$$= -1(2k + 8) - 1(3k - 7)$$
Replace $-$ with -1
$$= -2k - 8 - 3k + 7$$
Distributive property
$$= -5k - 1$$

127. $[(-2) + 7 - (-5)] + [-4 - (-10)]$
$$= \{[(-2) + 7] - (-5)\} + (-4 + 10)$$
$$= (5 + 5) + 6$$
$$= 10 + 6 = 16$$

128. $\left(-\dfrac{5}{6}\right)^2 = \left(-\dfrac{5}{6}\right)\left(-\dfrac{5}{6}\right)$
$$= \dfrac{25}{36}$$

129. $\dfrac{6(-4) + 2(-12)}{5(-3) + (-3)} = \dfrac{-24 + (-24)}{-15 + (-3)}$
$$= \dfrac{-48}{-18} = \dfrac{8 \cdot 6}{3 \cdot 6}$$
$$= \dfrac{8}{3}$$

130. $\dfrac{3}{8} - \dfrac{5}{12} = \dfrac{3 \cdot 3}{8 \cdot 3} - \dfrac{5 \cdot 2}{12 \cdot 2}$
$$= \dfrac{9}{24} - \dfrac{10}{24}$$
$$= \dfrac{9}{24} + \left(-\dfrac{10}{24}\right)$$
$$= -\dfrac{1}{24}$$

131. $\dfrac{8^2 + 6^2}{7^2 + 1^2} = \dfrac{64 + 36}{49 + 1}$
$$= \dfrac{100}{50} = 2$$

132. $-16(-3.5) - 7.2(-3)$
$$= 56 - [-(7.2)(3)]$$
$$= 56 - (-21.6)$$
$$= 56 + 21.6$$
$$= 77.6$$

133. $2\dfrac{5}{6} - 4\dfrac{1}{3} = \dfrac{17}{6} - \dfrac{13}{3}$
$$= \dfrac{17}{6} - \dfrac{13 \cdot 2}{3 \cdot 2}$$
$$= \dfrac{17}{6} - \dfrac{26}{6}$$
$$= \dfrac{17}{6} + \left(-\dfrac{26}{6}\right)$$
$$= -\dfrac{9}{6} = -\dfrac{3}{2} \text{ or } -1\dfrac{1}{2}$$

134. $-8 + [(-4 + 17) - (-3 - 3)]$
$$= -8 + \{(13) - [-3 + (-3)]\}$$
$$= -8 + [13 - (-6)]$$
$$= -8 + (13 + 6)$$
$$= -8 + 19 = 11$$

135. $-\dfrac{12}{5} \div \dfrac{9}{7} = -\dfrac{12}{5} \cdot \dfrac{7}{9}$
$$= -\dfrac{12 \cdot 7}{5 \cdot 9}$$
$$= -\dfrac{3 \cdot 4 \cdot 7}{5 \cdot 3 \cdot 3}$$
$$= -\dfrac{28}{15}$$

136. $(-8 - 3) - 5(2 - 9)$
$= [-8 + (-3)] - 5[2 + (-9)]$
$= -11 - 5(-7)$
$= -11 - (-35)$
$= -11 + 35 = 24$

137. $5x^2 - 12y^2 + 3x^2 - 9y^2$
$= (5x^2 + 3x^2) + (-12y^2 - 9y^2)$
$= (5 + 3)x^2 + (-12 - 9)y^2$
$= 8x^2 - 21y^2$

138. $-4(2t + 1) - 8(-3t + 4)$
$= -4(2t) - 4(1) - 8(-3t) - 8(4)$
$= -8t - 4 + 24t - 32$
$= 16t - 36$

139. Zero divided by any nonzero number is equal to zero. That is, $0 \div x = 0$ if $x \neq 0$.

Any number divided by zero is undefined. That is, $x \div 0$ is undefined.

140. The statement is not correct because it does not consider the operation involved. The product or quotient of two negative numbers is a positive number, but the sum of two negative numbers is a negative number.

141. "The product of 5 and the sum of a number and 7" is translated as
$$5(x + 7) = 5(x) + 5(7)$$
$$= 5x + 35.$$

142. $99 - 112 = 99 + (-112)$
$= -13$

The lowest temperature ever recorded in Albany was $-13°$F.

143. The value for the fourth quarter in 1994 was 61 and the value for the fourth quarter in 1995 was -18. The change is
$(-18) - (61) = -18 + (-61) = -79$.

144. The value for the third quarter in 1995 was 5 and the value for the third quarter in 1996 was 27. The change is $(27) - (5) = 22$.

Chapter 1 Test

1. $\dfrac{63}{99} = \dfrac{7 \cdot 9}{11 \cdot 9} = \dfrac{7}{11}$

2. The denominators are 8, 12, and 15; or equivalently, 2^3, $2^2 \cdot 3$, and $3 \cdot 5$. So the LCD is $2^3 \cdot 3 \cdot 5 = 120$.

$\dfrac{5}{8} + \dfrac{11}{12} + \dfrac{7}{15}$

$= \dfrac{5 \cdot 15}{8 \cdot 15} + \dfrac{11 \cdot 10}{12 \cdot 10} + \dfrac{7 \cdot 8}{15 \cdot 8}$

$= \dfrac{75}{120} + \dfrac{110}{120} + \dfrac{56}{120}$

$= \dfrac{241}{120}$

3. $\dfrac{19}{15} \div \dfrac{6}{5} = \dfrac{19}{15} \cdot \dfrac{5}{6} = \dfrac{19 \cdot 5}{3 \cdot 5 \cdot 6} = \dfrac{19}{18}$

4. **(a)** The number of passengers that used air travel is $\dfrac{2}{5}$ of 1230 million.
$$\dfrac{2}{5} \cdot 1230 = 492$$
So 492 million passengers used air travel.

(b) Since $\dfrac{3}{10}$ of the passengers used the bus, $\dfrac{7}{10}$ did not.
$$\dfrac{7}{10} \cdot 1230 = 861$$
So 861 million passengers did not use the bus.

5. $4[-20 + 7(-2)] = 4[-20 + (-14)]$
$= 4(-34) = -136$

Since $-136 \leq 135$, the statement "$4[-20 + 7(-2)] \leq 135$" is true.

6. $-1, -3, |-4|, |-1|$

Recall that $|-4| = 4$ and $|-1| = 1$. From smallest to largest, the numbers are $-3, -1, |-1|, |-4|$.

7. The number $-\dfrac{2}{3}$ can be written as a quotient of two integers with denominator not 0, so it is a *rational number*. Since all rational numbers are real numbers, it is also a *real number*.

8. If -8 and -1 are both graphed on a number line, we see that the point for -8 is to the *left* of the point for -1. This indicates that -8 is *less than* -1.

9. "The quotient of -6 and the sum of 2 and -8" is written $\dfrac{-6}{2 + (-8)}$,

and $\dfrac{-6}{2 + (-8)} = \dfrac{-6}{-6} = 1$.

10. **(a)** The subgroups that had less than $13 billion in revenues are: Japanese (12.6), Vietnamese (4.3), Filipino (4.8), Hawaiian (1.1), and Other Asian & Pacific Islander (7.7).

(b) The subgroups that had revenues greater than or equal to $16.2 billion are: Asian Indian (19.3), Chinese (30.2), and Korean (16.2).

11. $-2 - (5 - 17) + (-6)$
$= -2 - [5 + (-17)] + (-6)$
$= -2 - (-12) + (-6)$
$= (-2 + 12) + (-6)$
$= 10 + (-6) = 4$

12. $-5\dfrac{1}{2} + 2\dfrac{2}{3} = -\dfrac{11}{2} + \dfrac{8}{3}$
$= -\dfrac{11 \cdot 3}{2 \cdot 3} + \dfrac{8 \cdot 2}{3 \cdot 2}$
$= -\dfrac{33}{6} + \dfrac{16}{6}$
$= -\dfrac{17}{6}$ or $-2\dfrac{5}{6}$

13. $-6 - [-7 + (2 - 3)]$
$= -6 - [-7 + (-1)]$
$= -6 - (-8)$
$= -6 + 8 = 2$

14. $4^2 + (-8) - (2^3 - 6)$
$= 16 + (-8) - (8 - 6)$
$= [16 + (-8)] - 2$
$= 8 - 2 = 6$

15. $(-5)(-12) + 4(-4) + (-8)^2$
$= (-5)(-12) + 4(-4) + 64$
$= [60 + (-16)] + 64$
$= 44 + 64 = 108$

16. $\dfrac{-7 - (-6 + 2)}{-5 - (-4)} = \dfrac{-7 - (-4)}{-5 + 4}$
$= \dfrac{-7 + 4}{-1}$
$= \dfrac{-3}{-1} = 3$

17. $\dfrac{30(-1 - 2)}{-9[3 - (-2)] - 12(-2)}$
$= \dfrac{30(-3)}{-9(5) - (-24)}$
$= \dfrac{-90}{-45 + 24}$
$= \dfrac{-90}{-21}$
$= \dfrac{30 \cdot 3}{7 \cdot 3} = \dfrac{30}{7}$

18. $-x + 3 = -3$

If $x = 6$,
$$-6 + 3 = -3.$$
Therefore, the solution is 6.

19. $-3x = -12$

If $x = 4$,
$$-3x = -3(4) = -12.$$
Therefore, the solution is 4.

20. $3x - 4y^2$
$= 3(-2) - 4(4^2)$ *Let $x = -2$, $y = 4$*
$= 3(-2) - 4(16)$
$= -6 - 64 = -70$

21. $\dfrac{5x + 7y}{3(x + y)}$
$= \dfrac{5(-2) + 7(4)}{3(-2 + 4)}$ *Let $x = -2$, $y = 4$*
$= \dfrac{-10 + 28}{3(2)}$
$= \dfrac{18}{6} = 3$

22. **(a)** $\$144,100 - \$147,200 = -\$3100$

(b) $\$147,700 - \$144,100 = \$3600$

(c) $\$154,500 - \$147,700 = \$6800$

(d) $\$158,700 - \$154,500 = \$4200$

23. 4 saves (3 points per save)

+ 3 wins (3 points per win)

+ 2 losses (-2 points per loss)

+ 1 blown save (-2 points per blown save)

$= 4(3) + 3(3) + 2(-2) + 1(-2)$
$= 12 + 9 - 4 - 2$
$= 15$ points

24. **(a)** $83 - 97 = 83 + (-97) = -14$

The change from 1990 to 1991 was -14 name changes.

(b) $138 - 83 = 138 + (-83) = 55$

The change from 1991 to 1996 was 55 name changes.

(c) $138 - 107 = 138 + (-107) = 31$

The change from 1992 to 1996 was 31 name changes.

25. Commutative

$(5 + 2) + 8 = 8 + (5 + 2)$

illustrates a commutative property because the order of the numbers is changed, but not the grouping. The correct response is B.

26. Associative

$-5 + (3 + 2) = (-5 + 3) + 2$

illustrates an associative property because the grouping of the numbers is changed, but not the order. The correct response is D.

27. Inverse

$$-\frac{5}{3}\left(-\frac{3}{5}\right) = 1$$

illustrates an inverse property. The correct response is E.

28. Identity

$$3x + 0 = 3x$$

illustrates an identity property. The correct response is A.

29. Distributive

$$-3(x + y) = -3x + (-3y)$$

illustrates the distributive property. The correct response is C.

30. $3(x + 1) = 3 \cdot x + 3 \cdot 1$
$$= 3x + 3$$

The distributive property is used to rewrite $3(x + 1)$ as $3x + 3$.

31. **(a)** $-6[5 + (-2)] = -6(3) = -18$

(b) $-6[5 + (-2)] = -6(5) + (-6)(-2)$
$$= -30 + 12 = -18$$

(c) The above two answers must be the same because the distributive property states that $a(b + c) = ab + ac$ is true for all real numbers a, b, and c.

32. $8x + 4x - 6x + x + 14x$
$$= (8 + 4 - 6 + 1 + 14)x$$
$$= 21x$$

33. $5(2x - 1) - (x - 12) + 2(3x - 5)$
$$= 5(2x - 1) - 1(x - 12) + 2(3x - 5)$$
$$= 10x - 5 - x + 12 + 6x - 10$$
$$= (10 - 1 + 6)x + (-5 + 12 - 10)$$
$$= 15x - 3$$

CHAPTER 2   LINEAR EQUATIONS AND APPLICATIONS

Section 2.1

1. Equations that have exactly the same solution sets are **equivalent equations**.

   (a) $x + 2 = 6$

   $x + 2 - 2 = 6 - 2$ *Subtract 2*

   $x = 4$

   So $x + 2 = 6$ and $x = 4$ *are* equivalent equations.

   (b) $10 - x = 5$

   $10 - x - 10 = 5 - 10$ *Subtract 10*

   $-x = -5$

   $-1(-x) = -1(-5)$ *Multiply by -1*

   $x = 5$

   So $10 - x = 5$ and $x = -5$ *are not* equivalent equations.

   (c) As in part (a), subtract 3 from both sides to get $x = 6$, so $x + 3 = 9$ and $x = 6$ *are* equivalent equations.

   (d) Subtract 4 from both sides to get $x = 4$. The second equation is $x = -4$, so $4 + x = 8$ and $x = -4$ *are not* equivalent equations.

3. The addition property of equality says that the same number (or expression) added to both sides of an equation results in an equivalent equation. Example: $-x$ can be added to both sides of $2x + 3 = x - 5$ to get the equivalent equation $x + 3 = -5$. The multiplication property of equality says that the same nonzero number (or expression) multiplied on both sides of the equation results in an equivalent equation.

   Example: Multiplying both sides of $7x = 4$ by $\frac{1}{7}$ gives the equivalent equation $x = \frac{4}{7}$.

For Exercises 5–20, all solutions should be checked by substituting into the original equation. Checks will be shown here for only a few of the exercises.

5. $x + 6 = 12$

   $x + 6 - 6 = 12 - 6$

   $x = 6$

   Check this solution by replacing x with 6 in the original equation.

   $x + 6 = 12$

   $6 + 6 = 12$?   *Let x = 6*

   $12 = 12$ *True*

   Because the final statement is true, $\{6\}$ is the solution set.

7. $x - 8.4 = -2.1$

   $x - 8.4 + 8.4 = -2.1 + 8.4$

   $x = 6.3$

   Check $x = 6.3$:

   $6.3 - 8.4 = -2.1$?   *Let x = 6.3*

   $-2.1 = -2.1$ *True*

   Thus, $\{6.3\}$ is the solution set.

9. $\frac{2}{5}w - 6 = \frac{7}{5}w$

   $\frac{2}{5}w - 6 - \frac{2}{5}w = \frac{7}{5}w - \frac{2}{5}w$ *Subtract $\frac{2}{5}w$*

   $-6 = \frac{5}{5}w$

   $-6 = w$

   Checking yields a true statement, so $\{-6\}$ is the solution set.

11. $5.6x + 2 = 4.6x$

   $5.6x + 2 - 4.6x = 4.6x - 4.6x$

   $1.0x + 2 = 0$

   $x + 2 - 2 = 0 - 2$

   $x = -2$

   Checking yields a true statement, so $\{-2\}$ is the solution set.

13. $3p + 6 = 10 + 2p$

   $3p + 6 - 2p = 10 + 2p - 2p$

   $p + 6 = 10$

   $p + 6 - 6 = 10 - 6$

   $p = 4$

   $\{4\}$ is the solution set.

15. $1.2y - 4 = .2y - 4$

   $1.2y - 4 - .2y = .2y - 4 - .2y$

   $1.0y - 4 = -4$

   $y - 4 + 4 = -4 + 4$

   $y = 0$

   $\{0\}$ is the solution set.

17. $\frac{1}{2}x + 2 = -\frac{1}{2}x$

   $\frac{1}{2}x + \frac{1}{2}x + 2 = -\frac{1}{2}x + \frac{1}{2}x$

   $x + 2 = 0$

   $x + 2 - 2 = 0 - 2$

   $x = -2$

   $\{-2\}$ is the solution set.

19. $3x + 7 - 2x = 0$

   $x + 7 = 0$

   $x + 7 - 7 = 0 - 7$

   $x = -7$

   $\{-7\}$ is the solution set.

21. Equations (a) $x^2 - 5x + 6 = 0$ and (b) $x^3 = x$ are not linear equations in one variable because they cannot be written in the form $Ax + B = 0$. Note that in a linear equation the exponent on the variable must be 1.

A sample answer might be, "A linear equation in one variable is an equation that can be written using only one variable term with the variable to the first power."

23.
$$5t + 3 + 2t - 6t = 4 + 12$$
$$(5 + 2 - 6)t + 3 = 16$$
$$t + 3 - 3 = 16 - 3$$
$$t = 13$$

Check $t = 13$: $16 = 16$

$\{13\}$ is the solution set.

25.
$$6x + 5 + 7x + 3 = 12x + 4$$
$$13x + 8 = 12x + 4$$
$$13x + 8 - 12x = 12x + 4 - 12x$$
$$x + 8 = 4$$
$$x + 8 - 8 = 4 - 8$$
$$x = -4$$

Check $x = -4$: $-44 = -44$

$\{-4\}$ is the solution set.

27.
$$5.2q - 4.6 - 7.1q = -.9q - 4.6$$
$$-1.9q - 4.6 = -.9q - 4.6$$
$$-1.9q - 4.6 + .9q = -.9q - 4.6 + .9q$$
$$-1.0q - 4.6 = -4.6$$
$$-1.0q - 4.6 + 4.6 = -4.6 + 4.6$$
$$-1.0q = 0$$
$$\frac{-1.0q}{-1.0} = \frac{0}{-1.0}$$
$$q = 0$$

Check $q = 0$: $-4.6 = -4.6$

$\{0\}$ is the solution set.

29.
$$\frac{5}{7}x + \frac{1}{3} = \frac{2}{5} - \frac{2}{7}x + \frac{2}{5}$$
$$\frac{5}{7}x + \frac{1}{3} = \frac{4}{5} - \frac{2}{7}x$$
$$\frac{5}{7}x + \frac{2}{7}x + \frac{1}{3} = \frac{4}{5} - \frac{2}{7}x + \frac{2}{7}x \qquad \text{Add } \frac{2}{7}x$$
$$\frac{7}{7}x + \frac{1}{3} = \frac{4}{5} \qquad \text{Combine like terms}$$
$$1x + \frac{1}{3} - \frac{1}{3} = \frac{4}{5} - \frac{1}{3} \qquad \text{Subtract } \frac{1}{3}$$
$$x = \frac{12}{15} - \frac{5}{15} \qquad \text{LCD = 15}$$
$$x = \frac{7}{15}$$

Check $x = \frac{7}{15}$: $\frac{2}{3} = \frac{2}{3}$

$\left\{\dfrac{7}{15}\right\}$ is the solution set.

31.
$$(5y + 6) - (3 + 4y) = 10$$
$$5y + 6 - 3 - 4y = 10 \qquad \textit{Distributive property}$$
$$y + 3 = 10 \qquad \textit{Combine terms}$$
$$y + 3 - 3 = 10 - 3 \qquad \textit{Subtract 3}$$
$$y = 7$$

Check $y = 7$: $10 = 10$

$\{7\}$ is the solution set.

33.
$$2(p + 5) - (9 + p) = -3$$
$$2p + 10 - 9 - p = -3$$
$$p + 1 = -3$$
$$p + 1 - 1 = -3 - 1$$
$$p = -4$$

Check $p = -4$: $-3 = -3$

$\{-4\}$ is the solution set.

35.
$$-6(2b + 1) + (13b - 7) = 0$$
$$-12b - 6 + 13b - 7 = 0$$
$$b - 13 = 0$$
$$b - 13 + 13 = 0 + 13$$
$$b = 13$$

Check $b = 13$: $0 = 0$

$\{13\}$ is the solution set.

37.
$$10(-2x + 1) = -19(x + 1)$$
$$-20x + 10 = -19x - 19$$
$$-20x + 10 + 19x = -19x - 19 + 19x$$
$$-x + 10 = -19$$
$$-x + 10 - 10 = -19 - 10$$
$$-x = -29$$
$$x = 29$$

Check $x = 29$: $-570 = -570$

$\{29\}$ is the solution set.

39.
$$-2(8p + 2) - 3(2 - 7p) = 2(4 + 2p)$$
$$-16p - 4 - 6 + 21p = 8 + 4p$$
$$5p - 10 = 8 + 4p$$
$$5p - 10 - 4p = 8 + 4p - 4p$$
$$p - 10 = 8$$
$$p - 10 + 10 = 8 + 10$$
$$p = 18$$

Check $p = 18$: $80 = 80$

$\{18\}$ is the solution set.

41. If you multiply both sides of an equation by 0, you will obtain the equation $0 = 0$. While this equation is true, the variable has been lost and we are unable to solve the equation.

43. $5x = 30$

$$\frac{5x}{5} = \frac{30}{5} \quad \textit{Divide by 5}$$

$$1x = 6$$

$$x = 6$$

Check $x = 6$: $30 = 30$

$\{6\}$ is the solution set.

45. $3a = -15$

$$\frac{3a}{3} = \frac{-15}{3} \quad \textit{Divide by 3}$$

$$a = -5$$

Check $a = -5$: $-15 = -15$

$\{-5\}$ is the solution set.

47. $10t = -36$

$$\frac{10t}{10} = \frac{-36}{10} \quad \textit{Divide by 10}$$

$$t = -\frac{36}{10} = -\frac{18}{5} \quad \textit{Lowest terms}$$

Check $t = -\frac{18}{5}$: $-36 = -36$

$\left\{-\frac{18}{5}\right\}$ is the solution set.

49. $-6x = -72$

$$\frac{-6x}{-6} = \frac{-72}{-6} \quad \textit{Divide by -6}$$

$$x = 12$$

Check $x = 12$: $-72 = -72$

$\{12\}$ is the solution set.

51. $2r = 0$

$$\frac{2r}{2} = \frac{0}{2} \quad \textit{Divide by 2}$$

$$r = 0$$

Check $r = 0$: $0 = 0$

$\{0\}$ is the solution set.

53. $\frac{1}{4}y = -12$

$$4 \cdot \frac{1}{4}y = 4(-12) \quad \textit{Multiply by 4}$$

$$1y = -48$$

$$y = -48$$

Check $y = -48$: $-12 = -12$

$\{-48\}$ is the solution set.

55. $-y = 12$

$$-1 \cdot (-y) = -1 \cdot 12 \quad \textit{Multiply by -1}$$

$$y = -12$$

Check $y = -12$: $12 = 12$

$\{-12\}$ is the solution set.

57. $-x = -\frac{4}{7}$

$$-1 \cdot (-x) = -1 \cdot \left(-\frac{4}{7}\right)$$

$$x = \frac{4}{7}$$

Check $x = \frac{4}{7}$: $-\frac{4}{7} = -\frac{4}{7}$

$\left\{\frac{4}{7}\right\}$ is the solution set.

59. $.2t = 8$

$$\frac{.2t}{.2} = \frac{8}{.2}$$

$$t = 40$$

Check $t = 40$: $8 = 8$

$\{40\}$ is the solution set.

61. $4x + 3x = 21$

$$7x = 21$$

$$\frac{7x}{7} = \frac{21}{7}$$

$$x = 3$$

Check $x = 3$: $21 = 21$

$\{3\}$ is the solution set.

63. $5m + 6m - 2m = 63$

$$9m = 63$$

$$\frac{9m}{9} = \frac{63}{9}$$

$$m = 7$$

Check $m = 7$: $63 = 63$

$\{7\}$ is the solution set.

65. $\frac{x}{7} = -5$

$$\frac{1}{7}x = -5$$

$$7\left(\frac{1}{7}x\right) = 7(-5)$$

$$x = -35$$

Check $x = -35$: $-5 = -5$

$\{-35\}$ is the solution set.

67.
$$-\frac{2}{7}p = -5$$

$$-\frac{7}{2}\left(-\frac{2}{7}p\right) = -\frac{7}{2}(-5) \qquad \begin{array}{l}\textit{Multiply by} \\ \textit{the reciprocal} \\ \textit{of } -\frac{2}{7}\end{array}$$

$$p = \frac{35}{2}$$

Check $p = \dfrac{35}{2}$: $-5 = -5$

$\left\{\dfrac{35}{2}\right\}$ is the solution set.

69.
$$-\frac{7}{9}c = \frac{3}{5}$$

$$-\frac{9}{7}\left(-\frac{7}{9}c\right) = -\frac{9}{7} \cdot \frac{3}{5} \qquad \begin{array}{l}\textit{Multiply by} \\ \textit{the reciprocal} \\ \textit{of } -\frac{7}{9}\end{array}$$

$$c = -\frac{27}{35}$$

Check $c = -\dfrac{27}{35}$: $\dfrac{3}{5} = \dfrac{3}{5}$

$\left\{-\dfrac{27}{35}\right\}$ is the solution set.

71.
$$-2.1m = 25.62$$
$$\frac{-2.1m}{-2.1} = \frac{25.62}{-2.1}$$
$$m = -12.2$$

Check $m = -12.2$: $25.62 = 25.62$

$\{-12.2\}$ is the solution set.

73. Answers will vary. For example,

$$\frac{3}{2}x = -6.$$

75. "Three times a number is 17 more than twice the number."

$$3x = 2x + 17$$
$$3x - 2x = 2x + 17 - 2x$$
$$x = 17$$

The number is 17.

77. "If five times a number is added to three times the number, the result is the sum of seven times the number and 9."

$$5x + 3x = 7x + 9$$
$$8x = 7x + 9$$
$$8x - 7x = 7x + 9 - 7x$$
$$x = 9$$

The number is 9.

79. "When a number is divided by -5, the result is 2."

$$\frac{x}{-5} = 2$$

$$(-5)\left(-\frac{1}{5}x\right) = (-5)(2)$$

$$x = -10$$

The number is -10.

Section 2.2

1. In step 1, we use the distributive property to remove any parentheses on either side of the equation. Then we combine any like terms.

In step 2, we get the variable term on one side of the equation and a number on the other. To do this, add or subtract terms from both sides.

In step 3, we finish solving the equation so that we find the value of the variable that makes the equation true. To do this, we can multiply or divide both sides by any number except 0.

In step 4, we check our solution. Substitute the number you got in step 3 into the *original* equation. Examples will vary.

3.
$$2m + 8 = 16$$
$$2m = 8 \qquad \textit{Subtract 8}$$
$$m = 4 \qquad \textit{Divide by 2}$$

Check $m = 4$: $16 = 16$

$\{4\}$ is the solution set.

5.
$$-4x - 1 = -2x + 1$$
$$-4x = -2x + 2 \qquad \textit{Add 1}$$
$$-2x = 2 \qquad \textit{Add 2x}$$
$$x = -1 \qquad \textit{Divide by } -2$$

Check $x = -1$: $3 = 3$

$\{-1\}$ is the solution set.

7.
$$4x + 6 = -(x - 2)$$

$$4x + 6 = -x + 2 \qquad \begin{array}{l}\textit{Distributive} \\ \textit{property}\end{array}$$

$$4x = -x - 4 \qquad \textit{Subtract 6}$$
$$5x = -4 \qquad \textit{Add 1x}$$
$$x = -\frac{4}{5} \qquad \textit{Divide by 5}$$

Check $x = -\dfrac{4}{5}$: $\dfrac{14}{5} = \dfrac{14}{5}$

$\left\{-\dfrac{4}{5}\right\}$ is the solution set.

9. $5(2m + 3) = 3(4m + 9)$

$10m + 15 = 12m + 27$ *Distributive property*

$10m = 12m + 12$ *Subtract 15*

$-2m = 12$ *Subtract 12m*

$m = -6$ *Divide by -2*

Check $m = -6$: $-45 = -45$

$\{-6\}$ is the solution set.

11. $6(4x - 1) = 12(2x + 3)$

$24x - 6 = 24x + 36$

$-6 = 36$ *Subtract 24x*

The variable has "disappeared," and the resulting equation is false. Therefore, the equation has no solution set, symbolized by $\emptyset$.

13. $3(2x - 4) = 6(x - 2)$

$6x - 12 = 6x - 12$

$-12 = -12$ *Subtract 6x*

$0 = 0$ *Add 12*

The variable has "disappeared." Since the resulting statement is a *true* one, *any* real number is a solution. We indicate the solution set as {all real numbers}.

15. $7r - 5r + 2 = 5r - r$

$2r + 2 = 4r$

$2 = 2r$ *Subtract 2r*

$1 = r$ *Divide by 2*

Check $r = 1$: $4 = 4$

$\{1\}$ is the solution set.

17. $11x - 5(x + 3) - 6x = 0$

$11x - 5x - 15 - 6x = 0$

$6x - 15 - 6x = 0$

$-15 = 0$

The variable has "disappeared," and the resulting equation is false. Therefore, the equation has no solution set, symbolized by $\emptyset$.

19. The student's solution is not correct. We should never divide both sides by a variable because the value of the variable might be 0, and division by 0 is undefined. The correct solution follows.

$$7x = 3x$$
$$7x - 3x = 3x - 3x$$
$$4x = 0$$
$$\frac{4x}{4} = \frac{0}{4}$$
$$x = 0$$

The solution set is $\{0\}$.

21. If the equation has decimals as coefficients, multiply both sides of the equation by the power of 10 that makes all decimals into integers. If the equation has common fractions as coefficients, multiply both sides of the equation by the LCD of all fractions in the equation.

23. $\dfrac{3}{5}t - \dfrac{1}{10}t = t - \dfrac{5}{2}$

The least common denominator of all the fractions in the equation is 10.

$$10\left(\frac{3}{5}t - \frac{1}{10}t\right) = 10\left(t - \frac{5}{2}\right)$$

Multiply both sides by 10

$$10\left(\frac{3}{5}t\right) + 10\left(-\frac{1}{10}t\right) = 10t + 10\left(-\frac{5}{2}\right)$$

Distributive property

$6t - t = 10t - 25$

$5t = 10t - 25$

$-5t = -25$ *Subtract 10t*

$\dfrac{-5t}{-5} = \dfrac{-25}{-5}$ *Divide by -5*

$t = 5$

Check $t = 5$: $\dfrac{5}{2} = \dfrac{5}{2}$

$\{5\}$ is the solution set.

25. $-\dfrac{1}{4}(x - 12) + \dfrac{1}{2}(x + 2) = x + 4$

The LCD of all the fractions is 4.

$$4\left[-\frac{1}{4}(x - 12) + \frac{1}{2}(x + 2)\right] = 4(x + 4)$$

Multiply by 4

$$4\left(-\frac{1}{4}\right)(x - 12) + 4\left(\frac{1}{2}\right)(x + 2) = 4x + 16$$

Distributive property

$(-1)(x - 12) + 2(x + 2) = 4x + 16$

Multiply

$-x + 12 + 2x + 4 = 4x + 16$

Distributive property

$x + 16 = 4x + 16$

$-3x + 16 = 16$

$-3x = 0$

$\dfrac{-3x}{-3} = \dfrac{0}{-3}$

Divide by -3

$x = 0$

Check $x = 0$: $4 = 4$

$\{0\}$ is the solution set.

27. $\dfrac{2}{3}k - \left(k + \dfrac{1}{4}\right) = \dfrac{1}{12}(k + 4)$

The least common denominator of all the fractions in the equation is 12, so multiply both sides by 12 and solve for k.

$$12\left[\dfrac{2}{3}k - \left(k + \dfrac{1}{4}\right)\right] = 12\left[\dfrac{1}{12}(k + 4)\right]$$

$$12\left(\dfrac{2}{3}k\right) - 12\left(k + \dfrac{1}{4}\right) = 12\left[\dfrac{1}{12}(k + 4)\right]$$

Distributive property

$$8k - 12k - 12\left(\dfrac{1}{4}\right) = 1(k + 4)$$

$$8k - 12k - 3 = k + 4$$

$$-4k - 3 = k + 4$$

$$-5k - 3 = 4$$

$$-5k = 7$$

$$k = -\dfrac{7}{5}$$

Check $k = -\dfrac{7}{5}: \ \dfrac{13}{60} = \dfrac{13}{60}$

$\left\{-\dfrac{7}{5}\right\}$ is the solution set.

29. $.20(60) + .05x = .10(60 + x)$

To eliminate the decimal in .20 and .10, we need to multiply the equation by 10. But to eliminate the decimal in .05, we need to multiply by 100, so we choose 100.

$$100[.20(60) + .05x] = 100[.10(60 + x)]$$

Multiply by 100

$$100[.20(60)] + 100(.05x) = 100[.10(60 + x)]$$

Distributive property

$$20(60) + 5x = 10(60 + x)$$

Multiply

$$1200 + 5x = 600 + 10x$$

$$1200 - 5x = 600$$

$$-5x = -600$$

$$x = \dfrac{-600}{-5} = 120$$

Check $x = 120: \ 18 = 18$

$\{120\}$ is the solution set.

31. $1.00x + .05(12 - x) = .10(63)$

To clear the equation of decimals, we multiply both sides by 100.

$$100[1.00x + .05(12 - x)] = 100[.10(63)]$$

$$100(1.00x) + 100[.05(12 - x)] = (100)(.10)(.63)$$

$$100x + 5(12 - x) = 10(63)$$

$$100x + 60 - 5x = 630$$

$$95x + 60 = 630$$

$$95x = 570$$

$$x = \dfrac{570}{95} = 6$$

Check $x = 6: \ 6.3 = 6.3$

$\{6\}$ is the solution set.

33. $.06(10,000) + .08x = .072(10,000 + x)$

$$1000[.06(10,000)] + 1000(.08x) =$$
$$1000[.072(10,000 + x)]$$

Multiply by 1000, not 100

$$60(10,000) + 80x = 72(10,000 + x)$$

$$600,000 + 80x = 720,000 + 72x$$

$$600,000 + 8x = 720,000$$

$$8x = 120,000$$

$$x = \dfrac{120,000}{8} = 15,000$$

Check $x = 15,000: \ 1800 = 1800$

$\{15,000\}$ is the solution set.

35. If $a = 2$ and $b = 4$,

$$100ab = 100(2 \cdot 4)$$
$$= 100 \cdot 8$$
$$= 800.$$

36. If $a = 2$ and $b = 4$,

$$(100a)b = (100 \cdot 2) \cdot 4$$
$$= 200 \cdot 4$$
$$= 800.$$

The answers in Exercises 35 and 36 are the same because of the associative property of multiplication.

37. $(100a)(100b) = 100 \cdot 100 \cdot a \cdot b$
$$= 10,000ab$$
$$\neq 100ab$$

No, the term $(100a)(100b)$ is not equivalent to $100ab$. Note that $(100a)(100b)$ contains two factors of 100, while $100ab$ contains only one.

38. The distributive property does not apply because it involves the operation of *addition* as well as multiplication. In Exercise 37, we have only multiplication.

39. Yes, the expressions

$$100 \cdot .05(x+2)$$

and

$$\big[100 \cdot .05\big](x+2)$$

are equivalent by the associative property of multiplication.

40. No, it is not correct to "distribute" the 100 to both .05 and $(x+2)$. This would result in having two factors of 100, that is, in multiplying by 10,000 rather than 100, as we observed in Exercise 37.

41.
$$10(2x-1) = 8(2x+1) + 14$$
$$20x - 10 = 16x + 8 + 14$$
$$20x - 10 = 16x + 22$$
$$4x - 10 = 22$$
$$4x = 32$$
$$x = 8$$

Check $x = 8$: $\ 150 = 150$

The solution set is $\{8\}$.

43.
$$-(4y+2) - (-3y-5) = 3$$
$$-1(4y+2) - 1(-3y-5) = 3$$
$$-4y - 2 + 3y + 5 = 3$$
$$-y + 3 = 3$$
$$-y = 0$$
$$y = 0$$

Check $y = 0$: $\ 3 = 3$

The solution set is $\{0\}$.

45.
$$\frac{1}{2}(x+2) + \frac{3}{4}(x+4) = x + 5$$

To clear fractions, multiply both sides by the LCD, which is 4.

$$4\left[\frac{1}{2}(x+2) + \frac{3}{4}(x+4)\right] = 4(x+5)$$
$$4\left(\frac{1}{2}\right)(x+2) + 4\left(\frac{3}{4}\right)(x+4) = 4x + 20$$
$$2(x+2) + 3(x+4) = 4x + 20$$
$$2x + 4 + 3x + 12 = 4x + 20$$
$$5x + 16 = 4x + 20$$
$$x + 16 = 20$$
$$x = 4$$

Check $x = 4$: $\ 9 = 9$

The solution set is $\{4\}$.

47.
$$.10(x+80) + .20x = 14$$

To eliminate the decimals, multiply both sides by 10.
$$10[.10(x+80) + .20x] = 10(14)$$
$$1(x+80) + 2x = 140$$
$$x + 80 + 2x = 140$$
$$3x + 80 = 140$$
$$3x = 60$$
$$x = 20$$

Check $x = 20$: $\ 14 = 14$

The solution set is $\{20\}$.

49.
$$4(x+8) = 2(2x+6) + 20$$
$$4x + 32 = 4x + 12 + 20$$
$$4x + 32 = 4x + 32$$
$$4x = 4x$$
$$0 = 0$$

Since $0 = 0$ is a true statement, the solution set is {all real numbers}.

51.
$$9(v+1) - 3v = 2(3v+1) - 8$$
$$9v + 9 - 3v = 6v + 2 - 8$$
$$6v + 9 = 6v - 6$$
$$9 = -6 \ \textit{False}$$

Because this is a false statement, the equation has no solution set, symbolized by $\emptyset$.

53. The sum of q and the other number is 11. To find the other number, you would subtract q from 11, so the other number is $11 - q$.

55. The total number of yards is $x + 7$.

57. If Mary is a years old now, in 12 years she will be $a + 12$ years old. Five years ago she was $a - 5$ old.

59. Since each bill is worth 5 dollars, the number of bills is $\dfrac{t}{5}$.

Section 2.3

1. Choice (c), $6\frac{2}{3}$, is not a reasonable answer in an applied problem that requires finding the number of coins in a jar, since you cannot have $\frac{2}{3}$ of a coin. The number of coins must be a whole number.

3. It is important to read the problem carefully before you write anything down. You have to pick a variable to stand for the unknown number and then express any other unknown quantities in terms of the same variable. Be sure to write these things down. Sometimes a figure or diagram will help you to write an equation for the problem. Once you've written the equation, solve it by the methods covered earlier in this chapter. Use your solution to answer the question that the problem asked. Finally, check your solution in the words of the original problem and make sure your answer makes sense.

5. Step 1   Let $x =$ the unknown number.

Step 2

$x + 1 =$ one is added to a number

$2(x + 1) =$ this sum is doubled

$x + 5 = 5$ more than the number

Step 3 $2(x + 1) = x + 5$

Step 4 $2x + 2 = x + 5$

$x + 2 = 5$

$x = 3$

Step 5   The number is 3.

Step 6   Check that 3 is the correct answer by substituting this result into the words of the original problem. One added to the number is 4, double this value is 8. Five more than 3 is also 8, so the number is 3.

7. Step 1   Let $x =$ the unknown number.

Step 2

$2x + 3 = 3$ is added to twice the number

$4(2x + 3) =$ the sum multiplied by 4

$7x + 8 =$ the number is multiplied by 7 and 8 is added to the product

Step 3 $4(2x + 3) = 7x + 8$

Step 4 $8x + 12 = 7x + 8$

$x + 12 = 8$

$x = -4$

Step 5   The number is -4.

Step 6   Check that -4 is the correct answer by substituting this result into the words of the original problem.

Twice the number is $2(-4) = -8$.

Three added to twice the number is

$-8 + 3 = -5$.

This sum multiplied by 4 is

$4(-5) = -20$.

The number multiplied by 7 is

$7(-4) = -28$.

Eight added to this product is

$-28 + 8 = -20$.

Because both results are -20, the answer, -4, checks.

9. Let $x =$ the number of Democrats;

$x + 7 =$ the number of Republicans.

The number of Democrats	plus	the number of Republicans
↓	↓	↓
x	$+$	$(x + 7)$

equals	the number of members of the Senate.
↓	↓
$=$	99 (one seat vacant)

Solve the equation.

$x + (x + 7) = 99$

$2x + 7 = 99$

$2x = 92$

$x = 46$

There were 46 Democrats and

$46 + 7 = 53$ Republicans.

11. Let $x =$ the number of women.

Then $x + 2783 =$ the number of men.

Since the total number of competitors was $10,341$, we can write the equation

$x + (x + 2783) = 10,341$.

Solve this equation.

$2x + 2783 = 10,341$

$2x = 7558$

$x = 3779$

$x + 2783 = 6562$

There were 3779 women and 6562 men. Since $3779 + 6562 = 10,341$, this answer checks.

13. Let $x =$ Fernandez's score.

Then $x + 1 =$ Irwin's score.

Since the total of the two scores was 571, we can write the equation

$x + (x + 1) = 571$.

Solve this equation.

$2x + 1 = 571$

$2x = 570$

$x = 285$

$x + 1 = 286$

continued

Fernandez's score was 285 and Irwin's score was 286. Since $285 + 286 = 571$, this answer checks.

15. Let $x =$ the number of packages delivered by Airborne Express.

Then $3x =$ the number of packages delivered by Federal Express,

and $x - 2 =$ the number of packages delivered by United Parcel Service.

$$x + 3x + (x - 2) = 13$$
$$5x - 2 = 13$$
$$5x = 15$$
$$x = 3$$

One package was delivered by United Parcel Service, 3 were delivered by Airborne Express, and 9 were delivered by Federal Express.

17. Let $x =$ the distance of Mercury from the sun (in millions of miles);

$x + 31.2 =$ the distance of Venus from the sun;

$x + 57 =$ the distance of Earth from the sun.

Mercury's distance $\downarrow$ x + Venus's distance $\downarrow$ $x + 31.2$

Earth's distance $\downarrow$ $x + 57$ = Total distance $\downarrow$ 196.2

$$x + (x + 31.2) + (x + 57) = 196.2$$
Solve this equation.
$$3x + 88.2 = 196.2$$
$$3x = 108$$
$$x = 36$$

Mercury is 36 million miles from the sun.

19. Let $x =$ the number of innings pitched by Maddux.

Then $x + 8\frac{2}{3} =$ the number of innings pitched by Smoltz;

$x - 167\frac{2}{3} =$ the number of innings pitched by Wohlers.

The total number of innings pitched was 576, so
$$x + \left(x + 8\frac{2}{3}\right) + \left(x - 167\frac{2}{3}\right) = 576.$$
Solve this equation.
$$3x - 159 = 576$$

$$3x = 735$$
$$x = 245$$
$$x + 8\frac{2}{3} = 253\frac{2}{3}$$
$$x - 167\frac{2}{3} = 77\frac{1}{3}$$

Smoltz pitched $253\frac{2}{3}$ innings, Maddux pitched 245 innings, and Wohlers pitched $77\frac{1}{3}$ innings.

21. Let $x =$ the measure of angle A;

$x =$ the measure of angle B (since both A and B have the same measure);

$x + 60 =$ measure of angle C.

The sum of the measures of the angles of any triangle is 180°, so
$$x + x + (x + 60) = 180$$
Solve this equation.
$$3x + 60 = 180$$
$$3x = 120$$
$$x = 40$$

Angles A and B have measures of 40 degrees, and angle C has a measure of $40 + 60 = 100$ degrees. Since
$$40 + 40 + 100 = 180,$$
the answer checks.

23. Let $x =$ the number of calories burned in regulating body temperature;

$\left(5\frac{3}{8}\right)x =$ the number of calories burned in exertion.

Since the total number of calories burned is $11,200$, we can write the equation
$$x + \left(5\frac{3}{8}\right)x = 11,200.$$
Solve this equation.
$$(1)x + \left(5\frac{3}{8}\right)x = 11,200$$
$$\left(6\frac{3}{8}\right)x = 11,200$$
$$\frac{51}{8}x = 11,200$$
$$x = \frac{11,200}{\frac{51}{8}}$$
$$\approx 1756.86$$
$$\left(5\frac{3}{8}\right)x \approx 9443.14$$

Rounding our answers, we see that 1757 calories are burned in regulating body temperature and 9443 calories are burned in exertion (per day).

25. Let $x = $ the number of prescriptions for tranquilizers.

Then $\frac{4}{3}x = $ the number of prescriptions for antibiotics.

The total number of prescriptions is 42, so

$$x + \frac{4}{3}x = 42.$$

Solve this equation.

$$3\left(x + \frac{4}{3}x\right) = 3(42)$$
$$3x + 4x = 126$$
$$7x = 126$$
$$x = 18$$

There were 18 prescriptions for tranquilizers. The number of prescriptions for antibiotics was $\frac{4}{3}(18) = 24$. The total number of prescriptions was $18 + 24 = 42$.

27. Let $x = $ the number of ounces of cashews;

$5x = $ the number of ounces of peanuts.

There is a total of 27 ounces, so

$$x + 5x = 27.$$

Solve this equation.

$$6x = 27$$
$$x = \frac{27}{6} = \frac{9}{2} = 4\frac{1}{2}$$

There are $4\frac{1}{2}$ ounces of cashews and

$5\left(\frac{9}{2}\right) = \frac{45}{2} = 22\frac{1}{2}$ ounces of peanuts. The total

number of ounces was $4\frac{1}{2} + 22\frac{1}{2} = 27$.

29. Subtract one of the numbers (m) from the sum (k) to express the other number. The other number is $k - m$.

31. An angle cannot have its supplement equal to its complement. The sum of an angle and its supplement equals 180°, while the sum of an angle and its complement equals 90°. If we try to solve the equation

$$90 - x = 180 - x,$$

we will get

$$90 - x + x = 180 - x + x$$
$$90 = 180 \quad \textit{False}$$

so this equation has no solution.

33. The next smaller consecutive integer is less than a number. Thus, if x represents an integer, the next smaller consecutive integer is $x - 1$.

35. Step 1   Let $x = $ the measure of the angle.

Step 2
$90 - x = $ the measure of its complement;
$180 - x = $ the measure of its supplement.

Step 3   Write an equation.

The measure of the supplement	is	10 times	the measure of the complement.
↓	↓	↓	↓
$180 - x$	$=$	10 $\cdot$	$90 - x$

Step 4   Solve this equation.

$$180 - x = 10(90 - x)$$
$$180 - x = 900 - 10x$$
$$-x = 720 - 10x$$
$$9x = 720$$
$$x = 80$$

Step 5   The measure of the angle is 80°.

Step 6   The measure of the supplement $(180 - 80 = 100)$ is 10 times the complement $(90 - 80 = 10)$.

37. Let $x = $ the measure of the angle;
$180 - x = $ the measure of the angle's supplement;
$90 - x = $ the measure of the angle's complement.

Angle's supplement	three times	angle's complement.	less 38°
↓	↓	↓	↓
$180 - x =$	3 $\cdot$	$(90 - x)$	-38

$$180 - x = 270 - 3x - 38$$
$$180 - x = 232 - 3x$$
$$180 + 2x = 232$$
$$2x = 52$$
$$x = 26$$

The angle measures 26°.

39. Step 1   Let $x = $ the measure of the angle.

Step 2
$90 - x = $ the measure of its complement;
$180 - x = $ the measure of its supplement.

Step 3   Write an equation.

measure of complement		measure of supplement		sum
$\downarrow$		$\downarrow$		$\downarrow$
$90 - x$	$+$	$180 - x$	$=$	160

Step 4   Solve the equation.

$$(90 - x) + (180 - x) = 160$$
$$-2x + 270 = 160$$
$$-2x = -110$$
$$x = 55$$

Step 5   The measure of the angle is $55°$.

Step 6   The sum of the measures of its complement ($90° - 55° = 35°$) and its supplement ($180° - 55° = 125°$) is $160°$ ($35° + 125° = 160°$).

41.   Let $x =$ the smaller integer;
$x + 1 =$ the larger integer.

$$x + (x + 1) = 137$$
$$2x + 1 = 137$$
$$2x = 136$$
$$x = 68$$

If $x = 68$, $x + 1 = 69$.

The integers are 68 and 69.

43.   Let $x =$ the smaller even integer;
$x + 2 =$ the larger even integer.

$$x + 3(x + 2) = 46$$
$$x + 3x + 6 = 46$$
$$4x + 6 = 46$$
$$4x = 40$$
$$x = 10$$

If $x = 10$, $x + 2 = 12$.

The integers are 10 and 12.

45.   Because the two pages are back-to-back, they must have page numbers that are consecutive integers.

Let $x =$ the smaller page number;
$x + 1 =$ the larger page number.

$$x + (x + 1) = 203$$
$$2x + 1 = 203$$
$$2x = 202$$
$$x = 101$$

If $x = 101$, $x + 1 = 102$.

The page numbers are 101 and 102.

47.   Let $x =$ the smaller integer;
$x + 1 =$ the larger integer.

$$x + 3(x + 1) = 43$$
$$x + 3x + 3 = 43$$
$$4x + 3 = 43$$
$$4x = 40$$
$$x = 10$$

If $x = 10$, $x + 1 = 11$.

The integers are 10 and 11.

49.   Let $x =$ the smallest odd integer;
$x + 2 =$ the middle odd integer;
$x + 4 =$ the largest odd integer.

$$2[(x + 4) - 6] = [x + 2(x + 2)] - 23$$
$$2(x - 2) = x + 2x + 4 - 23$$
$$2x - 4 = 3x - 19$$
$$-4 = x - 19$$
$$15 = x$$

If $x = 15$, $x + 2 = 17$, and $x + 4 = 19$.

The integers are 15, 17, and 19.

51.   Let $x =$ the amount of Head Start funding in 1993 (in billions of dollars).

Then $x + .55 =$ the amount of funding in 1994,

and $(x + .55) + .20 = x + .75$
$=$ the amount of funding in 1995.

The total funding was 9.64 billion dollars, so

$$x + (x + .55) + (x + .75) = 9.64.$$

Solve this equation.

$$3x + 1.30 = 9.64$$
$$3x = 8.34$$
$$x = 2.78$$

If $x = 2.78$, $x + .55 = 3.33$, and $x + .75 = 3.53$. The Head Start funding was 2.78 billion dollars in 1993, 3.33 billion dollars in 1994, and 3.53 billion dollars in 1995.

Section 2.4

1.   The perimeter of a geometric figure is the distance around the figure. It can be found by adding up the lengths of all the sides. Perimeter is a one-dimensional (linear) measurement, so it is given in linear units (inches, centimeters, feet, etc.).

3.   Carpeting for a bedroom covers the surface of the bedroom floor, so area would be used.

5. To measure fencing for a yard, use perimeter since you would need to measure the lengths of the sides of the yard.

7. Tile for a bathroom covers the surface of the bathroom floor, so area would be used.

9. To determine the cost for replacing a linoleum floor with a wood floor, use area since you need to know the measure of the surface covered by the wood.

In Exercises 11–26, substitute the given values into the formula and then solve for the remaining variable.

11. $P = 2L + 2W$; $L = 6$; $W = 4$

$$P = 2L + 2W$$
$$= 2(6) + 2(4)$$
$$= 12 + 8$$
$$P = 20$$

13. $P = 4s$; $s = 6$

$$P = 4s$$
$$= 4(6)$$
$$P = 24$$

15. $A = \frac{1}{2}bh$; $b = 10$; $h = 14$

$$A = \frac{1}{2}bh$$
$$A = \frac{1}{2}(10)(14)$$
$$= (5)(14)$$
$$A = 70$$

17. $d = rt$; $d = 100$, $t = 2.5$

$$d = rt$$
$$100 = r(2.5)$$
$$100 = 2.5r$$
$$\frac{100}{2.5} = \frac{2.5r}{2.5}$$
$$40 = r$$

19. $I = prt$; $p = 5000$, $r = .025$, $t = 7$

$$I = prt$$
$$= (5000)(.025)(7)$$
$$I = 875$$

21. $A = \frac{1}{2}h(b + B)$; $h = 7$; $b = 12$; $B = 14$

$$A = \frac{1}{2}h(b + B)$$
$$= \frac{1}{2}(7)(12 + 14)$$
$$= \frac{1}{2}(7)(26)$$
$$= \frac{1}{2}(182)$$
$$A = 91$$

23. $C = 2\pi r$; $C = 8.164$, $\pi = 3.14$

$$C = 2\pi r$$
$$8.164 = 2(3.14)r$$
$$8.164 = 6.28r$$
$$1.3 = r$$

25. $A = \pi r^2$; $r = 12$, $\pi = 3.14$

$$A = \pi r^2$$
$$A = 3.14(12)^2$$
$$= 3.14(144)$$
$$A = 452.16$$

27. You would need to be given 4 values in a formula with 5 variables to find the value of any one variable.

In Exercises 29–34, substitute the given values into the formula and then solve for V.

29. $V = LWH$; $L = 12$, $W = 8$, $H = 4$

$$V = LWH$$
$$= (12)(8)(4)$$
$$V = 384$$

31. $V = \frac{1}{3}Bh$; $B = 36$, $h = 4$

$$V = \frac{1}{3}Bh$$
$$= \frac{1}{3}(36)(4)$$
$$= (12)(4)$$
$$V = 48$$

33. $V = \frac{4}{3}\pi r^3$; $r = 6$, $\pi = 3.14$

$$V = \frac{4}{3}\pi r^3$$
$$= \frac{4}{3}(3.14)(6)^3$$
$$= \frac{4}{3}(3.14)(216)$$
$$V = 904.32$$

35. The diameter of the circle is 443 feet, so its radius is $\frac{443}{2} = 221.5$ ft. Use the area of a circle formula to find the enclosed area.

$$A = \pi r^2$$
$$= \pi(221.5)^2$$
$$\approx 154,133.6 \text{ ft}^2,$$

or about $154,000$ ft^2. (If 3.14 is used for π, the value is $154,055.465$.)

37. The diameter of the circular dome is 630 feet, so its radius is $\dfrac{630}{2} = 315$ ft. Use the circumference of a circle formula.

$$C = 2\pi r$$
$$= 2\pi(315)$$
$$\approx 1979.2 \text{ ft}$$

If we use 3.14 for π, the answer is 1978.2 ft.

39. A page of the newspaper is a rectangle with length 51 inches and width 35 inches, so use the formulas for the perimeter and area of a rectangle.

$$P = 2L + 2W$$
$$= 2(51) + 2(35)$$
$$= 102 + 70$$
$$P = 172$$

The perimeter was 172 inches.

$$A = LW$$
$$= (51)(35)$$
$$A = 1785$$

The area was 1785 square inches.

41. Substitute the values given in the problem into the formula $V = LWH$ to find the volume of the rectangular set.

$$V = LWH$$
$$= \left(6\frac{3}{4}\right)(3)\left(1\frac{1}{8}\right) \quad \begin{array}{l} \text{Let } L = 6\frac{3}{4}, \\ W = 3, H = 1\frac{1}{8} \end{array}$$
$$= \left(\frac{27}{4}\right)(3)\left(\frac{9}{8}\right)$$
$$V = \frac{729}{32} = 22\frac{25}{32}$$

The volume of the television set is $\dfrac{729}{32}$ or $22\dfrac{25}{32}$ cubic inches.

43. Use the formula for the area of a trapezoid with $B = 115.80$, $b = 171.00$, and $h = 165.97$.

$$A = \frac{1}{2}(B + b)h$$
$$= \frac{1}{2}(115.80 + 171.00)(165.97)$$
$$= \frac{1}{2}(286.80)(165.97)$$
$$= 23,800.098$$

To the nearest hundredth of a square foot, the combined area of the two lots is $23,800.10$ square feet.

45. In the figure, the two angles are supplementary; thus their sum is 180°.

$$(10x + 7) + (7x + 3) = 180$$
$$17x + 10 = 180$$
$$17x = 170$$
$$x = 10$$

To find the measures of the angles, replace x with 10 in the two expressions.

$$10x + 7 = 10(10) + 7$$
$$= 100 + 7 = 107$$
$$7x + 3 = 7(10) + 3$$
$$= 70 + 3 = 73$$

The two angle measures are 107° and 73°.

47. The two angles are vertical angles, which have equal measures. Set their measures equal to each other and solve for x.

$$7x + 5 = 3x + 45$$
$$4x + 5 = 45$$
$$4x = 40$$
$$x = 10$$

The measure of the first angle is $7(10) + 5 = 75°$; the measure of the second angle is $3(10) + 45$, which is also 75°.

49. The angles are vertical angles, which have equal measures. Set $11x - 37$ equal to $7x + 27$ and solve.

$$11x - 37 = 7x + 27$$
$$4x - 37 = 27$$
$$4x = 64$$
$$x = 16$$

Replacing x with 16 gives us

$$11(16) - 37 = 139 \text{ and}$$
$$7(16) + 27 = 139.$$

The angles both measure 139°.

51. $d = rt$ for r

$$\frac{d}{t} = \frac{rt}{t} \quad \text{Divide by } t$$
$$\frac{d}{t} = r \text{ or } r = \frac{d}{t}$$

53. $I = prt$ for p

$$I = prt$$
$$\frac{I}{rt} = \frac{prt}{rt} \quad \text{Divide by } rt$$
$$\frac{I}{rt} = p \text{ or } p = \frac{I}{rt}$$

55. $P = a + b + c$ for a

$$P - b - c = a + b + c - b - c$$

$$\text{Subtract } b \text{ and } c$$

$$P - b - c = a \text{ or } a = P - b - c$$

57. $A = \dfrac{1}{2}bh$ for b

$$2A = 2\left(\frac{1}{2}bh\right) \quad \textit{Multiply by 2}$$

$$2A = bh$$

$$\frac{2A}{h} = \frac{bh}{h} \qquad \textit{Divide by } h$$

$$\frac{2A}{h} = b \text{ or } b = \frac{2A}{h}$$

59. $A = p + prt$ for r

$$A - p = p + prt - p \quad \textit{Subtract } p$$

$$A - p = prt$$

$$\frac{A - p}{pt} = \frac{prt}{pt} \qquad \textit{Divide by } pt$$

$$\frac{A - p}{pt} = r \text{ or } r = \frac{A - p}{pt}$$

61. $V = \pi r^2 h$ for h

$$\frac{V}{\pi r^2} = \frac{\pi r^2 h}{\pi r^2} \quad \textit{Divide by } \pi r^2$$

$$\frac{V}{\pi r^2} = h \text{ or } h = \frac{V}{\pi r^2}$$

63. $y = mx + b$ for m

$$y - b = mx + b - b \quad \textit{Subtract } b$$

$$y - b = mx$$

$$\frac{y - b}{x} = \frac{mx}{x} \qquad \textit{Divide by } x$$

$$\frac{y - b}{x} = m$$

65. $P = 2L + 2W$

(a) $P - 2L = 2L + 2W - 2L$

$\quad P - 2L = 2W$

(b) $\dfrac{P - 2L}{2} = \dfrac{2W}{2}$

$\quad \dfrac{P - 2L}{2} = W$

66. $P = 2L + 2W$

(a) $\dfrac{P}{2} = \dfrac{2L}{2} + \dfrac{2W}{2}$

$\quad \dfrac{P}{2} = L + W$

(b) $\dfrac{P}{2} - L = W$

67. (a) Multiplicative identity property

(b) An expression divided by 1 is equal to itself.

(c) Rule for multiplication of fractions

(d) Rule for subtraction of fractions

68.
$$\frac{5T + 4}{4} = \frac{5T}{4} + \frac{4}{4}$$
$$= \frac{5T}{4} + 1$$

is another valid answer.

Section 2.5

1. The ratio of 25 feet to 40 feet is

$$\frac{25 \text{ feet}}{40 \text{ feet}} = \frac{25}{40} = \frac{5}{8}.$$

3. The ratio of 18 dollars to 72 dollars is

$$\frac{18 \text{ dollars}}{72 \text{ dollars}} = \frac{18}{72} = \frac{1}{4}.$$

5. First convert 6 feet to inches.

$$6 \text{ feet} = 6 \cdot 12 = 72 \text{ inches}$$

The ratio of 144 inches to 6 feet is then

$$\frac{144}{72} = \frac{2 \cdot 72}{72} = \frac{2}{1}.$$

7. 5 days $= 5 \cdot 24 = 120$ hours

The ratio of 5 days to 40 hours is

$$\frac{120 \text{ hours}}{40 \text{ hours}} = \frac{3 \cdot 40}{40} = \frac{3}{1}.$$

9. (a) $.4 = \dfrac{4}{10} = \dfrac{2 \cdot 2}{2 \cdot 5} = \dfrac{2}{5}$

(b) 4 to 10 $= \dfrac{4}{10} = \dfrac{2 \cdot 2}{2 \cdot 5} = \dfrac{2}{5}$

(c) 20 to 50 $= \dfrac{20}{50} = \dfrac{2 \cdot 10}{5 \cdot 10} = \dfrac{2}{5}$

(d) 5 to 2 $= \dfrac{5}{2}$

Choice (d) is not the same as the ratio 2 to 5.

11. A ratio is a comparison, while a proportion is a statement that two ratios are equal. For example, $\dfrac{2}{3}$ is a ratio and $\dfrac{2}{3} = \dfrac{8}{12}$ is a proportion.

13. $\dfrac{5}{35} = \dfrac{8}{56}$

Check to see whether the cross products are equal.

$$5 \cdot 56 = 280$$
$$35 \cdot 8 = 280$$

The cross products are equal, so the proportion is true.

15. $\dfrac{120}{82} = \dfrac{7}{10}$

Compare the cross products.

$$120 \cdot 10 = 1200$$
$$82 \cdot 7 = 574$$

The cross products are different, so the proportion is false.

17. $\dfrac{\frac{1}{2}}{5} = \dfrac{1}{10}$

Compare the cross products.

$$\frac{1}{2} \cdot 10 = 5$$
$$5 \cdot 1 = 5$$

The cross products are equal, so the proportion is true.

19. $\dfrac{k}{4} = \dfrac{175}{20}$

$20k = 4(175)$ *Cross products are equal*

$20k = 700$

$\dfrac{20k}{20} = \dfrac{700}{20}$ *Divide by 20*

$k = 35$

The solution set is $\{35\}$.

21. $\dfrac{x}{6} = \dfrac{18}{4}$

$x \cdot 4 = 6 \cdot 18$ *Cross products are equal*

$4x = 108$

$\dfrac{4x}{4} = \dfrac{108}{4}$ *Divide by 4*

$x = 27$

The solution set is $\{27\}$.

23. $\dfrac{3y - 2}{5} = \dfrac{6y - 5}{11}$

$11(3y - 2) = 5(6y - 5)$ *Cross products are equal*

$33y - 22 = 30y - 25$ *Distributive property*

$3y - 22 = -25$ *Subtract 30y*

$3y = -3$ *Add 22*

$y = -1$ *Divide by 3*

The solution set is $\{-1\}$.

25. Let $x =$ the number of fluid ounces of oil required to fill the tank.

Set up a proportion with one ratio involving the number of ounces of oil and the other involving the number of gallons of gasoline.

$$\frac{2.5 \text{ ounces}}{x \text{ ounces}} = \frac{1 \text{ gallon}}{2.75 \text{ gallons}}$$

$$\frac{2.5}{x} = \frac{1}{2.75}$$

$$x \cdot 1 = 2.5(2.75)$$

$$x = 6.875$$

To fill the tank of the chain saw, 6.875 fluid ounces of oil are required.

27. Let $x =$ the number of U.S. dollars Margaret exchanged.

Set up a proportion.

$$\frac{\$1.6762}{x \text{ dollars}} = \frac{1 \text{ pound}}{400 \text{ pounds}}$$

$$\frac{1.6762}{x} = \frac{1}{400}$$

$$x \cdot 1 = 1.6762(400)$$

$$x = 670.48$$

Margaret exchanged $670.48.

29. Let $x =$ the cost for filling a 15–gallon tank.

Set up a proportion.

$$\frac{x \text{ dollars}}{\$3.72} = \frac{15 \text{ gallons}}{6 \text{ gallons}}$$

$$\frac{x}{3.72} = \frac{15}{6}$$

$$6x = 15(3.72)$$

$$6x = 55.80$$

$$x = 9.30$$

It would cost $9.30 to completely fill a 15–gallon tank.

31. Let $x =$ the distance between Memphis and Philadelphia on the map (in feet).

Set up a proportion with one ratio involving map distances and the other involving actual distances.

$$\frac{x \text{ feet}}{2.4 \text{ feet}} = \frac{1000 \text{ miles}}{600 \text{ miles}}$$

$$\frac{x}{2.4} = \frac{1000}{600}$$

$$600x = (2.4)(1000)$$

$$600x = 2400$$

$$x = 4$$

The distance on the map between Memphis and Philadelphia would be 4 feet.

33. Let $x = $ the number of fish in Willow Lake.

Set up a proportion with one ratio involving the sample and the other involving the total number of fish.

$$\frac{7 \text{ fish}}{350 \text{ fish}} = \frac{250 \text{ fish}}{x \text{ fish}}$$

$$7x = 350 \cdot 250$$
$$7x = 87,500$$
$$x = 12,500$$

We estimate that there are $12,500$ fish in Willow Lake.

35. **(a)** Set up a proportion with one ratio involving percentages and the other involving revenues.

$$\frac{26\%}{100\%} = \frac{\$x}{\$350 \text{ million}}$$
$$100x = 26(350) = 9100$$
$$x = 91$$

The amount of revenue provided by tickets was $91 million.

(b) For sponsors:

$$\frac{32\%}{100\%} = \frac{\$x}{\$350 \text{ million}}$$
$$100x = 32(350) = 11,200$$
$$x = \$112 \text{ million}$$

If the 10 sponsors contributed equally, then each sponsor would have provided $11.2 million.

(c) For TV rights:

$$\frac{34\%}{100\%} = \frac{\$x}{\$350 \text{ million}}$$
$$100x = 34(350) = 11,900$$
$$x = \$119 \text{ million}$$

In Exercises 37–42, to find the best buy, divide the price by the number of units to get the unit cost. Each result was found by using a calculator and rounding the answer to three decimal places.

37.

Size	Unit Cost (dollars per bag)
20–count	$\frac{\$3.09}{20} = \$.155$
30–count	$\frac{\$4.59}{30} = \$.153$

Since $\$.153 < \$.155$, the 30–count size is the best buy.

39.

Size	Unit Cost (dollars per ounce)
15–ounce	$\frac{\$2.99}{15} = \$.199$
25–ounce	$\frac{\$4.49}{25} = \$.180$
31–ounce	$\frac{\$5.49}{31} = \$.177$

The 31–ounce size is the best buy.

41.

Size	Unit Cost (dollars per ounce)
14–ounce	$\frac{\$.89}{14} = \$.064$
32–ounce	$\frac{\$1.19}{32} = \$.037$
64–ounce	$\frac{\$2.95}{64} = \$.046$

The 32–ounce size is the best buy.

43. $\dfrac{x}{12} = \dfrac{3}{9}$

$$9x = 12 \cdot 3 = 36$$
$$x = 4$$

Other possibilities for the proportion are:

$$\frac{12}{x} = \frac{9}{3}, \quad \frac{x}{12} = \frac{5}{15}, \quad \frac{12}{x} = \frac{15}{5}$$

45. $\dfrac{x}{3} = \dfrac{2}{6}$

$$6x = 3 \cdot 2 = 6$$
$$x = 1$$

47. **(a)**

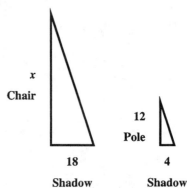

(b) These two triangles are similar, so their sides are proportional.

$$\frac{x}{12} = \frac{18}{4}$$
$$4x = 18(12)$$
$$4x = 216$$
$$x = 54$$

The chair is 54 feet tall.

49. Let $x =$ the 1992 price of electricity.

$$\frac{1990\text{ price}}{1990\text{ index}} = \frac{1992\text{ price}}{1992\text{ index}}$$

$$\frac{225}{130.7} = \frac{x}{140.3}$$

$$130.7x = 225(140.3)$$

$$x = \frac{225(140.3)}{130.7} \approx 241.53$$

The 1992 price would be about $242.

51. Let $x =$ the 1994 price of electricity.

$$\frac{1990\text{ price}}{1990\text{ index}} = \frac{1994\text{ price}}{1994\text{ index}}$$

$$\frac{225}{130.7} = \frac{x}{148.2}$$

$$130.7x = 225(148.2)$$

$$x = \frac{225(148.2)}{130.7} \approx 255.13$$

The 1994 price would be about $255.

53. Let $x =$ the 1991 cost of shelter.

$$\frac{1981\text{ cost}}{1981\text{ index}} = \frac{1991\text{ cost}}{1991\text{ index}}$$

$$\frac{3000}{90.5} = \frac{x}{146.3}$$

$$90.5x = 3000(146.3)$$

$$x = \frac{3000(146.3)}{90.5} \approx 4849.72$$

The shelter cost in 1991 would be about $4850.

55. $\dfrac{x}{6} = \dfrac{2}{5}$

The least common denominator of the two fractions is $6 \cdot 5 = 30$.

56. **(a)** $\dfrac{x}{6} = \dfrac{2}{5}$

$$30\left(\frac{x}{6}\right) = 30\left(\frac{2}{5}\right)$$

$$5x = 12$$

(b) $\dfrac{5x}{5} = \dfrac{12}{5}$

$$x = \frac{12}{5}$$

The solution is $\dfrac{12}{5}$.

57. **(a)** $\dfrac{x}{6} = \dfrac{2}{5}$

$$5x = 2 \cdot 6$$

$$5x = 12$$

(b) $\dfrac{5x}{5} = \dfrac{12}{5}$

$$x = \frac{12}{5}$$

58. The results are the same. Solving by cross products yields the same solution as multiplying by the LCD.

59. Let I denote the interest and r the rate. The interest varies directly as the rate and can be written as

$$I = kr$$

$$48 = k(5) \quad \textit{Substitute}$$

$$\frac{48}{5} = k \quad \textit{Solve for } k$$

So the formula is

$$I = \frac{48}{5}r$$

$$= (9.6)(4.2)$$

$$= 40.32$$

The interest is $40.32.

61. Let D denote the distance the spring stretches and F the force applied.

$$D = kF$$

$$16 = k(30) \quad \textit{Substitute}$$

$$\frac{16}{30} = k \quad \textit{Solve for } k$$

So the formula is

$$D = \frac{16}{30}F$$

$$= \frac{16}{30}(50)$$

$$= \frac{80}{3} \text{ or } 26\frac{2}{3}$$

The spring will be stretched $26\frac{2}{3}$ inches.

63. Let D denote the distance traveled and t the time spent. Distance varies directly as time, so

$$D = kt$$

$$2226 = k(103) \quad \textit{Substitute}$$

$$\frac{2226}{103} = k \quad \textit{Solve for } k$$

So the formula is

$$D = \frac{2226}{103}t$$

$$= \frac{2226}{103}(120)$$

$$\approx 2593.4$$

They would have gone about 2593 miles in 120 days.

Section 2.6

1. The amount of pure acid in 250 milliliters of a 14% acid solution is

    $$250 \quad \times \quad .14 \quad = 35 \text{ milliliters.}$$
    $$\uparrow \qquad\qquad \uparrow \qquad\qquad \uparrow$$

    | Amount of solution | Rate of concentration | Amount of pure acid |

3. If $10,000 is invested for one year at 3.5% simple interest, the amount of interest earned is

    $$\$10,000 \quad \times \quad .035 \quad = \$350.$$
    $$\uparrow \qquad\qquad \uparrow \qquad\qquad \uparrow$$

    Principal | Interest Rate | Interest earned

5. The monetary value of 283 nickels is

    $$283 \quad \times \quad \$.05 \quad = \$14.15.$$
    $$\uparrow \qquad\qquad \uparrow \qquad\qquad \uparrow$$

    Number of coins | Denomination | Monetary value

7. Solving $d = rt$ for t gives us $t = \dfrac{d}{r}$.

    $$\frac{\text{Distance}}{\text{Rate}} = \frac{500 \text{ miles}}{149.956 \text{ mph}} =$$

    3.334 hours, rounded to the nearest thousandth.

9. Solving $d = rt$ for r gives us $r = \dfrac{d}{t}$.

    $$\frac{\text{Distance}}{\text{Time}} = \frac{100 \text{ meters}}{12.58 \text{ seconds}} =$$

    7.95 meters per second, rounded to the nearest hundredth.

11. The problem may be stated as follows: "What is 16.4% of $3250?"

    $$16.4\% \text{ of } 3250 = (.164)(3250)$$
    $$= 533$$

    Thus, 16.4% of $3250 is $533.

13. The increase in value is

    $$\$2400 - \$625 = \$1775$$

    Let x represent the percent increase in decimal form.

    $$(x)(625) = 1775$$
    $$\frac{625x}{625} = \frac{1775}{625}$$
    $$x = 2.84$$

    $2.84 = 284\%$, so the percent increase is 284%.

15. **(a)** 18 to 24

    $$5\% \text{ of } \$15,691,000,000$$
    $$= .05(15,691,000,000)$$
    $$= 784,550,000$$

    The amount spent by the 18 to 24 age group was about $785 million.

    (b) 25 to 34

    $$11\% \text{ of } \$15,691,000,000$$
    $$= .11(15,691,000,000)$$
    $$= 1,726,010,000$$

    The amount spent by the 25 to 34 age group was about $1726 million.

    (c) 45 to 64

    $$3\% \text{ of } \$15,691,000,000$$
    $$= .03(15,691,000,000)$$
    $$= 470,730,000$$

    The amount spent by the 45 to 64 age group was about $471 million.

17. There would be no alcohol in pure water $(r \cdot 0 = 0)$, so the amount is 0 liters.

19. Step 1

    Let $x =$ the number of gallons of 50% solution needed;

    $x + 80 =$ the number of gallons of 40% solution.

    Step 2 and 3

    Use the box diagram in the textbook to write the equation.

    | Pure antifreeze in 50% solution | plus | pure antifreeze in 20% solution |
    | $\downarrow$ | | $\downarrow$ $\downarrow$ |
    | $.50x$ | $+$ | $.20(80)$ |

    is | pure antifreeze in 40% solution.

    $\downarrow$ $\downarrow$
    $=$ $.40(x + 80)$

    Step 4   Solve the equation.

    $$.50x + .20(80) = .40(x + 80)$$

    Multiply by 10 to clear decimals.
    $$5x + 2(80) = 4(x + 80)$$
    $$5x + 160 = 4x + 320$$
    $$x + 160 = 320$$
    $$x = 160$$

    Step 5   160 gallons of 50% antifreeze are needed.

Step 6   50% of 160 gallons plus 20% of 80 gallons is 80 gallons plus 16 gallons, or 96 gallons, of pure antifreeze; which is equal to 40% of (160 + 80) gallons. [.40(240) = 96]

In Exercises 21–26, the steps are similar and are not identified.

21.   Let $x =$ the number of kilograms of 20% tin;

$x + 80 =$ the number of kilograms of 50% tin.

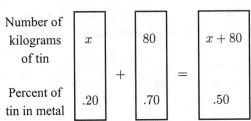

Tin in 20% metal   plus   tin in 70% metal

↓   ↓   ↓

.20x   +   .70(80)

is   tin in 50% metal.

↓   ↓

=   .50(x + 80)

Solve the equation.

$.20x + .70(80) = .50x + 40$

Multiply by 10 to clear decimals.

$2x + 7(80) = 5x + 400$
$2x + 560 = 5x + 400$
$560 = 3x + 400$
$160 = 3x$
$x = \dfrac{160}{3} = 53\dfrac{1}{3}$

$53\dfrac{1}{3}$ kilograms of 20% tin are needed.

Check $x = 53\dfrac{1}{3}$:

LS and RS refer to the left side and right side of the original equation.

LS: $.20\left(53\dfrac{1}{3}\right) + .70(80) = 66\dfrac{2}{3}$

RS: $.50\left(53\dfrac{1}{3} + 80\right) = 66\dfrac{2}{3}$

23.   Let $x =$ the number of liters of 60% acid solution;

$20 - x =$ the number of liters of 75% acid solution.

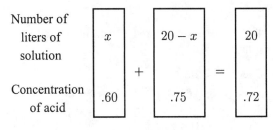

Pure acid in 60% solution   plus   pure acid in 75% solution

↓   ↓   ↓

.60x   +   .75(20 − x)

is   pure acid in 72% solution.

↓   ↓

=   .72(20)

Solve the equation.

$.60x + .75(20 - x) = .72(20)$
$60x + 75(20 - x) = 72(20)$
$60x + 1500 - 75x = 1440$
$1500 - 15x = 1440$
$-15x = -60$
$x = 4$

4 liters of 60% acid solution must be mixed with 16 liters of 75% solution.

Check $x = 4$:

LS: $.60(4) + .75(20 - 4) = 14.4$
RS: $.72(20) = 14.4$

25.   Let $x =$ the number of milliliters of 4% solution.

Milliliters of solution	50		x		$50 + x$
Concentration of minoxidil	.01	+	.04	=	.02

Pure minoxidil in 1%   pure minoxidil in 4%   pure minoxidil in 2%

↓   ↓   ↓

$.01(50) +$ $.04x$ $= .02(50 + x)$

continued

$$1(50) + 4x = 2(50 + x)$$
$$50 + 4x = 100 + 2x$$
$$50 + 2x = 100$$
$$2x = 50$$
$$x = 25$$

The pharmacist must add 25 milliliters of 4% solution.

Check $x = 25$:

LS: $.01(50) + .04(25) = 1.5$

RS: $.02(50 + 25) = 1.5$

27. The concentration of the new solution could not be more than the strength of the stronger of the original solutions, so the correct answer is (d), since 32% is stronger than both 20% and 30%.

29. Let $x =$ the amount invested at 3% (in dollars);

$x - 4000 =$ the amount invested at 5% (in dollars).

Amount invested (in dollars)	Rate of interest	Interest for one year
x	.03	.03x
$x - 4000$	.05	$.05(x - 4000)$

Since the total annual interest was $200, the equation is

$$.03x + .05(x - 4000) = 200.$$
$$3x + 5(x - 4000) = 100(200)$$
$$3x + 5x - 20,000 = 20,000$$
$$8x = 40,000$$
$$x = 5000$$

If $x = 5000$, $x - 4000 = 1000$.

Li invested $5000 at 3% and $1000 at 5%.

31.   Let $x =$ the amount invested at 3%;

$2x + 30,000 =$ the amount invested at 4%.

$$.03x + .04(2x + 30,000) = 5600$$
$$3x + 4(2x + 30,000) = 100(5600)$$
$$3x + 8x + 120,000 = 560,000$$
$$11x + 120,000 = 560,000$$
$$11x = 440,000$$
$$x = 40,000$$

If $x = 40,000$, $2x + 30,000 = 110,000$.

The actor invested $40,000 at 3% and $110,000 at 4%.

33. Let $x =$ the number of five-dollar bills;

$x + 5 =$ the number of twenty-dollar bills.

The value of fives plus the value of twenties is $725

$$\downarrow \qquad \downarrow \qquad \qquad \downarrow \qquad \downarrow \qquad \downarrow$$
$$5x \quad + \quad 20(x + 5) = \quad 725$$

$$5x + 20x + 100 = 725$$
$$25x + 100 = 725$$
$$25x = 625$$
$$x = 25$$

The teller has 25 five-dollar bills.

35.   Let $x =$ the number of fives;

$126 - x =$ the number of tens.

The value of fives plus the value of tens is total value.

$$\downarrow \qquad \downarrow \qquad \downarrow \qquad \qquad \downarrow \quad \downarrow$$
$$5 \cdot x \quad + \quad 10(126 - x) = \quad 840$$

$$5x + 1260 - 10x = 840$$
$$-5x + 1260 = 840$$
$$-5x = -420$$
$$x = 84$$
and $126 - x = 42$

The cashier has 84 fives and 42 tens.

37. Let $x =$ the number of pounds of candy worth $5/lb;

$x + 40 =$ the number of pounds of candy worth $3/lb.

The value of $5 candy and the value of $2 candy is the value of $3 candy.

$$\downarrow \qquad \downarrow \qquad \downarrow \qquad \downarrow \qquad \downarrow$$
$$5(x) \quad + \quad 2(40) \quad = \quad 3(x + 40)$$

$$5x + 80 = 3x + 120$$
$$2x + 80 = 120$$
$$2x = 40$$
$$x = 20$$

Twenty pounds of the candy worth $5 per pound should be used in the mixture.

39. The number of liters of a solution does not have to be a whole number, so a problem like the one in Example 2 can have a fraction (or decimal) as an answer. However, a number of bills (or coins, stamps, etc.) must be a whole number, so a problem like the one in Example 4 cannot have a fraction as an answer.

40. (a) Let x = the number of nickels;

$3400 - x$ = the number of dimes.

$$.05x + .10(3400 - x) = 290$$

(b) $5x + 10(3400 - x) = 100(290)$

$$5x + 34,000 - 10x = 29,000$$

$$-5x + 34,000 = 29,000$$

$$-5x = -5000$$

$$x = 1000$$

and $\quad 3400 - x = 2400$

There are 1000 nickels and 2400 dimes.

41. (a) Let x = the amount invested at 5%;

$3400 - x$ = the amount invested at 10%.

$$.05x + .10(3400 - x) = 290$$

(b) $5x + 10(3400 - x) = 100(290)$

$$5x + 34,000 - 10x = 29,000$$

$$-5x + 34,000 = 29,000$$

$$-5x = -5000$$

$$x = 1000$$

and $\quad 3400 - x = 2400$

She invested $1000 at 5% and $2400 at 10%.

42. The equations in Exercises 40(a) and 41(a) are the same.

43. In either of the problems in Exercises 40 and 41, if you let x represent the other unknown quantity, you will get a different solution to the equation, but you *will* get the same answers to the problem.

45. Use the formula $d = rt$ with $r = 53$ and $t = 10$.

$$d = rt$$

$$= (53)(10)$$

$$= 530$$

The distance between Memphis and Chicago is 530 miles.

47. The distance traveled cannot be found by multiplying 45 and 30 because the rate is given in miles per hour, while the time is given in minutes. To find the correct distance, start by converting the time to hours.

$$30 \text{ minutes} = \frac{1}{2} \text{ hour}$$

$$d = rt$$

$$= 45\left(\frac{1}{2}\right)$$

$$= 22.5$$

The car traveled 22.5 $\left(\text{or } 22\frac{1}{2}\right)$ miles.

49. Let t = the time each plane travels.

Use the chart in the text to help write the equation.

Distance of plane leaving Portland	plus	distance of plane leaving St. Louis	is	distance between Portland and St. Louis.
↓	↓	↓ ↓		↓
$90t$	+	$116t$	=	2060

$$206t = 2060$$

$$t = \frac{2060}{206} = 10$$

It will take the planes 10 hours to meet.

51. Let x = the rate of the northbound train;

$x + 20$ = the rate of the southbound train.

Using the formula $d = rt$ and the chart in the text, we see that

$$d_{\text{north}} + d_{\text{south}} = d_{\text{total}}$$

$$x(2) + (x + 20)(2) = 280$$

$$2x + 2x + 40 = 280$$

$$4x = 240$$

$$x = 60$$

and $\quad x + 20 = 80$

The speed of the northbound train is 60 mph and the speed of the southbound train is 80 mph.

53. Let t = the number of hours until the steamboats will be 35 miles apart.

Make a chart, using the formula $d = rt$.

	r	t	d
Slower boat	18	t	$18t$
Faster boat	25	t	$25t$

Distance traveled by faster boat	minus	Distance traveled by slower boat	is 35.
↓	↓	↓	↓ ↓
$25t$	−	$18t$	= 35

$$25t - 18t = 35$$

$$7t = 35$$

$$t = 5$$

In 5 hours, the steamboats will be 35 miles apart.

55. Let x = the average amount spent by a single woman;

$x + 10.87$ = the average amount spent by a single man.

$$x + (x + 10.87) = 92.29$$
$$2x + 10.87 = 92.29$$
$$2x = 81.42$$
$$x = 40.71$$

If $x = 40.71$, $x + 10.87 = 51.58$.

The average expenditures were \$40.71 for a single woman and \$51.58 for a single man.

57. Let $T = 21,782,000$.

Business/Government:

28% of $T = .28T$
$$= 6,098,960,$$

or about 6.1 million.

K – 12:

23% of $T = .23T$
$$= 5,009,860,$$

or about 5 million.

HED:

4% of $T = .04T$
$$= 871,280,$$

or about .9 million.

Homes:

45% of $T = .45T$
$$= 9,801,900,$$

or about 9.8 million.

59. Let r = the speed of each driver.

	r	t	d
First driver	r	3	$3r$
Second driver	r	2.5	$2.5r$

Since the first car was 240 miles from the finish line after 2.5 hours, while the other car was 300 miles from the finish line after 3 hours, the difference between the distances traveled by the two cars in these amounts of time is $300 - 240 = 60$ miles.

$$3r - 2.5r = 60$$
$$.5r = 60$$
$$r = \frac{60}{.5} = 120$$

The speed of each driver was 120 miles per hour.

Chapter 2 Review Exercises

1. $m - 5 = 1$

$\quad m = 6 \quad$ *Add 5*

The solution set is $\{6\}$.

2. $y + 8 = -4$

$\quad y = -12 \quad$ *Subtract 8*

The solution set is $\{-12\}$.

3. $3k + 1 = 2k + 8$

$\quad k + 1 = 8 \qquad$ *Subtract 2k*

$\quad k = 7 \qquad$ *Subtract 1*

The solution set is $\{7\}$.

4. $5k = 4k + \dfrac{2}{3}$

$\quad k = \dfrac{2}{3} \qquad$ *Subtract 4k*

The solution set is $\left\{\dfrac{2}{3}\right\}$.

5. $(4r - 2) - (3r + 1) = 8$

$(4r - 2) - 1(3r + 1) = 8 \quad$ *Replace – with –1*

$\quad 4r - 2 - 3r - 1 = 8 \quad$ *Distributive property*

$\quad r - 3 = 8$

$\quad r = 11 \quad$ *Add 3*

The solution set is $\{11\}$.

6. $3(2y - 5) = 2 + 5y$

$\quad 6y - 15 = 2 + 5y \quad$ *Distributive property*

$\quad y - 15 = 2 \qquad$ *Subtract 5y*

$\quad y = 17 \qquad$ *Add 15*

The solution set is $\{17\}$.

7. $7k = 35$

$\quad k = 5 \quad$ *Divide by 7*

The solution set is $\{5\}$.

8. $12r = -48$

$\quad r = -4 \quad$ *Divide by 12*

The solution set is $\{-4\}$.

9. $2p - 7p + 8p = 15$

$\quad 3p = 15$

$\quad p = 5 \quad$ *Divide by 3*

The solution set is $\{5\}$.

10. $\dfrac{m}{12} = -1$

$\quad m = -12 \quad$ *Multiply by 12*

The solution set is $\{-12\}$.

11.
$$\frac{5}{8}k = 8$$

$$\frac{8}{5}\left(\frac{5}{8}k\right) = \frac{8}{5}(8) \quad \textit{Multiply by } \frac{8}{5}$$

$$k = \frac{64}{5}$$

The solution set is $\left\{\frac{64}{5}\right\}$.

12. $12m + 11 = 59$

$\quad 12m = 48 \quad \textit{Subtract 11}$

$\quad\quad m = 4 \quad \textit{Divide by 12}$

The solution set is $\{4\}$.

13. $3(2x + 6) - 5(x + 8) = x - 22$

$\quad 6x + 18 - 5x - 40 = x - 22$

$\quad\quad\quad\quad x - 22 = x - 22$

This is a true statement, so the solution set is $\{$all real numbers$\}$.

14. $5x + 9 - (2x - 3) = 2x - 7$

$\quad 5x + 9 - 2x + 3 = 2x - 7$

$\quad\quad\quad 3x + 12 = 2x - 7$

$\quad\quad\quad\quad x + 12 = -7$

$\quad\quad\quad\quad\quad\quad x = -19$

The solution set is $\{-19\}$.

15.
$$\frac{1}{2}r - \frac{r}{3} = \frac{r}{6}$$

$$6\left(\frac{1}{2}r\right) - 6\left(\frac{r}{3}\right) = 6\left(\frac{r}{6}\right) \quad \textit{Multiply by 6}$$

$$3r - 2r = r$$

$$r = r$$

This is a true statement, so the solution set is $\{$all real numbers$\}$.

16. $.10(x + 80) + .20x = 14$

$$10[.10(x + 80) + .20x] = 10(14) \quad \begin{array}{l}\textit{Multiply}\\\textit{by 10}\end{array}$$

$$(x + 80) + 2x = 140 \quad \begin{array}{l}\textit{Distributive}\\\textit{property}\end{array}$$

$$3x + 80 = 140$$

$$3x = 60$$

$$x = 20$$

The solution set is $\{20\}$.

17. $3x - (-2x + 6) = 4(x - 4) + x$

$\quad 3x + 2x - 6 = 4x - 16 + x$

$\quad\quad\quad 5x - 6 = 5x - 16$

$\quad\quad\quad\quad\quad -6 = -16$

This statement is false, so there is no solution set, symbolized by $\emptyset$.

18. $2(y - 3) - 4(y + 12) = -2(y + 27)$

$\quad 2y - 6 - 4y - 48 = -2y - 54$

$\quad\quad\quad\quad -2y - 54 = -2y - 54$

This is a true statement, so the solution set is $\{$all real numbers$\}$.

19. Step 1:   Let $x =$ the number of nations participating in 1986.

Step 2: $x - 6 =$ number in 1990

$\quad\quad\quad\quad x + 28 =$ number in 1994

Step 3: $x + (x - 6) + (x + 28) = 358$

Step 4: $3x + 22 = 358$

$\quad\quad\quad\quad\quad 3x = 336$

$\quad\quad\quad\quad\quad\quad x = 112$

Step 5:   There were 112 nations participating in the 1986 World Cup.

Step 6:   112 nations in 1986, 106 in 1990, and 140 in 1994 add up to 358 in all three years.

20. Step 1:   Let $x =$ the number of Republicans.

Step 2: $x + 30 =$ the number of Democrats

Step 3: $x + (x + 30) = 120$

Step 4: $2x + 30 = 120$

$\quad\quad\quad\quad\quad 2x = 90$

$\quad\quad\quad\quad\quad\quad x = 45$

Step 5:   If $x = 45$, $x + 30 = 75$. There were 75 Democrats and 45 Republicans.

Step 6:   There are 30 more Democrats than Republicans and the total is 120.

21. Step 1:   Let $x =$ the land area of Rhode Island.

Step 2: $x + 5213 =$ land area of Hawaii

Step 3:   The areas total 7637 square miles, so $x + (x + 5213) = 7637$.

Step 4: $2x + 5213 = 7637$

$\quad\quad\quad\quad\quad 2x = 2424$

$\quad\quad\quad\quad\quad\quad x = 1212$

Step 5:   If $x = 1212$, $x + 5213 = 6425$. The land area of Rhode Island is 1212 square miles and that of Hawaii is 6425 square miles.

Step 6:   The land area of Hawaii is 5213 square miles greater than the land area of Rhode Island and the total is 7637 square miles.

22. Step 1: Let x = the height of Twin Falls.

Step 2: $\dfrac{5}{2}x$ = the height of Seven Falls

Step 3: The sum of the heights is 420 feet, so
$$x + \dfrac{5}{2}x = 420.$$

Step 4:
$$2\left(x + \dfrac{5}{2}x\right) = 2(420)$$
$$2x + 5x = 840$$
$$7x = 840$$
$$x = 120$$

Step 5: If $x = 120$, $\dfrac{5}{2}x = \dfrac{5}{2}(120) = 300$.

The height of Twin Falls is 120 feet and that of Seven Falls is 300 feet.

Step 6: The height of Seven Falls is $\dfrac{5}{2}$ the height of Twin Falls and the sum is 420.

23. Step 1: Let x = the measure of the angle.

Step 2: $90 - x =$ the measure of its complement

$180 - x =$ the measure of its supplement.

Step 3: $180 - x = 10(90 - x)$

Step 4:
$$180 - x = 900 - 10x$$
$$9x + 180 = 900$$
$$9x = 720$$
$$x = 80$$

Step 5: The measure of the angle is 80°. Its complement measures $90° - 80° = 10°$, and its supplement measures $180° - 80° = 100°$.

Step 6: The measure of the supplement is 10 times the measure of the complement.

In Exercises 24–27, substitute the given values into the given formula and then solve for the remaining variable.

24. $A = \dfrac{1}{2}bh$; $A = 44$, $b = 8$

$$A = \dfrac{1}{2}bh$$
$$44 = \dfrac{1}{2}(8)h$$
$$44 = 4h$$
$$11 = h$$

25. $A = \dfrac{1}{2}h(b + B)$; $b = 3$, $B = 4$, $h = 8$

$$A = \dfrac{1}{2}h(b + B)$$
$$A = \dfrac{1}{2}(8)(3 + 4)$$
$$= \dfrac{1}{2}(8)(7)$$
$$= (4)(7)$$
$$A = 28$$

26. $C = 2\pi r$; $C = 29.83$, $\pi = 3.14$

$$C = 2\pi r$$
$$29.83 = 2(3.14)r$$
$$29.83 = 6.28r$$
$$\dfrac{29.83}{6.28} = \dfrac{6.28r}{6.28}$$
$$4.75 = r$$

27. $V = \dfrac{4}{3}\pi r^3$; $r = 6$, $\pi = 3.14$

$$V = \dfrac{4}{3}\pi r^3$$
$$= \dfrac{4}{3}(3.14)(6)^3$$
$$= \dfrac{4}{3}(3.14)(216)$$
$$= \dfrac{4}{3}(678.24)$$
$$V = 904.32$$

28. $A = LW$ for L

$$\dfrac{A}{W} = \dfrac{LW}{W} \quad \text{Divide by } W$$
$$\dfrac{A}{W} = L \quad \text{or} \quad L = \dfrac{A}{W}$$

29. $A = \dfrac{1}{2}h(b + B)$ for h

$$2A = 2\left[\dfrac{1}{2}h(b + B)\right] \quad \text{Multiply by 2}$$
$$2A = h(b + B)$$
$$\dfrac{2A}{(b + B)} = \dfrac{h(b + B)}{(b + B)} \quad \text{Divide by } b + B$$
$$\dfrac{2A}{b + B} = h \quad \text{or} \quad h = \dfrac{2A}{b + B}$$

30. Because the two angles are supplementary,

$$(8x - 1) + (3x - 6) = 180.$$
$$11x - 7 = 180$$
$$11x = 187$$
$$x = 17$$

so $8x - 1 = 135$

and $3x - 6 = 45.$

The measures of the two angles are 135° and 45°.

31. The angles are vertical angles, so their measures are equal.

$$3x + 10 = 4x - 20$$
$$10 = x - 20$$
$$30 = x$$

so $3x + 10 = 100$

and $4x - 20 = 100$.

Each angle has a measure of 100°.

32. First, use the formula for the circumference of a circle to find the value of r.

$$C = 2\pi r$$
$$62.5 = 2(3.14)(r) \quad \text{Let } C = 62.5, \pi = 3.14$$
$$62.5 = 6.28r$$
$$\frac{62.5}{6.28} = \frac{6.28r}{6.28}$$
$$9.95 \approx r$$

The radius of the turntable is approximately 9.95 feet. The diameter is twice the radius, so the diameter is approximately 19.9 feet.

Now use the formula for the area of a circle.

$$A = \pi r^2$$
$$= (3.14)(9.95)^2 \quad \text{Let } \pi = 3.14, r = 9.95$$
$$= (3.14)(99.0025)$$
$$A \approx 311$$

The area of the turntable is approximately 311 square feet.

33. The sum of the three marked angles in the triangle is 180°.

$$45° + (x + 12.2)° + (3x + 2.8)° = 180°$$
$$4x + 60 = 180$$
$$4x = 120$$
$$x = 30$$

So $(x + 12.2)° = 42.2°$

and $(3x + 2.8)° = 92.8°$.

34. Knowing the values of h and b is not enough information to find the value of A. We would also need to know the value of B. Note that B and b are different variables. In general, to find the numerical value of one variable in a formula, we need to know the values of all the other variables.

35. The ratio of 60 centimeters to 40 centimeters is

$$\frac{60}{40} = \frac{3 \cdot 20}{2 \cdot 20} = \frac{3}{2}.$$

36. To find the ratio of 5 days to 2 weeks, first convert 2 weeks to days.

$$2 \text{ weeks} = 2 \cdot 7 = 14 \text{ days}$$

Thus, the ratio of 5 days to 2 weeks is $\frac{5}{14}$.

37. To find the ratio of 90 inches to 10 feet, first convert 10 feet to inches.

$$10 \text{ feet} = 10 \cdot 12 = 120 \text{ inches}$$

Thus, the ratio of 90 inches to 10 feet is

$$\frac{90}{120} = \frac{3 \cdot 30}{4 \cdot 30} = \frac{3}{4}.$$

38. To find the ratio of 3 months to 3 years, first convert 3 years to months.

$$3 \text{ years} = 3 \cdot 12 = 36 \text{ months}$$

Thus, the ratio of 3 months to 3 years is

$$\frac{3}{36} = \frac{1 \cdot 3}{12 \cdot 3} = \frac{1}{12}.$$

39.
$$\frac{p}{21} = \frac{5}{30}$$
$$30p = 105 \quad \textit{Cross products are equal}$$
$$\frac{30p}{30} = \frac{105}{30} \quad \textit{Divide by 30}$$
$$p = \frac{105}{30} = \frac{7 \cdot 15}{2 \cdot 15} = \frac{7}{2}$$

The solution set is $\left\{ \frac{7}{2} \right\}$.

40.
$$\frac{5 + x}{3} = \frac{2 - x}{6}$$
$$6(5 + x) = 3(2 - x) \quad \textit{Cross products are equal}$$
$$30 + 6x = 6 - 3x \quad \textit{Distributive property}$$
$$30 + 9x = 6 \quad \textit{Add 3x}$$
$$9x = -24 \quad \textit{Subtract 30}$$
$$x = \frac{-24}{9} = -\frac{8}{3}$$

The solution set is $\left\{ -\frac{8}{3} \right\}$.

41.
$$\frac{y}{5} = \frac{6y - 5}{11}$$
$$11y = 5(6y - 5)$$
$$11y = 30y - 25$$
$$-19y = -25$$
$$y = \frac{-25}{-19} = \frac{25}{19}$$

The solution set is $\left\{ \frac{25}{19} \right\}$.

42. Let x = the number of pounds of fertilizer needed to cover 500 square feet.

$$\frac{x \text{ pounds}}{2 \text{ pounds}} = \frac{500 \text{ square feet}}{150 \text{ square feet}}$$

$$150x = 2(500)$$

$$x = \frac{1000}{150} = \frac{20 \cdot 50}{3 \cdot 50}$$

$$= \frac{20}{3} = 6\frac{2}{3}$$

$6\frac{2}{3}$ pounds of fertilizer will cover 500 square feet.

43. Let x = distance from the lens to the object.

$$\frac{\text{length of the film image}}{\text{length of the object}} = \frac{\substack{\text{distance from the lens} \\ \text{to the film image}}}{\substack{\text{distance from the} \\ \text{lens to the object}}}$$

$$\frac{35 \text{ millimeters}}{1 \text{ meter}} = \frac{7 \text{ millimeters}}{x \text{ meter(s)}}$$

$$35x = 1(7)$$

$$x = \frac{7}{35} = \frac{1}{5} \text{ or } 0.2$$

The child should stand 0.2 meter (200 millimeters) from the lens.

44. Let x = the tax on a $36.00 item.

Set up a proportion with one ratio involving sales tax and the other involving the costs of the items.

$$\frac{x \text{ dollars}}{\$2.04} = \frac{\$36}{\$24}$$

$$24x = (2.04)(36) = 73.44$$

$$x = \frac{73.44}{24} = 3.06$$

The sales tax on a $36.00 item is $3.06.

45. Let x = the actual distance between the second pair of cities (in kilometers).

Set up a proportion with one ratio involving map distances and the other involving actual distances.

$$\frac{x \text{ kilometers}}{150 \text{ kilometers}} = \frac{80 \text{ centimeters}}{32 \text{ centimeters}}$$

$$32x = (150)(80) = 12,000$$

$$x = \frac{12,000}{32} = 375$$

The cities are 375 kilometers apart.

46. Let x = the number of gold medals earned by Romania.

$$\frac{x \text{ gold medals}}{20 \text{ medals}} = \frac{2 \text{ gold medals}}{10 \text{ medals}}$$

$$10x = 2(20) = 40$$

$$x = 4$$

At the 1996 Olympic Games, 4 gold medals were earned by Romania.

47. Unit costs are rounded to three decimal places.

Size	Unit Cost (dollars per ounce)	
15–ounce	$\dfrac{\$2.69}{15} = \$.179$	(most expensive)
20–ounce	$\dfrac{\$3.29}{20} = \$.165$	
25.5–ounce	$\dfrac{\$3.49}{25.5} = \$.137$	(least expensive)

The 25.5–ounce size is the best buy.

48. $160 million is what percent of $290 million?

$$\frac{160}{290} \approx .5517$$

Approximately 55.2% of the cost of Pacific Bell Park was borrowed.

49. The cost of the interest is 9% of $160 million.

$$(.09)(160) = \$14.4 \text{ million}$$

The cash flow is

$$(1.07)(14.4) = \$15.408 \text{ million},$$

or approximately $15.4 million.

50. Let x = the number of liters of the 60% solution to be used;

$x + 15$ = the number of liters of the 20% mixture.

Number of liters	15		x		$x + 15$	
Strength of solution	.10	+	.60	=	.20	

Drug amount in 10% solution	plus	Drug amount in 60% solution	is	Drug amount in 20% solution
$\downarrow$	$\downarrow$	$\downarrow$	$\downarrow$	$\downarrow$
.10(15)	+	.60(x)	=	.20(x + 15)

Multiply by 10 to clear decimals.

$$1(15) + 6x = 2(x + 15)$$

$$15 + 6x = 2x + 30$$

$$15 + 4x = 30$$

$$4x = 15$$

$$x = \frac{15}{4} = 3.75$$

3.75 liters of 60% solution are needed.

51. Let $x =$ the amount invested at 5%;
$10,000 - x =$ the amount invested at 6%.

Interest at 5%	plus	Interest at 6%		equals $550.
↓	↓	↓	↓	↓
.05x	+	.06(10,000 − x)	=	550

$$5x + 6(10,000 - x) = 100(550)$$
$$5x + 60,000 - 6x = 55,000$$
$$-x = -5000$$
$$x = 5000$$

Todd invested $5000 at 5% and
$10,000 - 5000 = \$5000$ at 6%.

52. Use the formula $d = rt$ or $r = \dfrac{d}{t}$.

$$r = \frac{d}{t} = \frac{3150}{384} \approx 8.203$$

Rounded to the nearest tenth, the Yorkshire's average speed was 8.2 mph.

53. Use the formula $d = rt$ or $t = \dfrac{d}{r}$.

$$t = \frac{d}{r} = \frac{819}{63} = 13$$

Sue drove for 13 hours.

54. Let $t =$ the number of hours until the planes are 1925 miles apart.

Use $d = rt$.

The distance one plane flies north	plus	the distance the other plane flies south	is	the distance between the planes.
↓	↓	↓	↓	↓
350t	+	420t	=	1925

$$770t = 1925$$
$$t = \frac{1925}{770} = \frac{5}{2} = 2\frac{1}{2}$$

The planes will be 1925 miles apart in $2\frac{1}{2}$ hours.

55. Let $x =$ the number of hours that Annie biked before she caught up with Jim.

Use the formula $d = rt$.

	r	t	d
Annie	8	x	$8x$
Jim	5	$x + \dfrac{1}{2}$	$5\left(x + \dfrac{1}{2}\right)$

Since they traveled the same distance, set the two distance expressions equal to each other and solve for x.

$$8x = 5\left(x + \frac{1}{2}\right)$$
$$8x = 5x + \frac{5}{2}$$
$$3x = \frac{5}{2}$$
$$\frac{1}{3}(3x) = \frac{1}{3}\left(\frac{5}{2}\right) \quad \textit{Multiply by } \frac{1}{3}$$
$$x = \frac{5}{6}$$

Annie will take $\dfrac{5}{6}$ of an hour or 50 minutes to catch up with Jim.

56. Let C denote the circumference and r the radius.

$$C = kr$$
$$5 = k(.796) \quad \textit{Substitute}$$
$$\frac{5}{.796} = k \quad \textit{Solve for } k$$

So the formula is

$$C = \frac{5}{.796}r$$
$$17.5 = \frac{5}{.796}r \quad \textit{C=17.5}$$
$$\frac{17.5(.796)}{5} = r \quad \textit{Solve for } r$$

Thus, $r = 2.786$ inches.

57. Let S denote the supply and p the price.

$$S = kp$$
$$40 = k(50) \quad \textit{Substitute}$$
$$k = \frac{40}{50} = \frac{4}{5} \quad \textit{Solve for } k$$

So the formula is

$$S = \frac{4}{5}p$$
$$= \frac{4}{5}(35) = 28$$

When the price is $35, the supply should be 28 games.

58. Let p denote the pressure exerted by a liquid and d the depth of the point beneath the surface of the liquid.

$$p = kd$$
$$\frac{80}{3} = k(10) \quad Substitute$$
$$\frac{8}{3} = k \quad\quad Solve\ for\ k$$

So $\quad p = \frac{8}{3}d$
$$= \frac{8}{3}(30) = 80$$

The pressure at 30 meters is 80 newtons per square centimeter.

59. Answers will vary.

60. Answers will vary.

61. $\dfrac{y}{7} = \dfrac{y-5}{2}$

$2y = 7(y-5) \quad Cross\ products\ are\ equal$
$2y = 7y - 35$
$-5y = -35$
$y = 7$

The solution set is $\{7\}$.

62. $I = prt$ for r

$\dfrac{I}{pt} = \dfrac{prt}{pt} \quad Divide\ by\ pt$

$\dfrac{I}{pt} = r$ or $r = \dfrac{I}{pt}$

63. $-2x = -4 - 2(2-x)$
$-2x = -4 - 4 + 2x$
$-4x = -8$
$x = 2$
The solution set is $\{2\}$.

64. $2k - 5 = 4k + 13$
$-2k - 5 = 13 \quad\quad Subtract\ 4k$
$-2k = 18 \quad\quad Add\ 5$
$k = -9 \quad\quad Divide\ by\ -2$

The solution set is $\{-9\}$.

65. $.05x + .02x = 4.9$

To clear decimals, multiply both sides by 100.

$100(.05x + .02x) = 100(4.9)$
$5x + 2x = 490$
$7x = 490$
$x = 70$

The solution set is $\{70\}$.

66. $2 - 3(y-5) = 4 + y$
$2 - 3y + 15 = 4 + y$
$17 - 3y = 4 + y$
$17 - 4y = 4$
$-4y = -13$
$y = \dfrac{-13}{-4} = \dfrac{13}{4}$

The solution set is $\left\{\dfrac{13}{4}\right\}$.

67. $9x - (7x + 2) = 3x + (2 - x)$
$9x - 7x - 2 = 3x + 2 - x$
$2x - 2 = 2x + 2$
$-2 = 2$

Because $-2 = 2$ is a false statement, the given equation has no solution, symbolized by $\emptyset$.

68. $\dfrac{1}{3}s + \dfrac{1}{2}s + 7 = \dfrac{5}{6}s + 5 + 2$

$\dfrac{1}{3}s + \dfrac{1}{2}s = \dfrac{5}{6}s \quad\quad Subtract\ 7$

The least common denominator is 6.

$6\left(\dfrac{1}{3}s + \dfrac{1}{2}s\right) = 6\left(\dfrac{5}{6}s\right)$
$2s + 3s = 5s$
$5s = 5s$

Because $5s = 5s$ is a true statement, the solution set is {all real numbers}.

69. Let $x = 6$ in the equation.

$$3 - (8 + 4x) = 2x + 7$$
$$3 - [8 + 4(6)] = 2(6) + 7$$
$$-29 = 19$$

This is false, so $x = 6$ is not a solution of the equation.

Solve the equation.

$$3 - 8 - 4x = 2x + 7$$
$$-5 - 4x = 2x + 7$$
$$-6x = 12$$
$$x = -2$$

The solution set is $\{-2\}$. The student probably got the incorrect answer by writing

$$3 - (8 + 4x) = 3 - 8 + 4x$$

and then solving the equation, which *does* have solution set $\{6\}$.

70. Let $x =$ the number of cups of crumbs needed for 30 servings.

Set up a proportion with one ratio involving number of cups of crumbs and the other involving number of servings.

$$\frac{x \text{ cups}}{\frac{2}{3} \text{ cups}} = \frac{30 \text{ servings}}{8 \text{ servings}}$$

$$(x)(8) = \left(\frac{2}{3}\right)(30)$$

$$8x = 20$$

$$x = \frac{20}{8} = \frac{5}{2} \text{ or } 2\frac{1}{2}$$

$2\frac{1}{2}$ cups of crumbs would be needed for 30 servings.

71. Let $\quad x =$ the length of the Brooklyn Bridge;

$x + 2605 =$ the length of the Golden Gate Bridge.

$$x + (x + 2605) = 5795$$
$$2x + 2605 = 5795$$
$$2x = 3190$$
$$x = 1595$$
$$x + 2605 = 4200$$

The length of the Brooklyn Bridge is 1595 feet and that of the Golden Gate Bridge is 4200 feet.

72. Substituting 0 for x in the equation makes the fractions undefined since we cannot divide by 0.

73. The unit costs are rounded to four decimal places.

Size	Unit Cost (dollars per ounce)	
32–ounce size	$\frac{\$1.19}{32} = \$.0372$	
48–ounce size	$\frac{\$1.79}{48} = \$.0373$	(most expensive)
64–ounce size	$\frac{\$1.99}{64} = \$.0311$	(least expensive)

The 64-ounce size is the best buy.

74. Let $x =$ the number of quarts of oil needed for 192 quarts of gasoline.

Set up a proportion with one ratio involving oil and the other involving gasoline.

$$\frac{x \text{ quarts}}{1 \text{ quart}} = \frac{192 \text{ quarts}}{24 \text{ quarts}}$$

$$x \cdot 24 = 1 \cdot 192 \qquad \textit{Cross products}$$
$$x = 8 \qquad \textit{Divide by 24}$$

The amount of oil needed is 8 quarts.

75. Let $\quad x =$ the speed of the slower train;

$x + 30 =$ the speed of the faster train.

	r	t	d
Slower train	x	3	$3x$
Faster train	$x + 30$	3	$3(x + 30)$

The sum of the distances traveled by the two trains is 390 miles, so

$$3x + 3(x + 30) = 390.$$
$$3x + 3x + 90 = 390$$
$$6x + 90 = 390$$
$$6x = 300$$
$$x = 50$$

and $\quad x + 30 = 80$

The speed of the slower train is 50 miles per hour and the speed of the faster train is 80 miles per hour.

76. Let $x =$ the length of the shortest side of the triangle;

$x + 3 =$ the length of the second side;

$2x =$ the length of the third side.

Use $P = a + b + c$, with $a = x$, $b = x + 3$, and $c = 2x$.

The perimeter is 39.

$$x + (x + 3) + 2x = 39$$

$$x + (x + 3) + 2x = 39$$
$$4x + 3 = 39$$
$$4x = 36$$
$$x = 9$$

The shortest side is 9 centimeters.

77. The formula for the area of a trapezoid is $A = \frac{1}{2}(b + B)h$, where h is the height of the trapezoid.

$$A = \frac{1}{2}(b + B)h$$

Let $A = 360$
$b = 42$
$B = 48$

$$360 = \frac{1}{2}(42 + 48)h$$

$$360 = \frac{1}{2}(90)h$$
$$360 = 45h$$
$$8 = h$$

The height is 8 centimeters.

78. Let $s =$ the length of a side of the square.

The formula for the perimeter of a square is $P = 4s$.

The perimeter is 200.

$$4s = 200$$

$$4s = 200$$
$$s = 50$$

The length of a side is 50 meters.

79. No. Only equations that have one fractional term on each side can be solved by cross multiplication. This equation has two terms on the left side.

Chapter 2 Test

1. $5x + 9 = 7x + 21$

$\qquad -2x + 9 = 21$ *Subtract 7x*

$\qquad\quad -2x = 12$ *Subtract 9*

$\qquad\qquad x = -6$ *Divide by –2*

The solution set is $\{-6\}$.

2. $\qquad\qquad -\dfrac{4}{7}x = -12$

$\left(-\dfrac{7}{4}\right)\left(-\dfrac{4}{7}x\right) = \left(-\dfrac{7}{4}\right)(-12)$

$\qquad\qquad\quad x = 21$

The solution set is $\{21\}$.

3. $7 - (m - 4) = -3m + 2(m + 1)$

$\qquad 7 - m + 4 = -3m + 2m + 2$

$\qquad -m + 11 = -m + 2$

Because the last statement is false, the equation has no solution set, symbolized by $\emptyset$.

4. $.06(x + 20) + .08(x - 10) = 4.6$

To clear decimals, multiply both sides by 100.

$100[.06(x + 20) + .08(x - 10)] = 100(4.6)$

$\qquad 6(x + 20) + 8(x - 10) = 460$

$\qquad 6x + 120 + 8x - 80 = 460$

$\qquad\qquad\quad 14x + 40 = 460$

$\qquad\qquad\qquad\quad 14x = 420$

$\qquad\qquad\qquad\qquad x = 30$

The solution set is $\{30\}$.

5. $-8(2x + 4) = -4(4x + 8)$

$\qquad -16x - 32 = -16x - 32$

Because the last statement is true, the solution set is {all real numbers}.

6.   Let $x =$ the area of Kauai (in square miles);

$x + 177 =$ the area of Maui (in square miles);

$(x + 177) + 3293 = x + 3470$

$\qquad\qquad\qquad = $ the area of Hawaii.

$x + (x + 177) + (x + 3470) = 5300$

$\qquad\qquad\quad 3x + 3647 = 5300$

$\qquad\qquad\qquad\quad 3x = 1653$

$\qquad\qquad\qquad\qquad x = 551$

If $x = 551$, $x + 177 = 728$, and $x + 3470 = 4021$.

The area of Hawaii is 4021 square miles, the area of Maui is 728 square miles, and the area of Kauai is 551 square miles.

7.   **(a)** Solve $P = 2L + 2W$ for W.

$\qquad P - 2L = 2W$

$\qquad \dfrac{P - 2L}{2} = W$

$\qquad W = \dfrac{P - 2L}{2}$ or $W = \dfrac{P}{2} - L$

(b) Substitute 116 for P and 40 for L in either form of the formula obtained in (a).

$\qquad W = \dfrac{P - 2L}{2}$

$\qquad\quad = \dfrac{116 - 2(40)}{2}$

$\qquad\quad = \dfrac{116 - 80}{2}$

$\qquad\quad = \dfrac{36}{2} = 18$

8.   The angles are vertical angles, so their measures are equal.

$\qquad 3x + 15 = 4x - 5$

$\qquad\qquad 15 = x - 5$

$\qquad\qquad 20 = x$

So $3x + 15 = 75$ and $4x - 5 = 75$. Both angles have measure 75°.

9.   Let $\qquad x =$ the measure of the angle.

Then $90 - x =$ the measure of its complement

and $180 - x =$ the measure of its supplement.

$\qquad 180 - x = 3(90 - x) + 10$

$\qquad 180 - x = 270 - 3x + 10$

$\qquad 180 - x = 280 - 3x$

$\qquad 180 + 2x = 280$

$\qquad\qquad 2x = 100$

$\qquad\qquad\quad x = 50$

The measure of the angle is 50°. The measure of its supplement, 130°, is 10° more than three times its complement, 40°.

10. $\dfrac{y + 5}{3} = \dfrac{y - 3}{4}$

$\qquad 4(y + 5) = 3(y - 3)$

$\qquad 4y + 20 = 3y - 9$

$\qquad\quad y + 20 = -9$

$\qquad\qquad\quad y = -29$

The solution set is $\{-29\}$.

11. Let C denote the circumference of a circle and r its radius.

$$C = kr$$
$$37.68 = k(6) \quad \textit{Substitute}$$
$$6.28 = k \quad \textit{Solve for } k$$

So the formula is
$$C = 6.28r$$
$$= 6.28(10)$$
$$= 62.8 \text{ centimeters}$$

12. Let $x =$ the actual distance between Seattle and Cincinnati.

$$\frac{x \text{ miles}}{1050 \text{ miles}} = \frac{46 \text{ inches}}{21 \text{ inches}}$$
$$21x = 46(1050)$$
$$x = \frac{48,300}{21} = 2300$$

The actual distance between Seattle and Cincinnati is 2300 miles.

13. Let $x =$ the amount invested at 3%;
$x + 6000 =$ the amount invested at 4.5%.

Amount invested (in dollars)	Rate of interest	Interest for one year
x	.03	$.03x$
$x + 6000$	.045	$.045(x + 6000)$

$$.03x + .045(x + 6000) = 870$$
$$1000(.03x) + 1000[.045(x + 6000)] = 1000(870)$$
$$30x + 45(x + 6000) = 870,000$$
$$30x + 45x + 270,000 = 870,000$$
$$75x + 270,000 = 870,000$$
$$75x = 600,000$$
$$x = 8000$$

and $\qquad\qquad x + 6000 = 14,000$

Laura invested \$8000 at 3% and \$14,000 at 4.5%.

14. What percent of 75 is 177?

$$\frac{177}{75} = 2.36$$

So the debt is 236% of the franchise value.

15. Use the formula $d = rt$ and let t be the number of hours they traveled.

	r	t	d
First car	50	t	$50t$
Second car	65	t	$65t$

First car's distance	and	second car's distance
↓	↓	↓
$50t$	+	$65t$

is	total distance.
↓	↓
=	460

$$50t + 65t = 460$$
$$115t = 460$$
$$t = 4$$

The two cars will be 460 miles apart in 4 hours.

16. Let $x =$ West's score;
$x + 12 =$ East's score.

$$x + (x + 12) = 252$$
$$2x + 12 = 252$$
$$2x = 240$$
$$x = 120$$

The score of the game was East 132, West 120.

17. Let $x =$ Sprewell's points;
$x + 7 =$ Rice's points.

$$x + (x + 7) = 45$$
$$2x + 7 = 45$$
$$2x = 38$$
$$x = 19$$

Sprewell scored 19 points and Rice scored 26 points.

Cumulative Review Exercises Chapters 1–2

1. $\dfrac{108}{144} = \dfrac{3 \cdot 36}{4 \cdot 36} = \dfrac{3}{4}$

2. $\dfrac{5}{6} + \dfrac{1}{4} - \dfrac{7}{15} = \dfrac{50}{60} + \dfrac{15}{60} - \dfrac{28}{60}$

$\qquad = \dfrac{65 - 28}{60}$

$\qquad = \dfrac{37}{60}$

3. $\dfrac{9}{8} \cdot \dfrac{16}{3} \div \dfrac{5}{8} = \dfrac{9}{8} \cdot \dfrac{16}{3} \cdot \dfrac{8}{5}$

$\qquad = \dfrac{3 \cdot 3 \cdot 16 \cdot 8}{8 \cdot 3 \cdot 5}$

$\qquad = \dfrac{48}{5}$

4. "The difference between half a number and 18" is written

$$\frac{1}{2}x - 18.$$

5. "The quotient of 6 and 12 more than a number is 2" is written

$$\frac{6}{x + 12} = 2.$$

6. $\dfrac{8(7) - 5(6 + 2)}{3 \cdot 5 + 1} \geq 1$

$\dfrac{8(7) - 5(8)}{3 \cdot 5 + 1} \geq 1$

$\dfrac{56 - 40}{15 + 1} \geq 1$

$\dfrac{16}{16} \geq 1$

$1 \geq 1$

The statement is true.

7. $9 - (-4) + (-2) = (9 + 4) + (-2)$
$= 13 - 2$
$= 11$

8. $\dfrac{-4(9)(-2)}{-3^2} = \dfrac{-36(-2)}{-1 \cdot 3^2}$
$= \dfrac{72}{-9}$
$= -8$

9. $(-7 - 1)(-4) + (-4) = (-8)(-4) + (-4)$
$= 32 + (-4)$
$= 28$

10. $\dfrac{3x^2 - y^3}{-4z} = \dfrac{3(-2)^2 - (-4)^3}{-4(3)}$ Let $x = -2$,
$y = -4, z = 3$
$= \dfrac{3(4) - (-64)}{-12}$
$= \dfrac{12 + 64}{-12}$
$= \dfrac{76}{-12}$
$= -\dfrac{19}{3}$

11. $7(k + m) = 7k + 7m$

The multiplication of 7 is distributed over the sum, which illustrates the distributive property.

12. $3 + (5 + 2) = 3 + (2 + 5)$

The order of the numbers added in the parentheses is changed, which illustrates the commutative property.

13. $-4(k + 2) + 3(2k - 1)$
$= (-4)(k) + (-4)(2) + (3)(2k) + (3)(-1)$
$= -4k - 8 + 6k - 3$
$= -4k + 6k - 8 - 3$
$= 2k - 11$

14. $2r - 6 = 8r$
$-6 = 6r$
$-1 = r$

Check $r = -1$: $-8 = -8$

The solution set is $\{-1\}$.

15. $4 - 5(a + 2) = 3(a + 1) - 1$
$4 - 5a - 10 = 3a + 3 - 1$
$-5a - 6 = 3a + 2$
$-8a - 6 = 2$
$-8a = 8$
$a = -1$

Check $a = -1$: $-1 = -1$

The solution set is $\{-1\}$.

16. $\dfrac{2}{3}y + \dfrac{3}{4}y = -17$
$12\left(\dfrac{2}{3}y + \dfrac{3}{4}y\right) = 12(-17)$ $LCD = 12$
$8y + 9y = -204$
$17y = -204$
$y = -12$

Check $y = -12$: $-17 = -17$

The solution set is $\{-12\}$.

17. $-(m - 1) = 3 - 2m$
$-m + 1 = 3 - 2m$
$m = 2$

Check $m = 2$: $-1 = -1$

The solution set is $\{2\}$.

18. $\dfrac{2x + 3}{5} = \dfrac{x - 4}{2}$
$(2x + 3)(2) = (5)(x - 4)$
$4x + 6 = 5x - 20$
$6 = x - 20$
$26 = x$

Check $x = 26$: $11 = 11$

The solution set is $\{26\}$.

19. $\dfrac{y - 2}{3} = \dfrac{2y + 1}{5}$
$(y - 2)(5) = 3(2y + 1)$
$5y - 10 = 6y + 3$
$-y = 13$
$y = -13$

Check $y = -13$: $-5 = -5$

The solution set is $\{-13\}$.

20. $3x + 4y = 24$ for y
$4y = 24 - 3x$
$y = \dfrac{24 - 3x}{4}$

21. $A = P(1 + ni)$ for n

$A = P + Pni$
$A - P = Pni$
$\dfrac{A - P}{Pi} = n$

22. Let $x =$ the length of the middle-sized piece;

$3x =$ the length of the longest piece;

$x - 5 =$ the length of the shortest piece.

$$x + 3x + (x - 5) = 40$$
$$5x - 5 = 40$$
$$5x = 45$$
$$x = 9$$

The length of the middle-sized piece is 9 centimeters, of the longest piece is 27 centimeters, and of the shortest piece is 4 centimeters.

23. The radius r is one-half the diameter, so

$$r = \frac{1}{2}(328) = 164 \text{ feet.}$$

The formula for the circumference is $C = 2\pi r$.

$$C = 2\pi r$$
$$= 2(3.14)(164)$$
$$= 1029.92$$

The circumference is about 1030 feet. Note that the formula $C = \pi d$ could be used here.

24.
$$\frac{x \text{ cups}}{1\frac{1}{4} \text{ cups}} = \frac{20 \text{ people}}{6 \text{ people}}$$

$$6x = \left(1\frac{1}{4}\right)(20)$$
$$6x = 25$$
$$x = \frac{25}{6} \text{ or } 4\frac{1}{6} \text{ cups}$$

25. Let $x =$ speed of slower car;

$x + 20 =$ speed of faster car.

Use the formula $d = rt$,

$$d_{\text{slower}} + d_{\text{faster}} = d_{\text{total}}$$
$$(x)(4) + (x + 20)(4) = 400$$
$$4x + 4x + 80 = 400$$
$$8x + 80 = 400$$
$$8x = 320$$
$$x = 40$$

The speeds are 40 mph and 60 mph.

CHAPTER 3   LINEAR INEQUALITIES AND ABSOLUTE VALUE

Section 3.1

1. $x \le 3$ can be written in interval notation as $(-\infty, 3]$. Choice D

3. $x < 3$ represents all numbers less than 3, which is graphed in choice B.

5. $-3 \le x \le 3$ can be written in interval notation as $[-3, 3]$. Choice F

7. If an endpoint is an element of the solution set, then this is shown on the graph using a bracket. If an endpoint is *not* an element of the solution set, then this is shown on the graph using a parenthesis.

9. $$4x + 1 \ge 21$$
    Subtract 1 from both sides.
    $$4x + 1 - 1 \ge 21 - 1$$
    $$4x \ge 20$$
    Divide both sides by 4.
    $$\frac{4x}{4} \ge \frac{20}{4}$$
    $$x \ge 5$$
    Solution set: $[5, \infty)$

11. $$\frac{3k - 1}{4} > 5$$
    $$4\left(\frac{3k - 1}{4}\right) > 4(5) \qquad \textit{Multiply by 4.}$$
    $$3k - 1 > 20$$
    $$3k - 1 + 1 > 20 + 1 \quad \textit{Add 1.}$$
    $$3k > 21$$
    $$\frac{3k}{3} > \frac{21}{3} \qquad \textit{Divide by 3.}$$
    $$k > 7$$
    Solution set: $(7, \infty)$

    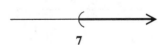

13. $-4x < 16$
    Divide both sides by -4, and reverse the inequality sign.
    $$\frac{-4x}{-4} > \frac{16}{-4}$$
    $$x > -4$$

    Solution set: $(-4, \infty)$

    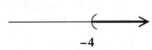

15. $$-\frac{3}{4}r \ge 30$$
    Multiply both sides by $-\frac{4}{3}$, and reverse the inequality sign.
    $$-\frac{4}{3}\left(-\frac{3}{4}r\right) \le -\frac{4}{3}(30)$$
    $$r \le -40$$
    Solution set: $(-\infty, -40]$

17. $-1.3m \ge -5.2$
    Divide both sides by -1.3, and reverse the inequality sign.
    $$\frac{-1.3m}{-1.3} \le \frac{-5.2}{-1.3}$$
    $$m \le 4$$

    Solution set: $(-\infty, 4]$

19. $$\frac{2k - 5}{-4} > 5$$
    Multiply both sides by -4, and reverse the inequality sign.
    $$-4\left(\frac{2k - 5}{-4}\right) < -4(5)$$
    $$2k - 5 < -20$$
    $$2k < -15 \qquad \textit{Add 5.}$$
    $$k < -\frac{15}{2} \qquad \textit{Divide by 2.}$$
    Solution set: $\left(-\infty, -\frac{15}{2}\right)$

    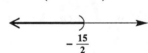

21. $y + 4(2y - 1) \ge y$
    $$y + 8y - 4 \ge y \qquad \textit{Clear parentheses.}$$
    $$9y - 4 \ge y \qquad \textit{Combine terms.}$$
    $$8y \ge 4 \qquad \textit{Add 4; Subtract y.}$$
    $$y \ge \frac{4}{8} = \frac{1}{2} \qquad \textit{Divide by 8.}$$
    Solution set: $\left[\frac{1}{2}, \infty\right)$

    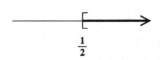

23. $-(4 + r) + 2 - 3r < -14$

$-4 - r + 2 - 3r < -14$ *Clear parentheses.*

$-4r - 2 < -14$ *Combine terms.*

$-4r < -12$ *Add 2.*

Divide by -4, and reverse the inequality sign.

$r > 3$

Solution set: $(3, \infty)$

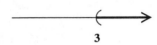

25. $-3(z - 6) > 2z - 2$

$-3z + 18 > 2z - 2$ *Clear parentheses.*

$-5z > -20$ *Subtract 2z; subtract 18.*

Divide by -5, and reverse the inequality sign.

$z < 4$

Solution set: $(-\infty, 4)$

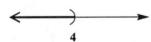

27. $\dfrac{2}{3}(3k - 1) \geq \dfrac{3}{2}(2k - 3)$

Multiply both sides by 6 to clear the fractions.

$6 \cdot \dfrac{2}{3}(3k - 1) \geq 6 \cdot \dfrac{3}{2}(2k - 3)$

$4(3k - 1) \geq 9(2k - 3)$

$12k - 4 \geq 18k - 27$ *Clear parentheses.*

$-6k \geq -23$ *Subtract 18k; add 4.*

Divide by -6, and reverse the inequality sign.

$k \leq \dfrac{23}{6}$

Solution set: $\left(-\infty, \dfrac{23}{6}\right]$

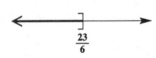

29. $-\dfrac{1}{4}(p + 6) + \dfrac{3}{2}(2p - 5) < 10$

Multiply each term by 4 to clear the fractions.

$-1(p + 6) + 6(2p - 5) < 40$

$-p - 6 + 12p - 30 < 40$

$11p - 36 < 40$

$11p < 76$

$p < \dfrac{76}{11}$

Solution set: $\left(-\infty, \dfrac{76}{11}\right)$

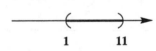

31. $3(2x - 4) - 4x < 2x + 3$

$6x - 12 - 4x < 2x + 3$

$2x - 12 < 2x + 3$

$-12 < 3$ *True*

The statement is true for all values of x. Therefore, the original inequality is true for any real number.

Solution set: $(-\infty, \infty)$

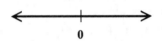

33. $8\left(\dfrac{1}{2}x + 3\right) < 8\left(\dfrac{1}{2}x - 1\right)$

$4x + 24 < 4x - 8$

$24 < -8$ *False*

This is a false statement, so the inequality is a contradiction.

Solution set: $\emptyset$

35. It is incorrect. The inequality symbol should be reversed only when multiplying or dividing by a negative number. Since 5 is positive, the inequality symbol should not be reversed.

37. $-4 < x - 5 < 6$

Add 5 to each part of the inequality to isolate the variable x.

$-4 + 5 < x - 5 + 5 < 6 + 5$

$1 < x < 11$

Solution set: $(1, 11)$

39. $-9 \leq k + 5 \leq 15$

Subtract 5 from each part.

$-9 - 5 \leq k + 5 - 5 \leq 15 - 5$

$-14 \leq k \leq 10$

Solution set: $[-14, 10]$

41. $-6 \leq 2z + 4 \leq 16$

   $-10 \leq 2z \leq 12$ *Subtract 4.*

   $-5 \leq z \leq 6$ *Divide by 2.*

   Solution set: $[-5, 6]$

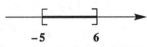

43. $-19 \leq 3x - 5 \leq 1$

   $-14 \leq 3x \leq 6$ *Add 5.*

   $-\dfrac{14}{3} \leq x \leq 2$ *Divide by 3.*

   Solution set: $\left[-\dfrac{14}{3}, 2\right]$

45. $-1 \leq \dfrac{2x - 5}{6} \leq 5$

   $-6 \leq 2x - 5 \leq 30$ *Multiply by 6.*

   $-1 \leq 2x \leq 35$ *Add 5.*

   $-\dfrac{1}{2} \leq x \leq \dfrac{35}{2}$ *Divide by 2.*

   Solution set: $\left[-\dfrac{1}{2}, \dfrac{35}{2}\right]$

47. $4 \leq 5 - 9x < 8$

   $-1 \leq -9x < 3$ *Subtract 5.*

   Divide each part by -9; reverse the inequalities.

   $\dfrac{1}{9} \geq x > -\dfrac{1}{3}$

   This inequality may be written

   $-\dfrac{1}{3} < x \leq \dfrac{1}{9}.$

   Solution set: $\left(-\dfrac{1}{3}, \dfrac{1}{9}\right]$

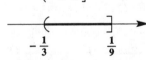

49.   "Exceed" means greater than. From the graph, the percent of tornadoes was greater than 7.7% in April, May, June, and July.

51.   A total of $17,252$ tornadoes were reported. To find the months in which fewer than (or less than) 1500 were reported, find what percent 1500 is of $17,252$.

$\dfrac{1500}{17,252} \approx .087 \approx 8.7\%$

The months where less then 8.7% of tornadoes were reported were January, February, March, August, September, October, November, and December.

53.   From the graph, the seasons in which ticket sales were less than or equal to 8 million are: 1990–1991, 1991–1992, 1992–1993, 1993–1994.

55.   Let $x =$ the number of tickets (in millions) that would need to be sold during the 1993–1994 season to average 7.75 million for the 4-year period.

$\dfrac{7.3 + 7.5 + 7.9 + x}{4} \geq 7.75$

$7.3 + 7.5 + 7.9 + x \geq 4(7.75)$

$x + 22.7 \geq 31$

$x \geq 8.3$

At least 8.3 million tickets must be sold.

57.   Notice from the problem that the taxicab rates are assessed per $\dfrac{1}{5}$ mi. Therefore, let

$x =$ the number of $\dfrac{1}{5}$–mi distances

Dantrell can travel.
He must pay \$1.50 plus $.25x$, and this amount must be no more than \$3.75.

$1.50 + .25x \leq 3.75$

$.25x \leq 2.25$

$x \leq 9$

Thus, Dantrell can travel the first $\dfrac{1}{5}$ mi plus $\dfrac{9}{5}$ additional miles or

$\dfrac{1}{5} + \dfrac{9}{5} = \dfrac{10}{5} = 2$ mi.

59.   Let $x =$ her score on the third test.
Her average must be at least 84 (≥ 84). To find the average of three numbers, add them and divide by 3.

$\dfrac{90 + 82 + x}{3} \geq 84$

$\dfrac{172 + x}{3} \geq 84$ *Add.*

$172 + x \geq 252$ *Multiply by 3.*

$x \geq 80$ *Subtract 172.*

She must score at least 80 on her third test.

61. Let $x =$ the number of miles driven.
The cost of renting from Ford is \$35 plus the mileage cost of $.14x$, while the cost of renting from Chevrolet is \$34 plus the mileage cost of $.16x$.

$$\text{Cost from Chevrolet} > \text{Cost from Ford}$$
$$34 + .16x > 35 + .14x$$
$$3400 + 16x > 3500 + 14x$$
Multiply by 100.
$$2x > 100$$
$$x > 50$$

After 50 mi, the price to rent the Chevrolet exceeds the price to rent the Ford.

63. Cost $C = 20x + 100$; Revenue $R = 24x$
The business will show a profit only when $R > C$. Substitute the given expressions for R and C.

$$R > C$$
$$24x > 20x + 100$$
$$4x > 100$$
$$x > 25$$

The company will show a profit upon selling 26 tapes.

65. $5(x + 3) - 2(x - 4) = 2(x + 7)$
$$5x + 15 - 2x + 8 = 2x + 14$$
$$3x + 23 = 2x + 14$$
$$x = -9$$
Solution set: $\{-9\}$

The graph is the point -9 on a number line.

66. $5(x + 3) - 2(x - 4) > 2(x + 7)$
$$5x + 15 - 2x + 8 > 2x + 14$$
$$3x + 23 > 2x + 14$$
$$x > -9$$
Solution set: $(-9, \infty)$

The graph extends from -9 to the right on a number line; -9 is not included in the graph.

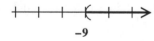

67. $5(x + 3) - 2(x - 4) < 2(x + 7)$
$$5x + 15 - 2x + 8 < 2x + 14$$
$$3x + 23 < 2x + 14$$
$$x < -9$$
Solution set: $(-\infty, -9)$

The graph extends from -9 to the left on a number line; -9 is not included in the graph.

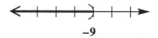

68. If we graph all the solution sets from Exercises 65–67; that is, $\{-9\}$, $(-9, \infty)$, and $(-\infty, -9)$, on the same number line, we will have graphed the set of all real numbers.

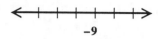

69. The solution set of the given equation is the point -3 on a number line. The solution set of the first inequality extends from -3 to the right (toward ∞) on the same number line. Based on Exercises 65–67, the solution set of the second inequality should then extend from -3 to the left (toward $-\infty$) on the number line. Complete the statement with $\underline{(-\infty, -3)}$.

71. $4 < y < 1$ is an improper statement since $4 \not< 1$. There is no such number y.

Section 3.2

1. This statement is true. The solution set of $x + 1 = 5$ is $\{4\}$. The solution set of $x + 1 > 5$ is $(4, \infty)$. The solution set of $x + 1 < 5$ is $(-\infty, 4)$. Taken together we have the set of real numbers. (See Section 3.1, Exercises 65–69, for a discussion of this concept.)

3. This statement is false. The union is $(-\infty, 8) \cup (8, \infty)$. The only real number that is *not* in the union is 8.

5. This statement is false since 0 is a rational number but not an irrational number. The sets of rational numbers and irrational numbers have no common elements so their intersection is $\emptyset$.

In Exercises 7–20, let
$$A = \{1, 2, 3, 4, 5, 6\},\ B = \{1, 3, 5\},\ C = \{1, 6\},$$
and $D = \{4\}$.

7. The intersection of sets B and A contains only those elements in both sets B and A.
$$B \cap A = \{1, 3, 5\} \text{ or set } B$$

9. The intersection of sets A and D is the set of all elements in both set A and D. Therefore,
$$A \cap D = \{4\} \text{ or set } D.$$

11. The intersection of set B and the set of no elements (empty set), $B \cap \emptyset$, is the set of no elements or $\emptyset$.

13. The union of sets A and B is the set of all elements that are in either set A or set B or both sets A and B. Since all numbers in set B are also in set A, the set $A \cup B$ will be the same as set A.
$$A \cup B = \{1, 2, 3, 4, 5, 6\} \text{ or set } A$$

15. The union of sets B and C contains all elements in either set B or set C or both sets B and C.

$$B \cup C = \{1, 3, 5, 6\}$$

17. The union of sets C and D is the set of all elements that are in either set C or set D or both sets C and D.

$$C \cup D = \{1, 4, 6\}$$

19. $B \cap C =$ the set of elements common to both B and $C = \{1\}$.

$$A \cap (B \cap C) = A \cap \{1\} = \{1\}.$$
$$A \cap B = \{1, 3, 5\}.$$
$$(A \cap B) \cap C = \{1, 3, 5\} \cap C = \{1\}.$$

Therefore,

$$A \cap (B \cap C) = (A \cap B) \cap C.$$

This illustrates the associative property of set intersection.

21. One example of how the concept of intersection can be applied to a real-life situation is in the grocery store. The set of red apples is the intersection of the set of red fruit and the set of apples.

23. The first graph represents the set $(-\infty, 2)$. The second graph represents the set $(-3, \infty)$. The intersection includes the elements common to both sets, that is, $(-3, 2)$.

25. The first graph represents the set $(-\infty, 5]$. The second graph represents the set $(-\infty, 2]$. The intersection includes the elements common to both sets, that is, $(-\infty, 2]$.

27. $x - 3 \le 6$ and $x + 2 \ge 7$
$\quad\quad x \le 9$ and $\quad\quad x \ge 5$
The graph of the solution set is all numbers that are both less than or equal to 9 and greater than or equal to 5. This is the intersection. The elements common to both sets are the numbers between 5 and 9, including the endpoints. The solution set is $[5, 9]$.

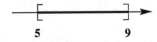

29. $-3x > 3$ and $x + 3 > 0$
$\quad\quad x < -1$ and $\quad\quad x > -3$
The graph of the solution set is all numbers that are both less than -1 and greater than -3. This is the intersection. The elements common to both

sets are the numbers between -3 and -1, not including the endpoints. The solution set is $(-3, -1)$.

31. $3x - 4 \le 8$ and $-4x + 1 \ge -15$
$\quad\quad 3x \le 12$ and $\quad\quad -4x \ge -16$
$\quad\quad\; x \le 4$ and $\quad\quad\quad\; x \le 4$
Since both inequalities are identical, the graph of the solution set is the same as the graph of one of the inequalities. The solution set is $(-\infty, 4]$.

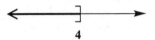

33. The first graph represents the set $(-\infty, 2]$. The second graph represents the set $[4, \infty)$. The union includes all elements in either set, or in both, that is, $(-\infty, 2] \cup [4, \infty)$.

35. The first graph represents the set $(-\infty, 1]$. The second graph represents the set $(-\infty, 8]$. The union includes all elements in either set, or in both, that is, $(-\infty, 8]$.

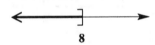

37. $x + 2 > 7$ or $1 - x > 6$
$\quad\quad\quad\quad\quad\quad\quad\quad -x > 5$
$\quad\quad x > 5$ or $\quad\quad x < -5$
The graph of the solution set is all numbers either greater than 5 or less than -5. This is the union. The solution set is

$$(-\infty, -5) \cup (5, \infty).$$

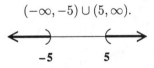

39. $x + 1 > 3$ or $-4x + 1 > 5$
$\quad\quad\quad\quad\quad\quad\quad\quad -4x > 4$
$\quad\quad x > 2$ or $\quad\quad x < -1$
The graph of the solution set is all numbers either less than -1 or greater than 2. This is the union. The solution set is $(-\infty, -1) \cup (2, \infty)$.

41. $(-\infty, -1] \cap [-4, \infty)$
The intersection is the set of numbers less than or equal to -1 and greater than or equal to -4. The numbers common to both original sets are between, and including, -4 and -1. The solution set is $[-4, -1]$.

43. $(-\infty, -6] \cap [-9, \infty)$

The intersection is the set of numbers less than or equal to -6 and greater than or equal to -9. The numbers common to both original sets are between, and including, -9 and -6. The solution set is $[-9, -6]$.

45. $(-\infty, 3) \cup (-\infty, -2)$

The union is the set of numbers that are either less than 3 or less than -2, or both. This is all numbers less than 3. The solution set is $(-\infty, 3)$.

47. $[3, 6] \cup (4, 9)$

The union is the set of numbers between, and including, 3 and 6, or between, but not including, 4 and 9. This is the set of numbers greater than or equal to 3 and less than 9. The solution set is $[3, 9)$.

49. $x < -1$ and $x > -5$

The word "and" means to take the intersection of both sets. $x < -1$ and $x > -5$ is true only when

$$-5 < x < -1.$$

The graph of the solution set is all numbers greater than -5 *and* less than -1. This is all numbers between -5 and -1, not including -5 or -1. The solution set is $(-5, -1)$.

51. $x < 4$ or $x < -2$

The word "or" means to take the union of both sets. The graph of the solution set is all numbers that are either less than 4 *or* less than -2, or both. This is all numbers less than 4. The solution set is $(-\infty, 4)$.

53. $-3x \le -6$ or $-3x \ge 0$

$\quad\quad x \ge 2$ or $\quad x \le 0$

The word "or" means to take the union of both sets. The graph of the solution set is all numbers that are either greater than or equal to 2 *or* less than or equal to 0. The solution set is $(-\infty, 0] \cup [2, \infty)$.

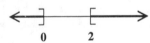

55. $x + 1 \ge 5$ and $x - 2 \le 10$

$\quad\quad x \ge 4$ and $\quad x \le 12$

The word "and" means to take the intersection of both sets. The graph of the solution set is all numbers that are both greater than or equal to 4 *and* less than or equal to 12. This is all numbers between, and including, 4 and 12. The solution set is $[4, 12]$.

57. From the graph, the number of stations sold exceeded 1200 in the years 1993, 1994, 1995, and 1996.

The value of transactions exceeded 3000 (in millions of dollars) in 1995 and 1996. Because of the word "and," we want the intersection of these two sets of years, which is 1995 and 1996.

For Exercises 59–64, find the area and perimeter of each of the given yards.

For Luigi's, Maria's, and Than's yards, use the formulas $A = LW$ and $P = 2L + 2W$.

Luigi's yard
$A = 50(30) = 1500 \text{ ft}^2$
$P = 2(50) + 2(30) = 160 \text{ ft}$

Maria's yard
$A = 40(35) = 1400 \text{ ft}^2$
$P = 2(40) + 2(35) = 150 \text{ ft}$

Than's yard
$A = 60(50) = 3000 \text{ ft}^2$
$P = 2(60) + 2(50) = 220 \text{ ft}$

For Joe's yard, use the formulas $A = \dfrac{1}{2}bh$ and $P = a + b + c$.

Joe's yard
$A = \dfrac{1}{2}(40)(30) = 600 \text{ ft}^2$
$P = 30 + 40 + 50 = 120 \text{ ft}$

To be fenced, a yard must have a perimeter $P \le 150$ ft. To be sodded, a yard must have an area $A \le 1400 \text{ ft}^2$.

59. Find "the yard can be fenced *and* the yard can be sodded."

A yard that can be fenced has $P \le 150$. Maria and Joe qualify.

A yard that can be sodded has $A \le 1400$. Again, Maria and Joe qualify.

Find the intersection. Maria's and Joe's yards are common to both sets, so Maria and Joe can have their yards both fenced and sodded.

60. Find "the yard can be fenced *and* the yard cannot be sodded."

A yard that can be fenced has $P \le 150$. Maria and Joe qualify.

A yard that cannot be sodded has $A > 1400$. Luigi and Than qualify.

Find the intersection. There are no yards common to both sets, so none of them qualify.

61. Find "the yard cannot be fenced *and* the yard can be sodded."

A yard that cannot be fenced has $P > 150$. Luigi and Than qualify.

A yard that can be sodded has $A \leq 1400$. Maria and Joe qualify.

Find the intersection. There are no yards common to both sets, so none of the qualify.

62. Find "the yard cannot be fenced *and* the yard cannot be sodded."

A yard that cannot be fenced has $P > 150$. Luigi and Than qualify.

A yard that cannot be sodded has $A > 1400$. Again, Luigi and Than qualify.

Find the intersection. Luigi's and Than's yards are common to both sets, so Luigi and Than qualify.

63. Find "the yard can be fenced *or* the yard can be sodded." From Exercise 59, Maria's and Joe's yards qualify for both conditions, so the union is Maria and Joe.

64. Find "the yard cannot be fenced *or* the yard can be sodded." From Exercise 61, Luigi's and Than's yards cannot be fenced, and Maria's and Joe's yards can be sodded. The union includes all of them.

Section 3.3

1. **(a)** $|x| = 5$ has two solutions, $x = 5$ or $x = -5$. The graph is Choice E.

(b) $|x| < 5$ is written $-5 < x < 5$. Notice that -5 and 5 are not included. The graph is Choice C, which uses parentheses.

(c) $|x| > 5$ is written $x < -5$ or $x > 5$. The graph is Choice D, which uses parentheses.

(d) $|x| \leq 5$ is written $-5 \leq x \leq 5$. This time -5 and 5 are included. The graph is Choice B, which uses brackets.

(e) $|x| \geq 5$ is written $x \leq -5$ or $x \geq 5$. The graph is Choice A, which uses brackets.

3. **(a)** $|ax + b| = k, k = 0$
This means the distance from $ax + b$ to 0 is 0, so $ax + b = 0$, which has one solution.

(b) $|ax + b| = k, k > 0$
This means the distance from $ax + b$ to 0 is a positive number, so $ax + b = k$ or $ax + b = -k$. There are two solutions.

(c) $|ax + b| = k, k < 0$
This means the distance from $ax + b$ to 0 is a negative number, which is impossible because distance is always positive. There are no solutions.

5. $|x| = 12$
$x = 12$ or $x = -12$
Solution set: $\{-12, 12\}$

7. $|4x| = 20$
$4x = 20$ or $4x = -20$
$x = 5$ or $x = -5$
Solution set: $\{-5, 5\}$

9. $|y - 3| = 9$
$y - 3 = 9$ or $y - 3 = -9$
$y = 12$ or $y = -6$
Solution set: $\{-6, 12\}$

11. $|2x + 1| = 7$
$2x + 1 = 7$ or $2x + 1 = -7$
$2x = 6$ $2x = -8$
$x = 3$ or $x = -4$
Solution set: $\{-4, 3\}$

13. $|4r - 5| = 17$
$4r - 5 = 17$ or $4r - 5 = -17$
$4r = 22$ $4r = -12$
$r = \dfrac{22}{4} = \dfrac{11}{2}$ or $r = -3$
Solution set: $\left\{-3, \dfrac{11}{2}\right\}$

15. $|2y + 5| = 14$
$2y + 5 = 14$ or $2y + 5 = -14$
$2y = 9$ $2y = -19$
$y = \dfrac{9}{2}$ or $y = -\dfrac{19}{2}$
Solution set: $\left\{-\dfrac{19}{2}, \dfrac{9}{2}\right\}$

17. $\left|\dfrac{1}{2}x + 3\right| = 2$

$\dfrac{1}{2}x + 3 = 2$ or $\dfrac{1}{2}x + 3 = -2$
$\dfrac{1}{2}x = -1$ $\dfrac{1}{2}x = -5$
$x = -2$ or $x = -10$
Solution set: $\{-10, -2\}$

19. $\left|1 - \dfrac{3}{4}k\right| = 7$

$1 - \dfrac{3}{4}k = 7 \quad \text{or} \quad 1 - \dfrac{3}{4}k = -7$

Multiply all sides by 4.

$4 - 3k = 28 \quad \text{or} \quad 4 - 3k = -28$

$-3k = 24 \qquad\qquad -3k = -32$

$k = -8 \quad \text{or} \qquad k = \dfrac{32}{3}$

Solution set: $\left\{-8, \dfrac{32}{3}\right\}$

21. When solving an absolute value equation or inequality of the form

(a) $|ax + b| = k$

(b) $|ax + b| < k$, or

(c) $|ax + b| > k$,

where k is a positive number, use

(a) *or* for the $=$ case,

(b) *and* for the $<$ case, and

(c) *or* for the $>$ case.

23. $|x| > 3$

$x > 3 \quad \text{or} \quad x < -3$

Solution set: $(-\infty, -3) \cup (3, \infty)$

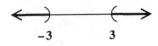

25. $|k| \ge 4$

$k \ge 4 \quad \text{or} \quad k \le -4$

Solution set: $(-\infty, -4] \cup [4, \infty)$

27. $|t + 2| > 10$

$t + 2 > 10 \quad \text{or} \quad t + 2 < -10$

$t > 8 \quad \text{or} \qquad t < -12$

Solution set: $(-\infty, -12) \cup (8, \infty)$

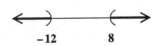

29. $|3 - x| > 5$

$3 - x > 5 \quad \text{or} \quad 3 - x < -5$

$-x > 2 \quad \text{or} \qquad -x < -8$

Multiply by -1,

and reverse the inequality signs.

$x < -2 \quad \text{or} \qquad x > 8$

Solution set: $(-\infty, -2) \cup (8, \infty)$

31. $|x| \le 3$

$-3 \le x \le 3$

Solution set: $[-3, 3]$

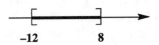

33. $|k| < 4$

$-4 < k < 4$

Solution set: $(-4, 4)$

35. $|t + 2| \le 10$

$-10 \le t + 2 \le 10$

$-12 \le t \le 8$

Solution set: $[-12, 8]$

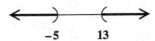

37. $|3 - x| \le 5$

$-5 \le 3 - x \le 5$

$-8 \le -x \le 2$

Multiply by -1, and reverse the inequality signs.

$8 \ge x \ge -2 \text{ or } -2 \le x \le 8$

Solution set: $[-2, 8]$

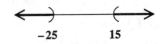

39. $|-4 + k| > 9$

$-4 + k > 9 \quad \text{or} \quad -4 + k < -9$

$k > 13 \quad \text{or} \qquad k < -5$

Solution set: $(-\infty, -5) \cup (13, \infty)$

41. $|r + 5| > 20$

$r + 5 > 20 \quad \text{or} \quad r + 5 < -20$

$r > 15 \quad \text{or} \qquad r < -25$

Solution set: $(-\infty, -25) \cup (15, \infty)$

43. $|7 + 2z| = 5$

$7 + 2z = 5$ or $\quad 7 + 2z = -5$

$\qquad 2z = -2 \qquad\qquad 2z = -12$

$\qquad z = -1$ or $\qquad\qquad z = -6$.

Solution set: $\{-6, -1\}$

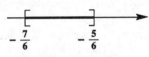

45. $|3r - 1| \le 11$

$-11 \le 3r - 1 \le 11$

$-10 \le 3r \le 12$

$-\dfrac{10}{3} \le r \le 4$

Solution set: $\left[-\dfrac{10}{3}, 4\right]$

47. $|-6x - 6| \le 1$

$-1 \le -6x - 6 \le 1$

$5 \le -6x \le 7$

Divide by -6. and reverse
the inequality signs.

$-\dfrac{5}{6} \ge x \ge -\dfrac{7}{6}$ or $-\dfrac{7}{6} \le x \le -\dfrac{5}{6}$

Solution set: $\left[-\dfrac{7}{6}, -\dfrac{5}{6}\right]$

49. $|3x - 1| \ge 8$

$3x - 1 \ge 8$ or $\quad 3x - 1 \le -8$

$\quad 3x \ge 9 \qquad\qquad 3x \le -7$

$\quad x \ge 3$ or $\qquad x \le -\dfrac{7}{3}$

Solution set: $\left(-\infty, -\dfrac{7}{3}\right] \cup [3, \infty)$

51. The distance between x and 4 equals 9 can be
written

$$|x - 4| = 9 \quad \text{or} \quad |4 - x| = 9.$$

53. $|x + 4| + 1 = 2$

$\quad |x + 4| = 1$

$x + 4 = 1$ or $\quad x + 4 = -1$

$\quad x = -3$ or $\qquad x = -5$

Solution set: $\{-5, -3\}$

55. $|2x + 1| + 3 > 8$

$\quad |2x + 1| > 5$

$2x + 1 > 5$ or $\quad 2x + 1 < -5$

$\quad 2x > 4 \qquad\qquad 2x < -6$

$\quad x > 2$ or $\qquad x < -3$

Solution set: $(-\infty, -3) \cup (2, \infty)$

57. $|x + 5| - 6 \le -1$

$\quad |x + 5| \le 5$

$\quad -5 \le x + 5 \le 5$

$\quad -10 \le x \le 0$

Solution set: $[-10, 0]$

59. $|3x + 1| = |2x + 4|$

$3x + 1 = 2x + 4$ or $\quad 3x + 1 = -(2x + 4)$

$\qquad\qquad\qquad\qquad 3x + 1 = -2x - 4$

$\qquad\qquad\qquad\qquad 5x = -5$

$\quad x = 3 \qquad$ or $\qquad x = -1$

Solution set: $\{-1, 3\}$

61. $\left| m - \dfrac{1}{2} \right| = \left| \dfrac{1}{2}m - 2 \right|$

$m - \dfrac{1}{2} = \dfrac{1}{2}m - 2$ or $m - \dfrac{1}{2} = -\left(\dfrac{1}{2}m - 2\right)$

Multiply by 2. $\qquad m - \dfrac{1}{2} = -\dfrac{1}{2}m + 2$

$2m - 1 = m - 4 \qquad 2m - 1 = -m + 4$

$\qquad\qquad\qquad\qquad\qquad 3m = 5$

$\quad m = -3 \qquad$ or $\qquad m = \dfrac{5}{3}$

Solution set: $\left\{-3, \dfrac{5}{3}\right\}$

63. $|6x| = |9x + 1|$

$6x = 9x + 1$ or $\quad 6x = -(9x + 1)$

$-3x = 1 \qquad\qquad 6x = -9x - 1$

$\qquad\qquad\qquad\qquad 15x = -1$

$\quad x = -\dfrac{1}{3} \qquad$ or $\qquad x = -\dfrac{1}{15}$

Solution set: $\left\{-\dfrac{1}{3}, -\dfrac{1}{15}\right\}$

65. $|2p - 6| = |2p + 11|$

$2p - 6 = 2p + 11$ or $\quad 2p - 6 = -(2p + 11)$

$\quad -6 = 11 \quad$ *False* $\qquad 2p - 6 = -2p - 11$

$\qquad\qquad\qquad\qquad\qquad 4p = -5$

$\quad$ *No solution* $\quad$ or $\qquad p = -\dfrac{5}{4}$

Solution set: $\left\{-\dfrac{5}{4}\right\}$

67. $|12t - 3| = -8$

Since the absolute value of an expression can never be negative, there are no solutions for this equation.

Solution set: $\emptyset$

69. $|4x + 1| = 0$

The expression $4x + 1$ will equal 0 *only* for the solution of the equation

$$4x + 1 = 0.$$
$$4x = -1$$
$$x = \frac{-1}{4} \text{ or } -\frac{1}{4}$$

Solution set: $\left\{-\frac{1}{4}\right\}$

71. $|2q - 1| < -6$

There are no numbers whose absolute value is negative, so this inequality has no solution.

Solution set: $\emptyset$

73. $|x + 5| > -9$

Since the absolute value of an expression is always nonnegative (positive or zero), the inequality is true for any real number x.

Solution set: $(-\infty, \infty)$

75. $|7x + 3| \leq 0$

The absolute value of an expression is always nonnegative (positive or zero), so this inequality is true only when

$$7x + 3 = 0$$
$$7x = -3$$
$$x = -\frac{3}{7}.$$

Solution set: $\left\{-\frac{3}{7}\right\}$

77. $|5x - 2| \geq 0$

The absolute value of an expression is always nonnegative, so the inequality is true for any real number x.

Solution set: $(-\infty, \infty)$

79. $|10z + 7| > 0$

Since an absolute value expression is always nonnegative and $|10z + 7| \neq 0$, there is only one possible value of z that makes this statement false. The equation $10z + 7 = 0$ will give that value.

$$10z + 7 = 0$$
$$10z = -7$$
$$z = -\frac{7}{10}$$

Solution set: $\left(-\infty, -\frac{7}{10}\right) \cup \left(-\frac{7}{10}, \infty\right)$

81. Add the given heights with a calculator to get 4602. There are 10 numbers, so divide the sum by 10.

$$\frac{4602}{10} = 460.2$$

The average height is 460.2 ft.

82. $|x - k| < 50$

Substitute 460.2 for k and solve the inequality.

$$|x - 460.2| < 50$$
$$-50 < x - 460.2 < 50$$
$$410.2 < x < 510.2$$

The buildings with heights between 410.2 ft and 510.2 ft are the Federal Office Building, City Hall, Kansas City Power and Light, and the Hyatt Regency.

83. $|x - k| < 75$

Substitute 460.2 for k and solve the inequality.

$$|x - 460.2| < 75$$
$$-75 < x - 460.2 < 75$$
$$385.2 < x < 535.2$$

The buildings with heights between 385.2 ft and 535.2 ft are Southwest Bell Telephone, City Center Square, Commerce Tower, the Federal Office Building, City Hall, Kansas City Power and Light, and the Hyatt Regency.

84. **(a)** This would be the opposite of the inequality in Exercise 83, that is,

$$|x - 460.2| \geq 75.$$

(b) $|x - 460.2| \geq 75$

$$x - 460.2 \geq 75 \qquad \text{or} \qquad x - 460.2 \leq -75$$
$$x \geq 535.2 \quad \text{or} \qquad\qquad x \leq 385.2$$

(c) The buildings that are not within 75 ft of the average have height less than or equal to 385.2 or greater than or equal to 535.2. This would include Pershing Road Associates, AT&T Town Pavilion, and One Kansas City Place.

(d) The answer makes sense because it includes all the buildings *not* listed earlier which had heights within 75 ft of the average.

85. **(a)** For the weight x of the box to be within .5 ounce of 16 ounces, we must have

$$|x - 16| \leq .5.$$

(b) $\quad -.5 \leq x - 16 \leq .5$

$\qquad 15.5 \leq x \leq 16.5$

(c) The process is out of control if $|x - 16| \geq .5$, which is the same as $x \leq 15.5$ or $x \geq 16.5$.

Summary: Exercises on Solving Linear and Absolute Value Equations and Inequalities

1. $\quad 4z + 1 = 49$

$\qquad 4z = 48$

$\qquad z = 12$

Solution set: $\{12\}$

3. $\quad 6q - 9 = 12 + 3q$

$\qquad 3q = 21$

$\qquad q = 7$

Solution set: $\{7\}$

5. $\quad |a + 3| = -4$

Since the absolute value of an expression is always nonnegative, there is no number that makes this statement true. Therefore, the solution set is $\emptyset$.

7. $\quad 8r + 2 \geq 5r$

$\qquad 3r \geq -2$

$\qquad r \geq -\dfrac{2}{3}$

Solution set: $\left[-\dfrac{2}{3}, \infty \right)$

9. $\quad 2q - 1 = -7$

$\qquad 2q = -6$

$\qquad q = -3$

Solution set: $\{-3\}$

11. $\quad 6z - 5 \leq 3z + 10$

$\qquad 3z \leq 15$

$\qquad z \leq 5$

Solution set: $(-\infty, 5]$

13. $\quad 9y - 3(y + 1) = 8y - 7$

$\qquad 9y - 3y - 3 = 8y - 7$

$\qquad 6y - 3 = 8y - 7$

$\qquad 4 = 2y$

$\qquad 2 = y$

Solution set: $\{2\}$

15. $\quad 9y - 5 \geq 9y + 3$

$\qquad -5 \geq 3 \quad$ *False*

This is a false statement, so the inequality is a contradiction.

Solution set: $\emptyset$

17. $\quad |q| < 5.5$

$\qquad -5.5 < q < 5.5$

Solution set: $(-5.5, 5.5)$

19. $\quad \dfrac{2}{3}y + 8 = \dfrac{1}{4}y$

$\qquad 8y + 96 = 3y \qquad$ *Multiply by 12.*

$\qquad 5y = -96$

$\qquad y = -\dfrac{96}{5}$

Solution set: $\left\{ -\dfrac{96}{5} \right\}$

21. $\quad \dfrac{1}{4}p < -6$

$\qquad 4\left(\dfrac{1}{4}p \right) < 4(-6)$

$\qquad p < -24$

Solution set: $(-\infty, -24)$

23. $\quad \dfrac{3}{5}q - \dfrac{1}{10} = 2$

$\qquad 6q - 1 = 20 \qquad$ *Multiply by 10.*

$\qquad 6q = 21$

$\qquad q = \dfrac{21}{6} \text{ or } \dfrac{7}{2}$

Solution set: $\left\{ \dfrac{7}{2} \right\}$

25. $\quad r + 9 + 7r = 4(3 + 2r) - 3$

$\qquad 8r + 9 = 12 + 8r - 3$

$\qquad 8r + 9 = 8r + 9$

$\qquad 0 = 0 \quad$ *True*

The last statement is true for any real number r.

Solution set: $(-\infty, \infty)$

27. $\quad |2p - 3| > 11$

$\qquad 2p - 3 > 11 \quad$ or $\quad 2p - 3 < -11$

$\qquad 2p > 14 \qquad\qquad 2p < -8$

$\qquad p > 7 \quad$ or $\qquad p < -4$

Solution set: $(-\infty, -4) \cup (7, \infty)$

29. $\quad |5a + 1| \leq 0$

The expression $|5a + 1|$ is never less than 0 since an absolute value expression must be nonnegative. However, $|5a + 1| = 0$ if $5a + 1 = 0$.

$$5a = -1$$

$$a = \dfrac{-1}{5} = -\dfrac{1}{5}$$

Solution set: $\left\{ -\dfrac{1}{5} \right\}$

31. $-2 \le 3x - 1 \le 8$

$-1 \le 3x \le 9$

$-\dfrac{1}{3} \le x \le 3$

Solution set: $\left[-\dfrac{1}{3}, 3\right]$

33. $|7z - 1| = |5z + 3|$

$7z - 1 = 5z + 3$ or $7z - 1 = -(5z + 3)$

$2z = 4 \qquad\qquad 7z - 1 = -5z - 3$

$12z = -2$

$z = 2 \qquad$ or $\qquad z = \dfrac{-2}{12} = -\dfrac{1}{6}$

Solution set: $\left\{-\dfrac{1}{6}, 2\right\}$

35. $|1 - 3x| \ge 4$

$1 - 3x \ge 4 \qquad$ or $\qquad 1 - 3x \le -4$

$-3x \ge 3 \qquad\qquad\qquad -3x \le -5$

$x \le -1 \qquad$ or $\qquad x \ge \dfrac{5}{3}$

Solution set: $(-\infty, -1] \cup \left[\dfrac{5}{3}, \infty\right)$

37. $-(m + 4) + 2 = 3m + 8$

$-m - 4 + 2 = 3m + 8$

$-m - 2 = 3m + 8$

$-10 = 4m$

$m = \dfrac{-10}{4} = -\dfrac{5}{2}$

Solution set: $\left\{-\dfrac{5}{2}\right\}$

39. $-6 \le \dfrac{3}{2} - x \le 6$

$-12 \le 3 - 2x \le 12$

$-15 \le -2x \le 9$

$\dfrac{15}{2} \ge x \ge -\dfrac{9}{2}$ or $-\dfrac{9}{2} \le x \le \dfrac{15}{2}$

Solution set: $\left[-\dfrac{9}{2}, \dfrac{15}{2}\right]$

41. $|y - 1| \ge -6$

The absolute value of an expression is always nonnegative, so the inequality is true for any real number x.

Solution set: $(-\infty, \infty)$

43. $8q - (1 - q) = 3(1 + 3q) - 4$

$8q - 1 + q = 3 + 9q - 4$

$9q - 1 = 9q - 1$ *True*

This is an identity.

Solution set: $(-\infty, \infty)$

45. $|r - 5| = |r + 9|$

$r - 5 = r + 9 \qquad$ or $\qquad r - 5 = -(r + 9)$

$-5 = 9$ *False* $\qquad\qquad r - 5 = -r - 9$

$2r = -4$

No solution $\qquad$ or $\qquad r = -2$

Solution set: $\{-2\}$

47. $2x + 1 > 5 \quad$ or $\quad 3x + 4 < 1$

$2x > 4 \qquad\qquad 3x < -3$

$x > 2 \quad$ or $\quad x < -1$

Solution set: $(-\infty, -1) \cup (2, \infty)$

Chapter 3 Review Exercises

1. $-\dfrac{2}{3}k < 6$

$-2k < 18 \qquad\qquad$ *Multiply by 3.*

Divide by -2; reverse the inequality sign.

$k > -9$

Solution set: $(-9, \infty)$

2. $-5x - 4 \ge 11$

$-5x \ge 15$

Divide by -5; reverse the inequality sign.

$x \le -3$

Solution set: $(-\infty, -3]$

3. $\dfrac{6a + 3}{-4} < -3$

Multiply by -4; reverse the inequality sign.

$6a + 3 > 12$

$6a > 9$

$a > \dfrac{9}{6}$ or $\dfrac{3}{2}$

Solution set: $\left(\dfrac{3}{2}, \infty\right)$

4. $5 - (6 - 4k) \ge 2k - 7$

$5 - 6 + 4k \ge 2k - 7$

$4k - 1 \ge 2k - 7$

$2k \ge -6$

$k \ge -3$

Solution set: $[-3, \infty)$

5. $8 \le 3y - 1 < 14$

$9 \le 3y < 15$

$3 \le y < 5$

Solution set: $[3, 5)$

6. $\dfrac{5}{3}(m-2) + \dfrac{2}{5}(m+1) > 1$

$25(m-2) + 6(m+1) > 15$

Multiply by 15.

$25m - 50 + 6m + 6 > 15$

$31m - 44 > 15$

$31m > 59$

$m > \dfrac{59}{31}$

Solution set: $\left(\dfrac{59}{31}, \infty\right)$

7.   Let $x =$ the student's grade on the fifth test. The average of the five test grades must be at least 70. The inequality is

$$\dfrac{75 + 79 + 64 + 71 + x}{5} \ge 70.$$

$75 + 79 + 64 + 71 + x \ge 350$

$289 + x \ge 350$

$x \ge 61$

The student will pass algebra if at least 61% on the fifth test is achieved.

8.   Let $x =$ the length of the playground.

Since the perimeter, twice the length x plus twice the width 22, must be less than or equal to 120 m, solve the inequality

$$2x + 2(22) \le 120$$

$2x + 44 \le 120$

$2x \le 76$

$x \le 38.$

The length of the playground must be 38 m or less.

9.   Let $x =$ the number of tickets that can be purchased.

We can add the $50 discount to the $1000 that is available to purchase tickets, and use $1050 as the available amount. The number of $10.59 tickets must have a value of no more than $1050.

$$10.59x \le 1050$$

$x \le \dfrac{1050}{10.59} \approx 99.2$

The group can purchase 99 or fewer tickets.

10.   The result, $-8 < -13$, is a false statement. There are no real numbers that make this inequality true. The solution set is $\emptyset$.

For Exercises 11–14, let $A = \{a, b, c, d\}$, $B = \{a, c, e, f\}$, and $C = \{a, e, f, g\}$.

11. $A \cap B = \{a, b, c, d\} \cap \{a, c, e, f\}$

$= \{a, c\}$

12. $A \cap C = \{a, b, c, d\} \cap \{a, e, f, g\}$

$= \{a\}$

13. $B \cup C = \{a, c, e, f\} \cup \{a, e, f, g\}$

$= \{a, c, e, f, g\}$

14. $A \cup C = \{a, b, c, d\} \cup \{a, e, f, g\}$

$= \{a, b, c, d, e, f, g\}$

15. $x \le 4$ and $x < 3$

The graph of the solution set is all numbers both less than or equal to 4 and less than 3. This is the intersection. The elements common to both sets are the numbers less than 3. The solution set is $(-\infty, 3)$.

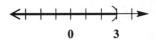

16. $x + 4 > 12$ and $x - 2 < 1$

$x > 8$ and $x < 3$

The graph of the solution set is all numbers both greater than 8 and less than 3. This is the intersection. Since there are no numbers that are both greater than 8 and less than 3, the solution set is $\emptyset$.

17. $x > 5$ or $x \le -1$

The graph of the solution set is all numbers either greater than 5 or less than or equal to -1. This is the union. The solution set is $(-\infty, -1] \cup (5, \infty)$.

18. $x - 4 > 6$ or $x + 3 \le 18$

$x > 10$ or $x \le 15$

The graph of the solution set is all numbers either greater than 10 or less than or equal to 15, or both. This is the union. The solution set is the set of all real numbers, or $(-\infty, \infty)$.

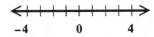

19. $(-3, \infty) \cap (-\infty, 4)$

$(-3, \infty)$ includes all real numbers greater than -3.

$(-\infty, 4)$ includes all real numbers less than 4. Find the intersection. The numbers common to both sets are greater than -3 and less than 4.

$$-3 < x < 4$$

Solution set: $(-3, 4)$

20. $(-\infty, 6) \cap (-\infty, 2)$

$(-\infty, 6)$ includes all real numbers less than 6.

$(-\infty, 2)$ includes all real numbers less than 2. Find the intersection. The numbers common to both sets are less than 2.

Solution set: $(-\infty, 2)$

21. $(4, \infty) \cup (9, \infty)$

$(4, \infty)$ includes all real numbers greater than 4.
$(9, \infty)$ includes all real numbers greater than 9.
Find the union. The numbers in the first set, the second set, or in both sets are all the real numbers that are greater than 4.

Solution set: $(4, \infty)$

22. $(1, 2) \cup (1, \infty)$

$(1, 2)$ includes the real numbers between 1 and 2, not including 1 and 2.
$(1, \infty)$ includes all real numbers greater than 1.
Find the union. The numbers in the first set, the second set, or in both sets are all real numbers greater than 1.

Solution set: $(1, \infty)$

23. **(a)** The median earnings for men are less than $900 includes managerial and professional specialty, waiters, and bus drivers.
The median earnings for women are greater than $500 includes managerial and professional specialty and mathematical and computer scientists.
Find the intersection. The only occupation in both groups is managerial and professional specialty.

(b) The median earnings for men are greater than $900 includes mathematical and computer sciences.
The median earnings for women are greater than $600 includes mathematical and computer sciences and managerial and professional specialty, which is also the union.

24. $|x| = 7$

$x = 7$ or $x = -7$

Solution set: $\{-7, 7\}$

25. $|3k - 7| = 8$

$3k - 7 = 8$ or $3k - 7 = -8$
    $3k = 15$ $3k = -1$

     $k = 5$ or $k = -\dfrac{1}{3}$

Solution set: $\left\{-\dfrac{1}{3}, 5\right\}$

26. $|z - 4| = -12$

Since the absolute value of an expression can never be negative, there are no solutions for this equation.

Solution set: $\emptyset$

27. $|4a + 2| - 7 = -3$

$|4a + 2| = 4$

$4a + 2 = 4$ or $4a + 2 = -4$
   $4a = 2$ $4a = -6$

     $a = \dfrac{2}{4}$ $a = -\dfrac{6}{4}$

     $a = \dfrac{1}{2}$ or $a = -\dfrac{3}{2}$

Solution set: $\left\{-\dfrac{3}{2}, \dfrac{1}{2}\right\}$

28. $|3p + 1| = |p + 2|$

$3p + 1 = p + 2$ or $3p + 1 = -(p + 2)$
    $2p = 1$ $3p + 1 = -p - 2$
                        $4p = -3$

     $p = \dfrac{1}{2}$ or $p = -\dfrac{3}{4}$

Solution set: $\left\{-\dfrac{3}{4}, \dfrac{1}{2}\right\}$

29. $|2m - 1| = |2m + 3|$

$2m - 1 = 2m + 3$ or $2m - 1 = -(2m + 3)$
    $0 = 4$ *False* $2m - 1 = -2m - 3$
                        $4m = -2$

  No solution   or $m = -\dfrac{2}{4} = -\dfrac{1}{2}$

Solution set: $\left\{-\dfrac{1}{2}\right\}$

30. $|-y + 6| \leq 7$

$-7 \leq -y + 6 \leq 7$
$-13 \leq -y \leq 1$

Multiply by -1; reverse the inequality signs.

$13 \geq y \geq -1$ or $-1 \leq y \leq 13$

Solution set: $[-1, 13]$

31. $|2p + 5| \leq 1$

$-1 \leq 2p + 5 \leq 1$
$-6 \leq 2p \leq -4$
$-3 \leq p \leq -2$

Solution set: $[-3, -2]$

32. $|x + 1| \geq -3$

Since the absolute value of an expression is always nonnegative (positive or zero), the inequality is true for any real number x.

Solution set: $(-\infty, \infty)$

33. $|5r - 1| > 9$

$5r - 1 > 9$ or $5r - 1 < -9$

$5r > 10$ $\qquad$ $5r < -8$

$r > 2$ or $\qquad r < -\dfrac{8}{5}$

Solution set: $\left(-\infty, -\dfrac{8}{5}\right) \cup (2, \infty)$

34. $|11x - 3| \leq -2$

There are no numbers whose absolute value is negative, so this inequality has no solution.

Solution set: $\emptyset$

35. $|11x - 3| \leq 0$

The absolute value of an expression is always nonnegative (positive or zero), so this inequality is true only when

$11x - 3 = 0$

$11x = 3$

$x = \dfrac{3}{11}.$

Solution set: $\left\{\dfrac{3}{11}\right\}$

36. The distance between x and 14 is greater than 12 can be written

$|x - 14| > 12$ or $|14 - x| > 12.$

37. $5 - (6 - 4k) > 2k - 5$

$5 - 6 + 4k > 2k - 5$

$-1 + 4k > 2k - 5$

$2k > -4$

$k > -2$

Solution set: $(-2, \infty)$

38. $x < 3$ and $x \geq -2$

The real numbers that are common to both sets are the numbers greater than or equal to -2 and less than 3.

$-2 \leq x < 3$

Solution set: $[-2, 3)$

39. $|3k + 6| \geq 0$

The absolute value of an expression is always nonnegative, so the inequality is true for any real number k.

Solution set: $(-\infty, \infty)$

40. $|2k - 7| + 4 = 11$

$|2k - 7| = 7$

$2k - 7 = 7$ or $2k - 7 = -7$

$2k = 14$ $\qquad$ $2k = 0$

$k = 7$ or $\qquad k = 0$

Solution set: $\{0, 7\}$

41. Let $x =$ the amount earned during the fifth month.

Since the employee must average at least \$1000 per month over the five-month period, solve the inequality

$\dfrac{900 + 1200 + 1040 + 760 + x}{5} \geq 1000$

$\dfrac{3900 + x}{5} \geq 1000$

$3900 + x \geq 5000$

$x \geq 1100.$

Any amount greater than or equal to \$1100 will qualify the employee.

42. $|p| < 14$

$-14 < p < 14$

Solution set: $(-14, 14)$

43. $\dfrac{3}{4}(a - 2) - \dfrac{1}{3}(5 - 2a) < -2$

$9(a - 2) - 4(5 - 2a) < -24$

Multiply by 12.

$9a - 18 - 20 + 8a < -24$

$17a - 38 < -24$

$17a < 14$

$a < \dfrac{14}{17}$

Solution set: $\left(-\infty, \dfrac{14}{17}\right)$

44. $-4 < 3 - 2k < 9$

$-7 < -2k < 6$

Divide by -2; reverse the inequality signs.

$\dfrac{7}{2} > k > -3$ or $-3 < k < \dfrac{7}{2}$

Solution set: $\left(-3, \dfrac{7}{2}\right)$

45. $-.3x + 2.1(x - 4) \leq -6.6$

$-3x + 21(x - 4) \leq -66$

Multiply by 10.

$-3x + 21x - 84 \leq -66$

$18x - 84 \leq -66$

$18x \leq 18$

$x \leq 1$

Solution set: $(-\infty, 1]$

46. $|5r - 1| > 14$

$5r - 1 > 14$ or $5r - 1 < -14$

$5r > 15$ $5r < -13$

$r > 3$ or $r < -\dfrac{13}{5}$

Solution set: $\left(-\infty, -\dfrac{13}{5}\right) \cup (3, \infty)$

47. $x \geq -2$ or $x < 4$

The solution set includes all numbers either greater than or equal to -2 or all numbers less than 4. This is the union and is the set of all real numbers. The solution set is $(-\infty, \infty)$.

48. $|m - 1| = |2m + 3|$

$m - 1 = 2m + 3$ or $m - 1 = -(2m + 3)$

$m - 1 = -2m - 3$

$3m = -2$

$-4 = m$ or $m = -\dfrac{2}{3}$

Solution set: $\left\{-4, -\dfrac{2}{3}\right\}$

49. $|m + 3| \leq 1$

$-1 \leq m + 3 \leq 1$

$-4 \leq m \leq -2$

Solution set: $[-4, -2]$

50. $x > 6$ and $x < 8$

The graph of the solution set is all numbers both greater than 6 *and* less than 8. This is the intersection. The elements common to both sets are the numbers between 6 and 8, not including the endpoints. The solution set is $(6, 8)$.

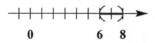

51. $-5x + 1 \geq 11$ or $3x + 5 \geq 26$

$-5x \geq 10$ $3x \geq 21$

$x \leq -2$ or $x \geq 7$

The graph of the solution set is all numbers either less than or equal to -2 *or* greater than or equal to 7. This is the union. The solution set is $(-\infty, -2] \cup [7, \infty)$.

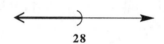

52. (a) The endpoints on the number line are $-\dfrac{11}{3}$ and 1. The solution set of $|3x + 4| \geq 7$ would be

$\left(-\infty, -\dfrac{11}{3}\right] \cup [1, \infty)$.

(b) The solution set of $|3x + 4| \leq 7$ would be

$\left[-\dfrac{11}{3}, 1\right]$.

Chapter 3 Test

1. When multiplying or dividing both sides of an inequality by a negative number, remember to reverse the direction of the inequality symbol.

2. $4 - 6(x + 3) \leq -2 - 3(x + 6) + 3x$

$4 - 6x - 18 \leq -2 - 3x - 18 + 3x$

$-6x - 14 \leq -20$

$-6x \leq -6$

Divide by -6, and reverse the inequality sign.

$x \geq 1$

Solution set: $[1, \infty)$

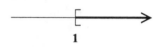

3. $-\dfrac{4}{7}x > -16$

$-4x > -112$ *Multiply by 7.*

Divide by -4, and reverse the inequality sign.

$x < 28$

Solution set: $(-\infty, 28)$

4. $-6 \leq \dfrac{4}{3}x - 2 \leq 2$

$-18 \leq 4x - 6 \leq 6$ *Multiply by 3.*

$-12 \leq 4x \leq 12$ *Add 6.*

$-3 \leq x \leq 3$ *Divide by 4.*

Solution set: $[-3, 3]$

5. To find the inequalities equivalent to $x < -3$, solve each inequality for x.

(a) $-3x < 9$

$x > -3$ *Not equivalent*

(b) $-3x > -9$

$x < 3$ *Not equivalent*

(c) $-3x > 9$

$x < -3$ *Equivalent*

(d) $-3x < -9$

$\qquad x > 3 \qquad$ *Not equivalent*

Of these four choices, only (c) is equivalent to $x < -3$.

6.   From the graph, the number of departures to Europe increased the most between 1992 and 1993 (about 1 million). The graph is flat between 1991 and 1992, so that is when the departures stayed the same.

7.   Let $x =$ the student's grade on the fourth test.

The average of the four exams equals

$$\frac{83 + 76 + 79 + x}{4} = \frac{238 + x}{4}.$$

To get a B the student must have an average of at least 80%. The inequality is

$$\frac{238 + x}{4} \geq 80$$
$$238 + x \geq 320$$
$$x \geq 82.$$

The minimum grade is 82%.

8.   Given $C = 50x + 5000$ and $R = 60x$.

To find the values of x for which R is at least equal to C, solve the inequality

$$R \geq C$$
$$60x \geq 50x + 5000$$
$$10x \geq 5000$$
$$x \geq 500.$$

The values of x for which R is at least equal to C are in the interval $[500, \infty)$.

9.   **(a)** $A \cap B = \{1, 2, 5, 7\} \cap \{1, 5, 9, 12\}$
$\qquad\qquad = \{1, 5\}$

(b) $A \cup B = \{1, 2, 5, 7\} \cup \{1, 5, 9, 12\}$
$\qquad\qquad = \{1, 2, 5, 7, 9, 12\}$

10. **(a)** $3k \geq 6 \quad$ and $\quad k - 4 < 5$
$\qquad\quad k \geq 2 \quad$ and $\qquad k < 9$
The solution set is all numbers both greater than or equal to 2 *and* less than 9. This is the intersection. The numbers common to both sets are between 2 and 9, including 2 but not 9. The solution set is $[2, 9)$.

(b) $-4x \leq -24 \quad$ or $\quad 4x - 2 < 10$
$\qquad\qquad\qquad\qquad\qquad\quad 4x < 12$
$\qquad x \geq 6 \qquad$ or $\qquad x < 3$
The solution set is all numbers less than 3 or greater than or equal to 6. This is the union. The solution set is $(-\infty, 3) \cup [6, \infty)$.

11. $|4x - 3| = 7$

$4x - 3 = 7 \qquad$ or $\qquad 4x - 3 = -7$
$\quad 4x = 10 \qquad\qquad\qquad\qquad 4x = -4$
$\quad\ x = \dfrac{10}{4} = \dfrac{5}{2} \quad$ or $\qquad\qquad x = -1$

Solution set: $\left\{-1, \dfrac{5}{2}\right\}$

12. $|4x - 3| > 7$
$4x - 3 > 7 \qquad$ or $\qquad 4x - 3 < -7$
$\quad 4x > 10 \qquad\qquad\qquad\qquad 4x < -4$
$\quad\ x > \dfrac{10}{4}$
$\quad\ x > \dfrac{5}{2} \qquad$ or $\qquad\qquad x < -1$

Solution set: $(-\infty, -1) \cup \left(\dfrac{5}{2}, \infty\right)$

Note that the solution sets for Exercises 12–13 could be obtained from the solution set in Exercise 11.

13. $|4x - 3| < 7$
$-7 < 4x - 3 < 7$
$-4 < 4x < 10$
$-1 < x < \dfrac{10}{4}$

Solution set: $\left(-1, \dfrac{5}{2}\right)$

14. $|3 - 5x| = |2x + 8|$
$3 - 5x = 2x + 8 \quad$ or $\quad 3 - 5x = -(2x + 8)$
$\quad -7x = 5 \qquad\qquad\qquad 3 - 5x = -2x - 8$
$\qquad\qquad\qquad\qquad\qquad\qquad -3x = -11$
$\quad\ x = -\dfrac{5}{7} \qquad$ or $\qquad\quad x = \dfrac{11}{3}$

Solution set: $\left\{-\dfrac{5}{7}, \dfrac{11}{3}\right\}$

15. $|-3x + 4| - 4 < -1$
$\qquad |-3x + 4| < 3$
$-3 < -3x + 4 < 3$
$-7 < -3x < -1$
$\dfrac{7}{3} > x > \dfrac{1}{3} \quad$ or $\quad \dfrac{1}{3} < x < \dfrac{7}{3}$

Solution set: $\left(\dfrac{1}{3}, \dfrac{7}{3}\right)$

16. **(a)** $|5x + 3| < k$
If $k < 0$, then $|5x + 3|$ would be less than a negative number. Since the absolute value of an expression is always nonnegative (positive or zero), the solution set is $\emptyset$.

(b) $|5x + 3| > k$

If $k < 0$, then $|5x + 3|$ would be greater than a negative number. Since the absolute value of an expression is always nonnegative (positive or zero), the solution set is the set of all real numbers, $(-\infty, \infty)$.

(c) $|5x + 3| = k$

If $k < 0$, then $|5x + 3|$ would be equal to a negative number. Since the absolute value of an expression is always nonnegative (positive or zero), the solution set is $\emptyset$.

Cumulative Review Exercises Chapters 1–3

Exercises 1–6 refer to set A.

Let $A = \left\{ -8, -\dfrac{2}{3}, -\sqrt{6}, 0, \dfrac{4}{5}, 9, \sqrt{36} \right\}$.

Simplify $\sqrt{36} = 6$.

1. The elements 9 and 6 are natural numbers.

2. The elements 0, 9, and 6 are whole numbers.

3. The elements -8, 0, 9, and 6 are integers.

4. The elements -8, $-\dfrac{2}{3}$, 0, $\dfrac{4}{5}$, 9, and 6 are rational numbers.

5. The element $-\sqrt{6}$ is an irrational number.

6. All the elements in set A are real numbers.

7. $-\dfrac{4}{3} - \left(-\dfrac{2}{7} \right) = -\dfrac{4}{3} + \dfrac{2}{7}$

    $= -\dfrac{28}{21} + \dfrac{6}{21}$

    $= -\dfrac{22}{21}$

8. $|-4| - |2| + |-6| = 4 - 2 + 6$

    $= 2 + 6$

    $= 8$

9. $(-3)^5 = (-3)(-3)(-3)(-3)(-3) = -243$

10. $\left(\dfrac{6}{7} \right)^3 = \dfrac{6}{7} \cdot \dfrac{6}{7} \cdot \dfrac{6}{7} = \dfrac{216}{343}$

11. $-\sqrt{36} = -(6) = -6$

    $\sqrt{-36}$ is not a real number.

12. $\dfrac{4 - 4}{4 + 4} = \dfrac{0}{8} = 0$

    $\dfrac{4 + 4}{4 - 4} = \dfrac{8}{0}$, which is undefined.

For Exercises 13–16, let $a = 2$, $b = -3$, and $c = 4$.

13. $-3a + 2b - c = -3(2) + 2(-3) - 4$

    $= -6 - 6 - 4$

    $= -16$

14. $-2b^2 - 4c = -2(-3)^2 - 4(4)$

    $= -2(9) - 4(4)$

    $= -18 - 16$

    $= -34$

15. $-8(a^2 + b^3) = -8[2^2 + (-3)^3]$

    $= -8[4 + (-27)]$

    $= -8(-23)$

    $= 184$

16. $\dfrac{3a^3 - b}{4 + 3c} = \dfrac{3(2)^3 - (-3)}{4 + 3(4)}$

    $= \dfrac{3(8) - (-3)}{4 + 3(4)}$

    $= \dfrac{24 + 3}{4 + 12}$

    $= \dfrac{27}{16}$

17. $-7r + 5 - 13r + 12$

    $= -7r - 13r + 5 + 12$

    $= (-7 - 13)r + (5 + 12)$

    $= -20r + 17$

18. $-(3k + 8) - 2(4k - 7) + 3(8k + 12)$

    $= -3k - 8 - 8k + 14 + 24k + 36$

    $= -3k - 8k + 24k - 8 + 14 + 36$

    $= 13k + 42$

19. $(a + b) + 4 = 4 + (a + b)$

    The order of the terms $(a + b)$ and 4 have been reversed. This is an illustration of the commutative property.

20. $4x + 12x = (4 + 12)x$

    The common variable, x, has been removed from each term. This is an illustration of the distributive property.

21. $-9 + 9 = 0$

    The sum of a number and its opposite is equal to 0. This is an illustration of the inverse property.

22. The product of a number and its reciprocal must equal 1. Given the number $-\dfrac{2}{3}$, its reciprocal is $-\dfrac{3}{2}$, since

    $$-\dfrac{2}{3}\left(-\dfrac{3}{2} \right) = 1.$$

23. $-4x + 7(2x + 3) = 7x + 36$

    $-4x + 14x + 21 = 7x + 36$

    $10x + 21 = 7x + 36$

    $3x = 15$

    $x = 5$

    Solution set: $\{5\}$

24. $-\dfrac{3}{5}x + \dfrac{2}{3}x = 2$

$3(-3x) + 5(2x) = 15(2)$ *Multiply by 15.*

$-9x + 10x = 30$

$x = 30$

Solution set: $\{30\}$

25. $.06x + .03(100 + x) = 4.35$

$6x + 3(100 + x) = 435$ *Multiply by 100.*

$6x + 300 + 3x = 435$

$9x = 135$

$x = 15$

Solution set: $\{15\}$

26. Solve $P = a + b + c$ for b.

$P - a - c = b$ or

$b = P - a - c$

27. $3 - 2(x + 7) \le -x + 3$

$3 - 2x - 14 \le -x + 3$

$-2x - 11 \le -x + 3$

$-x \le 14$

Multiply by -1, and reverse the inequality sign.

$x \ge -14$

Solution set: $[-14, \infty)$

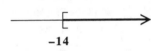

-14

28. $-4 < 5 - 3x \le 0$

$-9 < -3x \le -5$

Divide by -3, and reverse the inequality sign.

$3 > x \ge \dfrac{5}{3}$ or $\dfrac{5}{3} \le x < 3$

Solution set: $\left[\dfrac{5}{3}, 3\right)$

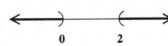

$\dfrac{5}{3}$ 3

29. $2x + 1 > 5$ or $2 - x > 2$

$2x > 4$ $-x > 0$

$x > 2$ or $x < 0$

Solution set: $(-\infty, 0) \cup (2, \infty)$

0 2

30. $|-7k + 3| \ge 4$

$-7k + 3 \ge 4$ or $-7k + 3 \le -4$

$-7k \ge 1$ $-7k \le -7$

$k \le -\dfrac{1}{7}$ or $k \ge 1$

Solution set: $\left(-\infty, -\dfrac{1}{7}\right] \cup [1, \infty)$

$-\dfrac{1}{7}$ 1

31. Let $x =$ the amount invested at 7% and at 10%. The total amount invested is $2x$, and the total interest is

$.1(2x) - 150 = .2x - 150.$

Interest at 7%	+	interest at 10%	=	total interest.
$.07x$	$+$	$.10x$	$=$	$.2x - 150$

$.17x = .20x - 150$

$150 = .03x$

$5000 = x$

She invested $5000 at each rate.

32. Let $x =$ the amount of food C.

$2x =$ the amount of food A.

$5 =$ the amount of food B.

The total is at most 24 grams.

$x + 2x + 5 \le 24$

$3x \le 19$

$x \le \dfrac{19}{3}$ or $6\dfrac{1}{3}$

He may use at most $6\dfrac{1}{3}$ g of food C.

33. Let $x =$ the grade the student must make on the third test.

To find the average of the three tests, add them and divide by 3. This average must be at least 80.

$\dfrac{88 + 78 + x}{3} \ge 80$

$\dfrac{166 + x}{3} \ge 80$

$166 + x \ge 240$

$x \ge 74$

She must earn at least 74 points on her third test.

34. Let $x =$ the time it takes for Jack to be $\dfrac{1}{4}$ mi ahead of Jill.

Use the formula $d = rt$. Make a table.

	r	t	d
Jack	7	x	$7x$
Jill	5	x	$5x$

Jack's distance must be $\dfrac{1}{4}$ mi more than Jill's distance.

continued

$$7x = 5x + \frac{1}{4}$$

$$28x = 20x + 1 \quad \textit{Multiply by 4.}$$

$$8x = 1$$

$$x = \frac{1}{8}$$

It will take Jack $\frac{1}{8}$ hr.

35. Let $x =$ the amount of pure alcohol that should be added.

Strength	Liters of Solution	Liters of Pure Alcohol
100%	x	$1.00x$
10%	7	$.10(7)$
30%	$x + 7$	$.30(x + 7)$

From the last column:

$$1.00x + .10(7) = .30(x + 7)$$

$$10x + 1(7) = 3(x + 7) \quad \textit{Multiply by 10.}$$

$$10x + 7 = 3x + 21$$

$$7x = 14$$

$$x = 2$$

2 L of pure alcohol should be added to the solution.

36. Let $x =$ the number of nickels.

$x - 4 =$ the number of quarters.
The collection contains 29 coins, so the number of pennies is

$$29 - x - (x - 4) = 33 - 2x.$$

	Number of coins	Denomination	Value
Pennies	$33 - 2x$	.01	$.01(33 - 2x)$
Nickels	x	.05	$.05x$
Quarters	$x - 4$	.25	$.25(x - 4)$
Total	29		$2.69

From the last column:

$$.01(33 - 2x) + .05x + .25(x - 4) = 2.69$$

$$1(33 - 2x) + 5x + 25(x - 4) = 269$$

$$\textit{Multiply by 100.}$$

$$33 - 2x + 5x + 25x - 100 = 269$$

$$28x - 67 = 269$$

$$28x = 336$$

$$x = 12$$

There are $33 - 2(12) = 9$ pennies, 12 nickels, and $12 - 4 = 8$ quarters.

In Exercises 37 and 38, use Clark's rule.

$$\frac{\text{Weight of child in pounds}}{150} \times \frac{\text{adult}}{\text{dose}} = \frac{\text{child's}}{\text{dose}}$$

37. If the child weighs 55 lb and the adult dosage is 120 mg, then

$$\frac{55}{150} \times 120 = 44.$$

The child's dosage is 44 mg.

38. If the child weighs 75 lb and the adult dosage is 40 drops, then

$$\frac{75}{150} \times 40 = 20.$$

The child's dosage is 20 drops.

39. **(a)** From 1990 to 1995, the number of daily newspapers decreased by $1611 - 1532 = 79$.

(b) $\dfrac{79}{1611} \approx .049 = 4.9\%$

40. Use the BMI formula.

$$\text{BMI} = \frac{704 \text{ (weight in pounds)}}{\text{(height in inches)}^2}$$

Ken Griffey, Jr., weighs 205 lb and is 6 ft, 3 in or 75 in (1 ft = 12 in) tall, so his

$$\text{BMI} = \frac{704(205)}{75^2} = \frac{144{,}320}{5625} \approx 25.7$$

CHAPTER 4 LINEAR EQUATIONS IN TWO VARIABLES

Section 4.1

1. The symbol (x, y) *does* represent an ordered pair, while the symbols $[x, y]$ and $\{x, y\}$ *do not* represent ordered pairs. (Note that only parentheses are used to write ordered pairs.)

3. All points having coordinates in the form

    (negative, positive)

    are in quadrant II, so the point whose graph has coordinates $(-4, 2)$ is in quadrant II.

5. All ordered pairs that are solutions of the equation $y = 3$ have y-coordinates equal to 3, so the ordered pair $(4, 3)$ is a solution of the equation $y = 3$.

7. From 1990 to 1991, the winnings increased from \$205,000 to \$226,000; an increase of \$21,000. From 1992 to 1993, the winnings increased from \$174,000 to \$296,000; an increase of \$122,000.

9. The winnings were the greatest in 1993 (\$296,000).

11. In 1989, the number of shipments of Dot Matrix printers was 5.8 million, Laser printers was 1.5 million, and of Inkjet printers was .2 million.

    $$5.8 + 1.5 + .2 = 7.5$$

    The total number of units was 7.5 million, which is choice (d).

13. Since 1992, the graph which rises most rapidly is the one for Inkjet printers. Therefore, the most rapid increase occurs in Inkjet printer shipments.

15. The number of new products decreased from 1994 to 1995 (where the line graph goes down). The number of new products increased the greatest from 1993 to 1994 (where the line graph is steepest).

17. Using the estimates from Exercise 16, the increase from 1993 to 1994 was $22,000 - 17,200 = 4800$

    and the increase from 1995 to 1996 was

    $24,300 - 20,900 = 3400$.

    Since $4800 > 3400$, this confirms our answer to Exercise 15.

19. $x + y = 9$; $(0, 9)$

    To determine whether $(0, 9)$ is a solution of the given equation, substitute 0 for x and 9 for y.

    $$\begin{aligned} x + y &= 9 \\ 0 + 9 &= 9 \quad ? \quad \textit{Let x = 0, y = 9} \\ 9 &= 9 \quad \textit{True} \end{aligned}$$

    The result is true, so $(0, 9)$ is a solution of the given equation $x + y = 9$.

21. $2p - q = 6$; $(4, 2)$

    Substitute 4 for p and 2 for q.

    $$\begin{aligned} 2p - q &= 6 \\ 2(4) - 2 &= 6 \quad ? \quad \textit{Let p = 4, q = 2} \\ 8 - 2 &= 6 \quad ? \\ 6 &= 6 \quad \textit{True} \end{aligned}$$

    The result is true, so $(4, 2)$ is a solution of $2p - q = 6$.

23. $4x - 3y = 6$; $(2, 1)$

    Substitute 2 for x and 1 for y.

    $$\begin{aligned} 4x - 3y &= 6 \\ 4(2) - 3(1) &= 6 \quad ? \quad \textit{Let x = 2, y = 1} \\ 8 - 3 &= 6 \quad ? \\ 5 &= 6 \quad \textit{False} \end{aligned}$$

    The result is false, so $(2, 1)$ is not a solution of $4x - 3y = 6$.

25. $y = 3x$: $(2, 6)$

    Substitute 2 for x and 6 for y.

    $$\begin{aligned} y &= 3x \\ 6 &= 3(2) \quad ? \quad \textit{Let x = 2, y = 6} \\ 6 &= 6 \quad \textit{True} \end{aligned}$$

    The result is true, so $(2, 6)$ is a solution of $y = 3x$.

27. $x = -6$; $(-6, 5)$

    Since y does not appear in the equation, we just substitute -6 for x.

    $$\begin{aligned} x &= -6 \\ -6 &= -6 \quad \textit{Let x = -6; true} \end{aligned}$$

    The result is true, so $(-6, 5)$ is a solution of $x = -6$.

29. $x + 4 = 0$; $(-6, 2)$

    Since y does not appear in the equation, we just substitute -6 for x.

    $$\begin{aligned} x + 4 &= 0 \\ -6 + 4 &= 0 \quad \textit{Let x = -6} \\ -2 &= 0 \quad \textit{False} \end{aligned}$$

    The result is false, so $(-6, 2)$ is not a solution of $x + 4 = 0$.

31. Because there is an infinite number of real numbers that can replace either variable in a linear equation in two variables, there is an infinite number of solutions. A linear equation in one variable may have 0, 1, or an infinite number of solutions.

33. The ordered pairs $(4, -1)$ and $(-1, 4)$ are different ordered pairs because the order of the x- and y- coordinates has been reversed. Two *ordered* pairs are equal only if they have the same numbers in the same *order*.

35. $y = 2x + 7;\ (2,\ \ )$

In this ordered pair, $x = 2$. Find the corresponding value of y by replacing x with 2 in the given equation.

$$y = 2x + 7$$
$$y = 2(2) + 7 \quad \textit{Let x = 2}$$
$$y = 4 + 7$$
$$y = 11$$

The ordered pair is $(2, 11)$.

37. $y = 2x + 7;\ (\ ,0)$

In this ordered pair, $y = 0$. Find the corresponding value of x by replacing y with 0 in the given equation.

$$y = 2x + 7$$
$$0 = 2x + 7 \quad \textit{Let y = 0}$$
$$-7 = 2x$$
$$\frac{-7}{2} = x$$

The ordered pair is $\left(-\dfrac{7}{2}, 0\right)$.

39. $y = -4x - 4;\ (0,\ \ )$

$$y = -4x - 4$$
$$y = -4(0) - 4 \quad \textit{Let x = 0}$$
$$y = 0 - 4$$
$$y = -4$$

The ordered pair is $(0, -4)$.

41. $y = -4x - 4;\ (\ ,16)$

$$y = -4x - 4$$
$$16 = -4x - 4 \quad \textit{Let y = 16}$$
$$20 = -4x$$
$$-5 = x$$

The ordered pair is $(-5, 16)$.

43. $y = 6x + 2$

Since $6\left(\dfrac{1}{3}\right) = 2$, while $6\left(\dfrac{1}{7}\right) = \dfrac{6}{7}$, finding the corresponding y-value for $x = \dfrac{1}{3}$ would not give us a fraction, while finding the y-value for $x = \dfrac{1}{7}$ would.

45. $2x + 3y = 12$
If $x = 0$,

$$2(0) + 3y = 12$$
$$0 + 3y = 12$$
$$3y = 12$$
$$y = 4.$$

If $y = 0$,

$$2x + 3(0) = 12$$
$$2x + 0 = 12$$
$$2x = 12$$
$$x = 6.$$

If $y = 8$,

$$2x + 3(8) = 12$$
$$2x + 24 = 12$$
$$2x = -12$$
$$x = -6.$$

The completed table of values is shown below.

x	y
0	4
6	0
−6	8

47. $3x - 5y = -15$
If $x = 0$,

$$3(0) - 5y = -15$$
$$0 - 5y = -15$$
$$-5y = -15$$
$$y = 3.$$

If $y = 0$,

$$3x - 5(0) = -15$$
$$3x - 0 = -15$$
$$3x = -15$$
$$x = -5.$$

If $y = -6$,

$$3x - 5(-6) = -15$$
$$3x + 30 = -15$$
$$3x = -45$$
$$x = -15.$$

The completed table of values is shown below.

x	y
0	3
−5	0
−15	−6

49. $x = -9$

No matter which value of y is chosen, the value of x will always be -9. Each ordered pair can be completed by placing -9 in the first position.

x	y
-9	6
-9	2
-9	-3

51. $y = -6$

No matter which value of x is chosen, the value of y will always be -6. Each ordered pair can be completed by placing -6 in the second position.

x	y
8	-6
4	-6
-2	-6

For Exercises 53–60, the ordered pairs are plotted on the graph following the solution for Exercise 60.

53. To plot $(6, 2)$, start at the origin, go 6 units to the right, and then go up 2 units.

55. To plot $(-4, 2)$, start at the origin, go 4 units to the left, and then go up 2 units.

57. To plot $\left(-\dfrac{4}{5}, -1\right)$, start at the origin, go $\dfrac{4}{5}$

units to the left, and then go down 1 unit.

59. To plot $(0, 4)$, start at the origin and go up 4 units. The point lies on the y-axis.

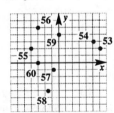

61. The point with coordinates (x, y) is in quadrant III if x is *negative* and y is *negative*.

63. The point with coordinates (x, y) is in quadrant IV if x is *positive* and y is *negative*.

65. $x - 2y = 6$

x	y
0	
	0
2	
	-1

Substitute the given values to complete the ordered pairs.

$$x - 2y = 6$$
$$0 - 2y = 6 \quad \textit{Let x = 0}$$
$$-2y = 6$$
$$y = -3$$

$$x - 2y = 6$$
$$x - 2(0) = 6 \quad \textit{Let y = 0}$$
$$x - 0 = 6$$
$$x = 6$$

$$x - 2y = 6$$
$$2 - 2y = 6 \quad \textit{Let x = 2}$$
$$-2y = 4$$
$$y = -2$$

$$x - 2y = 6$$
$$x - 2(-1) = 6 \quad \textit{Let y = -1}$$
$$x + 2 = 6$$
$$x = 4$$

The completed table of values follows.

x	y
0	-3
6	0
2	-2
4	-1

Plot the points $(0, -3)$, $(6, 0)$, $(2, -2)$, and $(4, -1)$ on a coordinate system.

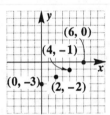

67. $3x - 4y = 12$

x	y
0	
	0
-4	
	-4

Substitute the given values to complete the ordered pairs.

$$3x - 4y = 12$$
$$3(0) - 4y = 12 \quad \textit{Let x = 0}$$
$$0 - 4y = 12$$
$$-4y = 12$$
$$y = -3$$

$$3x - 4y = 12$$
$$3x - 4(0) = 12 \quad \textit{Let y = 0}$$
$$3x - 0 = 12$$
$$3x = 12$$
$$x = 4$$

$$3x - 4y = 12$$
$$3(-4) - 4y = 12 \qquad Let\ x = -4$$
$$-12 - 4y = 12$$
$$-4y = 24$$
$$y = -6$$

$$3x - 4y = 12$$
$$3x - 4(-4) = 12 \qquad Let\ y = -4$$
$$3x + 16 = 12$$
$$3x = -4$$
$$x = -\frac{4}{3}$$

The completed table is as follows.

x	y
0	-3
4	0
-4	-6
$-\dfrac{4}{3}$	-4

Plot the points $(0, -3)$, $(4, 0)$, $(-4, -6)$, and

$\left(-\dfrac{4}{3}, -4\right)$ on a coordinate system.

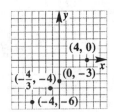

69. The given equation, $y + 4 = 0$, can be written as $y = -4$. So regardless of the value of x, the value of y is -4.

x	y
0	-4
5	-4
-2	-4
-3	-4

Plot the points $(0, -4)$, $(5, -4)$, $(-2, -4)$, and

$(-3, -4)$ on a coordinate system.

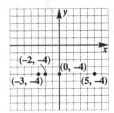

71. (a) When $x = 100$, $y = 5050$. Thus, the ordered pair is $(100, 5050)$.

(b) When $y = 6000$, $x = 2000$. Thus, the ordered pair is $(2000, 6000)$.

73. (a) Yes, there is an approximate linear relationship between the years since 1980 and the minutes beyond two hours to complete a game.

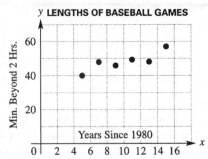

(b) Yes, a prediction could be made by determining the linear equation that describes the linear relationship. However, only input values for the years included in the given data, 1980–1996, should be used.

(c) In this exercise, we can think of the input values as the x-values; that is, the number of years since 1980. The output values are the y-values; that is, the number of minutes beyond two hours.

75. (a) Estimating the y-values from the graph gives us the ordered pairs $(30, 133)$, $(40, 126)$, $(60, 112)$, and $(70, 105)$.

(b) The inputs are the ages between 20 and 80. The outputs are the *lower* limits of the target heart rate zone.

77. No. To go beyond the given data at either end assumes that the graph continues in the same way. This may not be true.

Section 4.2

1. To determine which equation has x-intercept 4, set y equal to 0 and see which equation is equivalent to $x = 4$. Choice (c) is correct since

$$2x - 5y = 8$$
$$2x - 5(0) = 8$$
$$2x = 8$$
$$x = 4$$

3. If the graph of the equation goes through the origin, then substituting 0 for x and 0 for y will result in a true statement. Choice (d) is correct since

$$x + 4y = 0$$
$$(0) + 4(0) = 0$$
$$0 = 0$$

is a true statement.

5. Choice (b) is correct since the graph of $y = -3$ is a horizontal line.

7. $y = -x + 5$

(0,), (,0), (2,)

If $x = 0$,   If $y = 0$,
 $y = -0 + 5$ $0 = -x + 5$
 $y = 5.$ $x = 5.$

If $x = 2$,
 $y = -2 + 5$
 $y = 3.$

The ordered pairs are $(0, 5)$, $(5, 0)$, and $(2, 3)$. Plot the corresponding points and draw a line through them.

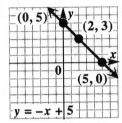

9. $y = \dfrac{2}{3}x + 1$

(0,), (3,), (-3,)

If $x = 0$,   If $x = 3$,
 $y = \dfrac{2}{3}(0) + 1$ $y = \dfrac{2}{3}(3) + 1$
 $y = 0 + 1$ $y = 2 + 1$
 $y = 1.$ $y = 3.$

If $x = -3$,
 $y = \dfrac{2}{3}(-3) + 1$
 $y = -2 + 1$
 $y = -1.$

The ordered pairs are $(0, 1)$, $(3, 3)$, and

$(-3, -1)$. Plot the corresponding points and draw a line through them.

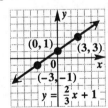

11. $3x = -y - 6$

$(0, \ ), (\ , 0), \left(-\dfrac{1}{3}, \quad \right)$

If $x = 0$,   If $y = 0$,
 $3(0) = -y - 6$ $3x = -0 - 6$
 $0 = -y - 6$ $3x = -6$
 $y = -6.$ $x = -2.$

If $x = -\dfrac{1}{3}$,

$3\left(-\dfrac{1}{3}\right) = -y - 6$

 $-1 = -y - 6$
 $y - 1 = -6$
 $y = -5.$

The ordered pairs are $(0, -6)$, $(-2, 0)$, and

$\left(-\dfrac{1}{3}, -5\right)$. Plot the corresponding points and draw a line through them.

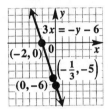

13. To find the x-intercept, let $y = 0$.

$$2x - 3y = 24$$
$$2x - 3(0) = 24$$
$$2x - 0 = 24$$
$$2x = 24$$
$$x = 12$$

The x-intercept is $(12, 0)$.

To find the y-intercept, let $x = 0$.

$$2x - 3y = 24$$
$$2(0) - 3y = 24$$
$$0 - 3y = 24$$
$$-3y = 24$$
$$y = -8$$

The y-intercept is $(0, -8)$.

15. To find the x-intercept, let $y = 0$.

$$x + 6y = 0$$
$$x + 6(0) = 0$$
$$x + 0 = 0$$
$$x = 0$$

The x-intercept is $(0, 0)$. Since we have found the point with x equal to 0, this is also the y-intercept.

17. The equation of the x-axis is $y = 0$.

19. Her next step would be to let x be some number other than 0 and find the corresponding value of y. This would give coordinates of a second point on the line.

21. Begin by finding the intercepts.

$$x = y + 2$$
$$x = 0 + 2 \quad Let \ y = 0$$
$$x = 2$$

$$x = y + 2$$
$$0 = y + 2 \quad Let \ x = 0$$
$$-2 = y$$

The x-intercept is $(2, 0)$ and the y-intercept is $(0, -2)$. To find a third point, choose $y = 1$.

$$x = y + 2$$
$$x = 1 + 2 \quad Let \ y = 1$$
$$x = 3$$

This gives the ordered pair $(3, 1)$. Plot $(2, 0)$, $(0, -2)$, and $(3, 1)$ and draw a line through them.

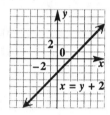

23. Find the intercepts.

$$x - y = 4$$
$$x - 0 = 4 \quad Let \ y = 0$$
$$x = 4$$

$$x - y = 4$$
$$0 - y = 4 \quad Let \ x = 0$$
$$y = -4$$

The x-intercept is $(4, 0)$ and the y-intercept is $(0, -4)$. To find a third point, choose $y = 1$.

$$x - y = 4$$
$$x - 1 = 4 \quad Let \ y = 1$$
$$x = 5$$

This gives the ordered pair $(5, 1)$. Plot $(4, 0)$, $(0, -4)$, and $(5, 1)$ and draw a line through them.

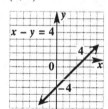

25. Find the intercepts.

$$2x + y = 6$$
$$2x + 0 = 6 \quad Let \ y = 0$$
$$2x = 6$$
$$x = 3$$

$$2x + y = 6$$
$$2(0) + y = 6 \quad Let \ x = 0$$
$$0 + y = 6$$
$$y = 6$$

The x-intercept is $(3, 0)$ and the y-intercept is $(0, 6)$. To find a third point, choose $x = 1$.

$$2x + y = 6$$
$$2(1) + y = 6 \quad Let \ x = 1$$
$$2 + y = 6$$
$$y = 4$$

This gives the point $(1, 4)$. Plot $(3, 0)$, $(0, 6)$, and $(1, 4)$ and draw a line through them.

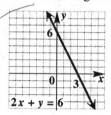

27. Find the intercepts.

$$3x + 7y = 14$$
$$3x + 7(0) = 14 \quad Let \ y = 0$$
$$3x = 14$$
$$x = \frac{14}{3}$$

$$3x + 7y = 14$$
$$3(0) + 7y = 14 \quad Let \ x = 0$$
$$0 + 7y = 14$$
$$y = 2$$

The x-intercept is $\left(\dfrac{14}{3}, 0\right)$ and the y-intercept is $(0, 2)$. To find a third point, choose $x = 2$.

$$3x + 7y = 14$$
$$3(2) + 7y = 14 \quad Let \ x = 2$$
$$6 + 7y = 14$$
$$7y = 8$$
$$y = \frac{8}{7}$$

This gives the ordered pair $\left(2, \dfrac{8}{7}\right)$. Plot $\left(\dfrac{14}{3}, 0\right)$, $(0, 2)$, and $\left(2, \dfrac{8}{7}\right)$. Writing $\dfrac{14}{3}$ as the mixed number $4\dfrac{2}{3}$ and $\dfrac{8}{7}$ as $1\dfrac{1}{7}$ will be helpful for plotting. Draw a line through these three points.

continued

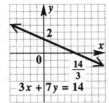

$3x + 7y = 14$

29. $y - 2x = 0$

If $y = 0$, $x = 0$. Both intercepts are the origin, $(0, 0)$. Find two additional points.

$$y - 2x = 0$$
$$y - 2(1) = 0 \quad \textit{Let } x = 1$$
$$y - 2 = 0$$
$$y = 2$$

$$y - 2x = 0$$
$$y - 2(-3) = 0 \quad \textit{Let } x = -3$$
$$y + 6 = 0$$
$$y = -6$$

Plot $(0, 0)$, $(1, 2)$, and $(-3, -6)$ and draw a line through them.

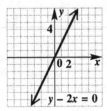

$y - 2x = 0$

31. $y = -6x$

Find three points on the line.

If $x = 0$, $y = -6(0) = 0$.

If $x = 1$, $y = -6(1) = -6$.

If $x = -1$, $y = -6(-1) = 6$.

Plot $(0, 0)$, $(1, -6)$, and $(-1, 6)$ and draw a line through these points.

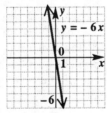

$y = -6x$

33. $y + 1 = 0$
$$y = -1$$

For any value of x, the value of y is -1. Three ordered pairs are $(-4, -1)$, $(0, -1)$, and $(3, -1)$. Plot these points and draw a line through them. The graph is a horizontal line.

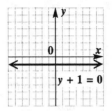

$y + 1 = 0$

35. $x = -2$

For any value of y, the value of x is -2. Three ordered pairs are $(-2, 3)$, $(-2, 0)$, and $(-2, -4)$. Plot these points and draw a line through them. The graph is a vertical line.

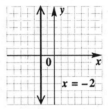

$x = -2$

37. $8 - 2(3x - 4) - 2x = 0$
$$8 - 6x + 8 - 2x = 0$$
$$-8x + 16 = 0$$
$$-8x = -16$$
$$x = \frac{-16}{-8} = 2$$

This is the same as the x-intercept on the calculator screen.

39. $.6x - .1x - x + 2.5 = 0$
$$(.6 - .1 - 1)x = -2.5$$
$$-.5x = -2.5$$
$$x = \frac{-2.5}{-.5} = 5$$

This is the same as the x-intercept on the calculator screen.

41. The x-intercept of the graph is the x-value when y is 0. This is the same as the solution of the equation with one side equal to 0.

43. **(a)** $y = 2.81x + 24.1$

Year	x-value	y-value
1980	10	$2.81(10) + 24.1 = 52.2$
1985	15	$2.81(15) + 24.1 = 66.25$
1990	20	$2.81(20) + 24.1 = 80.3$
1994	24	$2.81(24) + 24.1 = 91.54$

Note that all the y-values are measured in thousands.

(b) From the graph, we get the following estimates: $(1980, 54)$, $(1985, 67)$, $(1990, 77)$, and $(1994, 93)$.

(c) Yes, they are quite close.

45. $y = 73.5 + 3.9x$

(a) $y = 73.5 + 3.9(23)$ *Let x = 23*
$y = 163.2$

The height is 163.2 centimeters.

(b) $y = 73.5 + 3.9(25)$ *Let x = 25*
$y = 171$

The height is 171 centimeters.

(c) $y = 73.5 + 3.9(20)$ *Let x = 20*
$y = 151.5$

The height is 151.5 centimeters.

(d) Graph the ordered pairs $(23, 163.2)$, $(25, 171)$, and $(20, 151.5)$ and draw the line through them.

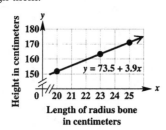

47. Note the break in the y-axis near the origin. This indicates that you should count the units only between 10,000 and 15,000. Since there are 4 units between these numbers, each unit is equivalent to $1250.

(a) – (c) As the graph shows, the value goes down by 1 unit or $1250 each year.

(d) The total depreciation over the 5-year period is $5(\$1250) = \$6250$.

49. $\{(0, 5), (2, 3), (4, 1), (6, -1), (8, -3)\}$

The input 0 is paired with the output 5, the input 2 is paired with the output 3, and so on. Every input value corresponds to exactly one output value, so the set defines a function.

51. Given any particular value for x, we could find the corresponding *single* value for y. Thus, the ordered pairs that satisfy $-x + 2y = 9$ defines a function.

53. The only restriction for the equation $x = 8$ is that all ordered pairs have 8 for their x-value. Some ordered pairs that satisfy the equation are $(8, 0)$, $(8, 2)$, and $(8, -3)$. Since the x-value 8 corresponds to more than one y-value, the ordered pairs that satisfy $x = 8$ do not define a function.

55. Each year corresponds to exactly one number of master's degrees, so the relationship defines a function.

Section 4.3

1. Rise is the vertical change between two different points on a line.

Run is the horizontal change between two different points on a line.

3. The indicated points have coordinates $(-1, -3)$ and $(1, 5)$. Use the definition of slope with $(-1, -3) = (x_1, y_1)$ and $(1, 5) = (x_2, y_2)$.

$$\text{slope} = m = \frac{y_2 - y_1}{x_2 - x_1}$$
$$= \frac{5 - (-3)}{1 - (-1)}$$
$$= \frac{8}{2} = 4$$

5. The indicated points have coordinates $(-3, 2)$ and $(5, -2)$. Use the definition of slope with $(-3, 2) = (x_1, y_1)$ and $(5, -2) = (x_2, y_2)$.

$$\text{slope} = m = \frac{y_2 - y_1}{x_2 - x_1}$$
$$= \frac{-2 - 2}{5 - (-3)}$$
$$= \frac{-4}{8} = -\frac{1}{2}$$

7. The indicated points have coordinates $(-2, -4)$ and $(4, -4)$. Use the definition of slope with $(-2, -4) = (x_1, y_1)$ and $(4, -4) = (x_2, y_2)$.

$$\text{slope} = m = \frac{y_2 - y_1}{x_2 - x_1}$$
$$= \frac{-4 - (-4)}{4 - (-2)}$$
$$= \frac{-4 + 4}{4 + 2} = \frac{0}{6} = 0$$

9. Yes, the slope will be the same. If we *start* at $(-1, -4)$ and *end* at $(3, 2)$, the vertical change will be 6 (6 units up) and the horizontal change will be 4 (4 units to the right), giving a slope of

$$m = \frac{6}{4} = \frac{3}{2}.$$

If we *start* at $(3, 2)$ and *end* at $(-1, -4)$, the vertical change will be -6 (6 units down) and the horizontal change will be -4 (4 units to the left), giving a slope of

$$m = \frac{-6}{-4} = \frac{3}{2}.$$

11. positive slope

Sketches will vary. The line must rise from left to right. One such line is shown in the following graph.

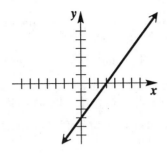

13. zero slope

Sketches will vary. The line must be horizontal. One such line is shown in the following graph.

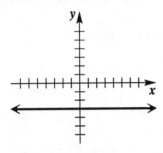

15. Because he found the difference $3 - 5 = -2$ in the numerator, he should have subtracted in the same order in the denominator to get $-1 - 2 = -3$. The correct slope is $\dfrac{-2}{-3} = \dfrac{2}{3}$.

Note that these slopes are opposites of one another.

17. Let $(1, -2) = (x_1, y_1)$ and $(-3, -7) = (x_2, y_2)$.

$$\text{slope} = m = \frac{y_2 - y_1}{x_2 - x_1}$$
$$= \frac{-7 - (-2)}{-3 - 1}$$
$$= \frac{-5}{-4} = \frac{5}{4}$$

19. Let $(0, 3) = (x_1, y_1)$ and $(-2, 0) = (x_2, y_2)$.

$$\text{slope} = m = \frac{y_2 - y_1}{x_2 - x_1}$$
$$= \frac{0 - 3}{-2 - 0}$$
$$= \frac{-3}{-2} = \frac{3}{2}$$

21. Let $(4, 3) = (x_1, y_1)$ and $(-6, 3) = (x_2, y_2)$.

$$\text{slope} = m = \frac{y_2 - y_1}{x_2 - x_1}$$
$$= \frac{3 - 3}{-6 - 4}$$
$$= \frac{0}{-10} = 0$$

23. Let $(-12, 3) = (x_1, y_1)$ and $(-12, -7) = (x_2, y_2)$.

$$\text{slope} = m = \frac{y_2 - y_1}{x_2 - x_1}$$
$$= \frac{-7 - 3}{-12 - (-12)}$$
$$= \frac{-10}{0},$$

which is undefined.

25. Let $\left(-\dfrac{7}{5}, \dfrac{3}{10}\right) = (x_1, y_1)$ and $\left(\dfrac{1}{5}, -\dfrac{1}{2}\right) = (x_2, y_2)$.

$$\text{slope} = m = \frac{y_2 - y_1}{x_2 - x_1}$$
$$= \frac{-\dfrac{1}{2} - \dfrac{3}{10}}{\dfrac{1}{5} - \left(-\dfrac{7}{5}\right)}$$
$$= \frac{-\dfrac{5}{10} - \dfrac{3}{10}}{\dfrac{1}{5} + \dfrac{7}{5}}$$
$$= \frac{-\dfrac{8}{10}}{\dfrac{8}{5}}$$
$$= \left(-\dfrac{8}{10}\right)\left(\dfrac{5}{8}\right)$$
$$= -\dfrac{1}{2}$$

27. $y = 5x + 12$

Since the equation is already solved for y, the slope is given by the coefficient of x, which is 5. Thus, the slope of the line is 5.

29. Solve the equation for y.

$$4y = x + 1$$
$$y = \frac{1}{4}x + \frac{1}{4} \quad \text{Divide by 4}$$

The slope of the line is given by the coefficient of x, so the slope is $\dfrac{1}{4}$.

31. Solve the equation for y.

$$3x - 2y = 3$$
$$-2y = -3x + 3 \quad \text{Subtract } 3x$$
$$y = \frac{3}{2}x - \frac{3}{2} \quad \text{Divide by } -2$$

The slope of the line is given by the coefficient of x, so the slope is $\dfrac{3}{2}$.

33. $x = 6$

This is an equation of a vertical line. Its slope is *undefined*.

35. **(a)** Because the line *falls* from left to right, its slope is *negative*.

(b) Because the line intersects the y-axis *at the* origin, the y-value of its y-intercept is *zero*.

37. **(a)** Because the line *rises* from left to right, its slope is *positive*.

(b) Because the line intersects the y-axis *below* the origin, the y-value of its y-intercept is *negative*.

39. **(a)** The line is *horizontal*, so its slope is *zero*.

(b) The line intersects the y-axis *below* the origin, so the y-value of its y-intercept is *negative*.

41. If two lines are both vertical or both horizontal, they are *parallel*. Choice (a) is correct.

43. Find the slope of each line by solving the equations for y.

$$2x + 5y = 4$$
$$5y = -2x + 4 \quad \textit{Subtract 2x}$$
$$y = -\frac{2}{5}x + \frac{4}{5} \quad \textit{Divide by 5}$$

The slope of the first line is $-\dfrac{2}{5}$.

$$4x + 10y = 1$$
$$10y = -4x + 1 \quad \textit{Subtract 4x}$$
$$y = -\frac{4}{10}x + \frac{1}{4} \quad \textit{Divide by 10}$$
$$y = -\frac{2}{5}x + \frac{1}{4} \quad \textit{Lowest terms}$$

The slope of the second line is $-\dfrac{2}{5}$.

The slopes are equal, so the lines are *parallel*.

45. Find the slope of each line by solving the equations for y.

$$8x - 9y = 6$$
$$-9y = -8x + 6 \quad \textit{Subtract 8x}$$
$$y = \frac{8}{9}x - \frac{2}{3} \quad \textit{Divide by -9}$$

The slope of the first line is $\dfrac{8}{9}$.

$$8x + 6y = -5$$
$$6y = -8x - 5 \quad \textit{Subtract 8x}$$
$$y = -\frac{4}{3}x - \frac{5}{6} \quad \textit{Divide by 6}$$

The slope of the second line is $-\dfrac{4}{3}$.

The slopes are not equal, so the lines are not parallel. The slopes are not negative reciprocals $\left(\text{the negative reciprocal of } \dfrac{8}{9} \text{ is } -\dfrac{9}{8}\right)$, so the lines are not perpendicular. Thus, the lines are *neither* parallel nor perpendicular.

47. Find the slope of each line by solving the equations for y.

$$3x - 2y = 6$$
$$-2y = -3x + 6 \quad \textit{Subtract 3x}$$
$$y = \frac{3}{2}x - 3 \quad \textit{Divide by -2}$$

The slope of the first line is $\dfrac{3}{2}$.

$$2x + 3y = 3$$
$$3y = -2x + 3 \quad \textit{Subtract 2x}$$
$$y = -\frac{2}{3}x + 1 \quad \textit{Divide by 3}$$

The slope of the second line is $-\dfrac{2}{3}$.

The product of the slopes is

$$\frac{3}{2}\left(-\frac{2}{3}\right) = -1,$$

so the lines are *perpendicular*.

49. Since the slope is the ratio of the vertical rise to the horizontal run, the slope is

$$\frac{6}{20} = \frac{3}{10}.$$

51. We use the points with coordinates

$(1990, 11{,}338)$ and $(2005, 14{,}818)$.

$$m = \frac{14{,}818 \text{ thousand } - \ 11{,}338 \text{ thousand}}{2005 - 1990}$$
$$= \frac{3480 \text{ thousand}}{15}$$
$$= 232 \text{ thousand}$$

or 232,000.

52. The slope of the line in Figure A is *positive*. This means that during the period represented, enrollment *increased* in grades 9 – 12.

53. The increase is approximately 232,000 students per year.

54. We use the points with coordinates $(1990, 22)$ and $(1996, 10)$.

$$m = \frac{10 - 22}{1996 - 1990}$$
$$= \frac{-12}{6}$$
$$= -2 \text{ students per computer}$$

55. The slope of the line in Figure B is *negative*. This means that during the period represented, the number of students per computer *decreased*.

56. The decrease is 2 students per computer per year.

57. **(a)** The input numbers are the years 1991 to 1996. Substituting these years into $y = .797x - 1549$ gives us the output numbers (in thousands): 37.8, 38.6, 39.4, 40.2, 41.0, 41.8 (rounded to one decimal place). Possible ordered pairs are (1991, 37.8), (1992, 38.6), (1993, 39.4), (1994, 40.2), (1995, 41.0), and (1996, 41.8).

(b) Subtract the output numbers in part (a) from the data given in the table.

$$38.0 - 37.8 = .2$$
$$39.0 - 38.6 = .4$$
$$39.6 - 39.4 = .2$$
$$40.4 - 40.2 = .2$$
$$41.2 - 41.0 = .2$$
$$42.1 - 41.8 = .3$$

So the approximations are reasonably close — all within .4 thousand (400).

59. The *y*-values change .1 billion (or 100,000,000) square feet each year. Since the change each year is the same, the graph is a straight line.

61. The two points have coordinates $(-4, -5)$ and $(2, 7)$. Therefore, the slope is

$$m = \frac{7 - (-5)}{2 - (-4)} = \frac{12}{6} = 2.$$

63. The two points we use have coordinates $(-12, -.8)$ and $(0, 4)$. Therefore, the slope is

$$m = \frac{4 - (-.8)}{0 - (-12)} = \frac{4.8}{12} = .4 = \frac{2}{5}.$$

65. From the table, when $x = 0$, $Y_1 = 4$. Therefore, the *y*-intercept is $(0, 4)$.

Chapter 4 Review Exercises

1. $y = 3x + 2$ $(-1, \ )$ $(0, \ )$ $(\ , 5)$

$$y = 3x + 2$$
$$y = 3(-1) + 2 \quad Let\ x = -1$$
$$y = -3 + 2$$
$$y = -1$$

$$y = 3x + 2$$
$$y = 3(0) + 2 \quad Let\ x = 0$$
$$y = 0 + 2$$
$$y = 2$$

$$y = 3x + 2$$
$$5 = 3x + 2 \quad Let\ y = 5$$
$$3 = 3x$$
$$1 = x$$

The ordered pairs are $(-1, -1)$, $(0, 2)$, and $(1, 5)$.

2. $4x + 3y = 6$ $(0, \ )$ $(\ , 0)$ $(-2, \ )$

$$4x + 3y = 6$$
$$4(0) + 3y = 6 \quad Let\ x = 0$$
$$3y = 6$$
$$y = 2$$

$$4x + 3y = 6$$
$$4x + 3(0) = 6 \quad Let\ y = 0$$
$$4x = 6$$
$$x = \frac{6}{4} = \frac{3}{2}$$

$$4x + 3y = 6$$
$$4(-2) + 3y = 6 \quad Let\ x = -2$$
$$-8 + 3y = 6$$
$$3y = 14$$
$$y = \frac{14}{3}$$

The ordered pairs are $(0, 2)$, $\left(\frac{3}{2}, 0\right)$, and $\left(-2, \frac{14}{3}\right)$.

3. $x - 7 = 0$ $(\ , -3)$ $(\ , 0)$ $(\ , 5)$

The given equation may be written $x = 7$. For any value of y, the value of x will always be 7. The ordered pairs are $(7, -3)$, $(7, 0)$, and $(7, 5)$.

4. $x + y = 7$; $(2, 5)$

Substitute 2 for x and 5 for y in the given equation.

$$x + y = 7$$
$$2 + 5 = 7 \ ?$$
$$7 = 7 \quad True$$

Yes, $(2, 5)$ is a solution of the given equation.

5. $2x + y = 5$; $(-1, 3)$

Substitute -1 for x and 3 for y in the given equation.

$$2x + y = 5$$
$$2(-1) + 3 = 5 \ ?$$
$$-2 + 3 = 5 \ ?$$
$$1 = 5 \quad False$$

No, $(-1, 3)$ is not a solution of the given equation.

6. $3x - y = 4$; $\left(\dfrac{1}{3}, -3\right)$

Substitute $\dfrac{1}{3}$ for x and -3 for y in the given equation.

$$3x - y = 4$$
$$3\left(\dfrac{1}{3}\right) - (-3) = 4 \quad ?$$
$$1 + 3 = 4 \quad ?$$
$$4 = 4 \quad True$$

Yes, $\left(\dfrac{1}{3}, -3\right)$ is a solution of the given equation.

Graph for Exercises 7–10

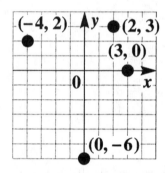

7. To plot $(2, 3)$, start at the origin, go 2 units to the right, and then go up 3 units. (See above graph.)

8. To plot $(-4, 2)$, start at the origin, go 4 units to the left, and then go up 2 units.

9. To plot $(3, 0)$, start at the origin, go 3 units to the right. The point lies on the x-axis.

10. To plot $(0, -6)$, start at the origin, go down 6 units. The point lies on the y-axis.

11. The product of two numbers is positive whenever the two numbers have the same sign. If $xy > 0$, either $x > 0$ and $y > 0$, so that (x, y) lies in quadrant I, or $x < 0$ and $y < 0$, so that (x, y) lies in quadrant III.

12. The y-coordinate is 0 for any point on the x-axis. Thus, for any real value of k, the point $(k, 0)$ lies on the x-axis.

13. To find the x-intercept, let $y = 0$.

$$y = 2x + 5$$
$$0 = 2x + 5$$
$$-2x = 5$$
$$x = -\dfrac{5}{2}$$

The x-intercept is $\left(-\dfrac{5}{2}, 0\right)$.

To find the y-intercept, let $x = 0$.

$$y = 2x + 5$$
$$y = 2(0) + 5$$
$$y = 5$$

The y-intercept is $(0, 5)$.

To find a third point, choose $x = -1$.

$$y = 2x + 5$$
$$y = 2(-1) + 5$$
$$y = 3$$

This gives the ordered pair $(-1, 3)$. Plot $\left(-\dfrac{5}{2}, 0\right)$, $(0, 5)$, and $(-1, 3)$ and draw a line through them.

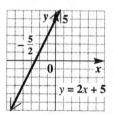

14. To find the x-intercept, let $y = 0$.

$$3x + 2y = 8$$
$$3x + 2(0) = 8$$
$$3x = 8$$
$$x = \dfrac{8}{3}$$

The x-intercept is $\left(\dfrac{8}{3}, 0\right)$.

To find the y-intercept, let $x = 0$.

$$3x + 2y = 8$$
$$3(0) + 2y = 8$$
$$2y = 8$$
$$y = 4$$

The y-intercept is $(0, 4)$.

To find a third point, choose $x = 1$.

$$3x + 2y = 8$$
$$3(1) + 2y = 8$$
$$2y = 5$$
$$y = \dfrac{5}{2}$$

This gives the ordered pair $\left(1, \dfrac{5}{2}\right)$. Plot $\left(\dfrac{8}{3}, 0\right)$, $(0, 4)$, and $\left(1, \dfrac{5}{2}\right)$ and draw a line through them.

continued

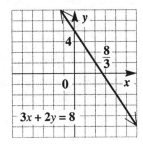

15. $x + 2y = -4$

Find the intercepts.

If $y = 0$, $x = -4$, so the x-intercept is

$(-4, 0)$.

If $x = 0$, $y = -2$, so the y-intercept is

$(0, -2)$.

To find a third point, choose $x = 2$.

$$x + 2y = -4$$
$$2 + 2y = -4 \quad \textit{Let x = 2}$$
$$2y = -6$$
$$y = -3$$

This gives the ordered pair $(2, -3)$. Plot $(-4, 0)$, $(0, -2)$, and $(2, -3)$ and draw a line through them.

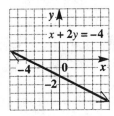

16. $x + y = 0$

This line goes through the origin, so both intercepts are $(0, 0)$. Two other points on the line are $(2, -2)$ and $(-2, 2)$.

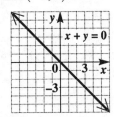

17. $\{(5, -3), (-4, -3), (-1, 8), (2, 8)\}$

The input 5 is paired with the output -3, the input -4 is paired with the output -3, and so on. Every input value corresponds to exactly one output value, so the set defines a function.

18. $3x - 9 = 0$

$$3x = 9 \quad \textit{Add 9}$$
$$x = 3 \quad \textit{Divide by 3}$$

The only restriction for the equation $x = 3$ is that all ordered pairs have 3 for their x-value. Some ordered pairs that satisfy the equation are $(3, 0)$, $(3, 2)$, and $(3, -3)$. Since the x-value 3 corresponds to more than one y-value, the ordered pairs that satisfy $x = 3$ (and hence, $3x - 9 = 0$) do not define a function.

19. Given any year, the value of x is found by subtracting 1980 from the year. This value of x is then substituted into the equation

$$y = .425x - 1.225,$$

which gives exactly one number representing the sponsorship spending in North America (in billions of dollars). Thus, the relationship defines a function.

20. The input values are the years since 1980; that is, the numbers 1, 2, 3, ... The output values are the numbers representing the sponsorship spending in billions of dollars.

21. Let $(2, 3) = (x_1, y_1)$ and $(-4, 6) = (x_2, y_2)$.

$$\text{slope} = m = \frac{y_2 - y_1}{x_2 - x_1} = \frac{6 - 3}{-4 - 2}$$
$$= \frac{3}{-6} = -\frac{1}{2}$$

22. Let $(0, 6) = (x_1, y_1)$ and $(1, 6) = (x_2, y_2)$.

$$\text{slope} = m = \frac{6 - 6}{1 - 0}$$
$$= \frac{0}{1} = 0$$

23. Let $(2, 5) = (x_1, y_1)$ and $(2, 8) = (x_2, y_2)$.

$$\text{slope} = m = \frac{8 - 5}{2 - 2}$$
$$= \frac{3}{0}, \text{ which is undefined.}$$

24. $y = 3x - 4$

The equation is already solved for y, so the slope of the line is given by the coefficient of x. Thus, the slope is 3.

25. $y = \frac{2}{3}x + 1$

The equation is already solved for y, so the slope of the line is given by the coefficient of x. Thus, the slope is $\frac{2}{3}$.

26. From the table, we choose the two points $(0, 1)$ and $(2, 4)$. Therefore,

$$\text{slope} = m = \frac{4 - 1}{2 - 0} = \frac{3}{2}.$$

27. The points on the graph are $(-1, 2)$ and $(6, -2)$. The slope of the line through the points is

$$m = \frac{-2 - 2}{6 - (-1)} = \frac{-4}{7} = -\frac{4}{7}.$$

28. $y = 5$ is the equation of a horizontal line. Its slope is 0.

29. Because perpendicular lines have slopes which are negative reciprocals of each other and the slope of the graph of $y = -3x + 3$ is -3, the slope of a line perpendicular to it will be

$$-\frac{1}{-3} = \frac{1}{3}.$$

30. If the product of two numbers is negative, the numbers must have opposite signs. Because the product of the slopes of two perpendicular lines (neither of which is vertical) is -1, one line must have a positive slope and the other a negative slope. Therefore, the signs of the slopes cannot be the same.

31. Find the slope of each line by solving the equations for y.

$$3x + 2y = 6$$
$$2y = -3x + 6 \quad \textit{Subtract 3x}$$
$$y = -\frac{3}{2}x + 3 \quad \textit{Divide by 2}$$

The slope of the first line is $-\dfrac{3}{2}$.

$$6x + 4y = 8$$
$$4y = -6x + 8 \quad \textit{Subtract 6x}$$
$$y = -\frac{6}{4}x + 2 \quad \textit{Divide by 4}$$
$$y = -\frac{3}{2}x + 2 \quad \textit{Lowest terms}$$

The slope of the second line is $-\dfrac{3}{2}$. The slopes are equal so the lines are *parallel*.

32. Find the slope of each line by solving the equations for y.

$$x - 3y = 1$$
$$-3y = -x + 1 \quad \textit{Subtract x}$$
$$y = \frac{1}{3}x - \frac{1}{3} \quad \textit{Divide by –3}$$

The slope of the first line is $\dfrac{1}{3}$.

$$3x + y = 4$$
$$y = -3x + 4 \quad \textit{Subtract 3x}$$

The slope of the second line is -3.

The product of the slopes is

$$\frac{1}{3}(-3) = -1,$$

so the lines are *perpendicular*.

33. Find the slope of each line by solving the equations for y.

$$x - 2y = 8$$
$$-2y = -x + 8 \quad \textit{Subtract x}$$
$$y = \frac{1}{2}x - 4 \quad \textit{Divide by –2}$$

The slope of the first line is $\dfrac{1}{2}$.

$$x + 2y = 8$$
$$2y = -x + 8 \quad \textit{Subtract x}$$
$$y = -\frac{1}{2}x + 4 \quad \textit{Divide by 2}$$

The slopes are not equal and their product is

$$\left(\frac{1}{2}\right)\left(-\frac{1}{2}\right) = -\frac{1}{4} \neq -1,$$

so the lines are *neither* parallel nor perpendicular.

34. The growth rates are:

$$1992 \rightarrow 4.9\%$$
$$1993 \rightarrow 9.6\%$$
$$1994 \rightarrow 6.4\%$$

To estimate the sales in 1994, we'll start with the sales in 1993 and multiply by the 1994 growth rate.

$$456(.064) = 29.184 \approx 29$$

So the sales in 1994 are about $456 + 29 = \$485$ billion.

35. $x = 3y \quad (0, \ ) \ (8, \ ) \ (\ , -3)$

$$x = 3y$$
$$0 = 3y \quad \textit{Let x = 0}$$
$$0 = y$$

$$x = 3y$$
$$8 = 3y \quad \textit{Let x = 8}$$
$$\frac{8}{3} = y$$

$$x = 3y$$
$$x = 3(-3) \quad \textit{Let y = –3}$$
$$x = -9$$

The ordered pairs are $(0, 0)$, $\left(8, \dfrac{8}{3}\right)$, and $(-9, -3)$.

36. $x = -2$ is an equation of a vertical line. The x-intercept is $(-2, 0)$ and there is no y-intercept. Its slope is undefined.

37.

$$y = 2x + 3$$
$$0 = 2x + 3 \quad \textit{Let } y = 0$$
$$-2x = 3$$
$$x = -\frac{3}{2}$$

The x-intercept is $\left(-\frac{3}{2}, 0\right)$.

$$y = 2x + 3$$
$$y = 2(0) + 3 \quad \textit{Let } x = 0$$
$$y = 3$$

The y-intercept is $(0, 3)$.

Since the equation is solved for y, the slope is the coefficient of x, which is 2.

38. $11x - 3y = 4$

x-intercept:

$$11x - 3(0) = 4 \quad \textit{Let } y = 0$$
$$11x = 4$$
$$x = \frac{4}{11}$$

The x-intercept is $\left(\frac{4}{11}, 0\right)$.

y-intercept:

$$11(0) - 3y = 4 \quad \textit{Let } x = 0$$
$$-3y = 4$$
$$y = -\frac{4}{3}$$

The y-intercept is $\left(0, -\frac{4}{3}\right)$.

To find the slope, solve the equation for y.

$$-3y = -11x + 4$$
$$y = \frac{11}{3}x - \frac{4}{3}$$

The slope is the coefficient of x, so $m = \frac{11}{3}$.

39.   Let $(4, -1) = (x_1, y_1)$ and $(-2, -3) = (x_2, y_2)$.

$$\begin{aligned}
\text{slope} = m &= \frac{y_2 - y_1}{x_2 - x_1} \\
&= \frac{-3 - (-1)}{-2 - 4} \\
&= \frac{-3 + 1}{-6} \\
&= \frac{-2}{-6} = \frac{1}{3}
\end{aligned}$$

40.   A line with undefined slope is vertical. Any line perpendicular to a vertical line is a horizontal line, which has a slope of 0.

41. $5x - 3y = 16;$ $\left(0, -\frac{16}{3}\right)$

Substitute 0 for x and $-\frac{16}{3}$ for y in the given equation.

$$5x - 3y = 16$$
$$5(0) - 3\left(-\frac{16}{3}\right) = 16 \quad ?$$
$$0 + 16 = 16 \quad ?$$
$$16 = 16 \quad \textit{True}$$

Yes, $\left(0, -\frac{16}{3}\right)$ is a solution of the given equation.

42. $x + 3y = 9$

If $x = 0$, $y = 3$, and the y-intercept is $(0, 3)$.
If $y = 0$, $x = 9$, and the x-intercept is $(9, 0)$.
For a third point, choose $x = 3$.

$$3 + 3y = 9$$
$$3y = 6$$
$$y = 2$$

So the point is $(3, 2)$. Plot the points and draw the line through them.

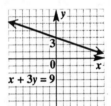

43.   The equation $x - 5 = 0$ can be written $x = 5$. This is the equation of a vertical line with x-intercept $(5, 0)$.

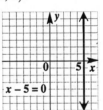

44. $2x - y = 3$

If $y = 0$, $x = \frac{3}{2}$, so the x-intercept is $(\frac{3}{2}, 0)$.
If $x = 0$, $y = -3$, so the y-intercept is $(0, -3)$.

Draw a line through these two points. A third point may be used as a check.

continued

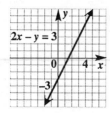

45. A third point serves as a check. If all three points lie on a line, we can be reasonably sure that our line is correct.

46. Since each city corresponds to exactly one population, the set of ordered pairs defines a function. The inputs are the five cities and the outputs are the corresponding populations.

Chapter 4 Test

1. The dot above 1993 appears to correspond to a cost of \$850,000. The best estimate is choice (b), \$848,000.

2. The increase from 1995 to 1996 is $1100 - 950 = 150$ (in thousands of dollars). Thus, the percent increase is

$$\frac{150}{950} \cdot 100 \approx 15.8 \approx 16,$$

to the nearest percent.

3. $3x + 5y = -30 \quad (0, \ ), (\ , 0), (\ , 3)$

$$3x + 5y = -30$$
$$3(0) + 5y = -30 \quad Let \ x = 0$$
$$5y = -30$$
$$y = -6$$

$$3x + 5y = -30$$
$$3x + 5(0) = -30 \quad Let \ y = 0$$
$$3x = -30$$
$$x = -10$$

$$3x + 5y = -30$$
$$3x + 5(3) = -30 \quad Let \ y = 3$$
$$3x + 15 = -30$$
$$3x = -45$$
$$x = -15$$

The ordered pairs are $(0, -6)$, $(-10, 0)$, and $(-15, 3)$.

4. $y + 12 = 0 \quad (0, \ ), (-4, \ ), \left(\dfrac{5}{2}, \ \right)$

The equation can also be written as $y = -12$. For every value of x, the value of y will always be -12. Complete the ordered pairs by placing -12 in the second position in each pair.

The ordered pairs are $(0, -12)$, $(-4, -12)$, and $\left(\dfrac{5}{2}, -12\right)$.

5.
$$4x - 7y = 9 \quad ?$$
$$Let \ x = 4 \ and \ y = -1$$
$$4(4) - 7(-1) = 9 \quad ?$$
$$16 + 7 = 9 \quad ?$$
$$23 = 9 \quad ? \quad No$$

So $(4, -1)$ is not a solution of $4x - 7y = 9$.

6. To find the x-intercept, let $y = 0$ and solve for x.

To find the y-intercept, let $x = 0$ and solve for y.

7. $3x + y = 6$

If $y = 0$, $x = 2$, so the x-intercept is $(2, 0)$.
If $x = 0$, $y = 6$, so the y-intercept is $(0, 6)$.

A third point, such as $(1, 3)$, can be used as a check. Draw a line through $(0, 6)$, $(1, 3)$, and $(2, 0)$.

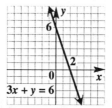

8. $y + 3 = 0$ can also be written as $y = -3$. Its graph is a horizontal line with y-intercept $(0, -3)$. There is no x-intercept.

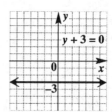

9. Yes, because each input determines exactly one output.

10. The inputs are the years from 1990 through 1996, and the outputs are the corresponding Super Bowl advertising costs in thousands of dollars. The outputs are 650, 780, 830, 850, 880, 950, and 1100.

11. Through $(-4, 6)$ and $(-1, -2)$

Use the definition of slope with $(x_1, y_1) = (-4, 6)$ and $(x_2, y_2) = (-1, -2)$.

$$slope = m = \frac{y_2 - y_1}{x_2 - x_1}$$
$$= \frac{-2 - 6}{-1 - (-4)}$$
$$= \frac{-8}{3} = -\frac{8}{3}$$

12. $2x + y = 10$

To find the slope, solve the given equation for y.

$$2x + y = 10$$
$$y = -2x + 10$$

The equation is now written in $y = mx + b$ form, so the slope is given by the coefficient of x, which is -2.

13. The indicated points are $(0, -4)$ and $(2, 1)$. Use the definition of slope with $(x_1, y_1) = (0, -4)$ and $(x_2, y_2) = (2, 1)$.

$$\text{slope} = m = \frac{y_2 - y_1}{x_2 - x_1}$$
$$= \frac{1 - (-4)}{2 - 0}$$
$$= \frac{5}{2}$$

14. Use the points $(0, 4)$ and $(-8, 0)$.

$$\text{slope} = m = \frac{0 - 4}{-8 - 0} = \frac{-4}{-8} = \frac{1}{2}$$

15. For the graph of $y = -4x + 6$, the slope of the line is -4.

A line parallel to the graph of $y = -4x + 6$ has the same slope. Thus, the slope is -4.

16. For the graph of $y = -4x + 6$, the slope of the line is -4.

A line perpendicular to the graph of $y = -4x + 6$ has a slope which is the negative reciprocal of -4. Thus, the slope is

$$-\frac{1}{-4} = \frac{1}{4}.$$

Cumulative Review Exercises Chapters 1–4

1.
$$16\frac{7}{8} - 3\frac{1}{10} = \frac{135}{8} - \frac{31}{10}$$
$$= \frac{675}{40} - \frac{124}{40}$$
$$= \frac{551}{40} \text{ or } 13\frac{31}{40}$$

2.
$$\frac{3}{4} \div \frac{5}{8} = \frac{3}{4} \cdot \frac{8}{5} = \frac{3 \cdot 2 \cdot 4}{4 \cdot 5} = \frac{6}{5}$$

3. $-11 + 20 + (-2) = 9 + (-2) = 7$

4.
$$\frac{(-3)^2 - (-4)(2^4)}{5 \cdot 2 - (-2)^3}$$
$$= \frac{9 - (-4)(16)}{10 - (-8)} \quad \textit{Do exponents first}$$
$$= \frac{9 - (-64)}{10 - (-8)} \quad \textit{Multiply}$$
$$= \frac{9 + 64}{10 + 8} = \frac{73}{18} \text{ or } 4\frac{1}{18}$$

5. First, find how much of the room was painted on the two days.

$$\frac{1}{4} + \frac{1}{3} = \frac{1 \cdot 3}{4 \cdot 3} + \frac{1 \cdot 4}{3 \cdot 4} \quad \textit{Common denominator of 12}$$
$$= \frac{3}{12} + \frac{4}{12}$$
$$= \frac{7}{12}$$

Next, subtract from the whole (1) how much of the room is unpainted.

$$1 - \frac{7}{12} = \frac{12}{12} - \frac{7}{12}$$
$$= \frac{5}{12}$$

$\frac{5}{12}$ of the room is still unpainted.

6.
$$\frac{4(3 - 9)}{2 - 6} \geq 6 \ ?$$
$$\frac{4(-6)}{-4} \geq 6 \ ?$$
$$\frac{-24}{-4} \geq 6 \ ?$$
$$6 \geq 6$$

The statement is *true* since $6 = 6$.

7. $xz^3 - 5y^2 = (-2)(3)^3 - 5(-4)^2$

Let $x = -2$, $y = -4$, $z = 3$
$$= (-2)(27) + (-5)(16)$$
$$= -54 + (-80)$$
$$= -134$$

8. $3(-2 + x) = 3 \cdot (-2) + 3(x)$
$$= -6 + 3x$$

illustrates the *distributive property*.

9. $-4p - 6 + 3p + 8$
$$= (-4p + 3p) + (-6 + 8)$$
$$= -p + 2$$

10.
$$V = \frac{1}{3}\pi r^2 h$$
$$3V = \pi r^2 h \quad \textit{Multiply by 3}$$
$$\frac{3V}{\pi r^2} = h \quad \textit{Divide by } \pi r^2$$

11. $6 - 3(1 + a) = 2(a + 5) - 2$
$$6 - 3 - 3a = 2a + 10 - 2$$
$$3 - 3a = 2a + 8$$
$$-5a = 5$$
$$a = -1$$

The solution set is $\{-1\}$.

12. $-(m-1) = 3 - 2m$

$\quad -m + 1 = 3 - 2m$ *Distributive Prop.*

$\quad\quad m + 1 = 3$ *Add 2m*

$\quad\quad\quad m = 2$ *Subtract 1*

The solution set is $\{2\}$.

13. $\dfrac{y-2}{3} = \dfrac{2y+1}{5}$

$(y-2)(5) = (3)(2y+1)$ *Cross products*

$\quad 5y - 10 = 6y + 3$

$\quad\quad -10 = y + 3$

$\quad\quad -13 = y$

The solution set is $\{-13\}$.

14. $\quad -5z \geq 4z - 18$

$-5z - 4z \geq 4z - 18 - 4z$

$\quad\quad -9z \geq -18$

$\quad\quad \dfrac{-9z}{-9} \leq \dfrac{-18}{-9}$ *Reverse sign*

$\quad\quad\quad z \leq 2$

The solution set is $(-\infty, 2]$.

15. $2 < -6(z+1) < 10$

$2 < -6z - 6 < 10$

$8 < -6z < 16$

$\dfrac{8}{-6} > \dfrac{-6z}{-6} > \dfrac{16}{-6}$

$-\dfrac{4}{3} > z > -\dfrac{8}{3}$

or $-\dfrac{8}{3} < z < -\dfrac{4}{3}$

The solution set is $\left(-\dfrac{8}{3}, -\dfrac{4}{3}\right)$.

16. Let x = amount earned in October.

$\dfrac{200 + 375 + 325 + x}{4} \geq 300$

Average is at least 300

$\dfrac{900 + x}{4} \geq 300$

$900 + x \geq 1200$

$x \geq 300$

She must earn $300 or more to average at least $300 for the four months.

17. Use the formula for the circumference of a circle.

$C = 2\pi r$

Let C = 39

$39 = 2\pi r$

$\dfrac{39}{2\pi} = r$

Thus, r is about 6.2 miles. To the nearest mile, the radius of the circular base is 6 miles.

18. Let $x = $ the number of liters of 20% chemical solution.

Number of liters of solution	x	$+$	30	$=$	$x + 30$
Strength of solution	$.20$		$.60$		$.50$

$.20x + .60(30) = .50(x+30)$

$2x + 6(30) = 5(x+30)$ *Multiply by 10*

$2x + 180 = 5x + 150$

$-3x + 180 = 150$

$-3x = -30$

$x = 10$

Use 10 liters of 20% solution.

19. If $x = 12$, then

$y = -.4685(12) + 95.07$

$= 89.448 \approx 89.45$

If $x = 28$, then

$y = -.4685(28) + 95.07$

$= 81.952 \approx 81.95$

If $x = 36$, then

$y = -.4685(36) + 95.07$

$= 78.204 \approx 78.20$

20. For the purchase of a home, we find 14% of $50,000.

$.14(\$50,000) = \$7000$

For retirement, we find 20% of $50,000.

$.20(\$50,000) = \$10,000$

21. To find the x-intercept, let $y = 0$.

$3x + 2y = 12$

$3x + 2(0) = 12$

$3x = 12$

$x = 4$

The x-intercept is $(4, 0)$.

To find the y-intercept, let $x = 0$.

$3x + 2y = 12$

$3(0) + 2y = 12$

$2y = 12$

$y = 6$

The y-intercept is $(0, 6)$.

22. To find a third point, let $x = 2$.

$$3x + 2y = 12$$
$$3(2) + 2y = 12$$
$$2y = 6$$
$$y = 3$$

Plot the points $(4, 0)$, $(0, 6)$, and $(2, 3)$ and draw a line through them.

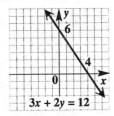

23. To find the slope of the line, solve the equation for y.

$$3x + 2y = 12$$
$$2y = -3x + 12$$
$$y = -\frac{3}{2}x + 6$$

The slope is the coefficient of x, $-\dfrac{3}{2}$.

24. $x + 5y = -6$

$$5y = -x - 6$$
$$y = -\frac{1}{5}x - \frac{6}{5}$$

The slope of the first line is $-\dfrac{1}{5}$.

The slope of the second line, $y = 5x - 8$, is 5.

Since $-\dfrac{1}{5}$ is the negative reciprocal of 5, the lines are *perpendicular*.

25. The set of ordered pairs represents a function since each importer corresponds to exactly one number (% change). The smallest change corresponds to the ordered pair (Taiwan, -6.6) and the largest is (South Korea, -40.5). To determine the smallest and largest *change*, in the usual sense, we need to consider the absolute value of the changes. Algebraically, the smallest change (leftmost number on a number line) is -40.5 and the largest change (rightmost number on a number line) is 28.7.

CHAPTER 5 POLYNOMIALS AND EXPONENTS

Section 5.1

1. In the term $7x^5$, the coefficient is 7 and the exponent is 5.

3. The degree of the term $-4x^8$ is 8, the exponent.

5. When $x^2 + 10$ is evaluated for $x = 4$, the result is
$$4^2 + 10 = 16 + 10 = 26.$$

7. $3xy + 2xy - 5xy = (3 + 2 - 5)xy$
$$= (0)xy$$
$$= 0$$

9. The polynomial $6x^4$ has one term. The coefficient of this term is 6.

11. The polynomial t^4 has one term. The coefficient of this term is 1.

13. The polynomial $-19r^2 - r$ has two terms. The coefficients are -19 and -1.

15. The polynomial $x + 8x^2 + 5x^3$ has three terms. The coefficient of x is 1, the coefficient of x^2 is 8, and the coefficient of x^3 is 5.

In Exercises 17–28, use the distributive property to add like terms.

17. $-3m^5 + 5m^5 = (-3 + 5)m^5 = 2m^5$

19. $2r^5 + (-3r^5) = [2 + (-3)]r^5$
$$= -1r^5 = -r^5$$

21. The polynomial $.2m^5 - .5m^2$ cannot be simplified. The two terms are unlike because the exponents on the variables are different.

23. $-3x^5 + 2x^5 - 4x^5 = (-3 + 2 - 4)x^5$
$$= -5x^5$$

25. $-4p^7 + 8p^7 + 5p^9 = (-4 + 8)p^7 + 5p^9$
$$= 4p^7 + 5p^9$$

In descending powers of the variable, this polynomial is written $5p^9 + 4p^7$.

27. $-4y^2 + 3y^2 - 2y^2 + y^2$
$$= (-4 + 3 - 2 + 1)y^2$$
$$= -2y^2$$

29. $6x^4 - 9x$

This polynomial has no like terms, so it is already simplified. It is already written in descending powers of the variable x. The highest degree of any nonzero term is 4, so the degree of the polynomial is 4. There are two terms, so this is a binomial.

31. $5m^4 - 3m^2 + 6m^4 - 7m^3$
$$= (5m^4 + 6m^4) + (-7m^3) + (-3m^2)$$
$$= 11m^4 - 7m^3 - 3m^2$$

The degree of the simplified polynomial is 4. The simplified polynomial is a trinomial.

33. $\frac{5}{3}x^4 - \frac{2}{3}x^4 = \left(\frac{5}{3} - \frac{2}{3}\right)x^4$
$$= \frac{3}{3}x^4 = x^4$$

The degree of the simplified polynomial is 4. The simplified polynomial is a monomial.

35. $.8x^4 - .3x^4 - .5x^4 + 7$
$$= (.8 - .3 - .5)x^4 + 7$$
$$= 0x^4 + 7 = 7$$

Since 7 can be written as $7x^0$, the degree is 0. The simplified polynomial has one term, so it is a monomial.

37. **(a)** $2x^5 - 4x^4 + 5x^3 - x^2$
$$= 2(2)^5 - 4(2)^4 + 5(2)^3 - (2)^2 \quad \textit{Let } x = 2$$
$$= 2(32) - 4(16) + 5(8) - 4$$
$$= 64 - 64 + 40 - 4$$
$$= 36$$

(b) $2x^5 - 4x^4 + 5x^3 - x^2$
$$= 2(-1)^5 - 4(-1)^4 + 5(-1)^3 - (-1)^2$$
$$\textit{Let } x = -1$$
$$= 2(-1) - 4(1) + 5(-1) - 1$$
$$= -2 - 4 - 5 - 1$$
$$= -12$$

39. **(a)** $-3x^2 + 14x - 2$
$$= -3(2)^2 + 14(2) - 2 \quad \textit{Let } x = 2$$
$$= -3(4) + 28 - 2$$
$$= -12 + 28 - 2$$
$$= 14$$

(b) $-3x^2 + 14x - 2$
$$= -3(-1)^2 + 14(-1) - 2 \quad \textit{Let } x = -1$$
$$= -3(1) - 14 - 2$$
$$= -3 - 14 - 2$$
$$= -19$$

41. $2x^2 - 3x - 5$
$$= 2(2)^2 - 3(2) - 5 \quad \textit{Let } x = 2$$
$$= 2(4) - 6 - 5$$
$$= 8 - 11$$
$$= -3$$

43. If $x = 4$,
$$1.25x = 1.25(4) = 5.00.$$

If $\underline{4}$ gallons are purchased, the cost is $\underline{\$5.00}$.

44. If $x = 6$,

$$2x + 15 = 2(6) + 15 = 12 + 15 = 27.$$

If the saw is rented for $\underline{6}$ days, the cost is $\underline{\$27}$.

45. If $x = 2.5$,

$$-16x^2 + 60x + 80$$
$$= -16(2.5)^2 + 60(2.5) + 80$$
$$= -16(6.25) + 150 + 80$$
$$= -100 + 230$$
$$= 130.$$

If $\underline{2.5}$ seconds have elapsed, the height of the object is $\underline{130}$ feet.

46. Using the hint that any power of 1 is equal to 1, we add the coefficients and the constant:

$$2.69 + 4.75 + 452.43 = 459.87$$

So the number of revenue passenger miles is approximately 460 billion.

47. Add, column by column.

$$\begin{array}{r} 3m^2 + 5m \\ 2m^2 - 2m \\ \hline 5m^2 + 3m \end{array}$$

49. Subtract.

$$\begin{array}{r} 12x^4 - x^2 \\ 8x^4 + 3x^2 \\ \hline \end{array}$$

Change all the signs in the second row, and then add,

$$\begin{array}{r} 12x^4 - x^2 \\ -8x^4 - 3x^2 \\ \hline 4x^4 - 4x^2 \end{array}$$

51. Add.

$$\begin{array}{r} \frac{2}{3}x^2 + \frac{1}{5}x + \frac{1}{6} \\ \frac{1}{2}x^2 - \frac{1}{3}x + \frac{2}{3} \\ \hline \end{array}$$

Rewrite the fractions so that the fractions in each column have a common denominator; then add column by column.

$$\begin{array}{r} \frac{4}{6}x^2 + \frac{3}{15}x + \frac{1}{6} \\ \frac{3}{6}x^2 - \frac{5}{15}x + \frac{4}{6} \\ \hline \frac{7}{6}x^2 - \frac{2}{15}x + \frac{5}{6} \end{array}$$

53. Add.

$$\begin{array}{r} 9m^3 - 5m^2 + 4m - 8 \\ -3m^3 + 6m^2 + 8m - 6 \\ \hline 6m^3 + m^2 + 12m - 14 \end{array}$$

55. Subtract.

$$\begin{array}{r} 12m^3 - 8m^2 + 6m + 7 \\ -3m^3 + 5m^2 - 2m - 4 \\ \hline \end{array}$$

Change all the signs in the second row, and then add.

$$\begin{array}{r} 12m^3 - 8m^2 + 6m + 7 \\ 3m^3 - 5m^2 + 2m + 4 \\ \hline 15m^3 - 13m^2 + 8m + 11 \end{array}$$

57. Vertical addition and subtraction of polynomials is preferable because like terms are arranged in columns.

59. $(8m^2 - 7m) - (3m^2 + 7m - 6)$
$$= (8m^2 - 7m) + (-3m^2 - 7m + 6)$$
$$= (8 - 3)m^2 + (-7 - 7)m + 6$$
$$= 5m^2 - 14m + 6$$

61. $(16x^3 - x^2 + 3x) + (-12x^3 + 3x^2 + 2x)$
$$= 16x^3 - x^2 + 3x - 12x^3 + 3x^2 + 2x$$
$$= (16 - 12)x^3 + (-1 + 3)x^2 + (3 + 2)x$$
$$= 4x^3 + 2x^2 + 5x$$

63. $(7y^4 + 3y^2 + 2y) - (18y^4 - 5y^2 + y)$
$$= (7y^4 + 3y^2 + 2y) + (-18y^4 + 5y^2 - y)$$
$$= (7 - 18)y^4 + (3 + 5)y^2 + (2 - 1)y$$
$$= -11y^4 + 8y^2 + y$$

65. $(9a^4 - 3a^2 + 2) + (4a^4 - 4a^2 + 2)$
$$+ (-12a^4 + 6a^2 - 3)$$
$$= (9a^4 + 4a^4 - 12a^4)$$
$$\quad + (-3a^2 - 4a^2 + 6a^2) + (2 + 2 - 3)$$
$$= a^4 - a^2 + 1$$

67. $[(8m^2 + 4m - 7) - (2m^2 - 5m + 2)]$
$$- (m^2 + m + 1)$$
$$= (8m^2 + 4m - 7) + (-2m^2 + 5m - 2)$$
$$\quad + (-m^2 - m - 1)$$
$$= (8 - 2 - 1)m^2 + (4 + 5 - 1)m$$
$$\quad + (-7 - 2 - 1)$$
$$= 5m^2 + 8m - 10$$

69. $[(3x^2 - 2x + 7) - (4x^2 + 2x - 3)]$
$$- [(9x^2 + 4x - 6) + (-4x^2 + 4x + 4)]$$
$$= [(3 - 4)x^2 + (-2 - 2)x + (7 + 3)]$$
$$\quad - [(9 - 4)x^2 + (4 + 4)x + (-6 + 4)]$$
$$= (-x^2 - 4x + 10) - (5x^2 + 8x - 2)$$
$$= -x^2 - 4x + 10 - 5x^2 - 8x + 2$$
$$= -6x^2 - 12x + 12$$

71. The coefficients of the x^2 terms are -4, $-(-2)$, and -8. The sum of these numbers is

$$-4 + 2 - 8 = -10.$$

73. $(6b + 3c) + (-2b - 8c)$
$= (6b - 2b) + (3c - 8c)$
$= 4b - 5c$

75. $(4x + 2xy - 3) - (-2x + 3xy + 4)$
$= (4x + 2xy - 3) + (2x - 3xy - 4)$
$= (4x + 2x) + (2xy - 3xy) + (-3 - 4)$
$= 6x - xy - 7$

77. $\left(5x^2y - 2xy + 9xy^2\right)$
$\quad - \left(8x^2y + 13xy + 12xy^2\right)$
$= \left(5x^2y - 2xy + 9xy^2\right)$
$\quad + \left(-8x^2y - 13xy - 12xy^2\right)$
$= \left(5x^2y - 8x^2y\right) + \left(-2xy - 13xy\right)$
$\quad + \left(9xy^2 - 12xy^2\right)$
$= -3x^2y - 15xy - 3xy^2$

79. The perimeter of a rectangle of length
$L = 4x^2 + 3x + 1$ and width $W = x + 2$ is

$P = 2L + 2W$
$\quad = 2\left(4x^2 + 3x + 1\right) + 2(x + 2)$
$\quad = 8x^2 + 6x + 2 + 2x + 4$
$\quad = 8x^2 + 8x + 6.$

81. (a) Use the formula for the perimeter of a
triangle, $P = a + b + c$, with $a = 2y - 3t$,
$b = 5y + 3t$, and $c = 16y + 5t$.

$P = (2y - 3t) + (5y + 3t) + (16y + 5t)$
$\quad = (2y + 5y + 16y) + (-3t + 3t + 5t)$
$\quad = 23y + 5t$

The perimeter of the triangle is $23y + 5t$.

(b) Use the fact that the sum of the angles of any
triangle is 180°.

$(10x + 3)° + (8x + 2)° + (5x - 1)° = 180°$
$(10x + 8x + 5x) + (3 + 2 - 1) = 180$
$23x + 4 = 180$
$23x = 176$
$x = \dfrac{176}{23}$

If $x = \dfrac{176}{23}$,

$10x + 3 = 10\left(\dfrac{176}{23}\right) + 3 \approx 79.52,$

$8x + 2 = 8\left(\dfrac{176}{23}\right) + 2 \approx 63.22,$

and $\quad 5x - 1 = 5\left(\dfrac{176}{23}\right) - 1 \approx 37.26.$

The measures of the angles are approximately
79.52°, 63.22°, and 37.26°.

83. $\left(3x^2 - 2\right) - \left(9x^2 - 6x + 5\right)$
$= \left(3x^2 - 2\right) + \left(-9x^2 + 6x - 5\right)$
$= \left(3x^2 - 9x^2\right) + 6x + (-2 - 5)$
$= -6x^2 + 6x - 7$

85. $\left[\left(5x^2 + 2x - 3\right) + \left(x^2 - 8x + 2\right)\right]$
$\quad - \left[\left(7x^2 - 3x + 6\right) + \left(-x^2 + 4x - 6\right)\right]$
$= \left(6x^2 - 6x - 1\right) - \left(6x^2 + x\right)$
$= \left(6x^2 - 6x - 1\right) + \left(-6x^2 - x\right)$
$= \left(6x^2 - 6x^2\right) + (-6x - x) + (-1)$
$= -7x - 1$

87. $y = x^2 - 4$
$x = -2 : y = (-2)^2 - 4 = 4 - 4 = 0$
$x = -1 : y = (-1)^2 - 4 = 1 - 4 = -3$
$x = 0 : y = (0)^2 - 4 = 0 - 4 = -4$
$x = 1 : y = (1)^2 - 4 = 1 - 4 = -3$
$x = 2 : y = (2)^2 - 4 = 4 - 4 = 0$

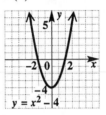

89. $y = 2x^2 - 1$
$x = -2 : y = 2(-2)^2 - 1 = 2 \cdot 4 - 1 = 7$
$x = -1 : y = 2(-1)^2 - 1 = 2 \cdot 1 - 1 = 1$
$x = 0 : y = 2(0)^2 - 1 = 2 \cdot 0 - 1 = -1$
$x = 1 : y = 2(1)^2 - 1 = 2 \cdot 1 - 1 = 1$
$x = 2 : y = 2(2)^2 - 1 = 2 \cdot 4 - 1 = 7$

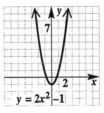

91. $y = 4 - x^2$
$x = -2 : y = 4 - (-2)^2 = 4 - 4 = 0$
$x = -1 : y = 4 - (-1)^2 = 4 - 1 = 3$
$x = 0 : y = 4 - (0)^2 = 4 - 0 = 4$
$x = 1 : y = 4 - (1)^2 = 4 - 1 = 3$
$x = 2 : y = 4 - (2)^2 = 4 - 4 = 0$

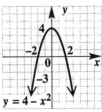

93. $y = (x+3)^2$

$x = -5 : y = (-5+3)^2 = (-2)^2 = 4$

$x = -4 : y = (-4+3)^2 = (-1)^2 = 1$

$x = -3 : y = (-3+3)^2 = (0)^2 = 0$

$x = -2 : y = (-2+3)^2 = (1)^2 = 1$

$x = -1 : y = (-1+3)^2 = (2)^2 = 4$

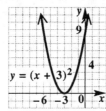

Section 5.2

1. $3^3 = 3 \cdot 3 \cdot 3 = 27$, so the statement $3^3 = 9$ is *false*.

3. $\left(a^2\right)^3 = a^{2(3)} = a^6$, so the statement $\left(a^2\right)^3 = a^5$ is *false*.

5. $w \cdot w \cdot w \cdot w \cdot w \cdot w = w^6$

7. $\dfrac{1}{4 \cdot 4 \cdot 4 \cdot 4} = \dfrac{1}{4^4}$

9. $(-7x)(-7x)(-7x)(-7x) = (-7x)^4$

11. $\left(\dfrac{1}{2}\right)\left(\dfrac{1}{2}\right)\left(\dfrac{1}{2}\right)\left(\dfrac{1}{2}\right)\left(\dfrac{1}{2}\right)\left(\dfrac{1}{2}\right) = \left(\dfrac{1}{2}\right)^6$

13. In $(-3)^4$, -3 is the base.

In -3^4, 3 is the base.

$(-3)^4 = (-3)(-3)(-3)(-3) = 81$

$-3^4 = -(3 \cdot 3 \cdot 3 \cdot 3) = -81$

15. In the exponential expression 3^5, the base is 3 and the exponent is 5.

$3^5 = 3 \cdot 3 \cdot 3 \cdot 3 \cdot 3 = 243$

17. In the expression $(-3)^5$, the base is -3 and the exponent is 5.

$(-3)^5 = (-3)(-3)(-3)(-3)(-3) = -243$

19. In the expression $(-6x)^4$, the base is $-6x$ and the exponent is 4.

21. In the expression $-6x^4$, -6 is not part of the base. The base is x and the exponent is 4.

23. The product rule does not apply to $5^2 + 5^3$ because the expression is a sum, not a product. The product rule would apply if we had $5^2 \cdot 5^3$.

$5^2 + 5^3 = 25 + 125$

$= 150$

25. $5^2 \cdot 5^6 = 5^{2+6} = 5^8$

27. $4^2 \cdot 4^7 \cdot 4^3 = 4^{2+7+3} = 4^{12}$

29. $(-7)^3(-7)^6 = (-7)^{3+6} = (-7)^9$

31. $t^3 \cdot t^8 \cdot t^{13} = t^{3+8+13} = t^{24}$

33. $\left(-8r^4\right)\left(7r^3\right) = -8 \cdot 7 \cdot r^4 \cdot r^3$

$= -56r^{4+3}$

$= -56r^7$

35. $\left(-6p^5\right)\left(-7p^5\right) = (-6)(-7)p^5 \cdot p^5$

$= (-6)(-7)p^{5+5}$

$= 42p^{10}$

37. The product rule does not apply to $3^2 \cdot 4^3$ because the bases are different. One base is 3 and the other base is 4. To use the product rule, both bases must be the same.

$3^2 \cdot 4^3 = 9 \cdot 64$

$= 576$

39. $(-2)\wedge 5 = (-2)(-2)(-2)(-2)(-2)$

$= -32$

$-2\wedge 5 = -1 \cdot 2 \cdot 2 \cdot 2 \cdot 2 \cdot 2$

$= -32$

Since the left side is equal to the right side, the statement is true and the calculator will return a 1.

41. $(-3)\wedge 6 = (-3)(-3)(-3)(-3)(-3)(-3)$

$= 729$

$-3\wedge 6 = -1 \cdot 3 \cdot 3 \cdot 3 \cdot 3 \cdot 3 \cdot 3$

$= -729$

Since the left side is not equal to the right side, the statement is false and the calculator will return a 0.

43. $\left(4^3\right)^2 = 4^{3 \cdot 2}$ *Power rule (a)*

$= 4^6$

45. $\left(t^4\right)^5 = t^{4 \cdot 5} = t^{20}$ *Power rule (a)*

47. $(7r)^3 = 7^3 r^3$ *Power rule (b)*

49. $(5xy)^5 = 5^5 x^5 y^5$ *Power rule (b)*

51. $\left(-5^2\right)^6 = \left(-1 \cdot 5^2\right)^6$

$= (-1)^6 \cdot \left(5^2\right)^6$ *Power rule (b)*

$= 1 \cdot 5^{2 \cdot 6}$ *Power rule (a)*

$= 1 \cdot 5^{12} = 5^{12}$

53. $\left(-8^3\right)^5 = \left(-1 \cdot 8^3\right)^5$

$= (-1)^5 \cdot \left(8^3\right)^5$ *Power rule (b)*

$= -1 \cdot 8^{3 \cdot 5}$ *Power rule (a)*

$= -8^{15}$

55. $8(qr)^3 = 8q^3 r^3$ *Power rule (b)*

57. $\left(\dfrac{1}{2}\right)^3 = \dfrac{1^3}{2^3} = \dfrac{1}{2^3}$ *Power rule (c)*

59. $\left(\dfrac{a}{b}\right)^3 (b \neq 0) = \dfrac{a^3}{b^3}$ *Power rule (c)*

61. $\left(\dfrac{9}{5}\right)^8 = \dfrac{9^8}{5^8}$ *Power rule (c)*

63. $\left(\dfrac{5}{2}\right)^3 \cdot \left(\dfrac{5}{2}\right)^2 = \left(\dfrac{5}{2}\right)^{3+2}$ *Product rule*

$= \left(\dfrac{5}{2}\right)^5$

$= \dfrac{5^5}{2^5}$ *Power rule (c)*

65. $\left(\dfrac{9}{8}\right)^3 \cdot 9^2 = \dfrac{9^3}{8^3} \cdot \dfrac{9^2}{1}$ *Power rule (c)*

$= \dfrac{9^3 \cdot 9^2}{8^3 \cdot 1}$ *Multiply fractions*

$= \dfrac{9^{3+2}}{8^3}$ *Product rule*

$= \dfrac{9^5}{8^3}$

67. $(2x)^9 (2x)^3 = (2x)^{9+3}$ *Product rule*

$= (2x)^{12}$

$= 2^{12} x^{12}$ *Product rule (b)*

69. $(-6p)^4 (-6p)$

$= (-6p)^4 (-6p)^1$

$= (-6p)^5$ *Product rule*

$= (-1)^5 6^5 p^5$ *Power rule (b)*

$= -6^5 p^5$

71. $(6x^2 y^3)^5 = 6^5 (x^2)^5 (y^3)^5$ *Power rule (b)*

$= 6^5 x^{2 \cdot 5} y^{3 \cdot 5}$ *Power rule (a)*

$= 6^5 x^{10} y^{15}$

73. $(x^2)^3 (x^3)^5 = x^6 \cdot x^{15}$ *Power rule (a)*

$= x^{21}$ *Product rule*

75. $(2w^2 x^3 y)^2 (x^4 y)^5$

$= \left[2^2 (w^2)^2 (x^3)^2 y^2\right] \left[(x^4)^5 y^5\right]$ *Power rule (b)*

$= (2^2 w^4 x^6 y^2)(x^{20} y^5)$ *Power rule (a)*

$= 2^2 w^4 (x^6 x^{20})(y^2 y^5)$

Commutative and associative properties

$= 2^2 w^4 x^{26} y^7$

77. $(-r^4 s)^2 (-r^2 s^3)^5$

$= \left[(-1)r^4 s\right]^2 \left[(-1)r^2 s^3\right]^5$

$= \left[(-1)^2 (r^4)^2 s^2\right] \left[(-1)^5 (r^2)^5 (s^3)^5\right]$

Power rule (b)

$= \left[(-1)^2 r^8 s^2\right] \left[(-1)^5 r^{10} s^{15}\right]$ *Power rule (a)*

$= (-1)^7 r^{18} s^{17}$ *Product rule*

$= -r^{18} s^{17}$

79. $\left(\dfrac{5a^2 b^5}{c^6}\right)^3$ $(c \neq 0)$

$= \dfrac{(5a^2 b^5)^3}{(c^6)^3}$ *Power rule (c)*

$= \dfrac{5^3 (a^2)^3 (b^5)^3}{(c^6)^3}$ *Power rule (b)*

$= \dfrac{5^3 a^6 b^{15}}{c^{18}}$ *Power rule (a)*

81. To simplify $(10^2)^3$ as 1000^6 is not correct. Using power rule (a) to simplify $(10^2)^3$, we obtain

$(10^2)^3 = 10^{2 \cdot 3}$

$= 10^6$

$= 10 \cdot 10 \cdot 10 \cdot 10 \cdot 10 \cdot 10$

$= 1,000,000.$

83. Use the formula for the area of a rectangle, $A = LW$, with $L = 4x^3$ and $W = 3x^2$.

$A = (4x^3)(3x^2)$

$= 4 \cdot 3 \cdot x^3 \cdot x^2$

$= 12x^5$

85. Use the formula for the area of a parallelogram, $A = bh$, with $b = 2p^5$ and $h = 3p^2$.

$A = (2p^5)(3p^2)$

$= 2 \cdot 3 \cdot p^5 \cdot p^2$

$= 6p^7$

87. Use the formula for the volume of a cube, $V = e^3$, with $e = 5x^2$.

$V = (5x^2)^3$

$= 5^3 x^6$

$= 125x^6$

89. If a is a positive number greater than 1, $a^4 > a^3$, $-(-a)^3$ is positive, $-a^3$ is negative, $(-a)^4$ is positive, and $-a^4$ is negative. Therefore, in order from smallest to largest, we have

$-a^4, \ -a^3, \ -(-a)^3, \ (-a)^4.$

Another way to determine the order is to choose a number greater than 1 and substitute it for a in each expression, and then arrange the terms from smallest to largest.

91. Use the formula $A = P(1 + r)^n$ with $P = 250$, $r = .04$, and $n = 5$.

$$A = 250(1 + .04)^5$$
$$= 250(1.04)^5$$
$$\approx 304.16$$

The amount of money in the account will be $304.16.

93. Use the formula $A = P(1 + r)^n$ with $P = 1500$, $r = .035$, and $n = 6$.

$$A = 1500(1 + .035)^6$$
$$= 1500(1.035)^6$$
$$\approx 1843.88$$

The amount of money in the account will be $1843.88.

Section 5.3

1. **(a)** $(5x^3)(6x^5)$

$$= 5 \cdot 6x^{3+5}$$
$$= 30x^8$$

Choice B is correct.

(b) $(-5x^5)(6x^3)$

$$= -5 \cdot 6x^{5+3}$$
$$= -30x^8$$

Choice D is correct.

(c) $(5x^5)^3 = (5)^3(x^5)^3$

$$= 125x^{5 \cdot 3}$$
$$= 125x^{15}$$

Choice A is correct.

(d) $(-6x^3)^3 = (-6)^3(x^3)^3$

$$= -216x^{3 \cdot 3}$$
$$= -216x^9$$

Choice C is correct.

3. $(-5a^9)(-8a^5) = (-5)(-8)a^{9+5}$
$$= 40a^{14}$$

5. $-2m(3m + 2) = (-2m)(3m) + (-2m)(2)$
$$= -6m^2 - 4m$$

7. $3p(8 - 6p + 12p^3)$

$$= (3p)(8) + (3p)(-6p) + (3p)(12p^3)$$
$$= 24p + (-18p^2) + 36p^4$$
$$= 24p - 18p^2 + 36p^4$$

9. $-8z(2z + 3z^2 + 3z^3)$

$$= (-8z)(2z) + (-8z)(3z^2)$$
$$+ (-8z)(3z^3)$$
$$= -16z^2 - 24z^3 - 24z^4$$

11. $7x^2y(2x^3y^2 + 3xy - 4y)$

$$= (7x^2y)(2x^3y^2) + (7x^2y)(3xy)$$
$$- (7x^2y)(4y)$$
$$= 14x^5y^3 + 21x^3y^2 - 28x^2y^2$$

In Exercises 13–20, we can multiply the polynomials horizontally or vertically. The following solutions illustrate these two methods.

13. $(6x + 1)(2x^2 + 4x + 1)$

$$= (6x)(2x^2) + (6x)(4x) + (6x)(1)$$
$$+ (1)(2x^2) + (1)(4x) + (1)(1)$$
$$= 12x^3 + 24x^2 + 6x + 2x^2 + 4x + 1$$
$$= 12x^3 + 26x^2 + 10x + 1$$

15. $(4m + 3)(5m^3 - 4m^2 + m - 5)$

Multiply vertically.

$$
\begin{array}{rrrrr}
5m^3 & - \ 4m^2 & + \ m & - \ 5 \\
& & 4m & + \ 3 \\
\hline
15m^3 & - \ 12m^2 & + \ 3m & - \ 15 \\
20m^4 & - \ 16m^3 & + \ 4m^2 & - \ 20m \\
\hline
20m^4 & - \ m^3 & - \ 8m^2 & - \ 17m & - \ 15
\end{array}
$$

17. $(2x - 1)(3x^5 - 2x^3 + x^2 - 2x + 3)$

Multiply vertically.

$$
\begin{array}{rrrrrr}
3x^5 & - \ 2x^3 & + \ x^2 & - \ 2x & + \ 3 \\
& & & 2x & - \ 1 \\
\hline
-3x^5 & + \ 2x^3 & - \ x^2 & + \ 2x & - \ 3 \\
6x^6 & - \ 4x^4 & + \ 2x^3 & - \ 4x^2 & + \ 6x \\
\hline
6x^6 - 3x^5 & - \ 4x^4 & + \ 4x^3 & - \ 5x^2 & + \ 8x & - \ 3
\end{array}
$$

19. $(5x^2 + 2x + 1)(x^2 - 3x + 5)$

Multiply vertically.

$$
\begin{array}{rrrrr}
5x^2 & + \ 2x & + \ 1 \\
x^2 & - \ 3x & + \ 5 \\
\hline
25x^2 & + \ 10x & + \ 5 \\
-15x^3 & - \ 6x^2 & - \ 3x \\
5x^4 & + \ 2x^3 & + \ x^2 \\
\hline
5x^4 & - \ 13x^3 & + \ 20x^2 & + \ 7x & + \ 5
\end{array}
$$

21. $(n - 2)(n + 3)$

$$
\begin{array}{cccc}
\mathbf{F} & \mathbf{O} & \mathbf{I} & \mathbf{L}
\end{array}
$$
$$= (n)(n) + (n)(3) + (-2)(n) + (-2)(3)$$
$$= n^2 + 3n - 2n - 6$$
$$= n^2 + n - 6$$

23. $(x+6)(x-6)$

$$\underset{\mathbf{F}}{}\quad\underset{\mathbf{O}}{}\quad\underset{\mathbf{I}}{}\quad\underset{\mathbf{L}}{}$$
$$= (x)(x) + (x)(-6) + (6)(x) + (6)(-6)$$
$$= x^2 - 6x + 6x - 36$$
$$= x^2 - 36$$

25. $(4r+1)(2r-3)$

$$\underset{\mathbf{F}}{}\quad\underset{\mathbf{O}}{}\quad\underset{\mathbf{I}}{}\quad\underset{\mathbf{L}}{}$$
$$= (4r)(2r) + (4r)(-3) + (1)(2r) + (1)(-3)$$
$$= 8r^2 - 12r + 2r - 3$$
$$= 8r^2 - 10r - 3$$

27. $(3x+2)(3x-2)$

$$= (3x)(3x) + (3x)(-2) + (2)(3x) + (2)(-2)$$
$$= 9x^2 - 6x + 6x - 4$$
$$= 9x^2 - 4$$

29. $(3q+1)(3q+1)$

$$= (3q)(3q) + (3q)(1) + (1)(3q) + (1)(1)$$
$$= 9q^2 + 3q + 3q + 1$$
$$= 9q^2 + 6q + 1$$

31. $(3x+y)(x-2y)$

$$= (3x)(x) + (3x)(-2y) + (y)(x) + (y)(-2y)$$
$$= 3x^2 - 6xy + xy - 2y^2$$
$$= 3x^2 - 5xy - 2y^2$$

33. $(-3t+4)(t+6)$

$$= (-3t)(t) + (-3t)(6) + (4)(t) + (4)(6)$$
$$= -3t^2 - 18t + 4t + 24$$
$$= -3t^2 - 14t + 24$$

35. Use the FOIL method to multiply the two binomials.

$$(2y+3)(y-5)$$
$$= (2y)(y) + (2y)(-5) + (3)(y) + (3)(-5)$$
$$= 2y^2 - 10y + 3y - 15$$
$$= 2y^2 - 7y - 15$$

Now multiply this result by $3y^3$.

$$3y^3\left(2y^2 - 7y - 15\right)$$
$$= \left(3y^3\right)\left(2y^2\right) + \left(3y^3\right)(-7y) + \left(3y^3\right)(-15)$$
$$= 6y^5 - 21y^4 - 45y^3$$

37. Use the formula for the area of a square, $A = s^2$, with $s = 6x + 2$.

$$A = (6x+2)^2$$
$$= (6x+2)(6x+2)$$
$$= (6x)(6x) + (6x)(2) + (2)(6x) + (2)(2)$$
$$= 36x^2 + 12x + 12x + 4$$
$$= 36x^2 + 24x + 4$$

39. $(x+4)(x-4)$
$$= (x)(x) + (x)(-4) + (4)(x) + (4)(-4)$$
$$= x^2 - 16$$
$$(y+2)(y-2)$$
$$= (y)(y) + (y)(-2) + (2)(y) + (2)(-2)$$
$$= y^2 - 4$$
$$(r+7)(r-7)$$
$$= (r)(r) + (r)(-7) + (7)(r) + (7)(-7)$$
$$= r^2 - 49$$

We observe that the product of the sum and difference of the same two numbers is the difference of their squares.

41. $\left(3p + \dfrac{5}{4}q\right)\left(2p - \dfrac{5}{3}q\right)$ *Use FOIL*

$$= (3p)(2p) + (3p)\left(-\frac{5}{3}q\right) + \left(\frac{5}{4}q\right)(2p)$$
$$\quad + \left(\frac{5}{4}q\right)\left(-\frac{5}{3}q\right)$$
$$= 6p^2 + (3)\left(-\frac{5}{3}\right)pq + \left(\frac{5}{4}\right)(2)pq$$
$$\quad + \left(\frac{5}{4}\right)\left(-\frac{5}{3}\right)q^2$$
$$= 6p^2 - 5pq + \frac{5}{2}pq - \frac{25}{12}q^2$$
$$= 6p^2 + \left(-\frac{10}{2} + \frac{5}{2}\right)pq - \frac{25}{12}q^2$$
$$= 6p^2 - \frac{5}{2}pq - \frac{25}{12}q^2$$

43. $\left(m^3 - 4\right)\left(2m^3 + 3\right)$ *Use FOIL*
$$= \left(m^3\right)\left(2m^3\right) + \left(m^3\right)(3) + (-4)\left(2m^3\right)$$
$$\quad + (-4)(3)$$
$$= 2m^6 + 3m^3 - 8m^3 - 12$$
$$= 2m^6 - 5m^3 - 12$$

45. $\left(2k^3 + h^2\right)\left(k^2 - 3h^2\right)$
$$= \left(2k^3\right)\left(k^2\right) + \left(2k^3\right)\left(-3h^2\right) + \left(h^2\right)\left(k^2\right)$$
$$\quad + \left(h^2\right)\left(-3h^2\right)$$
$$= 2k^5 - 6k^3h^2 + h^2k^2 - 3h^4$$

47. $3p^3(2p^2 + 5p)(p^3 + 2p + 1)$

$= [3p^3(2p^2) + 3p^3(5p)](p^3 + 2p + 1)$

$\qquad\qquad$ *Distributive property*

$= (6p^5 + 15p^4)(p^3 + 2p + 1)$

Now multiply vertically.

$$
\begin{array}{r}
p^3 + 2p + 1 \\
6p^5 + 15p^4 \\
\hline
15p^7 \qquad\quad + 30p^5 + 15p^4 \\
6p^8 \qquad + 12p^6 + 6p^5 \\
\hline
6p^8 + 15p^7 + 12p^6 + 36p^5 + 15p^4
\end{array}
$$

49. $-2x^5(3x^2 + 2x - 5)(4x + 2)$

$= [(-2x^5)(3x^2) + (-2x^5)(2x)$

$\quad + (-2x^5)(-5)](4x + 2)$

$\qquad\qquad$ *Distributive property*

$= (-6x^7 - 4x^6 + 10x^5)(4x + 2)$

Now multiply vertically.

$$
\begin{array}{r}
- 6x^7 - 4x^6 + 10x^5 \\
4x + 2 \\
\hline
- 12x^7 - 8x^6 + 20x^5 \\
- 24x^8 - 16x^7 + 40x^6 \\
\hline
- 24x^8 - 28x^7 + 32x^6 + 20x^5
\end{array}
$$

51. The area A of the shaded region is the difference between the area of the larger square, which has sides of length $x + 7$, and the area of the smaller square, which has sides of length x.

$A = (x + 7)^2 - (x)^2$

$= (x + 7)(x + 7) - x^2$

$= [(x)(x) + (x)(7) + (7)(x) + (7)(7)] - x^2$

$= [x^2 + 7x + 7x + 49] - x^2$

$= x^2 + 14x + 49 - x^2$

$= 14x + 49$

53. The area A of the shaded region is the difference between the area of the circle, which has radius x, and the area of the square, which has sides of length 3.

$A = \pi r^2 - s^2$

$= \pi(x)^2 - (3)^2$

$= \pi x^2 - 9$

55. Use the formula $A = LW$ with $L = 3x + 6$ and $W = 10$.

$A = (3x + 6)(10)$

$= (3x)(10) + (6)(10)$

$= 30x + 60$

56. If the area is 600 square yards, we have the equation

$$30x + 60 = 600.$$

57. $30x + 60 = 600$

$\qquad 30x = 540$

$\qquad\quad x = \dfrac{540}{30} = 18$

58. If $x = 18$,

$$3x + 6 = 3(18) + 6$$

$$= 54 + 6$$

$$= 60.$$

The dimensions of the rectangle are 10 yards by 60 yards.

59. To find the cost of covering the lawn with sod, we use area. Since the area is 600 square yards and the cost of sod is $3.50 per square yard, the total cost is given by

$$(600)(3.50) = 2100.$$

The total cost is $2100.

60. Use the formula $P = 2L + 2W$ with $L = 60$ and $W = 10$.

$$P = 2(60) + 2(10)$$

$$= 120 + 20$$

$$= 140$$

The perimeter is 140 yards.

61. To find the cost of constructing a fence around the lawn, we use perimeter. If the cost is $9.00 per yard to fence the lawn and 140 yards must be fenced, the total cost to fence the yard is given by

$$(9.00)(140) = 1260.$$

The total cost is $1260.

62. **(a)** From Exercise 55, the area of the lawn is given by the polynomial $30x + 60$. If it costs k dollars per square yard to sod the lawn, the cost to sod the entire lawn will be

$$(30x + 60)(k) = 30xk + 60k \text{ (dollars)}$$

(b) The total cost to fence the lawn is given by multiplying the perimeter by r.

Use $P = 2L + 2W$ with $L = 3x + 6$ and $W = 10$.

$$P = 2(3x + 6) + 2(10)$$

$$= 6x + 12 + 20$$

$$= 6x + 32$$

The total cost is given by

$$(6x + 32)(r) = 6xr + 32r.$$

The total cost is $6xr + 32r$ (dollars).

63. Suppose that we want to multiply $(x + 3)(2x + 5)$. The letter F stands for *first*. We multiply the two first terms to get $2x^2$. The letter O represents *outer*. We next multiply the two outer terms to get

$5x$. The letter I represents *inner*. The product of the two inner terms is $6x$. L stands for *last*. The product of the two last terms is 15. Very often the outer and inner products are like terms, as they are in this case. So we simplify $2x^2 + 5x + 6x + 15$ to get the final product, $2x^2 + 11x + 15$.

Section 5.4

1. $(2x + 3)^2$

   (a) The square of the first term is
   $$(2x)^2 = (2x)(2x) = 4x^2.$$

   (b) Twice the product of the two terms is
   $$2(2x)(3) = 12x.$$

   (c) The square of the last term is
   $$3^2 = 9.$$

   (d) The final product is the trinomial
   $$4x^2 + 12x + 9.$$

In Exercises 3–16, use one of the following formulas for the square of a binomial:
$$(a + b)^2 = a^2 + 2ab + b^2$$
$$(a - b)^2 = a^2 - 2ab + b^2$$

3. $(p + 2)^2 = p^2 + 2(p)(2) + 2^2$
   $$= p^2 + 4p + 4$$

5. $(a - c)^2 = a^2 - 2(a)(c) + c^2$
   $$= a^2 - 2ac + c^2$$

7. $(4x - 3)^2 = (4x)^2 - 2(4x)(3) + 3^2$
   $$= 16x^2 - 24x + 9$$

9. $(8t + 7s)^2 = (8t)^2 + 2(8t)(7s) + (7s)^2$
   $$= 64t^2 + 112ts + 49s^2$$

11. $\left(5x + \dfrac{2}{5}y\right)^2 = (5x)^2 + 2(5x)\left(\dfrac{2}{5}y\right) + \left(\dfrac{2}{5}y\right)^2$
    $$= 25x^2 + 4xy + \dfrac{4}{25}y^2$$

13. $(2x + 5)^2 = (2x)^2 + 2(2x)(5) + 5^2$
    $$= 4x^2 + 20x + 25$$

    Now multiply by x.
    $$x(4x^2 + 20x + 25) = 4x^3 + 20x^2 + 25x$$

15. $(4r - 2)^2 = (4r)^2 - 2(4r)(2) + 2^2$
    $$= 16r^2 - 16r + 4$$

    Now multiply by -1.
    $$-1(16r^2 - 16r + 4) = -16r^2 + 16r - 4$$

17. **(a)** $(7x)(7x) = 49x^2$

    (b) $(7x)(-3y) + (3y)(7x)$
    $$= -21xy + 21xy = 0$$

    (c) $(3y)(-3y) = -9y^2$

    (d) $49x^2 - 9y^2$

The sum found in (b) is omitted because it is 0. Adding 0, the identity element for addition, would not change the answer.

In Exercises 19–30, use the formula for the product of the sum and difference of two terms.
$$(a + b)(a - b) = a^2 - b^2$$

19. $(q + 2)(q - 2) = q^2 - 2^2$
    $$= q^2 - 4$$

21. $(2w + 5)(2w - 5) = (2w)^2 - 5^2$
    $$= 4w^2 - 25$$

23. $(10x + 3y)(10x - 3y) = (10x)^2 - (3y)^2$
    $$= 100x^2 - 9y^2$$

25. $(2x^2 - 5)(2x^2 + 5) = (2x^2)^2 - 5^2$
    $$= 4x^4 - 25$$

27. $\left(7x + \dfrac{3}{7}\right)\left(7x - \dfrac{3}{7}\right) = (7x)^2 - \left(\dfrac{3}{7}\right)^2$
    $$= 49x^2 - \dfrac{9}{49}$$

29. $(3p + 7)(3p - 7) = (3p)^2 - 7^2$
    $$= 9p^2 - 49$$

    Now multiply by p.
    $$p(9p^2 - 49) = 9p^3 - 49p$$

31. The large square has side of length $a + b$, so its area is $(a + b)^2$.

32. The red square has side of length a, so its area is a^2.

33. Each blue rectangle has length a and width b, so each has an area of ab. Thus, the sum of the areas of the blue rectangles is
    $$ab + ab = 2ab.$$

34. The yellow square has side b, so its area is b^2.

35. Sum $= a^2 + 2ab + b^2$

36. The area of the largest square equals the sum of the areas of the two smaller squares and the two rectangles. Therefore, $(a + b)^2$ must equal $a^2 + 2ab + b^2$.

37. $35^2 = (35)(35)$

$$
\begin{array}{r}
35 \\
35 \\
\hline
175 \\
105 \\
\hline
1225
\end{array}
$$

38. $(a+b)^2 = a^2 + 2ab + b^2$
$(30+5)^2 = 30^2 + 2(30)(5) + 5^2$

39. $30^2 + 2(30)(5) + 5^2$
$= 900 + 60(5) + 25$
$= 900 + 300 + 25$
$= 1225$

40. The answers are equal.

41. $101 \times 99 = (100+1)(100-1)$
$= 100^2 - 1^2$
$= 10,000 - 1$
$= 9999$

43. $201 \times 199 = (200+1)(200-1)$
$= 200^2 - 1^2$
$= 40,000 - 1$
$= 39,999$

45. $20\frac{1}{2} \times 19\frac{1}{2} = \left(20 + \frac{1}{2}\right)\left(20 - \frac{1}{2}\right)$
$= 20^2 - \left(\frac{1}{2}\right)^2$
$= 400 - \frac{1}{4}$
$= 399\frac{3}{4}$

47. $(m-5)^3$
$= (m-5)^2(m-5) \quad a^3 = a^2 \cdot a$
$= (m^2 - 10m + 25)(m-5)$ *Square the binomial*
$= m^3 - 10m^2 + 25m$
$\quad - 5m^2 + 50m - 125$ *Multiply polynomials*
$= m^3 - 15m^2 + 75m - 125$ *Combine like terms*

49. $(2a+1)^3$
$= (2a+1)^2(2a+1) \quad a^3 = a^2 \cdot a$
$= (4a^2 + 4a + 1)(2a+1)$ *Square the binomial*
$= 8a^3 + 8a^2 + 2a$
$\quad + 4a^2 + 4a + 1$ *Multiply polynomials*
$= 8a^3 + 12a^2 + 6a + 1$ *Combine like terms*

51. $(3r - 2t)^4$
$= (3r - 2t)^2(3r - 2t)^2 \quad a^4 = a^2 \cdot a^2$
$= (9r^2 - 12rt + 4t^2)(9r^2 - 12rt + 4t^2)$
Square each binomial
$= 81r^4 - 108r^3t + 36r^2t^2 - 108r^3t$
$\quad + 144r^2t^2 - 48rt^3 + 36r^2t^2 - 48rt^3 + 16t^4$
Multiply polynomials
$= 81r^4 - 216r^3t + 216r^2t^2 - 96rt^3 + 16t^4$
Combine like terms

53. $(x+y)^2$ is $x^2 + 2xy + y^2$, which is a trinomial. $x^2 + y^2$ is a binomial. Therefore,
$$x^2 + y^2 \neq (x+y)^2.$$

55. Use the formula for area of a triangle, $A = \frac{1}{2}bh$. Let $b = m + 2n$ and $h = m - 2n$.
$$A = \frac{1}{2} \cdot b \cdot h$$
$$= \frac{1}{2}(m+2n)(m-2n)$$
$$= \frac{1}{2}[m^2 - (2n)^2]$$
$$= \frac{1}{2}(m^2 - 4n^2)$$
$$= \frac{1}{2}m^2 - 2n^2$$

57. Use the formula for the area of a parallelogram, $A = bh$, with $b = 3a + 2$ and $h = 3a - 2$.
$$A = b \cdot h$$
$$= (3a+2)(3a-2)$$
$$= (3a)^2 - 2^2$$
$$= 9a^2 - 4$$

59. Use the formula for the area of a circle, $A = \pi r^2$, with $r = x + 2$.
$$A = \pi(x+2)^2$$
$$= \pi(x^2 + 4x + 4)$$
$$= \pi x^2 + 4\pi x + 4\pi$$

61. $V = (x+2)^3$
$= (x+2)^2(x+2)$
$= (x^2 + 4x + 4)(x+2)$
$= x^2(x) + x^2(2) + 4x(x)$
$\quad + 4x(2) + 4(x) + 4(2)$
$= x^3 + 2x^2 + 4x^2 + 8x + 4x + 8$
$= x^3 + 6x^2 + 12x + 8$

Section 5.5

1. $(-2)^{-4} = \dfrac{1}{(-2)^4}$ *Definition of negative exponent*

$= \dfrac{1}{16}$

Since $\dfrac{1}{16}$ is positive (greater than zero), the

statement $(-2)^{-4} < 0$ is *false*.

3. $(-5)^0 = 1$ *Definition of zero exponent*

The statement $(-5)^0 > 0$ is *true*.

5. $1 - 12^0 = 1 - 1$ *Definition of zero exponent*

$= 0$

The statement $1 - 12^0 = 1$ is *false*.

7. By definition, $a^0 = 1 \ (a \neq 0)$, so $9^0 = 1$.

9. $(-4)^0 = 1$ *Definition of zero exponent*

11. $-9^0 = -(9^0) = -(1) = -1$

13. $(-2)^0 - 2^0 = 1 - 1 = 0$

15. $\dfrac{0^{10}}{10^0} = \dfrac{0}{1} = 0$

17. $7^0 + 9^0 = 1 + 1 = 2$

19. $4^{-3} = \dfrac{1}{4^3}$ *Definition of negative exponent*

$= \dfrac{1}{64}$

21. When we evaluate a fraction raised to a negative exponent, we can use a shortcut. Note that

$$\left(\frac{a}{b}\right)^{-n} = \frac{1}{\left(\dfrac{a}{b}\right)^n} = \frac{1}{\dfrac{a^n}{b^n}} = \frac{b^n}{a^n} = \left(\frac{b}{a}\right)^n.$$

In words, a fraction raised to the negative of a number is equal to its reciprocal raised to the number. We will use the simple phrase "$\dfrac{a}{b}$ and $\dfrac{b}{a}$ are reciprocals" to indicate our use of this evaluation shortcut.

$\left(\dfrac{1}{2}\right)^{-4} = 2^4 = 16$ $\frac{1}{2}$ *and 2 are reciprocals*

23. $\left(\dfrac{6}{7}\right)^{-2} = \left(\dfrac{7}{6}\right)^2$ $\frac{6}{7}$ *and* $\frac{7}{6}$ *are reciprocals*

$= \dfrac{7^2}{6^2}$ *Power rule (c)*

$= \dfrac{49}{36}$

25. $(-3)^{-4} = \dfrac{1}{(-3)^4}$

$= \dfrac{1}{81}$

27. $5^{-1} + 3^{-1} = \dfrac{1}{5} + \dfrac{1}{3}$

$= \dfrac{3}{15} + \dfrac{5}{15} = \dfrac{8}{15}$

29. $\dfrac{5^8}{5^5} = 5^{8-5} = 5^3$ *Quotient rule*

31. $\dfrac{9^4}{9^5} = 9^{4-5}$ *Quotient rule*

$= 9^{-1}$

$= \dfrac{1}{9}$ *Definition of negative exponent*

33. $\dfrac{5}{5^{-1}} = \dfrac{5^1}{5^{-1}} = 5^1 \cdot 5^1$ *Changing from negative to positive exponents*

$= 5^{1+1} = 5^2$

35. $\dfrac{x^{12}}{x^{-3}} = x^{12} \cdot x^3$ *Changing from negative to positive exponents*

$= x^{12+3} = x^{15}$

37. $\dfrac{1}{6^{-3}} = 6^3$ *Changing from negative to positive exponents*

39. $\dfrac{2}{r^{-4}} = 2r^4$

41. $\dfrac{4^{-3}}{5^{-2}} = \dfrac{5^2}{4^3}$

43. $p^5 q^{-8} = \dfrac{p^5}{q^8}$

45. $\dfrac{r^5}{r^{-4}} = r^5 \cdot r^4 = r^{5+4} = r^9$

47. $\dfrac{6^4 x^8}{6^5 x^3} = 6^{4-5} x^{8-3} = 6^{-1} x^5 = \dfrac{x^5}{6}$

49. Treat the expression in parentheses as a single variable; that is, treat $(a + b)$ as you would treat x.

$$\dfrac{(a+b)^{-3}}{(a+b)^{-4}} = (a+b)^{-3-(-4)}$$

$$= (a+b)^{-3+4}$$

$$= (a+b)^1 = a + b$$

Another Method:

$$\dfrac{(a+b)^{-3}}{(a+b)^{-4}} = \dfrac{(a+b)^4}{(a+b)^3}$$

$$= (a+b)^{4-3}$$

$$= (a+b)^1 = a + b$$

51. $\dfrac{(x+2y)^{-3}}{(x+2y)^{-5}} = (x+2y)^{-3-(-5)}$

$\qquad\qquad\qquad = (x+2y)^{-3+5}$

$\qquad\qquad\qquad = (x+2y)^2$

53. In simplest form,

$$\dfrac{25}{25} = 1.$$

54. $\dfrac{25}{25} = \dfrac{5^2}{5^2}$

55. $\dfrac{5^2}{5^2} = 5^{2-2} = 5^0$

56. $1 = 5^0$

This supports the definition for zero as an exponent.

57. $\dfrac{\left(7^4\right)^3}{7^9} = \dfrac{7^{4\cdot 3}}{7^9}$ *Power rule (a)*

$\qquad\quad = \dfrac{7^{12}}{7^9}$

$\qquad\quad = 7^{12-9}$ *Quotient rule*

$\qquad\quad = 7^3$ or 343

59. $x^{-3} \cdot x^5 \cdot x^{-4}$

$\quad = x^{-3+5+(-4)}$ *Product rule*

$\quad = x^{-2}$

$\quad = \dfrac{1}{x^2}$ *Definition of negative exponent*

61. $\dfrac{(3x)^{-2}}{(4x)^{-3}} = \dfrac{(4x)^3}{(3x)^2}$ *Changing from negative to positive exponents*

$\qquad\qquad = \dfrac{4^3 x^3}{3^2 x^2}$ *Power rule (b)*

$\qquad\qquad = \dfrac{4^3 x^{3-2}}{3^2}$ *Quotient rule*

$\qquad\qquad = \dfrac{4^3 x}{3^2} = \dfrac{64x}{9}$

63. $\left(\dfrac{x^{-1}y}{z^2}\right)^{-2} = \dfrac{\left(x^{-1}y\right)^{-2}}{\left(z^2\right)^{-2}}$ *Power rule (c)*

$\qquad\qquad = \dfrac{\left(x^{-1}\right)^{-2}y^{-2}}{\left(z^2\right)^{-2}}$ *Power rule (b)*

$\qquad\qquad = \dfrac{x^2 y^{-2}}{z^{-4}}$ *Power rule (a)*

$\qquad\qquad = \dfrac{x^2 z^4}{y^2}$ *Definition of negative exponent*

65. $(6x)^4 (6x)^{-3} = (6x)^{4+(-3)}$ *Product rule*

$\qquad\qquad\quad = (6x)^1 = 6x$

67. $\dfrac{\left(m^7 n\right)^{-2}}{m^{-4}n^3} = \dfrac{\left(m^7\right)^{-2}n^{-2}}{m^{-4}n^3}$

$\qquad\qquad = \dfrac{m^{7(-2)}n^{-2}}{m^{-4}n^3}$

$\qquad\qquad = \dfrac{m^{-14}n^{-2}}{m^{-4}n^3}$

$\qquad\qquad = m^{-14-(-4)}n^{-2-3}$

$\qquad\qquad = m^{-10}n^{-5}$

$\qquad\qquad = \dfrac{1}{m^{10}n^5}$

69. $\dfrac{\left(x^{-1}y^2 z\right)^{-2}}{\left(x^{-3}y^3 z\right)^{-1}} = \dfrac{\left(x^{-1}\right)^{-2}\left(y^2\right)^{-2}z^{-2}}{\left(x^{-3}\right)^{-1}\left(y^3\right)^{-1}z^{-1}}$

$\qquad\qquad = \dfrac{x^2 y^{-4} z^{-2}}{x^3 y^{-3} z^{-1}}$

$\qquad\qquad = \dfrac{x^2 y^3 z^1}{x^3 y^4 z^2}$

$\qquad\qquad = \dfrac{1}{xyz}$

71. $\left(\dfrac{xy^{-2}}{x^2 y}\right)^{-3} = \dfrac{x^{-3}\left(y^{-2}\right)^{-3}}{\left(x^2\right)^{-3}y^{-3}}$

$\qquad\qquad = \dfrac{x^{-3}y^6}{x^{-6}y^{-3}}$

$\qquad\qquad = \dfrac{x^6 y^6 y^3}{x^3}$

$\qquad\qquad = x^3 y^9$

73. The student attempted to use the quotient rule with unequal bases. The correct way to simplify this expression is

$$\dfrac{16^3}{2^2} = \dfrac{\left(2^4\right)^3}{2^2} = \dfrac{2^{12}}{2^2} = 2^{10} = 1024.$$

Section 5.6

1. In the statement $\dfrac{6x^2 + 8}{2} = 3x^2 + 4$, $6x^2 + 8$ is the dividend, 2 is the divisor, and $3x^2 + 4$ is the quotient.

3. To check the division shown in Exercise 1, multiply $3x^2 + 4$ by 2 (or 2 by $3x^2 + 4$) and show that the product is $6x^2 + 8$.

5. In this section, we are dividing a polynomial by a monomial. The problem

$$\dfrac{16m^3 - 12m^2}{4m}$$

is an example of such a division. However, in the problem

$$\dfrac{4m}{16m^3 - 12m^2},$$

we are dividing a polynomial by a binomial. Therefore, the methods of this section do not apply.

7. $\dfrac{60x^4 - 20x^2 + 10x}{2x}$

$= \dfrac{60x^4}{2x} - \dfrac{20x^2}{2x} + \dfrac{10x}{2x}$

$= \dfrac{60}{2}x^{4-1} - \dfrac{20}{2}x^{2-1} + \dfrac{10}{2}$

$= 30x^3 - 10x + 5$

9. $\dfrac{20m^5 - 10m^4 + 5m^2}{5m^2}$

$= \dfrac{20m^5}{5m^2} - \dfrac{10m^4}{5m^2} + \dfrac{5m^2}{5m^2}$

$= \dfrac{20}{5}m^{5-2} - \dfrac{10}{5}m^{4-2} + \dfrac{5}{5}$

$= 4m^3 - 2m^2 + 1$

11. $\dfrac{8t^5 - 4t^3 + 4t^2}{2t}$

$= \dfrac{8t^5}{2t} - \dfrac{4t^3}{2t} + \dfrac{4t^2}{2t}$

$= 4t^4 - 2t^2 + 2t$

13. $\dfrac{4a^5 - 4a^2 + 8}{4a}$

$= \dfrac{4a^5}{4a} - \dfrac{4a^2}{4a} + \dfrac{8}{4a}$

$= a^4 - a + \dfrac{2}{a}$

15. $\dfrac{12x^5 - 9x^4 + 6x^3}{3x^2}$

$= \dfrac{12x^5}{3x^2} - \dfrac{9x^4}{3x^2} + \dfrac{6x^3}{3x^2}$

$= 4x^3 - 3x^2 + 2x$

17. $\dfrac{3x^2 + 15x^3 - 27x^4}{3x^2}$

$= \dfrac{3x^2}{3x^2} + \dfrac{15x^3}{3x^2} - \dfrac{27x^4}{3x^2}$

$= 1 + 5x - 9x^2$

19. $\dfrac{36x + 24x^2 + 6x^3}{3x^2}$

$= \dfrac{36x}{3x^2} + \dfrac{24x^2}{3x^2} + \dfrac{6x^3}{3x^2}$

$= \dfrac{12}{x} + 8 + 2x$

21. $\dfrac{4x^4 + 3x^3 + 2x}{3x^2}$

$= \dfrac{4x^4}{3x^2} + \dfrac{3x^3}{3x^2} + \dfrac{2x}{3x^2}$

$= \dfrac{4x^2}{3} + x + \dfrac{2}{3x}$

23. $\dfrac{-27r^4 + 36r^3 - 6r^2 - 26r + 2}{-3r}$

$= \dfrac{-27r^4}{-3r} + \dfrac{36r^3}{-3r} - \dfrac{6r^2}{-3r} - \dfrac{26r}{-3r} + \dfrac{2}{-3r}$

$= 9r^3 - 12r^2 + 2r + \dfrac{26}{3} - \dfrac{2}{3r}$

25. $\dfrac{2m^5 - 6m^4 + 8m^2}{-2m^3}$

$= \dfrac{2m^5}{-2m^3} - \dfrac{6m^4}{-2m^3} + \dfrac{8m^2}{-2m^3}$

$= -m^2 + 3m - \dfrac{4}{m}$

27. $\left(20a^4 - 15a^5 + 25a^3\right) \div \left(5a^4\right)$

$= \dfrac{20a^4 - 15a^5 + 25a^3}{5a^4}$

$= \dfrac{20a^4}{5a^4} - \dfrac{15a^5}{5a^4} + \dfrac{25a^3}{5a^4}$

$= 4 - 3a + \dfrac{5}{a}$

29. $\left(120x^{11} - 60x^{10} + 140x^9 - 100x^8\right) \div \left(10x^{12}\right)$

$= \dfrac{120x^{11} - 60x^{10} + 140x^9 - 100x^8}{10x^{12}}$

$= \dfrac{120x^{11}}{10x^{12}} - \dfrac{60x^{10}}{10x^{12}} + \dfrac{140x^9}{10x^{12}} - \dfrac{100x^8}{10x^{12}}$

$= \dfrac{12}{x} - \dfrac{6}{x^2} + \dfrac{14}{x^3} - \dfrac{10}{x^4}$

31. $\dfrac{2}{3x}$ is not the same as $\dfrac{2}{3}x$.

$$\dfrac{2}{3}x = \dfrac{2}{3} \cdot \dfrac{x}{1} = \dfrac{2x}{3}$$

Note that in $\dfrac{2}{3x}$, x is in the denominator, while in $\dfrac{2}{3}x = \dfrac{2x}{3}$, x is in the numerator.

33. We use $A = LW$ with

$$A = 15x^3 + 12x^2 - 9x + 3$$

and $W = 3$.

$$15x^3 + 12x^2 - 9x + 3 = L(3)$$

$$\dfrac{15x^3 + 12x^2 - 9x + 3}{3} = L$$

$$L = \dfrac{15x^3}{3} + \dfrac{12x^2}{3} - \dfrac{9x}{3} + \dfrac{3}{3}$$

$$L = 5x^3 + 4x^2 - 3x + 1$$

35. "The quotient of a certain polynomial and $-7m^2$ is $9m^2 + 3m + 5 - \dfrac{2}{m}$" means that the polynomial is the product of the factors $-7m^2$ and $9m^2 + 3m + 5 - \dfrac{2}{m}$.

Multiply these factors to determine the polynomial.

$$\left(-7m^2\right)\left(9m^2 + 3m + 5 - \dfrac{2}{m}\right)$$

$$= \left(-7m^2\right)\left(9m^2\right) + \left(-7m^2\right)(3m)$$

$$+ \left(-7m^2\right)(5) + \left(-7m^2\right)\left(-\dfrac{2}{m}\right)$$

$$= -63m^4 - 21m^3 - 35m^2 + 14m$$

37.
$$2 \overline{\smash{\big)}\,2846} \quad 1423$$

38.
$$1423 = \left(1 \times 10^3\right) + \left(4 \times 10^2\right)$$
$$+ \left(2 \times 10^1\right) + \left(3 \times 10^0\right)$$

39.
$$\frac{2x^3 + 8x^2 + 4x + 6}{2}$$
$$= \frac{2x^3}{2} + \frac{8x^2}{2} + \frac{4x}{2} + \frac{6}{2}$$
$$= x^3 + 4x^2 + 2x + 3$$

40. They are similar in that the coefficients of the powers of ten are equal to the coefficients of the powers of x. They are different in that one is a constant, while the other is a polynomial. They are equal if $x = 10$ (the base of our decimal system).

41.
$$\frac{x^2 - x - 6}{x - 3}$$

$$\begin{array}{r} x + 2 \\ x - 3 \overline{\smash{\big)}\, x^2 - x - 6} \\ \underline{x^2 - 3x} \\ 2x - 6 \\ \underline{2x - 6} \\ 0 \end{array}$$

The remainder is 0. The answer is the quotient, $x + 2$.

43.
$$\frac{2y^2 + 9y - 35}{y + 7}$$

$$\begin{array}{r} 2y - 5 \\ y + 7 \overline{\smash{\big)}\, 2y^2 + 9y - 35} \\ \underline{2y^2 + 14y} \\ -5y - 35 \\ \underline{-5y - 35} \\ 0 \end{array}$$

The remainder is 0. The answer is the quotient, $2y - 5$.

45.
$$\frac{p^2 + 2p + 20}{p + 6}$$

$$\begin{array}{r} p - 4 \\ p + 6 \overline{\smash{\big)}\, p^2 + 2p + 20} \\ \underline{p^2 + 6p} \\ -4p + 20 \\ \underline{-4p - 24} \\ 44 \end{array}$$

The remainder is 44. Write the remainder as the numerator of a fraction that has the divisor $p + 6$ as its denominator. The answer is

$$p - 4 + \frac{44}{p + 6}.$$

47. $\left(r^2 - 8r + 15\right) \div (r - 3)$

$$\begin{array}{r} r - 5 \\ r - 3 \overline{\smash{\big)}\, r^2 - 8r + 15} \\ \underline{r^2 - 3r} \\ -5r + 15 \\ \underline{-5r + 15} \\ 0 \end{array}$$

The remainder is 0. The answer is the quotient, $r - 5$.

49.
$$\frac{12m^2 - 20m + 3}{2m - 3}$$

$$\begin{array}{r} 6m - 1 \\ 2m - 3 \overline{\smash{\big)}\, 12m^2 - 20m + 3} \\ \underline{12m^2 - 18m} \\ -2m + 3 \\ \underline{-2m + 3} \\ 0 \end{array}$$

The remainder is 0. The answer is the quotient, $6m - 1$.

51.
$$\frac{4a^2 - 22a + 32}{2a + 3}$$

$$\begin{array}{r} 2a - 14 \\ 2a + 3 \overline{\smash{\big)}\, 4a^2 - 22a + 32} \\ \underline{4a^2 + 6a} \\ -28a + 32 \\ \underline{-28a - 42} \\ 74 \end{array}$$

The remainder is 74. The answer is

$$2a - 14 + \frac{74}{2a + 3}.$$

53.
$$\frac{8x^3 - 10x^2 - x + 3}{2x + 1}$$

$$\begin{array}{r} 4x^2 - 7x + 3 \\ 2x + 1 \overline{\smash{\big)}\, 8x^3 - 10x^2 - x + 3} \\ \underline{8x^3 + 4x^2} \\ -14x^2 - x \\ \underline{-14x^2 - 7x} \\ 6x + 3 \\ \underline{6x + 3} \\ 0 \end{array}$$

The remainder is 0. The answer is the quotient,

$$4x^2 - 7x + 3.$$

55. $\dfrac{8k^4 - 12k^3 - 2k^2 + 7k - 6}{2k - 3}$

$$
\begin{array}{r}
4k^3 \qquad\quad - k + 2 \\
2k - 3 \,\big|\, 8k^4 - 12k^3 - 2k^2 + 7k - 6 \\
\underline{8k^4 - 12k^3} \\
-2k^2 + 7k \\
\underline{-2k^2 + 3k} \\
+4k - 6 \\
\underline{+4k - 6} \\
0
\end{array}
$$

The remainder is 0.

The quotient is $4k^3 - k + 2$.

57.
$$
\begin{array}{r}
5y^3 \qquad\quad + 2y - 3 \\
y + 1 \,\big|\, 5y^4 + 5y^3 + 2y^2 - y - 3 \\
\underline{5y^4 + 5y^3} \\
2y^2 - y \\
\underline{2y^2 + 2y} \\
-3y - 3 \\
\underline{-3y - 3} \\
0
\end{array}
$$

The remainder is 0.

The quotient is $5y^3 + 2y - 3$.

59. $\dfrac{3k^3 - 4k^2 - 6k + 10}{k - 2}$

$$
\begin{array}{r}
3k^2 + 2k - 2 \\
k - 2 \,\big|\, 3k^3 - 4k^2 - 6k + 10 \\
\underline{3k^3 - 6k^2} \\
2k^2 - 6k \\
\underline{2k^2 - 4k} \\
-2k + 10 \\
\underline{-2k + 4} \\
6
\end{array}
$$

The remainder is 6.

The quotient is $3k^2 + 2k - 2$.

The answer is $3k^2 + 2k - 2 + \dfrac{6}{k - 2}$.

61. $\dfrac{6p^4 - 16p^3 + 15p^2 - 5p + 10}{3p + 1}$

$$
\begin{array}{r}
2p^3 - 6p^2 + 7p - 4 \\
3p + 1 \,\big|\, 6p^4 - 16p^3 + 15p^2 - 5p + 10 \\
\underline{6p^4 + 2p^3} \\
-18p^3 + 15p^2 \\
\underline{-18p^3 - 6p^2} \\
21p^2 - 5p \\
\underline{21p^2 + 7p} \\
-12p + 10 \\
\underline{-12p - 4} \\
14
\end{array}
$$

The remainder is 14.

The quotient is $2p^3 - 6p^2 + 7p - 4$.

The answer is $2p^3 - 6p^2 + 7p - 4 + \dfrac{14}{3p + 1}$.

63. $\dfrac{5 - 2r^2 + r^4}{r^2 - 1}$

Insert $0r$ in the divisor for the missing term. Rearrange terms of the dividend in descending powers and insert missing terms.

$$
\begin{array}{r}
r^2 \qquad\quad - 1 \\
r^2 + 0r - 1 \,\big|\, r^4 + 0r^3 - 2r^2 + 0r + 5 \\
\underline{r^4 + 0r^3 - r^2} \\
-r^2 + 0r + 5 \\
\underline{-r^2 + 0r + 1} \\
4
\end{array}
$$

The remainder is 4.

The quotient is $r^2 - 1$.

The answer is $r^2 - 1 + \dfrac{4}{r^2 - 1}$.

65. $\dfrac{y^3 + 1}{y + 1}$

$$
\begin{array}{r}
y^2 - y + 1 \\
y + 1 \,\big|\, y^3 + 0y^2 + 0y + 1 \\
\underline{y^3 + y^2} \\
-y^2 + 0y \\
\underline{-y^2 - y} \\
y + 1 \\
\underline{y + 1} \\
0
\end{array}
$$

The remainder is 0. The answer is the quotient, $y^2 - y + 1$.

67. $\dfrac{a^4 - 1}{a^2 - 1}$

$$
\begin{array}{r}
a^2 \qquad\qquad\quad +1 \\
a^2 + 0a - 1 \overline{\smash{)}\; a^4 \;\; + 0a^3 \;\; + 0a^2 \;\; + 0a \;\; - 1} \\
\underline{a^4 \;\; + 0a^3 \;\; - \;\; a^2} \\
a^2 \;\; + 0a \;\; - 1 \\
\underline{a^2 \;\; + 0a \;\; - 1} \\
0
\end{array}
$$

The remainder is 0. The answer is the quotient, $a^2 + 1$.

69. $\dfrac{x^4 - 4x^3 + 5x^2 - 3x + 2}{x^2 + 3}$

$$
\begin{array}{r}
x^2 \;\; - \;\; 4x \;\; + 2 \\
x^2 + 0x + 3 \overline{\smash{)}\; x^4 \;\; - 4x^3 \;\; + 5x^2 \;\; - \;\; 3x \;\; + 2} \\
\underline{x^4 \;\; + 0x^3 \;\; + 3x^2} \\
- 4x^3 \;\; + 2x^2 \;\; - \;\; 3x \\
\underline{- 4x^3 \;\; + 0x^2 \;\; - 12x} \\
2x^2 \;\; + \;\; 9x \;\; + 2 \\
\underline{2x^2 \;\; + \;\; 0x \;\; + 6} \\
9x \;\; - 4
\end{array}
$$

$$\frac{x^4 - 4x^3 + 5x^2 - 3x + 2}{x^2 + 3}$$

$$= x^2 - 4x + 2 + \frac{9x - 4}{x^2 + 3}$$

71. $\dfrac{2x^5 + 9x^4 + 8x^3 + 10x^2 + 14x + 5}{2x^2 + 3x + 1}$

$$
\begin{array}{r}
x^3 \;\; + \;\; 3x^2 \;\; - \;\; x \;\; + 5 \\
2x^2 + 3x + 1 \overline{\smash{)}\; 2x^5 \;\; + 9x^4 \;\; + 8x^3 \;\; + 10x^2 \;\; + 14x \;\; + 5} \\
\underline{2x^5 \;\; + 3x^4 \;\; + \;\; x^3} \\
6x^4 \;\; + 7x^3 \;\; + 10x^2 \\
\underline{6x^4 \;\; + 9x^3 \;\; + \;\; 3x^2} \\
- 2x^3 \;\; + \;\; 7x^2 \;\; + 14x \\
\underline{- 2x^3 \;\; - \;\; 3x^2 \;\; - \;\; x} \\
10x^2 \;\; + 15x \;\; + 5 \\
\underline{10x^2 \;\; + 15x \;\; + 5} \\
0
\end{array}
$$

The remainder is 0. The answer is the quotient, $x^3 + 3x^2 - x + 5$.

73. $\left(3a^2 - 11a + 17\right) \div (2a + 6)$

$$
\begin{array}{r}
\frac{3}{2}a \;\; - 10 \\
2a + 6 \overline{\smash{)}\; 3a^2 \;\; - 11a \;\; + 17} \\
\underline{3a^2 \;\; + \;\; 9a} \\
- 20a \;\; + 17 \\
\underline{- 20a \;\; - 60} \\
77
\end{array}
$$

The remainder is 77.

The quotient is $\dfrac{3}{2}a - 10$.

The answer is $\dfrac{3}{2}a - 10 + \dfrac{77}{2a + 6}$.

75. The division process stops when the degree of the remainder is less than the degree of the divisor.

77. Use $A = LW$ with

$$A = 5x^3 + 7x^2 - 13x - 6$$

and $W = 5x + 2$.

$$5x^3 + 7x^2 - 13x - 6 = L(5x + 2)$$

$$L = \frac{5x^3 + 7x^2 - 13x - 6}{5x + 2}$$

$$
\begin{array}{r}
x^2 \;\; + \;\; x \;\; - 3 \\
5x + 2 \overline{\smash{)}\; 5x^3 \;\; + 7x^2 \;\; - 13x \;\; - 6} \\
\underline{5x^3 \;\; + 2x^2} \\
5x^2 \;\; - 13x \\
\underline{5x^2 \;\; + \;\; 2x} \\
- 15x \;\; - 6 \\
\underline{- 15x \;\; - 6} \\
0
\end{array}
$$

The length is $x^2 + x - 3$ units.

79. Use the distance formula, $d = rt$, with $d = 5x^3 - 6x^2 + 3x + 14$ miles and $r = x + 1$ miles per hour.

$$5x^3 - 6x^2 + 3x + 14 = (x + 1)t$$

$$t = \frac{5x^3 - 6x^2 + 3x + 14}{x + 1}$$

$$
\begin{array}{r}
5x^2 \;\; - 11x \;\; + 14 \\
x + 1 \overline{\smash{)}\; 5x^3 \;\; - \;\; 6x^2 \;\; + \;\; 3x \;\; + 14} \\
\underline{5x^3 \;\; + \;\; 5x^2} \\
- 11x^2 \;\; + \;\; 3x \\
\underline{- 11x^2 \;\; - 11x} \\
14x \;\; + 14 \\
\underline{14x \;\; + 14} \\
0
\end{array}
$$

The time is $5x^2 - 11x + 14$ hours.

81. (a) $2x + 7$

(b) $2x - 7$

If $x = 1$,

$$2x + 7 = 2(1) + 7 = 9.$$

If $x = 1$,

$$2x - 7 = 2(1) - 7 = -5.$$

If $x = 1$,

$$\frac{2x^2 + 3x - 14}{x - 2} = \frac{2(1)^2 + 3(1) - 14}{1 - 2}$$
$$= \frac{-9}{-1} = 9.$$

Therefore, (a) is correct, and (b) is incorrect.

82. (a) $x^2 - 4x - 2$

(b) $x^2 + 4x - 2$

If $x = 1$,

$x^2 - 4x - 2 = (1)^2 - 4(1) - 2 = -5.$

If $x = 1$,

$x^2 + 4x - 2 = (1)^2 + 4(1) - 2 = 3.$

If $x = 1$,

$$\frac{x^4 + 4x^3 - 5x^2 - 12x + 6}{x^2 - 3}$$
$$= \frac{(1)^4 + 4(1)^3 - 5(1)^2 - 12(1) + 6}{(1)^2 - 3}$$
$$= \frac{-6}{-2} = 3.$$

Therefore, (a) is incorrect, and (b) is correct.

83. (a) $y^2 + 5y + 1$

(b) $y^2 - 5y + 1$

If $y = 1$,

$y^2 + 5y + 1 = (1)^2 + 5(1) + 1 = 7.$

If $y = 1$,

$y^2 - 5y + 1 = (1)^2 - 5(1) + 1 = -3.$

If $y = 1$,

$$\frac{2y^3 + 17y^2 + 37y + 7}{2y + 7}$$
$$= \frac{2(1)^3 + 17(1)^2 + 37(1) + 7}{2(1) + 7}$$
$$= \frac{63}{9} = 7.$$

Therefore, (a) is correct, and (b) is incorrect.

84. A polynomial is easy to evaluate for 1 because the value is simply the sum of the coefficients of the polynomial. We would not be able to use 1 if the divisor is $x - 1$ because the quotient would then have 0 as the value of its denominator and would thus be undefined.

Section 5.7

1. Move the decimal point to the left 4 places due to the exponent, -4.

$4.6 \times 10^{-4} = .00046$

Choice A is correct.

3. Move the decimal point to the right 5 places due to the exponent, 5.

$4.6 \times 10^5 = 460,000$

Choice C is correct.

5. 4.56×10^3 is written in scientific notation because 4.56 is between 1 and 10, and 10^3 is a power of 10.

7. $5,600,000$ is not in scientific notation. It can be written in scientific notation as 5.6×10^6.

9. $.8 \times 10^2$ is not in scientific notation because $|.8| = .8$ is not greater than or equal to 1 and less than 10. It can be written in scientific notation as 8×10^1.

11. $.004$ is not in scientific notation because $|.004| = .004$ is not between 1 and 10. It can be written in scientific notation as 4×10^{-3}.

13. A number is written in scientific notation if it is written as a product of two numbers, the first of which has absolute value less than 10 and greater than or equal to 1 and the second of which is a power of 10. Some examples are 2.3×10^{-4} and 6.02×10^{23}.

15. $5,876,000,000$

Move the decimal point to the right of the first nonzero digit and count the number of places the decimal point was moved.

$5.876,000,000$ *9 places*

Because moving the decimal point to the *left* made the number *smaller*, we must multiply by a *positive* power of 10 so that the product 5.876×10^n will equal the larger number. Thus, $n = 9$, and

$5,876,000,000 = 5.876 \times 10^9.$

17. $82,350$

Move the decimal point left 4 places so it is to the right of the first nonzero digit.

8.2350 *4 places*

Since the number got smaller, multiply by a positive power of 10.

$82,350 = 8.2350 \times 10^4 = 8.235 \times 10^4$

(Note that the final zero need not be written.)

19. .000007

Move the decimal point to the right of the first nonzero digit.

$$.000007. \quad \textit{6 places}$$

Since moving the decimal point to the *right* made the number *larger*, we must multiply by a *negative* power of 10 so that the product 7×10^n will equal the smaller number. Thus, $n = -6$, and

$$.000007 = 7 \times 10^{-6}.$$

21. .00203

We move the decimal point to the right of the first nonzero digit. We move it 3 places. Since 2.03 is greater than .00203, the exponent on 10 must be negative.

$$.00203 = 2.03 \times 10^{-3}$$

23. 7.5×10^5

Because the exponent is positive, make 7.5 larger by moving the decimal point 5 places to the right.

$$7.5 \times 10^5 = 750,000$$

25. 5.677×10^{12}

Since the exponent is positive, make 5.677 larger by moving the decimal point 12 places to the right. We need to add 9 zeros.

$$5.677 \times 10^{12} = 5,677,000,000,000$$

27. 6.21×10^0

Because the exponent is 0, the decimal point should not be moved.

$$6.21 \times 10^0 = 6.21$$

We know this result is correct because $10^0 = 1$.

29. 7.8×10^{-4}

Since the exponent is negative, move the decimal point 4 places to the left.

$$7.8 \times 10^{-4} = .00078$$

31. 5.134×10^{-9}

Because the exponent is negative, make 5.134 smaller by moving the decimal point 9 places to the left.

$$5.134 \times 10^{-9} = .000000005134$$

33. $\left(2 \times 10^8\right) \times \left(3 \times 10^3\right)$

$$
\begin{aligned}
&= (2 \times 3)\left(10^8 \times 10^3\right) \quad \textit{Commutative and associative properties} \\
&= 6 \times 10^{11} \quad \textit{Product rule for exponents} \\
&= 600,000,000,000
\end{aligned}
$$

35. $\left(5 \times 10^4\right) \times \left(3 \times 10^2\right)$

$$
\begin{aligned}
&= (5 \times 3)\left(10^4 \times 10^2\right) \\
&= 15 \times 10^6 \\
&= 15,000,000
\end{aligned}
$$

37. $\left(3 \times 10^{-4}\right) \times \left(2 \times 10^8\right)$

$$
\begin{aligned}
&= (3 \times 2)\left(10^{-4} \times 10^8\right) \\
&= 6 \times 10^4 = 60,000
\end{aligned}
$$

39. $\dfrac{9 \times 10^{-5}}{3 \times 10^{-1}} = \dfrac{9}{3} \times \dfrac{10^{-5}}{10^{-1}}$

$$
\begin{aligned}
&= 3 \times 10^{-5-(-1)} \\
&= 3 \times 10^{-4} \\
&= .0003
\end{aligned}
$$

41. $\dfrac{8 \times 10^3}{2 \times 10^2} = \dfrac{8}{2} \times \dfrac{10^3}{10^2}$

$$
\begin{aligned}
&= 4 \times 10^1 \\
&= 40
\end{aligned}
$$

43. $\dfrac{2.6 \times 10^{-3}}{2 \times 10^2} = \dfrac{2.6}{2} \times \dfrac{10^{-3}}{10^2}$

$$
\begin{aligned}
&= 1.3 \times 10^{-5} \\
&= .000013
\end{aligned}
$$

45. To work in scientific mode on the TI-83, press MODE and then change from "Normal" to "Sci."

$$.00000047 = 4.7 \times 10^{-7}$$

Prediction: 4.7 E -7

47. $(8\,\text{E}5)/(4\,\text{E}\,{-}2) = \dfrac{8 \times 10^5}{4 \times 10^{-2}}$

$$
\begin{aligned}
&= \frac{8}{4} \times 10^{5-(-2)} \\
&= 2 \times 10^7
\end{aligned}
$$

Prediction: 2 E7

49. $(2\,\text{E}6)*(2\,\text{E}\,{-}3)/(4\,\text{E}2)$

$$
\begin{aligned}
&= \frac{(2 \times 10^6)(2 \times 10^{-3})}{4 \times 10^2} \\
&= \frac{(2 \times 2) \times (10^6 \times 10^{-3})}{4 \times 10^2} \\
&= \frac{4 \times 10^3}{4 \times 10^2} \\
&= \frac{4}{4} \times 10^{3-2} \\
&= 1 \times 10^1
\end{aligned}
$$

Prediction: 1 E1

51. 10 billion $= 10,000,000,000$
$$= 1 \times 10^{10}$$

53. $2 \times 10^9 = 2,000,000,000$

55. $\$7,326,000,000 = \$7.326 \times 10^9$

57. $\$30,262,000,000 = \$3.0262 \times 10^{10}$

59. Solve the formula $d = rt$ for t.

$$t = \frac{d}{r}$$
$$= \frac{4.58 \times 10^9}{3.00 \times 10^5}$$
$$= \frac{4.58}{3.00} \times 10^{9-5}$$
$$\approx 1.5267 \times 10^4$$
$$\approx 15,300 \text{ seconds}$$

It took about $15,300$ seconds for the signals to reach Earth.

61. **(a)** For the year 1995, 106.3 billion can be written as $106,300,000,000$. In scientific notation, this is 1.063×10^{11}.

1996: 1.124×10^{11}

1997: 1.250×10^{11}

(b) The slope of the line segment between the tops of the bars for 1995 and 1997 is

$$m = \frac{125.0 - 106.3}{1997 - 1995}$$
$$= \frac{18.7}{2}$$
$$= 9.35 \text{ in } \frac{\text{billion dollars}}{\text{yr}}$$

or $9.35 \times 10^9 \dfrac{\text{dollars}}{\text{yr}}$

Chapter 5 Review Exercises

1. $9m^2 + 11m^2 + 2m^2 = (9 + 11 + 12)m^2$
$$= 22m^2$$

The degree is 2.

To determine if the polynomial is a monomial, binomial, or trinomial, count the number of terms in the final expression.

There is one term, so this is a monomial.

2. $-4p + p^3 - p^2 + 8p + 2$
$$= p^3 - p^2 + (-4 + 8)p + 2$$
$$= p^3 - p^2 + 4p + 2$$

The degree is 3.

To determine if the polynomial is a monomial, binomial, or trinomial, count the number of terms in the final expression. Since there are four terms, it is none of these.

3. $12a^5 - 9a^4 + 8a^3 + 2a^2 - a + 3$ cannot be simplified further and is already written in descending powers.

The degree is 5.

This polynomial has 6 terms, so it is none of the names listed.

4. $-7y^5 - 8y^4 - y^5 + y^4 + 9y$
$$= -7y^5 - 1y^5 - 8y^4 + 1y^4 + 9y$$
$$= (-7 - 1)y^5 + (-8 + 1)y^4 + 9y$$
$$= -8y^5 - 7y^4 + 9y$$

The degree is 5.

There are three terms, so the polynomial is a trinomial.

5. $(12r^4 - 7r^3 + 2r^2) - (5r^4 - 3r^3 + 2r^2 - 1)$
$$= (12r^4 - 7r^3 + 2r^2)$$
$$\quad + (-5r^4 + 3r^3 - 2r^2 + 1)$$

Change signs in the second polynomial and add
$$= 12r^4 - 7r^3 + 2r^2 - 5r^4 + 3r^3 - 2r^2 + 1$$
$$= 7r^4 - 4r^3 + 1$$

The degree is 4.

The polynomial is a trinomial.

6. $(5x^3y^2 - 3xy^5 + 12x^2)$
$$\quad - (-9x^2 - 8x^3y^2 + 2xy^5)$$
$$= (5x^3y^2 - 3xy^5 + 12x^2)$$
$$\quad + (9x^2 + 8x^3y^2 - 2xy^5)$$
$$= (5x^3y^2 + 8x^3y^2) + (-3xy^5 - 2xy^5)$$
$$\quad + (12x^2 + 9x^2)$$
$$= 13x^3y^2 - 5xy^5 + 21x^2$$

7. Add.

$$\begin{array}{r} -2a^3 + 5a^2 \\ 3a^3 - a^2 \\ \hline a^3 + 4a^2 \end{array}$$

8. Subtract.

$$\begin{array}{r} 6y^2 - 8y + 2 \\ 5y^2 + 2y - 7 \end{array}$$

Change all signs in the second row and then add.

$$\begin{array}{r} 6y^2 - 8y + 2 \\ -5y^2 - 2y + 7 \\ \hline y^2 - 10y + 9 \end{array}$$

9. Subtract.

$$-12k^4 - 8k^2 + 7k$$
$$\underline{\quad k^4 + 7k^2 - 11k}$$

Change all signs in the second row and then add.

$$-12k^4 - 8k^2 + 7k$$
$$\underline{-k^4 - 7k^2 + 11k}$$
$$-13k^4 - 15k^2 + 18k$$

10. $y = -x^2 + 5$

If $x = -2$, then

$$y = -(-2)^2 + 5$$
$$= -(4) + 5 = 1.$$

Other values can be found in a similar fashion.

x	-2	-1	0	1	2
y	1	4	5	4	1

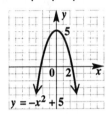

11. $y = 3x^2 - 2$

If $x = -2$, then

$$y = 3(-2)^2 - 2$$
$$= 3(4) - 2$$
$$= 12 - 2 = 10.$$

x	-2	-1	0	1	2
y	10	1	-2	1	10

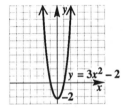

12. The product rule does not apply to $7^2 + 7^4$ because you are adding powers of 7, not multiplying them.

13. $4^3 \cdot 4^8 = 4^{3+8} = 4^{11}$

14. $(-5)^6(-5)^5 = (-5)^{6+5} = (-5)^{11}$

15. $(-8x^4)(9x^3) = (-8)(9)(x^4)(x^3)$
$$= -72x^{4+3} = -72x^7$$

16. $(2x^2)(5x^3)(x^9) = (2)(5)(x^2)(x^3)(x^9)$
$$= 10x^{2+3+9} = 10x^{14}$$

17. $(19x)^5 = 19^5 x^5$

18. $(-4y)^7 = (-4)^7 y^7$

19. $5(pt)^4 = 5p^4 t^4$

20. $\left(\dfrac{7}{5}\right)^6 = \dfrac{7^6}{5^6}$

21. $(6x^2 z^4)^2 (x^3 y z^2)^4$
$$= 6^2 (x^2)^2 (z^4)^2 (x^3)^4 (y)^4 (z^2)^4$$
$$= 6^2 x^4 z^8 x^{12} y^4 z^8$$
$$= 6^2 x^{4+12} y^4 z^{8+8}$$
$$= 6^2 x^{16} y^4 z^{16}$$

22. $\left(\dfrac{2m^3 n}{p^2}\right)^3 = \dfrac{2^3 (m^3)^3 n^3}{(p^2)^3}$
$$= \dfrac{2^3 m^9 n^3}{p^6}$$

23. $(a + 2)(a^2 - 4a + 1)$

Multiply vertically.

$$a^2 - 4a + 1$$
$$\underline{\qquad\qquad a + 2}$$
$$2a^2 - 8a + 2$$
$$\underline{a^3 - 4a^2 + \quad a}$$
$$a^3 - 2a^2 - 7a + 2$$

24. $(3r - 2)(2r^2 + 4r - 3)$

Multiply vertically.

$$2r^2 + 4r - 3$$
$$\underline{\qquad\qquad 3r - 2}$$
$$-\quad 4r^2 - 8r + 6$$
$$\underline{6r^3 + 12r^2 - 9r}$$
$$6r^3 + 8r^2 - 17r + 6$$

25. $(5p^2 + 3p)(p^3 - p^2 + 5)$
$$= 5p^2(p^3) + 5p^2(-p^2) + 5p^2(5)$$
$$\quad + 3p(p^3) + 3p(-p^2) + 3p(5)$$
$$= 5p^5 - 5p^4 + 25p^2 + 3p^4 - 3p^3 + 15p$$
$$= 5p^5 - 2p^4 - 3p^3 + 25p^2 + 15p$$

26. $(m - 9)(m + 2)$

$$\begin{array}{cccc} \mathbf{F} & \mathbf{O} & \mathbf{I} & \mathbf{L} \end{array}$$
$$= (m)(m) + (m)(2) + (-9)(m) + (-9)(2)$$
$$= m^2 + 2m - 9m - 18$$
$$= m^2 - 7m - 18$$

27. $(3k - 6)(2k + 1)$

$$\begin{array}{cccc} \mathbf{F} & \mathbf{O} & \mathbf{I} & \mathbf{L} \end{array}$$
$$= (3k)(2k) + (3k)(1) + (-6)(2k) + (-6)(1)$$
$$= 6k^2 + 3k - 12k - 6$$
$$= 6k^2 - 9k - 6$$

28. $(a + 3b)(2a - b)$

$$\overset{\mathbf{F}}{} \quad \overset{\mathbf{O}}{} \quad \overset{\mathbf{I}}{} \quad \overset{\mathbf{L}}{}$$
$= (a)(2a) + (a)(-b) + (3b)(2a) + (3b)(-b)$
$= 2a^2 - ab + 6ab - 3b^2$
$= 2a^2 + 5ab - 3b^2$

29. $(6k + 5q)(2k - 7q)$

$$\overset{\mathbf{F}}{} \quad \overset{\mathbf{O}}{} \quad \overset{\mathbf{I}}{} \quad \overset{\mathbf{L}}{}$$
$= (6k)(2k) + (6k)(-7q) + (5q)(2k) + (5q)(-7q)$
$= 12k^2 - 42kq + 10kq - 35q^2$
$= 12k^2 - 32kq - 35q^2$

30. $(s - 1)^3 = (s - 1)(s - 1)(s - 1)$
$ = (s^2 - 2s + 1)(s - 1)$

Now, we will use vertical multiplication.

$$
\begin{array}{r}
s^2 - 2s + 1 \\
s - 1 \\
\hline
-\ s^2 + 2s - 1 \\
s^3 - 2s^2 + s \\
\hline
s^3 - 3s^2 + 3s - 1
\end{array}
$$

$(s - 1)^3 = s^3 - 3s^2 + 3s - 1$

31. Use the formula $A = LW$ with $L = 2x - 3$ and $W = x + 2$.

$A = (2x - 3)(x + 2)$
$ = (2x)(x) + (2x)(2) + (-3)(x) + (-3)(2)$
$ = 2x^2 + 4x - 3x - 6$
$ = 2x^2 + x - 6$

The area of the rectangle is

$2x^2 + x - 6$ square units.

32. Use the formula $A = s^2$ with

$$s = 5x^4 + 2x^2$$

$A = (5x^4 + 2x^2)^2$
$ = (5x^4)^2 + 2(5x^4)(2x^2) + (2x^2)^2$
$ = 25x^8 + 20x^6 + 4x^4$

The area of the square is

$$25x^8 + 20x^6 + 4x^4$$

square units.

33. $(a + 4)^2 = a^2 + 2(a)(4) + 4^2$
$ = a^2 + 8a + 16$

34. $(2r + 5t)^2 = (2r)^2 + 2(2r)(5t) + (5t)^2$
$ = 4r^2 + 20rt + 25t^2$

35. $(6m - 5)(6m + 5) = (6m)^2 - 5^2$
$ = 36m^2 - 25$

36. $(5a + 6b)(5a - 6b) = (5a)^2 - (6b)^2$
$ = 25a^2 - 36b^2$

37. $(r + 2)^3 = (r + 2)^2(r + 2)$
$ = (r^2 + 4r + 4)(r + 2)$
$ = r^3 + 4r^2 + 4r + 2r^2 + 8r + 8$
$ = r^3 + 6r^2 + 12r + 8$

38. $t(5t - 3)^2 = t(25t^2 - 30t + 9)$
$ = 25t^3 - 30t^2 + 9t$

39. Answers will vary. One example is given here.

(a) $(x + y)^2 \neq x^2 + y^2$
Let $x = 2$ and $y = 3$.

$(x + y)^2 = (2 + 3)^2 = 5^2 = 25$
$x^2 + y^2 = 2^2 + 3^2 = 4 + 9 = 13$

Since $25 \neq 13$,

$$(x + y)^2 \neq x^2 + y^2.$$

(b) $(x + y)^3 \neq x^3 + y^3$
Let $x = 2$ and $y = 3$.

$(x + y)^3 = (2 + 3)^3 = 5^3 = 125$
$x^3 + y^3 = 2^3 + 3^3 = 8 + 27 = 35$

Since $125 \neq 35$,

$$(x + y)^3 \neq x^3 + y^3.$$

40. To find the third power of a binomial, such as $(a + b)^3$, first square the binomial and then multiply that result by the binomial:

$(a + b)^3 = (a + b)^2(a + b)$
$ = (a^2 + 2ab + b^2)(a + b)$
$ = (a^3 + 2a^2b + ab^2)$
$ + (a^2b + 2ab^2 + b^3)$
$ = a^3 + 3a^2b + 3ab^2 + b^3$

41. If we chose to let $x = 0$ and $y = 1$, we would get the true equation $1 = 1$ for both (a) and (b). These results would not be sufficient to illustrate the truth, in general, of the inequalities. The next step in working the exercise would be to use two other values instead of $x = 0$ and $y = 1$.

42. Use the formula for the volume of a cube, $V = e^3$ with $e = x^2 + 2$ centimeters.

$V = (x^2 + 2)^3$
$ = (x^2 + 2)^2(x^2 + 2)$
$ = (x^4 + 4x^2 + 4)(x^2 + 2)$

Now we use vertical multiplication.

$$
\begin{array}{r}
x^4 + 4x^2 + 4 \\
x^2 + 2 \\
\hline
2x^4 + 8x^2 + 8 \\
x^6 + 4x^4 + 4x^2 \\
\hline
x^6 + 6x^4 + 12x^2 + 8
\end{array}
$$

continued

The volume of the cube is

$$x^6 + 6x^4 + 12x^2 + 8$$

cubic centimeters.

43. Use the formula for the volume of a sphere, $V = \dfrac{4}{3}\pi r^3$, with $r = x + 1$ inches.

$$V = \frac{4}{3}\pi(x+1)^3$$

$$= \frac{4}{3}\pi(x+1)^2(x+1)$$

$$= \frac{4}{3}\pi\left(x^2 + 2x + 1\right)(x+1)$$

Now use vertical multiplication.

$$
\begin{array}{r}
x^2 + 2x + 1 \\
x + 1 \\
\hline
x^2 + 2x + 1 \\
x^3 + 2x^2 + x \\
\hline
x^3 + 3x^2 + 3x + 1
\end{array}
$$

$$V = \frac{4}{3}\pi\left(x^3 + 3x^2 + 3x + 1\right)$$

$$= \frac{4}{3}\pi x^3 + 4\pi x^2 + 4\pi x + \frac{4}{3}\pi$$

The volume of the sphere is

$$\frac{4}{3}\pi x^3 + 4\pi x^2 + 4\pi x + \frac{4}{3}\pi$$

cubic inches.

44. $6^0 + (-6)^0 = 1 + 1 = 2$

45. $(-23)^0 - (-23)^0 = 1 - 1 = 0$

46. $-10^0 = -(10^0) = -(1) = -1$

47. $-7^{-2} = -\dfrac{1}{7^2} = -\dfrac{1}{49}$

48. $\left(\dfrac{5}{8}\right)^{-2} = \left(\dfrac{8}{5}\right)^2 = \dfrac{64}{25}$

49. $\left(5^{-2}\right)^{-4} = 5^{(-2)(-4)}$ *Power rule*
$= 5^8$

50. $9^3 \cdot 9^{-5} = 9^{3+(-5)} = 9^{-2} = \dfrac{1}{9^2} = \dfrac{1}{81}$

51. $2^{-1} + 4^{-1} = \dfrac{1}{2^1} + \dfrac{1}{4^1}$

$$= \frac{1}{2} + \frac{1}{4}$$

$$= \frac{2}{4} + \frac{1}{4} = \frac{3}{4}$$

52. $\dfrac{6^{-5}}{6^{-3}} = \dfrac{6^3}{6^5} = \dfrac{1}{6^2} = \dfrac{1}{36}$

53. $\dfrac{x^{-7}}{x^{-9}} = \dfrac{x^9}{x^7}$

$= x^2$ *Quotient rule*

54. $\dfrac{y^4 \cdot y^{-2}}{y^{-5}} = \dfrac{y^4 \cdot y^5}{y^2}$

$$= \frac{y^9}{y^2} = y^7$$

55. $\left(6r^{-2}\right)^{-1} = (6)^{-1}\left(r^{-2}\right)^{-1}$

$$= \left(6^{-1}\right)\left(r^{-2(-1)}\right)$$

$$= 6^{-1}r^2$$

$$= \frac{r^2}{6}$$

56. $(3p)^4\left(3p^{-7}\right) = \left(3^4 p^4\right)\left(3p^{-7}\right)$

$$= 3^{4+1} \cdot p^{4-7}$$

$$= 3^5 p^{-3}$$

$$= \frac{3^5}{p^3}$$

57. $\dfrac{ab^{-3}}{a^4 b^2} = \dfrac{a}{a^4 b^2 b^3} = \dfrac{1}{a^3 b^5}$

58. $\dfrac{\left(6r^{-1}\right)^2\left(2r^{-4}\right)}{r^{-5}\left(r^2\right)^{-3}} = \dfrac{\left(6^2 r^{-2}\right)\left(2r^{-4}\right)}{r^{-5}r^{-6}}$

$$= \frac{72r^{-6}}{r^{-11}}$$

$$= \frac{72r^{11}}{r^6}$$

$$= 72r^5$$

59. $\dfrac{-15y^4}{9y^2} = \dfrac{-15y^{4-2}}{9} = \dfrac{-5y^2}{3}$

60. $\dfrac{6y^4 - 12y^2 + 18y}{6y} = \dfrac{6y4}{6y} - \dfrac{12y^2}{6y} + \dfrac{18y}{6y}$

$$= y^3 - 2y + 3$$

61. $\left(-10m^4n^2 + 5m^3n^2 + 6m^2n^4\right) \div \left(5m^2n\right)$

$$= \frac{-10m^4n^2}{5m^2n} + \frac{5m^3n^2}{5m^2n} + \frac{6m^2n^4}{5m^2n}$$

$$= -2m^2n + mn + \frac{6n^3}{5}$$

62. Let P be the polynomial that when multiplied by $6m^2n$ gives the product $12m^3n^2 + 18m^6n^3 - 24m^2n^2$.

$$(P)\left(6m^2n\right) = 12m^3n^2 + 18m^6n^3 - 24m^2n^2$$

$$P = \frac{12m^3n^2 + 18m^6n^3 - 24m^2n^2}{6m^2n}$$

$$= \frac{12m^3n^2}{6m^2n} + \frac{18m^6n^3}{6m^2n} - \frac{24m^2n^2}{6m^2n}$$

$$= 2mn + 3m^4n^2 - 4n$$

63. $\dfrac{6x^2 - 12x}{6}$ is not $x^2 - 12x$. The error made was not dividing both terms in the numerator by 6. The correct method is as follows:

$$\frac{6x^2 - 12x}{6} = \frac{6x^2}{6} - \frac{12x}{6}$$
$$= x^2 - 2x.$$

64. $\dfrac{2r^2 + 3r - 14}{r - 2}$

$$
\begin{array}{r}
2r + 7 \\
r - 2 \enclose{longdiv}{2r^2 + 3r - 14} \\
\underline{2r^2 - 4r} \\
7r - 14 \\
\underline{7r - 14} \\
0
\end{array}
$$

The remainder is 0.

The answer is the quotient, $2r + 7$.

65. $\dfrac{10a^3 + 9a^2 - 14a + 9}{5a - 3}$

$$
\begin{array}{r}
2a^2 + 3a - 1 \\
5a - 3 \enclose{longdiv}{10a^3 + 9a^2 - 14a + 9} \\
\underline{10a^3 - 6a^2} \\
15a^2 - 14a \\
\underline{15a^2 - 9a} \\
- 5a + 9 \\
\underline{- 5a + 3} \\
6
\end{array}
$$

The answer is

$$2a^2 + 3a - 1 + \frac{6}{5a - 3}.$$

66. $\dfrac{x^4 + 3x^3 - 5x^2 - 3x + 4}{x^2 - 1}$

We use 0 as the coefficient of the missing term.

$$
\begin{array}{r}
x^2 + 3x - 4 \\
x^2 + 0x - 1 \enclose{longdiv}{x^4 + 3x^3 - 5x^2 - 3x + 4} \\
\underline{x^4 + 0x^3 - x^2} \\
3x^3 - 4x^2 - 3x \\
\underline{3x^3 + 0x^2 - 3x} \\
- 4x^2 + 4 \\
\underline{- 4x^2 + 4} \\
0
\end{array}
$$

The remainder is 0. The answer is the quotient, $x^2 + 3x - 4$.

67. $\dfrac{m^4 + 4m^3 - 5m^2 - 12m + 6}{m^2 - 3}$

$$
\begin{array}{r}
m^2 + 4m - 2 \\
m^2 + 0m - 3 \enclose{longdiv}{m^4 + 4m^3 - 5m^2 - 12m + 6} \\
\underline{m^4 + 0m^3 - 3m^2} \\
4m^3 - 2m^2 - 12m \\
\underline{4m^3 + 0m^2 - 12m} \\
- 2m^2 + 0m + 6 \\
\underline{- 2m^2 + 0m + 6} \\
0
\end{array}
$$

The remainder is 0. The answer is the quotient, $m^2 + 4m - 2$.

68. $48,000,000 = 4.8 \times 10^7$

69. $28,988,000,000 = 2.8988 \times 10^{10}$

70. $.0000000824 = 8.24 \times 10^{-8}$

71. $2.4 \times 10^4 = 24,000$

Move the decimal point 4 places to the right.

72. $7.83 \times 10^7 = 78,300,000$

Move the decimal point 7 places to the right.

73. $8.97 \times 10^{-7} = .000000897$

Move the decimal point 7 places to the left.

74. $\left(2 \times 10^{-3}\right) \times \left(4 \times 10^5\right)$
$$= (2 \times 4)\left(10^{-3} \times 10^5\right)$$
$$= 8 \times 10^{-3+5} = 8 \times 10^2$$
$$= 800$$

75. $\dfrac{8 \times 10^4}{2 \times 10^{-2}} = \dfrac{8}{2} \times \dfrac{10^4}{10^{-2}} = 4 \times 10^{4-(-2)}$
$$= 4 \times 10^6 = 4,000,000$$

76. $\dfrac{12 \times 10^{-5} \times 5 \times 10^4}{4 \times 10^3 \times 6 \times 10^{-2}}$
$$= \frac{12 \times 5}{4 \times 6} \times \frac{10^{-5} \times 10^4}{10^3 \times 10^{-2}}$$
$$= \frac{60}{24} \times \frac{10^{-1}}{10^1}$$
$$= \frac{5}{2} \times 10^{-1-1}$$
$$= 2.5 \times 10^{-2}$$
$$= .025$$

77. $2 \times 10^{-6} = .000002$

78. $1.6 \times 10^{-12} = .0000000000016$

79. There are 41 zeros so the number is 4.2×10^{42}.

80. $747 = 7.47 \times 10^2$

200 billion $= 200,000,000,000$
$$= 2 \times 10^{11}$$

81. **(a)** 2.82 billion $= 2,820,000,000$
$$= 2.82 \times 10^9$$

(b) 578.4 billion $= 578,400,000,000$
$$= 5.784 \times 10^{11}$$

(c) 17.7 billion $= 17,700,000,000$
$$= 1.77 \times 10^{10}$$

82. **(a)** $1000 = 1 \times 10^3$

(b) $2000 = 2 \times 10^3$

(c) $50,000 = 5 \times 10^4$

(d) $100,000 = 1 \times 10^5$

In Exercises 83–88, use the following polynomials:
$$P : x - 3$$
$$Q : x^3 + x^2 - 11x - 3.$$

83. **(a)** When $x = 5$,
$$P = 5 - 3 = 2.$$

(b) When $x = 5$,
$$Q = 5^3 + 5^2 - 11(5) - 3$$
$$= 125 + 25 - 55 - 3$$
$$= 92.$$

(c) $P + Q$
$$= (x - 3) + (x^3 + x^2 - 11x - 3)$$
$$= x^3 + x^2 - 10x - 6$$

(d) If $x = 5$,
$$P + Q = 5^3 + 5^2 - 10(5) - 6$$
$$= 125 + 25 - 50 - 6$$
$$= 94.$$

This is the sum of the values from parts (a) and (b).

84. **(a)** If $x = 4$,
$$P = 4 - 3 = 1.$$

(b) If $x = 4$,
$$Q = 4^3 + 4^2 - 11(4) - 3$$
$$= 64 + 16 - 44 - 3$$
$$= 33.$$

(c) $P - Q$
$$= (x - 3) - (x^3 + x^2 - 11x - 3)$$
$$= (x - 3) + (-x^3 - x^2 + 11x + 3)$$
$$= -x^3 - x^2 + 12x$$

(d) If $x = 4$,
$$P - Q = -(4)^3 - (4)^2 + 12(4)$$
$$= -64 - 16 + 48$$
$$= -32.$$

This is the difference of the values from parts (a) and (b).

85. **(a)** If $x = -3$,
$$P = -3 - 3 = -6.$$

(b) If $x = -3$,
$$Q = (-3)^3 + (-3)^2 - 11(-3) - 3$$
$$= -27 + 9 + 33 - 3$$
$$= 12.$$

(c) $P \cdot Q$
$$= (x - 3)(x^3 + x^2 - 11x - 3)$$

Use vertical multiplication.

$$
\begin{array}{r}
x^3 + x^2 - 11x - 3 \\
x - 3 \\
\hline
-3x^3 - 3x^2 + 33x + 9 \\
x^4 + x^3 - 11x^2 - 3x \\
\hline
x^4 - 2x^3 - 14x^2 + 30x + 9
\end{array}
$$

$$P \cdot Q = x^4 - 2x^3 - 14x^2 + 30x + 9$$

(d) If $x = -3$,
$$P \cdot Q = (-3)^4 - 2(-3)^3 - 14(-3)^2 + 30(-3) + 9$$
$$= 81 - 2(-27) - 14(9) + 30(-3) + 9$$
$$= 81 + 54 - 126 - 90 + 9$$
$$= -72.$$

This is the product of the values from parts (a) and (b).

86. **(a)** If $x = 6$,
$$P = 6 - 3 = 3.$$

(b) If $x = 6$,
$$Q = 6^3 + 6^2 - 11(6) - 3$$
$$= 216 + 36 - 66 - 3$$
$$= 183.$$

(c) $\dfrac{Q}{P} = \dfrac{x^3 + x^2 - 11x - 3}{x - 3}$

$$
\begin{array}{r}
x^2 + 4x + 1 \\
x - 3 \overline{\smash{)}\, x^3 + x^2 - 11x - 3} \\
\underline{x^3 - 3x^2} \\
4x^2 - 11x \\
\underline{4x^2 - 12x} \\
x - 3 \\
\underline{x - 3} \\
0
\end{array}
$$

$$\frac{Q}{P} = x^2 + 4x + 1$$

(d) If $x = 6$,

$$\frac{Q}{P} = (6)^2 + 4(6) + 1$$
$$= 36 + 24 + 1$$
$$= 61.$$

This is the quotient of the value from part (b) divided by the value from part (a).

87. The answers will vary depending upon the value chosen for x.

88. We could not choose 3 as a replacement for x because it would make the denominator equal to zero. It is the only invalid replacement for x.

89. $5^0 + 7^0 = 1 + 1 = 2$

90. $\left(\dfrac{6r^2p}{5}\right)^3 = \dfrac{6^3 r^{2 \cdot 3} p^3}{5^3}$

$$= \dfrac{6^3 r^6 p^3}{5^3}$$

91. $(12a + 1)(12a - 1) = (12a)^2 - 1^2$
$$= 144a^2 - 1$$

92. $2^{-4} = \dfrac{1}{2^4} = \dfrac{1}{16}$

93. $\left(8^{-3}\right)^4 = 8^{(-3)(4)}$
$$= 8^{-12}$$
$$= \dfrac{1}{8^{12}}$$

94. $\dfrac{2p^3 - 6p^2 + 5p}{2p^2}$

$$= \dfrac{2p^3}{2p^2} - \dfrac{6p^2}{2p^2} + \dfrac{5p}{2p^2}$$

$$= p - 3 + \dfrac{5}{2p}$$

95. $\dfrac{(2m^{-5})(3m^2)^{-1}}{m^{-2}(m^{-1})^2}$

$$= \dfrac{(2m^{-5})(3^{-1}m^{-2})}{m^{-2}(m^{-2})}$$

$$= \dfrac{2}{3} \cdot \dfrac{m^{-5-2}}{m^{-2-2}}$$

$$= \dfrac{2}{3} \cdot \dfrac{m^{-7}}{m^{-4}}$$

$$= \dfrac{2}{3} \cdot \dfrac{m^4}{m^7}$$

$$= \dfrac{2}{3m^3}$$

96. $(3k - 6)(2k^2 + 4k + 1)$

Multiply vertically.

$$
\begin{array}{rrrrr}
& 2k^2 & + & 4k & + 1 \\
& & & 3k & - 6 \\
\hline
- & 12k^2 & - & 24k & - 6 \\
6k^3 & + & 12k^2 & + & 3k \\
\hline
6k^3 & & & - 21k & - 6
\end{array}
$$

97. $\dfrac{r^9 \cdot r^{-5}}{r^{-2} \cdot r^{-7}} = \dfrac{r^9 r^2 r^7}{r^5}$

$$= \dfrac{r^{18}}{r^5}$$
$$= r^{13}$$

98. $(2r + 5s)^2$
$$= (2r)^2 + 2(2r)(5s) + (5s)^2$$
$$= 4r^2 + 20rs + 25s^2$$

99. $(-5y^2 + 3y - 11) + (4y^2 - 7y + 15)$
$$= -5y^2 + 4y^2 + 3y - 7y - 11 + 15$$
$$= -y^2 - 4y + 4$$

100. $(2r + 5)(5r - 2)$
$$= 2r(5r) + 2r(-2) + 5(5r) + 5(-2)$$
$$= 10r^2 - 4r + 25r - 10$$
$$= 10r^2 + 21r - 10$$

101. $\dfrac{2y^3 + 17y^2 + 37y + 7}{2y + 7}$

$$
\begin{array}{r}
y^2 + 5y + 1 \\
2y + 7 \overline{\smash{\big)}\, 2y^3 + 17y^2 + 37y + 7} \\
\underline{2y^3 + 7y^2} \\
10y^2 + 37y \\
\underline{10y^2 + 35y} \\
2y + 7 \\
\underline{2y + 7} \\
0
\end{array}
$$

The remainder is 0. The answer is the quotient, $y^2 + 5y + 1$.

102. $(25x^2y^3 - 8xy^2 + 15x^3y) \div (10x^2y^3)$

$$= \dfrac{25x^2y^3 - 8xy^2 + 15x^3y}{10x^2y^3}$$

$$= \dfrac{25x^2y^3}{10x^2y^3} - \dfrac{8xy^2}{10x^2y^3} + \dfrac{15x^3y}{10x^2y^3}$$

$$= \dfrac{5}{2} - \dfrac{4}{5xy} + \dfrac{3x}{2y^2}$$

103. $(6p^2 - p - 8) - (-4p^2 + 2p - 3)$
$$= (6p^2 - p - 8) + (4p^2 - 2p + 3)$$
$$= 10p^2 - 3p - 5$$

104. $\dfrac{5^8}{5^{19}} = 5^{8-19}$

$$= 5^{-11} = \dfrac{1}{5^{11}}$$

105. $(-7 + 2k)^2$
$$= (-7)^2 + 2(-7)(2k) + (2k)^2$$
$$= 49 - 28k + 4k^2$$

106. $\left(\dfrac{x}{y^{-3}}\right)^{-4} = \dfrac{x^{-4}}{(y^{-3})^{-4}}$
$$= \dfrac{x^{-4}}{y^{12}}$$
$$= \dfrac{1}{x^4 y^{12}}$$

Chapter 5 Test

1. $5x^2 + 8x - 12x^2 = 5x^2 - 12x^2 + 8x$
$$= -7x^2 + 8x$$

degree 2; binomial (2 terms)

2. $13n^3 - n^2 + n^4 + 3n^4 - 9n^2$
$$= n^4 + 3n^4 + 13n^3 - n^2 - 9n^2$$
$$= 4n^4 + 13n^3 - 10n^2$$

degree 4; trinomial (3 terms)

3. $y = 2x^2 - 4$
If $x = -2$, then
$$y = 2(-2)^2 - 4$$
$$= 2 \cdot 4 - 4$$
$$= 8 - 4 = 4.$$

x	-2	-1	0	1	2
y	4	-2	-4	-2	4

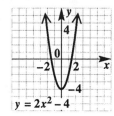

$y = 2x^2 - 4$

4. $\left(2y^2 - 8y + 8\right) + \left(-3y^2 + 2y + 3\right)$
$\quad - \left(y^2 + 3y - 6\right)$
$$= \left(2y^2 - 8y + 8\right) + \left(-3y^2 + 2y + 3\right)$$
$$+ \left(-y^2 - 3y + 6\right)$$
$$= 2y^2 - 3y^2 - y^2 - 8y + 2y - 3y + 8 + 3 + 6$$
$$= -2y^2 - 9y + 17$$

5. $\left(-9a^3 b^2 + 13ab^5 + 5a^2 b^2\right)$
$\quad - \left(6ab^5 + 12a^3 b^2 + 10a^2 b^2\right)$
$$= \left(-9a^3 b^2 + 13ab^5 + 5a^2 b^2\right)$$
$$+ \left(-6ab^5 - 12a^3 b^2 - 10a^2 b^2\right)$$
$$= \left(-9a^3 b^2 - 12a^3 b^2\right) + \left(13ab^5 - 6ab^5\right)$$
$$+ \left(5a^2 b^2 - 10a^2 b^2\right)$$
$$= -21a^3 b^2 + 7ab^5 - 5a^2 b^2$$

6. Subtract.

$$9t^3 - 4t^2 + 2t + 2$$
$$\underline{9t^3 + 8t^2 - 3t - 6}$$

Change all the signs in the second row; then add.

$$9t^3 - \quad 4t^2 + 2t + 2$$
$$\underline{-9t^3 - \quad 8t^2 + 3t + 6}$$
$$\quad\quad -12t^2 + 5t + 8$$

7. $3x^2\left(-9x^3 + 6x^2 - 2x + 1\right)$
$$= \left(3x^2\right)\left(-9x^3\right) + \left(3x^2\right)\left(6x^2\right)$$
$$+ \left(3x^2\right)(-2x) + \left(3x^2\right)(1)$$
$$= -27x^5 + 18x^4 - 6x^3 + 3x^2$$

8.
$$\quad\quad\quad\quad\textbf{F}\quad\textbf{O}\quad\textbf{I}\quad\textbf{L}$$
$$(t - 8)(t + 3) = t^2 + 3t - 8t - 24$$
$$= t^2 - 5t - 24$$

9. $(4x + 3y)(2x - y)$
$$\quad\quad\textbf{F}\quad\quad\textbf{O}\quad\quad\textbf{I}\quad\quad\textbf{L}$$
$$= 8x^2 - 4xy + 6xy - 3y^2$$
$$= 8x^2 + 2xy - 3y^2$$

10. $(5x - 2y)^2 = (5x)^2 - 2(5x)(2y) + (2y)^2$
$$= 25x^2 - 20xy + 4y^2$$

11. $(10v + 3w)(10v - 3w)$
$$= (10v)^2 - (3w)^2$$
$$= 100v^2 - 9w^2$$

12. $(2r - 3)\left(r^2 + 2r - 5\right)$

Multiply vertically.

$$r^2 + \quad 2r - 5$$
$$\underline{\quad\quad\quad\quad 2r - 3}$$
$$-3r^2 - \quad 6r + 15$$
$$\underline{2r^3 + 4r^2 - 10r}$$
$$2r^3 + \quad r^2 - 16r + 15$$

13. Use the formula for the area of a square, $A = s^2$, with $s = 3x + 9$.

$$A = s^2$$
$$A = (3x + 9)^2$$
$$= (3x)^2 + 2(3x)(9) + 9^2$$
$$= 9x^2 + 54x + 81$$

14. $5^{-4} = \dfrac{1}{5^4} = \dfrac{1}{625}$

15. $(-3)^0 + 4^0 = 1 + 1 = 2$

16. $4^{-1} + 3^{-1} = \dfrac{1}{4} + \dfrac{1}{3} = \dfrac{3}{12} + \dfrac{4}{12} = \dfrac{7}{12}$

17. $\dfrac{(3x^2y)^2(xy^3)^2}{(xy)^3} = \dfrac{3^2(x^2)^2y^2x^2(y^3)^2}{x^3y^3}$

$= \dfrac{9x^4y^2x^2y^6}{x^3y^3}$

$= \dfrac{9x^6y^8}{x^3y^3}$

$= 9x^3y^5$

18. $\dfrac{8^{-1}\cdot 8^4}{8^{-2}} = \dfrac{8^{(-1)+4}}{8^{-2}} = \dfrac{8^3}{8^{-2}} = 8^{3-(-2)} = 8^5$

19. $\dfrac{(x^{-3})^{-2}(x^{-1}y)^2}{(xy^{-2})^2} = \dfrac{(x^{-3})^{-2}(x^{-1})^2(y)^2}{(x)^2(y^{-2})^2}$

$= \dfrac{x^6x^{-2}y^2}{x^2y^{-4}}$

$= \dfrac{x^4y^2}{x^2y^{-4}}$

$= x^{4-2}y^{2-(-4)}$

$= x^2y^6$

20. I disagree with the given statement because

$$3^{-4} = \dfrac{1}{3^4} = \dfrac{1}{81},$$

which is positive. A negative exponent indicates a reciprocal, not a negative number.

21. $\dfrac{8y^3 - 6y^2 + 4y + 10}{2y}$

$= \dfrac{8y^3}{2y} - \dfrac{6y^2}{2y} + \dfrac{4y}{2y} + \dfrac{10}{2y}$

$= 4y^2 - 3y + 2 + \dfrac{5}{y}$

22. $(-9x^2y^3 + 6x^4y^3 + 12xy^3) \div (3xy)$

$= \dfrac{-9x^2y^3}{3xy} + \dfrac{6x^4y^3}{3xy} + \dfrac{12xy^3}{3xy}$

$= -3xy^2 + 2x^3y^2 + 4y^2$

23. $(3x^3 - x + 4) \div (x - 2)$

$$
\begin{array}{r}
3x^2 + 6x + 11 \\
x-2\,\overline{\big)\,3x^3 + 0x^2 - x + 4} \\
\underline{3x^3 - 6x^2} \\
6x^2 - x \\
\underline{6x^2 - 12x} \\
11x + 4 \\
\underline{11x - 22} \\
26
\end{array}
$$

The result is

$$3x^2 + 6x + 11 + \dfrac{26}{x-2}.$$

24. **(a)** $45{,}000{,}000{,}000 = 4.5 \times 10^{10}$

(b) $3.6 \times 10^{-6} = .0000036$

(c) $\dfrac{9.5 \times 10^{-1}}{5 \times 10^3} = \dfrac{9.5}{5} \times \dfrac{10^{-1}}{10^3}$

$= 1.9 \times 10^{-1-3}$

$= 1.9 \times 10^{-4}$

$= .00019$

25. **(a)** $170{,}000 = 1.7 \times 10^5$

$1000 = 1 \times 10^3$

(b) $(1.7 \times 10^5)(1 \times 10^3)$

$= 1.7 \times 10^{5+3}$

$= 1.7 \times 10^8$

The total payload is more than 1.7×10^8 pounds.

Cumulative Review Exercises Chapters 1–5

1. $\dfrac{28}{16} = \dfrac{7\cdot 4}{4\cdot 4} = \dfrac{7}{4}$

2. $\dfrac{55}{11} = \dfrac{5\cdot 11}{1\cdot 11} = 5$

3. $\dfrac{2}{3} + \dfrac{1}{8} = \dfrac{16}{24} + \dfrac{3}{24} = \dfrac{19}{24}$

4. $\dfrac{7}{4} - \dfrac{9}{5} = \dfrac{35}{20} - \dfrac{36}{20} = -\dfrac{1}{20}$

5. Each shed requires $1\frac{1}{4}$ cubic yards of concrete, so the total amount of concrete needed for 25 sheds would be

$$25 \times 1\dfrac{1}{4} = 25 \times \dfrac{5}{4}$$

$$= \dfrac{125}{4}$$

$$= 31\dfrac{1}{4} \text{ cubic yards.}$$

6. Use the formula for simple interest, $I = Prt$, with $P = \$34{,}000$, $r = 5.4\%$, and $t = 1$.

$I = Prt$

$= (34{,}000)(.054)(1)$

$= 1836$

She earned \$1836 in interest.

7. The positive integer factors of 45 are 1, 3, 5, 9, 15, and 45.

8. $b - a = b + (-a)$

b is positive.

Since a is negative, $-a$ is positive.

The sum of two positive numbers is positive, so $b - a$ is positive.

9. $\dfrac{4x - 2y}{x + y} = \dfrac{4(-2) - 2(4)}{(-2) + 4}$ *Let $x = -2$, $y = 4$*

$= \dfrac{-8 - 8}{2} = \dfrac{-16}{2} = -8$

10. $x^3 - 4xy = (-2)^3 - 4(-2)(4)$ *Let x = -2,*
$$y = 4$$
$$= -8 + 32 = 24$$

11. $\dfrac{(-13 + 15) - (3 + 2)}{6 - 12} = \dfrac{2 - 5}{-6} = \dfrac{-3}{-6} = \dfrac{1}{2}$

12. $-7 - 3[2 + (5 - 8)] = -7 - 3[2 + (-3)]$
$$= -7 - 3[-1]$$
$$= -7 + 3 = -4$$

13. $(9 + 2) + 3 = 9 + (2 + 3)$

The numbers are in the same order but grouped differently, so this is an example of the associative property of addition.

14. $6(4 + 2) = 6(4) + 6(2)$

The number 6 outside the parentheses is "distributed" over the 4 and the 2. This is an example of the distributive property.

15. **(a)** The change in the number of airline passenger deaths from 1994 to 1995 is

$$557 - 732 = -175;$$

that is, the number of deaths *decreased* 175.

(b) From 1995 to 1996, we have

$$1132 - 557 = 575;$$

an *increase* of 575.

16. $-3\left(2x^2 - 8x + 9\right) - \left(4x^2 + 3x + 2\right)$
$$= -6x^2 + 24x - 27 - 4x^2 - 3x - 2$$
$$= -10x^2 + 21x - 29$$

17. $2 - 3(t - 5) = 4 + t$
$$2 - 3t + 15 = 4 + t$$
$$-3t + 17 = 4 + t$$
$$-4t + 17 = 4$$
$$-4t = -13$$
$$t = \dfrac{13}{4}$$

The solution set is $\left\{\dfrac{13}{4}\right\}$.

18. $2(5h + 1) = 10h + 4$
$$10h + 2 = 10h + 4$$
$$2 = 4 \qquad \textit{False}$$

The false statement indicates that the equation has no solution.

19. $d = rt$ for r

$$\dfrac{d}{t} = \dfrac{rt}{t}$$
$$\dfrac{d}{t} = r$$

20. $\dfrac{x}{5} = \dfrac{x - 2}{7}$

$7x = 5(x - 2)$ *Cross products are equal*

$$7x = 5x - 10$$
$$2x = -10$$
$$x = -5$$

The solution set is $\{-5\}$.

21. $\dfrac{1}{3}p - \dfrac{1}{6}p = -2$

To clear fractions, multiply both sides of the equation by the least common denominator, which is 6.

$$6\left(\dfrac{1}{3}p - \dfrac{1}{6}p\right) = (6)(-2)$$
$$6\left(\dfrac{1}{3}p\right) - 6\left(\dfrac{1}{6}p\right) = -12$$
$$2p - p = -12$$
$$p = -12$$

The solution set is $\{-12\}$.

22. $.05x + .15(50 - x) = 5.50$

To clear decimals, multiply both sides of the equation by 100.

$$100[.05x + .15(50 - x)] = 100(5.50)$$
$$100(.05x) + 100[.15(50 - x)] = 100(5.50)$$
$$5x + 15(50 - x) = 550$$
$$5x + 750 - 15x = 550$$
$$-10x + 750 = 550$$
$$-10x = -200$$
$$x = 20$$

23. $4 - (3x + 12) = (2x - 9) - (5x - 1)$
$$4 - 3x - 12 = 2x - 9 - 5x + 1$$
$$-3x - 8 = -3x - 8 \qquad \textit{True}$$

The true statement indicates that the solution set is {all real numbers} or $(-\infty, \infty)$.

24. Let $x = $ the number of breaths per minute taken by the elephant;

$16x = $ the number for the mouse.
$$x + 16x = 170$$
$$17x = 170$$
$$x = 10$$

The elephant takes 10 breaths per minute and the mouse takes 160 breaths per minute.

25.   Let $x =$ the unknown number.

$$3(8 - x) = 3x$$
$$24 - 3x = 3x$$
$$24 = 6x$$
$$x = 4$$

The unknown number is 4.

26.   Let $x = $ the number of accidents in 1995 and 1996;

$x + 2 = $ the number in 1994.

$$(x + 2) + x + x = 68$$
$$3x + 2 = 68$$
$$3x = 66$$
$$x = 22$$

There were 22 aircraft accidents involving passenger fatalities in 1995 and 1996, and 24 accidents in 1994.

27. $-8x \leq -80$

$$\frac{-8x}{-8} \geq \frac{-80}{-8} \qquad \textit{Divide by –8;}$$
$$\qquad\qquad\qquad \textit{reverse the symbol}$$
$$x \geq 10$$

The solution set is $[10, \infty)$.

28. $-2(x + 4) > 3x + 6$

$$-2x - 8 > 3x + 6$$
$$-2x > 3x + 14$$
$$-5x > 14$$
$$\frac{-5x}{-5} < \frac{14}{-5}$$
$$x < -\frac{14}{5}$$

The solution set is $\left(-\infty, -\frac{14}{5}\right)$.

29. $-3 \leq 2x + 5 < 9$

$$-8 \leq 2x < 4 \qquad \textit{Subtract 5}$$
$$\frac{-8}{2} \leq \frac{2x}{2} < \frac{4}{2} \qquad \textit{Divide by 2}$$
$$-4 \leq x < 2$$

The solution set is $[-4, 2)$.

30.   Let $x = $ one side of the triangle;

$2x = $ the second side of the triangle.

The perimeter of the triangle cannot be more than 50 feet. This is equivalent to stating that the sum of the lengths of the sides must be less than or equal to 50 feet. Write this statement as an inequality and solve.

$$x + 2x + 17 \leq 50$$
$$3x + 17 \leq 50$$
$$3x \leq 33$$
$$x \leq 11$$

One side cannot be more than 11 feet. The other side cannot be more than $2 \cdot 11 = 22$ feet.

31. $-2x + 4y = 8$

$x = 0:\ -2(0) + 4y = 8$
$$4y = 8$$
$$y = 2$$

$x = 2:\ -2(2) + 4y = 8$
$$4y = 12$$
$$y = 3$$

$x = 4:\ -2(4) + 4y = 8$
$$4y = 16$$
$$y = 4$$

$y = 0:\ -2x + 4(0) = 8$
$$-2x = 8$$
$$x = -4$$

$y = 1:\ -2x + 4(1) = 8$
$$-2x = 4$$
$$x = -2$$

x	0	-4	2	-2	4
y	2	0	3	1	4

32.   We recognize $y = -3x + 6$ as the equation of a line with y-intercept 6 and slope -3.

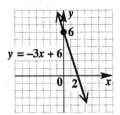

33. $y = (x+4)^2$

If $x = -6$, then

$$y = (-6+4)^2$$
$$= (-2)^2 = 4.$$

x	-6	-5	-4	-3	-2
y	4	1	0	1	4

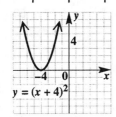

$y = (x+4)^2$

34. $(7x^3 - 12x^2 - 3x + 8) + (6x^2 + 4)$
$\quad - (-4x^3 + 8x^2 - 2x - 2)$
$\quad = (7x^3 - 12x^2 - 3x + 8) + (6x^2 + 4)$
$\qquad + (4x^3 - 8x^2 + 2x + 2)$
$\quad = 7x^3 + 4x^3 - 12x^2 + 6x^2 - 8x^2$
$\qquad - 3x + 2x + 8 + 4 + 2$
$\quad = 11x^3 - 14x^2 - x + 14$

35. $6x^5(3x^2 - 9x + 10)$
$\quad = (6x^5)(3x^2) + (6x^5)(-9x)$
$\qquad + (6x^5)(10)$
$\quad = 18x^7 - 54x^6 + 60x^5$

36. $(7x + 4)(9x + 3)$
$\quad = 63x^2 + 21x + 36x + 12 \quad FOIL$
$\quad = 63x^2 + 57x + 12$

37. $(5x + 8)^2 = (5x)^2 + 2(5x)(8) + (8)^2$
$\qquad\qquad = 25x^2 + 80x + 64$

38. $\dfrac{14x^3 - 21x^2 + 7x}{7x}$
$\quad = \dfrac{14x^3}{7x} - \dfrac{21x^2}{7x} + \dfrac{7x}{7x}$
$\quad = 2x^2 - 3x + 1$

39. $\dfrac{y^3 - 3y^2 + 8y - 6}{y - 1}$

$$
\begin{array}{r}
y^2 - 2y + 6 \\
y-1 \overline{\smash{\big)}\, y^3 - 3y^2 + 8y - 6} \\
\underline{y^3 - y^2} \\
-2y^2 + 8y \\
\underline{-2y^2 + 2y} \\
6y - 6 \\
\underline{6y - 6} \\
0
\end{array}
$$

The remainder is 0. The answer is the quotient,
$y^2 - 2y + 6$.

40. $4^{-1} + 3^0 = \dfrac{1}{4} + 1 = 1\dfrac{1}{4}$ or $\dfrac{5}{4}$

41. $2^{-4} \cdot 2^5 = 2^{-4+5} = 2^1 = 2$

42. $\dfrac{8^{-5} \cdot 8^7}{8^2} = \dfrac{8^{-5+7}}{8^2} = \dfrac{8^2}{8^2} = 1$

43. $\dfrac{(a^{-3}b^2)^2}{(2a^{-4}b^{-3})^{-1}} = \dfrac{(a^{-3})^2(b^2)^2}{2^{-1}(a^{-4})^{-1}(b^{-3})^{-1}}$

$$= \dfrac{a^{-6}b^4}{2^{-1}a^4b^3} = \dfrac{2b^4}{a^6a^4b^3}$$

$$= \dfrac{2b}{a^{10}}$$

44. $34,500 = 3.45 \times 10^4$

45. $5.36 \times 10^{-7} = .000000536$

CHAPTER 6   FACTORING AND APPLICATIONS

Section 6.1

1. 18, 24, 42

    Find the prime factored form of each number

    $$18 = 2^1 \cdot 3^2$$
    $$24 = 2^3 \cdot 3^1$$
    $$42 = 2^1 \cdot 3^1 \cdot 7^1$$

    Since 2 and 3 are factors of each number, they must appear in the greatest common factor (GCF). The smallest exponent on 2 and 3 is 1, so the GCF is $2^1 \cdot 3^1 = 6$.

3. Factoring is the opposite of multiplication.

5. Answers will vary. One example of three numbers whose GCF is 5 is 5, 10, 15. Another is 5, 25, 125.

7. Write each term in prime factored form.

    $$16y = 2^4 \cdot y$$
    $$24 = 2^3 \cdot 3$$

    There is no y in the second term, so y will not appear in the GCF. Thus, the GCF of $16y$ and 24 is

    $$2^3 = 8.$$

9. $$30x^3 = 2 \cdot 3 \cdot 5 \cdot x^3$$
    $$40x^6 = 2^3 \cdot 5 \cdot x^6$$
    $$50x^7 = 2 \cdot 5^2 \cdot x^7$$

    The GCF of the coefficients, 30, 40, and 50, is $2^1 \cdot 5^1 = 10$. The smallest exponent on the variable x is 3. Thus the GCF of the given terms is $10x^3$.

11. $$12m^3n^2 = 2^2 \cdot 3 \cdot m^3 \cdot n^2$$
    $$18m^5n^4 = 2 \cdot 3^2 \cdot m^5 \cdot n^4$$
    $$36m^8n^3 = 2^2 \cdot 3^2 \cdot m^8 \cdot n^3$$

    The GCF is $2 \cdot 3 \cdot m^3 \cdot n^2 = 6m^3n^2$.

13. $$-x^4y^3 = -1 \cdot x^4 \cdot y^3$$
    $$-xy^2 = -1 \cdot x \cdot y^2$$

    The GCF is xy^2.

15. $$42ab^3 = 2 \cdot 3 \cdot 7 \cdot a \cdot b^3$$
    $$-36a = -1 \cdot 2^2 \cdot 3^2 \cdot a$$
    $$90b = 2 \cdot 3^2 \cdot 5 \cdot b$$
    $$-48ab = -1 \cdot 2^4 \cdot 3 \cdot a \cdot b$$

    The GCF is $2 \cdot 3 = 6$.

17. $2k^2(5k)$ is written as a product of $2k^2$ and $5k$ and hence, it is *factored*.

19. $2k^2 + (5k + 1)$ is written as a sum of $2k^2$ and $(5k + 1)$ and hence, it is *not factored*.

21. Yes, $-xy$ is a common factor of $-x^4y^3$ and $-xy^2$. When $-xy$ is multiplied by x^3y^2, the result is $-x^4y^3$.

23. $$12 = 6(2)$$

    Factor out 6 from 12 to obtain 2.

25. $$3x^2 = 3x(x)$$

    Factor out $3x$ from $3x^2$ to obtain x.

27. $$9m^4 = 3m^2(3m^2)$$

    Factor out $3m^2$ from $9m^4$ to obtain $3m^2$.

29. $$-8z^9 = -4z^5(2z^4)$$

    Factor out $-4z^5$ from $-8z^9$ to obtain $2z^4$.

31. $$6m^4n^5 = 3m^3n(2mn^4)$$

    Factor out $3m^3n$ from $6m^4n^5$ to obtain $2mn^4$.

33. $$-14x^4y^3 = 2xy(-7x^3y^2)$$

    Factor out $2xy$ from $-14x^4y^3$ to obtain $-7x^3y^2$.

35. The GCF of $12y$ and -24 is 12.

    $$12y - 24 = 12 \cdot y - 12 \cdot 2$$
    $$= 12(y - 2)$$

37. $10a^2 - 20a$

    The GCF is $10a$.

    $$10a^2 - 20a = 10a(a) - 10a(2)$$
    $$= 10a(a - 2)$$

39. $65y^{10} + 35y^6$

    The GCF is $5y^6$.

    $$65y^{10} + 35y^6 = \left(5y^6\right)\left(13y^4\right) + \left(5y^6\right)(7)$$
    $$= 5y^6(13y^4 + 7)$$

41. $11w^3 - 100$

    The two terms of this expression have no common factor (except 1).

43. $8m^2n^3 + 24m^2n^2$

    The GCF is $8m^2n^2$.

    $$8m^2n^3 + 24m^2n^2 = \left(8m^2n^2\right)(n) + \left(8m^2n^2\right)(3)$$
    $$= 8m^2n^2(n + 3)$$

45. $13y^8 + 26y^4 - 39y^2$

    The GCF is $13y^2$.

    $$13y^8 + 26y^4 - 39y^2$$
    $$= 13y^2\left(y^6\right) + 13y^2\left(2y^2\right) + 13y^2(-3)$$
    $$= 13y^2\left(y^6 + 2y^2 - 3\right)$$

47. $45q^4p^5 + 36qp^6 + 81q^2p^3$

The GCF is $9qp^3$.

$$45q^4p^5 + 36qp^6 + 81q^2p^3$$
$$= 9qp^3(5q^3p^2) + 9qp^3(4p^3)$$
$$\quad + 9qp^3(9q)$$
$$= 9qp^3(5q^3p^2 + 4p^3 + 9q)$$

49. $a^5 + 2a^3b^2 - 3a^5b^2 + 4a^4b^3$

The GCF is a^3.

$$a^5 + 2a^3b^2 - 3a^5b^2 + 4a^4b^3$$
$$= a^3(a^2) + a^3(2b^2) + a^3(-3a^2b^2)$$
$$\quad + a^3(4ab^3)$$
$$= a^3(a^2 + 2b^2 - 3a^2b^2 + 4ab^3)$$

51. The GCF of the terms $c(x+2) - d(x+2)$ is the binomial $x + 2$.

$$c(x+2) - d(x+2)$$
$$= (x+2)(c) - (x+2)(d)$$
$$= (x+2)(c-d)$$

53. The GCF of the terms of
$m(m+2n) + n(m+2n)$ is the binomial $m + 2n$.

$$m(m+2n) + n(m+2n)$$
$$= (m+2n)(m) + (m+2n)(n)$$
$$= (m+2n)(m+n)$$

55. $8(7t+4) + x(7t+4)$

This expression is the *sum* of two terms, $8(7t+4)$ and $x(7t+4)$, so it is not in factored form. We can factor out $7t + 4$.

$$8(7t+4) + x(7t+4)$$
$$= (7t+4)(8) + (7t+4)(x)$$
$$= (7t+4)(8+x)$$

57. $(8+x)(7t+4)$

This expression is the *product* of two factors, $8 + x$ and $7t + 4$, so it is in factored form.

59. $18x^2(y+4) + 7(y+4)$

This expression is the *sum* of two terms, $18x^2(y+4)$ and $7(y+4)$, so it is not in factored form. We can factor out $y + 4$.

$$18x^2(y+4) + 7(y+4)$$
$$= (y+4)(18x^2) + (y+4)(7)$$
$$= (y+4)(18x^2+7)$$

61. It is not possible to factor the expression in Exercise 60 because the two terms, $12k^3(s-3)$ and $7(s+3)$, do not have a common factor.

63. $p^2 + 4p + 3p + 12$

The first two terms have a common factor of p, and the last two terms have a common factor of 3. Thus,

$$p^2 + 4p + 3p + 12$$
$$= (p^2 + 4p) + (3p + 12)$$
$$= p(p+4) + 3(p+4).$$

Now we have two terms which have a common binomial factor of $p + 4$. Thus,

$$p^2 + 4p + 3p + 12$$
$$= p(p+4) + 3(p+4)$$
$$= (p+4)(p+3).$$

65. $a^2 - 2a + 5a - 10$

$$= (a^2 - 2a) + (5a - 10) \quad \text{\textit{Group the terms}}$$
$$= a(a-2) + 5(a-2) \quad \text{\textit{Factor each group}}$$
$$= (a-2)(a+5) \quad \text{\textit{Factor out } a-2}$$

67. $7z^2 + 14z - az - 2a$

$$= (7z^2 + 14z) + (-az - 2a) \quad \text{\textit{Group the terms}}$$
$$= 7z(z+2) - a(z+2) \quad \text{\textit{Factor each group}}$$
$$= (z+2)(7z-a) \quad \text{\textit{Factor out } z+2}$$

69. $18r^2 + 12ry - 3xr - 2xy$

$$= (18r^2 + 12ry) + (-3xr - 2xy) \quad \text{\textit{Group the terms}}$$
$$= 6r(3r+2y) - x(3r+2y) \quad \text{\textit{Factor each group}}$$
$$= (3r+2y)(6r-x) \quad \text{\textit{Factor out } 3r+2y}$$

71. $3a^3 + 3ab^2 + 2a^2b + 2b^3$

$$= (3a^3 + 3ab^2) + (2a^2b + 2b^3) \quad \text{\textit{Group the terms}}$$
$$= 3a(a^2+b^2) + 2b(a^2+b^2) \quad \text{\textit{Factor each group}}$$
$$= (a^2+b^2)(3a+2b) \quad \text{\textit{Factor out } a^2+b^2}$$

73. $1 - a + ab - b$

$= (1 - a) + (ab - b)$ *Group the terms*

$= 1(1 - a) - b(-a + 1)$ *Factor each group*

$= 1(1 - a) - b(1 - a)$

$= (1 - a)(1 - b)$ *Factor out $1 - a$*

75. $16m^3 - 4m^2p^2 - 4mp + p^3$

$= (16m^3 - 4m^2p^2) + (-4mp + p^3)$

$= 4m^2(4m - p^2) - p(4m - p^2)$

$= (4m - p^2)(4m^2 - p)$

77. $5m + 15 - 2mp - 6p$

$= (5m + 15) + (-2mp - 6p)$

$= 5(m + 3) - 2p(m + 3)$

$= (m + 3)(5 - 2p)$

79. $18r^2 + 12ry - 3ry - 2y^2$

$= (18r^2 + 12ry) + (-3ry - 2y^2)$

$= 6r(3r + 2y) - y(3r + 2y)$

$= (3r + 2y)(6r - y)$

81. $a^5 + 2a^5b - 3 - 6b$

$= (a^5 + 2a^5b) + (-3 - 6b)$

$= a^5(1 + 2b) - 3(1 + 2b)$

$= (1 + 2b)(a^5 - 3)$

83. In order to rewrite

$$2xy + 12 - 3y - 8x$$

as

$$2xy - 8x + (-3y) + 12,$$

we must change the order of the terms and also the grouping of the terms. The properties that allow us to do this are the commutative and associative properties of addition.

84. If we group the first pair of terms, the polynomial becomes

$$(2xy - 8x) + (-3y) + 12.$$

The greatest common factor for the first pair of terms is $2x$.

85. After we group both pairs of terms in the rearranged polynomial, we have

$$(2xy - 8x) + (-3y + 12).$$

The second pair does have -3 as a common factor.

86. $(2xy - 8x) + (-3y + 12)$

$= 2x(y - 4) - 3(y - 4)$

87. The expression obtained in Exercise 86 is the *difference* between two terms, $2x(y - 4)$ and $3(y - 4)$, so it is *not* in factored form.

88. $2x(y - 4) - 3(y - 4)$

$= (y - 4)(2x - 3)$

or $= (2x - 3)(y - 4)$

Yes, this is the same result as the one shown in Example 5(b), even though the terms were grouped in a different way.

89. (a) To determine whether the result $(a - 1)(b - 1)$ is correct, multiply the factors by using FOIL.

$$(a - 1)(b - 1) = ab - a - b + 1$$

Rearranging the terms of this product, we obtain

$$ab - a - b + 1 = 1 - a + ab - b,$$

which is the polynomial given in Exercise 73, so the student's answer is correct.

(b) Both answers are acceptable because in each case the product of the factors is the given polynomial. We can also see the two factored forms are equivalent in the following way:

$$1 - a = -1(a - 1)$$
$$1 - b = -1(b - 1).$$

Thus,

$$(1 - a)(1 - b)$$
$$= [-1(a - 1)] \cdot [-1(b - 1)]$$
$$= (-1)(-1)(a - 1)(b - 1)$$
$$= 1(a - 1)(b - 1)$$
$$= (a - 1)(b - 1).$$

Section 6.2

1. If the coefficient of the last term of the trinomial is negative, then a and b must have different signs, one positive and one negative.

3. A *prime polynomial* is one that cannot be factored into polynomials with integer coefficients.

5. Product: 48   Sum: -19

Factors of 48	Sum of factors
$1, 48$	$1 + 48 = 49$
$-1, -48$	$-1 + (-48) = -49$
$2, 24$	$2 + 24 = 26$
$-2, -24$	$-2 + (-24) = -26$
$3, 16$	$3 + 16 = 19$
$-3, -16$	$-3 + (-16) = -19$ ←
$4, 12$	$4 + 12 = 16$
$-4, -12$	$-4 + (-12) = -16$
$6, 8$	$6 + 8 = 14$
$-6, -8$	$-6 + (-8) = -14$

The pair of integers whose product is 48 and whose sum is -19 is -3 and -16.

7. Product: -24 Sum: -5

Factors of -24	Sum of factors
$1, -24$	$1 + (-24) = -23$
$-1, 24$	$-1 + 24 = 23$
$2, -12$	$2 + (-12) = -10$
$-2, 12$	$-2 + 12 = 10$
$3, -8$	$3 + (-8) = -5$ ←
$-3, 8$	$-3 + 8 = 5$
$4, -6$	$4 + (-6) = -2$
$-4, 6$	$-4 + 6 = 2$

The pair of integers whose product is -24 and whose sum is -5 is 3 and -8.

9. $x^2 - 12x + 32$

Multiply each of the given pairs of factors to determine which one gives the required product.

(a) $(x - 8)(x + 4) = x^2 - 4x - 32$

(b) $(x + 8)(x - 4) = x^2 + 4x - 32$

(c) $(x - 8)(x - 4) = x^2 - 12x + 32$

(d) $(x + 8)(x + 4) = x^2 + 12x + 32$

Choice (c) is the correct factored form.

11. $p^2 + 11p + 30 = (p + 5)(\quad)$

Look for an integer whose product with 5 is 30 and whose sum with 5 is 11. That integer is 6.

$$p^2 + 11p + 30 = (p + 5)(p + 6)$$

13. $x^2 + 15x + 44 = (x + 4)(\quad)$

Look for an integer whose product with 4 is 44 and whose sum with 4 is 15. That integer is 11.

$$x^2 + 15x + 44 = (x + 4)(x + 11)$$

15. $x^2 - 9x + 8 = (x - 1)(\quad)$

Look for an integer whose product with -1 is 8 and whose sum with -1 is -9. That integer is -8.

$$x^2 - 9x + 8 = (x - 1)(x - 8)$$

17. $y^2 - 2y - 15 = (y + 3)(\quad)$

Look for an integer whose product with 3 is -15 and whose sum with 3 is -2. That integer is -5.

$$y^2 - 2y - 15 = (y + 3)(y - 5)$$

19. $x^2 + 9x - 22 = (x - 2)(\quad)$

Look for an integer whose product with -2 is -22 and whose sum with -2 is 9. That integer is 11.

$$x^2 + 9x - 22 = (x - 2)(x + 11)$$

21. $y^2 - 7y - 18 = (y + 2)(\quad)$

Look for an integer whose product with 2 is -18 and whose sum with 2 is -7. That integer is -9.

$$y^2 - 7y - 18 = (y + 2)(y - 9)$$

23. $-3 + 4 = 1$

24. $3 + (-4) = -1$

25. The sums obtained in Exercises 23 and 24 are opposites as well.

26. $(x + 3)(x - 4) = x^2 - 4x + 3x - 12$
$$= x^2 - x - 12$$
$$\neq x^2 + x - 12$$

$(x + 3)(x - 4)$ is not the correct factored form of $x^2 + x - 12$ because the middle term is incorrect.

27. $(x - 3)(x + 4) = x^2 + 4x - 3x - 12$
$$= x^2 + x - 12$$

The product of the given factors is $x^2 + x - 12$. This is the correct factored form of $x^2 + x - 12$ because we obtain the exact trinomial we were given to factor.

28. In the incorrect factored form, the sign of the middle term is the opposite of what it should be.

29. When I factor a trinomial into a product of binomials, and the middle term of the product is different only in sign, I should *reverse the signs of the two second terms of the binomials* in order to obtain the correct factored form.

30. We will obtain the correct factored from by reversing the signs of the two second terms of the binomials. Thus, the correct factored form would be $(x - 5)(x + 3)$.

31. $y^2 + 9y + 8$

Look for two integers whose product is 8 and whose sum is 9. Both integers must be positive because b and c are both positive.

Factors of 8	Sum of factors
$1, 8$	9 ←
$2, 4$	6

Thus,
$$y^2 + 9y + 8 = (y + 8)(y + 1).$$

33. $b^2 + 8b + 15$

Look for two integers whose product is 15 and whose sum is 8. Both integers must be positive because b and c are both positive.

Factors of 15	Sum of factors
$1, 15$	16
$3, 5$	8 ←

continued

Thus,
$$b^2 + 8b + 15 = (b + 3)(b + 5).$$

35. $m^2 + m - 20$

Look for two integers whose product is -20 and whose sum is 1. Since c is negative, one integer must be positive and one must be negative.

Factors of -20	Sum of factors
$-1, 20$	19
$1, -20$	-19
$-2, 10$	8
$2, -10$	-8
$-4, 5$	1 ←
$4, -5$	-1

Thus,
$$m^2 + m - 20 = (m - 4)(m + 5).$$

37. $y^2 - 8y + 15$

Find two integers whose product is 15 and whose sum is -8. Since c is positive and b is negative, both integers must be negative.

Factors of 15	Sum of factors
$-1, -15$	-16
$-3, -5$	-8 ←

Thus,
$$y^2 - 8y + 15 = (y - 5)(y - 3).$$

39. $x^2 + 4x + 5$

Look for two integers whose product is 5 and whose sum is 4. Both integers must be positive since b and c are both positive.

Product	Sum
$5 \cdot 1 = 5$	$5 + 1 = 6$

There is no other pair of positive integers whose product is 5. Since there is no pair of integers whose product is 5 and whose sum is 4, $x^2 + 4x + 5$ is a *prime* polynomial.

41. $t^2 - 8t + 16$

Find two integers whose product is 16 and whose sum is -8. Since c is positive and b is negative, both integers must be negative.

Factors of 16	Sum of factors
$-1, -16$	-17
$-2, -8$	-10
$-4, -4$	-8 ←

Thus,
$$t^2 - 8t + 16 = (t - 4)(t - 4)$$
$$\text{or } (t - 4)^2.$$

43. $r^2 - r - 30$

Look for two integers whose product is -30 and whose sum is -1. Because c is negative, one integer must be positive and the other must be negative.

Factors of -30	Sum of factors
$-1, 30$	29
$1, -30$	-29
$-2, 15$	13
$2, -15$	-13
$-3, 10$	7
$3, -10$	-7
$-5, 6$	1
$5, -6$	-1 ←

Thus,
$$r^2 - r - 30 = (r + 5)(r - 6).$$

45. $n^2 - 12n - 35$

Look for two integers whose product is -35 and whose sum is -12. Because c is negative, one integer must be positive and one must be negative.

Factors of -35	Sum of factors
$-1, 35$	34
$1, -35$	-34
$-5, 7$	2
$5, -7$	-2

This list does not produce the required integers, and there are no other possibilities to try. Therefore, $n^2 - 12n - 35$ is *prime*.

47. $r^2 + 3ra + 2a^2$

Look for two expressions whose product is $2a^2$ and whose sum is $3a$. They are $2a$ and a, so
$$r^2 + 3ra + 2a^2 = (r + 2a)(r + a).$$

49. $t^2 - tz - 6z^2$

Look for two expressions whose product is $-6z^2$ and whose sum is $-z$. They are $2z$ and $-3z$, so
$$t^2 - tz - 6z^2 = (t + 2z)(t - 3z).$$

51. $x^2 + 4xy + 3y^2$

Look for two expressions whose product is $3y^2$ and whose sum is $4y$. The expressions are $3y$ and y, so
$$x^2 + 4xy + 3y^2 = (x + 3y)(x + y).$$

53. $v^2 - 11vw + 30w^2$

Factors of $30w^2$	Sum of factors
$-30w, -w$	$-31w$
$-15w, -2w$	$-17w$
$-10w, -3w$	$-13w$
$-5w, -6w$	$-11w$

The complete factored form is

$$v^2 - 11vw + 30w^2 = (v - 5w)(v - 6w).$$

55. $4x^2 + 12x - 40$

First, factor out the GCF, 4.

$$4x^2 + 12x - 40 = 4(x^2 + 3x - 10)$$

Now factor $x^2 + 3x - 10$.

Factors of -10	Sum of factors
$-1, 10$	9
$1, -10$	-9
$2, -5$	-3
$-2, 5$	$3 \leftarrow$

Thus,

$$x^2 + 3x - 10 = (x - 2)(x + 5).$$

The complete factored form is

$$4x^2 + 12x - 40 = 4(x - 2)(x + 5).$$

57. $2t^3 + 8t^2 + 6t$

First, factor out the GCF, $2t$.

$$2t^3 + 8t^2 + 6t = 2t(t^2 + 4t + 3)$$

Then factor $t^2 + 4t + 3$.

$$t^2 + 4t + 3 = (t + 1)(t + 3)$$

The complete factored form is

$$2t^3 + 8t^2 + 6t = 2t(t + 1)(t + 3).$$

59. $2x^6 + 8x^5 - 42x^4$

First, factor out the GCF, $2x^4$.

$$2x^6 + 8x^5 - 42x^4 = 2x^4(x^2 + 4x - 21)$$

Now factor $x^2 + 4x - 21$.

Factors of -21	Sum of factors
$1, -21$	-20
$-1, 21$	20
$3, -7$	-4
$-3, 7$	$4 \leftarrow$

Thus,

$$x^2 + 4x - 21 = (x - 3)(x + 7).$$

The complete factored form is

$$2x^6 + 8x^5 - 42x^4 = 2x^4(x - 3)(x + 7).$$

61. $m^3n - 10m^2n^2 + 24mn^3$

First, factor out the GCF, mn.
$$m^3n - 10m^2n^2 + 24mn^3$$
$$= mn(m^2 - 10mn + 24n^2)$$

The expressions $-6n$ and $-4n$ have a product of $24n^2$ and a sum of $-10n$. The complete factored form is
$$m^3n - 10m^2n^2 + 24mn^3$$
$$= mn(m - 6n)(m - 4n).$$

63. $(2x + 4)(x - 3)$

	F	**O**	**I**	**L**
$= (2x)(x) + (2x)(-3) + (4)(x) + (4)(-3)$
$= 2x^2 - 6x + 4x - 12$
$= 2x^2 - 2x - 12$

It is incorrect to completely factor $2x^2 - 2x - 12$ as $(2x + 4)(x - 3)$ because $2x + 4$ can be factored further as $2(2x + 2)$. The first step should be to factor out the GCF, 2. The correct factorization is

$$2x^2 - 2x - 12 = 2(x^2 - x - 6)$$
$$= 2(x + 2)(x - 3).$$

65. $a^5 + 3a^4b - 4a^3b^2$

The GCF is a^3, so

$$a^5 + 3a^4b - 4a^3b^2$$
$$= a^3(a^2 + 3ab - 4b^2).$$

Now factor $a^2 + 3ab - 4b^2$. The expressions $4b$ and $-b$ have a product of $-4b^2$ and a sum of $3b$. The complete factored form is

$$a^5 + 3a^4b - 4a^3b^2 = a^3(a + 4b)(a - b).$$

67. $y^3z + y^2z^2 - 6yz^3$

The GCF is yz, so

$$y^3z + y^2z^2 - 6yz^3 = yz(y^2 + yz - 6z^2).$$
Now factor $y^2 + yz - 6z^2$. The expressions $3z$ and $-2z$ have a product of $-6z^2$ and a sum of z. The complete factored form is

$$y^3z + y^2z^2 - 6yz^3 = yz(y + 3z)(y - 2z).$$

69. $z^{10} - 4z^9y - 21z^8y^2$
$$= z^8(z^2 - 4zy - 21y^2) \quad \textit{GCF is } z^8$$
$$= z^8(z - 7y)(z + 3y)$$

71. The GCF is $(a + b)$, so

$$(a + b)x^2 + (a + b)x - 12(a + b)$$
$$= (a + b)(x^2 + x - 12).$$

Now factor $x^2 + x - 12$.

$$x^2 + x - 12 = (x + 4)(x - 3)$$

continued

The complete factored form is

$$(a+b)x^2 + (a+b)x - 12(a+b)$$
$$= (a+b)(x+4)(x-3).$$

73. The GCF is $(2p+q)$, so

$$(2p+q)r^2 - 12(2p+q)r + 27(2p+q)$$
$$= (2p+q)(r^2 - 12r + 27).$$

Now factor $r^2 - 12r + 27$.

$$r^2 - 12r + 27 = (r-9)(r-3)$$

The complete factored form is

$$(2p+q)r^2 - 12(2p+q)r + 27(2p+q)$$
$$= (2p+q)(r-9)(r-3).$$

75. Multiply the factors using FOIL to determine the polynomial.

$$(a+9)(a+4)$$

$$\begin{array}{cccc} \textbf{F} & \textbf{O} & \textbf{I} & \textbf{L} \end{array}$$
$$= (a)(a) + (a)(4) + (9)(a) + (9)(4)$$
$$= a^2 + 4a + 9a + 36$$
$$= a^2 + 13a + 36$$

Section 6.3

1. Find the product of _2_ and _−21_. This is −42.

3. Write the middle term x as _−6x_ + _7x_, or as _7x_ + _−6x_.

5. $2x^2 - 6x + 7x - 21$
$$= (2x^2 - 6x)(7x - 21)$$
$$= 2x(x-3) + 7(x-3)$$

Factor out the common factor of _x − 3_.

7. $6a^2 + 7ab - 20b^2 = (3a\underline{})(\underline{} + 5b)$

We know that the missing term in the first expression must be $-4b$ since

$$(-4b)(5b) = -20b^2.$$

The missing term in the second expression must be $2a$ since

$$(3a)(2a) = 6a^2.$$

Checking our answer by multiplying, we see that $(3a-4b)(2a+5b) = 6a^2 + 7ab - 20b^2$, as desired.

9. $3x^2 - 9x - 30 = 3(x^2\underline{})$

Factoring 3 out of $-9x - 30$ gives us the missing expression $-3x - 10$. Factoring $x^2 - 3x - 10$ gives us

$$3(x\underline{-5})(x\underline{+2}).$$

11. $4y^2 + 17y - 15$

Multiply the factors in the choices together to see which ones give the correct product. Since

$$(y+5)(4y-3) = 4y^2 + 17y - 15$$

and

$$(2y-5)(2y+3) = 4y^2 - 4y - 15,$$

the correct factored form is choice (a), $(y+5)(4y-3)$.

13. $4k^2 + 13mk + 3m^2$

Since

$$(4k+m)(k+3m) = 4k^2 + 13km + 3m^2$$

and

$$(4k+3m)(k+m) = 4k^2 + 7km + 3m^2,$$

the correct factored form is choice (a), $(4k+m)(k+3m)$.

15. Since 2 is not a factor of $12x^2 + 7x - 12$, it cannot be a factor of any factor of $12x^2 + 7x - 12$. Since 2 is a factor of $2x - 6$, this means that $2x - 6$ cannot be a factor of $12x^2 + 7x - 12$.

Note: In Exercises 17–56, either the trial and error method (which uses FOIL in reverse) or the grouping method can be used to factor each polynomial.

17. $3a^2 + 10a + 7$

Factor by the grouping method. Look for two integers whose product is $3(7) = 21$ and whose sum is 10. The integers are 3 and 7. Use these integers to rewrite the middle term, $10a$, as $3a + 7a$, and then factor the resulting four-term polynomial by grouping.

$$3a^2 + 10a + 7$$
$$= 3a^2 + 3a + 7a + 7 \quad 10a = 3a + 7a$$
$$= (3a^2 + 3a) + (7a + 7) \quad \begin{array}{l}\textit{Group}\\ \textit{the terms}\end{array}$$
$$= 3a(a+1) + 7(a+1) \quad \begin{array}{l}\textit{Factor}\\ \textit{each group}\end{array}$$
$$= (a+1)(3a+7) \quad \begin{array}{l}\textit{Factor out}\\ a+1\end{array}$$

19. $4r^2 + r - 3$

Factor by trial and error.

Possible factors of $4r^2$ are $4r$ and r, $2r$ and $2r$, or r and $4r$.

Possible factors of -3 are -3 and 1 or 3 and -1.

$$(4r+3)(r-1) = 4r^2 - r - 3 \quad \textit{Incorrect}$$

The middle term differs only in sign from the correct product, so change the signs of the two factors.

$$(4r-3)(r+1) = 4r^2 + r - 3 \quad \textit{Correct}$$

21. $15m^2 + m - 2$

Factor by the grouping method. Look for two integers whose product is $15(-2) = -30$ and whose sum is 1. The integers are 6 and -5.

$15m^2 + m - 2$
$= 15m^2 + 6m - 5m - 2$ $m = 6m - 5m$
$= (15m^2 + 6m) + (-5m - 2)$ *Group the terms*
$= 3m(5m + 2) - 1(5m + 2)$ *Factor each group*
$= (5m + 2)(3m - 1)$ *Factor out $5m + 2$*

23. $8m^2 - 10m - 3$

Factor by the grouping method. Look for two integers whose product is $8(-3) = -24$ and whose sum is -10. The integers are -12 and 2.

$8m^2 - 10m - 3$
$= 8m^2 - 12m + 2m - 3$ $-10m = -12m + 2m$
$= (8m^2 - 12m) + (2m - 3)$ *Group the terms*
$= 4m(2m - 3) + 1(2m - 3)$ *Factor each group*
$= (2m - 3)(4m + 1)$ *Factor out $2m - 3$*

25. $20x^2 + 11x - 3$

Factor by the grouping method. Look for two integers whose product is $20(-3) = -60$ and whose sum is 11. The integers are 15 and -4.

$20x^2 + 11x - 3$
$= 20x^2 + 15x - 4x - 3$ $11x = 15x - 4x$
$= (20x^2 + 15x) + (-4x - 3)$ *Group the terms*
$= 5x(4x + 3) - 1(4x + 3)$ *Factor each group*
$= (4x + 3)(5x - 1)$ *Factor out $4x + 3$*

27. $21m^2 + 13m + 2$

Factor by the grouping method. Look for two integers whose product is $21(2) = 42$ and whose sum is 13. The integers are 7 and 6.

$21m^2 + 13m + 2$
$= 21m^2 + 7m + 6m + 2$ $13m = 7m + 6m$
$= (21m^2 + 7m) + (6m + 2)$ *Group the terms*
$= 7m(3m + 1) + 2(3m + 1)$ *Factor each group*
$= (3m + 1)(7m + 2)$ *Factor out $3m + 1$*

29. $20y^2 + 39y - 11$

Factor by the grouping method. Look for two integers whose product is $20(-11) = -220$ and whose sum is 39. The integers are 44 and -5.

$20y^2 + 39y - 11$
$= 20y^2 + 44y - 5y - 11$ $39y = 44y - 5y$
$= (20y^2 + 44y) + (-5y - 11)$ *Group the terms*
$= 4y(5y + 11) - 1(5y + 11)$ *Factor each group*
$= (5y + 11)(4y - 1)$ *Factor out $5y + 11$*

31. $6b^2 + 7b + 2$

Factor by the grouping method. Look for two integers whose product is $6(2) = 12$ and whose sum is 7. The integers are 3 and 4.

$6b^2 + 7b + 2$
$= 6b^2 + 3b + 4b + 2$ $7b = 3b + 4b$
$= (6b^2 + 3b) + (4b + 2)$ *Group the terms*
$= 3b(2b + 1) + 2(2b + 1)$ *Factor each group*
$= (2b + 1)(3b + 2)$ *Factor out $2b + 1$*

33. Factor out the GCF, 3.
$24x^2 - 42x + 9 = 3(8x^2 - 14x + 3)$

Use the grouping method to factor $8x^2 - 14x + 3$. Look for two integers whose product is $8(3) = 24$ and whose sum is -14. The integers are -12 and -2.

$24x^2 - 42x + 9$
$= 3(8x^2 - 12x - 2x + 3)$ $-14x = -12x - 2x$
$= 3[(8x^2 - 12x) + (-2x + 3)]$ *Group the terms*
$= 3[4x(2x - 3) - 1(2x - 3)]$ *Factor each group*
$= 3(2x - 3)(4x - 1)$ *Factor out $2x - 3$*

35. $40m^2q + mq - 6q$

First, factor out the greatest common factor, q.

$$40m^2q + mq - 6q = q(40m^2 + m - 6)$$

Now factor $40m^2 + m - 6$ by trial and error to obtain

$$40m^2 + m - 6 = (5m + 2)(8m - 3).$$

The complete factorization is

$$40m^2q + mq - 6q = q(5m + 2)(8m - 3).$$

37. $2m^3 + 2m^2 - 40m$

$$= 2m(m^2 + m - 20) \quad \begin{array}{l} \textit{Factor out} \\ \textit{the GCF, 2m} \end{array}$$

$$= 2m(m + 5)(m - 4) \quad \begin{array}{l} \textit{5(-4) = -20;} \\ \textit{5 + (-4) = 1} \end{array}$$

39. Factor out the GCF, $3n^2$.

$$15n^4 - 39n^3 + 18n^2 = 3n^2(5n^2 - 13n + 6)$$

Factor $5n^2 - 13n + 6$ by the trial and error method. Possible factors of $5n^2$ are $5n$ and n.

Possible factors of 6 are -6 and -1, -3 and -2, -2 and -3, or -1 and -6.

$(5n - 6)(n - 1) = 5n^2 - 11n + 6$ *Incorrect*
$(5n - 3)(n - 2) = 5n^2 - 13n + 6$ *Correct*

The complete factored form is
$$15n^4 - 39n^3 + 18n^2 = 3n^2(5n - 3)(n - 2).$$

41. $18x^5 + 15x^4 - 75x^3$

$$= 3x^3(6x^2 + 5x - 25) \quad \textit{Factor out } 3x^3$$

$$= 3x^3(6x^2 + 15x - 10x - 25) \quad \textit{5x = 15x - 10x}$$

$$= 3x^3[(6x^2 + 15x) + (-10x - 25)] \quad \begin{array}{l} \textit{Group} \\ \textit{the terms} \end{array}$$

$$= 3x^3[3x(2x + 5) - 5(2x + 5)] \quad \begin{array}{l} \textit{Factor} \\ \textit{each group} \end{array}$$

$$= 3x^3(2x + 5)(3x - 5) \quad \begin{array}{l} \textit{Factor out} \\ \textit{2x + 5} \end{array}$$

43. Factor out the GCF, y^2.
$$15x^2y^2 - 7xy^2 - 4y^2 = y^2(15x^2 - 7x - 4)$$

Factor $15x^2 - 7x - 4$ by the grouping method. Look for two integers whose product is $15(-4) = -60$ and whose sum is -7. The integers are -12 and 5.

$$15x^2y^2 - 7xy^2 - 4y^2$$
$$= y^2(15x^2 - 12x + 5x - 4)$$
$$= y^2[3x(5x - 4) + 1(5x - 4)]$$
$$= y^2(5x - 4)(3x + 1)$$

45. $12p^2 + 7pq - 12q^2$

$$= 12p^2 + 16pq - 9pq - 12q^2$$

$$ \quad \textit{7pq = 16pq - 9pq}$$

$$= (12p^2 + 16pq) + (-9pq - 12q^2)$$

$$\quad \textit{Group the terms}$$

$$= 4p(3p + 4q) - 3q(3p + 4q)$$

$$\quad \textit{Factor each group}$$

$$= (3p + 4q)(4p - 3q)$$

$$\quad \textit{Factor out 3p + 4q}$$

47. $25a^2 + 25ab + 6b^2$

Use the grouping method. Find two integers whose product is $(25)(6) = 150$ and whose sum is 25. The numbers are 15 and 10.

$$25a^2 + 25ab + 6b^2$$
$$= 25a^2 + 15ab + 10ab + 6b^2$$
$$= 5a(5a + 3b) + 2b(5a + 3b)$$
$$= (5a + 3b)(5a + 2b)$$

49. $6a^2 - 7ab - 5b^2$

$$= 6a^2 - 10ab + 3ab - 5b^2$$

$$\quad \textit{-7ab = -10ab + 3ab}$$

$$= (6a^2 - 10ab) + (3ab - 5b^2)$$

$$\quad \textit{Group the terms}$$

$$= 2a(3a - 5b) + b(3a - 5b)$$

$$\quad \textit{Factor each group}$$

$$= (3a - 5b)(2a + b)$$

$$\quad \textit{Factor out 3a - 5b}$$

51. Factor out the GCF, m^4n.

$$6m^6n + 7m^5n^2 + 2m^4n^3$$
$$= m^4n(6m^2 + 7mn + 2n^2)$$

Now factor $6m^2 + 7mn + 2n^2$ by trial and error.

Possible factors of $6m^2$ are $6m$ and m or $3m$ and $2m$. Possible factors of $2n^2$ are $2n$ and n.

$$(3m + 2n)(2m + n) = 6m^2 + 7mn + 2n^2$$

$$\quad \textit{Correct}$$

Thus,

$$6m^6n + 7m^5n^2 + 2m^4n^3$$
$$= m^4n(3m + 2n)(2m + n).$$

53. $5 - 6x + x^2$

$$= 5 - 5x - x + x^2 \quad \textit{-6x = -5x - x}$$

$$= (5 - 5x) + (-x + x^2) \quad \begin{array}{l} \textit{Group} \\ \textit{the terms} \end{array}$$

$$= 5(1 - x) - x(1 - x) \quad \begin{array}{l} \textit{Factor} \\ \textit{each group} \end{array}$$

$$= (1 - x)(5 - x) \quad \begin{array}{l} \textit{Factor out} \\ \textit{1 - x} \end{array}$$

55. $16 + 16x + 3x^2$

Factor by the grouping method. Find two integers whose product is $(16)(3) = 48$ and whose sum is 16. The numbers are 4 and 12.

$$16 + 16x + 3x^2$$
$$= 16 + 4x + 12x + 3x^2$$
$$= 4(4 + x) + 3x(4 + x)$$
$$= (4 + x)(4 + 3x)$$

57. $-10x^3 + 5x^2 + 140x$

First, factor out $-5x$; then complete the factoring by trial and error, using FOIL to test various possibilities until the correct one is found.

$$-10x^3 + 5x^2 + 140x$$
$$= -5x(2x^2 - x - 28)$$
$$= -5x(2x + 7)(x - 4)$$

59. $-x^2 - 4x + 21 = -1(x^2 + 4x - 21)$
$$= -1(x + 7)(x - 3)$$

61. $-3x^2 - x + 4 = -1(3x^2 + x - 4)$
$$= -1(3x + 4)(x - 1)$$

63. $-2a^2 - 5ab - 2b^2$
$$= -1(2a^2 + 5ab + 2b^2)$$

Factor out –1

$$= -1(2a^2 + 4ab + ab + 2b^2)$$

5ab = 4ab + ab

$$= -1[(2a^2 + 4ab) + (ab + 2b^2)]$$

Group the terms

$$= -1[2a(a + 2b) + b(a + 2b)]$$

Factor each group

$$= -1(a + 2b)(2a + b)$$

Factor out a + 2b

65. Yes, $(x + 7)(3 - x)$ is equivalent to $-1(x + 7)(x - 3)$ because $-1(x - 3) = -x + 3 = 3 - x$.

67. First, factor out the GCF, $(m + 1)^3$; then factor the resulting trinomial by trial and error.

$$25q^2(m + 1)^3 - 5q(m + 1)^3 - 2(m + 1)^3$$
$$= (m + 1)^3(25q^2 - 5q - 2)$$
$$= (m + 1)^3(5q - 2)(5q + 1)$$

69. $15x^2(r + 3)^3 - 34xy(r + 3)^3 - 16y^2(r + 3)^3$
$$= (r + 3)^3(15x^2 - 34xy - 16y^2)$$
$$= (r + 3)^3(5x + 2y)(3x - 8y)$$

71. $5x^2 + kx - 1$

Look for two integers whose product is $5(-1) = -5$ and whose sum is k.

Factors of -5	Sum of factors (k)
$-5, 1$	-4
$5, -1$	4

Thus, there are two possible integer values for k: -4 and 4.

73. $2m^2 + km + 5$

Look for two integers whose product is $2(5) = 10$ and whose sum is k.

Factors of 10	Sum of factors (k)
$-10, -1$	-11
$10, 1$	11
$-5, -2$	-7
$5, 2$	7

Thus, there are four possible integer values for k: $-11, -7, 7,$ and 11.

75. $35 = 5 \cdot 7$

76. $35 = (-5)(-7)$

77. Verify the given factored form by multiplying the two binomials by FOIL.

$$(3x - 4)(2x - 1) = 6x^2 - 3x - 8x + 4$$
$$= 6x^2 - 11x + 4$$

Thus, the product of $3x - 4$ and $2x - 1$ is $6x^2 - 11x + 4$.

78. Verify the given factored form by multiplying the two binomials by FOIL.

$$(4 - 3x)(1 - 2x) = 4 - 8x - 3x + 6x^2$$
$$= 6x^2 - 11x + 4$$

Thus, the product of $4 - 3x$ and $1 - 2x$ is $6x^2 - 11x + 4$.

79. The factors in Exercise 78 are the opposites of the factors in Exercise 77.

80. We can form another valid factored form by taking the opposites of the two given factors. The opposite of $7t - 3$ is $-1(7t - 3) = -7t + 3 = 3 - 7t$, and the opposite of $2t - 5$ is $5 - 2t$, so we obtain the factored form $(3 - 7t)(5 - 2t)$.

81. If an integer factors as the product of a and b, then it also factors as the product of $\underline{-a}$ and $\underline{-b}$.

82. If a trinomial factors as the product of the binomials P and Q, then it also factors as the product of the binomials $\underline{-P}$ and $\underline{-Q}$.

Section 6.4

1.
$$1^2 = \underline{1} \qquad 2^2 = \underline{4} \qquad 3^2 = \underline{9}$$
$$4^2 = \underline{16} \qquad 5^2 = \underline{25} \qquad 6^2 = \underline{36}$$
$$7^2 = \underline{49} \qquad 8^2 = \underline{64} \qquad 9^2 = \underline{81}$$
$$10^2 = \underline{100} \quad 11^2 = \underline{121} \quad 12^2 = \underline{144}$$
$$13^2 = \underline{169} \quad 14^2 = \underline{196} \quad 15^2 = \underline{225}$$
$$16^2 = \underline{256} \quad 17^2 = \underline{289} \quad 18^2 = \underline{324}$$
$$19^2 = \underline{361} \quad 20^2 = \underline{400}$$

3.
$$1^3 = \underline{1} \qquad 2^3 = \underline{8} \qquad 3^3 = \underline{27}$$
$$4^3 = \underline{64} \qquad 5^3 = \underline{125} \qquad 6^3 = \underline{216}$$
$$7^3 = \underline{343} \qquad 8^3 = \underline{512} \qquad 9^3 = \underline{729}$$
$$10^3 = \underline{1000}$$

5. **(a)** $64x^6y^{12} = (8x^3y^6)^2$, so $64x^6y^{12}$ is a perfect square.
$64x^6y^{12} = (4x^2y^4)^3$, so $64x^6y^{12}$ is also a perfect cube.
Therefore, the answer is "both of these."

(b) $125t^6 = (5t^2)^3$, so $125t^6$ is a perfect cube. Since 125 is not a perfect square, $125t^6$ is not a perfect square.

(c) $49x^{12} = (7x^6)^2$, so $49x^{12}$ is a perfect square. Since 49 is not a perfect cube, $49x^{12}$ is not a perfect cube.

(d) $81r^{10} = (9r^5)^2$, so $81r^{10}$ is a perfect square. It is not a perfect cube.

7. $y^2 - 25$

To factor this binomial, use the rule for factoring a difference of two squares.
$$a^2 - b^2 = (a + b)(a - b)$$
$$\downarrow \quad \downarrow \qquad \downarrow \quad \downarrow \quad \downarrow \quad \downarrow$$
$$y^2 - 25 = y^2 - 5^2 = (y + 5)(y - 5)$$

9. $9r^2 - 4 = (3r)^2 - 2^2$
$$= (3r + 2)(3r - 2)$$

11. $36m^2 - \dfrac{16}{25} = (6m)^2 - \left(\dfrac{4}{5}\right)^2$
$$= \left(6m + \dfrac{4}{5}\right)\left(6m - \dfrac{4}{5}\right)$$

13. $36x^2 - 16$

First factor out the GCF, 4; then use the rule for factoring the difference of two squares.
$$36x^2 - 16 = 4(9x^2 - 4)$$
$$= 4[(3x)^2 - 2^2]$$
$$= 4(3x + 2)(3x - 2)$$

15. $196p^2 - 225 = (14p)^2 - 15^2$
$$= (14p + 15)(14p - 15)$$

17. $16r^2 - 25a^2 = (4r)^2 - (5a)^2$
$$= (4r + 5a)(4r - 5a)$$

19. $100x^2 + 49$

This binomial is the *sum* of two squares and the terms have no common factor. Unlike the *difference* of two squares, it cannot be factored. It is a prime polynomial.

21. $p^4 - 49 = (p^2)^2 - 7^2$
$$= (p^2 + 7)(p^2 - 7)$$

23. $x^4 - 1$

To factor this binomial completely, factor the difference of two squares twice.
$$x^4 - 1 = (x^2)^2 - 1^2$$
$$= (x^2 + 1)(x^2 - 1)$$
$$= (x^2 + 1)(x^2 - 1^2)$$
$$= (x^2 + 1)(x + 1)(x - 1)$$

25. $p^4 - 256$

To factor this binomial completely, factor the difference of two squares twice.
$$p^4 - 256 = (p^2)^2 - 16^2$$
$$= (p^2 + 16)(p^2 - 16)$$
$$= (p^2 + 16)(p^2 - 4^2)$$
$$= (p^2 + 16)(p + 4)(p - 4)$$

27. The student's answer is not a complete factorization because $x^2 - 9$ can be factored further. The correct complete factorization is
$$x^4 - 81 = (x^2 + 9)(x + 3)(x - 3).$$
Because the teacher had directed the student to factor the polynomial *completely*, she was justified in her grading of the item.

In Exercises 29–46, use the rules for factoring perfect square trinomials:
$$a^2 + 2ab + b^2 = (a + b)^2$$
$$a^2 - 2ab + b^2 = (a - b)^2.$$

29. $w^2 + 2w + 1$

The first and last terms are perfect squares, w^2 and 1^2. This trinomial is a perfect square, since the middle term is twice the product of w and 1, or
$$2 \cdot w \cdot 1 = 2w.$$
Therefore,
$$w^2 + 2w + 1 = (w + 1)^2.$$

31. $x^2 - 8x + 16$

The first and last terms are perfect squares, x^2 and $(-4)^2$. This trinomial is a perfect square, since the middle term is twice the product of x and -4, or

$$2 \cdot x \cdot (-4) = -8x.$$

Therefore,

$$x^2 - 8x + 16 = (x - 4)^2.$$

33. $t^2 + t + \dfrac{1}{4}$

t^2 is a perfect square, and $\dfrac{1}{4}$ is a perfect square since $\dfrac{1}{2} \cdot \dfrac{1}{2} = \dfrac{1}{4}$. The middle term is twice the product of the first and last terms:

$$t = 2(t)\left(\dfrac{1}{2}\right).$$

Therefore,

$$t^2 + t + \dfrac{1}{4} = \left(t + \dfrac{1}{2}\right)^2.$$

35. $x^2 - 1.0x + .25$

The first and last terms are perfect squares, x^2 and $(-.5)^2$. The trinomial is a perfect square, since the middle term is

$$2 \cdot x \cdot (-.5) = -1.0x.$$

Therefore,

$$x^2 - 1.0x + .25 = (x - .5)^2.$$

37. $2x^2 + 24x + 72$

First, factor out the GCF, 2.

$$2x^2 + 24x + 72 = 2\left(x^2 + 12x + 36\right)$$

Now factor $x^2 + 12x + 36$ as a perfect square trinomial.

$$x^2 + 12x + 36 = (x + 6)^2$$

The final factored form is

$$2x^2 + 24x + 72 = 2(x + 6)^2.$$

39. $16x^2 - 40x + 25$

The first and last terms are perfect squares, $(4x)^2$ and $(-5)^2$. The middle term is

$$2(4x)(-5) = -40x.$$

Therefore,

$$16x^2 - 40x + 25 = (4x - 5)^2.$$

41. $49x^2 - 28xy + 4y^2$

The first and last terms are perfect squares, $(7x)^2$ and $(-2y)^2$. The middle term is

$$2(7x)(-2y) = -28xy.$$

Therefore,

$$49x^2 - 28xy + 4y^2 = (7x - 2y)^2.$$

43. $64x^2 + 48xy + 9y^2$
$$= (8x)^2 + 2(8x)(3y) + (3y)^2$$
$$= (8x + 3y)^2$$

45. $-50h^2 + 40hy - 8y^2$
$$= -2\left(25h^2 - 20hy + 4y^2\right)$$
$$= -2\left[(5h)^2 - 2(5h)(2y) + (2y)^2\right]$$
$$= -2(5h - 2y)^2$$

47. $a^3 + 1$

Let $x = a$ and $y = 1$ in the pattern for the sum of two cubes.

$$x^3 + y^3 = (x + y)\left(x^2 - xy + y^2\right)$$
$$a^3 + 1 = a^3 + 1^3 = (a + 1)\left(a^2 - a \cdot 1 + 1^2\right)$$
$$= (a + 1)\left(a^2 - a + 1\right)$$

49. $a^3 - 1$

Let $x = a$ and $y = 1$ in the pattern for the difference of two cubes.

$$x^3 - y^3 = (x - y)\left(x^2 + xy + y^2\right)$$
$$a^3 - 1 = a^3 - 1^3 = (a - 1)\left(a^2 + a \cdot 1 + 1^2\right)$$
$$= (a - 1)\left(a^2 + a + 1\right)$$

51. $p^3 + q^3 = (p + q)\left(p^2 - pq + q^2\right)$ *Sum of two cubes*

53. Factor $27x^3 - 1$ as the difference of two cubes.
$$27x^3 - 1 = (3x)^3 - 1^3$$
$$= (3x - 1)\left[(3x)^2 + 3x \cdot 1 + 1^2\right]$$
$$= (3x - 1)\left(9x^2 + 3x + 1\right)$$

55. Factor $8p^3 + 729q^3$ as the sum of two cubes.
$$8p^3 + 729q^3 = (2p)^3 + (9q)^3$$
$$= (2p + 9q)\left[(2p)^2 - 2p \cdot 9q + (9q)^2\right]$$
$$= (2p + 9q)\left(4p^2 - 18pq + 81q^2\right)$$

57. Factor $y^3 - 8x^3$ as the difference of two cubes.
$$y^3 - 8x^3 = y^3 - (2x)^3$$
$$= (y - 2x)\left[y^2 + y \cdot 2x + (2x)^2\right]$$
$$= (y - 2x)\left(y^2 + 2yx + 4x^2\right)$$

59. Factor $27a^3 - 64b^3$ as the difference of two cubes.
$$27a^3 - 64b^3 = (3a)^3 - (4b)^3$$
$$= (3a - 4b)\left[(3a)^2 + 3a \cdot 4b + (4b)^2\right]$$
$$= (3a - 4b)\left(9a^2 + 12ab + 16b^2\right)$$

61. Factor $125t^3 + 8s^3$ as the sum of two cubes.

$$125t^3 + 8s^3 = (5t)^3 + (2s)^3$$
$$= (5t + 2s)\left[(5t)^2 - 5t \cdot 2s + (2s)^2\right]$$
$$= (5t + 2s)(25t^2 - 10ts + 4s^2)$$

63. $x^2 - y^2 = (x + y)(x - y)$
Difference of Two Squares;
Examples are Exercises 7–18 and 21–26.

$x^2 + 2xy + y^2 = (x + y)^2$
Perfect Square Trinomial;
Examples are Exercises 29, 30, 33, 34, 37, 43, 44.

$x^2 - 2xy + y^2 = (x - y)^2$
Perfect Square Trinomial;
Examples are Exercises 31, 32, 35, 36, 38–42, 45, 46.

$x^3 - y^3 = (x - y)(x^2 + xy + y^2)$
Difference of Two Cubes;
Examples are Exercises 49, 50, 53, 54, 57–60.

$x^3 + y^3 = (x + y)(x^2 - xy + y^2)$
Sum of Two Cubes;
Examples are Exercises 47, 48, 51, 52, 55, 56, 61, 62.

64. Factor by trial and error to obtain
$$10x^2 + 11x - 6 = (5x - 2)(2x + 3).$$

65.
$$
\begin{array}{r}
5x \;\; - 2 \\
2x + 3 \overline{\big)\, 10x^2 + 11x - 6} \\
\underline{10x^2 + 15x} \\
-\,4x - 6 \\
\underline{-\,4x - 6} \\
0
\end{array}
$$

Thus,
$$\frac{10x^2 + 11x - 6}{2x + 3} = 5x - 2.$$

66. Yes. If $10x^2 + 11x - 6$ factors as $(5x - 2)(2x + 3)$, then when $10x^2 + 11x - 6$ is divided by $2x + 3$, the quotient should be $5x - 2$.

67.
$$
\begin{array}{r}
x^2 + x + 1 \\
x - 1 \overline{\big)\, x^3 + 0x^2 + 0x - 1} \\
\underline{x^3 - x^2} \\
x^2 + 0x \\
\underline{x^2 - x} \\
x - 1 \\
\underline{x - 1} \\
0
\end{array}
$$

The quotient is $x^2 + x + 1$, so
$$x^3 - 1 = (x - 1)(x^2 + x + 1).$$

69. $(a - b)^3 - (a + b)^3$

Factor as the difference of two cubes. Substitute into the rule using $x = a - b$ and $y = a + b$.

$(a - b)^3 - (a + b)^3$
$$= [(a - b) - (a + b)]$$
$$\cdot \left[(a - b)^2 + (a - b)(a + b) + (a + b)^2\right]$$
$$= (a - b - a - b)$$
$$\cdot \left[(a^2 - 2ab + b^2) + (a^2 - b^2)\right.$$
$$\left. + (a^2 + 2ab + b^2)\right]$$
$$= -2b(3a^2 + b^2) \quad \textit{Combine like terms}$$

71. $3r - 3k + 3r^2 - 3k^2$

First factor out the GCF, 3.
$$3r - 3k + 3r^2 - 3k^2$$
$$= 3(r - k + r^2 - k^2)$$

Now factor $r - k + r^2 - k^2$ by grouping, noting that $r^2 - k^2$ is the difference of two squares.

$r - k + r^2 - k^2$
$$= 1(r - k) + (r + k)(r - k) \quad \begin{array}{l}\textit{Factor}\\ \textit{each group}\end{array}$$
$$= (r - k)[1 + (r + k)] \quad \begin{array}{l}\textit{Factor out}\\ r - k\end{array}$$
$$= (r - k)(1 + r + k)$$

Therefore,
$$3r - 3k + 3r^2 - 3k^2$$
$$= 3(r - k)(1 + r + k).$$

72. $(x^2 + 2x + 1) - 4 = (x + 1)^2 - 4$

73. Using the rule for factoring the difference of two squares, we obtain
$$(x + 1)^2 - 4 = (x + 1)^2 - 2^2$$
$$= [(x + 1) - 2][(x + 1) + 2].$$

74. Combining the constants within the factors from Exercise 73, we obtain
$$[(x + 1) - 2][(x + 1) + 2]$$
$$= (x + 1 - 2)(x + 1 + 2)$$
$$= (x - 1)(x + 3).$$

75. $(x^2 + 2x + 1) - 4 = x^2 + 2x - 3$

76. To factor $x^2 + 2x - 3$, look for two integers whose product is -3 and whose sum is 2. The integers are -1 and 3. Thus,
$$x^2 + 2x - 3 = (x - 1)(x + 3).$$

77. The results are the same. (Preference is a matter of individual choice.)

79. Find b so that
$$x^2 + bx + 25 = (x + 5)^2.$$

Since $(x + 5)^2 = x^2 + 10x + 25$, $b = 10$.

81. Find a so that
$$ay^2 - 12y + 4 = (3y - 2)^2.$$
Since $(3y - 2)^2 = 9y^2 - 12y + 4$, $a = 9$.

Summary: Exercises on Factoring

1. $a^2 - 4a - 12 = (a - 6)(a + 2)$

3. $6y^2 - 6y - 12 = 6(y^2 - y - 2)$
$$= 6(y - 2)(y + 1)$$

5. $6a + 12b + 18c$
$$= 6(a + 2b + 3c)$$

7. $p^2 - 17p + 66 = (p - 11)(p - 6)$

9. $10z^2 - 7z - 6$

Use the grouping method.

Look for two integers whose product is $10(-6) = -60$ and whose sum is -7. The integers are -12 and 5.

$10z^2 - 7z - 6$
$$= 10z^2 - 12z + 5z - 6$$
$$= 2z(5z - 6) + 1(5z - 6) \quad \textit{Factor each group}$$
$$= (5z - 6)(2z + 1) \quad \textit{Factor out } 5z - 6$$

11. $m^2 - n^2 + 5m - 5n$

Factor by grouping.

$m^2 - n^2 + 5m - 5n$
$$= (m + n)(m - n) + 5(m - n) \quad \textit{Factor each group}$$
$$= (m - n)(m + n + 5) \quad \textit{Factor out } m - n$$

13. $8a^5 - 8a^4 - 48a^3$
$$= 8a^3(a^2 - a - 6)$$
$$= 8a^3(a - 3)(a + 2)$$

15. $z^2 - 3za - 10a^2$
$$= (z - 5a)(z + 2a)$$

17. $x^2 - 4x - 5x + 20$
$$= x(x - 4) - 5(x - 4) \quad \textit{Factor each group}$$
$$= (x - 4)(x - 5) \quad \textit{Factor out } x - 4$$

19. $6n^2 - 19n + 10$
$$= (3n - 2)(2n - 5)$$

21. $16x + 20 = 4(4x + 5)$

23. $6y^2 - 5y - 4$

Factor by grouping. Find two integers whose product is $6(-4) = -24$ and whose sum is -5. The integers are -8 and 3.

$6y^2 - 5y - 4$
$$= 6y^2 - 8y + 3y - 4$$
$$= 2y(3y - 4) + 1(3y - 4)$$
$$= (3y - 4)(2y + 1)$$

25. $6z^2 + 31z + 5 = (6z + 1)(z + 5)$

27. $4k^2 - 12k + 9$
$$= (2k)^2 - 2 \cdot 2k \cdot 3 + 3^2$$
$$= (2k - 3)^2 \quad \textit{Perfect square trinomial}$$

29. $54m^2 - 24z^2$
$$= 6(9m^2 - 4z^2)$$
$$= 6[(3m)^2 - (2z)^2]$$
$$= 6(3m + 2z)(3m - 2z)$$

31. $3k^2 + 4k - 4 = (3k - 2)(k + 2)$

33. $14k^3 + 7k^2 - 70k$
$$= 7k(2k^2 + k - 10)$$
$$= 7k(2k + 5)(k - 2)$$

35. $y^4 - 16$
$$= (y^2)^2 - 4^2$$
$$= (y^2 + 4)(y^2 - 4) \quad \textit{Difference of two squares}$$
$$= (y^2 + 4)(y + 2)(y - 2) \quad \textit{Difference of two squares}$$

37. $8m - 16m^2 = 8m(1 - 2m)$

39. Factor $z^3 - 8$ as the difference of two cubes.
$$z^3 - 8 = z^3 - 2^3$$
$$= (z - 2)(z^2 + 2z + 2^2)$$
$$= (z - 2)(z^2 + 2z + 4)$$

41. $k^2 + 9$ cannot be factored because it is the sum of two squares. The expression is prime.

43. $32m^9 + 16m^5 + 24m^3$
$$= 8m^3(4m^6 + 2m^2 + 3)$$

45. $16r^2 + 24rm + 9m^2$
$$= (4r)^2 + 2 \cdot 4r \cdot 3m + (3m)^2$$
$$= (4r + 3m)^2 \quad \textit{Perfect square trinomial}$$

47. $15h^2 + 11hg - 14g^2$

Factor by grouping. Look for two integers whose product is $15(-14) = -210$ and whose sum is 11. The integers are 21 and -10.

$$15h^2 + 11hg - 14g^2$$
$$= 15h^2 + 21hg - 10hg - 14g^2$$
$$= 3h(5h + 7g) - 2g(5h + 7g) \quad \text{\textit{Factor each group}}$$
$$= (5h + 7g)(3h - 2g) \quad \text{\textit{Factor out } 5h + 7g}$$

49. $k^2 - 11k + 30$
$$= (k - 5)(k - 6)$$

51. $3k^3 - 12k^2 - 15k$
$$= 3k(k^2 - 4k - 5)$$
$$= 3k(k - 5)(k + 1)$$

53. Factor $1000p^3 + 27$ as the sum of two cubes.

$$1000p^3 + 27$$
$$= (10p)^3 + 3^3$$
$$= (10p + 3)\left[(10p)^2 - 10p \cdot 3 + 3^2\right]$$
$$= (10p + 3)\left(100p^2 - 30p + 9\right)$$

55. $6 + 3m + 2p + mp$
$$= 3(2 + m) + p(2 + m)$$
$$= (2 + m)(3 + p)$$

57. $16z^2 - 8z + 1 = (4z - 1)^2$

Perfect square trinomial

59. $108m^2 - 36m + 3$
$$= 3\left(36m^2 - 12m + 1\right)$$
$$= 3(6m - 1)^2 \quad \text{\textit{Perfect square trinomial}}$$

61. $64m^2 - 40mn + 25n^2$ is prime. The middle term would have to be $+80mn$ or $-80mn$ in order to make this a perfect square trinomial.

63. $32z^3 + 56z^2 - 16z$
$$= 8z\left(4z^2 + 7z - 2\right)$$
$$= 8z(4z - 1)(z + 2)$$

65. $20 + 5m + 12n + 3mn$
$$= 5(4 + m) + 3n(4 + m)$$
$$= (4 + m)(5 + 3n)$$

67. $6a^2 + 10a - 4$
$$= 2\left(3a^2 + 5a - 2\right)$$
$$= 2(3a - 1)(a + 2)$$

69. $a^3 - b^3 + 2a - 2b$

Factor by grouping. The first two terms form a difference of cubes.

$$a^3 - b^3 + 2a - 2b$$
$$= \left(a^3 - b^3\right) + (2a - 2b)$$
$$= (a - b)\left(a^2 + ab + b^2\right) + 2(a - b)$$
$$= (a - b)\left(a^2 + ab + b^2 + 2\right) \quad \text{\textit{Factor out } (a - b)}$$

71. $64m^2 - 80mn + 25n^2$
$$= (8m)^2 - 2(8m)(5n) + (5n)^2$$
$$= (8m - 5n)^2 \quad \text{\textit{Perfect square trinomial}}$$

73. $8k^2 - 2kh - 3h^2$
$$= 8k^2 - 6kh + 4kh - 3h^2$$
$$= 2k(4k - 3h) + h(4k - 3h) \quad \text{\textit{Factor each group}}$$
$$= (4k - 3h)(2k + h) \quad \text{\textit{Factor out } 4k - 3h}$$

75. $(m + 1)^3 + 1$

Use the pattern for the sum of two cubes with $x = m + 1$ and $y = 1$.

$$(m + 1)^3 + 1$$
$$= (m + 1)^3 + 1^3$$
$$= [(m + 1) + 1]$$
$$\quad \cdot \left[(m + 1)^2 - 1(m + 1) + 1^2\right]$$
$$= (m + 2)\left(m^2 + 2m + 1 - m - 1 + 1\right)$$
$$= (m + 2)\left(m^2 + m + 1\right)$$

77. $10y^2 - 7yz - 6z^2$
$$= 10y^2 - 12yz + 5yz - 6z^2$$
$$= 2y(5y - 6z) + z(5y - 6z) \quad \text{\textit{Factor each group}}$$
$$= (5y - 6z)(2y + z) \quad \text{\textit{Factor out } 5y - 6z}$$

79. $8a^2 + 23ab - 3b^2$
$$= 8a^2 + 24ab - ab - 3b^2$$
$$= 8a(a + 3b) - b(a + 3b) \quad \text{\textit{Factor each group}}$$
$$= (a + 3b)(8a - b) \quad \text{\textit{Factor out } a + 3b}$$

81. $x^6 - 1 = \left(x^3\right)^2 - 1^2$
$$= \left(x^3 + 1\right)\left(x^3 - 1\right)$$
$$\text{or} \quad \left(x^3 - 1\right)\left(x^3 + 1\right)$$

82. From Exercise 81, we have
$$x^6 - 1 = \left(x^3 - 1\right)\left(x^3 + 1\right).$$

Use the rules for the difference and sum of two cubes to factor further.

Since
$$x^3 - 1 = (x - 1)\left(x^2 + x + 1\right)$$
and
$$x^3 + 1 = (x + 1)\left(x^2 - x + 1\right),$$
we obtain the factorization
$$x^6 - 1 = (x - 1)\left(x^2 + x + 1\right)$$
$$\quad \cdot (x + 1)\left(x^2 - x + 1\right).$$

83. $x^6 - 1 = \left(x^2\right)^3 - 1^3$

$$= \left(x^2 - 1\right)\left[\left(x^2\right)^2 + x^2 \bullet 1 + 1^2\right]$$

$$= \left(x^2 - 1\right)\left(x^4 + x^2 + 1\right)$$

84. From Exercise 83, we have

$$x^6 - 1 = \left(x^2 - 1\right)\left(x^4 + x^2 + 1\right).$$

Use the rule for the difference of two squares to factor the binomial.

$$x^2 - 1 = (x - 1)(x + 1)$$

Thus, we obtain the factorization

$$x^6 - 1 = (x - 1)(x + 1)\left(x^4 + x^2 + 1\right).$$

85. The result in Exercise 82 is the completely factored form.

86. Multiply the trinomials from the factored form in Exercise 82 vertically.

$$
\begin{array}{r}
x^2 \quad + x \quad + 1 \\
x^2 \quad - x \quad + 1 \\
\hline
x^2 \quad + x \quad + 1 \\
- x^3 \quad - x^2 \quad - x \\
x^4 \quad + x^3 \quad + x^2 \\
\hline
x^4 \qquad\qquad + x^2 \qquad\qquad + 1
\end{array}
$$

87. In general, if I must choose between factoring first using the method for difference of squares or the method for difference of cubes, I should choose the *difference of squares* method to eventually obtain the complete factored form.

88. $x^6 - 729 = \left(x^3\right)^2 - 27^2$

$$= \left(x^3 - 27\right)\left(x^3 + 27\right)$$

$$= (x - 3)\left(x^2 + 3x + 9\right)$$

$$\bullet\, (x + 3)\left(x^2 - 3x + 9\right)$$

Section 6.5

1. A quadratic equation is an equation that can be put in the form $ax^2 + bx + c = 0.$ $(a \neq 0)$

3. If a quadratic equation is in standard form, to solve the equation we should begin by *factoring* the polynomial.

For all equations in this section, answers should be checked by substituting into the original equation. These checks will be shown here for only a few of the exercises.

5. $(3k + 8)(k + 7) = 0$

By the zero-factor property, the only way that the product of these two factors can be zero is if at least one of the factors is zero.

$$3k + 8 = 0 \quad \text{or} \quad k + 7 = 0$$

Solve each of these linear equations.

$$3k = -8 \quad \text{or} \quad k = -7$$

$$k = -\frac{8}{3}$$

The solution set is $\left\{-\dfrac{8}{3}, -7\right\}.$

7. $t(t + 4) = 0$

$$t = 0 \text{ or } t + 4 = 0 \quad \begin{array}{l}\textit{Zero-factor}\\ \textit{property}\end{array}$$

Solve each equation.

$$t = 0 \quad \text{or} \quad t = -4$$

The solution set is $\{0, -4\}.$

9. We can consider the factored form as $2 \bullet x(3x - 4)$, the product of three factors, $2, x,$ and $3x - 4$. Applying the zero-factor property yields three equations,

$$2 = 0 \quad \text{or} \quad x = 0 \quad \text{or} \quad 3x - 4 = 0.$$

Since "$2 = 0$" is impossible and thus has no solution, we end up with the two solutions, $x = 0$ and $x = \dfrac{4}{3}.$

We conclude that multiplying a polynomial by a constant does not affect the solutions of the corresponding equation.

11. 9 is called a *double solution* for $(x - 9)^2 = 0$ because it occurs twice when the equation is solved.

$$(x - 9)^2 = 0$$
$$(x - 9)(x - 9) = 0$$
$$x - 9 = 0 \quad \text{or} \quad x - 9 = 0$$
$$x = 9 \quad \text{or} \quad x = 9$$

13. $y^2 + 3y + 2 = 0$

Factor the polynomial.

$$(y + 2)(y + 1) = 0$$

Set each factor equal to 0.

$$y + 2 = 0 \quad \text{or} \quad y + 1 = 0$$

Solve each equation.

$$y = -2 \quad \text{or} \quad y = -1$$

Check these solutions by substituting -2 for y and then -1 for y in the original equation.

$$y^2 + 3y + 2 = 0$$
$$(-2)^2 + 3(-2) + 2 = 0 ? \; \textit{Let } y = -2$$
$$4 - 6 + 2 = 0 ?$$
$$-2 + 2 = 0 \quad \textit{True}$$

$$y^2 + 3y + 2 = 0$$
$$(-1)^2 + 3(-1) + 2 = 0 \; ? \; \textit{Let } y = -1$$
$$1 - 3 + 2 = 0 \; ?$$
$$-2 + 2 = 0 \quad \textit{True}$$

The solution set is $\{-2, -1\}$.

15. $y^2 - 3y + 2 = 0$

Factor the polynomial.

$$(y - 1)(y - 2) = 0$$

Set each factor equal to 0.

$$y - 1 = 0 \quad \text{or} \quad y - 2 = 0$$

Solve each equation.

$$y = 1 \quad \text{or} \quad y = 2$$

The solution set is $\{1, 2\}$.

17. $x^2 = 24 - 5x$

Write the equation in standard form.

$$x^2 + 5x - 24 = 0 \quad \text{Factor the polynomial.}$$
$$(x + 8)(x - 3) = 0$$

Set each factor equal to 0.

$$x + 8 = 0 \quad \text{or} \quad x - 3 = 0$$

Solve each equation.

$$x = -8 \quad \text{or} \quad x = 3$$

The solution set is $\{-8, 3\}$.

19.
$$5x^2 = 15 + 10x$$
$$5x^2 - 10x - 15 = 0$$
$$5(x^2 - 2x - 3) = 0$$
$$5(x + 1)(x - 3) = 0$$
$$x + 1 = 0 \quad \text{or} \quad x - 3 = 0$$
$$x = -1 \quad \text{or} \quad x = 3$$

The solution set is $\{-1, 3\}$.

21. $z^2 = -2 - 3z$

Write the equation in standard form.

$$z^2 + 3z + 2 = 0$$

Factor the polynomial.

$$(z + 2)(z + 1) = 0$$

Set each factor equal to 0.

$$z + 2 = 0 \quad \text{or} \quad z + 1 = 0$$
$$z = -2 \quad \text{or} \quad z = -1$$

The solution set is $\{-2, -1\}$.

23. $m^2 + 8m + 16 = 0$

Factor $m^2 + 8m + 16$ as a perfect square trinomial.

$$(m + 4)^2 = 0 \cdot$$

Set the factor $m + 4$ equal to 0 and solve.

$$m + 4 = 0$$
$$m = -4$$

The solution set is $\{-4\}$.

25. $3x^2 + 5x - 2 = 0$

Factor the polynomial.

$$(3x - 1)(x + 2) = 0$$

Set each factor equal to 0.

$$3x - 1 = 0 \quad \text{or} \quad x + 2 = 0$$
$$3x = 1$$
$$x = \frac{1}{3} \quad \text{or} \quad x = -2$$

The solution set is $\left\{ \frac{1}{3}, -2 \right\}$.

27.
$$6p^2 = 4 - 5p$$
$$6p^2 + 5p - 4 = 0$$
$$(3p + 4)(2p - 1) = 0$$
$$3p + 4 = 0 \quad \text{or} \quad 2p - 1 = 0$$
$$3p = -4 \qquad\qquad 2p = 1$$
$$p = -\frac{4}{3} \quad \text{or} \quad p = \frac{1}{2}$$

The solution set is $\left\{ -\frac{4}{3}, \frac{1}{2} \right\}$.

29. $9s^2 + 12s = -4$
$$9s^2 + 12s + 4 = 0$$
$$(3s + 2)^2 = 0$$

Set the factor $3s + 2$ equal to 0 and solve.

$$3s + 2 = 0$$
$$3s = -2$$
$$s = -\frac{2}{3}$$

The solution set is $\left\{ -\frac{2}{3} \right\}$.

31.
$$2y^2 - 18 = 0$$
$$2(y^2 - 9) = 0$$
$$2(y + 3)(y - 3) = 0$$
$$y + 3 = 0 \quad \text{or} \quad y - 3 = 0$$
$$y = -3 \quad \text{or} \quad y = 3$$

The solution set is $\{-3, 3\}$.

33.
$$16k^2 - 49 = 0$$
$$(4k + 7)(4k - 7) = 0$$

$$4k + 7 = 0 \quad \text{or} \quad 4k - 7 = 0$$
$$4k = -7 \qquad\qquad 4k = 7$$
$$k = -\frac{7}{4} \quad \text{or} \quad k = \frac{7}{4}$$

The solution set is $\left\{-\frac{7}{4}, \frac{7}{4}\right\}$.

35.
$$n^2 = 121$$
$$n^2 - 121 = 0$$
$$(n + 11)(n - 11) = 0$$

$$n + 11 = 0 \quad \text{or} \quad n - 11 = 0$$
$$n = -11 \quad \text{or} \quad n = 11$$

The solution set is $\{-11, 11\}$.

37. Because $(-11)^2 = 121$, -11 is another solution. The correct way to solve the equation is as follows.

$$n^2 = 121$$
$$n^2 - 121 = 0$$
$$(n + 11)(n - 11) = 0$$

$$n + 11 = 0 \quad \text{or} \quad n - 11 = 0$$
$$n = -11 \quad \text{or} \quad n = 11$$

39.
$$x^2 = 7x$$
$$x^2 - 7x = 0$$
$$x(x - 7) = 0$$

$$x = 0 \quad \text{or} \quad x - 7 = 0$$
$$x = 0 \quad \text{or} \quad x = 7$$

The solution set is $\{0, 7\}$.

41.
$$6r^2 = 3r$$
$$6r^2 - 3r = 0$$
$$3r(2r - 1) = 0$$

$$3r = 0 \quad \text{or} \quad 2r - 1 = 0$$
$$2r = 1$$
$$r = 0 \quad \text{or} \quad r = \frac{1}{2}$$

The solution set is $\left\{0, \frac{1}{2}\right\}$.

43.
$$g(g - 7) = -10$$
$$g^2 - 7g = -10$$
$$g^2 - 7g + 10 = 0$$
$$(g - 2)(g - 5) = 0$$

$$g - 2 = 0 \quad \text{or} \quad g - 5 = 0$$
$$g = 2 \quad \text{or} \quad g = 5$$

The solution set is $\{2, 5\}$.

45.
$$3z(2z + 7) = 12$$
$$6z^2 + 21z = 12$$
$$6z^2 + 21z - 12 = 0$$
$$3(2z^2 + 7z - 4) = 0$$
$$3(2z - 1)(z + 4) = 0$$

$$2z - 1 = 0 \quad \text{or} \quad z + 4 = 0$$
$$2z = 1$$
$$z = \frac{1}{2} \quad \text{or} \quad z = -4$$

The solution set is $\left\{\frac{1}{2}, -4\right\}$.

47.
$$2(y^2 - 66) = -13y$$
$$2y^2 - 132 = -13y$$
$$2y^2 + 13y - 132 = 0$$
$$(2y - 11)(y + 12) = 0$$

$$2y - 11 = 0 \quad \text{or} \quad y + 12 = 0$$
$$2y = 11$$
$$y = \frac{11}{2} \quad \text{or} \quad y = -12$$

Check

$$2(y^2 - 66) = -13y$$
$$2\left[\left(\frac{11}{2}\right)^2 - 66\right] = -13\left(\frac{11}{2}\right) \text{ ? } \textit{Let } y = \frac{11}{2}$$
$$2\left(\frac{121}{4} - 66\right) = -\frac{143}{2} \text{ ?}$$
$$2\left(\frac{121}{4} - \frac{264}{4}\right) = -\frac{143}{2} \text{ ?}$$
$$2\left(-\frac{143}{4}\right) = -\frac{143}{2} \text{ ?}$$
$$-\frac{143}{2} = -\frac{143}{2} \quad \textit{True}$$
$$2(y^2 - 66) = -13y$$
$$2[(-12)^2 - 66] = -13(-12) \text{ ? } \textit{Let } y = -12$$
$$2[144 - 66] = 156 \text{ ?}$$
$$2(78) = 156 \text{ ?}$$
$$156 = 156 \quad \textit{True}$$

The solution set is $\left\{-12, \frac{11}{2}\right\}$.

49.
$$3x(x + 1) = (2x + 3)(x + 1)$$
$$3x^2 + 3x = 2x^2 + 5x + 3$$
$$x^2 - 2x - 3 = 0$$
$$(x - 3)(x + 1) = 0$$

$$x - 3 = 0 \quad \text{or} \quad x + 1 = 0$$
$$x = 3 \quad \text{or} \quad x = -1$$

The solution set is $\{3, -1\}$.

Alternatively, we could begin by moving all the terms to the left side and then factoring out $x + 1$.

$$3x(x+1) - (2x+3)(x+1) = 0$$
$$(x+1)[3x - (2x+3)] = 0$$
$$(x+1)(x-3) = 0$$

The rest of the solution is the same.

51. $(2r+5)(3r^2 - 16r + 5) = 0$

Begin by factoring $3r^2 - 16r + 5$.

$$(2r+5)(3r-1)(r-5) = 0$$

Set each of the three factors equal to 0 and solve the resulting equations.

$2r + 5 = 0$ or $3r - 1 = 0$ or $r - 5 = 0$
$2r = -5$ $\qquad$ $3r = 1$
$r = -\dfrac{5}{2}$ or $r = \dfrac{1}{3}$ or $r = 5$

The solution set is $\left\{-\dfrac{5}{2}, \dfrac{1}{3}, 5\right\}$.

53. $(2x+7)(x^2 + 2x - 3) = 0$
$(2x+7)(x+3)(x-1) = 0$

$2x + 7 = 0$ or $x + 3 = 0$ or $x - 1 = 0$
$2x = -7$
$x = -\dfrac{7}{2}$ or $x = -3$ or $x = 1$

The solution set is $\left\{-\dfrac{7}{2}, -3, 1\right\}$.

55. $9y^3 - 49y^2 = 0$

To factor the polynomial, begin by factoring out the greatest common factor.

$$y(9y^2 - 49) = 0$$

Now factor $9y^2 - 49$ as the difference of two squares.

$$y(3y+7)(3y-7) = 0$$

Set each of the three factors equal to 0 and solve.

$y = 0$ or $3y + 7 = 0$ or $3y - 7 = 0$
$\qquad$ $3y = -7$ $\qquad$ $3y = 7$
$y = 0$ or $y = -\dfrac{7}{3}$ or $y = \dfrac{7}{3}$

The solution set is $\left\{-\dfrac{7}{3}, 0, \dfrac{7}{3}\right\}$.

57. $r^3 - 2r^2 - 8r = 0$
$r(r^2 - 2r - 8) = 0$ *Factor out r*
$r(r-4)(r+2) = 0$ *Factor*

Set each factor equal to zero and solve.

$r = 0$ or $r - 4 = 0$ or $r + 2 = 0$
$r = 0$ or $r = 4$ or $r = -2$

The solution set is $\{0, 4, -2\}$.

59. $a^3 + a^2 - 20a = 0$
$a(a^2 + a - 20) = 0$ *Factor out a*
$a(a+5)(a-4) = 0$ *Factor*

Set each factor equal to zero and solve.

$a = 0$ or $a + 5 = 0$ or $a - 4 = 0$
$a = 0$ or $a = -5$ or $a = 4$

The solution set is $\{0, -5, 4\}$.

61. $r^4 = 2r^3 + 15r^2$

Rewrite with all terms on the left side.

$$r^4 - 2r^3 - 15r^2 = 0$$
$$r^2(r^2 - 2r - 15) = 0 \quad \text{\textit{Factor out }} r^2$$
$$r^2(r-5)(r+3) = 0 \quad \text{\textit{Factor}}$$

Set each factor equal to zero and solve.

$r^2 = 0$ or $r - 5 = 0$ or $r + 3 = 0$
$r = 0$ or $r = 5$ or $r = -3$

The solution set is $\{0, 5, -3\}$.

63. $6p^2(p+1) = 4(p+1) - 5p(p+1)$
$6p^2(p+1) + 5p(p+1) - 4(p+1) = 0$
Rewrite with all terms on the left
$(p+1)(6p^2 + 5p - 4) = 0$
Factor out p + 1
$(p+1)(2p-1)(3p+4) = 0$
Factor

Set each factor equal to zero and solve.

$p + 1 = 0$ or $2p - 1 = 0$ or $3p + 4 = 0$
$\qquad$ $2p = 1$ $\qquad$ $3p = -4$
$p = -1$ or $p = \dfrac{1}{2}$ or $p = -\dfrac{4}{3}$

The solution set is $\left\{-1, \dfrac{1}{2}, -\dfrac{4}{3}\right\}$.

65. $(k+3)^2 - (2k-1)^2 = 0$
$(k^2 + 6k + 9) - (4k^2 - 4k + 1) = 0$
Square the binomials
$k^2 + 6k + 9 - 4k^2 + 4k - 1 = 0$
$-3k^2 + 10k + 8 = 0$
Combine like terms
$-1(3k^2 - 10k - 8) = 0$
Factor out –1
$-1(3k+2)(k-4) = 0$
Factor

Set each factor equal to zero and solve.

continued

$-1 = 0$ or $3k + 2 = 0$ or $k - 4 = 0$

$$3k = -2$$

$$k = -\frac{2}{3} \quad \text{or} \quad k = 4$$

The solution set is $\left\{-\dfrac{2}{3}, 4\right\}$.

Alternatively we could begin by factoring the left side as the difference of two squares.

$$(k + 3)^2 - (2k - 1)^2 = 0$$
$$[(k + 3) + (2k - 1)][(k + 3) - (2k - 1)] = 0$$
$$[3k + 2][-k + 4] = 0$$

The same solution set is obtained.

67. To solve a quadratic equation by factoring we use the zero-factor property, which states that if a and b are real numbers and if $ab = 0$, then $a = 0$ or $b = 0$. This property only applies when the product of two factors is *zero*. There is no similar property that applies when the product of two factors is 1. To solve the equation $(x - 3)(x + 2) = 1$ (†), it is necessary to multiply the factors on the left, put the resulting equation in standard form, factor the new polynomial, and then use the zero-factor property.

In this case (†), the solutions of the two equations

$$x - 3 = 1 \quad \text{and} \quad x + 2 = 1$$

are $x = 4$ and $x = -1$. Substituting 4 for x in the equation $(x - 3)(x + 2) = 1$ gives us $(4 - 3)(4 + 2) = 1$ or $6 = 1$, which is a false statement. Similarly, $x = -1$ is not a solution of the given quadratic equation.

69. From the calculator screens, we see that the solution set of

$$x^2 + .4x - .05 = 0$$

is $\{-.5, .1\}$. Substituting $-.5$ for x gives us

$$(-.5)^2 + (.4)(-.5) - .05 = 0 \ ?$$
$$.25 - .20 - .05 = 0 \ ?$$
$$.25 - .25 = 0 \quad \textit{True}$$

This shows that $-.5$ is a solution of the given quadratic equation.

A similar check shows that .1 is also a solution.

71. From the calculator screens, we see that the solution set of

$$2x^2 + 7.2x + 5.5 = 0$$

is $\{-2.5, -1.1\}$. Verify as in Exercise 69.

Section 6.6

1. Read the problem carefully, choose *a variable* to represent *the unknown* and write it down.

3. Translate the problem into *an equation*.

5. $A = LW; \ A = 80, \ L = x + 8, \ W = x - 8$

(a) $A = LW$

$$80 = (x + 8)(x - 8)$$

(b) $80 = x^2 - 64$

$$0 = x^2 - 144$$
$$0 = (x + 12)(x - 12)$$

$x + 12 = 0 \qquad$ or $\qquad x - 12 = 0$

$\quad x = -12 \quad$ or $\qquad \quad x = 12$

(c) The solution cannot be $x = -12$ since, when substituted, $-12 + 8$ and $-12 - 8$ are negative numbers and length and width cannot be negative. Thus, $x = 12$ and

$$L = x + 8 = 12 + 8 = 20$$
$$W = x - 8 = 12 - 8 = 4.$$

The length is 20 units, and the width is 4 units.

(d) $LW = 20 \cdot 4 = 80$, the desired value of A.

7. $A = \dfrac{1}{2}bh; \ A = 60, \ b = 3x + 6, \ h = x + 5$

(a) $A = \dfrac{1}{2}bh$

$$60 = \frac{1}{2}(3x + 6)(x + 5)$$

(b) Solve the equation.

$120 = (3x + 6)(x + 5) \quad$ *Multiply by 2*

$120 = 3x^2 + 21x + 30 \quad$ *FOIL*

$0 = 3x^2 + 21x - 90 \quad$ *Standard form*

$0 = 3(x^2 + 7x - 30) \quad$ *Factor out 3*

$0 = 3(x - 3)(x + 10) \quad$ *Factor the trinomial*

$3 = 0 \qquad$ or $\quad x - 3 = 0$ or $\quad x + 10 = 0$

No solution $\qquad \qquad x = 3 \quad$ or $\qquad x = -10$

(c) Substitute these solutions for x in the expressions $3x + 6$ and $x + 5$ to find the values of b and h.

$$b = 3x + 6 = 3(3) + 6 = 15$$
or $\quad b = 3x + 6 = 3(-10) + 6 = -24$

We must discard the second solution since the base cannot have a negative length. Since $x = -10$ will not give a realistic answer for the base, we only need to substitute 3 for x to compute the height.

$$h = x + 5 = 3 + 5 = 8$$

The base is 15 units and the height is 8 units.

(d) $\dfrac{1}{2}bh = \dfrac{1}{2} \cdot 15 \cdot 8 = 60$, the desired value of A.

9. Let x = the width of the shell.

Then $x + 3$ = the length of the shell.

Substitute 28 for the area, x for the width, and $x + 3$ for the length in the formula for the area of a rectangle.

$$A = LW$$
$$28 = (x + 3)x$$
$$28 = x^2 + 3x$$
$$0 = x^2 + 3x - 28$$
$$0 = (x + 7)(x - 4)$$

$$x + 7 = 0 \quad \text{or} \quad x - 4 = 0$$
$$x = -7 \quad \text{or} \quad x = 4$$

The width of a rectangle cannot be negative, so we reject -7. The width of the shell is 4 inches, and the length is $4 + 3 = 7$ inches.

11. Let w = the width of the original screen.

Then $w + 3$ = the length of the original screen

and $w + 4$ = the length of the new screen.

Length × width (area) of new screen	equals	length × width (area) of original screen
↓	↓	↓
$w(w + 4)$	$=$	$w(w + 3)$

increased by	8.
↓	↓
$+$	8

Simplify this equation and solve it.

$$w^2 + 4w = w^2 + 3w + 8$$
$$w = 8$$

The width of the original screen is 8 inches, and the length is $w + 3 = 8 + 3 = 11$ inches.

13. Let h = the height of the triangle;

$2h + 2$ = the base of the triangle.

The area of the triangle is 30 square inches.

$$A = \frac{1}{2}bh$$
$$30 = \frac{1}{2}(2h + 2) \cdot h$$
$$60 = (2h + 2)h$$
$$60 = 2h^2 + 2h$$
$$0 = 2h^2 + 2h - 60$$
$$0 = 2(h^2 + h - 30)$$
$$0 = 2(h + 6)(h - 5)$$

$$h + 6 = 0 \quad \text{or} \quad h - 5 = 0$$
$$h = -6 \quad \text{or} \quad h = 5$$

The solution $h = -6$ must be discarded since a triangle cannot have a negative height. Thus,

$$h = 5 \text{ and } 2h + 2 = 2(5) + 2 = 12.$$

The height is 5 inches, and the base is 12 inches.

15. Let x = the width of the aquarium.

Then $x + 3$ = the height of the aquarium.

Use the formula for the volume of a rectangular box.

$$V = LWH$$
$$2730 = 21x(x + 3)$$
$$130 = x(x + 3) \qquad \textit{Divide by 21}$$
$$130 = x^2 + 3x$$
$$0 = x^2 + 3x - 130$$
$$0 = (x + 13)(x - 10)$$

$$x + 13 = 0 \quad \text{or} \quad x - 10 = 0$$
$$x = -13 \quad \text{or} \quad x = 10$$

We discard -13 because the width cannot be negative. The width is 10 inches. The height is $10 + 3 = 13$ inches.

17. Let $x = \dfrac{\text{the length of the longer leg}}{\text{of the right triangle}}$;

$x + 1$ = the length of the hypotenuse;

$x - 7$ = the length of the shorter leg.

Refer to the figure in the text. Use the Pythagorean formula with $a = x$, $b = x - 7$, and $c = x + 1$.

$$a^2 + b^2 = c^2$$
$$x^2 + (x - 7)^2 = (x + 1)^2$$
$$x^2 + (x^2 - 14x + 49) = x^2 + 2x + 1$$
$$2x^2 - 14x + 49 = x^2 + 2x + 1$$
$$x^2 - 16x + 48 = 0$$
$$(x - 12)(x - 4) = 0$$

$$x - 12 = 0 \quad \text{or} \quad x - 4 = 0$$
$$x = 12 \quad \text{or} \quad x = 4$$

Discard 4 because if the length of the longer leg is 4 centimeters, by the conditions of the problem, the length of the shorter leg would be $4 - 7 = -3$ centimeters, which is impossible. The length of the longer leg is 12 centimeters.

19.

Let $x = $ the distance from the bottom of the ladder to the bottom of the wall;

$2x + 1 = $ the length of the ladder.

Use the Pythagorean formula.

$$a^2 + b^2 = c^2$$
$$x^2 + 15^2 = (2x + 1)^2$$
$$x^2 + 225 = 4x^2 + 4x + 1$$
$$0 = 3x^2 + 4x - 224$$
$$0 = (3x + 28)(x - 8)$$

$$3x + 28 = 0 \quad \text{or} \quad x - 8 = 0$$
$$3x = -28$$
$$x = -\frac{28}{3} \quad \text{or} \quad x = 8$$

Discard $-\frac{28}{3}$ because the length cannot be negative. Thus, the distance from the wall to the bottom of the ladder is 8 feet.

21.

Let $x = $ the length of the shorter leg;

$x + 2 = $ the length of the longer leg;

$2x - 2 = $ the length of the hypotenuse.

Use the Pythagorean formula with $a = x$, $b = x + 2$, and $c = 2x - 2$.

$$a^2 + b^2 = c^2$$
$$x^2 + (x + 2)^2 = (2x - 2)^2$$
$$x^2 + (x^2 + 4x + 4) = 4x^2 - 8x + 4$$
$$2x^2 + 4x + 4 = 4x^2 - 8x + 4$$
$$0 = 2x^2 - 12x$$
$$0 = 2x(x - 6)$$

$$2x = 0 \quad \text{or} \quad x - 6 = 0$$
$$x = 0 \quad \text{or} \quad x = 6$$

Discard 0 because the length cannot be zero. The length of the shorter leg is 6 meters.

23.

Let $h = 64$ in the given formula and solve for t.

$$h = -16t^2 + 32t + 48$$
$$64 = -16t^2 + 32t + 48$$
$$16t^2 - 32t + 16 = 0$$
$$16(t^2 - 2t + 1) = 0$$
$$16(t - 1)^2 = 0$$
$$t - 1 = 0$$
$$t = 1$$

The height of the object will be 64 feet after 1 second.

25.

To find the time when the object hits the ground, let $h = 0$ and solve for t.

$$h = -16t^2 + 32t + 48$$
$$0 = -16t^2 + 32t + 48$$
$$16t^2 - 32t - 48 = 0$$
$$16(t^2 - 2t - 3) = 0$$
$$16(t + 1)(t - 3) = 0$$

$$t + 1 = 0 \quad \text{or} \quad t - 3 = 0$$
$$t = -1 \quad \text{or} \quad t = 3$$

We discard -1 because time cannot be negative. The object will hit the ground after 3 seconds.

27.

Let $h = 48$ in the given equation and solve for t.

$$h = -16t^2 + 64t$$
$$48 = -16t^2 + 64t$$
$$16t^2 - 64t + 48 = 0$$
$$16(t^2 - 4t + 3) = 0$$
$$16(t - 1)(t - 3) = 0$$

$$t - 1 = 0 \quad \text{or} \quad t - 3 = 0$$
$$t = 1 \quad \text{or} \quad t = 3$$

The height of the object will be 48 feet after 1 second (on the way up) and after 3 seconds (on the way down).

29.

Let $h = 0$ and solve for t.

$$h = -16t^2 + 64t$$
$$0 = -16t^2 + 64t$$
$$0 = -16t(t - 4)$$

$$-16t = 0 \quad \text{or} \quad t - 4 = 0$$
$$t = 0 \quad \text{or} \quad t = 4$$

The first solution, $t = 0$, represents the time when the object is propelled upward from the ground. The object will hit (return to) the ground after 4 seconds.

31.

Let $x = $ the length of the rectangular base;

$x - 3 = $ the width.

Use $A = LW$ to find the area of the base B. This gives

$$B = x(x - 3).$$

Now use the formula for the volume of a pyramid with $V = 144$, $B = x(x - 3)$, and $h = 8$.

$$V = \frac{1}{3}Bh$$
$$144 = \frac{1}{3}x(x - 3) \cdot 8$$
$$18 = \frac{1}{3}x(x - 3) \quad \textit{Divide by 8}$$
$$54 = x^2 - 3x \quad \textit{Multiply by 3}$$

$$0 = x^2 - 3x - 54$$
$$0 = (x - 9)(x + 6)$$

$x - 9 = 0$	or	$x + 6 = 0$
$x = 9$	or	$x = -6$

Discard -6 since length cannot be negative. Thus, the length of the base is 9 centimeters.

33. $d = \dfrac{1}{2}gt^2$

$d = \dfrac{1}{2}(32)(4)^2$ *Let g = 32 and t = 4*

$= \dfrac{1}{2}(32)(16)$

$= 256$

In 4 seconds, the object would fall 256 feet.

35. Navarre Building, New York City   512 feet

To find the amount of time it would take an object to fall from the top of this building, let $d = 512$ and $g = 32$ in the given formula and solve for t.

$$d = \frac{1}{2}gt^2$$
$$512 = \frac{1}{2}(32)t^2$$
$$512 = 16t^2$$
$$32 = t^2$$

Using trial and error, $(5.6)^2 = 31.36$ and $(5.7)^2 = 32.49$. Since 32.49 is closer to 32 than 31.36 is, $t \approx 5.7$.

Note that time cannot be negative. Note also that the formula gives the time in seconds.

It would take about 5.7 seconds for an object to fall from the top of the Navarre Building.

37. Central Plaza, Hong Kong   1028 feet

Let $d = 1028$ and $g = 32$ in the given formula and solve for t.

$$d = \frac{1}{2}gt^2$$
$$1028 = \frac{1}{2}(32)t^2$$
$$1028 = 16t^2$$
$$64.25 = t^2$$

Using trial and error, $(8.0)^2 = 64$ and $(8.1)^2 = 65.61$. Since 64 is closer to 64.25 than 65.61 is, $t \approx 8.0$.

It would take about 8.0 seconds for an object to fall from the top of Central Plaza.

39. Let n = the first integer;

$n + 1$ = the next integer.

The product	is 11	more than	their sum.
↓	↓ ↓	↓	↓

$$n(n + 1) = 11 + [n + (n + 1)]$$
$$n^2 + n = 11 + n + n + 1$$
$$n^2 + n = 2n + 12$$
$$n^2 - n - 12 = 0$$
$$(n - 4)(n + 3) = 0$$

$n - 4 = 0$	or	$n + 3 = 0$
$n = 4$	or	$n = -3$

If $n = 4$, then $n + 1 = 5$.

If $n = -3$ then $n + 1 = -2$.

The two integers are 4 and 5, or -3 and -2.

41. Let n = the smallest odd integer;

$n + 2$ = the next odd integer;

$n + 4$ = the largest odd integer.

$$3[n + (n + 2) + (n + 4)] = n(n + 2) + 18$$
$$3(3n + 6) = n^2 + 2n + 18$$
$$9n + 18 = n^2 + 2n + 18$$
$$0 = n^2 - 7n$$
$$0 = n(n - 7)$$

$n = 0$	or	$n - 7 = 0$
$n = 0$	or	$n = 7$

We must discard 0 because it is even and the problem requires the integers to be odd. If $n = 7$, $n + 2 = 9$, and $n + 4 = 11$. The three integers are 7, 9, and 11.

43. Let n = the smallest even integer;

$n + 2$ = the next even integer;

$n + 4$ = the largest even integer.

$$n^2 + (n + 2)^2 = (n + 4)^2$$
$$n^2 + n^2 + 4n + 4 = n^2 + 8n + 16$$
$$n^2 - 4n - 12 = 0$$
$$(n - 6)(n + 2) = 0$$

$n - 6 = 0$	or	$n + 2 = 0$
$n = 6$	or	$n = -2$

If $n = 6$, $n + 2 = 8$, and $n + 4 = 10$.

If $n = -2$, $n + 2 = 0$, and $n + 4 = 2$.

The three integers are 6, 8, and 10 or -2, 0, and 2.

45. **(a)** Transit ridership peaked in 1990 at 9.0 billion passengers.

(b) $y = -.012x^2 + .381x + 5.96$

For the year 1990, let $x = 20$.

$$y = -.012(20)^2 + .381(20) + 5.96$$
$$= 8.78 \text{ billion passengers.}$$

This approximation is .22 billion less than the value in the table, 9.0.

(c) x corresponds to the number of years after 1970, so 1995 corresponds to 25.

(d) $y = -.012(25)^2 + .381(25) + 5.96$
$$= 7.985 \text{ billion passengers.}$$

This approximation is .015 billion less than the value in the table, 8.0.

(e) Since the difference in the approximation and the table value is smaller in part (d), the 1995 result is the best approximation.

47. **(a)** $y = -416.72x^2 + 4416.7x + 8500$

Let $x = 8$

$$y = -416.72(8)^2 + 4416.7(8) + 8500$$
$$= 17,163.52 \text{ or about}$$

17,000 injuries in taxi accidents.

(b) Using the previous answers, we have the ratio

$$\frac{12,235.2 \text{ accidents}}{17,163.52 \text{ injuries}}$$
$$\approx .7 \text{ or 7 to 10.}$$

The average number of people injured per accident is the reciprocal of this value, about 1.4.

49. The dark square labeled ③ in Figure A has sides

of length c, so its area is c^2.

50. The dark square labeled ① in Figure B has sides

of length b, so its area is b^2.

51. The dark square labeled ② in Figure B has sides

of length a, so its area is a^2.

52. The sum of the areas of the dark regions in Figure B is $a^2 + b^2$. The area of the dark region in Figure A is c^2. Thus, the given statement is represented by the equation

$$a^2 + b^2 = c^2.$$

This equation represents the Pythagorean formula.

Chapter 6 Review Exercises

1. $7t + 14 = 7 \cdot t + 7 \cdot 2 = 7(t + 2)$

2. $60z^3 + 30z = 30z \cdot 2z^2 + 30z \cdot 1$
$$= 30z(2z^2 + 1)$$

3. $2xy - 8y + 3x - 12$
$$= (2xy - 8y) + (3x - 12)$$
$$= 2y(x - 4) + 3(x - 4)$$
$$= (x - 4)(2y + 3)$$

4. $6y^2 + 9y + 4y + 6$
$$= (6y^2 + 9y) + (4y + 6)$$
$$= 3y(2y + 3) + 2(2y + 3)$$
$$= (2y + 3)(3y + 2)$$

5. $x^2 + 5x + 6$

Find two integers whose product is 6 and whose sum is 5. The integers are 3 and 2. Thus,

$$x^2 + 5x + 6 = (x + 3)(x + 2).$$

6. $y^2 - 13y + 40$

Find two integers whose product is -40 and whose sum is -13.

Factors of 40	Sum of factors
$-1, -40$	-41
$-2, -20$	-22
$-4, -10$	-14
$-5, -8$	-13

The integers are -5 and -8, so

$$y^2 - 13y + 40 = (y - 5)(y - 8).$$

7. $q^2 + 6p - 27$

Find two integers whose product is -27 and whose sum is 6. The integers are -3 and 9, so

$$q^2 + 6q - 27 = (q - 3)(q + 9).$$

8. $r^2 - r - 56$

Find two integers whose product is -56 and whose sum is -1. The integers are 7 and -8, so

$$r^2 - r - 56 = (r + 7)(r - 8).$$

9. $r^2 - 4rs - 96s^2$

Find two expressions whose product is $-96s^2$ and whose sum is $-4s$. The expressions are $8s$ and $-12s$, so

$$r^2 - 4rs - 96s^2 = (r + 8s)(r - 12s).$$

10. $p^2 + 2pq - 120q^2$

Find two expressions whose product is $-120q^2$ and whose sum is $2q$. The expressions are $12q$ and $-10q$, so

$$p^2 + 2pq - 120q^2 = (p + 12q)(p - 10q).$$

11. $8p^3 - 24p^2 - 80p$

First, factor out the GCF, $8p$.

$$8p^3 - 24p^2 - 80p = 8p(p^2 - 3p - 10)$$

Now factor $p^2 - 3p - 10$.

$$p^2 - 3p - 10 = (p + 2)(p - 5)$$

The complete factored form is

$$8p^3 - 24p^2 - 80p$$
$$= 8p(p + 2)(p - 5).$$

12. $3x^4 + 30x^3 + 48x^2$
$$= 3x^2(x^2 + 10x + 16)$$
$$= 3x^2(x + 2)(x + 8)$$

13. $p^7 - p^6q - 2p^5q^2 = p^5(p^2 - pq - 2q^2)$
$$= p^5(p + q)(p - 2q)$$

14. $3r^5 - 6r^4s - 45r^3s^2$
$$= 3r^3(r^2 - 2rs - 15s^2)$$
$$= 3r^3(r + 3s)(r - 5s)$$

15. To begin factoring $6r^2 - 5r - 6$, the possible first terms of the two binomial factors are r and $6r$, or $2r$ and $3r$, if we consider only positive integer coefficients.

16. When factoring $2z^3 + 9z^2 - 5z$, the first step is to factor out the GCF, z.

In Exercises 17–24, either the trial and error method or the grouping method can be used to factor each polynomial.

17. $2k^2 - 5k + 2$

Factor by trial and error.

$$2k^2 - 5k + 2 = (2k - 1)(k - 2)$$

18. Factor $3r^2 + 11r - 4$ by grouping. Look for two integers whose product is $3(-4) = -12$ and whose sum is 11. The integers are 12 and -1.

$$3r^2 + 11r - 4 = 3r^2 + 12r - r - 4$$
$$= (3r^2 + 12r) + (-r - 4)$$
$$= 3r(r + 4) - 1(r + 4)$$
$$= (r + 4)(3r - 1)$$

19. Factor $6r^2 - 5r - 6$ by grouping. Find two integers whose product is $6(-6) = -36$ and whose sum is -5. The integers are -9 and 4.

$$6r^2 - 5r - 6 = 6r^2 - 9r + 4r - 6$$
$$= (6r^2 - 9r) + (4r - 6)$$
$$= 3r(2r - 3) + 2(2r - 3)$$
$$= (2r - 3)(3r + 2)$$

20. $10z^2 - 3z - 1$

Factor by trial and error.

$$10z^2 - 3z - 1 = (5z + 1)(2z - 1)$$

21. Factor $8v^2 + 17v - 21$ by grouping. Look for two integers whose product is $8(-21) = -168$ and whose sum is 17. The integers are 24 and -7.

$$8v^2 + 17v - 21 = 8v^2 + 24v - 7v - 21$$
$$= (8v^2 + 24v) + (-7v - 21)$$
$$= 8v(v + 3) - 7(v + 3)$$
$$= (v + 3)(8v - 7)$$

22. $24x^5 - 20x^4 + 4x^3$

Factor out the GCF, $4x^3$. Then complete the factoring by trial and error.

$$24x^5 - 20x^4 + 4x^3$$
$$= 4x^3(6x^2 - 5x + 1)$$
$$= 4x^3(3x - 1)(2x - 1)$$

23. $-6x^2 + 3x + 30 = -3(2x^2 - x - 10)$
$$= -3(2x - 5)(x + 2)$$

24. $10r^3s + 17r^2s^2 + 6rs^3$
$$= rs(10r^2 + 17rs + 6s^2)$$
$$= rs(5r + 6s)(2r + s)$$

25. Only (b) $4x^2y^2 - 25z^2$ is the difference of two squares. In (a), 32 is not a perfect square. In (c), we have a sum, not a difference. In (d), y^3 is not a square. The correct choice is (b).

26. Only (d) $x^2 - 20x + 100$ is a perfect square trinomial because $x^2 = x \cdot x$, $100 = 10 \cdot 10$, and $-20x = -2(x)(10)$.

In Exercises 27–30, use the rule for factoring a difference of two squares.

27. $n^2 - 49 = n^2 - 7^2 = (n + 7)(n - 7)$

28. $25b^2 - 121 = (5b)^2 - 11^2$
$$= (5b + 11)(5b - 11)$$

29. $49y^2 - 25w^2 = (7y)^2 - (5w)^2$
$$= (7y + 5w)(7y - 5w)$$

30. $144p^2 - 36q^2 = 36(4p^2 - q^2)$
$$= 36[(2p)^2 - q^2]$$
$$= 36(2p + q)(2p - q)$$

31. $x^2 + 100$

This polynomial is prime because it is the sum of two squares and the two terms have no common factor.

In Exercises 32–33, use the rules for factoring a perfect square trinomial.

32. $r^2 - 12r + 36 = r^2 - 2(6)(r) + 6^2$
$$= (r - 6)^2$$

33. $9t^2 - 42t + 49 = (3t)^2 - 2(3t)(7) + 7^2$
$$= (3t - 7)^2$$

In Exercises 34 – 35, use the rule for factoring a sum of two cubes.

34. $m^3 + 1000$

$$= m^3 + 10^3$$
$$= (m + 10)(m^2 - 10 \cdot m + 10^2)$$
$$= (m + 10)(m^2 - 10m + 100)$$

35. $125k^3 + 64x^3$

$$= (5k)^3 + (4x)^3$$
$$= (5k + 4x)[(5k)^2 - 5k \cdot 4x + (4x)^2]$$
$$= (5k + 4x)(25k^2 - 20kx + 16x^2)$$

36. $343x^3 - 64$

Use the rule for factoring a difference of two cubes.

$$343x^3 - 64$$
$$= (7x)^3 - 4^3$$
$$= (7x - 4)[(7x)^2 + 7x \cdot 4 + 4^2]$$
$$= (7x - 4)(49x^2 + 28x + 16)$$

In Exercises 37–46, all solutions should be checked by substituting in the original equations. The checks will not be shown here.

37. $z^2 + 4z + 3 = 0$

$$(z + 3)(z + 1) = 0$$

$z + 3 = 0$ or $z + 1 = 0$

$z = -3$ or $z = -1$

The solution set is $\{-3, -1\}$.

38. $2m^2 - 10m + 8 = 0$

$$2(m^2 - 5m + 4) = 0$$
$$2(m - 1)(m - 4) = 0$$

$m - 1 = 0$ or $m - 4 = 0$

$m = 1$ or $m = 4$

The solution set is $\{1, 4\}$.

39. $x(x - 8) = -15$

$$x^2 - 8x = -15$$
$$x^2 - 8x + 15 = 0$$
$$(x - 3)(x - 5) = 0$$

$x - 3 = 0$ or $x - 5 = 0$

$x = 3$ or $x = 5$

The solution set is $\{3, 5\}$.

40. $3z^2 - 11z - 20 = 0$

$$(3z + 4)(z - 5) = 0$$

$3z + 4 = 0$ or $z - 5 = 0$

$3z = -4$

$z = -\dfrac{4}{3}$ or $z = 5$

The solution set is $\left\{-\dfrac{4}{3}, 5\right\}$.

41. $81t^2 - 64 = 0$

$$(9t + 8)(9t - 8) = 0$$

$9t + 8 = 0$ or $9t - 8 = 0$

$9t = -8$ $9t = 8$

$t = -\dfrac{8}{9}$ or $t = \dfrac{8}{9}$

The solution set is $\left\{-\dfrac{8}{9}, \dfrac{8}{9}\right\}$.

42. $y^2 = 8y$

$$y^2 - 8y = 0$$
$$y(y - 8) = 0$$

$y = 0$ or $y - 8 = 0$

$y = 0$ or $y = 8$

The solution set is $\{0, 8\}$.

43. $3n(n - 5) = 18$

$$3n^2 - 15n = 18$$
$$3n^2 - 15n - 18 = 0$$
$$3(n^2 - 5n - 6) = 0$$
$$3(n + 1)(n - 6) = 0$$

$n + 1 = 0$ or $n - 6 = 0$

$n = -1$ or $n = 6$

The solution set is $\{-1, 6\}$.

44. $t^2 - 14t + 49 = 0$

$$(t - 7)^2 = 0$$
$$t - 7 = 0$$
$$t = 7$$

The solution set is $\{7\}$.

45. $t^2 = 12(t - 3)$

$$t^2 = 12t - 36$$
$$t^2 - 12t + 36 = 0$$
$$(t - 6)^2 = 0$$
$$t - 6 = 0$$
$$t = 6$$

The solution set is $\{6\}$.

46. $(5z + 2)(z^2 + 3z + 2) = 0$

$$(5z + 2)(z + 2)(z + 1) = 0$$

$5z + 2 = 0$ or $z + 2 = 0$ or $z + 1 = 0$

$5z = -2$

$z = -\dfrac{5}{2}$ or $z = -2$ or $z = -1$

The solution set is $\left\{-\dfrac{2}{5}, -2, -1\right\}$.

47.   Let x = the width of the rectangle.

Then $x + 6$ = the length.

$$A = LW$$
$$40 = (x + 6)x$$
$$40 = x^2 + 6x$$
$$x^2 + 6x - 40 = 0$$
$$(x + 10)(x - 4) = 0$$

$$x + 10 = 0 \qquad \text{or} \qquad x - 4 = 0$$
$$x = -10 \qquad \text{or} \qquad x = 4$$

Reject -10 since the width cannot be negative. The width of the rectangle is 4 meters and the length is $4 + 6$ or 10 meters.

48.   Let x = the width of the rectangle.

Then $3x$ = the length.

Use $A = LW$.

The width increased by 3	times	the same length	would be	an area of 30.
↓	↓	↓	↓	↓
$(x + 3)$	$\bullet$	$3x$	$=$	30

Solve the equation.

$$(x + 3)(3x) = 30$$
$$3x^2 + 9x = 30$$
$$3x^2 + 9x - 30 = 0$$
$$3(x^2 + 3x - 10) = 0$$
$$3(x + 5)(x - 2) = 0$$

$$x + 5 = 0 \qquad \text{or} \qquad x - 2 = 0$$
$$x = -5 \qquad \text{or} \qquad x = 2$$

Reject -5. The width of the original rectangle is 2 meters and the length is $3(2)$ or 6 meters.

49.   Let x = the length of the box.

Then $x - 1$ = the height.

$$V = LWH$$
$$120 = x(4)(x - 1)$$
$$120 = 4x^2 - 4x$$
$$4x^2 - 4x - 120 = 0$$
$$4(x^2 - x - 30) = 0$$
$$4(x - 6)(x + 5) = 0$$

$$x - 6 = 0 \qquad \text{or} \qquad x + 5 = 0$$
$$x = 6 \qquad \text{or} \qquad x = -5$$

Reject -5. The length of the box is 6 meters and the height is $6 - 1 = 5$ meters.

50.   Let x = the length of the shortest side.

Then $x + 1$ = the length of the middle side and $x + 2$ = the length of the hypotenuse.

Use the Pythagorean formula with $a = x$, $b = x + 1$, and $c = x + 2$.

$$x^2 + (x + 1)^2 = (x + 2)^2$$
$$x^2 + x^2 + 2x + 1 = x^2 + 4x + 4$$
$$x^2 - 2x - 3 = 0$$
$$(x - 3)(x + 1) = 0$$

$$x - 3 = 0 \qquad \text{or} \qquad x + 1 = 0$$
$$x = 3 \qquad \text{or} \qquad x = -1$$

Reject -1 because the length of a side cannot be negative. If $x = 3$, $x + 1 = 4$, and $x + 2 = 5$. The length of the sides are 3 feet, 4 feet, and 5 feet.

51. $h = 128t - 16t^2$
$$h = 128(1) - 16(1)^2 \quad \textit{Let } t = 1$$
$$= 128 - 16$$
$$= 112$$

After 1 second, the height is 112 feet.

52. $h = 128t - 16t^2$
$$h = 128(2) - 16(2)^2 \quad \textit{Let } t = 2$$
$$= 256 - 16(4)$$
$$= 256 - 64$$
$$= 192$$

After 2 seconds, the height is 192 feet.

53. $h = 128t - 16t^2$
$$h = 128(4) - 16(4)^2 \quad \textit{Let } t = 4$$
$$= 512 - 256$$
$$= 256$$

After 4 seconds, the height is 256 feet.

54.   The object hits the ground when $h = 0$.

$$h = 128t - 16t^2$$
$$0 = 128t - 16t^2 \quad \textit{Let } h = 0$$
$$0 = 16t(8 - t)$$

$$16t = 0 \qquad \text{or} \qquad 8 - t = 0$$
$$t = 0 \qquad \text{or} \qquad 8 = t$$

The solution $t = 0$ represents the time before the object is thrown. The object returns to the ground after 8 seconds.

55. Draw a sketch.

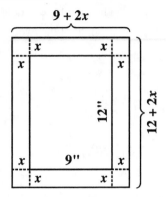

Let x represent the width of the border.

Then $9 + 2x$ represents the width of the finished mat and picture, and $12 + 2x$ represents the length.

Use $A = LW$ with $A = 208$, $L = 12 + 2x$, and $W = 9 + 2x$.

$$A = LW$$
$$208 = (12 + 2x)(9 + 2x)$$
$$208 = 108 + 42x + 4x^2$$
$$4x^2 + 42x - 100 = 0$$
$$2(2x^2 + 21x - 50) = 0$$
$$2(2x + 25)(x - 2) = 0$$

$$2x + 25 = 0 \quad \text{or} \quad x - 2 = 0$$
$$2x = -25$$
$$x = -\frac{25}{2} \quad \text{or} \quad x = 2$$

Reject $-\dfrac{25}{2}$ since the width cannot be negative.
The width of the border is 2 inches.

56. Draw a sketch.

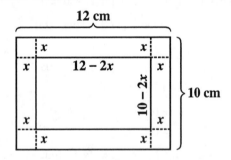

Let x represent the length of a side of each cutout square.

Then $12 - 2x$ represents the length of the bottom of the box and $10 - 2x$ represents the width.

Use $A = LW$ with $A = 48$, $L = 12 - 2x$, and $W = 10 - 2x$.

$$A = LW$$
$$48 = (12 - 2x)(10 - 2x)$$
$$48 = 120 - 44x + 4x^2$$
$$4x^2 - 44x + 72 = 0$$
$$4(x^2 - 11x + 18) = 0$$
$$4(x - 9)(x - 2) = 0$$

$$x - 9 = 0 \quad \text{or} \quad x - 2 = 0$$
$$x = 9 \quad \text{or} \quad x = 2$$

Reject 9 because if $x = 9$, then $12 - 2x = -6$, and a length cannot be negative. The length of a side of the cutout squares is 2 centimeters.

57. $y = -23.4x^2 + 285x - 831$

(a) If $x = 7$ (for 1997), then $y = 17.4$ (in millions of dollars).

(b) If $x = 8$, then $y = -48.6$.

(c) The equation is based on the data for 1994 to 1996. To use it to predict much beyond 1996 may lead to incorrect results, since it is possible that the conditions the equation was based on will change.

58. $z^2 - 11zx + 10x^2 = (z - x)(z - 10x)$

59. $3k^2 + 11k + 10$

Two integers with product $3(10) = 30$ and sum 11 are 5 and 6.

$$3k^2 + 11k + 10$$
$$= 3k^2 + 5k + 6k + 10$$
$$= (3k^2 + 5k) + (6k + 10)$$
$$= k(3k + 5) + 2(3k + 5)$$
$$= (3k + 5)(k + 2)$$

60. $15m^2 + 20mp - 12mp - 16p^2$

$$= 5m(3m + 4p) - 4p(3m + 4p) \qquad \textit{Factor by grouping}$$

$$= (3m + 4p)(5m - 4p)$$

61. $y^4 - 625$

$$= (y^2)^2 - 25^2$$

$$= (y^2 + 25)(y^2 - 25) \qquad \textit{Difference of two squares}$$

$$= (y^2 + 25)(y + 5)(y - 5) \qquad \textit{Difference of two squares}$$

62. $6m^3 - 21m^2 - 45m$

$$= 3m(2m^2 - 7m - 15)$$

$$= 3m\left[(2m^2 - 10m) + (3m - 15)\right] \qquad \textit{Factor by grouping}$$

$$= 3m[2m(m - 5) + 3(m - 5)]$$

$$= 3m(m - 5)(2m + 3)$$

63. $24ab^3c^2 - 56a^2bc^3 + 72a^2b^2c$

$$= 8abc(3b^2c - 7ac^2 + 9ab)$$

64. $25a^2 + 15ab + 9b^2$ is a prime polynomial.

65. $12x^2yz^3 + 12xy^2z - 30x^3y^2z^4$
$= 6xyz\left(2xz^2 + 2y - 5x^2yz^3\right)$

66. $2a^5 - 8a^4 - 24a^3$
$= 2a^3\left(a^2 - 4a - 12\right)$
$= 2a^3(a - 6)(a + 2)$

67. $12r^2 + 18rq - 10rq - 15q^2$
$= 6r(2r + 3q) - 5q(2r + 3q)$ *Factor by grouping*
$= (2r + 3q)(6r - 5q)$

68. $1000a^3 + 27$
$= (10a)^3 + 3^3$
$= (10a + 3)\left[(10a)^2 - 10a \cdot 3 + 3^2\right]$
$= (10a + 3)\left(100a^2 - 30a + 9\right)$

69. $49t^2 + 56t + 16$
$= (7t)^2 + 2(7t)(4) + 4^2$
$= (7t + 4)^2$

70. $t(t - 7) = 0$

$t = 0$ or $t - 7 = 0$
$t = 0$ or $t = 7$

The solution set is $\{0, 7\}$.

71. $x(x + 3) = 10$
       $x^2 + 3x = 10$
   $x^2 + 3x - 10 = 0$
  $(x + 5)(x - 2) = 0$

$x + 5 = 0$ or $x - 2 = 0$
    $x = -5$ or $x = 2$

The solution set is $\{-5, 2\}$.

72. $4x^2 - x - 3 = 0$
  $(4x + 3)(x - 1) = 0$

$4x + 3 = 0$ or $x - 1 = 0$
     $x = -\dfrac{3}{4}$ or $x = 1$

The solution set is $\left\{-\dfrac{3}{4}, 1\right\}$.

73. $y = 2.5x^2 - 453.5x + 20,850$

(a) If $x = 98$ (for 1998), then $y = 417$ (417,000 vehicles).

(b) The estimate may be unreliable because the conditions that prevailed in the years 1995–1997 may have changed, causing either a greater increase or a greater decrease in the numbers of alternative-fueled vehicles.

74.   Let $x =$ the first even integer;
     $x + 2 =$ the next even integer.
$x + (x + 2) = x(x + 2) - 34$
   $x + x + 2 = x^2 + 2x - 34$
     $2x + 2 = x^2 + 2x - 34$
         $0 = x^2 - 36$
         $0 = (x + 6)(x - 6)$

$x = -6$ or $x = 6$

If $x = -6$, then $x + 2 = -4$.

If $x = 6$, then $x + 2 = 8$.

The integers are -6 and -4, or 6 and 8.

75.   Let $x =$ the width of the house;
     $x + 7 =$ the length of the house.

Use $A = LW$ with 170 for A, $x + 7$ for L, and x for W.

$$170 = (x + 7)(x)$$
$$170 = x^2 + 7x$$
$$0 = x^2 + 7x - 170$$
$$0 = (x + 17)(x - 10)$$

$x = -17$ or $x = 10$

Discard -17 because the width cannot be negative.

$$W = x = 10$$
$$L = x + 7 = 10 + 7 = 17$$

The width is 10 meters and the length is 17 meters.

76.   Let $b =$ the base of the sail;
     $b + 4 =$ the height of the sail.

Use the formula for the area of a triangle.

$$A = \frac{1}{2}bh$$
$$30 = \frac{1}{2}(b)(b + 4) \quad \textit{Let } A = 30$$
$$60 = b^2 + 4b$$
$$0 = b^2 + 4b - 60$$
$$0 = (b + 10)(b - 6)$$

$b = -10$ or $b = 6$

Discard -10 since the base of a triangle cannot be negative. The base of the triangular sail is 6 meters.

77.

$$\text{Let } x = \begin{array}{l}\text{the distance}\\\text{traveled north;}\end{array}$$

$$x + 14 = \begin{array}{l}\text{the distance}\\\text{traveled east;}\end{array}$$

$$(x + 14) + 4 = x + 18$$

$$= \begin{array}{l}\text{the distance}\\\text{between the cars.}\end{array}$$

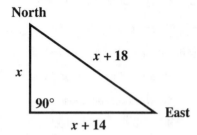

Use the Pythagorean formula.

$$a^2 + b^2 = c^2$$

Substitute x for a, $x + 14$ for b, and $x + 18$ for c.

$$x^2 + (x + 14)^2 = (x + 18)^2$$
$$x^2 + \left(x^2 + 28x + 196\right) = x^2 + 36x + 324$$
$$2x^2 + 28x + 196 = x^2 + 36x + 324$$
$$x^2 - 8x - 128 = 0$$
$$(x - 16)(x + 8) = 0$$
$$x = 16 \quad \text{or} \quad x = -8$$

Discard -8 since distance cannot be negative.

$$x + 18 = 16 + 18 = 34$$

The cars were 34 miles apart.

78. Let $x =$ the length of the ladder;

$$x - 4 = \begin{array}{l}\text{the distance from the bottom of}\\\text{the ladder to the building;}\end{array}$$

$$x - 2 = \begin{array}{l}\text{the distance up the side of the}\\\text{building that the ladder reaches.}\end{array}$$

Use the Pythagorean formula with the length of the ladder as the hypotenuse.

$$a^2 + b^2 = c^2$$
$$(x - 4)^2 + (x - 2)^2 = x^2$$
$$\left(x^2 - 8x + 16\right) + \left(x^2 - 4x + 4\right) = x^2$$
$$x^2 - 12x + 20 = 0$$
$$(x - 10)(x - 2) = 0$$
$$x = 10 \quad \text{or} \quad x = 2$$

Discard 2 since $x - 4$ be a negative distance. The ladder is $10 - 2$ or 8 feet up the side of the building.

79.

$$\text{Let } x = \begin{array}{l}\text{the distance traveled}\\\text{by the bicyclist;}\end{array}$$

$$x + 17 = \begin{array}{l}\text{the distance traveled}\\\text{by the motorist;}\end{array}$$

$$(x + 17) + 1 \text{ or } x + 18$$

$$= \text{the distance between them.}$$

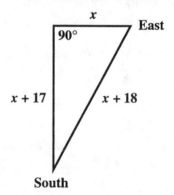

Use the Pythagorean formula.

$$a^2 + b^2 = c^2$$

Substitute x for a, $x + 17$ for b, and $x + 18$ for c.

$$x^2 + (x + 17)^2 = (x + 18)^2$$
$$x^2 + \left(x^2 + 34x + 289\right) = x^2 + 36x + 324$$
$$x^2 - 2x - 35 = 0$$
$$(x - 7)(x + 5) = 0$$
$$x = 7 \quad \text{or} \quad x = -5$$

Discard -5 since distance cannot be negative.

If $x = 7$,

$$x + 18 = 7 + 18 = 25.$$

The distance between them was 25 miles.

80. The given polynomial is not factored completely because the binomial $2x + 8$ contains a common factor of 2. If we factor out this common factor, we obtain the correct completely factored form:

$$6x^2 + 16x - 32 = 2(x + 4)(3x - 4)$$

The most efficient way to factor the given polynomial is to factor out the GCF of 2 as the first step and then factor the resulting trinomial:

$$6x^2 + 16x - 32 = 2\left(3x^2 + 8x - 16\right)$$
$$= 2(x + 4)(3x - 4)$$

Chapter 6 Test

1. $$2x^2 - 2x - 24 = 2\left(x^2 - x - 12\right)$$
$$= 2(x + 3)(x - 4)$$

The correct completely factored form is choice (d). Note that the factored forms (a) $(2x + 6)(x - 4)$ and (b) $(x + 3)(2x - 8)$ also can be multiplied to give a product of $2x^2 - 2x - 24$, but neither of these is completely factored because $2x + 6$ and $2x - 8$ both contain a common factor of 2.

2. $2m^3n^2 + 3m^3n - 5m^2n^2$
$= m^2n(2mn + 3m - 5n)$

3. $x^2 - 5x - 24$

Find two integers whose product is -24 and whose sum is -5. The integers are 3 and -8.

$$x^2 - 5x - 24 = (x + 3)(x - 8)$$

4. $2x^2 + x - 3$

Factor by trial and error.

$$2x^2 + x - 3 = (2x + 3)(x - 1)$$

5. $10z^2 - 17z + 3$

Factor by trial and error.

$$10z^2 - 17z + 3 = (2z - 3)(5z - 1)$$

6. $t^2 + 2t + 3$

We cannot find two integers whose product is 3 and whose sum is 2. This polynomial is prime.

7. $x^2 + 36$

This polynomial is prime because the sum of two squares cannot be factored and the two terms have no common factor.

8. $12 - 6a + 2b - ab$
$= (12 - 6a) + (2b - ab)$
$= 6(2 - a) + b(2 - a)$
$= (2 - a)(6 + b)$

9. $9y^2 - 64 = (3y)^2 - 8^2$
$= (3y + 8)(3y - 8)$

10. $4x^2 - 28xy + 49y^2$
$= (2x)^2 - 2(2x)(7y) + (7y)^2$
$= (2x - 7y)^2$

11. $-2x^2 - 4x - 2$
$= -2(x^2 + 2x + 1)$
$= -2(x^2 + 2 \cdot x \cdot 1 + 1^2)$
$= -2(x + 1)^2$

12. $6t^4 + 3t^3 - 108t^2$
$= 3t^2(2t^2 + t - 36)$
$= 3t^2(2t + 9)(t - 4)$

13. $r^3 - 125 = r^3 - 5^3$
$= (r - 5)(r^2 + 5 \cdot r + 5^2)$
$= (r - 5)(r^2 + 5r + 25)$

14. $8k^3 + 64 = 8(k^3 + 8)$
$= 8(k^3 + 2^3)$
$= 8(k + 2)[k^2 - 2 \cdot k + 2^2]$
$= 8(k + 2)(k^2 - 2k + 4)$

15. $(p + 3)(p + 3)$ is not the correct factored form of $p^2 + 9$ because

$$(p + 3)(p + 3) = p^2 + 6p + 9$$
$$\neq p^2 + 9.$$

The binomial $p^2 + 9$ is a prime polynomial.

16. $2r^2 - 13r + 6 = 0$
$(2r - 1)(r - 6) = 0$

$2r - 1 = 0 \qquad \text{or} \qquad r - 6 = 0$
$2r = 1$
$r = \dfrac{1}{2} \qquad \text{or} \qquad r = 6$

The solution set is $\left\{ \dfrac{1}{2}, 6 \right\}$.

17. $25x^2 - 4 = 0$
$(5x + 2)(5x - 2) = 0$

$5x + 2 = 0 \qquad \text{or} \qquad 5x - 2 = 0$
$5x = -2 \qquad\qquad\qquad 5x = 2$
$x = -\dfrac{2}{5} \qquad \text{or} \qquad x = \dfrac{2}{5}$

The solution set is $\left\{ -\dfrac{2}{5}, \dfrac{2}{5} \right\}$.

18. $x(x - 20) = -100$
$x^2 - 20x = -100$
$x^2 - 20x + 100 = 0$
$(x - 10)^2 = 0$
$x - 10 = 0$
$x = 10$

The solution set is $\{10\}$.

19. $t^3 = 9t$
$t^3 - 9t = 0$
$t(t^2 - 9) = 0$
$t(t + 3)(t - 3) = 0$

$t = 0 \quad \text{or} \quad t + 3 = 0 \quad \text{or} \quad t - 3 = 0$
$t = 0 \quad \text{or} \qquad t = -3 \quad \text{or} \qquad t = 3$

The solution set is $\{0, -3, 3\}$.

20. Let $h = 108$ and solve the given formula for t.

$$h = -16t^2 + 96t$$
$$108 = -16t^2 + 96t$$
$$16t^2 - 96t + 108 = 0$$
$$4(4t^2 - 24t + 27) = 0$$
$$4(2t - 3)(2t - 9) = 0$$

$2t - 3 = 0 \qquad \text{or} \qquad 2t - 9 = 0$
$2t = 3 \qquad\qquad\qquad 2t = 9$
$t = \dfrac{3}{2} \qquad \text{or} \qquad t = \dfrac{9}{2}$

The height of the object will be 108 feet after $\dfrac{3}{2}$ $\left(\text{or } 1\dfrac{1}{2} \right)$ seconds and after $\dfrac{9}{2}$ $\left(\text{or } 4\dfrac{1}{2} \right)$ seconds.

21. (a) Let x = the length of the stud;

$3x - 7$ = the length of the brace.

(b) $2x - 1$ = the length of the floor

from the brace to the stud.

(c) Use the Pythagorean formula with the length of the brace as the hypotenuse.

$$a^2 + b^2 = c^2$$
$$(2x - 1)^2 + x^2 = (3x - 7)^2$$
$$(4x^2 - 4x + 1) + x^2 = 9x^2 - 42x + 49$$
$$4x^2 - 38x + 48 = 0$$
$$2(2x^2 - 19x + 24) = 0$$
$$2(x - 8)(2x - 3) = 0$$

$$x - 8 = 0 \quad \text{or} \quad 2x - 3 = 0$$
$$2x = 3$$

$$x = 8 \quad \text{or} \quad x = \frac{3}{2}$$

Discard $\dfrac{3}{2}$ since the length of the brace, $3x - 7$, would be negative. The length of the brace should be $3x - 7 = 3(8) - 7 = 17$ feet.

22. $y = 2.5x^2 - 453.5x + 20,850$

If $x = 95.5$ (for July 1, 1995), then $y = 341.375$. Rounded to the nearest whole number, we have $y = 341$, which represents $341,000$ alternative-fueled vehicles.

If $x = 96.5$ (for July 1, 1996), then $y = 367.875$, or rounded, 368 (in thousands).

23.
$$x^2 = \frac{4}{9}$$
$$x^2 - \frac{4}{9} = 0$$
$$\left(x + \frac{2}{3}\right)\left(x - \frac{2}{3}\right) = 0$$

$$x + \frac{2}{3} = 0 \quad \text{or} \quad x - \frac{2}{3} = 0$$
$$x = -\frac{2}{3} \quad \text{or} \quad x = \frac{2}{3}$$

"$x = \dfrac{2}{3}$" is not the correct response because $-\dfrac{2}{3}$ is also a solution of the given equation.

Cumulative Review Exercises Chapters 1–6

1.
$$3x + 2(x - 4) = 4(x - 2)$$
$$3x + 2x - 8 = 4x - 8$$
$$5x - 8 = 4x - 8$$
$$x - 8 = -8$$
$$x = 0$$

The solution set is $\{0\}$.

2. $.3x + .9x = .06$

Multiply both sides by 100 to clear decimals.

$$100(.3x + .9x) = 100(.06)$$
$$30x + 90x = 6$$
$$120x = 6$$
$$x = \frac{6}{120} = \frac{1}{20} = .05$$

The solution set is $\{.05\}$.

3. $\dfrac{2}{3}y - \dfrac{1}{2}(y - 4) = 3$

To clear fractions, multiply both sides by the least common denominator, which is 6.

$$6\left[\frac{2}{3}y - \frac{1}{2}(y - 4)\right] = 6(3)$$
$$4y - 3(y - 4) = 18$$
$$4y - 3y + 12 = 18$$
$$y + 12 = 18$$
$$y = 6$$

The solution set is $\{6\}$.

4.
$$A = P + Prt$$
$$A = P(1 + rt)$$
$$\frac{A}{1 + rt} = \frac{P(1 + rt)}{1 + rt}$$
$$\frac{A}{1 + rt} = P$$

5. The sum of the angles is $180°$.

$$(2x + 16) + (x + 23) = 180$$
$$3x + 39 = 180$$
$$3x = 141$$
$$x = 47$$
$$2x + 16 = 2(47) + 16 = 110$$
$$x + 23 = 47 + 23 = 70$$

The angles are $110°$ and $70°$.

6. Let x = the amount of imports (in millions of dollars);

$x + 315$ = the amount of exports (in millions of dollars).

imports	exports	total
↓	↓	↓
x	$+ (x + 315)$	$= 1167$

Solve the equation.

$$x + (x + 315) = 1167$$
$$2x + 315 = 1167$$
$$2x = 852$$
$$x = 426$$

The imports were $426 million; the exports were $426 + $315 = $741 million.

7. Let x = number of pounds of gravel;

$3x$ = number of pounds of cement.

$$x + 3x = 140$$
$$4x = 140$$
$$x = 35$$

There are 35 pounds of gravel.

8. The point with coordinates (a, b) is in

(a) quadrant II if a is *negative* and b is *positive*.

(b) quadrant III if a is *negative* and b is *negative*.

9. $y = 12x + 3$

If $x = 2$, then $y = 12(2) + 3 = 27$. Since the y-value is measured in thousands, an ordered pair representing the second year is $(2, 27{,}000)$.

If $x = 5$, then $y = 63$. $(5, 63{,}000)$

10. $y = 12x + 3$

Let $x = 0$ to find the y-intercept.

$$y = 12(0) + 3 = 3$$

The y-intercept is $(0, 3)$.

Let $y = 0$ to find the x-intercept.

$$0 = 12x + 3$$
$$-3 = 12x$$
$$-\frac{1}{4} = x$$

The x-intercept is $\left(-\frac{1}{4}, 0\right)$.

11. Besides the y-intercept, other points on the graph of $y = 12x + 3$ are $(1, 15)$ and $(2, 27)$.

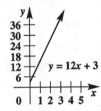

12. For the points $(95, 11{,}834)$ and $(90, 10{,}788)$, the slope m of the line through them is

$$m = \frac{11{,}834 - 10{,}788}{95 - 90} = \frac{1046}{5} = 209.2.$$

A slope of 209.2 means that the number of radio stations increased on the average by about 209 per year.

13. $2^{-3} \cdot 2^5 = 2^{-3+5} = 2^2 = 4$

14. $\left(\frac{3}{4}\right)^{-2} = \left(\frac{4}{3}\right)^2 = \frac{16}{9}$

15. $\left(\dfrac{4^{-3} \cdot 4^4}{4^5}\right)^{-1} = \left(\dfrac{4^5}{4^{-3} \cdot 4^4}\right)^1$

$$= \frac{4^5}{4^1} = 4^4 = 256$$

16. $\dfrac{\left(p^2\right)^3 p^{-4}}{\left(p^{-3}\right)^{-1} p} = \dfrac{p^{2 \cdot 3} p^{-4}}{p^{(-3)(-1)} p}$

$$= \frac{p^6 p^{-4}}{p^3 p^1}$$

$$= \frac{p^{6-4}}{p^{3+1}}$$

$$= \frac{p^2}{p^4} = \frac{1}{p^2}$$

17. $\left(2k^2 + 4k\right) - \left(5k^2 - 2\right) - \left(k^2 + 8k - 6\right)$

$$= \left(2k^2 + 4k\right) + \left(-5k^2 + 2\right)$$
$$+ \left(-k^2 - 8k + 6\right)$$
$$= 2k^2 + 4k - 5k^2 + 2 - k^2 - 8k + 6$$
$$= -4k^2 - 4k + 8$$

18. $3m^3\left(2m^5 - 5m^3 + m\right)$

$$= 3m^3 \cdot 2m^5 - 3m^3 \cdot 5m^3 + 3m^3 \cdot m$$
$$= 6m^8 - 15m^6 + 3m^4$$

19. $(y^2 + 3y + 5)(3y - 1)$

Multiply vertically.

$$
\begin{array}{r}
y^2 + 3y + 5 \\
3y - 1 \\
\hline
- y^2 - 3y - 5 \\
3y^3 + 9y^2 + 15y \\
\hline
3y^3 + 8y^2 + 12y - 5
\end{array}
$$

20. $(2p + 3q)(2p - 3q) = (2p)^2 - (3q)^2$

$$= 4p^2 - 9q^2$$

21. $\dfrac{8x^4 + 12x^3 - 6x^2 + 20x}{2x}$

$$= \frac{8x^4}{2x} + \frac{12x^3}{2x} - \frac{6x^2}{2x} + \frac{20x}{2x}$$
$$= 4x^3 + 6x^2 - 3x + 10$$

22. $(12p^3 + 2p^2 - 12p + 5) \div (2p - 2)$

$$
\begin{array}{r}
6p^2 + 7p + 1 \\
2p - 2 \enclose{longdiv}{12p^3 + 2p^2 - 12p + 5} \\
\underline{12p^3 - 2p^2} \\
14p^2 - 12p \\
\underline{14p^2 - 14p} \\
2p + 5 \\
\underline{2p - 2} \\
7
\end{array}
$$

The remainder is 7. We write the result as

$$6p^2 + 7p + 1 + \frac{7}{2p - 2}.$$

23. $2a^2 + 7a - 4$

Factor by trial and error.

$$2a^2 + 7a - 4 = (a + 4)(2a - 1)$$

24. $10m^2 + 19m + 6$

To factor by grouping, find two integers whose product is $10(6) = 60$ and whose sum is 19. The integers are 15 and 4.

$$\begin{aligned} 10m^2 + 19m + 6 &= 10m^2 + 15m + 4m + 6 \\ &= 5m(2m + 3) + 2(2m + 3) \\ &= (2m + 3)(5m + 2) \end{aligned}$$

25. $15x^2 - xy - 6y^2$
$$\begin{aligned} &= 15x^2 - 10xy + 9xy - 6y^2 \\ &= 5x(3x - 2y) + 3y(3x - 2y) \\ &= (3x - 2y)(5x + 3y) \end{aligned}$$

26. $9x^2 + 6x + 1 = (3x)^2 + 2(3x)(1) + 1^2$
$$= (3x + 1)^2$$

27. $-32t^2 - 112tz - 98z^2$
$$\begin{aligned} &= -2\left(16t^2 + 56tz + 49z^2\right) \\ &= -2\left[(4t)^2 + 2(4t)(7z) + (7z)^2\right] \\ &= -2(4t + 7z)^2 \end{aligned}$$

28. $25r^2 - 81t^2 = (5r)^2 - (9t)^2$
$$= (5r + 9t)(5r - 9t)$$

29. $100x^2 + 25 = 25\left(4x^2 + 1\right)$

30. $2pq + 6p^3q + 8p^2q$
$$\begin{aligned} &= 2pq\left(1 + 3p^2 + 4p\right) \\ &= 2pq\left(3p^2 + 4p + 1\right) \\ &= 2pq(3p + 1)(p + 1) \end{aligned}$$

31. $2ax - 2bx + ay - by$
$$\begin{aligned} &= (2ax - 2bx) + (ay - by) \\ &= 2x(a - b) + y(a - b) \\ &= (a - b)(2x + y) \end{aligned}$$

32. $(2p - 3)(p + 2)(p - 6) = 0$

Set each factor equal to 0 and solve the resulting linear equations.

$$2p - 3 = 0 \quad \text{or} \quad p + 2 = 0 \quad \text{or} \quad p - 6 = 0$$
$$2p = 3$$
$$p = \frac{3}{2} \quad \text{or} \qquad p = -2 \quad \text{or} \qquad p = 6$$

The solution set is $\left\{ \dfrac{3}{2}, -2, 6 \right\}$.

33. $6m^2 + m - 2 = 0$
$$(3m + 2)(2m - 1) = 0$$

$$3m + 2 = 0 \qquad \text{or} \qquad 2m - 1 = 0$$
$$3m = -2 \qquad\qquad\qquad 2m = 1$$
$$m = -\frac{2}{3} \quad \text{or} \qquad\quad m = \frac{1}{2}$$

The solution set is $\left\{ -\dfrac{2}{3}, \dfrac{1}{2} \right\}$.

34. Let $x = $ the smaller even integer;

$x + 2 = $ the larger even integer.

$$\begin{aligned} (x + 2)^2 - x^2 &= x^2 - 28 \\ x^2 + 4x + 4 - x^2 &= x^2 - 28 \\ 0 &= x^2 - 4x - 32 \\ 0 &= (x + 4)(x - 8) \\ x = -4 \quad &\text{or} \quad x = 8 \end{aligned}$$

If $x = -4$, $x + 2 = -2$.

If $x = 8$, $x + 2 = 10$.

The integers are -4 and -2, or 8 and 10.

35. Let $x = $ the length of the shorter leg;

$x + 7 = $ the length of the longer leg;

$2x + 3 = $ the length of the hypotenuse.

Use the Pythagorean formula.

$$\begin{aligned} x^2 + (x + 7)^2 &= (2x + 3)^2 \\ x^2 + \left(x^2 + 14x + 49\right) &= 4x^2 + 12x + 9 \\ 2x^2 + 14x + 49 &= 4x^2 + 12x + 9 \\ 0 &= 2x^2 - 2x - 40 \\ 0 &= 2\left(x^2 - x - 20\right) \\ 0 &= (x - 5)(x + 4) \\ x = 5 \quad &\text{or} \quad x = -4 \end{aligned}$$

Reject -4 because the length of a leg cannot be negative.

If $x = 5$, $x + 7 = 12$, and $2x + 3 = 2(5) + 3 = 13$.

The length of the sides are 5 meters, 12 meters, and 13 meters.

CHAPTER 7   RATIONAL EXPRESSIONS

Section 7.1

1. **(a)** The rational expression $\dfrac{x+5}{x-3}$ is undefined when its denominator is equal to 0; that is, when $x - 3 = 0$, or $x = 3$. It is equal to 0 when its numerator is equal to 0; that is, when $x + 5 = 0$, or $x = -5$.

(b) The rational expression $\dfrac{p-q}{q-p}$ is undefined when $q - p = 0$, or $p = q$. In all other cases,

$$\frac{p-q}{q-p} = \frac{-1(q-p)}{1(q-p)} = \frac{-1}{1} = -1.$$

3. A rational expression is a quotient of two polynomials, such as $\dfrac{x^2 + 3x - 6}{x + 4}$. One can think of this as an algebraic fraction.

5. $\dfrac{12}{5y}$

The denominator $5y$ will be zero when $y = 0$, so the given expression is undefined for $y = 0$.

7. $\dfrac{4x^2}{3x + 5}$

To find the values for which this expression is undefined, set the denominator equal to zero and solve for x.

$$3x + 5 = 0$$
$$3x = -5$$
$$x = -\frac{5}{3}$$

Because $x = -\dfrac{5}{3}$ will make the denominator zero, the given expression is undefined for $-\dfrac{5}{3}$.

9. $\dfrac{5m + 2}{m^2 + m - 6}$

To find the numbers that make the denominator 0, we must solve

$$m^2 + m - 6 = 0$$
$$(m + 3)(m - 2) = 0$$

$$m + 3 = 0 \quad \text{or} \quad m - 2 = 0$$
$$m = -3 \quad \text{or} \quad m = 2$$

The given expression is undefined for $m = -3$ and for $m = 2$.

11. $\dfrac{3x - 1}{x^2 + 2}$

This denominator cannot equal zero for any value of x because x^2 is always greater than or equal to zero, and adding 2 makes the sum greater than zero. Thus, the given rational expression is never undefined.

13. **(a)** $\dfrac{5x - 2}{4x} = \dfrac{5 \cdot 2 - 2}{4 \cdot 2}$ *Let x = 2*

$$= \frac{10 - 2}{8}$$
$$= \frac{8}{8} = 1$$

(b) $\dfrac{5x - 2}{4x} = \dfrac{5(-3) - 2}{4(-3)}$ *Let x = -3*

$$= \frac{-15 - 2}{-12}$$
$$= \frac{-17}{-12} = \frac{17}{12}$$

15. **(a)** $\dfrac{2x^2 - 4x}{3x} = \dfrac{2(2)^2 - 4(2)}{3(2)}$ *Let x = 2*

$$= \frac{2(4) - 4(2)}{3(2)}$$
$$= \frac{8 - 8}{6}$$
$$= \frac{0}{6} = 0$$

(b) $\dfrac{2x^2 - 4x}{3x} = \dfrac{2(-3)^2 - 4(-3)}{3(-3)}$ *Let x = -3*

$$= \frac{2(9) - (-12)}{-9}$$
$$= \frac{18 + 12}{-9}$$
$$= \frac{30}{-9} = -\frac{10}{3}$$

17. **(a)** $\dfrac{(-3x)^2}{4x + 12} = \dfrac{(-3 \cdot 2)^2}{4 \cdot 2 + 12}$ *Let x = 2*

$$= \frac{(-6)^2}{8 + 12}$$
$$= \frac{36}{20}$$
$$= \frac{9}{5}$$

(b) $\dfrac{(-3x)^2}{4x + 12} = \dfrac{[-3(-3)]^2}{4(-3) + 12}$ *Let x = -3*

$$= \frac{9^2}{-12 + 12}$$
$$= \frac{81}{0}$$

The expression is undefined when $x = -3$ because of the 0 in the denominator.

19. (a) $\dfrac{5x+2}{2x^2+11x+12}$

$= \dfrac{5(2)+2}{2(2)^2+11(2)+12}$ *Let x = 2*

$= \dfrac{10+2}{2(4)+11(2)+12}$

$= \dfrac{12}{8+22+12}$

$= \dfrac{12}{42} = \dfrac{2}{7}$

(b) $\dfrac{5x+2}{2x^2+11x+12}$

$= \dfrac{5(-3)+2}{2(-3)^2+11(-3)+12}$ *Let x = -3*

$= \dfrac{-15+2}{2(9)+11(-3)+12}$

$= \dfrac{-13}{18-33+12}$

$= \dfrac{-13}{-3} = \dfrac{13}{3}$

21. Any number divided by itself is 1, *provided the number is not 0*. This expression is equal to

$\dfrac{1}{x+2}$ for all values of x except -2 and 2, since

$\dfrac{x-2}{x^2-4} = \dfrac{x-2}{(x+2)(x-2)}$.

23. $\dfrac{18r^3}{6r} = \dfrac{3r^2(6r)}{1(6r)}$ *Factor*

$= 3r^2$ *Fundamental property*

25. $\dfrac{4(y-2)}{10(y-2)} = \dfrac{2 \cdot 2(y-2)}{5 \cdot 2(y-2)}$ *Factor*

$= \dfrac{2}{5}$ *Fundamental property*

27. $\dfrac{(x+1)(x-1)}{(x+1)^2} = \dfrac{(x+1)(x-1)}{(x+1)(x+1)}$

$= \dfrac{x-1}{x+1}$ *Fundamental property*

29. $\dfrac{7m+14}{5m+10} = \dfrac{7(m+2)}{5(m+2)}$ *Factor*

$= \dfrac{7}{5}$ *Fundamental property*

31. $\dfrac{m^2-n^2}{m+n} = \dfrac{(m+n)(m-n)}{m+n}$

$= m-n$

33. $\dfrac{12m^2-3}{8m-4} = \dfrac{3(4m^2-1)}{4(2m-1)}$

$= \dfrac{3(2m+1)(2m-1)}{4(2m-1)}$

$= \dfrac{3(2m+1)}{4}$

35. $\dfrac{3m^2-3m}{5m-5} = \dfrac{3m(m-1)}{5(m-1)}$

$= \dfrac{3m}{5}$

37. $\dfrac{9r^2-4s^2}{9r+6s} = \dfrac{(3r+2s)(3r-2s)}{3(3r+2s)}$

$= \dfrac{(3r-2s)}{3}$

39. Factor the numerator and denominator by grouping.

$\dfrac{zw+4z-3w-12}{zw+4z+5w+20}$

$= \dfrac{z(w+4)-3(w+4)}{z(w+4)+5(w+4)}$

$= \dfrac{(w+4)(z-3)}{(w+4)(z+5)}$

$= \dfrac{z-3}{z+5}$

41. $\dfrac{5k^2-13k-6}{5k+2} = \dfrac{(5k+2)(k-3)}{5k+2}$

$= k-3$

43. $\dfrac{2x^2-3x-5}{2x^2-7x+5} = \dfrac{(2x-5)(x+1)}{(2x-5)(x-1)}$

$= \dfrac{x+1}{x-1}$

45. $\dfrac{6-t}{t-6} = \dfrac{-1(t-6)}{1(t-6)} = \dfrac{-1}{1} = -1$

Note that $6-t$ and $t-6$ are opposites, so we know that their quotient will be -1.

47. $\dfrac{m^2-1}{1-m} = \dfrac{(m+1)(m-1)}{-1(m-1)}$

$= \dfrac{m+1}{-1}$

$= -(m+1)$

49. $\dfrac{q^2-4q}{4q-q^2} = \dfrac{q(q-4)}{q(4-q)}$

$= \dfrac{q-4}{4-q} = -1$

$q-4$ and $4-q$ are opposites.

51. $\dfrac{p+6}{p-6}$

This expression is already in lowest terms.

53. $L \cdot W = A$

$W = \dfrac{A}{L}$

$W = \dfrac{x^4+10x^2+21}{x^2+7}$

$= \dfrac{(x^2+7)(x^2+3)}{x^2+7}$

$= x^2+3$ *continued*

Note: If it is not apparent that we can factor A as $x^4 + 10x^2 + 21 = (x^2 + 7)(x^2 + 3)$, we may use "long division" to find the quotient $\dfrac{A}{L}$.

Remember to insert zeros for the coefficients of the missing terms in the dividend and divisor.

$$
\begin{array}{r}
x^2 \qquad\quad + 3 \\
x^2 + 0x + 7\,\overline{\big)\ x^4\ + 0x^3\ + 10x^2\ + 0x\ + 21} \\
\underline{x^4\ + 0x^3\ +\ 7x^2\qquad\qquad} \\
3x^2\ + 0x\ + 21 \\
\underline{3x^2\ + 0x\ + 21} \\
0
\end{array}
$$

The width of the rectangle is $x^2 + 3$.

55. To write four equivalent expressions for $-\dfrac{x + 4}{x - 3}$, we will follow the outline in Example 7. Applying the negative sign to the numerator we have

$$\frac{-(x + 4)}{x - 3}.$$

Distributing the negative sign gives us

$$\frac{-x - 4}{x - 3}.$$

Applying the negative sign to the denominator yields

$$\frac{x + 4}{-(x - 3)}.$$

Again, we distribute to get

$$\frac{x + 4}{-x + 3}.$$

57. $-\dfrac{2x - 3}{x + 3}$ is equivalent to each of the following:

$$\frac{-(2x - 3)}{x + 3}, \ \frac{-2x + 3}{x + 3},$$
$$\frac{2x - 3}{-(x + 3)}, \ \frac{2x - 3}{-x - 3}.$$

59. $\dfrac{-3x + 1}{5x - 6}$ is equivalent to each of the following:

$$\frac{-(3x - 1)}{5x - 6}, \ -\frac{3x - 1}{5x - 6},$$
$$\frac{3x - 1}{-5x + 6}, \ -\frac{-3x + 1}{-5x + 6}.$$

61. $\dfrac{m^2 - n^2 - 4m - 4n}{2m - 2n - 8}$

$= \dfrac{(m + n)(m - n) - 4(m + n)}{2(m - n - 4)}$ *Factor by grouping*

$= \dfrac{(m + n)(m - n - 4)}{2(m - n - 4)}$

$= \dfrac{m + n}{2}$ *Fundamental property*

63. The numerator is the difference of two cubes and the denominator is the difference of two squares.

$$\frac{b^3 - a^3}{a^2 - b^2}$$

$$= \frac{(b - a)(b^2 + ba + a^2)}{(a - b)(a + b)}$$

$$= (-1) \cdot \frac{(b^2 + ba + a^2)}{(a + b)} \quad \frac{(b - a)}{(a - b)} = -1$$

$$= -\frac{b^2 + ba + a^2}{a + b}$$

65. $\dfrac{z^3 + 27}{z^3 - 3z^2 + 9z}$

$= \dfrac{z^3 + 3^3}{z^3 - 3z^2 + 9z}$

$= \dfrac{(z + 3)(z^2 - 3z + 9)}{z(z^2 - 3z + 9)}$

$= \dfrac{z + 3}{z}$

67. Substituting -3 for x gives an error message for Y_1, so its denominator must be 0 when $x = -3$. Hence, we predict that the denominator of Y_1 is $x + 3$.

69. Substituting -5 for x gives an error message for Y_1, so its denominator must be 0 when $x = -5$. Hence, we predict that the denominator of Y_1 is $x + 5$.

71. Substituting 3 for x gives an error message for Y_1, so its denominator must be 0 when $x = 3$. Hence, we predict that the denominator of Y_1 is $x - 3$.

Section 7.2

1. (a) $\dfrac{5x^3}{10x^4} \cdot \dfrac{10x^7}{2x} = \dfrac{5 \cdot 10 \cdot x^3 \cdot x^7}{10 \cdot 2 \cdot x^4 \cdot x}$

$= \dfrac{5x^{10}}{2x^5}$

$= \dfrac{5x^5}{2}$ (B)

(b) $\dfrac{10x^4}{5x^3} \cdot \dfrac{10x^7}{2x} = \dfrac{10 \cdot 10 \cdot x^4 \cdot x^7}{5 \cdot 2 \cdot x^3 \cdot x}$

$= \dfrac{10x^{11}}{1x^4}$

$= 10x^7$ (D)

(c) $\dfrac{5x^3}{10x^4} \cdot \dfrac{2x}{10x^7} = \dfrac{5 \cdot 2 \cdot x^3 \cdot x}{10 \cdot 10 \cdot x^4 \cdot x^7}$

$= \dfrac{1x^4}{10x^{11}}$

$= \dfrac{1}{10x^7}$ (C)

(d) $\dfrac{10x^4}{5x^3} \cdot \dfrac{2x}{10x^7} = \dfrac{10 \cdot 2 \cdot x^4 \cdot x}{5 \cdot 10 \cdot x^3 \cdot x^7}$

$= \dfrac{2x^5}{5x^{10}}$

$= \dfrac{2}{5x^5}$ (A)

3. $\dfrac{15a^2}{14} \cdot \dfrac{7}{5a} = \dfrac{3 \cdot 5 \cdot a \cdot a \cdot 7}{2 \cdot 7 \cdot 5 \cdot a}$ *Multiply and factor*

$= \dfrac{3 \cdot a(5 \cdot 7 \cdot a)}{2(5 \cdot 7 \cdot a)}$

$= \dfrac{3a}{2}$ *Lowest terms*

5. $\dfrac{12x^4}{18x^3} \cdot \dfrac{-8x^5}{4x^2} = \dfrac{-96x^9}{72x^5}$ *Multiply numerators; multiply denominators*

$= \dfrac{-4x^4(24x^5)}{3(24x^5)}$ *Group common factors*

$= -\dfrac{4x^4}{3}$ *Lowest terms*

7. $\dfrac{2(c+d)}{3} \cdot \dfrac{18}{6(c+d)^2}$

$= \dfrac{3 \cdot 3 \cdot 2 \cdot 2(c+d)}{3 \cdot 3 \cdot 2(c+d)(c+d)}$ *Multiply and factor*

$= \dfrac{2}{c+d}$ *Lowest terms*

9. $\dfrac{9z^4}{3z^5} \div \dfrac{3z^2}{5z^3} = \dfrac{9z^4}{3z^5} \cdot \dfrac{5z^3}{3z^2}$

$= \dfrac{9 \cdot 5z^7}{3 \cdot 3z^7}$

$= 5$

11. $\dfrac{4t^4}{2t^5} \div \dfrac{(2t)^3}{-6} = \dfrac{4t^4}{2t^5} \cdot \dfrac{-6}{(2t)^3}$

$= \dfrac{4t^4}{2t^5} \cdot \dfrac{-6}{8t^3}$

$= \dfrac{-24t^4}{16t^8}$

$= \dfrac{-3(8t^4)}{2t^4(8t^4)}$

$= \dfrac{-3}{2t^4} = -\dfrac{3}{2t^4}$

13. $\dfrac{3}{2y-6} \div \dfrac{6}{y-3} = \dfrac{3}{2y-6} \cdot \dfrac{y-3}{6}$

$= \dfrac{3}{2(y-3)} \cdot \dfrac{y-3}{6}$

$= \dfrac{3(y-3)}{2 \cdot 2 \cdot 3(y-3)}$

$= \dfrac{1}{2 \cdot 2} = \dfrac{1}{4}$

15. If -4 is substituted for x in the expression, the value of the denominator in the first fraction will be zero.

16. If -5 is substituted for x in the expression, the value of the denominator in the second fraction will be zero.

17. If -7 is substituted for x in the second fraction, the numerator will be zero. This numerator will become a divisor when the reciprocal is formed to carry out the division process. Thus, we would be dividing by 0, which is undefined.

18. The expression $x - 6$ will never become a divisor, so the expression $x - 6$ may have a value of zero. We *are* allowed to divide 0 by a non-zero number.

19. Suppose I want to multiply $\dfrac{a^2-1}{6} \cdot \dfrac{9}{2a+2}$. I start by factoring where possible:

$$\dfrac{(a+1)(a-1)}{6} \cdot \dfrac{9}{2(a+1)}.$$

Next, I divide out common factors in the numerator and denominator to get $\dfrac{a-1}{2} \cdot \dfrac{3}{2}$. Finally, I multiply numerator times numerator and denominator times denominator to get the final product, $\dfrac{3(a-1)}{4}$.

21. $\dfrac{5x-15}{3x+9} \cdot \dfrac{4x+12}{6x-18}$

$= \dfrac{5(x-3)}{3(x+3)} \cdot \dfrac{4(x+3)}{6(x-3)}$

$= \dfrac{5 \cdot 4 \cdot (x-3)(x+3)}{3 \cdot 6 \cdot (x-3)(x+3)}$

$= \dfrac{10}{9}$

23. $\dfrac{2-t}{8} \div \dfrac{t-2}{6} = \dfrac{2-t}{8} \cdot \dfrac{6}{t-2}$ *Multiply by reciprocal*

$= \dfrac{6(2-t)}{8(t-2)}$ *Multiply numerators; multiply denominators*

$= \dfrac{6(-1)}{8}$ $\dfrac{2-t}{t-2} = -1$

$= -\dfrac{3}{4}$ *Lowest terms*

25. $\dfrac{27-3z}{4} \cdot \dfrac{12}{2z-18}$

$= \dfrac{3(9-z)}{4} \cdot \dfrac{3 \cdot 4}{2(z-9)}$

$= \dfrac{3 \cdot 3 \cdot 4(9-z)}{4 \cdot 2(z-9)}$

continued

$$= \frac{3 \cdot 3 \cdot (-1)}{2}$$

$$= -\frac{9}{2}$$

27. $\dfrac{p^2 + 4p - 5}{p^2 + 7p + 10} \div \dfrac{p - 1}{p + 4}$

$$= \frac{p^2 + 4p - 5}{p^2 + 7p + 10} \cdot \frac{p + 4}{p - 1}$$

$$= \frac{(p+5)(p-1) \cdot (p+4)}{(p+5)(p+2) \cdot (p-1)}$$

$$= \frac{p + 4}{p + 2}$$

29. $\dfrac{2k^2 - k - 1}{2k^2 + 5k + 3} \div \dfrac{4k^2 - 1}{2k^2 + k - 3}$

$$= \frac{2k^2 - k - 1}{2k^2 + 5k + 3} \cdot \frac{2k^2 + k - 3}{4k^2 - 1}$$

$$= \frac{(2k+1)(k-1)(2k+3)(k-1)}{(2k+3)(k+1)(2k+1)(2k-1)}$$

$$= \frac{(k-1)(k-1)}{(k+1)(2k-1)}$$

$$= \frac{(k-1)^2}{(k+1)(2k-1)}$$

31. $\dfrac{2k^2 + 3k - 2}{6k^2 - 7k + 2} \cdot \dfrac{4k^2 - 5k + 1}{k^2 + k - 2}$

$$= \frac{(2k-1)(k+2)}{(3k-2)(2k-1)} \cdot \frac{(4k-1)(k-1)}{(k+2)(k-1)}$$

$$= \frac{(2k-1)(k+2)(4k-1)(k-1)}{(3k-2)(2k-1)(k+2)(k-1)}$$

$$= \frac{4k - 1}{3k - 2}$$

33. $\dfrac{m^2 + 2mp - 3p^2}{m^2 - 3mp + 2p^2} \div \dfrac{m^2 + 4mp + 3p^2}{m^2 + 2mp - 8p^2}$

$$= \frac{m^2 + 2mp - 3p^2}{m^2 - 3mp + 2p^2} \cdot \frac{m^2 + 2mp - 8p^2}{m^2 + 4mp + 3p^2}$$

$$= \frac{(m+3p)(m-p)(m+4p)(m-2p)}{(m-2p)(m-p)(m+3p)(m+p)}$$

$$= \frac{m + 4p}{m + p}$$

35. $\dfrac{m^2 + 3m + 2}{m^2 + 5m + 4} \cdot \dfrac{m^2 + 10m + 24}{m^2 + 5m + 6}$

$$= \frac{(m+2)(m+1)}{(m+4)(m+1)} \cdot \frac{(m+6)(m+4)}{(m+3)(m+2)}$$

$$= \frac{m + 6}{m + 3} \quad \textit{Multiply and use}$$
$$\textit{fundamental property}$$

37. $\dfrac{y^2 + y - 2}{y^2 + 3y - 4} \div \dfrac{y + 2}{y + 3}$

$$= \frac{y^2 + y - 2}{y^2 + 3y - 4} \cdot \frac{y + 3}{y + 2}$$

$$= \frac{(y+2)(y-1)}{(y+4)(y-1)} \cdot \frac{y + 3}{y + 2}$$

$$= \frac{y + 3}{y + 4}$$

39. $\dfrac{2m^2 + 7m + 3}{m^2 - 9} \cdot \dfrac{m^2 - 3m}{2m^2 + 11m + 5}$

$$= \frac{(2m+1)(m+3)}{(m-3)(m+3)} \cdot \frac{m(m-3)}{(2m+1)(m+5)}$$

$$= \frac{(2m+1)(m+3)m(m-3)}{(m-3)(m+3)(2m+1)(m+5)}$$

$$= \frac{m}{m + 5}$$

41. $\dfrac{r^2 + rs - 12s^2}{r^2 - rs - 20s^2} \div \dfrac{r^2 - 2rs - 3s^2}{r^2 + rs - 30s^2}$

$$= \frac{r^2 + rs - 12s^2}{r^2 - rs - 20s^2} \cdot \frac{r^2 + rs - 30s^2}{r^2 - 2rs - 3s^2}$$

$$= \frac{(r-3s)(r+4s)(r+6s)(r-5s)}{(r-5s)(r+4s)(r-3s)(r+s)}$$

$$= \frac{r + 6s}{r + s}$$

43. $\dfrac{(q-3)^4(q+2)}{q^2 + 3q + 2} \div \dfrac{q^2 - 6q + 9}{q^2 + 4q + 4}$

$$= \frac{(q-3)^4(q+2)}{q^2 + 3q + 2} \cdot \frac{q^2 + 4q + 4}{q^2 - 6q + 9}$$

$$= \frac{(q-3)^4(q+2)(q+2)^2}{(q+2)(q+1)(q-3)^2}$$

$$= \frac{(q-3)^2(q+2)^2}{q + 1}$$

45. $\dfrac{x + 5}{x + 10} \div \left(\dfrac{x^2 + 10x + 25}{x^2 + 10x} \cdot \dfrac{10x}{x^2 + 15x + 50} \right)$

$$= \frac{x + 5}{x + 10} \div \left[\frac{(x+5)^2 \cdot 10x}{x(x+10)(x+5)(x+10)} \right]$$

$$= \frac{x + 5}{x + 10} \div \left[\frac{10(x+5)}{(x+10)^2} \right]$$

$$= \frac{x + 5}{x + 10} \cdot \frac{(x+10)^2}{10(x+5)}$$

$$= \frac{x + 10}{10}$$

47. $\dfrac{3a - 3b - a^2 + b^2}{4a^2 - 4ab + b^2} \cdot \dfrac{4a^2 - b^2}{2a^2 - ab - b^2}$

Factor $3a - 3b - a^2 + b^2$ by grouping.

$3a - 3b - a^2 + b^2$

$$= 3(a - b) - (a^2 - b^2)$$

$$= 3(a - b) - (a - b)(a + b)$$

$$= (a - b)[3 - (a + b)]$$

$$= (a - b)(3 - a - b)$$

continued

Thus,

$$\frac{3a - 3b - a^2 + b^2}{4a^2 - 4ab + b^2} \cdot \frac{4a^2 - b^2}{2a^2 - ab - b^2}$$

$$= \frac{(a-b)(3-a-b)}{(2a-b)(2a-b)} \cdot \frac{(2a-b)(2a+b)}{(2a+b)(a-b)}$$

$$= \frac{(a-b)(3-a-b)(2a-b)(2a+b)}{(2a-b)(2a-b)(2a+b)(a-b)}$$

$$= \frac{3-a-b}{2a-b}.$$

49. $\dfrac{-x^3 - y^3}{x^2 - 2xy + y^2} \div \dfrac{3y^2 - 3xy}{x^2 - y^2}$

$$= \frac{-1(x^3 + y^3)}{x^2 - 2xy + y^2} \cdot \frac{x^2 - y^2}{3y^2 - 3xy}$$

$$= \frac{-1(x+y)(x^2 - xy + y^2)}{(x-y)(x-y)}$$

$$\quad \cdot \frac{(x-y)(x+y)}{3y(y-x)}$$

$$= \frac{-1(x+y)(x^2 - xy + y^2)(x-y)(x+y)}{-1(x-y)(x-y)(3y)(x-y)}$$

$$= \frac{(x+y)^2(x^2 - xy + y^2)}{3y(x-y)^2}$$

If we had not changed $y - x$ to $-1(x - y)$ in the denominator, we would have obtained an alternate form of the answer,

$$-\frac{(x+y)^2(x^2 - xy + y^2)}{3y(y-x)(x-y)}.$$

51. Use the formula for the area of a rectangle with $A = \dfrac{5x^2 y^3}{2pq}$ and $L = \dfrac{2xy}{p}$ to solve for W.

$$A = L \cdot W$$

$$\frac{5x^2 y^3}{2pq} = \frac{2xy}{p} \cdot W$$

$$W = \frac{5x^2 y^3}{2pq} \div \frac{2xy}{p}$$

$$= \frac{5x^2 y^3}{2pq} \cdot \frac{p}{2xy}$$

$$= \frac{5x^2 y^3 p}{4pqxy}$$

$$= \frac{5xy^2}{4q}$$

Thus, the rational expression $\dfrac{5xy^2}{4q}$ represents the width of the rectangle.

Section 7.3

1. The factor a appears at most one time in any denominator as does the factor b. Thus, the LCD is the product of the two factors, ab. The correct response is (c).

3. Since $20 = 2^2 \cdot 5$, the LCD of $\dfrac{11}{20}$ and $\dfrac{1}{2}$ must have 5 as a factor and 2^2 as a factor. Because 2 appears twice in $2^2 \cdot 5$, we don't have to include another 2 in the LCD for the number $\dfrac{1}{2}$. Thus, the LCD is just $2^2 \cdot 5 = 20$. Note that this is a specific case of Exercise 2 since 2 is a factor of 20. The correct response is (c).

5. $\dfrac{-7}{15}, \dfrac{21}{20}$

Factor each denominator.

$$15 = 3 \cdot 5$$
$$20 = 2 \cdot 2 \cdot 5 = 2^2 \cdot 5$$

Take each factor the greatest number of times it appears as a factor in any one of the denominators.

$$LCD = 2^2 \cdot 3 \cdot 5 = 60$$

7. $\dfrac{17}{100}, \dfrac{23}{120}, \dfrac{43}{180}$

Factor each denominator.

$$100 = 2^2 \cdot 5^2$$
$$120 = 2^3 \cdot 3 \cdot 5$$
$$180 = 2^2 \cdot 3^2 \cdot 5$$

Take each factor the greatest number of times it appears as a factor in any one of the denominators.

$$LCD = 2^3 \cdot 3^2 \cdot 5^2 = 1800$$

9. $\dfrac{9}{x^2}, \dfrac{8}{x^5}$

The greatest number of times x appears as a factor in any denominator is the greatest exponent on x, which is 5.

$$LCD = x^5$$

11. $\dfrac{-2}{5p}, \dfrac{15}{6p}$

Factor each denominator.

$$5p = 5 \cdot p$$
$$6p = 2 \cdot 3 \cdot p$$

Take each factor the greatest number of times it appears; then multiply.

$$LCD = 2 \cdot 3 \cdot 5 \cdot p = 30p$$

13. $\dfrac{17}{15y^2}, \dfrac{55}{36y^4}$

Factor each denominator.

$$15y^2 = 3 \cdot 5 \cdot y^2$$
$$36y^4 = 2^2 \cdot 3^2 \cdot y^4$$

continued

Take each factor the greatest number of times it appears; then multiply.

$$LCD = 2^2 \cdot 3^2 \cdot 5 \cdot y^4 = 180y^4$$

15. $\dfrac{13}{5a^2b^3}, \dfrac{29}{15a^5b}$

Factor each denominator.

$$5a^2b^3 = 5 \cdot a^2 \cdot b^3$$
$$15a^5b = 3 \cdot 5 \cdot a^5 \cdot b$$

Take each factor the greatest number of times it appears; then multiply.

$$LCD = 3 \cdot 5 \cdot a^5 \cdot b^3 = 15a^5b^3$$

17. $\dfrac{7}{6p}, \dfrac{15}{4p-8}$

Factor each denominator.

$$6p = 2 \cdot 3 \cdot p$$
$$4p - 8 = 4(p-2) = 2^2(p-2)$$

Take each factor the greatest number of times it appears; then multiply.

$$LCD = 2^2 \cdot 3 \cdot p(p-2) = 12p(p-2)$$

19. $24 = 2^3 \cdot 3$
$20 = 2^2 \cdot 5$

To find the LCD, use each factor the greatest number of times it appears,

$$LCD = 2^3 \cdot 3 \cdot 5$$

20. $(t+4)^3(t-3)$
$(t+4)^2(t+8)$

Use $t + 4$ three times, $t - 3$ once, and $t + 8$ once as factors in the LCD.

$$LCD = (t+4)^3(t-3)(t+8)$$

21. The answers for Exercises 19 and 20 are similar in that $t + 4$ replaces 2, $t - 3$ replaces 3, and $t + 8$ replaces 5.

22. Common fractions contain only numbers, while algebraic fractions contain numbers and polynomials. The methods for finding the LCD for common fractions and algebraic fractions contain the same logical steps, but to learn how to find the LCD for algebraic fractions, one must be able to factor polynomials as well as numbers.

23. $\dfrac{37}{6r-12}, \dfrac{25}{9r-18}$

Factor each denominator.

$$6r - 12 = 6(r-2) = 2 \cdot 3(r-2)$$
$$9r - 18 = 9(r-2) = 3^2(r-2)$$

Take each factor the greatest number of times it appears; then multiply.

$$LCD = 2 \cdot 3^2(r-2) = 18(r-2)$$

25. $\dfrac{5}{12p+60}, \dfrac{17}{p^2+5p}, \dfrac{16}{p^2+10p+25}$

Factor each denominator.

$$12p + 60 = 12(p+5)$$
$$= 2^2 \cdot 3(p+5)$$
$$p^2 + 5p = p(p+5)$$
$$p^2 + 10p + 25 = (p+5)(p+5)$$
$$LCD = 2^2 \cdot 3 \cdot p(p+5)^2$$
$$= 12p(p+5)^2$$

27. $\dfrac{3}{8y+16}, \dfrac{22}{y^2+3y+2}$
$$8y + 16 = 8(y+2) = 2^3(y+2)$$
$$y^2 + 3y + 2 = (y+2)(y+1)$$
$$LCD = 8(y+2)(y+1)$$

29. $\dfrac{12}{m-3}, \dfrac{-4}{3-m}$

The expression $3 - m$ can be written as $-1(m-3)$, since

$$-1(m-3) = -m + 3 = 3 - m.$$

Because of this, either $m - 3$ or $3 - m$ can be used as the LCD.

31. $\dfrac{29}{p-q}, \dfrac{18}{q-p}$

The expression $q - p$ can be written as $-1(p-q)$, since

$$-1(p-q) = -p + q = q - p.$$

Because of this, either $p - q$ or $q - p$ can be used as the LCD.

33. $\dfrac{6}{a^2+6a}, \dfrac{-5}{a^2+3a-18}$
$$a^2 + 6a = a(a+6)$$
$$a^2 + 3a - 18 = (a+6)(a-3)$$
$$LCD = a(a+6)(a-3)$$

35. $\dfrac{-5}{k^2+2k-35}, \dfrac{-8}{k^2+3k-40}, \dfrac{9}{k^2-2k-15}$
$$k^2 + 2k - 35 = (k+7)(k-5)$$
$$k^2 + 3k - 40 = (k+8)(k-5)$$
$$k^2 - 2k - 15 = (k-5)(k+3)$$
$$LCD = (k+7)(k-5)(k+8)(k+3)$$

37. $(2x - 5)^2 = (2x)^2 - 2(2x)(5) + 5^2$
$$= 4x^2 - 20x + 25$$
$(5 - 2x)^2 = 5^2 - 2(5)(2x) + (2x)^2$
$$= 25 - 20x + 4x^2$$
$$= 4x^2 - 20x + 25$$

Yes, $(5 - 2x)^2$ is also acceptable as an LCD because

$$(2x - 5)^2 = (5 - 2x)^2.$$

39. $\dfrac{3}{4} = \dfrac{?}{28}$

To change 4 into 28, multiply by 7. If you multiply the denominator by 7, you must multiply the numerator by 7.

40. $\dfrac{3}{4} = \dfrac{3}{4} \cdot \dfrac{7}{7} = \dfrac{21}{28}$

Note that numerator and denominator are being multiplied by 7, so $\dfrac{3}{4}$ is being multiplied by the fraction $\dfrac{7}{7}$, which is equal to 1.

41. Since $\dfrac{7}{7}$ has a value of 1, the multiplier is 1. The identity property of multiplication is being used when we write a common fraction as an equivalent one with a larger denominator.

42. $\dfrac{2x + 5}{x - 4} = \dfrac{?}{7x - 28} = \dfrac{?}{7(x - 4)}$

The expression $7x - 28$ is factored as $7(x - 4)$, so the multiplier is 7.

43. $\dfrac{2x + 5}{x - 4} = \dfrac{}{7x - 28} = \dfrac{}{7(x - 4)}$

To form the new denominator, 7 must be used as the multiplier for the denominator. To form an equivalent fraction, the same multiplier must be used for numerator and denominator. Thus, the multiplier is $\dfrac{7}{7}$, which is equal to 1.

44. The identity property of multiplication is being used when we write an algebraic fraction as an equivalent one with a larger denominator.

45. $\dfrac{15m^2}{8k} = \dfrac{}{32k^4}$

$32k^4 = (8k)(4k^3)$, so we must multiply the numerator and the denominator by $4k^3$.

$\dfrac{15m^2}{8k} = \dfrac{15m^2}{8k} \cdot \dfrac{4k^3}{4k^3}$ *Multiplicative identity property*

$ = \dfrac{60m^2k^3}{32k^4}$

47. $\dfrac{19z}{2z - 6} = \dfrac{}{6z - 18}$

Begin by factoring each denominator.

$$2z - 6 = 2(z - 3)$$
$$6z - 18 = 6(z - 3)$$

The fractions may now be written as follows.

$$\dfrac{19z}{2(z - 3)} = \dfrac{}{6(z - 3)}$$

Comparing the two factored forms, we see that the denominator of the fraction on the left side must be multiplied by 3; the numerator must also be multiplied by 3.

$\dfrac{19z}{2z - 6}$

$= \dfrac{19z}{2(z - 3)} \cdot \dfrac{3}{3}$ *Multiplicative identity property*

$= \dfrac{19z(3)}{2(z - 3)(3)}$ *Multiplication of rational expressions*

$= \dfrac{57z}{6z - 18}$ *Multiply the factors*

49. $\dfrac{-2a}{9a - 18} = \dfrac{}{18a - 36}$

$\dfrac{-2a}{9(a - 2)} = \dfrac{}{18(a - 2)}$ *Factor each denominator*

$\dfrac{-2a}{9a - 18} = \dfrac{-2a}{9(a - 2)} \cdot \dfrac{2}{2}$ *Multiplicative identity property*

$\phantom{\dfrac{-2a}{9a - 18}} = \dfrac{-4a}{18a - 36}$ *Multiply*

51. $\dfrac{6}{k^2 - 4k} = \dfrac{}{k(k - 4)(k + 1)}$

$\dfrac{6}{k(k - 4)} = \dfrac{}{k(k - 4)(k + 1)}$ *Factor first denominator*

$\dfrac{6}{k^2 - 4k} = \dfrac{6}{k(k - 4)} \cdot \dfrac{(k + 1)}{(k + 1)}$ *Multiplicative identity property*

$\phantom{\dfrac{6}{k^2 - 4k}} = \dfrac{6(k + 1)}{k(k - 4)(k + 1)}$ *Multiply*

53. $\dfrac{36r}{r^2 - r - 6} = \dfrac{}{(r - 3)(r + 2)(r + 1)}$

$\dfrac{36r}{(r - 3)(r + 2)} = \dfrac{}{(r - 3)(r + 2)(r + 1)}$

Factor first denominator

$\dfrac{36r}{r^2 - r - 6} = \dfrac{36r}{(r - 3)(r + 2)} \cdot \dfrac{(r + 1)}{(r + 1)}$

Multiplicative identity property

$\phantom{\dfrac{36r}{r^2 - r - 6}} = \dfrac{36r(r + 1)}{(r - 3)(r + 2)(r + 1)}$

55. $\dfrac{a + 2b}{2a^2 + ab - b^2} = \dfrac{}{2a^3b + a^2b^2 - ab^3}$

$\dfrac{a + 2b}{2a^2 + ab - b^2} = \dfrac{}{ab(2a^2 + ab - b^2)}$

Factor second denominator

$\dfrac{a + 2b}{2a^2 + ab - b^2} = \dfrac{(a + 2b)}{(2a^2 + ab - b^2)} \cdot \dfrac{ab}{ab}$

Multiplicative identity property

$= \dfrac{ab(a + 2b)}{2a^3b + a^2b^2 - ab^3}$

57. $\dfrac{4r - t}{r^2 + rt + t^2} = \dfrac{}{t^3 - r^3}$

Factor the second denominator as the difference of two cubes.

$t^3 - r^3 = (t - r)(t^2 + rt + r^2)$

$\dfrac{4r - t}{r^2 + rt + t^2} = \dfrac{(4r - t)}{(r^2 + rt + t^2)} \cdot \dfrac{(t - r)}{(t - r)}$

Multiplicative identity property

$= \dfrac{(4r - t)(t - r)}{t^3 - r^3}$

Multiply the factors

59. $\dfrac{2(z - y)}{y^2 + yz + z^2} = \dfrac{}{y^4 - z^3y}$

Factor the second denominator.

$y^4 - z^3y = y(y^3 - z^3)$ *y is the GCF*

$= y(y - z)(y^2 + yz + z^2)$

Difference of two cubes

$\dfrac{2(z - y)}{y^2 + yz + z^2} = \dfrac{}{y(y - z)(y^2 + yz + z^2)}$

$\dfrac{2(z - y)}{y^2 + yz + z^2} = \dfrac{2(z - y)}{(y^2 + yz + z^2)} \cdot \dfrac{y(y - z)}{y(y - z)}$

$= \dfrac{2y(z - y)(y - z)}{y(y - z)(y^2 + yz + z^2)}$

Multiplicative identity property

$= \dfrac{2y(z - y)(y - z)}{y(y^3 - z^3)}$

$= \dfrac{2y(z - y)(y - z)}{y^4 - z^3y}$

or $\dfrac{-2y(y - z)^2}{y^4 - z^3y}$, since $z - y = -1(y - z)$

61. *Step 1:* Factor each denominator into prime factors.

Step 2: List each different denominator factor the greatest number of times it appears in any of the denominators.

Step 3: Multiply the factors in the list to get the LCD. For example, the least common denominator

for $\dfrac{1}{(x + y)^3}$ and $\dfrac{-2}{(x + y)^2(p + q)}$ is

$(x + y)^3(p + q)$.

Section 7.4

1. $\dfrac{5}{7 - 3} = \dfrac{5}{4}$

2. $\dfrac{5}{7 - 3} = \dfrac{(-1)(5)}{-1(7 - 3)}$

$= \dfrac{-5}{-7 + 3}$

3. $\dfrac{5}{7 - 3} = \dfrac{(-1)(5)}{-1(7 - 3)}$

$= \dfrac{-1}{-1} \cdot \dfrac{5}{7 - 3}$

The multiplier is $\dfrac{-1}{-1}$, which has a value of 1.

4. $\dfrac{-5}{-7 + 3} = \dfrac{-5}{-4} = \dfrac{5}{4}$

The two answers are equal.

5. Jill's answer is $\dfrac{5}{x - 3}$.

Jack's answer is

$\dfrac{-5}{3 - x} = \dfrac{(-1)(-5)}{-1(3 - x)} = \dfrac{5}{-3 + x} = \dfrac{5}{x - 3}$.

Jack's answer is also correct because his answer can be obtained from Jill's answer by multiplying

by $\dfrac{-1}{-1} = 1$, which is the identity element for multiplication.

6. Changing the sign of each term in the fraction is a way of multiplying by $\dfrac{-1}{-1}$ or 1, the identity element for multiplication.

7. Putting a negative sign in front of a fraction gives the opposite of the fraction. Changing the signs in either the numerator or the denominator also gives the opposite. Thus we have multiplied by $(-1)(-1) = 1$, the identity element for multiplication.

8. $\dfrac{y - 4}{3 - y}$

(a) $\dfrac{y - 4}{y - 3} = \dfrac{1(y - 4)}{-1(-y + 3)} = \dfrac{1(y - 4)}{-1(3 - y)}$

The multiplier is $\dfrac{1}{-1} = -1$, which is not the identity. This fraction is not equivalent to the original.

(b) $\dfrac{-y + 4}{-3 + y} = \dfrac{-1(-y + 4)}{-1(-3 + y)} = \dfrac{y - 4}{3 - y}$

The multiplier is $\dfrac{-1}{-1} = 1$, which is the identity element. This fraction is equivalent to the original.

(c) $-\dfrac{y-4}{y-3} = \dfrac{1}{-1} \cdot \dfrac{y-4}{y-3} = \dfrac{y-4}{-y+3}$

This fraction is equivalent to the original.

(d) $-\dfrac{y-4}{3-y} = \dfrac{-1}{1} \cdot \dfrac{y-4}{3-y} = \dfrac{-y+4}{3-y}$

Note that the denominator of the new fraction is the same as the original, but the numerator is the opposite. The fraction is not equivalent to the original.

(e) $-\dfrac{4-y}{y-3} = \dfrac{-1}{1} \cdot \dfrac{4-y}{y-3} = \dfrac{-4+y}{y-3} = \dfrac{y-4}{-3+y}$

Note that the numerator of the new fraction is the same as the original, but that the denominator is the opposite. The fraction is not equivalent to the original.

(f) $\dfrac{y+4}{3+y} = \dfrac{-1(y+4)}{-1(3+y)} = \dfrac{-y-4}{-3-y}$

After multiplying by $\dfrac{-1}{-1} = 1$, the new fraction is

not equivalent to the original.

9. $\dfrac{4}{m} + \dfrac{7}{m}$

The denominators are the same, so the sum is found by adding the two numerators and keeping the same (common) denominator.

$$\dfrac{4}{m} + \dfrac{7}{m} = \dfrac{4+7}{m} = \dfrac{11}{m}$$

11. $\dfrac{a+b}{2} - \dfrac{a-b}{2}$

The denominators are the same, so the difference is found by subtracting the second numerator from the first and keeping the same (common) denominator.

$$\dfrac{a+b}{2} - \dfrac{a-b}{2} = \dfrac{(a+b)-(a-b)}{2}$$
$$= \dfrac{a+b-a+b}{2}$$
$$= \dfrac{2b}{2} = b$$

13. $\dfrac{x^2}{x+5} + \dfrac{5x}{x+5} = \dfrac{x^2+5x}{x+5}$ *Add numerators*
$$= \dfrac{x(x+5)}{x+5}$$ *Factor numerator*
$$= x$$ *Lowest terms*

15. $\dfrac{y^2-3y}{y+3} + \dfrac{-18}{y+3} = \dfrac{y^2-3y-18}{y+3}$
$$= \dfrac{(y-6)(y+3)}{y+3}$$
$$= y-6$$

17. To add or subtract rational expressions with the same denominators, combine the numerators and keep the same denominator. For example, $\dfrac{3x+2}{x-6} + \dfrac{-2x-8}{x-6} = \dfrac{x-6}{x-6}$. Then write in lowest terms. In this example, the sum simplifies to 1.

19. $\dfrac{z}{5} + \dfrac{1}{3}$

The LCD is 15. Now rewrite each rational expression as a fraction with the LCD as its denominator.

$$\dfrac{z}{5} \cdot \dfrac{3}{3} = \dfrac{3z}{15}$$
$$\dfrac{1}{3} \cdot \dfrac{5}{5} = \dfrac{5}{15}$$

Since the fractions now have a common denominator, add the numerators and use the LCD as the denominator of the

sum. $\dfrac{z}{5} + \dfrac{1}{3} = \dfrac{3z}{15} + \dfrac{5}{15} = \dfrac{3z+5}{15}$

21. $\dfrac{5}{7} - \dfrac{r}{2} = \dfrac{5}{7} \cdot \dfrac{2}{2} - \dfrac{r}{2} \cdot \dfrac{7}{7}$ *LCD = 14*
$$= \dfrac{10}{14} - \dfrac{7r}{14}$$
$$= \dfrac{10-7r}{14}$$

23. $-\dfrac{3}{4} - \dfrac{1}{2x} = -\dfrac{3 \cdot x}{4 \cdot x} - \dfrac{1 \cdot 2}{2x \cdot 2}$ *LCD = 4x*
$$= \dfrac{-3x-2}{4x}$$

25. $\dfrac{x+1}{6} + \dfrac{3x+3}{9}$

First reduce the second fraction.

$$\dfrac{3x+3}{9} = \dfrac{3(x+1)}{9} = \dfrac{x+1}{3}$$

Now the LCD of $\dfrac{x+1}{6}$ and $\dfrac{x+1}{3}$ is 6. Thus,

$$\dfrac{x+1}{6} + \dfrac{x+1}{3} = \dfrac{x+1}{6} + \dfrac{x+1}{3} \cdot \dfrac{2}{2}$$
$$= \dfrac{x+1+2x+2}{6}$$
$$= \dfrac{3x+3}{6}$$
$$= \dfrac{3(x+1)}{6} = \dfrac{x+1}{2}.$$

27. $\dfrac{x+3}{3x} + \dfrac{2x+2}{4x} = \dfrac{x+3}{3x} + \dfrac{2(x+1)}{4x}$
$$= \dfrac{x+3}{3x} + \dfrac{x+1}{2x}$$ *Reduce*

continued

$$= \frac{x+3}{3x} \cdot \frac{2}{2} + \frac{x+1}{2x} \cdot \frac{3}{3}$$

$$\text{LCD} = 6x$$

$$= \frac{2x+6+3x+3}{6x}$$

$$= \frac{5x+9}{6x}$$

29. $\dfrac{7}{3p^2} - \dfrac{2}{p} = \dfrac{7}{3p^2} - \dfrac{2}{p} \cdot \dfrac{3p}{3p}$ $\quad \text{LCD} = 3p^2$

$$= \frac{7-6p}{3p^2}$$

31. $\dfrac{x}{x-2} + \dfrac{4}{x+2} - \dfrac{8}{x^2-4}$

$$= \frac{x}{x-2} + \frac{4}{x+2} - \frac{8}{(x+2)(x-2)}$$

$$= \frac{x}{x-2} \cdot \frac{x+2}{x+2} + \frac{4}{x+2} \cdot \frac{x-2}{x-2}$$

$$- \frac{8}{(x+2)(x-2)} \quad \text{LCD} = (x+2)(x-2)$$

$$= \frac{x(x+2) + 4(x-2) - 8}{(x+2)(x-2)}$$

$$= \frac{x^2 + 2x + 4x - 8 - 8}{(x+2)(x-2)}$$

$$= \frac{x^2 + 6x - 16}{(x+2)(x-2)}$$

$$= \frac{(x+8)(x-2)}{(x+2)(x-2)} = \frac{x+8}{x+2}$$

33. $\dfrac{t}{t+2} + \dfrac{5-t}{t} - \dfrac{4}{t^2+2t}$

$$= \frac{t}{t+2} + \frac{5-t}{t} - \frac{4}{t(t+2)}$$

$$= \frac{t}{t+2} \cdot \frac{t}{t} + \frac{5-t}{t} \cdot \frac{t+2}{t+2}$$

$$- \frac{4}{t(t+2)} \quad \text{LCD} = t(t+2)$$

$$= \frac{t \cdot t + (5-t)(t+2) - 4}{t(t+2)}$$

$$= \frac{t^2 + 5t + 10 - t^2 - 2t - 4}{t(t+2)}$$

$$= \frac{3t+6}{t(t+2)}$$

$$= \frac{3(t+2)}{t(t+2)} = \frac{3}{t}$$

35. $\dfrac{10}{m-2} + \dfrac{5}{2-m}$

Since

$$2 - m = -1(m-2),$$

either $m - 2$ or $2 - m$ could be used as the LCD.

37. $\dfrac{4}{x-5} + \dfrac{6}{5-x}$

The two denominators, $x - 5$ and $5 - x$, are opposites of each other, so either one may be used

as the common denominator. We will work the exercise both ways and compare the answers.

$$\frac{4}{x-5} + \frac{6}{5-x} = \frac{4}{x-5} + \frac{6(-1)}{(5-x)(-1)}$$

$$\text{LCD} = x - 5$$

$$= \frac{4}{x-5} + \frac{-6}{x-5}$$

$$= \frac{-2}{x-5}$$

$$\frac{4}{x-5} + \frac{6}{5-x} = \frac{4(-1)}{(x-5)(-1)} + \frac{6}{5-x}$$

$$\text{LCD} = 5 - x$$

$$= \frac{-4}{5-x} + \frac{6}{5-x}$$

$$= \frac{2}{5-x}$$

The two answers are equivalent, since

$$\frac{-2}{x-5} \cdot \frac{-1}{-1} = \frac{2}{5-x}.$$

39. $\dfrac{-1}{1-y} + \dfrac{3-4y}{y-1}$

The LCD is either $1 - y$ or $y - 1$. We'll use $y - 1$.

$$\frac{-1}{1-y} + \frac{3-4y}{y-1} = \frac{-1 \cdot -1}{-1 \cdot (1-y)} + \frac{3-4y}{y-1}$$

$$= \frac{1+3-4y}{y-1}$$

$$= \frac{4-4y}{y-1}$$

$$= \frac{4(1-y)}{y-1} = -4$$

41. $\dfrac{2}{x-y^2} + \dfrac{7}{y^2-x}$

$\text{LCD} = x - y^2$ or $y^2 - x$

We will use $x - y^2$.

$$\frac{2}{x-y^2} + \frac{7}{y^2-x}$$

$$= \frac{2}{x-y^2} + \frac{-1(7)}{-1(y^2-x)}$$

$$= \frac{2}{x-y^2} + \frac{-7}{-y^2+x}$$

$$= \frac{2}{x-y^2} + \frac{-7}{x-y^2}$$

$$= \frac{2+(-7)}{x-y^2} = \frac{-5}{x-y^2}$$

If $y^2 - x$ is used as the LCD, we will obtain the equivalent answer

$$\frac{5}{y^2-x}.$$

43. $\dfrac{x}{5x-3y} - \dfrac{y}{3y-5x}$

LCD $= 5x - 3y$ or $3y - 5x$

We will use $5x - 3y$.

$$\dfrac{x}{5x-3y} - \dfrac{y}{3y-5x}$$

$$= \dfrac{x}{5x-3y} - \dfrac{-1(y)}{-1(3y-5x)}$$

$$= \dfrac{x}{5x-3y} - \dfrac{-y}{-3y+5x}$$

$$= \dfrac{x}{5x-3y} - \dfrac{-y}{5x-3y}$$

$$= \dfrac{x-(-y)}{5x-3y} = \dfrac{x+y}{5x-3y}$$

If $3y - 5x$ is used as the LCD, we will obtain the equivalent answer

$$\dfrac{-x-y}{3y-5x}.$$

45. $\dfrac{3}{4p-5} + \dfrac{9}{5-4p}$

LCD $= 4p - 5$ or $5 - 4p$

We will use $4p - 5$.

$$\dfrac{3}{4p-5} + \dfrac{9}{5-4p}$$

$$= \dfrac{3}{4p-5} + \dfrac{-1(9)}{-1(5-4p)}$$

$$= \dfrac{3}{4p-5} + \dfrac{-9}{-5+4p}$$

$$= \dfrac{3}{4p-5} + \dfrac{-9}{4p-5}$$

$$= \dfrac{3+(-9)}{4p-5} = \dfrac{-6}{4p-5}$$

If $5 - 4p$ is used as the LCD, we will obtain the equivalent answer

$$\dfrac{6}{5-4p}.$$

47. $\dfrac{2m}{m-n} - \dfrac{5m+n}{2m-2n}$

$$= \dfrac{2m}{m-n} - \dfrac{5m+n}{2(m-n)} \quad \textit{Factor second}$$
$$\textit{denominator}$$

$$= \dfrac{2m}{m-n} \cdot \dfrac{2}{2} - \dfrac{5m+n}{2(m-n)} \quad \textit{LCD} = 2(m-n)$$

$$= \dfrac{4m-(5m+n)}{2(m-n)}$$

$$= \dfrac{4m-5m-n}{2(m-n)}$$

$$= \dfrac{-m-n}{2(m-n)}$$

$$= \dfrac{-(m+n)}{2(m-n)}$$

49. $\dfrac{5}{x^2-9} - \dfrac{x+2}{x^2+4x+3}$

To find the LCD, factor the denominators.

$$x^2 - 9 = (x+3)(x-3)$$
$$x^2 + 4x + 3 = (x+3)(x+1)$$

The LCD is $(x+3)(x-3)(x+1)$.

$$\dfrac{5}{x^2-9} - \dfrac{x+2}{x^2+4x+3}$$

$$= \dfrac{5\cdot(x+1)}{(x+3)(x-3)\cdot(x+1)}$$

$$- \dfrac{(x+2)\cdot(x-3)}{(x+3)(x+1)\cdot(x-3)}$$

$$= \dfrac{5x+5}{(x+3)(x-3)(x+1)}$$

$$- \dfrac{x^2-x-6}{(x+3)(x+1)(x-3)}$$

$$= \dfrac{(5x+5)-(x^2-x-6)}{(x+3)(x-3)(x+1)}$$

$$= \dfrac{5x+5-x^2+x+6}{(x+3)(x-3)(x+1)}$$

$$= \dfrac{-x^2+6x+11}{(x+3)(x-3)(x+1)}$$

51. $\dfrac{2q+1}{3q^2+10q-8} - \dfrac{3q+5}{2q^2+5q-12}$

$$= \dfrac{2q+1}{(3q-2)(q+4)} - \dfrac{3q+5}{(2q-3)(q+4)}$$

$$= \dfrac{(2q+1)\cdot(2q-3)}{(3q-2)(q+4)\cdot(2q-3)}$$

$$- \dfrac{(3q+5)\cdot(3q-2)}{(2q-3)(q+4)\cdot(3q-2)}$$

$$LCD = (3q-2)(q+4)(2q-3)$$

$$= \dfrac{(4q^2-4q-3)-(9q^2+9q-10)}{(3q-2)(q+4)(2q-3)}$$

$$= \dfrac{4q^2-4q-3-9q^2-9q+10}{(3q-2)(q+4)(2q-3)}$$

$$= \dfrac{-5q^2-13q+7}{(3q-2)(q+4)(2q-3)}$$

53. $\dfrac{4}{r^2-r} + \dfrac{6}{r^2+2r} - \dfrac{1}{r^2+r-2}$

$$= \dfrac{4}{r(r-1)} + \dfrac{6}{r(r+2)} - \dfrac{1}{(r+2)(r-1)}$$

$$= \dfrac{4\cdot(r+2)}{r(r-1)\cdot(r+2)} + \dfrac{6\cdot(r-1)}{r(r+2)\cdot(r-1)}$$

$$- \dfrac{1\cdot r}{r\cdot(r+2)(r-1)}$$

$$LCD = r(r+2)(r-1)$$

$$= \frac{4r + 8 + 6r - 6 - r}{r(r+2)(r-1)}$$

$$= \frac{9r + 2}{r(r+2)(r-1)}$$

55. $\dfrac{x + 3y}{x^2 + 2xy + y^2} + \dfrac{x - y}{x^2 + 4xy + 3y^2}$

$$= \frac{x + 3y}{(x+y)(x+y)} + \frac{x - y}{(x+3y)(x+y)}$$

$$= \frac{(x+3y) \cdot (x+3y)}{(x+y)(x+y) \cdot (x+3y)}$$

$$+ \frac{(x-y) \cdot (x+y)}{(x+3y)(x+y) \cdot (x+y)}$$

$LCD = (x+y)(x+y)(x+3y)$

$$= \frac{(x^2 + 6xy + 9y^2) + (x^2 - y^2)}{(x+y)(x+y)(x+3y)}$$

$$= \frac{2x^2 + 6xy + 8y^2}{(x+y)(x+y)(x+3y)}$$

or $\dfrac{2x^2 + 6xy + 8y^2}{(x+y)^2(x+3y)}$

57. $\dfrac{r + y}{18r^2 + 12ry - 3ry - 2y^2} + \dfrac{3r - y}{36r^2 - y^2}$

Factor $18r^2 + 12ry - 3ry - 2y^2$ by grouping.

$$18r^2 + 12ry - 3ry - 2y^2$$
$$= 6r(3r + 2y) - y(3r + 2y)$$
$$= (3r + 2y)(6r - y)$$

Factor the denominators.

$$\frac{r + y}{(3r + 2y)(6r - y)} + \frac{3r - y}{(6r - y)(6r + y)}$$

Rewrite fractions with the LCD,
$(3r + 2y)(6r - y)(6r + y)$.

$$= \frac{(r + y) \cdot (6r + y)}{(3r + 2y)(6r - y) \cdot (6r + y)}$$

$$+ \frac{(3r - y) \cdot (3r + 2y)}{(6r - y)(6r + y) \cdot (3r + 2y)}$$

$$= \frac{6r^2 + 7ry + y^2}{(3r + 2y)(6r - y)(6r + y)}$$

$$+ \frac{9r^2 + 3ry - 2y^2}{(3r + 2y)(6r - y)(6r + y)}$$

$$= \frac{6r^2 + 7ry + y^2 + 9r^2 + 3ry - 2y^2}{(3r + 2y)(6r - y)(6r + y)}$$

$$= \frac{15r^2 + 10ry - y^2}{(3r + 2y)(6r - y)(6r + y)}$$

59. $\left(\dfrac{-k}{2k^2 - 5k - 3} + \dfrac{3k - 2}{2k^2 - k - 1} \right) \dfrac{2k + 1}{k - 1}$

Factor the denominators.

$$= \left[\frac{-k}{(2k+1)(k-3)} + \frac{3k-2}{(2k+1)(k-1)} \right]$$
$$\cdot \frac{2k+1}{k-1}$$

$$= \left[\frac{-k(k-1)}{(2k+1)(k-3)(k-1)} \right.$$
$$\left. + \frac{(3k-2)(k-3)}{(2k+1)(k-1)(k-3)} \right] \cdot \frac{2k+1}{k-1}$$

$LCD = (2k+1)(k-3)(k-1)$

$$= \left[\frac{-k^2 + k}{(2k+1)(k-3)(k-1)} \right.$$
$$\left. + \frac{3k^2 - 11k + 6}{(2k+1)(k-3)(k-1)} \right] \cdot \frac{2k+1}{k-1}$$

$$= \left[\frac{2k^2 - 10k + 6}{(2k+1)(k-3)(k-1)} \right] \frac{2k+1}{k-1}$$

$$= \frac{(2k^2 - 10k + 6)(2k+1)}{(2k+1)(k-3)(k-1)^2}$$

$$= \frac{2k^2 - 10k + 6}{(k-3)(k-1)^2}$$

61. $\dfrac{k^2 + 4k + 16}{k + 4} \left(\dfrac{-5}{16 - k^2} + \dfrac{2k + 3}{k^3 - 64} \right)$

Factor $16 - k^2$ as the difference of two squares.

$$16 - k^2 = (4 + k)(4 - k)$$

Factor $k^3 - 64$ as the difference of two cubes.

$$k^3 - 64 = (k - 4)(k^2 + 4k + 16)$$

$$\frac{k^2 + 4k + 16}{k + 4} \left(\frac{-5}{16 - k^2} + \frac{2k + 3}{k^3 - 64} \right)$$

$$= \frac{k^2 + 4k + 16}{k + 4}$$
$$\cdot \left[\frac{-5}{(4 - k)(4 + k)} + \frac{2k + 3}{(k - 4)(k^2 + 4k + 16)} \right]$$

$$= \frac{k^2 + 4k + 16}{k + 4}$$
$$\cdot \left[\frac{-5}{-1(k - 4)(4 + k)} + \frac{2k + 3}{(k - 4)(k^2 + 4k + 16)} \right]$$

The least common denominator is
$(k - 4)(4 + k)(k^2 + 4k + 16)$.

$$= \frac{k^2 + 4k + 16}{k + 4}$$
$$\cdot \left[\frac{5(k^2 + 4k + 16)}{(k - 4)(4 + k)(k^2 + 4k + 16)} \right.$$
$$\left. + \frac{(2k + 3)(4 + k)}{(k - 4)(4 + k)(k^2 + 4k + 16)} \right]$$

continued

$$= \frac{k^2 + 4k + 16}{k + 4}$$
$$\cdot \left[\frac{5k^2 + 20k + 80 + 2k^2 + 11k + 12}{(k - 4)(4 + k)(k^2 + 4k + 16)} \right]$$

$$= \frac{k^2 + 4k + 16}{k + 4}$$
$$\cdot \left[\frac{7k^2 + 31k + 92}{(k - 4)(4 + k)(k^2 + 4k + 16)} \right]$$

$$= \frac{(k^2 + 4k + 16)(7k^2 + 31k + 92)}{(k + 4)(k - 4)(4 + k)(k^2 + 4k + 16)}$$

$$= \frac{7k^2 + 31k + 92}{(k - 4)(k + 4)^2}$$

63. (a) $P = 2L + 2W$

$$= 2\left(\frac{3k + 1}{10} \right) + 2\left(\frac{5}{6k + 2} \right)$$

$$= 2\left(\frac{3k + 1}{2 \cdot 5} \right) + 2\left(\frac{5}{2(3k + 1)} \right)$$

$$= \frac{3k + 1}{5} + \frac{5}{3k + 1}$$

To add the two fractions on the right, use $5(3k + 1)$ as the LCD.

$$P = \frac{(3k + 1)(3k + 1)}{5(3k + 1)} + \frac{(5)(5)}{5(3k + 1)}$$

$$= \frac{(3k + 1)(3k + 1) + (5)(5)}{5(3k + 1)}$$

$$= \frac{9k^2 + 6k + 1 + 25}{5(3k + 1)}$$

$$= \frac{9k^2 + 6k + 26}{5(3k + 1)}$$

(b) $A = L \cdot W$

$$A = \frac{3k + 1}{10} \cdot \frac{5}{6k + 2}$$

$$= \frac{3k + 1}{5 \cdot 2} \cdot \frac{5}{2(3k + 1)}$$

$$= \frac{1}{2 \cdot 2} = \frac{1}{4}$$

Section 7.5

1. (a) The LCD of $\frac{1}{2}$ and $\frac{1}{3}$ is $2 \cdot 3 = 6$. The simplified form of the numerator is

$$\frac{1}{2} - \frac{1}{3} = \frac{3}{6} - \frac{2}{6} = \frac{1}{6}.$$

(b) The LCD of $\frac{5}{6}$ and $\frac{1}{12}$ is 12 since 12 is a multiple of 6. The simplified form of the denominator is

$$\frac{5}{6} - \frac{1}{12} = \frac{10}{12} - \frac{1}{12} = \frac{9}{12} = \frac{3}{4}.$$

(c) $\dfrac{\dfrac{1}{6}}{\dfrac{3}{4}} = \dfrac{1}{6} \div \dfrac{3}{4}$

(d) $\dfrac{1}{6} \div \dfrac{3}{4} = \dfrac{1}{6} \cdot \dfrac{4}{3}$

$$= \frac{2 \cdot 2}{2 \cdot 3 \cdot 3} = \frac{2}{9}$$

3. $\dfrac{3 - \dfrac{1}{2}}{2 - \dfrac{1}{4}} = \dfrac{-3 + \dfrac{1}{2}}{-2 + \dfrac{1}{4}}$

Choice (d) is equivalent to the given fraction. Each term of the numerator and denominator has been multiplied by -1. Since $\dfrac{-1}{-1} = 1$, the fraction has been multiplied by the identity element, so its value is unchanged.

5. Method 1 indicates to write the complex fraction as a division problem, and then perform the division. For example, to simplify $\dfrac{\dfrac{1}{2}}{\dfrac{2}{3}}$, we write

$\dfrac{1}{2} \div \dfrac{2}{3}$. Then simplify as $\dfrac{1}{2} \cdot \dfrac{3}{2} = \dfrac{3}{4}$.

In Exercises 7–36, either Method 1 or Method 2 can be used to simplify each complex fraction. Only one method will be shown for each exercise.

7. To use Method 1, divide the numerator of the complex fraction by the denominator.

$$\frac{-\dfrac{4}{3}}{\dfrac{2}{9}} = \frac{-4}{3} \div \frac{2}{9} = -\frac{4}{3} \cdot \frac{9}{2}$$

$$= -\frac{36}{6} = -6$$

9. To use Method 2, multiply the numerator and denominator of the complex fraction by the LCD, y^2.

$$\frac{\dfrac{x}{y^2}}{\dfrac{x^2}{y}} = \frac{y^2\left(\dfrac{x}{y^2} \right)}{y^2\left(\dfrac{x^2}{y} \right)}$$

$$= \frac{x}{yx^2} = \frac{1}{xy}$$

11. $\dfrac{\dfrac{4a^4b^3}{3a}}{\dfrac{2ab^4}{b^2}} = \dfrac{4a^4b^3}{3a} \div \dfrac{2ab^4}{b^2}$ *Method 1*

$= \dfrac{4a^4b^3}{3a} \cdot \dfrac{b^2}{2ab^4}$

$= \dfrac{4a^4b^3 \cdot b^2}{3a \cdot 2ab^4}$

$= \dfrac{4a^4b^5}{6a^2b^4}$

$= \dfrac{2a^2b}{3}$

13. To use Method 2, multiply the numerator and denominator of the complex fraction by the LCD, $3m$.

$$\dfrac{\dfrac{m+2}{3}}{\dfrac{m-4}{m}} = \dfrac{3m\left(\dfrac{m+2}{3}\right)}{3m\left(\dfrac{m-4}{m}\right)}$$

$$= \dfrac{m(m+2)}{3(m-4)}$$

15. $\dfrac{\dfrac{2}{x} - 3}{\dfrac{2-3x}{2}} = \dfrac{2x\left(\dfrac{2}{x} - 3\right)}{2x\left(\dfrac{2-3x}{2}\right)}$ *Method 2; LCD = 2x*

$= \dfrac{2x\left(\dfrac{2}{x}\right) - 2x(3)}{x(2-3x)}$

$= \dfrac{4-6x}{x(2-3x)}$

$= \dfrac{2(2-3x)}{x(2-3x)}$ *Factor*

$= \dfrac{2}{x}$ *Lowest terms*

17. $\dfrac{\dfrac{1}{x} + x}{\dfrac{x^2+1}{8}} = \dfrac{8x\left(\dfrac{1}{x} + x\right)}{8x\left(\dfrac{x^2+1}{8}\right)}$ *Method 2; LCD = 8x*

$= \dfrac{8 + 8x^2}{x(x^2+1)}$ *Distributive property*

$= \dfrac{8(1+x^2)}{x(x^2+1)}$ *Factor*

$= \dfrac{8}{x}$ *Lowest terms*

19. $\dfrac{a - \dfrac{5}{a}}{a + \dfrac{1}{a}} = \dfrac{a\left(a - \dfrac{5}{a}\right)}{a\left(a + \dfrac{1}{a}\right)}$ *Method 2; LCD = a*

$= \dfrac{a^2 - 5}{a^2 + 1}$

21. $\dfrac{\dfrac{5}{8} + \dfrac{2}{3}}{\dfrac{7}{3} - \dfrac{1}{4}} = \dfrac{24\left(\dfrac{5}{8} + \dfrac{2}{3}\right)}{24\left(\dfrac{7}{3} - \dfrac{1}{4}\right)}$ *Method 2; LCD = 24*

$= \dfrac{24\left(\dfrac{5}{8}\right) + 24\left(\dfrac{2}{3}\right)}{24\left(\dfrac{7}{3}\right) - 24\left(\dfrac{1}{4}\right)}$

$= \dfrac{15 + 16}{56 - 6} = \dfrac{31}{50}$

23. $\dfrac{\dfrac{1}{x^2} + \dfrac{1}{y^2}}{\dfrac{1}{x} - \dfrac{1}{y}}$

$= \dfrac{x^2y^2\left(\dfrac{1}{x^2} + \dfrac{1}{y^2}\right)}{x^2y^2\left(\dfrac{1}{x} - \dfrac{1}{y}\right)}$ *Method 2; LCD = x²y²*

$= \dfrac{x^2y^2\left(\dfrac{1}{x^2}\right) + x^2y^2\left(\dfrac{1}{y^2}\right)}{x^2y^2\left(\dfrac{1}{x}\right) - x^2y^2\left(\dfrac{1}{y}\right)}$

$= \dfrac{y^2 + x^2}{xy^2 - x^2y} = \dfrac{y^2 + x^2}{xy(y-x)}$

25. $\dfrac{\dfrac{2}{p^2} - \dfrac{3}{5p}}{\dfrac{4}{p} + \dfrac{1}{4p}} = \dfrac{20p^2\left(\dfrac{2}{p^2} - \dfrac{3}{5p}\right)}{20p^2\left(\dfrac{4}{p} + \dfrac{1}{4p}\right)}$ *Method 2; LCD = 20p²*

$= \dfrac{20p^2\left(\dfrac{2}{p^2}\right) - 20p^2\left(\dfrac{3}{5p}\right)}{20p^2\left(\dfrac{4}{p}\right) + 20p^2\left(\dfrac{1}{4p}\right)}$

$= \dfrac{40 - 12p}{80p + 5p}$

$= \dfrac{40 - 12p}{85p}$

27. $\dfrac{\dfrac{5}{x^2y} - \dfrac{2}{xy^2}}{\dfrac{3}{x^2y^2} + \dfrac{4}{xy}}$

$= \dfrac{x^2y^2\left(\dfrac{5}{x^2y} - \dfrac{2}{xy^2}\right)}{x^2y^2\left(\dfrac{3}{x^2y^2} + \dfrac{4}{xy}\right)}$ *Method 2;*
LCD = x^2y^2

$= \dfrac{x^2y^2\left(\dfrac{5}{x^2y}\right) - x^2y^2\left(\dfrac{2}{xy^2}\right)}{x^2y^2\left(\dfrac{3}{x^2y^2}\right) + x^2y^2\left(\dfrac{4}{xy}\right)}$

$= \dfrac{5y - 2x}{3 + 4xy}$

29. $\dfrac{\dfrac{1}{4} - \dfrac{1}{a^2}}{\dfrac{1}{2} + \dfrac{1}{a}}$

$= \dfrac{4a^2\left(\dfrac{1}{4} - \dfrac{1}{a^2}\right)}{4a^2\left(\dfrac{1}{2} + \dfrac{1}{a}\right)}$ *Method 2;*
LCD = $4a^2$

$= \dfrac{a^2 - 4}{2a^2 + 4a}$ *Distributive*
property

$= \dfrac{(a-2)(a+2)}{2a(a+2)}$ *Factor numerator*
and denominator

$= \dfrac{a-2}{2a}$ *Use fundamental*
property

31. $\dfrac{\dfrac{1}{z+5}}{\dfrac{4}{z^2-25}}$

$= \dfrac{1}{z+5} \div \dfrac{4}{z^2-25}$ *Method 1*

$= \dfrac{1}{z+5} \cdot \dfrac{z^2-25}{4}$ *Multiply by reciprocal*

$= \dfrac{1 \cdot (z^2-25)}{(z+5) \cdot 4}$ *Multiply*

$= \dfrac{(z+5)(z-5)}{(z+5) \cdot 4}$ *Factor numerator*

$= \dfrac{z-5}{4}$ *Use fundamental*
property

33. $\dfrac{\dfrac{1}{m+1} - 1}{\dfrac{1}{m+1} + 1}$

$= \dfrac{(m+1)\left(\dfrac{1}{m+1} - 1\right)}{(m+1)\left(\dfrac{1}{m+1} + 1\right)}$ *Method 2;*
LCD = m + 1

$= \dfrac{1 - 1(m+1)}{1 + 1(m+1)}$ *Distributive*
property

$= \dfrac{1 - m - 1}{1 + m + 1}$ *Distributive*
property

$= \dfrac{-m}{m+2}$

35.

$\dfrac{\dfrac{1}{m-1} + \dfrac{2}{m+2}}{\dfrac{2}{m+2} - \dfrac{1}{m-3}}$

$= \dfrac{(m-1)(m+2)(m-3)\left(\dfrac{1}{m-1} + \dfrac{2}{m+2}\right)}{(m-1)(m+2)(m-3)\left(\dfrac{2}{m+2} - \dfrac{1}{m-3}\right)}$

Method 2;
LCD = (m – 1)(m + 2)(m – 3)

$= \dfrac{(m+2)(m-3) + 2(m-1)(m-3)}{2(m-1)(m-3) - (m-1)(m+2)}$

Distributive property

$= \dfrac{(m-3)[(m+2) + 2(m-1)]}{(m-1)[2(m-3) - (m+2)]}$

Factor out m – 3 in numerator
and m – 1 in denominator

$= \dfrac{(m-3)[m + 2 + 2m - 2]}{(m-1)[2m - 6 - m - 2]}$

Distributive property

$= \dfrac{3m(m-3)}{(m-1)(m-8)}$ *Combine like terms*

37. In a fraction, the fraction bar represents division. For example, $\dfrac{3}{5}$ can be read "3 divided by 5."

39. "The sum of $\dfrac{3}{8}$ and $\dfrac{5}{6}$ divided by 2" is written

$$\dfrac{\dfrac{3}{8} + \dfrac{5}{6}}{2}.$$

40. $\dfrac{\frac{3}{8}+\frac{5}{6}}{2}=\dfrac{\frac{9}{24}+\frac{20}{24}}{2}$ *Method 1*

$=\dfrac{\frac{29}{24}}{\frac{2}{1}}=\dfrac{29}{24}\cdot\dfrac{1}{2}=\dfrac{29}{48}$

41. $\dfrac{\frac{3}{8}+\frac{5}{6}}{2}=\dfrac{24\left(\frac{3}{8}+\frac{5}{6}\right)}{24(2)}$ *Method 2;*
LCD = 24

$=\dfrac{24\left(\frac{3}{8}\right)+24\left(\frac{5}{6}\right)}{24(2)}$

$=\dfrac{9+20}{48}=\dfrac{29}{48}$

42. Method 2 is usually shorter for more complex problems because the problem can be worked without adding and subtracting rational expressions, which can be complicated and time-consuming.

43. $1+\dfrac{1}{1+\frac{1}{1+1}}=1+\dfrac{1}{1+\frac{1}{2}}$

$=1+\dfrac{1}{\frac{2}{2}+\frac{1}{2}}$

$=1+\dfrac{1}{\frac{3}{2}}$

$=1+1\cdot\dfrac{2}{3}$

$=1+\dfrac{2}{3}$

$=\dfrac{3}{3}+\dfrac{2}{3}=\dfrac{5}{3}$

45. $7-\dfrac{3}{5+\frac{2}{4-2}}=7-\dfrac{3}{5+\frac{2}{2}}$

$=7-\dfrac{3}{5+1}$

$=7-\dfrac{3}{6}$

$=7-\dfrac{1}{2}$

$=\dfrac{14}{2}-\dfrac{1}{2}=\dfrac{13}{2}$

47. $r+\dfrac{r}{4-\frac{2}{6+2}}=r+\dfrac{r}{4-\frac{2}{8}}$

$=r+\dfrac{r}{4-\frac{1}{4}}$

$=r+\dfrac{r}{\frac{16}{4}-\frac{1}{4}}$

$=r+\dfrac{r}{\frac{15}{4}}$

$=r+r\cdot\dfrac{4}{15}$

$=r+\dfrac{4r}{15}$

$=\dfrac{15r}{15}+\dfrac{4r}{15}$

$=\dfrac{19r}{15}$

Section 7.6

1. $\dfrac{7}{8}x+\dfrac{1}{5}x$ is the sum of two terms, so it is an *expression* to be simplified. Simplify by finding the LCD, writing each coefficient with this LCD, and combining like terms.

$\dfrac{7}{8}x+\dfrac{1}{5}x=\dfrac{35}{40}x+\dfrac{8}{40}x$ *LCD = 40*

$=\dfrac{43}{40}x$ *Combine like terms*

3. $\dfrac{7}{8}x+\dfrac{1}{5}x=1$ has an equals sign, so this is an *equation* to be solved. Use the multiplication property of equality to clear fractions. The LCD is 40.

$\dfrac{7}{8}x+\dfrac{1}{5}x=1$

$40\left(\dfrac{7}{8}x+\dfrac{1}{5}x\right)=40\cdot 1$ *Multiply by 40*

$40\left(\dfrac{7}{8}x\right)+40\left(\dfrac{1}{5}x\right)=40\cdot 1$ *Distributive property*

continued

$$35x + 8x = 40 \quad \textit{Multiply}$$

$$43x = 40 \quad \begin{array}{l}\textit{Combine}\\\textit{like terms}\end{array}$$

$$x = \frac{40}{43} \quad \textit{Divide by 43}$$

The solution set is $\left\{\dfrac{40}{43}\right\}$.

5. $\dfrac{3}{5}y - \dfrac{7}{10}y$ is the difference of two terms, so it is an *expression* to be simplified.

$$\frac{3}{5}y - \frac{7}{10}y = \frac{6}{10}y - \frac{7}{10}y \quad \textit{LCD = 10}$$

$$= -\frac{1}{10}y \quad \begin{array}{l}\textit{Combine}\\\textit{like terms}\end{array}$$

7. $\dfrac{3}{5}y - \dfrac{7}{10}y = 1$ has an equals sign, so it is an *equation* to be solved.

$$\frac{3}{5}y - \frac{7}{10}y = 1$$

$$10\left(\frac{3}{5}y - \frac{7}{10}y\right) = 10 \cdot 1 \quad \textit{LCD = 10}$$

$$10\left(\frac{3}{5}y\right) - 10\left(\frac{7}{10}y\right) = 10 \cdot 1 \quad \begin{array}{l}\textit{Distributive}\\\textit{property}\end{array}$$

$$6y - 7y = 10 \quad \textit{Multiply}$$

$$-y = 10 \quad \begin{array}{l}\textit{Combine}\\\textit{like terms}\end{array}$$

$$y = -10 \quad \textit{Divide by -1}$$

The solution set is $\{-10\}$.

9. $\dfrac{3}{x+2} - \dfrac{5}{x} = 1$

The denominators, $x + 2$ and x, are equal to 0 for the values -2 and 0.

11. $\dfrac{-1}{(x+3)(x-4)} = \dfrac{1}{2x+1}$

The denominators, $(x+3)(x-4)$ and $2x+1$, are equal to 0 for the values -3, 4, and $-\dfrac{1}{2}$.

13. $\dfrac{4}{x^2 + 8x - 9} + \dfrac{1}{x^2 - 4} = 0$

The denominators, $x^2 + 8x - 9 = (x+9)(x-1)$ and $x^2 - 4 = (x+2)(x-2)$, are equal to 0 for the values $-9, 1, -2$, and 2.

15. When solving equations, the LCD is used as a multiplier for every term in the equation. As a result, the fractions are removed from the equation.

When adding and subtracting rational expressions, the LCD is used to make it possible to combine several separate rational expressions into one

rational expression. This does not necessarily eliminate fractions.

Note: In Exercises 17–62, all proposed solutions should be checked by substituting in the original equation. It is essential to determine whether a proposed solution will make any denominator in the original equation equal to zero. Checks will be shown here for only a few of the exercises.

17. $\dfrac{5}{m} - \dfrac{3}{m} = 8$

Multiply each side by the LCD, m.

$$m\left(\frac{5}{m} - \frac{3}{m}\right) = m \cdot 8$$

Use the distributive property to remove parentheses; then solve.

$$m\left(\frac{5}{m}\right) - m\left(\frac{3}{m}\right) = 8m$$

$$5 - 3 = 8m$$

$$2 = 8m$$

$$m = \frac{2}{8} = \frac{1}{4}$$

Check this proposed solution by replacing m with $\dfrac{1}{4}$ in the original equation.

$$\frac{5}{\frac{1}{4}} - \frac{3}{\frac{1}{4}} = 8 ? \quad \textit{Let m = } \frac{1}{4}$$

$$5 \cdot 4 - 3 \cdot 4 = 8 ? \quad \begin{array}{l}\textit{Multiply by}\\\textit{reciprocals}\end{array}$$

$$20 - 12 = 8 ?$$

$$8 = 8 \quad \textit{True}$$

Thus, the solution set is $\left\{\dfrac{1}{4}\right\}$.

19. $\dfrac{5}{y} + 4 = \dfrac{2}{y}$

$$y\left(\frac{5}{y} + 4\right) = y\left(\frac{2}{y}\right) \quad \begin{array}{l}\textit{Multiply by}\\\textit{LCD, y}\end{array}$$

$$y\left(\frac{5}{y}\right) + y(4) = y\left(\frac{2}{y}\right) \quad \begin{array}{l}\textit{Distributive}\\\textit{property}\end{array}$$

$$5 + 4y = 2$$

$$4y = -3$$

$$y = -\frac{3}{4}$$

Check $y = -\dfrac{3}{4}$: $-\dfrac{8}{3} = -\dfrac{8}{3}$

Thus, the solution set is $\left\{-\dfrac{3}{4}\right\}$.

21.
$$\frac{3x}{5} - 6 = x$$

$$5\left(\frac{3x}{5} - 6\right) = 5(x) \quad \text{\textit{Multiply by LCD, 5}}$$

$$5\left(\frac{3x}{5}\right) - 5(6) = 5x \quad \text{\textit{Distributive property}}$$

$$3x - 30 = 5x$$
$$-30 = 2x$$
$$-15 = x$$

Check $x = -15 : -15 = -15$

Thus, the solution set is $\{-15\}$.

23.
$$\frac{4m}{7} + m = 11$$

$$7\left(\frac{4m}{7} + m\right) = 7(11) \quad \text{\textit{Multiply by LCD, 7}}$$

$$7\left(\frac{4m}{7}\right) + 7(m) = 77 \quad \text{\textit{Distributive property}}$$

$$4m + 7m = 77$$
$$11m = 77$$
$$m = 7$$

Check $m = 7 : 11 = 11$

Thus, the solution set is $\{7\}$.

25.
$$\frac{z-1}{4} = \frac{z+3}{3}$$

$$12\left(\frac{z-1}{4}\right) = 12\left(\frac{z+3}{3}\right) \quad \text{\textit{Multiply by LCD, 12}}$$

$$3(z-1) = 4(z+3)$$

$$3z - 3 = 4z + 12 \quad \text{\textit{Distributive property}}$$

$$-15 = z$$

Check $z = -15 : -4 = -4$

Thus, the solution set is $\{-15\}$.

27.
$$\frac{3p+6}{8} = \frac{3p-3}{16}$$

$$16\left(\frac{3p+6}{8}\right) = 16\left(\frac{3p-3}{16}\right) \quad \text{\textit{Multiply by LCD, 16}}$$

$$2(3p+6) = 3p-3$$

$$6p + 12 = 3p - 3 \quad \text{\textit{Distributive property}}$$

$$3p = -15$$
$$p = -5$$

Check $p = -5 : -\frac{9}{8} = -\frac{9}{8}$

Thus, the solution set is $\{-5\}$.

29.
$$\frac{2x+3}{x} = \frac{3}{2}$$

$$2x\left(\frac{2x+3}{x}\right) = 2x\left(\frac{3}{2}\right) \quad \text{\textit{Multiply by LCD, 2x}}$$

$$2(2x+3) = 3x$$

$$4x + 6 = 3x \quad \text{\textit{Distributive property}}$$

$$x = -6$$

Check $x = -6 : \dfrac{3}{2} = \dfrac{3}{2}$

Thus, the solution set is $\{-6\}$.

31.
$$\frac{k}{k-4} - 5 = \frac{4}{k-4}$$

$$(k-4)\left(\frac{k}{k-4} - 5\right) = (k-4)\left(\frac{4}{k-4}\right)$$
$$\text{\textit{Multiply by LCD, k−4}}$$

$$(k-4)\left(\frac{k}{k-4}\right) - 5(k-4) = 4 \quad \text{\textit{Distributive property}}$$

$$k - 5k + 20 = 4$$
$$-4k = -16$$
$$k = 4$$

The proposed solution is 4. However, 4 cannot be a solution because it makes the denominator $k - 4$ equal 0. Therefore, the solution set is $\emptyset$.

33.
$$\frac{q+2}{3} + \frac{q-5}{5} = \frac{7}{3}$$

$$15\left(\frac{q+2}{3} + \frac{q-5}{5}\right) = 15\left(\frac{7}{3}\right) \quad \text{\textit{Multiply by LCD, 15}}$$

$$15\left(\frac{q+2}{3}\right) + 15\left(\frac{q-5}{5}\right) = 5 \cdot 7$$

$$5(q+2) + 3(q-5) = 35$$
$$5q + 10 + 3q - 15 = 35$$
$$8q - 5 = 35$$
$$8q = 40$$
$$q = 5$$

Check $q = 5 : \dfrac{7}{3} = \dfrac{7}{3}$

Thus, the solution set is $\{5\}$.

35.
$$\frac{x}{2} = \frac{5}{4} + \frac{x-1}{4}$$

$$4\left(\frac{x}{2}\right) = 4\left(\frac{5}{4} + \frac{x-1}{4}\right) \quad \text{\textit{Multiply by LCD, 4}}$$

$$2(x) = 4\left(\frac{5}{4}\right) + 4\left(\frac{x-1}{4}\right)$$

$$2x = 5 + x - 1$$
$$x = 4$$

Check $x = 4 : 2 = 2$

Thus, the solution set is $\{4\}$.

37.
$$\frac{a+7}{8} - \frac{a-2}{3} = \frac{4}{3}$$

$$24\left(\frac{a+7}{8} - \frac{a-2}{3}\right) = 24\left(\frac{4}{3}\right)$$

Multiply by LCD, 24

$$24\left(\frac{a+7}{8}\right) - 24\left(\frac{a-2}{3}\right) = 8(4)$$

$$3(a+7) - 8(a-2) = 32$$
$$3a + 21 - 8a + 16 = 32$$
$$-5a + 37 = 32$$
$$-5a = -5$$
$$a = 1$$

Check $a = 1 : \dfrac{4}{3} = \dfrac{4}{3}$

Thus, the solution set is $\{1\}$.

39.
$$\frac{p}{2} - \frac{p-1}{4} = \frac{5}{4}$$

$$4\left(\frac{p}{2} - \frac{p-1}{4}\right) = 4\left(\frac{5}{4}\right)$$

Multiply by LCD, 4

$$4\left(\frac{p}{2}\right) - 4\left(\frac{p-1}{4}\right) = 5$$

$$2p - 1(p-1) = 5$$
$$2p - p + 1 = 5$$
$$p = 4$$

Check $p = 4 : \dfrac{5}{4} = \dfrac{5}{4}$

Thus, the solution set is $\{4\}$.

41.
$$\frac{3x}{5} - \frac{x-5}{7} = 3$$

$$35\left(\frac{3x}{5} - \frac{x-5}{7}\right) = 35(3)$$

Multiply by LCD, 35

$$35\left(\frac{3x}{5}\right) - 35\left(\frac{x-5}{7}\right) = 105$$

$$7(3x) - 5(x-5) = 105$$
$$21x - 5x + 25 = 105$$
$$16x = 80$$
$$x = 5$$

Check $x = 5 : 3 = 3$

Thus, the solution set is $\{5\}$.

43.
$$\frac{2}{m} = \frac{m}{5m+12}$$

$$m(5m+12)\left(\frac{2}{m}\right) = m(5m+12)\left(\frac{m}{5m+12}\right)$$

Multiply by LCD, m(5m + 12)

$$(5m+12)(2) = m(m)$$

$$10m + 24 = m^2$$
$$-m^2 + 10m + 24 = 0$$
$$m^2 - 10m - 24 = 0 \qquad \textit{Multiply by } -1$$
$$(m-12)(m+2) = 0$$

$$m - 12 = 0 \qquad \text{or} \qquad m + 2 = 0$$
$$m = 12 \qquad \text{or} \qquad m = -2$$

Check $m = 12 : \dfrac{1}{6} = \dfrac{1}{6}$

Check $m = -2 : -1 = -1$

Thus, the solution set is $\{-2, 12\}$.

45.
$$\frac{-2}{z+5} + \frac{3}{z-5} = \frac{20}{z^2 - 25}$$

$$\frac{-2}{z+5} + \frac{3}{z-5} = \frac{20}{(z+5)(z-5)}$$

$$(z+5)(z-5)\left(\frac{-2}{z+5} + \frac{3}{z-5}\right)$$
$$= (z+5)(z-5)$$
$$\cdot \left(\frac{20}{(z+5)(z-5)}\right)$$

Multiply by LCD, (z + 5)(z − 5)

$$(z+5)(z-5)\left(\frac{-2}{z+5}\right)$$
$$+ (z+5)(z-5)\left(\frac{3}{z-5}\right) = 20$$
$$-2(z-5) + 3(z+5) = 20$$
$$-2z + 10 + 3z + 15 = 20$$
$$z + 25 = 20$$
$$z = -5$$

The proposed solution, −5, cannot be a solution because it would make the denominators $z + 5$ and $z^2 - 25$ equal 0 and the corresponding fractions undefined. Since −5 cannot be a solution, the solution set is $\emptyset$.

47.
$$\frac{3}{x-1} + \frac{2}{4x-4} = \frac{7}{4}$$

$$\frac{3}{x-1} + \frac{2}{4(x-1)} = \frac{7}{4}$$

$$4(x-1)\left(\frac{3}{x-1} + \frac{2}{4(x-1)}\right) = 4(x-1)\left(\frac{7}{4}\right)$$

Multiply by LCD, 4(x − 1)

$$4(3) + 2 = (x-1)(7)$$
$$14 = 7x - 7$$
$$21 = 7x$$
$$3 = x$$

Check $x = 3 : \dfrac{7}{4} = \dfrac{7}{4}$

Thus, the solution set is $\{3\}$.

49.

$$\frac{y}{3y+3} = \frac{2y-3}{y+1} - \frac{2y}{3y+3}$$

$$\frac{y}{3(y+1)} = \frac{2y-3}{y+1} - \frac{2y}{3(y+1)}$$

$$3(y+1)\left(\frac{y}{3(y+1)}\right) =$$

$$3(y+1)\left[\frac{2y-3}{y+1} - \frac{2y}{3(y+1)}\right]$$

Multiply by
LCD, 3(y + 1)

$$y = 3(y+1)\left(\frac{2y-3}{y+1}\right)$$

$$- 3(y+1)\left(\frac{2y}{3(y+1)}\right)$$

$$y = 3(2y-3) - 2y$$

$$y = 6y - 9 - 2y$$

$$y = 4y - 9$$

$$-3y = -9$$

$$y = 3$$

Check

$$\frac{y}{3y+3} = \frac{2y-3}{y+1} - \frac{2y}{3y+3}$$

$$\frac{3}{3(3)+3} = \frac{2(3)-3}{3+1} - \frac{2(3)}{3(3)+3} \ ? \ \textit{Let y = 3}$$

$$\frac{3}{9+3} = \frac{6-3}{4} - \frac{6}{9+3} \ ?$$

$$\frac{3}{12} = \frac{3}{4} - \frac{6}{12} \ ?$$

$$\frac{1}{4} = \frac{3}{4} - \frac{2}{4} \ ?$$

$$\frac{1}{4} = \frac{1}{4} \qquad \textit{True}$$

Thus, the solution set is $\{3\}$.

51.

$$\frac{5x}{14x+3} = \frac{1}{x}$$

$$x(14x+3)\left(\frac{5x}{14x+3}\right) = x(14x+3)\left(\frac{1}{x}\right)$$

Multiply by
LCD, x(14x + 3)

$$x(5x) = (14x+3)(1)$$

$$5x^2 = 14x + 3$$

$$5x^2 - 14x - 3 = 0$$

$$(5x+1)(x-3) = 0$$

$$x = -\frac{1}{5} \quad \text{or} \quad x = 3$$

Check $x = -\frac{1}{5} : -5 = -5$

Check $x = 3 : \frac{1}{3} = \frac{1}{3}$

Thus, the solution set is $\left\{-\frac{1}{5}, 3\right\}$.

53.

$$\frac{2}{x-1} - \frac{2}{3} = \frac{-1}{x+1}$$

$$3(x-1)(x+1)\left(\frac{2}{x-1} - \frac{2}{3}\right)$$

$$= 3(x-1)(x+1)\left(\frac{-1}{x+1}\right)$$

Multiply by
LCD, 3(x – 1)(x + 1)

$$3(x-1)(x+1)\left(\frac{2}{x-1}\right)$$

$$- 3(x-1)(x+1)\left(\frac{2}{3}\right)$$

$$= 3(x-1)(x+1)\left(\frac{-1}{x+1}\right)$$

$$3(x+1)(2) - (x-1)(x+1)(2)$$

$$= 3(x-1)(-1)$$

$$6(x+1) - 2(x^2 - 1) = -3(x-1)$$

$$6x + 6 - 2x^2 + 2 = -3x + 3$$

$$-2x^2 + 9x + 5 = 0$$

$$2x^2 - 9x - 5 = 0$$

$$(2x+1)(x-5) = 0$$

$$x = -\frac{1}{2} \quad \text{or} \quad x = 5$$

Check $x = -\frac{1}{2} : -2 = -2$

Check $x = 5 : -\frac{1}{6} = -\frac{1}{6}$

Thus, the solution set is $\left\{-\frac{1}{2}, 5\right\}$.

55.

$$\frac{x}{2x+2} = \frac{-2x}{4x+4} + \frac{2x-3}{x+1}$$

$$\frac{x}{2(x+1)} = \frac{-2x}{4(x+1)} + \frac{2x-3}{x+1}$$

$$4(x+1)\left(\frac{x}{2(x+1)}\right) = 4(x+1)\left(\frac{-2x}{4(x+1)}\right)$$

$$+ 4(x+1)\left(\frac{2x-3}{x+1}\right)$$

Multiply by
LCD, 4(x + 1)

$$2(x) = -2x + 4(2x-3)$$

$$2x = -2x + 8x - 12$$

$$-4x = -12$$

$$x = 3$$

Check $x = 3 : \frac{3}{8} = \frac{3}{8}$

Thus, the solution set is $\{3\}$.

57.
$$\frac{8x+3}{x} = 3x$$

$$x\left(\frac{8x+3}{x}\right) = x(3x) \quad \textit{Multiply by LCD, x}$$

$$8x+3 = 3x^2$$

$$0 = 3x^2 - 8x - 3$$

$$0 = (3x+1)(x-3)$$

$$x = -\frac{1}{3} \quad \text{or} \quad x = 3$$

Check $x = -\frac{1}{3} : -1 = -1$

Check $x = 3 : 9 = 9$

Thus, the solution set is $\left\{-\frac{1}{3}, 3\right\}$.

59.
$$\frac{3y}{y^2+5y+6}$$

$$= \frac{5y}{y^2+2y-3} - \frac{2}{y^2+y-2}$$

$$\frac{3y}{(y+2)(y+3)}$$

$$= \frac{5y}{(y+3)(y-1)} - \frac{2}{(y-1)(y+2)}$$

$$(y+2)(y+3)(y-1) \cdot \left[\frac{3y}{(y+2)(y+3)}\right]$$

$$= (y+2)(y+3)(y-1) \cdot \left[\frac{5y}{(y+3)(y-1)}\right]$$

$$- (y+2)(y+3)(y-1) \cdot \left[\frac{2}{(y-1)(y+2)}\right]$$

$$\textit{Multiply by LCD, } (y+2)(y+3)(y-1)$$

$$3y(y-1) = 5y(y+2) - 2(y+3)$$

$$3y^2 - 3y = 5y^2 + 10y - 2y - 6$$

$$0 = 2y^2 + 11y - 6$$

$$0 = (2y-1)(y+6)$$

$$y = \frac{1}{2} \quad \text{or} \quad y = -6$$

Check $y = \frac{1}{2} : \frac{6}{35} = \frac{6}{35}$

Check $y = -6 : -\frac{3}{2} = -\frac{3}{2}$

Thus, the solution set is $\left\{-6, \frac{1}{2}\right\}$.

61.
$$\frac{x+4}{x^2-3x+2} - \frac{5}{x^2-4x+3}$$

$$= \frac{x-4}{x^2-5x+6}$$

$$\frac{x+4}{(x-2)(x-1)} - \frac{5}{(x-3)(x-1)}$$

$$= \frac{x-4}{(x-3)(x-2)}$$

$$(x-2)(x-1)(x-3)$$
$$\cdot \left[\frac{x+4}{(x-2)(x-1)} - \frac{5}{(x-3)(x-1)}\right]$$

$$= (x-2)(x-1)(x-3)\left[\frac{x-4}{(x-3)(x-2)}\right]$$

$$\textit{Multiply by LCD, } (x-2)(x-1)(x-3)$$

$$(x+4)(x-3) - 5(x-2) = (x-1)(x-4)$$

$$x^2 + x - 12 - 5x + 10 = x^2 - 5x + 4$$

$$-4x - 2 = -5x + 4$$

$$x = 6$$

Check $x = 6 : \frac{1}{6} = \frac{1}{6}$

Thus, the solution set is $\{6\}$.

63. $kr - mr = km$

If you are solving for k, put both terms with k on one side and the remaining term on the other side.

$$kr - km = mr$$

65. $m = \frac{kF}{a}$ for F

We need to isolate F on one side of the equation.

$$m \cdot a = \left(\frac{kF}{a}\right)(a) \quad \textit{Multiply by a}$$

$$ma = kF$$

$$\frac{ma}{k} = \frac{kF}{k} \qquad \textit{Divide by k}$$

$$\frac{ma}{k} = F$$

67. $m = \frac{kF}{a}$ for a

$$m \cdot a = \left(\frac{kF}{a}\right)(a) \quad \textit{Multiply by a}$$

$$ma = kF$$

$$\frac{ma}{m} = \frac{kF}{m} \qquad \textit{Divide by m}$$

$$a = \frac{kF}{m}$$

69. $I = \frac{E}{R+r}$ for R

We need to isolate R on one side of the equation.

$$I(R+r) = \left(\frac{E}{R+r}\right)(R+r) \quad \textit{Multiply by } R+r$$

$$IR + Ir = E \qquad \textit{Distributive property}$$

$$IR = E - Ir \qquad \textit{Subtract Ir}$$

$$R = \frac{E-Ir}{I} \text{ or } R = \frac{E}{I} - r \quad \textit{Divide by I}$$

71.
$$h = \frac{2A}{B+b} \quad \text{for } A$$

$$(B+b)h = (B+b) \cdot \frac{2A}{B+b}$$
Multiply by B + b

$$h(B+b) = 2A$$

$$\frac{h(B+b)}{2} = A \qquad \textit{Divide by 2}$$

73.
$$d = \frac{2S}{n(a+L)} \quad \text{for } a$$

We need to isolate a on one side of the equation.

$$d \cdot n(a+L) = \frac{2S}{n(a+L)} \cdot n(a+L)$$
Multiply by n(a + L)

$$nd(a+L) = 2S$$
$$and + ndL = 2S$$

$$and = 2S - ndL \quad \textit{Subtract ndL}$$

$$a = \frac{2S - ndL}{nd} \quad \textit{Divide by nd}$$

$$\text{or} \quad a = \frac{2S}{nd} - L$$

75.
$$\frac{1}{x} = \frac{1}{y} - \frac{1}{z} \quad \text{for } y$$

The LCD of all the fractions in the equation is xyz, so multiply both sides by xyz.

$$xyz\left(\frac{1}{x}\right) = xyz\left(\frac{1}{y} - \frac{1}{z}\right)$$

$$xyz\left(\frac{1}{x}\right) = xyz\left(\frac{1}{y}\right) - xyz\left(\frac{1}{z}\right)$$
Distributive property

$$yz = xz - xy$$

Since we are solving for y, get all terms with y on one side of the equation.

$$yz + xy = xz \quad \textit{Add xy}$$

Factor out the common factor y on the left.

$$y(z + x) = xz$$

Finally, divide both sides by the coefficient of y, which is $z + x$.

$$y = \frac{xz}{z+x}$$

77.
$$9x + \frac{3}{z} = \frac{5}{y} \quad \text{for } z$$

$$yz\left(9x + \frac{3}{z}\right) = yz\left(\frac{5}{y}\right) \quad \begin{array}{l}\textit{Multiply by}\\ \textit{LCD, yz}\end{array}$$

$$yz(9x) + yz\left(\frac{3}{z}\right) = yz\left(\frac{5}{y}\right) \quad \begin{array}{l}\textit{Distributive}\\ \textit{property}\end{array}$$

$$9xyz + 3y = 5z$$

$$9xyz - 5z = -3y \quad \begin{array}{l}\textit{Get the z terms}\\ \textit{on one side}\end{array}$$

$$z(9xy - 5) = -3y \quad \textit{Factor out z}$$

$$z = \frac{-3y}{9xy - 5} \quad \textit{Divide by 9xy} - 5$$

$$\text{or} \quad z = \frac{3y}{5 - 9xy}$$

79.
$$\frac{x^2}{x-3} + \frac{2x-15}{x-3} = 0$$

$$(x-3)\left(\frac{x^2}{x-3}\right) + (x-3)\left(\frac{2x-15}{x-3}\right)$$
$$= (x-3)(0)$$

$$x^2 + 2x - 15 = 0$$
$$(x+5)(x-3) = 0$$
$$x = -5 \quad \text{or} \quad x = 3$$

A check will verify that -5 is a solution, but 3 must be rejected because it causes a denominator to become zero. The solution set is $\{-5\}$.

80.
$$\frac{x^2}{x-3} + \frac{2x-15}{x-3} = \frac{x^2 + (2x-15)}{x-3}$$
$$= \frac{x^2 + 2x - 15}{x-3}$$
$$= \frac{(x+5)(x-3)}{x-3}$$
$$= x + 5$$

(a) $x + 5 = 0$
$$x = -5$$

Since 3 is rejected, this solution is the same as the actual solution in Exercise 79.

(b) $x - 3 = 0$
$$x = 3$$

This solution and the rejected solution for Exercise 79 are the same.

81. The solution is -3. The number 1 must be rejected. Refer to Example 5 in the textbook for a complete solution.

82.

$$\frac{1}{x-1} + \frac{1}{2} - \frac{2}{x^2-1}$$

$$\frac{1}{x-1} + \frac{1}{2} - \frac{2}{(x+1)(x-1)}$$

$LCD = 2(x-1)(x+1)$

$$\frac{2(x+1)}{2(x-1)(x+1)} + \frac{(x-1)(x+1)}{2(x-1)(x+1)}$$

$$- \frac{2 \cdot 2}{2(x+1)(x-1)}$$

$$= \frac{2(x+1) + (x-1)(x+1) - 2 \cdot 2}{2(x+1)(x-1)}$$

$$= \frac{2x+2 + (x^2-1) - 4}{2(x+1)(x-1)}$$

$$= \frac{x^2 + 2x - 3}{2(x+1)(x-1)}$$

$$= \frac{(x+3)(x-1)}{2(x+1)(x-1)}$$

$$= \frac{x+3}{2(x+1)}$$

(a)

$$\frac{x+3}{2(x+1)} = 0$$

$$2(x+1)\left[\frac{x+3}{2(x+1)}\right] = 2(x+1) \cdot 0$$

$$x + 3 = 0$$

$$x = -3$$

This is the same as the actual solution.

(b) In the first part of the problem, $x - 1$ was the common factor.

$$x - 1 = 0$$

$$x = 1$$

This is the same as the rejected solution.

83. If an equation involving rational expressions is solved by using the LCD as a multiplier, each solution must be checked to make sure that one or more of the potential solutions does not make the value of a denominator zero.

If an equation involving rational expressions is solved by moving all terms to one side, combining over a common denominator, simplifying and reducing to lowest terms, only the actual solution will emerge. The problem solver must choose the method which yields the solution(s) most efficiently.

84. Move all terms to the left side of the equation, factor the denominators and find the LCD. Combine all fractions over the LCD, simplify, factor, and reduce to lowest terms. Set the expression equal to zero and solve for the variable. This procedure will give only the actual solution(s) and not any values that must be rejected.

85. If $x = 5$, then

$$\frac{-3}{x+5} = \frac{-3}{5+5} = -\frac{3}{10}.$$

87. If $x = 12$, then $\dfrac{-3}{12-x}$ has a zero denominator. An error message would occur.

89. If $x = 3$, then $\dfrac{1}{(x+4)(x-3)}$ has a zero denominator. An error message would occur.

Summary: Exercises on Operations and Equations with Rational Expressions

1. No equals sign appears so this is an *operation*.

$$\frac{4}{p} + \frac{6}{p} = \frac{4+6}{p} = \frac{10}{p}$$

3. No equals sign appears so this is an *operation*.

$$\frac{1}{x^2+x-2} \div \frac{4x^2}{2x-2}$$

$$= \frac{1}{x^2+x-2} \cdot \frac{2x-2}{4x^2}$$

$$= \frac{1}{(x+2)(x-1)} \cdot \frac{2(x-1)}{2 \cdot 2x^2}$$

$$= \frac{1}{2x^2(x+2)}$$

5. No equals sign appears so this is an *operation*.

$$\frac{2y^2+y-6}{2y^2-9y+9} \cdot \frac{y^2-2y-3}{y^2-1}$$

$$= \frac{(2y-3)(y+2)(y-3)(y+1)}{(2y-3)(y-3)(y+1)(y-1)}$$

$$= \frac{y+2}{y-1}$$

7.

$$\frac{x-4}{5} = \frac{x+3}{6}$$

There is an equals sign, so this is an *equation*.

$$30\left(\frac{x-4}{5}\right) = 30\left(\frac{x+3}{6}\right) \quad \textit{Multiply by LCD, 30}$$

$$6(x-4) = 5(x+3)$$

$$6x - 24 = 5x + 15$$

$$x = 39$$

Check $x = 39$: $7 = 7$

Thus, the solution set is $\{39\}$.

9. No equals sign appears so this is an *operation.*

$$\frac{4}{p+2} + \frac{1}{3p+6} = \frac{4}{p+2} + \frac{1}{3(p+2)}$$

$$= \frac{3 \cdot 4}{3(p+2)} + \frac{1}{3(p+2)} \quad LCD = 3(p+2)$$

$$= \frac{12+1}{3(p+2)}$$

$$= \frac{13}{3(p+2)}$$

11. $\dfrac{3}{t-1} + \dfrac{1}{t} = \dfrac{7}{2}$

There is an equals sign, so this is an *equation.*

$$2t(t-1)\left(\frac{3}{t-1} + \frac{1}{t}\right) = 2t(t-1)\left(\frac{7}{2}\right)$$

Multiply by
LCD, 2t(t − 1)

$$2t(t-1)\left(\frac{3}{t-1}\right) + 2t(t-1)\left(\frac{1}{t}\right) = 7t(t-1)$$

$$2t(3) + 2(t-1) = 7t(t-1)$$

$$6t + 2t - 2 = 7t^2 - 7t$$

$$0 = 7t^2 - 15t + 2$$

$$0 = (7t - 1)(t - 2)$$

$$t = \frac{1}{7} \quad \text{or} \quad y = 2$$

Check $t = \dfrac{1}{7}$: $\dfrac{7}{2} = \dfrac{7}{2}$

Check $t = 2$: $\dfrac{7}{2} = \dfrac{7}{2}$

Thus, the solution set is $\left\{\dfrac{1}{7}, 2\right\}$.

13. No equals sign appears so this is an *operation.*

$$\frac{5}{4z} - \frac{2}{3z} = \frac{3 \cdot 5}{3 \cdot 4z} - \frac{4 \cdot 2}{4 \cdot 3z} \quad LCD = 12z$$

$$= \frac{15}{12z} - \frac{8}{12z}$$

$$= \frac{15-8}{12z} = \frac{7}{12z}$$

15. No equals sign appears so this is an *operation.*

$$\frac{1}{m^2 + 5m + 6} + \frac{2}{m^2 + 4m + 3}$$

$$= \frac{1}{(m+2)(m+3)} + \frac{2}{(m+1)(m+3)}$$

$$= \frac{1(m+1)}{(m+1)(m+2)(m+3)}$$

$$+ \frac{2(m+2)}{(m+1)(m+2)(m+3)}$$

$$LCD = (m+1)(m+2)(m+3)$$

$$= \frac{(m+1) + (2m+4)}{(m+1)(m+2)(m+3)}$$

$$= \frac{3m+5}{(m+1)(m+2)(m+3)}$$

17. $\dfrac{2}{x+1} + \dfrac{5}{x-1} = \dfrac{10}{x^2-1}$

There is an equals sign, so this is an *equation.*

$$\frac{2}{x+1} + \frac{5}{x-1} = \frac{10}{(x+1)(x-1)}$$

$$(x+1)(x-1)\left(\frac{2}{x+1} + \frac{5}{x-1}\right)$$

$$= (x+1)(x-1)\left[\frac{10}{(x+1)(x-1)}\right]$$

Multiply by
LCD, (x + 1)(x − 1)

$$(x+1)(x-1)\left(\frac{2}{x+1}\right)$$

$$+ (x+1)(x-1)\left(\frac{5}{x-1}\right) = 10$$

Distributive
property

$$2(x-1) + 5(x+1) = 10$$

$$2x - 2 + 5x + 5 = 10$$

$$3 + 7x = 10$$

$$7x = 7$$

$$x = 1$$

Replacing x by 1 in the original equation makes the denominators $x - 1$ and $x^2 - 1$ equal to 0, so the solution set is $\emptyset$.

19. No equals sign appears so this is an *operation.*

$$\frac{4t^2 - t}{6t^2 + 10t} \div \frac{8t^2 + 2t - 1}{3t^2 + 11t + 10}$$

$$\frac{4t^2 - t}{6t^2 + 10t} \cdot \frac{3t^2 + 11t + 10}{8t^2 + 2t - 1} \quad \begin{array}{l}\textit{Multiply by} \\ \textit{reciprocal}\end{array}$$

$$= \frac{t(4t-1)}{2t(3t+5)} \cdot \frac{(3t+5)(t+2)}{(4t-1)(2t+1)}$$

Factor numerators
and denominators

$$= \frac{t+2}{2(2t+1)}$$

Section 7.7

1. **(a)** Let $x =$ <u>an amount</u>.

(b) An expression for "the numerator of the fraction $\dfrac{5}{6}$ is increased by an amount" is $\underline{5+x}$.

We could also use $\dfrac{5+x}{6}$.

(c) An equation that can be used to solve the problem is

$$\frac{5+x}{6} = \frac{13}{3}.$$

3. *Step 1* Let $x =$ the numerator of the original fraction.

Step 2 The $x + 6 =$ the denominator of the original fraction.

Step 3 $\dfrac{x+3}{(x+6)+3} = \dfrac{5}{7}$

Step 4 Since we have a fraction equal to another fraction, we can use cross multiplication.

$$7(x+3) = 5[(x+6)+3]$$
$$7x + 21 = 5x + 45$$
$$2x = 24$$
$$x = 12$$

Step 5 The original fraction is

$$\frac{x}{x+6} = \frac{12}{12+6} = \frac{12}{18}.$$

Step 6 Adding 3 to both the numerator and the denominator gives us

$$\frac{12+3}{18+3} = \frac{15}{21},$$

which is equivalent to $\dfrac{5}{7}$.

5. *Step 1* Let $x =$ the quantity.

Step 2 Then its $\dfrac{2}{3}$, its $\dfrac{1}{2}$, and its $\dfrac{1}{7}$ are

$$\frac{2}{3}x, \frac{1}{2}x, \text{ and } \frac{1}{7}x.$$

Step 3 $x + \dfrac{2}{3}x + \dfrac{1}{2}x + \dfrac{1}{7}x = 33$

Step 4 Multiply both sides by the LCD of 3, 2, and 7, which is 42.

$$42\left(x + \frac{2}{3}x + \frac{1}{2}x + \frac{1}{7}x\right) = 42(33)$$

$$42x + 42\left(\frac{2}{3}x\right) + 42\left(\frac{1}{2}x\right) + 42\left(\frac{1}{7}x\right) = 42(33)$$

$$42x + 28x + 21x + 6x = 1386$$
$$97x = 1386$$
$$x = \frac{1386}{97}$$

Step 5 The quantity is $\dfrac{1386}{97}$.

(Note that this fraction is already in lowest terms since 97 is a prime number and is not a factor of 1386.)

Step 6 Check $\dfrac{1386}{97}$ in the original problem.

$$x = \frac{1386}{97}, \frac{2}{3}x = \frac{924}{97},$$
$$\frac{1}{2}x = \frac{693}{97}, \frac{1}{7}x = \frac{198}{97}$$

Adding gives us

$$\frac{1386 + 924 + 693 + 198}{97}$$

$$= \frac{3201}{97} = 33, \text{ as desired.}$$

7. *Step 1* Let $x =$ the 1989 Medicare enrollment.

Step 2 Then $x + 624 =$ the 1990 Medicare enrollment.

Step 3 $\dfrac{x+624}{x} = \dfrac{877}{861}$

Step 4 Cross multiply to solve.

$$861(x+624) = 877(x)$$
$$861x + 537,264 = 877x$$
$$537,264 = 16x$$
$$33,579 = x$$

Step 5 The 1989 enrollment is $33,579$ (thousand) and the 1990 enrollment is $33,579 + 624 = 34,203$ (thousand).

Step 6 Check the ratio.

$$\frac{34,203}{33,579} = \frac{877 \cdot 39}{861 \cdot 39} = \frac{877}{861}$$

9. *Step 1* Let $x =$ the number of male physicians in the United States in 1995.

Step 2 Then $\dfrac{1}{4}x =$ the number of female physicians in the United States in 1995.

Step 3 $x + \dfrac{1}{4}x = 720,000$

Step 4 $1x + \dfrac{1}{4}x = 720,000$

$$\frac{5}{4}x = 720,000$$

$$x = \frac{4}{5}(720,000)$$

$$= 576,000$$

Step 5 In 1995 in the United States, there were $576,000$ male physicians and $\dfrac{1}{4}(576,000) = 144,000$ female physicians.

Step 6   The total number of physicians was $576,000 + 144,000 = 720,000$, as desired.

11. We are asked to find the average *rate*, so we'll use the distance, rate, and time relationship

$$r = \frac{d}{t}.$$

$$r = \frac{50 \text{ kilometers}}{2.07 \text{ hours}}$$

$$\approx 24.15 \text{ kilometers per hour}$$

13. We are asked to find the *time*, so we'll use the distance, rate, and time relationship

$$t = \frac{d}{r}.$$

$$t = \frac{500 \text{ miles}}{145.827 \text{ miles per hour}}$$

$$\approx 3.429 \text{ hours}$$

15. $r = \dfrac{d}{t}$

$$= \frac{400 \text{ meters}}{50.60 \text{ seconds}}$$

$$\approx 7.91 \text{ meters per second}$$

17. Let $x =$ the average speed of the plane in still air.

Then the speed against the wind is $x - 10$ and the speed with the wind is $x + 10$. The time flying against the wind is

$$t = \frac{d}{r} = \frac{500}{x - 10},$$

and the time flying with the wind is

$$t = \frac{d}{r} = \frac{600}{x + 10}.$$

Now complete the chart.

	d	r	t
Against the wind	500	$x - 10$	$\dfrac{500}{x - 10}$
With the wind	600	$x + 10$	$\dfrac{600}{x + 10}$

Since the problem states that the two times are equal, we have

$$\frac{500}{x - 10} = \frac{600}{x + 10}.$$

We would use this equation to solve the problem.

19. Use $\text{time} = \dfrac{\text{distance}}{\text{rate}}$ since we know that the times for Stephanie and Wally are the same.

continued

$$\text{time}_{\text{Stephanie}} = \text{time}_{\text{Wally}}$$

$$\frac{D}{R} = \frac{d}{r}$$

21. Let x represent the speed of the boat in still water. We fill in the chart as follows, realizing that the time column is filled in by using the formula $t = \dfrac{d}{r}.$

	d	r	t
Against the current	20	$x - 4$	$\dfrac{20}{x - 4}$
With the current	60	$x + 4$	$\dfrac{60}{x + 4}$

Since the times are equal, we get the following equation.

$$\frac{20}{x - 4} = \frac{60}{x + 4}$$

$$(x + 4)(x - 4)\frac{20}{x - 4} = (x + 4)(x - 4)\frac{60}{x + 4}$$

Multiply by LCD, $(x + 4)(x - 4)$

$$20(x + 4) = 60(x - 4)$$
$$20x + 80 = 60x - 240$$
$$320 = 40x$$
$$8 = x$$

The speed of the boat in still water is 8 miles per hour.

23. Let $x =$ the average speed of the ferry.

Use the formula $t = \dfrac{d}{r}$ to make a chart.

	d	r	t
Seattle-Victoria	148	x	$\dfrac{148}{x}$
Victoria-Vancouver	74	x	$\dfrac{74}{x}$

Since the time for the Victoria-Vancouver trip is 4 hours less than the time for the Seattle-Victoria trip, solve the equation

$$\frac{74}{x} = \frac{148}{x} - 4.$$

$$x\left(\frac{74}{x}\right) = x\left(\frac{148}{x} - 4\right) \quad \textit{Multiply by LCD, } x$$

$$74 = 148 - 4x$$
$$4x = 74$$
$$x = \frac{74}{4} = \frac{37}{2} \text{ or } 18\frac{1}{2}$$

The average speed of the ferry is $\dfrac{37}{2}$ or $18\dfrac{1}{2}$ miles per hour.

25.   Let $x =$ N'Deti's speed.

Then $.73x =$ McDermott's speed.

Use $t = \dfrac{d}{r}$ to complete the following table.

	d	r	t
N'Deti	26	x	$\dfrac{26}{x}$
McDermott	26	$.73x$	$\dfrac{26}{.73x}$

Since N'Deti ran the marathon in .8 of an hour less time than McDermott, we have

$$\frac{26}{x} = \frac{26}{.73x} - .8\,.$$

To solve this equation, multiply both sides of the equation by the LCD, $.73x$.

$$(.73x)\left(\frac{26}{x}\right) = (.73x)\left(\frac{26}{.73x} - .8\right)$$
$$18.98 = 26 - .584x$$
$$.584x = 7.02$$
$$x = \frac{7.02}{.584} \approx 12.02$$
$$.73x \approx 8.78$$

N'Deti's speed was about 12.02 miles per hour; McDermott's speed was about 8.78 miles per hour.

27.   If it takes Elayn 10 hours to do a job, her rate is

$$\frac{1}{10} \text{ job per hour.}$$

29.   Let $x =$ the number of hours it will take Jorge and Caterina to paint the room working together.

	Rate	Time working together	Fractional part of job done when working together
Jorge	$\dfrac{1}{8}$	x	$\dfrac{1}{8}x$
Caterina	$\dfrac{1}{6}$	x	$\dfrac{1}{6}x$

part done by Jorge	$+$	part done by Caterina	$=$	1 whole job
$\downarrow$	$\downarrow$	$\downarrow$	$\downarrow$	$\downarrow$
$\dfrac{1}{8}x$	$+$	$\dfrac{1}{6}x$	$=$	1

An equation that can be used to solve this problem is

$$\frac{1}{8}x + \frac{1}{6}x = 1.$$

Alternatively, we can compare the hourly rates of completion. In one hour, Jorge will complete $\dfrac{1}{8}$ of the job, Caterina will complete $\dfrac{1}{6}$ of the job, and together they will complete $\dfrac{1}{x}$ of the job. So another equation that can be used to solve this problem is

$$\frac{1}{8} + \frac{1}{6} = \frac{1}{x}.$$

31.   Let x represent the number of hours it will take for Geraldo and Luisa to do a day's laundry working together. Since Geraldo can do the laundry in 8 hours, his rate alone is $\dfrac{1}{8}$ job per hour. Also, since Luisa can do the job alone in 9 hours, her rate is $\dfrac{1}{9}$ job per hour.

	Rate	Time working together	Fractional part of the job done when working together
Geraldo	$\dfrac{1}{8}$	x	$\dfrac{1}{8}x$
Luisa	$\dfrac{1}{9}$	x	$\dfrac{1}{9}x$

Since together Geraldo and Luisa complete 1 whole job, we must add their individual fractional parts and set the sum equal to 1.

$$\frac{1}{8}x + \frac{1}{9}x = 1$$
$$72\left(\frac{1}{8}x\right) + 72\left(\frac{1}{9}x\right) = 72(1) \quad \textit{Multiply by LCD, 72}$$
$$9x + 8x = 72$$
$$17x = 72$$
$$x = \frac{72}{17} \text{ or } 4\frac{4}{17}$$

It will take Geraldo and Luisa $\dfrac{72}{17}$ or $4\dfrac{4}{17}$ hours to do a day's laundry if they work together.

33.   Let $x =$ the number of hours to pump the water using both pumps.

	Rate	Time working together	Fractional part of the job done when working together
Pump 1	$\dfrac{1}{10}$	x	$\dfrac{1}{10}x$
Pump 2	$\dfrac{1}{12}$	x	$\dfrac{1}{12}x$

Since together the two pumps complete 1 whole job, we must add their individual fractional parts and set the sum equal to 1.

$$\frac{1}{10}x + \frac{1}{12}x = 1$$

$$60\left(\frac{1}{10}x + \frac{1}{12}x\right) = 60(1) \quad \textit{Multiply by LCD, 60}$$

$$60\left(\frac{1}{10}x\right) + 60\left(\frac{1}{12}x\right) = 60$$

$$6x + 5x = 60$$

$$11x = 60$$

$$x = \frac{60}{11} \text{ or } 5\frac{5}{11}$$

It would take $\frac{60}{11}$ or $5\frac{5}{11}$ hours to pump out the basement if both pumps were used.

35. Let x represent the number of hours it will take the experienced employee to enter the data. Then $2x$ represents the number of hours it will take the new employee (the experienced employee takes less time). The experienced employee's rate is $\frac{1}{x}$ job per hour and the new employee's rate is $\frac{1}{2x}$ job per hour.

	Rate	Time working together	Fractional part of the job done when working together
Experienced employee	$\frac{1}{x}$	2	$\frac{1}{x} \cdot 2 = \frac{2}{x}$
New employee	$\frac{1}{2x}$	2	$\frac{1}{2x} \cdot 2 = \frac{1}{x}$

Since together the two employees complete the whole job, we must add their individual fractional parts and set the sum equal to 1.

$$\frac{2}{x} + \frac{1}{x} = 1$$

$$x\left(\frac{2}{x} + \frac{1}{x}\right) = x(1) \quad \textit{Multiply by LCD, x}$$

$$x\left(\frac{2}{x}\right) + x\left(\frac{1}{x}\right) = x$$

$$2 + 1 = x$$

$$3 = x$$

Working alone, it will take the experienced employee 3 hours to enter the data.

37. Let x = the number of hours to fill the pool $\frac{3}{4}$ full with both pipes working together.

	Rate	Time working together	Fractional part of the job done when working together
First pipe	$\frac{1}{6}$	x	$\frac{1}{6}x$
Second pipe	$\frac{1}{9}$	x	$\frac{1}{9}x$

$$\begin{array}{ccccc}
\text{Part done} & + & \text{Part done by} & = & \frac{3}{4}\text{ full} \\
\text{by first pipe} & & \text{second pipe} & & \\
\downarrow & \downarrow & \downarrow & \downarrow & \downarrow \\
\frac{1}{6}x & + & \frac{1}{9}x & = & \frac{3}{4}
\end{array}$$

$$36\left(\frac{1}{6}x + \frac{1}{9}x\right) = 36\left(\frac{3}{4}\right) \quad \textit{Multiply by LCD, 36}$$

$$36\left(\frac{1}{6}x\right) + 36\left(\frac{1}{9}x\right) = 36\left(\frac{3}{4}\right)$$

$$6x + 4x = 27$$

$$10x = 27$$

$$x = \frac{27}{10} \text{ or } 2\frac{7}{10}$$

It takes $\frac{27}{10}$ or $2\frac{7}{10}$ hours to fill the pool $\frac{3}{4}$ full using both pipes.

Alternatively, we could solve $\frac{1}{6}x + \frac{1}{9}x + = 1$ (filling the whole pool) and then multiply that answer by $\frac{3}{4}$.

39. Let x = the number of minutes it takes to fill the sink.

In 1 minute, the cold water faucet (alone) can fill $\frac{1}{12}$ of the sink. In the same time, the hot water faucet (alone) can fill $\frac{1}{15}$ of the sink. In 1 minute, the drain (alone) empties $\frac{1}{25}$ of the sink. Together, they fill $\frac{1}{x}$ of the sink in one minute, so solve the equation

$$\frac{1}{12} + \frac{1}{15} - \frac{1}{25} = \frac{1}{x}.$$

$$300x\left(\frac{1}{12} + \frac{1}{15} - \frac{1}{25}\right) = 300x\left(\frac{1}{x}\right)$$

Multiply by LCD, 300x

continued

$$25x + 20x - 12x = 300$$
$$33x = 300$$
$$x = \frac{300}{33} = \frac{100}{11} \text{ or } 9\frac{1}{11}$$

It will take $\frac{100}{11}$ or $9\frac{1}{11}$ minutes to fill the sink.

41. As the number of different lottery tickets you buy *increases*, the probability of winning that lottery *increases*. Thus, the variation between the quantities is *direct*.

43. As the amount of pressure put on the accelerator of a car *increases*, the speed of the car *increases*. Thus, the variation between the quantities is *direct*.

45. If the diameter of a balloon *increases*, then the surface area of the balloon *increases*. Thus, the variation between the quantities is *direct*.

47. As the number of days until the end of the baseball season *decreases*, the number of home runs that Sammy Sosa has *increases*. Thus, the variation between the quantities is *inverse*.

49. Since x varies directly as y, there is a constant k such that $x = ky$. First find the value of k.

$$27 = k(6) \quad Let\ x = 27,\ y = 6$$
$$k = \frac{27}{6} = \frac{9}{2}$$

When $k = \frac{9}{2}$, $x = ky$ becomes

$$x = \frac{9}{2}y.$$

Now find x when $y = 2$.

$$x = \frac{9}{2}(2) \quad Let\ y = 2$$
$$= 9$$

51. Since m varies inversely as p^2, there is a constant k such that $m = \frac{k}{p^2}$. First find the value of k.

$$20 = \frac{k}{2^2} \quad Let\ m = 20,\ p = 2$$
$$20 = \frac{k}{4}$$
$$80 = k$$

When $k = 80$, $m = \frac{k}{p^2}$ becomes

$$m = \frac{80}{p^2}.$$

Now find m when $p = 5$.

$$m = \frac{80}{5^2} \quad Let\ p = 5$$
$$m = \frac{80}{25} = \frac{16}{5} \text{ or } 3\frac{1}{5}$$

53. Since p varies inversely as q^2, there is a constant k such that $p = \frac{k}{q^2}$. First find the value of k.

$$4 = \frac{k}{\left(\frac{1}{2}\right)^2} \quad Let\ p = 4,\ q = \frac{1}{2}$$
$$4 = \frac{k}{\frac{1}{4}}$$
$$\frac{1}{4} \cdot 4 = k \qquad Multiply\ by\ \frac{1}{4}$$
$$1 = k$$

When $k = 1$, $p = \frac{k}{q^2}$ becomes

$$p = \frac{1}{q^2}.$$

Now find p when $q = \frac{3}{2}$.

$$p = \frac{1}{\left(\frac{3}{2}\right)^2} \quad Let\ q = \frac{3}{2}$$
$$p = \frac{1}{\frac{9}{4}} = \frac{4}{9}$$

55. If the constant of variation is positive and y varies directly as x, then as x increases, y *increases*.

57. The speed s varies inversely with time t, so there is a constant k such that $s = \frac{k}{t}$.

Find the value of k.

$$160 = \frac{k}{\frac{1}{2}} \quad Let\ s = 160,\ t = \frac{1}{2}$$
$$k = \frac{1}{2} \cdot 160 = 80$$

When $k = 80$, $s = \frac{k}{t}$ becomes

$$s = \frac{80}{t}.$$

Now find s when $t = \frac{3}{4}$.

$$s = \frac{80}{\frac{3}{4}} = 80 \cdot \frac{4}{3} = \frac{320}{3} \text{ or } 106\frac{2}{3}$$

A speed of $106\frac{2}{3}$ miles per hour is needed to go the same distance in three-fourths of a minute.

59. The rate of change r of the amount of raw sugar varies directly as the amount a of raw sugar remaining, so there is a constant k such that $r = ka$. Find the value of k.

$$200 = k(800) \quad \text{Let } r = 200, a = 800$$
$$k = \frac{200}{800} = \frac{1}{4}$$

When $k = \frac{1}{4}$, $r = ka$ becomes

$$r = \frac{1}{4}a.$$

Now find r when $a = 100$.

$$r = \frac{1}{4}(100) = 25$$

When only 100 kilograms of raw sugar are left, the rate of change is 25 kilograms per hour.

61. If the temperature is constant, the pressure P of a gas in a container varies inversely as the volume of the container, so

$$P = \frac{k}{V}$$
$$10 = \frac{k}{3} \quad \text{Let } P = 10, k = 3$$
$$k = 3 \cdot 10 = 30$$

So $P = \frac{30}{V}$ and when $V = 1.5$,

$$P = \frac{30}{1.5} = 20.$$

The pressure is 20 pounds per square foot.

63. For a constant area, the length L of a rectangle varies inversely as the width W, so

$$L = \frac{k}{W}.$$
$$27 = \frac{k}{10} \quad \text{Let } L = 27, W = 10$$
$$k = 27 \cdot 10 = 270$$

So $L = \frac{270}{W}$ and when $L = 18$,

$$18 = \frac{270}{W}$$
$$18W = 270$$
$$W = \frac{270}{18} = 15.$$

When the length is 18 feet, the width is 15 feet.

65. The distance d that a body falls varies directly as the square of the time t that it falls, so

$$d = kt^2.$$
$$400 = k(5^2) \quad \text{Let } d = 400, t = 5$$
$$k = \frac{400}{25} = 16$$

So $d = 16t^2$ and when $t = 3$,

$$d = 16 \cdot 3^2 = 144.$$

The body fell 144 feet in the first 3 seconds.

67. As x increases, y increases, so the variation is *direct*.

69. As x increases, y decreases, so the variation is *inverse*.

71. Let x represent the number of transplants in 1995. Then $3080 - x$ represents the number of transplants in 1985. "The ratio of the number in 1995 to the number in 1985 was approximately 33 to 10" can be written as

$$\frac{x}{3080 - x} = \frac{33}{10}.$$

Multiplying by the LCD (or cross multiplying) gives us

$$10x = 33(3080 - x)$$
$$10x = 101,640 - 33x$$
$$43x = 101,640$$
$$x = \frac{101,640}{43} \approx 2364$$
$$3080 - x = 716$$

There were approximately 2364 heart transplants in 1995 and 716 in 1985.

73. Answers will vary. Here is one possibility: I start with the fraction $\frac{11}{8}$. If I add -2 to both the numerator and denominator, I get $\frac{9}{6}$, which simplifies to $\frac{3}{2}$. The problem is stated as follows: "If a number is added to both the numerator and the denominator of $\frac{11}{8}$, the resulting fraction is equal to $\frac{3}{2}$. What is the number?" To solve, let $x =$ the number. The equation is $\frac{11 + x}{8 + x} = \frac{3}{2}$. This leads to $2(11 + x) = 3(8 + x)$, which leads to $22 + 2x = 24 + 3x$, or $-2 = x$. The number is -2.

Chapter 7 Review Exercises

1. $\dfrac{4}{x - 3}$

To find the values for which this expression is undefined, set the denominator equal to zero and solve for x.

$$x - 3 = 0$$
$$x = 3$$

Because $x = 3$ will make the denominator zero, the given expression is undefined for 3.

2. $\dfrac{y+3}{2y}$

   Set the denominator equal to zero and solve for y.

   $$2y = 0$$
   $$y = 0$$

   The given expression is undefined for 0.

3. $\dfrac{2k+1}{3k^2 + 17k + 10}$

   Set the denominator equal to zero and solve for k.

   $$3k^2 + 17k + 10 = 0$$
   $$(3k+2)(k+5) = 0$$
   $$k = -\frac{2}{3} \text{ or } k = -5$$

   The given expression is undefined for -5 and $-\dfrac{2}{3}$.

4. Set the denominator equal to 0 and solve the equation. Any solutions are values for which the rational expression is undefined.

5. (a) $\dfrac{4x-3}{5x+2} = \dfrac{4(-2)-3}{5(-2)+2}$ *Let x = -2*

   $$= \dfrac{-8-3}{-10+2} = \dfrac{-11}{-8} = \dfrac{11}{8}$$

   (b) $\dfrac{4x-3}{5x+2} = \dfrac{4(4)-3}{5(4)+2}$ *Let x = 4*

   $$= \dfrac{16-3}{20+2} = \dfrac{13}{22}$$

6. (a) $\dfrac{3x}{x^2-4} = \dfrac{3(-2)}{(-2)^2-4}$ *Let x = -2*

   $$= \dfrac{-6}{4-4} = \dfrac{-6}{0}$$

   Substituting -2 for x makes the denominator zero, so the given expression is undefined when $x = -2$.

   (b) $\dfrac{3x}{x^2-4} = \dfrac{3(4)}{(4)^2-4}$ *Let x = 4*

   $$= \dfrac{12}{16-4} = \dfrac{12}{12} = 1$$

7. $\dfrac{5a^3b^3}{15a^4b^2} = \dfrac{b \cdot 5a^3b^2}{3a \cdot 5a^3b^2} = \dfrac{b}{3a}$

8. $\dfrac{m-4}{4-m} = \dfrac{-1(4-m)}{4-m} = -1$

9. $\dfrac{4x^2-9}{6-4x} = \dfrac{(2x+3)(2x-3)}{-2(2x-3)}$

   $$= \dfrac{2x+3}{-2} = \dfrac{-1(2x+3)}{2}$$

   $$= \dfrac{-(2x+3)}{2}$$

10. $\dfrac{4p^2 + 8pq - 5q^2}{10p^2 - 3pq - q^2} = \dfrac{(2p-q)(2p+5q)}{(5p+q)(2p-q)}$

    $$= \dfrac{2p+5q}{5p+q}$$

11. $-\dfrac{4x-9}{2x+3}$

    Apply the negative sign to the numerator:

    $$\dfrac{-(4x-9)}{2x+3}$$

    Now distribute the sign:

    $$\dfrac{-4x+9}{2x+3}$$

    Apply it to the denominator:

    $$\dfrac{4x-9}{-(2x+3)}$$

    Again, distribute it:

    $$\dfrac{4x-9}{-2x-3}$$

12. $\dfrac{8-3x}{3+6x}$

    Four equivalent forms are:

    $$\dfrac{-8+3x}{-3-6x}, \dfrac{-(-8+3x)}{3+6x},$$
    $$\dfrac{8-3x}{-(-3-6x)}, -\dfrac{-8+3x}{3+6x}$$

13. $\dfrac{18p^3}{6} \cdot \dfrac{24}{p^4} = \dfrac{3 \cdot 24}{p} = \dfrac{72}{p}$

14. $\dfrac{8x^2}{12x^5} \cdot \dfrac{6x^4}{2x} = \dfrac{2}{3x^3} \cdot \dfrac{3x^3}{1} = 2$

15. $\dfrac{x-3}{4} \cdot \dfrac{5}{2x-6} = \dfrac{x-3}{4} \cdot \dfrac{5}{2(x-3)} = \dfrac{5}{8}$

16. $\dfrac{2r+3}{r-4} \cdot \dfrac{r^2-16}{6r+9}$

    $$= \dfrac{2r+3}{r-4} \cdot \dfrac{(r+4)(r-4)}{3(2r+3)}$$

    $$= \dfrac{r+4}{3}$$

17. $\dfrac{6a^2 + 7a - 3}{2a^2 - a - 6} \div \dfrac{a+5}{a-2}$

    $$= \dfrac{6a^2 + 7a - 3}{2a^2 - a - 6} \cdot \dfrac{a-2}{a+5}$$

    $$= \dfrac{(3a-1)(2a+3)}{(2a+3)(a-2)} \cdot \dfrac{a-2}{a+5}$$

    $$= \dfrac{3a-1}{a+5}$$

18. $\dfrac{y^2 - 6y + 8}{y^2 + 3y - 18} \div \dfrac{y - 4}{y + 6}$

$= \dfrac{y^2 - 6y + 8}{y^2 + 3y - 18} \cdot \dfrac{y + 6}{y - 4}$

$= \dfrac{(y - 4)(y - 2)}{(y + 6)(y - 3)} \cdot \dfrac{y + 6}{y - 4}$

$= \dfrac{y - 2}{y - 3}$

19. $\dfrac{2p^2 + 13p + 20}{p^2 + p - 12} \cdot \dfrac{p^2 + 2p - 15}{2p^2 + 7p + 5}$

$= \dfrac{(2p + 5)(p + 4)}{(p + 4)(p - 3)} \cdot \dfrac{(p + 5)(p - 3)}{(2p + 5)(p + 1)}$

$= \dfrac{p + 5}{p + 1}$

20. $\dfrac{3z^2 + 5z - 2}{9z^2 - 1} \cdot \dfrac{9z^2 + 6z + 1}{z^2 + 5z + 6}$

$= \dfrac{(3z - 1)(z + 2)}{(3z - 1)(3z + 1)} \cdot \dfrac{(3z + 1)^2}{(z + 3)(z + 2)}$

$= \dfrac{3z + 1}{z + 3}$

21. $\dfrac{4}{9y}, \dfrac{7}{12y^2}, \dfrac{5}{27y^4}$

Factor the denominators.

$$9y = 3^2 y$$
$$12y^2 = 2^2 \cdot 3 \cdot y^2$$
$$27y^4 = 3^3 \cdot y^4$$

LCD $= 2^2 \cdot 3^3 \cdot y^4 = 108y^4$

22. $\dfrac{3}{x^2 + 4x + 3}, \dfrac{5}{x^2 + 5x + 4}$

Factor the denominators.

$$x^2 + 4x + 3 = (x + 3)(x + 1)$$
$$x^2 + 5x + 4 = (x + 1)(x + 4)$$

LCD $= (x + 3)(x + 1)(x + 4)$

23. $\dfrac{3}{2a^3} = \dfrac{}{10a^4}$

$\dfrac{3}{2a^3} = \dfrac{3}{2a^3} \cdot \dfrac{5a}{5a} = \dfrac{15a}{10a^4}$

24. $\dfrac{9}{x - 3} = \dfrac{}{18 - 6x} = \dfrac{}{-6(x - 3)}$

$\dfrac{9}{x - 3} = \dfrac{9}{x - 3} \cdot \dfrac{-6}{-6}$

$= \dfrac{-54}{-6x + 18}$

$= \dfrac{-54}{18 - 6x}$

25. $\dfrac{-3y}{2y - 10} = \dfrac{}{50 - 10y} = \dfrac{}{-5(2y - 10)}$

$\dfrac{-3y}{2y - 10} = \dfrac{-3y}{2y - 10} \cdot \dfrac{-5}{-5}$

$= \dfrac{15y}{-10y + 50}$

$= \dfrac{15y}{50 - 10y}$

26. $\dfrac{4b}{b^2 + 2b - 3} = \dfrac{}{(b + 3)(b - 1)(b + 2)}$

$\dfrac{4b}{b^2 + 2b - 3} = \dfrac{4b}{(b + 3)(b - 1)}$

$= \dfrac{4b}{(b + 3)(b - 1)} \cdot \dfrac{b + 2}{b + 2}$

$= \dfrac{4b(b + 2)}{(b + 3)(b - 1)(b + 2)}$

27. $\dfrac{10}{x} + \dfrac{5}{x} = \dfrac{10 + 5}{x} = \dfrac{15}{x}$

28. $\dfrac{6}{3p} - \dfrac{12}{3p} = \dfrac{6 - 12}{3p} = \dfrac{-6}{3p} = -\dfrac{2}{p}$

29. $\dfrac{9}{k} - \dfrac{5}{k - 5} = \dfrac{9(k - 5)}{k(k - 5)} - \dfrac{5 \cdot k}{(k - 5)k}$

LCD = k(k − 5)

$= \dfrac{9(k - 5) - 5k}{k(k - 5)}$

$= \dfrac{9k - 45 - 5k}{k(k - 5)}$

$= \dfrac{4k - 45}{k(k - 5)}$

30. $\dfrac{4}{y} + \dfrac{7}{7 + y} = \dfrac{4(7 + y)}{y(7 + y)} + \dfrac{7 \cdot y}{(7 + y)y}$

LCD = y(7 + y)

$= \dfrac{28 + 4y + 7y}{y(7 + y)}$

$= \dfrac{28 + 11y}{y(7 + y)}$

31. $\dfrac{m}{3} - \dfrac{2 + 5m}{6} = \dfrac{m \cdot 2}{3 \cdot 2} - \dfrac{2 + 5m}{6}$ *LCD = 6*

$= \dfrac{2m - (2 + 5m)}{6}$

$= \dfrac{2m - 2 - 5m}{6}$

$= \dfrac{-2 - 3m}{6}$

32. $\dfrac{12}{x^2} - \dfrac{3}{4x} = \dfrac{12 \cdot 4}{x^2 \cdot 4} - \dfrac{3 \cdot x}{4x \cdot x}$ *LCD = 4x²*

$= \dfrac{48 - 3x}{4x^2}$

$= \dfrac{3(16 - x)}{4x^2}$

33. $\dfrac{5}{a-2b}+\dfrac{2}{a+2b}$

$=\dfrac{5(a+2b)}{(a-2b)(a+2b)}+\dfrac{2(a-2b)}{(a+2b)(a-2b)}$

$$LCD=(a-2b)(a+2b)$$

$=\dfrac{5(a+2b)+2(a-2b)}{(a-2b)(a+2b)}$

$=\dfrac{5a+10b+2a-4b}{(a-2b)(a+2b)}$

$=\dfrac{7a+6b}{(a-2b)(a+2b)}$

34. $\dfrac{4}{k^2-9}-\dfrac{k+3}{3k-9}$

$=\dfrac{4}{(k+3)(k-3)}-\dfrac{k+3}{3(k-3)}$

$$LCD=3(k+3)(k-3)$$

$=\dfrac{4\cdot 3}{(k+3)(k-3)\cdot 3}-\dfrac{(k+3)(k+3)}{3(k-3)(k+3)}$

$=\dfrac{12-(k+3)(k+3)}{3(k+3)(k-3)}$

$=\dfrac{12-(k^2+6k+9)}{3(k+3)(k-3)}$

$=\dfrac{12-k^2-6k-9}{3(k+3)(k-3)}$

$=\dfrac{-k^2-6k+3}{3(k+3)(k-3)}$

35. $\dfrac{8}{z^2+6z}-\dfrac{3}{z^2+4z-12}$

$=\dfrac{8}{z(z+6)}-\dfrac{3}{(z+6)(z-2)}$

$$LCD=z(z+6)(z-2)$$

$=\dfrac{8(z-2)}{z(z+6)(z-2)}-\dfrac{3\cdot z}{(z+6)(z-2)\cdot z}$

$=\dfrac{8(z-2)-3z}{z(z+6)(z-2)}$

$=\dfrac{8z-16-3z}{z(z+6)(z-2)}$

$=\dfrac{5z-16}{z(z+6)(z-2)}$

36. $\dfrac{11}{2p-p^2}-\dfrac{2}{p^2-5p+6}$

$=\dfrac{11}{p(2-p)}-\dfrac{2}{(p-3)(p-2)}$

$$LCD=p(p-3)(p-2)$$

$=\dfrac{11(-1)(p-3)}{p(2-p)(-1)(p-3)}$

$\quad-\dfrac{2\cdot p}{(p-3)(p-2)p}$

$=\dfrac{-11(p-3)-2p}{p(p-2)(p-3)}$

$=\dfrac{-11p+33-2p}{p(p-2)(p-3)}$

$=\dfrac{-13p+33}{p(p-2)(p-3)}$

37. (a) $\dfrac{\dfrac{a^4}{b^2}}{\dfrac{a^3}{b}}=\dfrac{a^4}{b^2}\div\dfrac{a^3}{b}$

$=\dfrac{a^4}{b^2}\cdot\dfrac{b}{a^3}$

$=\dfrac{a^4 b}{a^3 b^2}=\dfrac{a}{b}$

(b) $\dfrac{\dfrac{a^4}{b^2}}{\dfrac{a^3}{b}}=\dfrac{b^2\left(\dfrac{a^4}{b^2}\right)}{b^2\left(\dfrac{a^3}{b}\right)}$ *Multiply by LCD, b^2*

$=\dfrac{a^4}{ba^3}=\dfrac{a}{b}$

(c) For this problem, the difference in the methods is negligible. In general, Method 2 is preferable because it leads to quicker simplifications.

38. $\dfrac{\dfrac{2}{3}-\dfrac{1}{6}}{\dfrac{1}{4}+\dfrac{2}{5}}=\dfrac{60\left(\dfrac{2}{3}-\dfrac{1}{6}\right)}{60\left(\dfrac{1}{4}+\dfrac{2}{5}\right)}$ *Multiply by LCD, 60*

$\dfrac{60\cdot\dfrac{2}{3}-60\cdot\dfrac{1}{6}}{60\cdot\dfrac{1}{4}+60\cdot\dfrac{2}{5}}=\dfrac{40-10}{15+24}$

$=\dfrac{30}{39}=\dfrac{10}{13}$

39. $\dfrac{\dfrac{y-3}{y}}{\dfrac{y+3}{4y}}=\dfrac{y-3}{y}\cdot\dfrac{4y}{y+3}=\dfrac{4(y-3)}{y+3}$

40. $\dfrac{\dfrac{1}{p}-\dfrac{1}{q}}{\dfrac{1}{q-p}}=\dfrac{\left(\dfrac{1}{p}-\dfrac{1}{q}\right)pq(q-p)}{\left(\dfrac{1}{q-p}\right)pq(q-p)}$

Multiply by LCD, $pq(q-p)$

continued

$$= \frac{\frac{1}{p}[pq(q-p)] - \frac{1}{q}[pq(q-p)]}{pq}$$

$$= \frac{q(q-p) - p(q-p)}{pq}$$

$$= \frac{q^2 - pq - pq + p^2}{pq}$$

$$= \frac{q^2 - 2pq + p^2}{pq}$$

$$= \frac{(q-p)^2}{pq}$$

41.
$$\frac{x + \dfrac{1}{w}}{x - \dfrac{1}{w}}$$

$$= \frac{\left(x + \dfrac{1}{w}\right) \cdot w}{\left(x - \dfrac{1}{w}\right) \cdot w} \qquad \begin{array}{l}\textit{Multiply by}\\ \textit{LCD, } w\end{array}$$

$$= \frac{xw + \left(\dfrac{1}{w}\right)w}{xw - \left(\dfrac{1}{w}\right)w}$$

$$= \frac{xw + 1}{xw - 1}$$

42.
$$\frac{\dfrac{1}{r+t} - 1}{\dfrac{1}{r+t} + 1}$$

$$= \frac{\left(\dfrac{1}{r+t} - 1\right)(r+t)}{\left(\dfrac{1}{r+t} + 1\right)(r+t)} \qquad \begin{array}{l}\textit{Multiply by}\\ \textit{LCD, } r+t\end{array}$$

$$= \frac{\dfrac{1}{r+t}(r+t) - 1(r+t)}{\dfrac{1}{r+t}(r+t) + 1(r+t)}$$

$$= \frac{1 - r - t}{1 + r + t}$$

43. When 2 is substituted for m throughout the equation, the value of the denominator in the first and third expressions is zero.

44. $\dfrac{4-z}{z} + \dfrac{3}{2} = \dfrac{-4}{z}$

Multiply each side by the LCD, $2z$.

$$2z\left(\frac{4-z}{z} + \frac{3}{2}\right) = 2z\left(-\frac{4}{z}\right)$$

$$2z\left(\frac{4-z}{z}\right) + 2z\left(\frac{3}{2}\right) = -8$$

$$2(4-z) + 3z = -8$$

$$8 - 2z + 3z = -8$$

$$8 + z = -8$$

$$z = -16$$

Check $z = -16$: $\quad \dfrac{1}{4} = \dfrac{1}{4}$

Thus, the solution set is $\{-16\}$.

45.
$$\frac{3y-1}{y-2} = \frac{5}{y-2} + 1$$

$$(y-2)\left(\frac{3y-1}{y-2}\right) = (y-2)\left(\frac{5}{y-2} + 1\right)$$
$$\begin{array}{r}\textit{Multiply by}\\ \textit{LCD, } y-2\end{array}$$

$$(y-2)\left(\frac{3y-1}{y-2}\right) = (y-2)\left(\frac{5}{y-2}\right)$$
$$+ (y-2)(1)$$
$$\begin{array}{r}\textit{Distributive}\\ \textit{property}\end{array}$$

$$3y - 1 = 5 + y - 2$$

$$3y - 1 = 3 + y$$

$$2y = 4$$

$$y = 2$$

The solution set is $\emptyset$ because $y = 2$ makes the original denominators equal to zero.

46.
$$\frac{3}{m-2} + \frac{1}{m-1} = \frac{7}{m^2 - 3m + 2}$$

$$\frac{3}{m-2} + \frac{1}{m-1} = \frac{7}{(m-2)(m-1)}$$

$$(m-2)(m-1)\left(\frac{3}{m-2} + \frac{1}{m-1}\right)$$
$$= (m-2)(m-1) \cdot \frac{7}{(m-2)(m-1)}$$
$$\begin{array}{r}\textit{Multiply by}\\ \textit{LCD, } (m-2)(m-1)\end{array}$$

$$3(m-1) + 1(m-2) = 7$$

$$3m - 3 + m - 2 = 7$$

$$4m - 5 = 7$$

$$4m = 12$$

$$m = 3$$

Check $m = 3$: $\quad 3 + \dfrac{1}{2} = \dfrac{7}{2}$

Thus, the solution set is $\{3\}$.

47. $m = \dfrac{Ry}{t}$ for t

$$t \cdot m = t\left(\dfrac{Ry}{t}\right) \quad \text{Multiply by } t$$

$$tm = Ry$$

$$t = \dfrac{Ry}{m} \qquad \text{Divide by } m$$

48. $x = \dfrac{3y - 5}{4}$ for y

$$4x = 4\left(\dfrac{3y - 5}{4}\right)$$

$$4x = 3y - 5$$

$$4x + 5 = 3y$$

$$\dfrac{4x + 5}{3} = y$$

49. $p^2 = \dfrac{4}{3m - q}$ for m

$$(3m - q)p^2 = (3m - q)\left(\dfrac{4}{3m - q}\right)$$

$$3mp^2 - p^2q = 4$$

$$3mp^2 = 4 + p^2q$$

$$m = \dfrac{4 + p^2q}{3p^2}$$

50. *Step 1*   Let $x = $ the numerator of the original fraction.

Step 2   Then $x - 4 = $ the denominator of the original fraction.

Step 3 $\dfrac{x + 3}{(x - 4) + 3} = \dfrac{3}{2}$

Step 4 $\dfrac{x + 3}{x - 1} = \dfrac{3}{2}$

$$2(x - 1)\left(\dfrac{x + 3}{x - 1}\right) = 2(x - 1)\left(\dfrac{3}{2}\right)$$

$$2(x + 3) = 3(x - 1)$$

$$2x + 6 = 3x - 3$$

$$9 = x$$

Step 5   The original fraction was

$$\dfrac{9}{9 - 4} = \dfrac{9}{5}.$$

Step 6 $\dfrac{9 + 3}{5 + 3} = \dfrac{12}{8} = \dfrac{3}{2}$, as desired.

51. *Step 1*   Let $x = $ the numerator of the fraction.

Step 2   Then $3x = $ the denominator.

Step 3   If 2 is added to the numerator and 2 is subtracted from the denominator, the numerator would become $x + 2$ and the denominator would become $3x - 2$.

$$\dfrac{x + 2}{3x - 2} = 1$$

Step 4   Multiply both sides by $3x - 2$.

$$(3x - 2)\left(\dfrac{x + 2}{3x - 2}\right) = (3x - 2)(1)$$

$$x + 2 = 3x - 2$$

$$4 = 2x$$

$$2 = x$$

Step 5   The numerator of the original fraction is 2, and its denominator is $3 \cdot 2 = 6$, so the original fraction is $\dfrac{2}{6}$.

Step 6   This answer checks since

$$\dfrac{2 + 2}{3(2) - 2} = \dfrac{4}{4} = 1.$$

52. *Step 1*   Let $t = $ Scott's time.

Step 2   Use $d = rt$.

Step 3   Let $d = 200$ and $r = 130.934$.

$$200 = 130.934t$$

Step 4   Divide by 130.934

$$t = \dfrac{200}{130.934} \approx 1.527$$

Step 5   Scott's time was approximately 1.527 hours.

Step 6   Driving 1.527 hours with an average speed of 130.934 miles per hour gives us a distance of about 199.9 miles — slightly less than 200 due to rounding.

53. *Step 1*   Let $x = $ the number of hours it takes them to do the job working together.

Step 2

	Rate	Time working together	Fractional part of the job done when working together
Man	$\dfrac{1}{5}$	x	$\dfrac{1}{5}x$
Daughter	$\dfrac{1}{8}$	x	$\dfrac{1}{8}x$

Step 3   Working together, they do 1 whole job, so

$$\dfrac{1}{5}x + \dfrac{1}{8}x = 1.$$

Step 4   Solve this equation by multiplying both sides by the LCD, 40.

$$40\left(\frac{1}{5}x + \frac{1}{8}x\right) = 40(1)$$
$$8x + 5x = 40$$
$$13x = 40$$
$$x = \frac{40}{13} \text{ or } 3\frac{1}{13}$$

Step 5   Working together, it takes them $\frac{40}{13}$ or about $3\frac{1}{13}$ hours.

Step 6   The man does $\frac{1}{5}$ of the job per hour for $\frac{40}{13}$ hours:

$$\frac{1}{5} \cdot \frac{40}{13} = \frac{8}{13} \text{ of the job}$$

His daughter does $\frac{1}{8}$ of the job per hour for $\frac{40}{13}$ hours:

$$\frac{1}{8} \cdot \frac{40}{13} = \frac{5}{13} \text{ of the job}$$

Together, they have done
$$\frac{8}{13} + \frac{5}{13} = \frac{13}{13} = 1$$
total job.

54.   *Step 1*   Let $x =$ the time needed by the head gardener to mow the lawns.

Step 2   Then $2x =$ the time needed by the assistant to mow the lawns.

	Rate	Time working together	Fractional part of the job done when working together
Head gardener	$\frac{1}{x}$	$1\frac{1}{3} = \frac{4}{3}$	$\frac{1}{x} \cdot \frac{4}{3} = \frac{4}{3x}$
Assistant	$\frac{1}{2x}$	$1\frac{1}{3} = \frac{4}{3}$	$\frac{1}{2x} \cdot \frac{4}{3} = \frac{2}{3x}$

Step 3 $\dfrac{4}{3x} + \dfrac{2}{3x} = 1$

Step 4
$$\frac{4+2}{3x} = 1$$
$$\frac{6}{3x} = 1$$
$$3x\left(\frac{6}{3x}\right) = 3x(1)$$
$$6 = 3x$$
$$2 = x$$

Step 5   It takes the head gardener 2 hours to mow the lawns.

Step 6   The head gardener does $\frac{1}{2}$ of the job per hour for $\frac{4}{3}$ hours:

$$\frac{1}{2} \cdot \frac{4}{3} = \frac{2}{3} \text{ of the job}$$

The assistant does $\frac{1}{4}$ of the job per hour for $\frac{4}{3}$ hours:

$$\frac{1}{4} \cdot \frac{4}{3} = \frac{1}{3} \text{ of the job}$$

Together, they have done

$$\frac{2}{3} + \frac{1}{3} = \frac{3}{3} = 1$$

total job.

55.   Since a longer term is related to a lower rate per year, this is an *inverse* variation.

56.
Let $x =$ the number of male deaths due to firearms in 1994;

$38,505 - x =$ the number of female deaths due to firearms in 1994.

$$\frac{x}{38,505 - x} = \frac{63}{10}$$
$$10x = 63(38,505 - x)$$
$$10x = 2,425,815 - 63x$$
$$73x = 2,425,815$$
$$x = \frac{2,425,815}{73} \approx 33,230$$
$$38,505 - x = 5275$$

There were approximately $33,230$ male deaths and 5275 female deaths.

57.   Let $h =$ height of parallelogram;

$b =$ length of base of parallelogram.

The height varies inversely as the base, so

$$h = \frac{k}{b}.$$

Find k by replacing h with 8 and b with 12.

$$8 = \frac{k}{12}$$
$$k = 8 \cdot 12 = 96$$

So $h = \dfrac{96}{b}$ and when $b = 24$,

$$h = \frac{96}{24} = 4.$$

The height of the parallelogram is 4 centimeters.

58. Let $t =$ the number of hours until the distance between the boats is 35 miles.

	Rate	Time	Distance
Slower boat	18	t	$18t$
Faster boat	25	t	$25t$

$$25t - 18t = 35$$
$$7t = 35$$
$$t = 5$$

They will be 35 miles apart in 5 hours.

59. **(a)** $x + 3 = 0$
$$x = -3$$

This value of x makes the value of the denominator zero, so P will be undefined when $x = -3$.

(b) $x + 1 = 0$
$$x = -1$$

This value of x makes the value of the denominator zero, so Q will be undefined when $x = -1$.

(c) $x^2 + 4x + 3 = 0$
$$(x + 3)(x + 1) = 0$$
$$x = -3 \quad \text{or} \quad x = -1$$

These values of x make the value of the denominator zero, so the values for which R is undefined are -3 and -1.

60. $(P \cdot Q) \div R$

$$= \left(\frac{6}{x + 3} \cdot \frac{5}{x + 1} \right) \div \frac{4x}{x^2 + 4x + 3}$$

$$= \frac{30}{(x + 3)(x + 1)} \cdot \frac{x^2 + 4x + 3}{4x}$$

$$= \frac{30}{(x + 3)(x + 1)} \cdot \frac{(x + 3)(x + 1)}{4x}$$

$$= \frac{30}{4x} = \frac{15}{2x}$$

61. If $x = 0$, the divisor R is equal to zero. Division by zero is not permitted.

62. List the three denominators and factor if possible.

$x + 3$
$x + 1$
$x^2 + 4x + 3 = (x + 3)(x + 1)$

The LCD for P, Q, and R is $(x + 3)(x + 1)$.

63. $P + Q - R$

$$= \frac{6}{x + 3} + \frac{5}{x + 1} - \frac{4x}{x^2 + 4x + 3}$$

$$= \frac{6}{x + 3} + \frac{5}{x + 1} - \frac{4x}{(x + 3)(x + 1)}$$

$$LCD = (x + 3)(x + 1)$$

$$= \frac{6(x + 1)}{(x + 3)(x + 1)} + \frac{5(x + 3)}{(x + 3)(x + 1)}$$
$$- \frac{4x}{(x + 3)(x + 1)}$$

$$= \frac{6(x + 1) + 5(x + 3) - 4x}{(x + 3)(x + 1)}$$

$$= \frac{6x + 6 + 5x + 15 - 4x}{(x + 3)(x + 1)}$$

$$= \frac{7x + 21}{(x + 3)(x + 1)}$$

$$= \frac{7(x + 3)}{(x + 3)(x + 1)} = \frac{7}{x + 1}$$

64. $$\frac{P + Q}{R} = \frac{\dfrac{6}{x + 3} + \dfrac{5}{x + 1}}{\dfrac{4x}{x^2 + 4x + 3}} = \frac{\dfrac{6}{x + 3} + \dfrac{5}{x + 1}}{\dfrac{4x}{(x + 3)(x + 1)}}$$

To simplify this complex fraction, use Method 2. Multiply numerator and denominator by the LCD for all the fractions, $(x + 3)(x + 1)$.

$$\frac{(x + 3)(x + 1)\left(\dfrac{6}{x + 3} + \dfrac{5}{x + 1} \right)}{(x + 3)(x + 1)\left(\dfrac{4x}{(x + 3)(x + 1)} \right)}$$

$$= \frac{(x + 3)(x + 1)\left(\dfrac{6}{x + 3} \right) + (x + 3)(x + 1)\left(\dfrac{5}{x + 1} \right)}{4x}$$

$$= \frac{6(x + 1) + 5(x + 3)}{4x}$$

$$= \frac{6x + 6 + 5x + 15}{4x}$$

$$= \frac{11x + 21}{4x}$$

65. $$P + Q = R$$
$$\frac{6}{x + 3} + \frac{5}{x + 1} = \frac{4x}{x^2 + 4x + 3}$$
$$\frac{6}{x + 3} + \frac{5}{x + 1} = \frac{4x}{(x + 3)(x + 1)}$$

Multiply by
LCD, $(x + 3)(x + 1)$

$$(x+3)(x+1)\left(\frac{6}{x+3}\right) + (x+3)(x+1)\left(\frac{5}{x+1}\right)$$

$$= (x+3)(x+1)\left(\frac{4x}{(x+3)(x+1)}\right)$$

$$6(x+1) + 5(x+3) = 4x$$
$$6x + 6 + 5x + 15 = 4x$$
$$7x = -21$$
$$x = -3$$

To check, substitute -3 for x. The denominators of the first and third fractions become zero. Reject -3 as a solution. There is no solution to the equation, so the solution set is $\emptyset$.

66. We know that -3 is not allowed because P and R are undefined for $x = -3$.

67. $d = rt$

$$r = \frac{d}{t} = \frac{6}{x+3}$$

If $d = 6$ miles and $t = (x+3)$ minutes, then

$$r = \frac{6}{x+3}$$ miles per minute. Thus,

$$P = \frac{6}{x+3}$$

represents the rate of the car (in miles per minute).

68.
$$R = \frac{4x}{x^2 + 4x + 3}$$

$$\frac{40}{77} = \frac{4x}{x^2 + 4x + 3}$$

$$40(x^2 + 4x + 3) = (4x)(77) \quad \textit{Cross products}$$
$$\textit{are equal}$$

$$40x^2 + 160x + 120 = 308x$$
$$40x^2 - 148x + 120 = 0$$
$$10x^2 - 37x + 30 = 0$$
$$(5x - 6)(2x - 5) = 0$$

$$x = \frac{6}{5} \quad \text{or} \quad x = \frac{5}{2}$$

69.
$$\frac{\dfrac{5}{x-y} + 2}{3 - \dfrac{2}{x+y}}$$

To simplify this complex fraction, multiply numerator and denominator by the LCD of all the fractions, $(x - y)(x + y)$.

$$= \frac{\left(\dfrac{5}{x-y} + 2\right)(x-y)(x+y)}{\left(3 - \dfrac{2}{x+y}\right)(x-y)(x+y)}$$

$$= \frac{\left(\dfrac{5}{x-y}\right)(x-y)(x+y) + 2(x-y)(x+y)}{3(x-y)(x+y) - \left(\dfrac{2}{x+y}\right)(x-y)(x+y)}$$

$$= \frac{5(x+y) + 2(x-y)(x+y)}{3(x-y)(x+y) - 2(x-y)}$$

$$= \frac{(x+y)[5 + 2(x-y)]}{(x-y)[3(x+y) - 2]} \quad \begin{array}{l}\textit{Factor out} \\ (x+y) \textit{ and } (x-y)\end{array}$$

$$= \frac{(x+y)(5 + 2x - 2y)}{(x-y)(3x + 3y - 2)}$$

70. $\dfrac{4}{m-1} - \dfrac{3}{m+1}$

To perform the indicated subtraction, use $(m-1)(m+1)$ as the LCD.

$$\frac{4}{m-1} - \frac{3}{m+1}$$

$$= \frac{4(m+1)}{(m-1)(m+1)} - \frac{3(m-1)}{(m+1)(m-1)}$$

$$= \frac{4(m+1) - 3(m-1)}{(m-1)(m+1)}$$

$$= \frac{4m + 4 - 3m + 3}{(m-1)(m+1)}$$

$$= \frac{m+7}{(m-1)(m+1)}$$

71. $\dfrac{8p^5}{5} \div \dfrac{2p^3}{10}$

To perform the indicated division, multiply the first rational expression by the reciprocal of the second.

$$\frac{8p^5}{5} \div \frac{2p^3}{10} = \frac{8p^5}{5} \cdot \frac{10}{2p^3}$$

$$= \frac{80p^5}{10p^3}$$

$$= 8p^2$$

72. $\dfrac{r-3}{8} \div \dfrac{3r-9}{4} = \dfrac{r-3}{8} \cdot \dfrac{4}{3r-9}$

$$= \frac{r-3}{8} \cdot \frac{4}{3(r-3)}$$

$$= \frac{4}{24} = \frac{1}{6}$$

73.
$$\frac{\dfrac{5}{x}-1}{\dfrac{5-x}{3x}} = \frac{\left(\dfrac{5}{x}-1\right)3x}{\left(\dfrac{5-x}{3x}\right)3x}$$ *Multiply by LCD, 3x*

$$= \frac{\dfrac{5}{x}(3x)-1(3x)}{5-x}$$

$$= \frac{15-3x}{5-x}$$

$$= \frac{3(5-x)}{5-x} = 3$$

74.
$$\frac{4}{z^2-2z+1} - \frac{3}{z^2-1}$$

$$= \frac{4}{(z-1)^2} - \frac{3}{(z+1)(z-1)}$$

$$LCD = (z+1)(z-1)^2$$

$$= \frac{4(z+1)}{(z-1)^2(z+1)}$$
$$- \frac{3(z-1)}{(z+1)(z-1)(z-1)}$$

$$= \frac{4(z+1)-3(z-1)}{(z+1)(z-1)^2}$$

$$= \frac{4z+4-3z+3}{(z+1)(z-1)^2}$$

$$= \frac{z+7}{(z+1)(z-1)^2}$$

75. $\dfrac{1}{k}+\dfrac{3}{r}=\dfrac{5}{z}$ for r

The LCD of all the fractions is krz, so multiply both sides by krz.

$$krz\left(\frac{1}{k}+\frac{3}{r}\right) = krz\left(\frac{5}{z}\right)$$

$$krz\left(\frac{1}{k}\right)+krz\left(\frac{3}{r}\right) = krz\left(\frac{5}{z}\right)$$

$$rz+3kz = 5kr$$

Since we are solving for r, get all terms with r on one side.

$$3kz = 5kr - rz$$

Factor out the common factor r on the right.

$$3kz = r(5k-z)$$

Finally, divide both sides by the coefficient of r, which is $5k-z$.

$$\frac{3kz}{5k-z} = r$$

The answer may be written as

$$r = \frac{3kz}{5k-z} \text{ or } \frac{-3kz}{z-5k}.$$

76.
$$\frac{5+m}{m}+\frac{3}{4}=\frac{-2}{m}$$

$$4m\left(\frac{5+m}{m}+\frac{3}{4}\right) = 4m\left(\frac{-2}{m}\right)$$ *Multiply by LCD, 4m*

$$4(5+m)+3m = 4(-2)$$

$$20+4m+3m = -8$$

$$7m = -28$$

$$m = -4$$

Check $m=-4$: $\frac{1}{2}=\frac{1}{2}$

Thus, the solution set is $\{-4\}$.

77. Let $x =$ the number of people waiting for heart transplants in 1996;

$11,165-x =$ the number of people waiting for liver transplants in 1996.

$$\frac{x}{11,165-x}=\frac{2}{1}$$

$$x(1) = 2(11,165-x)$$

$$x = 22,330-2x$$

$$3x = 22,330$$

$$x = \frac{22,330}{3} \approx 7443$$

$$11,165-x = 3722$$

There were approximately 7443 people waiting for heart transplants in 1996 and 3722 waiting for liver transplants.

78. Let $x =$ the speed of the plane in still air. Then the speed of the plane with the wind is $x+50$, and the speed of the plane against the wind is $x-50$. Use $t=\dfrac{d}{r}$ to complete the chart.

	d	r	t
With wind	400	$x+50$	$\dfrac{400}{x+50}$
Against wind	200	$x-50$	$\dfrac{200}{x-50}$

The times are the same, so

$$\frac{400}{x+50}=\frac{200}{x-50}.$$

To solve this equation, multiply both sides by the LCD, $(x+50)(x-50)$.

$$(x+50)(x-50)\cdot\frac{400}{x+50}$$
$$= (x+50)(x-50)\cdot\frac{200}{x-50}$$

$$400(x-50) = 200(x+50)$$

$$400x-20,000 = 200x+10,000$$

$$200x = 30,000$$

$$x = 150$$

The speed of the plane is 150 kilometers per hour.

79. Since x varies directly as y, there is a constant k such that $x = ky$. First find the value of k.

$$x = ky$$
$$12 = k \cdot 5 \quad \textit{Let x = 12, y = 5}$$
$$\frac{12}{5} = k$$

Since $x = ky$ and $k = \frac{12}{5}$,

$$x = \frac{12}{5}y.$$

Now find x when $y = 3$.

$$x = \frac{12}{5}(3) \quad \textit{Let y = 3}$$
$$x = \frac{36}{5}$$

80. Let $x =$ the number of days it would take Luigi to do the job working alone.

	Rate	Time working together	Fractional part of the job done when working together
Mario	$\frac{1}{8}$	$3\frac{3}{7} = \frac{24}{7}$	$\frac{1}{8} \cdot \frac{24}{7} = \frac{3}{7}$
Luigi	$\frac{1}{x}$	$3\frac{3}{7} = \frac{24}{7}$	$\frac{1}{x} \cdot \frac{24}{7} = \frac{24}{7x}$

Working together, they do 1 whole job, so

$$\frac{3}{7} + \frac{24}{7x} = 1.$$
$$7x\left(\frac{3}{7} + \frac{24}{7x}\right) = 7x \cdot 1 \quad \begin{array}{l}\textit{Multiply by}\\ \textit{LCD, 7x}\end{array}$$
$$7x\left(\frac{3}{7}\right) + 7x\left(\frac{24}{7x}\right) = 7x$$
$$3x + 24 = 7x$$
$$24 = 4x$$
$$6 = x$$

It would take Luigi 6 days working alone.

Chapter 7 Test

1. $\dfrac{3x - 1}{x^2 - 2x - 8}$

Set the denominator equal to zero and solve for x.

$$x^2 - 2x - 8 = 0$$
$$(x + 2)(x - 4) = 0$$
$$x + 2 = 0 \quad \text{or} \quad x - 4 = 0$$
$$x = -2 \quad \text{or} \quad x = 4$$

The expression is undefined for -2 and 4.

2. (a) $\dfrac{6r + 1}{2r^2 - 3r - 20}$

$$= \frac{6(-2) + 1}{2(-2)^2 - 3(-2) - 20} \quad \textit{Let r = -2}$$
$$= \frac{-12 + 1}{2 \cdot 4 + 6 - 20}$$
$$= \frac{-11}{-6} = \frac{11}{6}$$

(b) $\dfrac{6r + 1}{2r^2 - 3r - 20}$

$$= \frac{6(4) + 1}{2(4)^2 - 3(4) - 20} \quad \textit{Let r = 4}$$
$$= \frac{24 + 1}{2 \cdot 16 - 12 - 20}$$
$$= \frac{25}{32 - 12 - 20}$$
$$= \frac{25}{20 - 20} = \frac{25}{0}$$

The expression is undefined when $r = 4$ because the denominator is 0.

3. $-\dfrac{6x - 5}{2x + 3}$

Apply the negative sign to the numerator:

$$\frac{-(6x - 5)}{2x + 3}$$

Now distribute the sign:

$$\frac{-6x + 5}{2x + 3}$$

Apply it to the denominator:

$$\frac{6x - 5}{-(2x + 3)}$$

Again, distribute it:

$$\frac{6x - 5}{-2x - 3}$$

4. $\dfrac{-15x^6 y^4}{5x^4 y} = \dfrac{(5x^4 y)(-3x^2 y^3)}{(5x^4 y)(1)}$

$$= \frac{5x^4 y}{5x^4 y} \cdot \frac{-3x^2 y^3}{1} = -3x^2 y^3$$

5. $\dfrac{6a^2 + a - 2}{2a^2 - 3a + 1} = \dfrac{(3a + 2)(2a - 1)}{(2a - 1)(a - 1)}$

$$= \frac{3a + 2}{a - 1}$$

6. $\dfrac{5(d - 2)}{9} \div \dfrac{3(d - 2)}{5}$

$$= \frac{5(d - 2)}{9} \cdot \frac{5}{3(d - 2)}$$
$$= \frac{5 \cdot 5}{9 \cdot 3} = \frac{25}{27}$$

7. $\dfrac{6k^2 - k - 2}{8k^2 + 10k + 3} \cdot \dfrac{4k^2 + 7k + 3}{3k^2 + 5k + 2}$

$= \dfrac{(3k - 2)(2k + 1)}{(4k + 3)(2k + 1)} \cdot \dfrac{(4k + 3)(k + 1)}{(3k + 2)(k + 1)}$

$= \dfrac{3k - 2}{3k + 2}$

8. $\dfrac{4a^2 + 9a + 2}{3a^2 + 11a + 10} \div \dfrac{4a^2 + 17a + 4}{3a^2 + 2a - 5}$

$= \dfrac{4a^2 + 9a + 2}{3a^2 + 11a + 10} \cdot \dfrac{3a^2 + 2a - 5}{4a^2 + 17a + 4}$

$= \dfrac{(4a + 1)(a + 2)}{(3a + 5)(a + 2)} \cdot \dfrac{(3a + 5)(a - 1)}{(4a + 1)(a + 4)}$

$= \dfrac{a - 1}{a + 4}$

9. $\dfrac{-3}{10p^2}, \dfrac{21}{25p^3}, \dfrac{-7}{30p^5}$

Factor each denominator.

$$10p^2 = 2 \cdot 5 \cdot p^2$$
$$25p^3 = 5^2 \cdot p^3$$
$$30p^5 = 2 \cdot 3 \cdot 5 \cdot p^5$$

$\text{LCD} = 2 \cdot 3 \cdot 5^2 \cdot p^5 = 150p^5$

10. $\dfrac{r + 1}{2r^2 + 7r + 6}, \dfrac{-2r + 1}{2r^2 - 7r - 15}$

Factor each denominator.

$$2r^2 + 7r + 6 = (2r + 3)(r + 2)$$
$$2r^2 - 7r - 15 = (2r + 3)(r - 5)$$

$\text{LCD} = (2r + 3)(r + 2)(r - 5)$

11. $\dfrac{15}{4p} = \dfrac{}{64p^3}$

$\dfrac{15}{4p} = \dfrac{15 \cdot 16p^2}{4p \cdot 16p^2} = \dfrac{240p^2}{64p^3}$

12. $\dfrac{3}{6m - 12} = \dfrac{}{42m - 84} = \dfrac{}{7(6m - 12)}$

$\dfrac{3}{6m - 12} = \dfrac{3 \cdot 7}{(6m - 12)7} = \dfrac{21}{42m - 84}$

13. $\dfrac{4x + 2}{x + 5} + \dfrac{-2x + 8}{x + 5}$

$= \dfrac{(4x + 2) + (-2x + 8)}{x + 5}$

$= \dfrac{2x + 10}{x + 5}$

$= \dfrac{2(x + 5)}{x + 5} = 2$

14. $\dfrac{-4}{y + 2} + \dfrac{6}{5y + 10}$

$= \dfrac{-4}{y + 2} + \dfrac{6}{5(y + 2)}$ *LCD = 5(y + 2)*

$= \dfrac{-4 \cdot 5}{(y + 2) \cdot 5} + \dfrac{6}{5(y + 2)}$

$= \dfrac{-20 + 6}{5(y + 2)} = \dfrac{-14}{5(y + 2)}$

15. Using LCD $= 3 - x$,

$\dfrac{x + 1}{3 - x} + \dfrac{x^2}{x - 3} = \dfrac{x + 1}{3 - x} + \dfrac{-1(x^2)}{-1(x - 3)}$

$= \dfrac{x + 1}{3 - x} + \dfrac{-x^2}{-x + 3}$

$= \dfrac{x + 1}{3 - x} + \dfrac{-x^2}{3 - x}$

$= \dfrac{(x + 1) + (-x^2)}{3 - x}$

$= \dfrac{-x^2 + x + 1}{3 - x}.$

If we use $x - 3$ for the LCD, we obtain the equivalent answer

$$\dfrac{x^2 - x - 1}{x - 3}.$$

16. $\dfrac{3}{2m^2 - 9m - 5} - \dfrac{m + 1}{2m^2 - m - 1}$

$= \dfrac{3}{(2m + 1)(m - 5)} - \dfrac{m + 1}{(2m + 1)(m - 1)}$

$\qquad$ *LCD = (2m + 1)(m - 5)(m - 1)*

$= \dfrac{3(m - 1)}{(2m + 1)(m - 5)(m - 1)}$

$\quad - \dfrac{(m + 1)(m - 5)}{(2m + 1)(m - 1)(m - 5)}$

$= \dfrac{3(m - 1) - (m + 1)(m - 5)}{(2m + 1)(m - 5)(m - 1)}$

$= \dfrac{(3m - 3) - (m^2 - 4m - 5)}{(2m + 1)(m - 5)(m - 1)}$

$= \dfrac{3m - 3 - m^2 + 4m + 5}{(2m + 1)(m - 5)(m - 1)}$

$= \dfrac{-m^2 + 7m + 2}{(2m + 1)(m - 5)(m - 1)}$

17. $\dfrac{\dfrac{2p}{k^2}}{\dfrac{3p^2}{k^3}} = \dfrac{2p}{k^2} \div \dfrac{3p^2}{k^3}$

$= \dfrac{2p}{k^2} \cdot \dfrac{k^3}{3p^2}$

$= \dfrac{2k^3 p}{3k^2 p^2} = \dfrac{2k}{3p}$

18.

$$\dfrac{\dfrac{1}{x+3}-1}{1+\dfrac{1}{x+3}}$$

$$=\dfrac{(x+3)\left(\dfrac{1}{x+3}-1\right)}{(x+3)\left(1+\dfrac{1}{x+3}\right)}\quad\begin{array}{l}\textit{Multiply by}\\\textit{LCD, x+3}\end{array}$$

$$=\dfrac{(x+3)\left(\dfrac{1}{x+3}\right)-(x+3)(1)}{(x+3)(1)+(x+3)\left(\dfrac{1}{x+3}\right)}$$

$$=\dfrac{1-(x+3)}{(x+3)+1}$$

$$=\dfrac{1-x-3}{x+4}$$

$$=\dfrac{-2-x}{x+4}$$

19. x cannot have a value that would make a denominator equal zero. Find these values.

$$\begin{array}{ll}x+1=0 & x-4=0\\ x=-1 & x=4\end{array}$$

x cannot be -1 or 4.

20.

$$\dfrac{2x}{x-3}+\dfrac{1}{x+3}=\dfrac{-6}{x^2-9}$$

$$\dfrac{2x}{x-3}+\dfrac{1}{x+3}=\dfrac{-6}{(x+3)(x-3)}$$

$$(x+3)(x-3)\left(\dfrac{2x}{x-3}+\dfrac{1}{x+3}\right)$$

$$=(x+3)(x-3)\left(\dfrac{-6}{(x+3)(x-3)}\right)$$

$$\begin{array}{l}\textit{Multiply by}\\\textit{LCD, (x + 3)(x − 3)}\end{array}$$

$$2x(x+3)+1(x-3)=-6$$

$$2x^2+6x+x-3=-6$$

$$2x^2+7x+3=0$$

$$(2x+1)(x+3)=0$$

$$x=-\dfrac{1}{2}\quad\text{or}\quad x=-3$$

x cannot equal -3 because the denominator $x+3$ would equal 0.

Check $x=-\dfrac{1}{2}:\ \dfrac{2}{7}+\dfrac{2}{5}=\dfrac{24}{35}$ *True*

Thus, the solution set is $\left\{-\dfrac{1}{2}\right\}$.

21. Solve $F=\dfrac{k}{d-D}$ for D.

$$(d-D)(F)=(d-D)\left(\dfrac{k}{d-D}\right)\quad\begin{array}{l}\textit{Multiply by}\\\textit{LCD, d − D}\end{array}$$

$$(d-D)(F)=k$$

$$dF-DF=k$$

$$-DF=k-dF$$

$$D=\dfrac{k-dF}{-F}\ \text{or}\ \dfrac{dF-k}{F}$$

22. Let $x=$ the speed of the current.

	r	t	d
Upstream	$7-x$	$\dfrac{20}{7-x}$	20
Downstream	$7+x$	$\dfrac{50}{7+x}$	50

The times are equal, so

$$\dfrac{20}{7-x}=\dfrac{50}{7+x}.$$

$$(7-x)(7+x)\left(\dfrac{20}{7-x}\right)=(7-x)(7+x)\left(\dfrac{50}{7+x}\right)$$

$$\begin{array}{l}\textit{Multiply by}\\\textit{LCD, (7 − x)(7 + x)}\end{array}$$

$$20(7+x)=50(7-x)$$

$$140+20x=350-50x$$

$$70x=210$$

$$x=3$$

The speed of the current is 3 miles per hour.

23. Let $x=$ the time required for the couple to paint the room working together.

	Rate	Time working together	Fractional part of the job done when working together
Husband	$\dfrac{1}{5}$	x	$\dfrac{1}{5}x$
Wife	$\dfrac{1}{4}$	x	$\dfrac{1}{4}x$

$$\dfrac{1}{5}x+\dfrac{1}{4}x=1$$

$$20\left(\dfrac{1}{5}x+\dfrac{1}{4}x\right)=20(1)\quad\begin{array}{l}\textit{Multiply by}\\\textit{LCD, 20}\end{array}$$

$$20\left(\dfrac{1}{5}x\right)+20\left(\dfrac{1}{4}x\right)=20$$

$$4x+5x=20$$

$$9x=20$$

$$x=\dfrac{20}{9}$$

The couple can paint the room in $\dfrac{20}{9}$ or $2\dfrac{2}{9}$ hours.

24. The length of time L that it takes for fruit to ripen during the growing season varies inversely as the average maximum temperature T during the season, so

$$L = \frac{k}{T}.$$

$$25 = \frac{k}{80} \quad \textit{Let L = 25, T = 80}$$

$$k = 25 \cdot 80 = 2000$$

So $L = \dfrac{2000}{T}$ and when $T = 75$,

$$L = \frac{2000}{75} = \frac{80}{3} \text{ or } 26\frac{2}{3}.$$

To the nearest whole number, L is 27 days.

Cumulative Review Exercises Chapters 1–7

1. $3 + 4\left(\dfrac{1}{2} - \dfrac{3}{4}\right)$

$$= 3 + 4\left(\frac{2}{4} - \frac{3}{4}\right)$$

$$= 3 + 4\left(-\frac{1}{4}\right)$$

$$= 3 + (-1)$$

$$= 2$$

2. $3(2y - 5) = 2 + 5y$

$$6y - 15 = 2 + 5y$$

$$y = 17$$

The solution set is $\{17\}$.

3. $A = \dfrac{1}{2}bh$ for b

$$2 \cdot A = 2 \cdot \frac{1}{2}bh$$

$$2A = bh$$

$$\frac{2A}{h} = b$$

4. $\dfrac{2 + m}{2 - m} = \dfrac{3}{4}$

$$4(2 + m) = 3(2 - m) \quad \textit{Cross multiply}$$

$$8 + 4m = 6 - 3m$$

$$7m = -2$$

$$m = -\frac{2}{7}$$

The solution set is $\left\{-\dfrac{2}{7}\right\}$.

5. $5y \le 6y + 8$

$$-y \le 8$$

$$y \ge -8 \quad \textit{Reverse the symbol}$$

The solution set is $[-8, \infty)$.

6. $5m - 9 > 2m + 3$

$$3m > 12$$

$$m > 4$$

The solution set is $(4, \infty)$.

7. $4x + 3y = -12$

(a) Let $y = 0$ to find the x-intercept.

$$4x + 3(0) = -12$$

$$4x = -12$$

$$x = -3$$

The x-intercept is $(-3, 0)$.

(b) Let $x = 0$ to find the y-intercept.

$$4(0) + 3y = -12$$

$$3y = -12$$

$$y = -4$$

The y-intercept is $(0, -4)$.

8. $y = -3x + 2$

This is an equation of a line.

If $x = 0$, $y = 2$; so the y-intercept is $(0, 2)$.

If $x = 1$, $y = -1$; and if $x = 2$, $y = -4$.

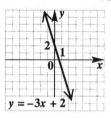

9. $y = -x^2 + 1$

This is the equation of a parabola opening downward with a y-intercept $(0, 1)$. If $x = +2$ or -2, $y = -3$.

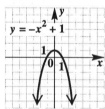

10. $\dfrac{(2x^3)^{-1}x}{2^3 x^5} = \dfrac{2^{-1}(x^3)^{-1}x}{2^3 x^5} = \dfrac{2^{-1}x^{-3}x}{2^3 x^5}$

$$= \frac{2^{-1}x^{-2}}{2^3 x^5} = \frac{1}{2^1 \cdot 2^3 \cdot x^2 \cdot x^5}$$

$$= \frac{1}{2^4 x^7}$$

11. $\dfrac{(m^{-2})^3 m}{m^5 m^{-4}} = \dfrac{m^{-6}m}{m^5 m^{-4}} = \dfrac{m \cdot m^4}{m^5 \cdot m^6}$

$$= \frac{m^5}{m^{11}} = \frac{1}{m^6}$$

12. $\dfrac{2p^3q^4}{8p^5q^3} = \dfrac{q \cdot 2p^3q^3}{4p^2 \cdot 2p^3q^3} = \dfrac{q}{4p^2}$

13. $\left(2k^2 + 3k\right) - \left(k^2 + k - 1\right)$
$= 2k^2 + 3k - k^2 - k + 1$
$= k^2 + 2k + 1$

14. $8x^2y^2\left(9x^4y^5\right) = 72x^6y^7$

15. $(2a - b)^2 = (2a)^2 - 2(2a)(b) + (b)^2$
$= 4a^2 - 4ab + b^2$

16. $\left(y^2 + 3y + 5\right)(3y - 1)$

Multiply vertically.

$$
\begin{array}{r}
y^2 + 3y + 5 \\
3y - 1 \\
\hline
-\ y^2 - 3y - 5 \\
3y^3 + 9y^2 + 15y \\
\hline
3y^3 + 8y^2 + 12y - 5
\end{array}
$$

17. $\dfrac{12p^3 + 2p^2 - 12p + 4}{2p - 2}$

$$
\begin{array}{r}
6p^2 + 7p + 1 \\
2p - 2\,\overline{)\,12p^3 + 2p^2 - 12p + 4} \\
\underline{12p^3 - 12p^2} \\
14p^2 - 12p \\
\underline{14p^2 - 14p} \\
2p + 4 \\
\underline{2p - 2} \\
6
\end{array}
$$

The result is

$$6p^2 + 7p + 1 + \frac{6}{2p - 2}$$

$$= 6p^2 + 7p + 1 + \frac{2 \cdot 3}{2(p - 1)}$$

$$= 6p^2 + 7p + 1 + \frac{3}{p - 1}.$$

18. If one operation can be done in 1.4×10^{-7} seconds, then one trillion operations will take

$$\left(1 \times 10^{12}\right)\left(1.4 \times 10^{-7}\right) = 1.4 \times 10^{12 + (-7)}$$
$$= 1.4 \times 10^5$$

or $140{,}000$ seconds.

19. $8t^2 + 10tv + 3v^2$

$= 8t^2 + 6tv + 4tv + 3v^2 \quad$ $6 \cdot 4 = 24;$
$ 6 + 4 = 10$

$= \left(8t^2 + 6tv\right) + \left(4tv + 3v^2\right)$

$= 2t(4t + 3v) + v(4t + 3v)$

$= (4t + 3v)(2t + v)$

20. $8r^2 - 9rs + 12s^2$

To factor this polynomial by the grouping method, we must find two integers whose product is $(8)(12) = 96$ and whose sum is -9. There is no pair of integers that satisfies both of these conditions, so the polynomial is prime.

21. $16x^4 - 1$
$= \left(4x^2\right)^2 - (1)^2$
$= \left(4x^2 + 1\right)\left(4x^2 - 1\right)$
$= \left(4x^2 + 1\right)\left[(2x)^2 - (1)^2\right]$
$= \left(4x^2 + 1\right)(2x + 1)(2x - 1)$

22. $\qquad r^2 = 2r + 15$
$r^2 - 2r - 15 = 0$
$(r + 3)(r - 5) = 0$

$r + 3 = 0 \qquad$ or $\qquad r - 5 = 0$
$r = -3 \qquad$ or $\qquad r = 5$

The solution set is $\{-3, 5\}$.

23. $(r - 5)(2r + 1)(3r - 2) = 0$

Note to reader: We may skip writing out the zero-factor property since this step can be easily performed mentally.

$r = 5 \quad$ or $\quad r = -\dfrac{1}{2} \quad$ or $\quad r = \dfrac{2}{3}$

The solution set is $\left\{5, -\dfrac{1}{2}, \dfrac{2}{3}\right\}$.

24. $\qquad$ Let x = the smaller number;
$x + 4$ = the larger number.
$x(x + 4) = x - 2$
$x^2 + 4x = x - 2$
$x^2 + 3x + 2 = 0$
$(x + 2)(x + 1) = 0$
$x = -2 \quad$ or $\quad x = -1$

The smaller number can be either -2 or -1.

25. $\quad$ Let w = the width of the rectangle;
$2w - 2$ = the length of the rectangle.

Use the formula $A = LW$

$60 = (2w - 2)w$
$60 = 2w^2 - 2w$
$2w^2 - 2w - 60 = 0$
$2\left(w^2 - w - 30\right) = 0$
$2(w - 6)(w + 5) = 0$
$w = 6 \quad$ or $\quad w = -5$

Discard -5 because the width cannot be negative. The width of the rectangle is 6 meters.

26.

$$\text{Let } x = \text{the number.}$$

$$\frac{x}{60} = \frac{15}{x} \quad \textit{Cross multiply}$$

$$x^2 = 60 \cdot 15 = 900$$

$$x^2 - 900 = 0$$

$$(x + 30)(x - 30) = 0$$

Since x is a positive number, it must be 30. Hence, the suggested maximum of one's diet for calories from fat is 30%.

27. All of the given expressions are equal to 1 for all real numbers for which they are defined. However, expressions (b), (c), and (d) all have one or more values for which the expression is undefined and therefore cannot be equal to 1. Since $k^2 + 2$ is *always* positive, the denominator in expression (a) is never equal to zero. This expression is defined and equal to 1 for all real numbers, so the correct choice is (a).

28. The appropriate choice is (d) since

$$\frac{-(3x + 4)}{7} = \frac{-3x - 4}{7}$$

$$\neq \frac{4 - 3x}{7}.$$

29. $\dfrac{5}{q} - \dfrac{1}{q} = \dfrac{5 - 1}{q} = \dfrac{4}{q}$

30. $\dfrac{3}{7} + \dfrac{4}{r} = \dfrac{3 \cdot r}{7 \cdot r} + \dfrac{4 \cdot 7}{r \cdot 7} \quad \textit{LCD = 7r}$

$$= \dfrac{3r + 28}{7r}$$

31. $\dfrac{4}{5q - 20} - \dfrac{1}{3q - 12}$

$$= \dfrac{4}{5(q - 4)} - \dfrac{1}{3(q - 4)}$$

$$= \dfrac{4 \cdot 3}{5(q - 4) \cdot 3} - \dfrac{1 \cdot 5}{3(q - 4) \cdot 5}$$

$$\textit{LCD} = 5 \cdot 3 \cdot (q - 4) = 15(q - 4)$$

$$= \dfrac{12 - 5}{15(q - 4)} = \dfrac{7}{15(q - 4)}$$

32. $\dfrac{2}{k^2 + k} - \dfrac{3}{k^2 - k}$

$$= \dfrac{2}{k(k + 1)} - \dfrac{3}{k(k - 1)}$$

$$= \dfrac{2(k - 1)}{k(k + 1)(k - 1)} - \dfrac{3(k + 1)}{k(k - 1)(k + 1)}$$

$$\textit{LCD} = k(k + 1)(k - 1)$$

$$= \dfrac{2(k - 1) - 3(k + 1)}{k(k + 1)(k - 1)}$$

$$= \dfrac{2k - 2 - 3k - 3}{k(k + 1)(k - 1)}$$

$$= \dfrac{-k - 5}{k(k + 1)(k - 1)}$$

33. $\dfrac{7z^2 + 49z + 70}{16z^2 + 72z - 40} \div \dfrac{3z + 6}{4z^2 - 1}$

$$= \dfrac{7z^2 + 49z + 70}{16z^2 + 72z - 40} \cdot \dfrac{4z^2 - 1}{3z + 6}$$

$$= \dfrac{7(z^2 + 7z + 10)}{8(2z^2 + 9z - 5)} \cdot \dfrac{(2z + 1)(2z - 1)}{3(z + 2)}$$

$$= \dfrac{7(z + 5)(z + 2)}{8(2z - 1)(z + 5)} \cdot \dfrac{(2z + 1)(2z - 1)}{3(z + 2)}$$

$$= \dfrac{7(2z + 1)}{8 \cdot 3} = \dfrac{7(2z + 1)}{24}$$

34. $\dfrac{\dfrac{4}{a} + \dfrac{5}{2a}}{\dfrac{7}{6a} - \dfrac{1}{5a}}$

$$= \dfrac{\left(\dfrac{4}{a} + \dfrac{5}{2a}\right) \cdot 30a}{\left(\dfrac{7}{6a} - \dfrac{1}{5a}\right) \cdot 30a} \quad \begin{array}{l}\textit{Multiply by}\\ \textit{LCD, 30a}\end{array}$$

$$= \dfrac{\dfrac{4}{a}(30a) + \dfrac{5}{2a}(30a)}{\dfrac{7}{6a}(30a) - \dfrac{1}{5a}(30a)}$$

$$= \dfrac{4 \cdot 30 + 5 \cdot 15}{7 \cdot 5 - 1 \cdot 6}$$

$$= \dfrac{120 + 75}{35 - 6} = \dfrac{195}{29}$$

35. $\dfrac{1}{x - 4} = \dfrac{3}{2x}$

To avoid zero denominators, x cannot equal 4 or 0.

36. $\dfrac{r + 2}{5} = \dfrac{r - 3}{3}$

$$15\left(\dfrac{r + 2}{5}\right) = 15\left(\dfrac{r - 3}{3}\right) \quad \begin{array}{l}\textit{Multiply by}\\ \textit{LCD, 15}\end{array}$$

$$3(r + 2) = 5(r - 3)$$

$$3r + 6 = 5r - 15$$

$$21 = 2r$$

$$\dfrac{21}{2} = r$$

The solution set is $\left\{\dfrac{21}{2}\right\}$.

37.

$$\frac{1}{x} = \frac{1}{x+1} + \frac{1}{2}$$

$$2x(x+1)\left(\frac{1}{x}\right) = 2x(x+1)\left(\frac{1}{x+1} + \frac{1}{2}\right)$$

Multiply by LCD, $2x(x+1)$

$$2(x+1) = 2x(x+1)\left(\frac{1}{x+1}\right)$$
$$+ 2x(x+1)\left(\frac{1}{2}\right)$$

$$2(x+1) = 2x + x(x+1)$$
$$2x + 2 = 2x + x^2 + x$$
$$0 = x^2 + x - 2$$
$$0 = (x+2)(x-1)$$
$$x = -2 \quad \text{or} \quad x = 1$$

Check $x = -2$: $\quad -\frac{1}{2} = -1 + \frac{1}{2}$ *True*

Check $x = 1$: $\quad 1 = \frac{1}{2} + \frac{1}{2}$ *True*

Thus, the solution set is $\{-2, 1\}$.

38. Let $x =$ amount of time for Arlene to reach her destination;

$x + \frac{1}{2} =$ amount of time for the return trip.

Use $d = rt$ to complete a table as follows.

	Rate	Time	Distance
To business	60	x	$60x$
Coming home	50	$x + \frac{1}{2}$	$50\left(x + \frac{1}{2}\right)$

Since the distances are equal,

$$60x = 50\left(x + \frac{1}{2}\right).$$

$$60x = 50x + 50\left(\frac{1}{2}\right)$$

$$10x = 25$$

$$x = \frac{25}{10} = \frac{5}{2}$$

Arlene's one way trip was

$$60 \cdot \frac{5}{2} = 30 \cdot 5 = 150 \text{ miles.}$$

39. Let $x =$ the number of hours it will take Juanita and Benito to weed the yard working together.

	Rate	Time working together	Fractional part of job done when working together
Juanita	$\frac{1}{3}$	x	$\frac{1}{3}x$
Benito	$\frac{1}{2}$	x	$\frac{1}{2}x$

$$\frac{1}{3}x + \frac{1}{2}x = 1$$

$$6\left(\frac{1}{3}x + \frac{1}{2}x\right) = 6 \cdot 1 \quad \begin{array}{l} \textit{Multiply by} \\ \textit{LCD, 6} \end{array}$$

$$6\left(\frac{1}{3}x\right) + 6\left(\frac{1}{2}x\right) = 6$$

$$2x + 3x = 6$$

$$5x = 6$$

$$x = \frac{6}{5}$$

If Juanita and Benito worked together, it would take them $\frac{6}{5}$ or $1\frac{1}{5}$ hours to weed the yard.

40. The force F required to compress a spring varies directly as the change c in the length of the spring, so

$$F = kc.$$
$$12 = k(3) \quad \textit{Let } F = 12, \; c = 3$$
$$4 = k$$

So $F = 4c$ and when $c = 5$,
$$F = 4(5) = 20.$$

A force of 20 pounds is required to compress the spring 5 inches.

CHAPTER 8   EQUATIONS OF LINES, INEQUALITIES, AND FUNCTIONS

Section 8.1

1. **(a)** The graph rises (so the numbers increase) in the consecutive years 1991–1992 and 1993–1994.

   (b) The biggest drop in the graph was between 1992 and 1993.

   (c) There were about 6000 solar collectors used for heating pools in 1993.

3. **(a)** The point $(1, 6)$ is located in quadrant I, since the x- and y-coordinates are both positive.

   (b) The point $(-4, -2)$ is located in quadrant III, since the x- and y-coordinates are both negative.

   (c) The point $(-3, 6)$ is located in quadrant II, since the x-coordinate is negative and the y-coordinate is positive.

   (d) The point $(7, -5)$ is located in quadrant IV, since the x-coordinate is positive and the y-coordinate is negative.

   (e) The point $(-3, 0)$ is located on the x-axis, so it does not belong to any quadrant.

For Exercises 5–14, see the rectangular coordinate system after Exercise 14.

5. To plot $(2, 3)$, go two units from zero to the right along the x-axis, and then go three units up parallel to the y-axis.

7. To plot $(-3, -2)$, go three units from zero to the left along the x-axis, and then go two units down parallel to the y-axis.

9. To plot $(0, 5)$, do not move along the x-axis at all since the x-coordinate is 0. Move five units up along the y-axis.

11. To plot $(-2, 4)$, go two units from zero to the left along the x-axis, and then go four units up parallel to the y-axis.

13. To plot $(-2, 0)$, go two units to the left along the x-axis. Do not move up or down since the y-coordinate is 0.

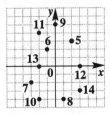

15. $x - y = 3$
    To complete the table, substitute the given values for x and y in the equation.
    For $x = 0$: $x - y = 3$
    $$0 - y = 3$$
    $$y = -3 \quad (0, -3)$$
    For $y = 0$: $x - y = 3$
    $$x - 0 = 3$$
    $$x = 3 \quad (3, 0)$$
    For $x = 5$: $x - y = 3$
    $$5 - y = 3$$
    $$-y = -2$$
    $$y = 2 \quad (5, 2)$$
    For $x = 2$: $x - y = 3$
    $$2 - y = 3$$
    $$-y = 1$$
    $$y = -1 \quad (2, -1)$$

    Plot the ordered pairs and draw the line through them.

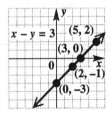

17. In quadrant III, both coordinates of the ordered pairs are negative. If $x + y = k$ and k is positive, then either x or y must be positive, because the sum of two negative numbers is negative.

19. $2x + 3y = 12$
    To find the x-intercept, let $y = 0$.
    $$2x + 3y = 12$$
    $$2x + 3(0) = 12$$
    $$2x = 12$$
    $$x = 6$$

    The x-intercept is $(6, 0)$.
    To find the y-intercept, let $x = 0$.
    $$2x + 3y = 12$$
    $$2(0) + 3y = 12$$
    $$3y = 12$$
    $$y = 4$$

    The y-intercept is $(0, 4)$.
    Plot the intercepts and draw the line through them.

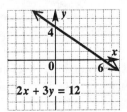

21. $x - 3y = 6$

To find the x-intercept, let $y = 0$.

$$x - 3y = 6$$
$$x - 3(0) = 6$$
$$x - 0 = 6$$
$$x = 6$$

The x-intercept is $(6, 0)$.
To find the y-intercept, let $x = 0$.

$$x - 3y = 6$$
$$0 - 3y = 6$$
$$-3y = 6$$
$$y = -2$$

The y-intercept is $(0, -2)$.
Plot the intercepts and draw the line through them.

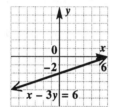

23. $\frac{2}{3}x - 3y = 7$

To find the x-intercept, let $y = 0$.

$$\frac{2}{3}x - 3(0) = 7$$
$$\frac{2}{3}x = 7$$
$$x = \frac{3}{2} \cdot 7 = \frac{21}{2}$$

The x-intercept is $\left(\frac{21}{2}, 0\right)$.

To find the y-intercept, let $x = 0$.

$$\frac{2}{3}(0) - 3y = 7$$
$$-3y = 7$$
$$y = -\frac{7}{3}$$

The y-intercept is $\left(0, -\frac{7}{3}\right)$.

Plot the intercepts and draw the line through them.

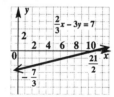

25. $y = 5$

This is a horizontal line. Every point has
y-coordinate 5, so no point has y-coordinate 0.
There is no x-intercept.
Since every point of the line has y-coordinate 5,

the y-intercept is $(0, 5)$. Draw the horizontal line
through $(0, 5)$.

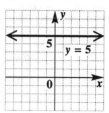

27. $x + 4 = 0 \quad (x = -4)$

This is a vertical line. Every point has x-coordinate
-4, so the x-intercept is $(-4, 0)$.
Since every point of the line has x-coordinate -4,
no point has x-coordinate 0. There is no
y-intercept. Draw the vertical line through $(-4, 0)$.

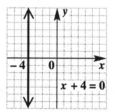

29. $x + 5y = 0$

To find the x-intercept, let $y = 0$.

$$x + 5y = 0$$
$$x + 5(0) = 0$$
$$x = 0$$

The x-intercept is $(0, 0)$, and since $x = 0$, this is
also the y-intercept. Since the intercepts are the
same, another point is needed to graph the line.
Choose any number for y, say $y = -1$, and solve
the equation for x.

$$x + 5y = 0$$
$$x + 5(-1) = 0$$
$$x = 5$$

This gives the ordered pair $(5, -1)$. Plot $(5, -1)$
and $(0, 0)$, and draw the line through them.

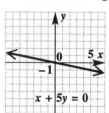

31. The graph goes through the point $(2, 6)$ which
satisfies only equation (c). The correct equation is
(c).

33. **(a)** From the table, the x-intercept is $(1.5, 0)$.

(b) The y-intercept is $(0, 3)$.

(c) The point $(0, 3)$ satisfies equations C and D.
The point $(.5, 2)$ satisfies only equation D.
Therefore, the correct equation is D.

35. R has x-coordinate 6 and y-coordinate -2, so the ordered pair is $(6, -2)$.

36. The midpoint of $\overline{PR}$ has x-coordinate halfway from 4 to 6, which is 5, and y-coordinate -2, so it's the ordered pair $S = (5, -2)$.

37. The midpoint of $\overline{QR}$ has x-coordinate 6 and y-coordinate halfway from -2 to 2, so it's the ordered pair $T = (6, 0)$.

38. $M = (x\text{-coordinate of } S, y\text{-coordinate of } T)$
$= (5, 0)$

39. $P = (4, -2)$ and $Q = (6, 2)$. The average of the x-coordinates of P and Q is $\dfrac{4+6}{2} = \dfrac{10}{2} = 5$.
The average of the y-coordinates of P and Q is
$\dfrac{-2+2}{2} = \dfrac{0}{2} = 0$.

40. The x-coordinate of M is the average of the x-coordinates of P and Q. The y-coordinate of M is the average of the y-coordinates of P and Q.

41. $\text{slope} = \dfrac{\text{change in vertical position}}{\text{change in horizontal position}}$

$= \dfrac{-30 \text{ feet}}{100 \text{ feet}}$

Choices (a) $-.3$, (b) $-\dfrac{3}{10}$, and (d) $-\dfrac{30}{100}$ are all correct.

43. To get to B from A, we must go up 2 units and move right 1 unit. Thus,

$$\text{slope of } AB = \dfrac{\text{rise}}{\text{run}} = \dfrac{2}{1} = 2.$$

45. $\text{slope of } CD = \dfrac{\text{rise}}{\text{run}} = \dfrac{-7}{0}$, which is undefined.

47. Let $(x_1, y_1) = (-2, -3)$ and $(x_2, y_2) = (-1, 5)$. Then

$$m = \dfrac{y_2 - y_1}{x_2 - x_1} = \dfrac{5 - (-3)}{-1 - (-2)} = \dfrac{8}{1} = 8.$$

The slope is 8.

49. Let $(x_1, y_1) = (-4, 1)$ and
$(x_2, y_2) = (2, 6)$. Then

$$m = \dfrac{y_2 - y_1}{x_2 - x_1} = \dfrac{6 - 1}{2 - (-4)} = \dfrac{5}{6}.$$

The slope is $\dfrac{5}{6}$.

51. Let $(x_1, y_1) = (2, 4)$ and
$(x_2, y_2) = (-4, 4)$. Then

$$m = \dfrac{y_2 - y_1}{x_2 - x_1} = \dfrac{4 - 4}{-4 - 2} = \dfrac{0}{-6} = 0.$$

The slope is 0.

53. "The line has positive slope" means that the line goes up from left to right. This is line B.

55. "The line has slope 0" means that there is no vertical change; that is, the line is horizontal. This is line A.

57. To find the slope of

$$x + 2y = 4,$$

first find the intercepts. Replace y with 0 to find that the x-intercept is $(4, 0)$; replace x with 0 to find that the y-intercept is $(0, 2)$. The slope is then

$$m = \dfrac{2 - 0}{0 - 4} = -\dfrac{2}{4} = -\dfrac{1}{2}.$$

To sketch the graph, plot the intercepts and draw the line through them.

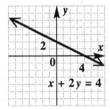

59. To find the slope of

$$-x + y = 4,$$

first find the intercepts. Replace y with 0 to find that the x-intercept is $(-4, 0)$; replace x with 0 to find that the y-intercept is $(0, 4)$. The slope is then

$$m = \dfrac{4 - 0}{0 - (-4)} = \dfrac{4}{4} = 1.$$

To sketch the graph, plot the intercepts and draw the line through them.

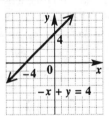

61. In the equation

$$y = 4x,$$

replace x with 0 and then x with 1 to get the ordered pairs $(0, 0)$ and $(1, 4)$, respectively. (There are other possibilities for ordered pairs.) The slope is then

$$m = \dfrac{4 - 0}{1 - 0} = \dfrac{4}{1} = 4.$$

To sketch the graph, plot the two points and draw the line through them.

continued

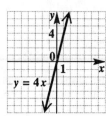

63. $x - 3 = 0 \quad (x = 3)$

The graph of $x = 3$ is the vertical line with
x-intercept $(3, 0)$. The slope of a vertical line is
undefined.

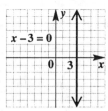

65. A vertical line has equation $\underline{x} = c$ for some
constant c; a horizontal line has equation $\underline{y} = d$
for some constant d.

67. Locate the point on coordinate system. Write the
slope as a fraction, if necessary. Move up or down
the number of units in the numerator and left or
right the number of units in the denominator to
determine a second point. Draw the line through
these two points.

69. To graph the line through $(-2, -3)$ with slope
$m = \dfrac{5}{4}$, locate $(-2, -3)$ on the graph. To find a
second point, use the definition of slope.

$$m = \frac{\text{change in } y}{\text{change in } x} = \frac{5}{4}$$

From $(-2, -3)$, go up 5 units. Then go 4 units to
the right to get to $(2, 2)$. Draw the line through
$(-2, -3)$ and $(2, 2)$.

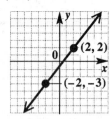

71. To graph the line through $(0, -4)$ with slope
$m = -\dfrac{3}{2}$, locate the point $(0, -4)$ on the graph.

To find a second point on the line, use the
definition of slope, writing $-\dfrac{3}{2}$ as $\dfrac{-3}{2}$.

$$m = \frac{\text{change in } y}{\text{change in } x} = \frac{-3}{2}$$

From $(0, -4)$, move 3 units down and then 2
units to the right. Draw a line through this second

point and $(0, -4)$. (Note that the slope could also
be written as $\dfrac{3}{-2}$. In this case, move 3 units up
and 2 units to the left to get another point on the
same line.)

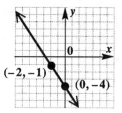

73. Locate $(-2, -4)$. Then use $m = 4 = \dfrac{4}{1}$ to go 4
units up and 1 unit right to $(-1, 0)$.

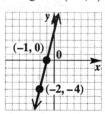

75. $3x = y$ and $2y - 6x = 5$

The slope of the first line is the coefficient of x,
namely 3. Solve the second equation for y.

$$2y = 6x + 5$$
$$y = 3x + \frac{5}{2}$$

So the slope of the second line is also 3, and the
lines are *parallel*.

77. $4x - 3y = 8$ and $4y + 3x = 12$

Solve the equations for y.

$$-3y = -4x + 8 \qquad\qquad 4y = -3x + 12$$
$$y = \frac{4}{3}x - \frac{8}{3} \qquad\qquad y = -\frac{3}{4}x + 3$$

The slopes, $\dfrac{4}{3}$ and $-\dfrac{3}{4}$, are negative reciprocals
of one another, so the lines are *perpendicular*.

79. $4x - 3y = 5$ and $3x - 4y = 2$

Solve the equations for y.

$$-3y = -4x + 5 \qquad\qquad -4y = -3x + 2$$
$$y = \frac{4}{3}x - \frac{5}{3} \qquad\qquad y = \frac{3}{4}x - \frac{1}{2}$$

The slopes are $\dfrac{4}{3}$ and $\dfrac{3}{4}$. The lines are *neither*
parallel nor perpendicular.

81. The slope of the line through $(4, 6)$ and $(-8, 7)$ is

$$m = \frac{7 - 6}{-8 - 4} = \frac{1}{-12} = -\frac{1}{12}.$$

The slope of the line through $(7, 4)$ and $(-5, 5)$ is

$$m = \frac{5 - 4}{-5 - 7} = \frac{1}{-12} = -\frac{1}{12}.$$

Since the slopes are equal, the two lines are *parallel*.

83. For 1991 to 1992:

$$m = \frac{15.15 - 15.64}{1992 - 1991} = \frac{-.49}{1} = -.49$$

For 1992 to 1994:

$$m = \frac{14.17 - 15.15}{1994 - 1992} = \frac{-.98}{2} = -.49$$

For 1993 to 1995:

$$m = \frac{13.68 - 14.66}{1995 - 1993} = \frac{-.98}{2} = -.49$$

These answers suggest that the average rate of change has remained constant at $-.49$

quadrillion $\dfrac{\text{BTUs}}{\text{yr}}$.

85. **(a)** For 1991 through 1996:

$$m = \frac{22.9 - 26.2}{1996 - 1991} = \frac{-3.3}{5} = -.66$$

The average rate of change is

$-.66$ million $\dfrac{\text{kilowatts}}{\text{year}}$.

(b) For 1993 through 1996:

$$m = \frac{22.9 - 26.2}{1996 - 1993} = \frac{-3.3}{3} = -1.1$$

The average rate of change is

-1.1 million $\dfrac{\text{kilowatts}}{\text{year}}$.

(c) The graph is not a straight line, so the average rate of change varies for different pairs of years.

Section 8.2

1. Choice (a) $3x - 2y = 5$ is in the form $Ax + By = C$ with $A \geq 0$ and integers A, B, and C having no common factor (except 1).

3. Choice (a) $y = 6x + 2$ is in the form $y = mx + b$.

5.
$$y = -3x + 10$$
$$3x + y = 10 \qquad \textit{Standard form}$$

7. $y = 2x + 3$
This line is in slope-intercept form with slope $m = 2$ and y-intercept $(0, b) = (0, 3)$. The only graph with positive slope and with a positive y-coordinate of its y-intercept is A.

9. $y = -2x - 3$
This line is in slope-intercept form with slope $m = -2$ and y-intercept $(0, b) = (0, -3)$. The

only graph with negative slope and with a negative y-coordinate of its y-intercept is C.

11. $y = 2x$
This line has slope $m = 2$ and y-intercept $(0, b) = (0, 0)$. The only graph with positive slope and with y-intercept $(0, 0)$ is H.

13. $y = 3$
This line is a horizontal line with y-intercept $(0, 3)$. Its y-coordinate is positive. The only graph that has these characteristics is B.

15. Through $(-2, 4)$; slope $= -\dfrac{3}{4}$
Use the point-slope form with $(x_1, y_1) = (-2, 4)$ and $m = -\dfrac{3}{4}$.

$$y - y_1 = m(x - x_1)$$
$$y - 4 = -\frac{3}{4}[x - (-2)]$$
$$y - 4 = -\frac{3}{4}(x + 2)$$
$$y - 4 = -\frac{3}{4}x - \frac{3}{2}$$
$$y = -\frac{3}{4}x - \frac{3}{2} + \frac{8}{2}$$
$$y = -\frac{3}{4}x + \frac{5}{2}$$

This slope-intercept form gives the y-intercept $\left(0, \dfrac{5}{2}\right)$.

17. Through $(5, 8)$; slope $= -2$
Use the point-slope form with $(x_1, y_1) = (5, 8)$ and $m = -2$.

$$y - y_1 = m(x - x_1)$$
$$y - 8 = -2(x - 5)$$
$$y - 8 = -2x + 10$$
$$y = -2x + 18$$

19. Through $(-5, 4)$; slope $= \dfrac{1}{2}$
Use the point-slope form with $(x_1, y_1) = (-5, 4)$ and $m = \dfrac{1}{2}$.

$$y - y_1 = m(x - x_1)$$
$$y - 4 = \frac{1}{2}[x - (-5)]$$
$$y - 4 = \frac{1}{2}(x + 5)$$
$$y - 4 = \frac{1}{2}x + \frac{5}{2}$$
$$y = \frac{1}{2}x + \frac{5}{2} + \frac{8}{2}$$
$$y = \frac{1}{2}x + \frac{13}{2}$$

21. Through $(3, 0)$; slope $= 4$

Use the point-slope form with $(x_1, y_1) = (3, 0)$ and $m = 4$.

$$y - y_1 = m(x - x_1)$$
$$y - 0 = 4(x - 3)$$
$$y = 4x - 12$$

23. The point-slope form, $y - y_1 = m(x - x_1)$, is used when the slope and one point on a line or two points on a line are known. The slope-intercept form, $y = mx + b$, is used when the slope and y-intercept are known. The standard form, $Ax + By = C$, is not useful for writing the equation, but can be found from other forms of the equation. The form $y = d$ is used for a horizontal line through the point (c, d). The form $x = c$ is used for a vertical line through the point (c, d).

25. Through $(9, 5)$; slope 0

A line with slope 0 is a horizontal line. A horizontal line through the point (x, k) has equation $y = k$. Here $k = 5$, so an equation is $y = 5$.

27. Through $(9, 10)$; undefined slope

A vertical line has undefined slope and equation $x = c$. Since the x-value in $(9, 10)$ is 9, the equation is $x = 9$.

29. Through $(.5, .2)$; vertical

A vertical line through the point (k, y) has equation $x = k$. Here $k = .5$, so the equation is $x = .5$.

31. Through $(-7, 8)$; horizontal

A horizontal line through the point (x, k) has equation $y = k$, so the equation is $y = 8$.

33. $(3, 4)$ and $(5, 8)$

Find the slope.

$$m = \frac{8 - 4}{5 - 3} = \frac{4}{2} = 2$$

Use the point-slope form with $(x_1, y_1) = (3, 4)$ and $m = 2$.

$$y - y_1 = m(x - x_1)$$
$$y - 4 = 2(x - 3)$$
$$y - 4 = 2x - 6$$
$$y = 2x - 2$$

35. $(6, 1)$ and $(-2, 5)$

Find the slope.

$$m = \frac{5 - 1}{-2 - 6} = \frac{4}{-8} = -\frac{1}{2}$$

Use the point-slope form with $(x_1, y_1) = (6, 1)$ and $m = -\frac{1}{2}$.

$$y - y_1 = m(x - x_1)$$
$$y - 1 = -\frac{1}{2}(x - 6)$$
$$y - 1 = -\frac{1}{2}x + 3$$
$$y = -\frac{1}{2}x + 4$$

37. $\left(-\frac{2}{5}, \frac{2}{5}\right)$ and $\left(\frac{4}{3}, \frac{2}{3}\right)$

Find the slope.

$$m = \frac{\dfrac{2}{3} - \dfrac{2}{5}}{\dfrac{4}{3} - \left(-\dfrac{2}{5}\right)} = \frac{\dfrac{10 - 6}{15}}{\dfrac{20 + 6}{15}}$$

$$= \frac{\dfrac{4}{15}}{\dfrac{26}{15}} = \frac{4}{26} = \frac{2}{13}$$

Use the point-slope form with $(x_1, y_1) = \left(-\dfrac{2}{5}, \dfrac{2}{5}\right)$ and $m = \dfrac{2}{13}$.

$$y - \frac{2}{5} = \frac{2}{13}\left[x - \left(-\frac{2}{5}\right)\right]$$
$$y - \frac{2}{5} = \frac{2}{13}\left(x + \frac{2}{5}\right)$$
$$y - \frac{2}{5} = \frac{2}{13}x + \frac{4}{65}$$
$$y = \frac{2}{13}x + \frac{4}{65} + \frac{26}{65}$$
$$y = \frac{2}{13}x + \frac{6}{13}\left(\frac{30}{65} = \frac{6}{13}\right)$$

39. $(2, 5)$ and $(1, 5)$

Find the slope.

$$m = \frac{5 - 5}{1 - 2} = \frac{0}{-1} = 0$$

A line with slope 0 is horizontal. A horizontal line through the point (x, k) has equation $y = k$, so the equation is $y = 5$.

41. $(7, 6)$ and $(7, -8)$

Find the slope.

$$m = \frac{-8 - 6}{7 - 7} = \frac{-14}{0} \quad \textit{Undefined}$$

A line with undefined slope is a vertical line. The equation of a vertical line is $x = k$, where k is the common x-value. So the equation is $x = 7$.

43. $(1, -3)$ and $(-1, -3)$

Find the slope.

$$m = \frac{-3 - (-3)}{-1 - 1} = \frac{0}{-2} = 0$$

A line with slope 0 is horizontal. A horizontal line through the point (x, k) has equation $y = k$, so the equation is $y = -3$.

45. $m = 5; b = 15$

Substitute these values in the slope-intercept form.

$$y = mx + b$$
$$y = 5x + 15$$

47. $m = -\dfrac{2}{3}; b = \dfrac{4}{5}$

Substitute these values in the slope-intercept form.

$$y = mx + b$$
$$y = -\dfrac{2}{3}x + \dfrac{4}{5}$$

49. Slope $\dfrac{2}{5}$; y-intercept $(0, 5)$

Here, $m = \dfrac{2}{5}$ and $b = 5$. Substitute these values in the slope-intercept form.

$$y = mx + b$$
$$y = \dfrac{2}{5}x + 5$$

51. To get to the point $(3, 3)$ from the y-intercept $(0, 1)$, we must go up 2 units and to the right 3 units, so the slope is $\dfrac{2}{3}$. The slope-intercept form is

$$y = \dfrac{2}{3}x + 1.$$

53. $x + y = 12$

(a) Solve for y to get the equation in slope-intercept form.

$$x + y = 12$$
$$y = -x + 12$$

(b) The slope is the coefficient of x, -1.

(c) The y-intercept is the point $(0, b)$ or $(0, 12)$.

55. $5x + 2y = 20$

(a) Solve for y.

$$2y = -5x + 20$$
$$y = -\dfrac{5}{2}x + 10$$

(b) The slope is $-\dfrac{5}{2}$.

(c) The y-intercept is $(0, 10)$.

57. $2x - 3y = 10$

(a) Solve for y.

$$-3y = -2x + 10$$
$$y = \dfrac{2}{3}x - \dfrac{10}{3}$$

(b) The slope is $\dfrac{2}{3}$.

(c) The y-intercept is $\left(0, -\dfrac{10}{3}\right)$.

59. Through $(7, 2)$; parallel to $3x - y = 8$

Find the slope of $3x - y = 8$.

$$-y = -3x + 8$$
$$y = 3x - 8$$

The slope is 3, so a line parallel to it also has slope 3. Use $m = 3$ and $(x_1, y_1) = (7, 2)$ in the point-slope form.

$$y - y_1 = m(x - x_1)$$
$$y - 2 = 3(x - 7)$$
$$y - 2 = 3x - 21$$
$$y = 3x - 19$$

61. Through $(-2, -2)$; parallel to $-x + 2y = 10$

Find the slope of $-x + 2y = 10$.

$$2y = x + 10$$
$$y = \dfrac{1}{2}x + 5$$

The slope is $\dfrac{1}{2}$, so a line parallel to it also has slope $\dfrac{1}{2}$. Use $m = \dfrac{1}{2}$ and $(x_1, y_1) = (-2, -2)$ in the point-slope form.

$$y - y_1 = m(x - x_1)$$
$$y - (-2) = \dfrac{1}{2}[x - (-2)]$$
$$y + 2 = \dfrac{1}{2}(x + 2)$$
$$y + 2 = \dfrac{1}{2}x + 1$$
$$y = \dfrac{1}{2}x - 1$$

63. Through $(8, 5)$; perpendicular to $2x - y = 7$

Find the slope of $2x - y = 7$.

$$-y = -2x + 7$$
$$y = 2x - 7$$

The slope of the line is 2. Therefore, the slope of the line perpendicular to it is $-\dfrac{1}{2}$ since

$$2\left(-\dfrac{1}{2}\right) = -1.$$ Use $m = -\dfrac{1}{2}$ and $(x_1, y_1) = (8, 5)$ in the point-slope form.

$$y - y_1 = m(x - x_1)$$
$$y - 5 = -\dfrac{1}{2}(x - 8)$$
$$y - 5 = -\dfrac{1}{2}x + 4$$
$$y = -\dfrac{1}{2}x + 9$$

65. Through $(-2, 7)$; perpendicular to $x = 9$
$x = 9$ is a vertical line so a line perpendicular to it will be a horizontal line. It goes through $(-2, 7)$ so its equation is

$$y = 7.$$

67. Since it costs $15 plus $3 per day to rent a chain saw, it costs $3x + 15$ dollars for x days. Thus, if y represents the charge to the user (in dollars), the equation is $y = 3x + 15$.
When $x = 1$, $y = 3(1) + 15 = 3 + 15 = 18$.

Ordered pair: $(1, 18)$

When $x = 5$, $y = 3(5) + 15 = 15 + 15 = 30$.

Ordered pair: $(5, 30)$

When $x = 10$, $y = 3(10) + 15 = 30 + 15 = 45$.

Ordered pair: $(10, 45)$

69. $(1, 80.1)$ and $(6, 89.4)$

$$m = \frac{89.4 - 80.1}{6 - 1} = \frac{9.3}{5} = 1.86$$

Use the point-slope form.

$$y - 80.1 = 1.86(x - 1)$$
$$y - 80.1 = 1.86x - 1.86$$
$$y = 1.86x + 78.24$$

71. **(a)** $(3, 21, 696)$ and $(7, 25, 050)$

$$m = \frac{25, 050 - 21, 696}{7 - 3} = \frac{3354}{4} = 838.5$$

Use the point-slope form.

$$y - 21, 696 = 838.5(x - 3)$$
$$y - 21, 696 = 838.5x - 2515.5$$
$$y = 838.5x + 19, 180.5$$

(b) $x = 5$ represents 1995

$$y = 838.5(5) + 19, 180.5$$
$$= 4192.5 + 19, 180.5$$
$$= 23, 373$$

This is close to the value $23, 583$ that is given.

73. **(a)** $2x + 7 - x = 4x - 2$
$$x + 7 = 4x - 2$$
$$-3x + 9 = 0$$
$$-3x + 9 = y$$

(b) From the screen, we see that $x = 3$ is the solution.

(c) $2x + 7 - x = 4x - 2$
$$x + 7 = 4x - 2$$
$$9 = 3x$$
$$3 = x$$
Solution set: $\{3\}$

75. **(a)** $3(2x + 1) - 2(x - 2) = 5$
$$6x + 3 - 2x + 4 - 5 = 0$$
$$4x + 2 = 0$$
$$4x + 2 = y$$

(b) From the screen, we see that $x = -.5$ is the solution.

(c) $3(2x + 1) - 2(x - 2) = 5$
$$6x + 3 - 2x + 4 = 5$$
$$4x + 7 = 5$$
$$4x = -2$$
$$x = -\frac{1}{2} \text{ or } -.5$$
Solution set: $\{-.5\}$

77. The solution to the equation $Y_1 = 0$ is the x-coordinate of the x-intercept. In this case, the x-intercept is greater than 10, so the correct choice must be (d).

79. When $C = 0°$, $F = \underline{32}°$, and when $C = 100°$, $F = \underline{212}°$. These are the freezing and boiling temperatures for water.

80. The two points of the form (C, F) would be $(0, 32)$ and $(100, 212)$.

81. $m = \dfrac{212 - 32}{100 - 0} = \dfrac{180}{100} = \dfrac{9}{5}$

82. Let $m = \dfrac{9}{5}$ and $(x_1, y_1) = (0, 32)$.

$$y - y_1 = m(x - x_1)$$
$$F - 32 = \frac{9}{5}(C - 0)$$
$$F - 32 = \frac{9}{5}C$$
$$F = \frac{9}{5}C + 32$$

83. $$F = \frac{9}{5}C + 32$$
$$F - 32 = \frac{9}{5}C$$
$$\frac{5}{9}(F - 32) = C$$

84. A temperature of $50°$ C corresponds to a temperature of $122°$ F.

Section 8.3

1. The boundary of the graph of $y \leq -x + 2$ will be a _solid_ line (since the inequality involves $\leq$), and the shading will be _below_ the line (since the inequality sign is $\leq$ or $<$).

3. The boundary of the graph of $y > -x + 2$ will be a _dashed_ line (since the inequality involves $>$), and the shading will be _above_ the line (since the inequality sign is $\geq$ or $>$).

5. Change the inequality to an equation and graph it. For example, to graph $3x + 2y > 5$, begin by graphing the line with equation $3x + 2y = 5$. The line should be dashed if the inequality involves $>$ or $<$. (This is the case for our example.) It should be solid if the inequality involves $\geq$ or $\leq$. Decide which side of the line to shade by solving for y. We get $y > -\dfrac{3}{2}x + \dfrac{5}{2}$. Because of the $>$ symbol, shade above the line. For the $<$ symbol, we would shade below the line. The shaded region shows the solutions.

7. $x + y \leq 2$

Graph the line $x + y = 2$ by drawing a solid line (since the inequality involves $\leq$) through the intercepts $(2, 0)$ and $(0, 2)$.

Test a point not on this line, such as $(0, 0)$.

$$x + y \leq 2$$
$$0 + 0 \leq 2$$
$$0 \leq 2 \quad \textit{True}$$

Shade that side of the line containing the test point $(0, 0)$.

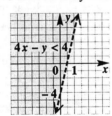

9. $4x - y < 4$

Graph the line $4x - y = 4$ by drawing a dashed line (since the inequality involves $<$) through the intercepts $(1, 0)$ and $(0, -4)$. Instead of using a test point, we will solve the inequality for y.

$$-y < -4x + 4$$
$$y > 4x - 4$$

Since we have "$y >$" in the last inequality, shade the region *above* the boundary line.

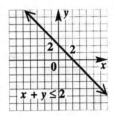

11. $x + 3y \geq -2$

Graph the solid line $x + 3y = -2$ (since the inequality involves $\geq$) through the intercepts $(-2, 0)$ and $\left(0, -\dfrac{2}{3}\right)$.

Test a point not on this line such as $(0, 0)$.

$$0 + 3(0) \geq -2$$
$$0 \geq -2 \quad \textit{True}$$

Shade that side of the line containing the test point $(0, 0)$.

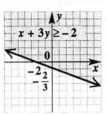

13. $x + y > 0$

Graph the line $x + y = 0$, which includes the points $(0, 0)$ and $(2, -2)$, as a dashed line (since the inequality involves $>$). Solving the inequality for y gives us

$$y > -x,$$

So shade the region above the boundary line.

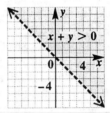

15. $x - 3y \leq 0$

Graph the solid line $x - 3y = 0$ through the points $(0, 0)$ and $(3, 1)$.

Solve the inequality for y.

$$-3y \leq -x$$
$$y \geq \frac{1}{3}x$$

Shade the region above the boundary line.

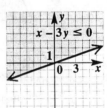

17. $y < x$

Graph the dashed line $y = x$ through $(0, 0)$ and $(2, 2)$. Since we have "$y <$" in the inequality, shade the region *below* the boundary line.

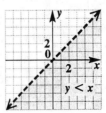

19. $x + y \leq 1$ and $x \geq 1$
Graph the solid line $x + y = 1$ through
$(0, 1)$ and $(1, 0)$. The inequality $x + y \leq 1$ can be
written as $y \leq -x + 1$, so shade the region below
the boundary line.
Graph the solid vertical line $x = 1$ through $(1, 0)$
and shade the region to the right. The required
graph is the common shaded area as well as the
portions of the lines that bound it.

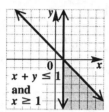

21. $2x - y \geq 2$ and $y < 4$
Graph the solid line $2x - y = 2$ through the
intercepts $(1, 0)$ and $(0, -2)$. Test $(0, 0)$ to get
$0 \geq 2$, a false statement. Shade that side of the
graph not containing $(0, 0)$. To graph $y < 4$ on the
same axes, graph the dashed horizontal line
through $(0, 4)$. Test $(0, 0)$ to get $0 < 4$, a true
statement. Shade that side of the dashed line
containing $(0, 0)$.
The word "and" indicates the intersection of the
two graphs. The final solution set consists of the
region where the two shaded regions overlap.

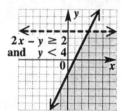

23. $x + y > -5$ and $y < -2$
Graph $x + y = -5$, which has intercepts $(-5, 0)$
and $(0, -5)$, as a dashed line. Test $(0, 0)$, which
yields $0 > -5$, a true statement. Shade the region
that includes $(0, 0)$.
Graph $y = -2$ as a dashed horizontal line. Shade
the region below $y = -2$. The required graph of
the intersection is the region common to both
graphs.

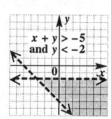

25. $|x| < 3$ can be rewritten as $-3 < x < 3$. The
boundaries are the dashed vertical lines $x = -3$
and $x = 3$. Since x is between -3 and 3, the graph
includes all points between the lines.

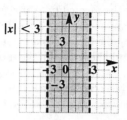

27. $|x + 1| < 2$ can be rewritten as
$$-2 < x + 1 < 2$$
$$-3 < x < 1.$$
The boundaries are the dashed vertical lines
$x = -3$ and $x = 1$. Since x is between -3 and 1,
the graph includes all points between the lines.

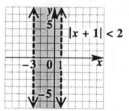

29. $x - y \geq 1$ or $y \geq 2$
Graph the solid line $x - y = 1$, which crosses the
y-axis at -1 and the x-axis at 1. Use $(0, 0)$ as a
test point, which yields $0 \geq 1$, a false statement.
Shade the region that does not include $(0, 0)$.
Now graph the solid line $y = 2$. Since the
inequality is $y \geq 2$, shade above this line.
The required graph of the union includes all the
shaded regions, that is, all the points that satisfy
either inequality.

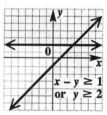

31. $x - 2 > y$ or $x < 1$
Graph $x - 2 = y$, which has intercepts
$(2, 0)$ and $(0, -2)$, as a dashed line. Test $(0, 0)$,
which yields $-2 > 0$, a false statement. Shade the
region that does not include $(0, 0)$.
Graph $x = 1$ as a dashed vertical line. Shade the
region to the left of $x = 1$.
The required graph of the union includes all the
shaded regions, that is, all the points that satisfy
either inequality.

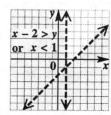

33. $3x + 2y < 6$ or $x - 2y > 2$
Graph $3x + 2y = 6$, which has intercepts
$(2, 0)$ and $(0, 3)$, as a dashed line. Test $(0, 0)$,
which yields $0 < 6$, a true statement. Shade the
region that includes $(0, 0)$.
Graph $x - 2y = 2$, which has intercepts
$(2, 0)$ and $(0, -1)$, as a dashed line. Test $(0, 0)$,
which yields $0 > 2$, a false statement. Shade the
region that does not include $(0, 0)$.
The required graph of the union includes all the
shaded regions, that is, all the points that satisfy
either inequality.

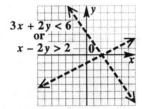

35. (a) The x-intercept is $(-4, 0)$, so the solution to
$y = 0$ is $\{-4\}$.

(b) The solution to $y < 0$ is $(-\infty, -4)$, since the
graph is below the x-axis for these values of x.

(c) The solution to $y > 0$ is $(-4, \infty)$, since the
graph is above the x-axis for these values of x.

37. (a) The x-intercept is $(3.5, 0)$, so the solution to
$y = 0$ is $\{3.5\}$.

(b) The solution to $y < 0$ is $(3.5, \infty)$, since the
graph is below the x-axis for these values of x.

(c) The solution to $y > 0$ is $(-\infty, 3.5)$, since the
graph is above the x-axis for these values of x.

39. $y \le 3x - 6$
The boundary line, $y = 3x - 6$, has slope 3 and
y-intercept -6. This would be graph B or graph C.
Since we want the region less than or equal to
$3x - 6$, we want the region on or below the
boundary line. The answer is graph C.

41. $y \le -3x - 6$
The slope of the boundary line $y = -3x - 6$ is
-3, and the y-intercept is -6. This would be
graph A or graph D. The inequality sign is $\le$, so
we want the region on or below the boundary line.
The answer is graph A.

43. (a) $5x + 3 = 0$
$$5x = -3$$
$$x = -\frac{3}{5} = -.6$$
Solution set: $\{-.6\}$

(b) $5x + 3 > 0$
$$5x > -3$$
$$x > -\frac{3}{5} \text{ or } -.6$$
Solution set: $(-.6, \infty)$

(c) $5x + 3 < 0$
$$5x < -3$$
$$x < -\frac{3}{5} \text{ or } -.6$$
Solution set: $(-\infty, -.6)$

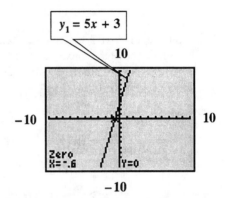

The x-intercept is $(-.6, 0)$, as in part (a). The
graph is above the x-axis for $x > -.6$, as in part
(b), and below the x-axis for $x < -.6$, as in part
(c).

45. (a) $-8x - (2x + 12) = 0$
$$-8x - 2x - 12 = 0$$
$$-10x - 12 = 0$$
$$-10x = 12$$
$$x = -1.2$$
Solution set: $\{-1.2\}$

(b) $-8x - (2x + 12) \ge 0$
$$-8x - 2x - 12 \ge 0$$
$$-10x - 12 \ge 0$$
$$-10x \ge 12$$
$$x \le -1.2$$
Solution set: $(-\infty, -1.2]$

(c) $-8x - (2x + 12) \le 0$
$$-8x - 2x - 12 \le 0$$
$$-10x - 12 \le 0$$
$$-10x \le 12$$
$$x \ge -1.2$$
Solution set: $[-1.2, \infty)$

continued

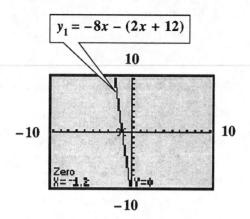

$$y_1 = -8x - (2x + 12)$$

The x-intercept is $(-1.2, 0)$, as in part (a). The graph is on or above the x-axis for $(-\infty, -1.2]$, as in part (b), and on or below the x-axis for $[-1.2, \infty)$, as in part (c).

47. The inequality we want to solve is

$$y \geq 85.$$
$$1.86x + 78.24 \geq 85$$
$$1.86x \geq 6.76$$
$$x \geq \frac{6.76}{1.86} \approx 3.6$$

Rounding 3.6 to 4, this inequality represents the years 1994–1996.

Section 8.4

1. We give one of many possible answers here. A function is a set of ordered pairs in which each first element determines exactly one second element. For example, $\{(0, 1), (1, 2), (2, 3), (3, 4) \dots\}$ is a function.

3. In an ordered pair of a relation, the first element is the independent variable.

5. $\{(2, 5), (3, 7), (4, 9), (5, 11)\}$
    The relation is a function since for each x-value, there is only one y-value. The domain is the set of x-values: $\{2, 3, 4, 5\}$. The range is the set of y-values: $\{5, 7, 9, 11\}$.

7. This relation is not a function since each input value corresponds to two output values. The domain is the set of positive real numbers; that is, $(0, \infty)$. The range is the set of positive real numbers along with their negatives; that is, $(-\infty, 0) \cup (0, \infty)$.

9. The relation is a function since each input value corresponds to exactly one output value. The domain is the set of inputs, {unleaded regular, unleaded premium, crude oil}. The range is the set of outputs, $\{1.22, 1.44, .21\}$.

11. Using the vertical line test, we find any vertical line will intersect the graph at most once. This indicates that the graph represents a function. This graph extends indefinitely to the left $(-\infty)$

and indefinitely to the right (∞). Therefore, the domain is $(-\infty, \infty)$. This graph extends indefinitely downward $(-\infty)$, and reaches a high point at $y = 4$. Therefore, the range is $(-\infty, 4]$.

13. Since a vertical line can intersect the graph of the relation in more than one point, the relation is not a function. The domain, the x-values of the points on the graph, is $[-4, 4]$. The range, the y-values of the points on the graph, is $[-3, 3]$.

15. $y = x^2$
    Each value of x corresponds to one y-value. For example, if $x = 3$, then $y = 3^2 = 9$. Therefore, $y = x^2$ defines y as a function of x.
    Since any x-value, positive, negative, or zero, can be squared, the domain is $(-\infty, \infty)$.

17. $x = y^2$
    The ordered pairs $(4, 2)$ and $(4, -2)$ both satisfy the equation. Since one value of x, 4, corresponds to two values of y, 2 and -2, the relation does not define a function. Because x is equal to the square of y, the values of x must always be nonnegative. The domain is $[0, \infty)$.

19. $x + y < 4$
    For a particular x-value, more than one y-value can be selected to satisfy $x + y < 4$. For example, if $x = 2$ and $y = 0$, then

$$2 + 0 < 4. \quad \textit{True}$$

Now, if $x = 2$ and $y = 1$, then

$$2 + 1 < 4. \quad \textit{Also true}$$

Therefore, $x + y < 4$ does not define y as a function of x.
    The graph of $x + y < 4$ consists of the shaded region below the dashed line $x + y = 4$, which extends indefinitely from left to right. Therefore, the domain is $(-\infty, \infty)$.

21. $y = \sqrt{x}$
    For any value of x, there is exactly one corresponding value for y, so this relation defines a function. Since the radicand must be a nonnegative number, x must always be nonnegative. The domain is $[0, \infty)$.

23. $xy = 1$
    Rewrite $xy = 1$ as $y = \dfrac{1}{x}$. Note that x can never equal 0, otherwise the denominator would equal 0. The domain is $(-\infty, 0) \cup (0, \infty)$.
    Each nonzero x-value gives exactly one y-value. Therefore, $xy = 1$ defines y as a function of x.

25. $y = \sqrt{4x + 2}$
    To determine the domain of $y = \sqrt{4x + 2}$, recall that the radicand must be nonnegative. Solve the inequality $4x + 2 \geq 0$, which gives us $x \geq -\dfrac{1}{2}$.

Therefore, the domain is $\left[-\dfrac{1}{2}, \infty\right)$.

Each x-value from the domain produces exactly one y-value. Therefore, $y = \sqrt{4x + 2}$ defines a function.

27. $y = \dfrac{2}{x - 9}$

Given any value of x, y is found by subtracting 9, then dividing the result into 2. This process produces exactly one value of y for each x-value, so the relation represents a function. The domain includes all real numbers except those that make the denominator 0, namely 9. The domain is $(-\infty, 9) \cup (9, \infty)$.

29. **(a)** The number of gallons of water varies from 0 to 3000, so the possible values are in the set $[0, 3000]$.

(b) The graph rises for the first 25 hours, so the water level increases for 25 hours. The graph falls for $t = 50$ to $t = 75$, so the water level decreases for 25 hours.

(c) There are 2000 gallons in the pool when $t = 90$.

(d) $f(0)$ is the number of gallons in the pool at time $t = 0$. Here, $f(0) = 0$.

31. The amount of income tax you pay depends on your taxable income, so income tax is a function of taxable income.

33. $f(x) = -3x + 4$
$f(0) = -3(0) + 4$
$\quad = 0 + 4$
$\quad = 4$

35. $f(x) = -3x + 4$
$f(-x) = -3(-x) + 4$
$\quad = 3x + 4$

37. $g(x) = -x^2 + 4x + 1$
$g(10) = -(10)^2 + 4(10) + 1$
$\quad = -100 + 40 + 1$
$\quad = -59$

39. $g(x) = -x^2 + 4x + 1$
$g\left(\dfrac{1}{2}\right) = -\left(\dfrac{1}{2}\right)^2 + 4\left(\dfrac{1}{2}\right) + 1$
$\quad = -\dfrac{1}{4} + 2 + 1$
$\quad = \dfrac{11}{4}$

41. $g(x) = -x^2 + 4x + 1$
$g(2p) = -(2p)^2 + 4(2p) + 1$
$\quad = -4p^2 + 8p + 1$

43. $g[f(1)]$
First find $f(1)$.
$$f(x) = -3x + 4$$
$$f(1) = -3(1) + 4 = 1$$
Since $f(1) = 1$, $g[f(1)] = g(1)$.
$$g(x) = -x^2 + 4x + 1$$
$$g(1) = -1^2 + 4(1) + 1 = 4$$
Thus, $g[f(1)] = 4$.

45. From Exercise 42, $f[g(1)] = -8$.
From Exercise 43, $g[f(1)] = 4$.
In general, $f[g(x)] \neq g[f(x)]$.

47. **(a)** Solve the equation for y.
$$x + 3y = 12$$
$$3y = 12 - x$$
$$y = \dfrac{12 - x}{3}$$
Since $y = f(x)$,
$$f(x) = \dfrac{12 - x}{3}.$$

(b) $f(3) = \dfrac{12 - 3}{3} = \dfrac{9}{3} = 3$

49. **(a)** Solve the equation for y.
$$y + 2x^2 = 3$$
$$y = 3 - 2x^2$$
Since $y = f(x)$,
$$f(x) = 3 - 2x^2.$$

(b) $f(3) = 3 - 2(3)^2$
$\quad = 3 - 2(9)$
$\quad = -15$

51. **(a)** Solve the equation for y.
$$4x - 3y = 8$$
$$-3y = 8 - 4x$$
$$y = \dfrac{8 - 4x}{-3}$$
Since $y = f(x)$,
$$f(x) = \dfrac{8 - 4x}{-3}.$$

(b) $f(3) = \dfrac{8 - 4(3)}{-3} = \dfrac{8 - 12}{-3}$
$\quad = \dfrac{-4}{-3} = \dfrac{4}{3}$

53. The equation $2x + y = 4$ has a straight line as its graph. To find y in $(3, y)$, let $x = 3$ in the equation.

$$2x + y = 4$$
$$2(3) + y = 4$$
$$6 + y = 4$$
$$y = -2$$

To use functional notation for $2x + y = 4$, solve for y to get

$$y = -2x + 4.$$

Replace y with $f(x)$ to get

$$f(x) = -2x + 4.$$
$$f(3) = -2(3) + 4 = -2$$

Because $y = -2$ when $x = 3$, the point $(3, -2)$ lies on the graph of the function.

55. $f(x) = -2x + 5$
The graph will be a line. The intercepts are $(0, 5)$ and $\left(\frac{5}{2}, 0\right)$.
The domain is $(-\infty, \infty)$.

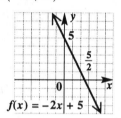

57. $h(x) = \frac{1}{2}x + 2$
The graph will be a line. The intercepts are $(0, 2)$ and $(-4, 0)$.
The domain is $(-\infty, \infty)$.

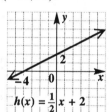

59. $G(x) = 2x$
This line includes the points $(0, 0), (1, 2)$, and $(2, 4)$. The domain is $(-\infty, \infty)$.

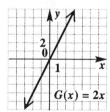

61. $g(x) = -4$
Using a y-intercept of $(0, -4)$ and a slope of

$m = 0$, we graph the horizontal line. From the graph we see that the domain is $(-\infty, \infty)$.

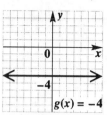

63. (a)

x	$f(x)$
0	0
1	\$1.50
2	\$3.00
3	\$4.50

(b) Since the charge equals the cost per mile, \$1.50, times the number of miles, x, $f(x) = \$1.50x$.

(c) To graph $f(x)$ for $x \in \{0, 1, 2, 3\}$, plot the points $(0, 0), (1, 1.50), (2, 3.00)$, and $(3, 4.50)$ from the chart.

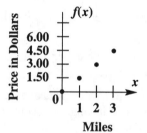

65. Since the length of a man's femur is given, use the formula
$$h(r) = 69.09 + 2.24r.$$
Let $r = 56$.
$$h(56) = 69.09 + 2.24(56)$$
$$= 194.53$$
The man is 194.53 cm tall.

67. Since the length of a woman's femur is given, use the formula
$$h(r) = 61.41 + 2.32r.$$
Let $r = 50$.
$$h(50) = 61.41 + 2.32(50)$$
$$= 177.41$$
The woman is 177.41 cm tall.

69. $f(x) = (.91)(3.14)x^2$
$$f(.8) = (.91)(3.14)(.8)^2$$
$$= 1.828736$$
To the nearest hundredth, the volume of the pool is 1.83 m^3.

71. $f(x) = (.91)(3.14)x^2$
$$f(1.2) = (.91)(3.14)(1.2)^2$$
$$= 4.114656$$
To the nearest hundredth, the volume of the pool is 4.11 m^3.

73. $f(x) = -183x + 40,034$

(a) $f(1) = -183(1) + 40,034 = 39,851$

(b) $f(3) = -183(3) + 40,034 = 39,485$

(c) $f(5) = -183(5) + 40,034 = 39,119$

(d) $x = 2$ corresponds to 1992. In 1992, there were $39,668$ post offices in the U.S.

75. The graph shows $x = 3$ and $y = 7$. In functional notation, this is

$$f(3) = 7.$$

77. Let $(x_1, y_1) = (-1, 8)$ and $(x_2, y_2) = (4, -7)$. Then

$$m = \frac{8 - (-7)}{-1 - 4} = \frac{15}{-5} = -3.$$

Use the point $(-1, 8)$ and $m = -3$ in the point-slope form.

$$y - y_1 = m(x - x_1)$$
$$y - 8 = -3[x - (-1)]$$
$$y - 8 = -3x - 3$$
$$y = -3x + 5$$
$$f(x) = -3x + 5$$

Chapter 8 Review Exercises

1. $3x + 2y = 10$

For $x = 0$:

$$3(0) + 2y = 10$$
$$2y = 10$$
$$y = 5 \quad (0, 5)$$

For $y = 0$:

$$3x + 2(0) = 10$$
$$3x = 10$$
$$x = \frac{10}{3} \quad \left(\frac{10}{3}, 0\right)$$

For $x = 2$:

$$3(2) + 2y = 10$$
$$6 + 2y = 10$$
$$2y = 4$$
$$y = 2 \quad (2, 2)$$

For $y = -2$:

$$3x + 2(-2) = 10$$
$$3x - 4 = 10$$
$$3x = 14$$
$$x = \frac{14}{3} \quad \left(\frac{14}{3}, -2\right)$$

Plot the ordered pairs, and draw the line through them.

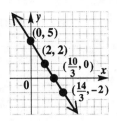

2. $x - y = 8$

For $x = 2$:

$$2 - y = 8$$
$$-y = 6$$
$$y = -6 \quad (2, -6)$$

For $y = -3$:

$$x - (-3) = 8$$
$$x + 3 = 8$$
$$x = 5 \quad (5, -3)$$

For $x = 3$:

$$3 - y = 8$$
$$-y = 5$$
$$y = -5 \quad (3, -5)$$

For $y = -2$:

$$x - (-2) = 8$$
$$x + 2 = 8$$
$$x = 6 \quad (6, -2)$$

Plot the ordered pairs, and draw the line through them.

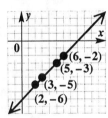

3. $4x - 3y = 12$

To find the x-intercept, let $y = 0$.

$$4x - 3y = 12$$
$$4x - 3(0) = 12$$
$$4x = 12$$
$$x = 3$$

The x-intercept is $(3, 0)$.
To find the y-intercept, let $x = 0$.

$$4x - 3y = 12$$
$$4(0) - 3y = 12$$
$$-3y = 12$$
$$y = -4$$

The y-intercept is $(0, -4)$.
Plot the intercepts and draw the line through them.

continued

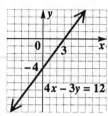

4. $5x + 7y = 28$
To find the x-intercept, let $y = 0$.

$$5x + 7y = 28$$
$$5x + 7(0) = 28$$
$$5x = 28$$
$$x = \frac{28}{5}$$

The x-intercept is $\left(\frac{28}{5}, 0\right)$.
To find the y-intercept, let $x = 0$.

$$5x + 7y = 28$$
$$5(0) + 7y = 28$$
$$7y = 28$$
$$y = 4$$

The y-intercept is $(0, 4)$.
Plot the intercepts and draw the line through them.

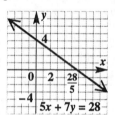

5. $2x + 5y = 20$
To find the x-intercept, let $y = 0$.

$$2x + 5y = 20$$
$$2x + 5(0) = 20$$
$$2x = 20$$
$$x = 10$$

The x-intercept is $(10, 0)$.
To find the y-intercept, let $x = 0$.

$$2x + 5y = 20$$
$$2(0) + 5y = 20$$
$$5y = 20$$
$$y = 4$$

The y-intercept is $(0, 4)$.
Plot the intercepts and draw the line through them.

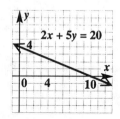

6. $x - 4y = 8$
To find the x-intercept, let $y = 0$.

$$x - 4y = 8$$
$$x - 4(0) = 8$$
$$x = 8$$

The x-intercept is $(8, 0)$.
To find the y-intercept, let $x = 0$.

$$0 - 4y = 8$$
$$-4y = 8$$
$$y = -2$$

The y-intercept is $(0, -2)$.
Plot the intercepts and draw the line through them.

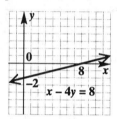

7. If both coordinates are positive, the point lies in quadrant I. If the first coordinate is negative and the second is positive, the point lies in quadrant II. To lie in quadrant III, the point must have both coordinates negative. To lie in quadrant IV, the first coordinate must be positive and the second must be negative.

8. Through $(-1, 2)$ and $(4, -5)$
$$m = \frac{\text{change in } y}{\text{change in } x} = \frac{2 - (-5)}{-1 - 4} = \frac{7}{-5} = -\frac{7}{5}$$

9. Through $(0, 3)$ and $(-2, 4)$
Let $(x_1, y_1) = (0, 3)$ and $(x_2, y_2) = (-2, 4)$.
$$m = \frac{y_2 - y_1}{x_2 - x_1} = \frac{4 - 3}{-2 - 0} = \frac{1}{-2} = -\frac{1}{2}$$

10. The slope of $y = 2x + 3$ is 2, the coefficient of x.

11. $3x - 4y = 5$
Write the equation in slope-intercept form.
$$-4y = -3x + 5$$
$$y = \frac{3}{4}x - \frac{5}{4}$$

The slope is $\frac{3}{4}$.

12. $x = 5$ is a vertical line and has undefined slope.

13. Parallel to $3y = 2x + 5$
Write the equation in slope-intercept form.
$$3y = 2x + 5$$
$$y = \frac{2}{3}x + \frac{5}{3}$$

The slope of $3y = 2x + 5$ is $\frac{2}{3}$; all lines parallel

to it will also have a slope of $\frac{2}{3}$.

14. Perpendicular to $3x - y = 4$
Solve for y.

$$y = 3x - 4$$

The slope is 3; the slope of a line perpendicular to

it is $-\frac{1}{3}$ since

$$3\left(-\frac{1}{3}\right) = -1.$$

15. Through $(-1, 5)$ and $(-1, -4)$

$$m = \frac{5 - (-4)}{-1 - (-1)} = \frac{9}{0} \quad \textit{Undefined}$$

This is a vertical line; it has undefined slope.

16. Find the slope using two points from the table, say $(0, 2)$ and $(6, -6)$.

$$m = \frac{2 - (-6)}{0 - 6} = \frac{2 + 6}{-6} = \frac{8}{-6} = -\frac{4}{3}$$

17. Find the slope using the two points from the screens; that is, $(-2, -6)$ and $(4, 9)$.

$$m = \frac{-6 - 9}{-2 - 4} = \frac{-15}{-6} = \frac{5}{2}$$

18. The line goes up from left to right, so it has positive slope.

19. The line goes down from left to right, so it has negative slope.

20. The line is horizontal, so it has zero slope.

21. The line is vertical, so it has undefined slope.

22. The slope is $\frac{2}{10}$ which can be written as

.2, 20%, $\frac{20}{100}$, or $\frac{1}{5}$.

The correct responses are (a), (b), (c), (d), and (f).

23. To rise 1 foot, we must move 4 feet in the horizontal direction. To rise 3 feet, we must move $3(4) = 12$ feet in the horizontal direction.

24. Let $(x_1, y_1) = (1970, 10,000)$ and $(x_2, y_2) = (1995, 41,000)$. Then

$$m = \frac{41,000 - 10,000}{1995 - 1970}$$
$$= \frac{31,000}{25} = 1240.$$

The average rate of change is $\dfrac{\$1240}{\text{yr}}$.

25. Slope $-\frac{1}{3}$, y-intercept $(0, -1)$

Use the slope-intercept form with $m = -\frac{1}{3}$ and $b = -1$.

$$y = mx + b$$
$$y = -\frac{1}{3}x - 1$$

26. Slope 0, y-intercept $(0, -2)$
Use the slope-intercept form with $m = 0$ and $b = -2$.

$$y = mx + b$$
$$y = (0)x - 2$$
$$y = -2$$

27. Slope $-\frac{4}{3}$, through $(2, 7)$

Use the point-slope form with $m = -\frac{4}{3}$ and $(x_1, y_1) = (2, 7)$.

$$y - y_1 = m(x - x_1)$$
$$y - 7 = -\frac{4}{3}(x - 2)$$
$$y - 7 = -\frac{4}{3}x + \frac{8}{3}$$
$$y = -\frac{4}{3}x + \frac{29}{3}$$

28. Slope 3, through $(-1, 4)$
Use the point-slope form with $m = 3$ and $(x_1, y_1) = (-1, 4)$.

$$y - y_1 = m(x - x_1)$$
$$y - 4 = 3[x - (-1)]$$
$$y - 4 = 3(x + 1)$$
$$y - 4 = 3x + 3$$
$$y = 3x + 7$$

29. Vertical, through $(2, 5)$
The equation of any vertical line is in the form $x = k$. Since the line goes through $(2, 5)$, the equation is $x = 2$. (Slope-intercept form is not possible.)

30. Through $(2, -5)$ and $(1, 4)$
Find the slope.

$$m = \frac{4 - (-5)}{1 - 2} = \frac{9}{-1} = -9$$

Use the point-slope form with $m = -9$ and $(x_1, y_1) = (2, -5)$.

$$y - y_1 = m(x - x_1)$$
$$y - (-5) = -9(x - 2)$$
$$y + 5 = -9x + 18$$
$$y = -9x + 13$$

31. Through $(-3, -1)$ and $(2, 6)$
Find the slope.

$$m = \frac{6 - (-1)}{2 - (-3)} = \frac{7}{5}$$

Use the point-slope form with $m = \frac{7}{5}$ and
$(x_1, y_1) = (2, 6)$.

$$y - y_1 = m(x - x_1)$$
$$y - 6 = \frac{7}{5}(x - 2)$$
$$y - 6 = \frac{7}{5}x - \frac{14}{5}$$
$$y = \frac{7}{5}x + \frac{16}{5}$$

32. Parallel to $4x - y = 3$ and through $(7, -1)$
$y = 4x - 3$ has slope 4. Lines parallel to it will
also have slope 4. The line with slope 4 through
$(7, -1)$ is :

$$y - y_1 = m(x - x_1)$$
$$y - (-1) = 4(x - 7)$$
$$y + 1 = 4x - 28$$
$$y = 4x - 29$$

33. Perpendicular to $2x - 5y = 7$ and through $(4, 3)$
Write the equation in slope-intercept form.

$$2x - 5y = 7$$
$$-5y = -2x + 7$$
$$y = \frac{2}{5}x - \frac{7}{5}$$

$y = \frac{2}{5}x - \frac{7}{5}$ has slope $\frac{2}{5}$ and is perpendicular to
lines with slope $-\frac{5}{2}$.
The line with slope $-\frac{5}{2}$ through $(4, 3)$ is

$$y - y_1 = m(x - x_1)$$
$$y - 3 = -\frac{5}{2}(x - 4)$$
$$y - 3 = -\frac{5}{2}x + 10$$
$$y = -\frac{5}{2}x + 13$$

34. From Exercise 16, $m = -\frac{4}{3}$.
From the table, the graph goes through $(0, 2)$, so
$b = 2$. Therefore, the equation is

$$y = -\frac{4}{3}x + 2.$$

35. From Exercise 17, $m = \frac{5}{2}$.
Use $(x_1, y_1) = (4, 9)$ and $m = \frac{5}{2}$ in the point-
slope form.

$$y - y_1 = m(x - x_1)$$
$$y - 9 = \frac{5}{2}(x - 4)$$
$$y - 9 = \frac{5}{2}x - 10$$
$$y = \frac{5}{2}x - 1$$

36. If $x = 0$ corresponds to 1980, then $x = 16$
corresponds to 1996. Substitute 16 for x in the
equation and solve for y.

$$y = 2.1x + 230$$
$$y = 2.1(16) + 230$$
$$= 33.6 + 230 = 263.6$$

The population in 1996 was about 264 million.

37. Substitute 247 for y in the equation and solve for
x.

$$y = 2.1x + 230$$
$$247 = 2.1x + 230$$
$$17 = 2.1x$$
$$x = \frac{17}{2.1} \approx 8$$

Since $x = 0$ corresponds to 1980, $x = 8$
corresponds to 1988. In 1988, the population
reached 247 million.

38. $3x - 2y \leq 12$
Graph $3x - 2y = 12$ as a solid line through
$(0, -6)$ and $(4, 0)$. Use $(0, 0)$ as a test point.
Since $(0, 0)$ satisfies the inequality, shade the
region on the side of the line containing $(0, 0)$.

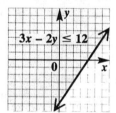

39. $5x - y > 6$
Graph $5x - y = 6$ as a dashed line through
$(0, -6)$ and $\left(\frac{6}{5}, 0\right)$. Use $(0, 0)$ as a test point.

Since $(0, 0)$ does not satisfy the inequality, shade
the region on the side of the line that does not
contain $(0, 0)$.

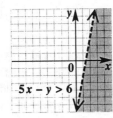

40. $x \geq 2$

Graph $x = 2$ as a solid vertical line, crossing the x-axis at 2. Since $(0, 0)$ does not satisfy the inequality, shade the region to the right of the line.

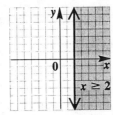

41. $2x + y \leq 1$ and $x \geq 2y$

Graph $2x + y = 1$ as a solid line through $\left(\dfrac{1}{2}, 0\right)$ and $(0, 1)$, and shade the region on the side containing $(0, 0)$ since it satisfies the inequality. Next, graph $x = 2y$ as a solid line through $(0, 0)$ and $(2, 1)$, and shade the region on the side containing $(2, 0)$ since $2 > 2(0)$ or $2 > 0$ is true. The intersection is the region where the graphs overlap.

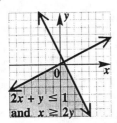

42. $\{(-4, 2), (-4, -2), (1, 5), (1, -5)\}$

The domain, the set of x-values, is $\{-4, 1\}$.
The range, the set of y-values, is $\{2, -2, 5, -5\}$.
Since each x-value has more than one y-value, the relation is not a function.

43. The domain, the x-values of the points on the graph, is $[-4, 4]$. The range, the y-values of the points on the graph, is $[0, 2]$. Since a vertical line intersects the graph of the relation in at most one point, the relation is a function.

44. The domain, the set of first components, is $\{$California, New York, Texas, Pennsylvania, Washington$\}$.
The range, the set of second components, is $\{71, 266, 50, 101, 48, 010, 42, 142, 38, 240\}$.
Since each state corresponds to one number of small offices/home offices, the relation is a function.

45. (a) The independent variable must be the country, because, for each country, there is exactly one amount of power. On the other hand, for a specific amount of power, there would be more than one country generating that amount.

(b) The domain, the set of countries, is $\{$United States, France, Japan, Germany, Canada, Russia$\}$. The range, the amounts of nuclear power generated (in billion kilowatt-hours), is $\{101, 154, 286, 377, 706\}$.

46. The line graph passes the vertical line test, so it is the graph of a function; in fact, it is a linear function.
The endpoints give us two ordered pairs on the line: $(1991, 80.1)$ and $(1996, 89.4)$.
The horizontal axis values vary from 1991 to 1996, so the domain is $[1991, 1996]$. The vertical axis values vary from 80.1 to 89.4, so the range is $[80.1, 89.4]$.

47. $y = 3x - 3$

For any value of x, there is exactly one value of y, so the equation defines a function, actually a linear function. The domain is the set of all real numbers $(-\infty, \infty)$.

48. $y < x + 2$

For any value of x, there are many values of y. For example, $(1, 0)$ and $(1, 1)$ are both solutions of the inequality that have the same x-value but different y-values. The inequality does not define a function. The domain is the set of all real numbers $(-\infty, \infty)$.

49. $y = |x - 4|$

For any value of x, there is exactly one value of y, so the equation defines a function. The domain is the set of all real numbers $(-\infty, \infty)$.

50. $y = \sqrt{4x + 7}$

Given any value of x, y is found by multiplying x by 4, adding 7, and taking the square root of the result. This process produces exactly one value of y for each x-value, so the equation defines a function. Since the radicand must be nonnegative,

$$4x + 7 \geq 0$$
$$4x \geq -7$$
$$x \geq -\frac{7}{4}.$$

The domain is $\left[-\dfrac{7}{4}, \infty\right)$.

51. $x = y^2$

The ordered pairs $(4, 2)$ and $(4, -2)$ both satisfy the equation. Since one value of x, 4, corresponds to two values of y, 2 and -2, the equation does not define a function. Because x is equal to the square of y, the values of x must always be nonnegative. The domain is $[0, \infty)$.

52. $y = \dfrac{7}{x-6}$

Given any value of x, y is found by subtracting 6, then dividing the result into 7. This process produces exactly one value of y for each x-value, so the equation defines a function. The domain includes all real numbers except those that make the denominator 0, namely 6. The domain is $(-\infty, 6) \cup (6, \infty)$.

53. If no vertical line intersects the graph in more than one point, then it is the graph of a function.

In Exercises 54–59, use

$$f(x) = -2x^2 + 3x - 6.$$

54. $f(0) = -2(0)^2 + 3(0) - 6 = -6$

55. $f(2.1) = -2(2.1)^2 + 3(2.1) - 6$
$$= -8.82 + 6.3 - 6 = -8.52$$

56. $f\left(-\dfrac{1}{2}\right) = -2\left(-\dfrac{1}{2}\right)^2 + 3\left(-\dfrac{1}{2}\right) - 6$
$$= -\dfrac{1}{2} - \dfrac{3}{2} - 6 = -8$$

57. $f(k) = -2k^2 + 3k - 6$

58. $f[f(0)]$
First find $f(0)$.
$$f(0) = -2(0)^2 + 3(0) - 6 = -6$$
Since $f(0) = -6$, $f[f(0)] = f(-6)$.
Find $f(-6)$.
$$f(-6) = -2(-6)^2 + 3(-6) - 6$$
$$= -72 - 18 - 6 = -96$$
So, $f[f(0)] = -96$.

59. $f(2p) = -2(2p)^2 + 3(2p) - 6$
$$= -2(4p^2) + 6p - 6$$
$$= -8p^2 + 6p - 6$$

60. $2x^2 - y = 0$
$$-y = -2x^2$$
$$y = 2x^2$$
Since $y = f(x)$,
$$f(x) = 2x^2,$$
and $f(3) = 2(3)^2 = 2(9) = 18$.

61. Solve for y in terms of x.
$$2x - 5y = 7$$
$$2x - 7 = 5y$$
$$\dfrac{2x - 7}{5} = y$$
This is the same as choice (c),

$$f(x) = \dfrac{-7 + 2x}{5}.$$

62. No, because the equation of a line with undefined slope is $x = c$, so the ordered pairs have the form (c, y), where c is a constant and y is a variable. Thus, the number c corresponds to an infinite number of values of y.

63. The slope is negative since the line falls from left to right.

64. Use the points $(-1, 5)$ and $(3, -1)$.
$$m = \dfrac{5 - (-1)}{-1 - 3} = \dfrac{5 + 1}{-1 - 3} = \dfrac{6}{-4} = -\dfrac{3}{2}$$

65. $2y = -3x + 7$
To find the x-intercept, let $y = 0$.
$$2(0) = -3x + 7$$
$$3x = 7$$
$$x = \dfrac{7}{3}$$
The x-intercept is $\left(\dfrac{7}{3}, 0\right)$.

66. $2y = -3x + 7$
To find the y-intercept, let $x = 0$.
$$2y = -3(0) + 7$$
$$2y = 7$$
$$y = \dfrac{7}{2}$$
The y-intercept is $\left(0, \dfrac{7}{2}\right)$.

67. Solve $2y = -3x + 7$ for y.
$$y = -\dfrac{3}{2}x + \dfrac{7}{2}$$
Since $y = f(x)$,
$$f(x) = -\dfrac{3}{2}x + \dfrac{7}{2}.$$

68. $f(x) = -\dfrac{3}{2}x + \dfrac{7}{2}$
$$f(8) = -\dfrac{3}{2}(8) + \dfrac{7}{2}$$
$$= -\dfrac{24}{2} + \dfrac{7}{2} = -\dfrac{17}{2}$$

69. $f(x) = -\dfrac{3}{2}x + \dfrac{7}{2}$
Let $f(x) = -8$.
$$-8 = -\dfrac{3}{2}x + \dfrac{7}{2}$$
$$-16 = -3x + 7 \qquad \textit{Multiply by 2.}$$
$$-23 = -3x$$
$$x = \dfrac{23}{3}$$

70.
$$f(x) \geq 0$$
$$-\frac{3}{2}x + \frac{7}{2} \geq 0$$
$$-\frac{3}{2}x \geq -\frac{7}{2}$$
$$x \leq \left(-\frac{7}{2}\right)\left(-\frac{2}{3}\right)$$
$$x \leq \frac{7}{3}$$

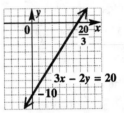

71. $f(x) = 0$ is equivalent to $y = 0$, which is the equation we solved in Exercise 65 to find the x-intercept.

Solution set: $\left\{\dfrac{7}{3}\right\}$

72. The graph is below the x-axis for $x > \dfrac{7}{3}$, so the solution set of $f(x) < 0$ is $\left(\dfrac{7}{3}, \infty\right)$.

73. The graph is above the x-axis for $x < \dfrac{7}{3}$, so the solution set of $f(x) > 0$ is $\left(-\infty, \dfrac{7}{3}\right)$.

74. Since $m = -\dfrac{3}{2}$, the slope of any line perpendicular to this line is $\dfrac{2}{3}$ since $\dfrac{2}{3}$ is the negative reciprocal of $-\dfrac{3}{2}$.

Chapter 8 Test

1. $2x - 3y = 12$
For $x = 1$:
$$2(1) - 3y = 12$$
$$2 - 3y = 12$$
$$-3y = 10$$
$$y = -\frac{10}{3} \quad \left(1, -\frac{10}{3}\right)$$

For $x = 3$:
$$2(3) - 3y = 12$$
$$6 - 3y = 12$$
$$-3y = 6$$
$$y = -2 \quad (3, -2)$$

For $y = -4$:
$$2x - 3(-4) = 12$$
$$2x + 12 = 12$$
$$2x = 0$$
$$x = 0 \quad (0, -4)$$

2. Through $(6, 4)$ and $(-4, -1)$
$$m = \frac{4 - (-1)}{6 - (-4)} = \frac{4 + 1}{6 + 4} = \frac{5}{10} = \frac{1}{2}$$
The slope of the line is $\dfrac{1}{2}$.

3. $3x - 2y = 20$
To find the x-intercept, let $y = 0$.
$$3x - 2(0) = 20$$
$$3x = 20$$
$$x = \frac{20}{3}$$
The x-intercept is $\left(\dfrac{20}{3}, 0\right)$.
To find the y-intercept, let $x = 0$.
$$3(0) - 2y = 20$$
$$-2y = 20$$
$$y = -10$$
The y-intercept is $(0, -10)$.
Draw the line through these two points.

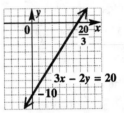

4. The graph of $y = 5$ is the horizontal line with slope 0 and y-intercept $(0, 5)$. There is no x-intercept.

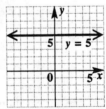

5. The graph of $x = 2$ is the vertical line with x-intercept at $(2, 0)$. There is no y-intercept.

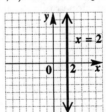

6. The graph of a line with undefined slope is the graph of a vertical line.

7.   Find the slope of each line.

$$5x - y = 8$$
$$-y = -5x + 8$$
$$y = 5x - 8$$

The slope is 5.

$$5y = -x + 3$$
$$y = -\frac{1}{5}x + \frac{3}{5}$$

The slope is $-\frac{1}{5}$.

Since $5\left(-\frac{1}{5}\right) = -1$, the two slopes are negative reciprocals and the lines are perpendicular.

8.   Find the slope of each line.

$$2y = 3x + 12$$
$$y = \frac{3}{2}x + 6$$

The slope is $\frac{3}{2}$.

$$3y = 2x - 5$$
$$y = \frac{2}{3}x - \frac{5}{3}$$

The slope is $\frac{2}{3}$.
The lines are neither parallel nor perpendicular.

9.   Through $(4, -1)$; $m = -5$
Let $m = -5$ and $(x_1, y_1) = (4, -1)$ in the point-slope form.

$$y - y_1 = m(x - x_1)$$
$$y - (-1) = -5(x - 4)$$
$$y + 1 = -5x + 20$$
$$y = -5x + 19$$

10.   Through $(-3, 14)$; horizontal
A horizontal line has equation $y = k$. Here $k = 14$, so the line has equation $y = 14$.

11.   Through $(-7, 2)$ and parallel to $3x + 5y = 6$
To find the slope of $3x + 5y = 6$, write the equation in slope-intercept form by solving for y.

$$3x + 5y = 6$$
$$5y = -3x + 6$$
$$y = -\frac{3}{5}x + \frac{6}{5}$$

The slope is $-\frac{3}{5}$, so a line parallel to it also has slope $-\frac{3}{5}$. Let $m = -\frac{3}{5}$ and $(x_1, y_1) = (-7, 2)$ in the point-slope form.

$$y - y_1 = m(x - x_1)$$
$$y - 2 = -\frac{3}{5}[x - (-7)]$$
$$y - 2 = -\frac{3}{5}(x + 7)$$
$$y - 2 = -\frac{3}{5}x - \frac{21}{5}$$
$$y = -\frac{3}{5}x - \frac{11}{5}$$

12.   Through $(-7, 2)$ and perpendicular to $y = 2x$
Since $y = 2x$ is in slope-intercept form ($b = 0$), the slope, m, of $y = 2x$ is 2. A line perpendicular to it has a slope that is the negative reciprocal of 2, that is, $-\frac{1}{2}$. Let $m = -\frac{1}{2}$ and $(x_1, y_1) = (-7, 2)$ in the point-slope form.

$$y - y_1 = m(x - x_1)$$
$$y - 2 = -\frac{1}{2}(x + 7)$$
$$y - 2 = -\frac{1}{2}x - \frac{7}{2}$$
$$y = -\frac{1}{2}x - \frac{3}{2}$$

13.   From the graphs, find the equation of the line through $(-2, 3)$ and $(6, -1)$ First find the slope.

$$m = \frac{3 - (-1)}{-2 - 6} = \frac{3 + 1}{-8} = \frac{4}{-8} = -\frac{1}{2}$$

Use $m = -\frac{1}{2}$ and $(x_1, y_1) = (-2, 3)$ in the point-slope form.

$$y - y_1 = m(x - x_1)$$
$$y - 3 = -\frac{1}{2}[x - (-2)]$$
$$y - 3 = -\frac{1}{2}(x + 2)$$
$$y - 3 = -\frac{1}{2}x - 1$$
$$y = -\frac{1}{2}x + 2$$

14.   Positive slope means that the line goes up from left to right. The only line that has positive slope and a negative y-coordinate for its y-intercept is choice (b).

15.   For 1994, $x = 4$ ($1994 - 1990 = 4$).

$$y = 1410x + 12,520$$
$$y = 1410(4) + 12,520 = 18,160$$

There were $18,160$ cases.

16.   (a) The number 1410 is the slope of the line.

(b) It is the annual rate of change in the number of cases served.

17. $3x - 2y > 6$

Graph the line $3x - 2y = 6$, which has intercepts $(2, 0)$ and $(0, -3)$, as a dashed line since the inequality involves $>$. Test $(0, 0)$, which yields $0 > 6$, a false statement. Shade the region that does not include $(0, 0)$.

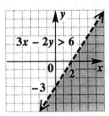

18. $y < 2x - 1$ and $x - y < 3$

First graph $y = 2x - 1$ as a dashed line through $(2, 3)$ and $(0, -1)$. Test $(0, 0)$, which yields $0 < -1$, a false statement. Shade the side of the line not containing $(0, 0)$.

Next, graph $x - y = 3$ as a dashed line through $(3, 0)$ and $(0, -3)$. Test $(0, 0)$, which yields $0 < 3$, a true statement.

Shade the side of the line containing $(0, 0)$. The intersection is the region where the graphs overlap.

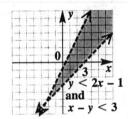

19. Choice (d) is the only graph that passes the vertical line test.

20. Choice (d) does not define a function, since its domain (input) element A is paired with two different range (output) elements, 1 and 2.

21. $f(x) = -x^2 + 2x - 1$

$f(1) = -(1)^2 + 2(1) - 1$

$= -1 + 2 - 1$

$= 0$

22. $f(x) = \dfrac{2}{3}x - 1$

This function represents a line with y-intercept $(0, -1)$ and x-intercept $\left(\dfrac{3}{2}, 0\right)$.

Draw the line through these two points. The domain is $(-\infty, \infty)$, and the range is $(-\infty, \infty)$.

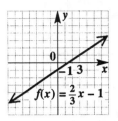

23. Choice (b) $\{(\text{Year, Death rate})\}$ defines a function since each year corresponds to one death rate. Choice (a) $\{(\text{Death rate, year})\}$ doesn't define a function since the values 8.6 and 8.8 would correspond to more than one year.

Cumulative Review Exercises Chapters 1–8

1. $-5(8 - 2z) + 4(7 - z) = 7(8 + z) - 3$

$-40 + 10z + 28 - 4z = 56 + 7z - 3$

> *Distributive property*

$6z - 12 = 7z + 53$ *Combine like terms*

$-65 = z$ *Subtract 6z; 53*

Solution set: $\{-65\}$

2.   Solve $A = p + prt$ for t.

$A = p + prt$

$A - p = prt$ *Subtract p*

$\dfrac{A - p}{pr} = t$ *Divide by pr*

3. $7x^2 + 8x + 1 = 0$

To solve this quadratic equation, begin by factoring the polynomial.

$$(7x + 1)(x + 1) = 0$$

Use the zero-factor property to set each factor equal to 0.

$$7x + 1 = 0 \text{ or } x + 1 = 0$$

Solve each of the linear equations.

$$7x = -1 \quad \text{or} \quad x = -1$$

$$x = -\dfrac{1}{7} \quad \text{or} \quad x = -1$$

The solution set of the original equation is

$$\left\{-1, -\dfrac{1}{7}\right\}.$$

4. $|2k - 7| + 4 = 11$

$|2k - 7| = 7$

$2k - 7 = 7$ or $2k - 7 = -7$

$2k = 14$ $2k = 0$

$k = 7$ or $k = 0$

Solution set: $\{0, 7\}$

5. $\dfrac{2}{x-1} = \dfrac{5}{x-1} - \dfrac{3}{4}$

Multiply both sides by the LCD, $4(x-1)$.

$$4(x-1)\left(\dfrac{2}{x-1}\right) = 4(x-1)\left(\dfrac{5}{x-1} - \dfrac{3}{4}\right)$$

$$4(x-1)\left(\dfrac{2}{x-1}\right) = 4(x-1)\left(\dfrac{5}{x-1}\right)$$
$$- 4(x-1)\left(\dfrac{3}{4}\right)$$

Distributive
property

$$4 \cdot 2 = 4 \cdot 5 - 3(x-1)$$
$$8 = 20 - 3x + 3$$
$$3x + 8 = 23$$
$$3x = 15$$
$$x = 5$$

Check $x = 5$: $\dfrac{2}{4} = \dfrac{5}{4} - \dfrac{3}{4}$ *True*

Thus, the solution set is $\{5\}$.

6. $-4 < 3 - 2k < 9$
$-7 < -2k < 6$

Divide by -2; reverse the inequality signs.

$\dfrac{7}{2} > k > -3$ or $-3 < k < \dfrac{7}{2}$

Solution set: $\left(-3, \dfrac{7}{2}\right)$

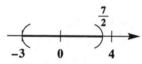

7. $-.3x + 2.1(x-4) \le -6.6$
$-3x + 21(x-4) \le -66$

Multiply by 10.

$-3x + 21x - 84 \le -66$
$18x - 84 \le -66$
$18x \le 18$
$x \le 1$

Solution set: $(-\infty, 1]$

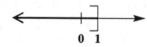

8. $-5x + 1 \ge 11$ or $3x + 5 > 26$
$\qquad -5x \ge 10 \qquad\qquad 3x > 21$
$\qquad\ x \le -2$ or $\qquad x > 7$

The graph of the solution set is all numbers either less than or equal to -2 *or* greater than 7. This is the union. The solution set is $(-\infty, -2] \cup (7, \infty)$.

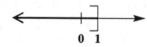

9. $\left(x^2 y^{-3}\right)\left(x^{-4} y^2\right)$

$= \left(x^2 x^{-4}\right)\left(y^{-3} y^2\right)$ *Commutative and associative properties*

$= x^{2+(-4)} y^{-3+2}$ *Product rule*

$= x^{-2} y^{-1}$ *Add*

$= \dfrac{1}{x^2 y}$ *Definition of negative exponent*

10. $\dfrac{x^{-6} y^3 z^{-1}}{x^7 y^{-4} z} = \dfrac{y^4 y^3}{x^6 x^7 z^1 z} = \dfrac{y^7}{x^{13} z^2}$

11. $\left(2m^{-2} n^3\right)^{-3}$

$= 2^{-3} \left(m^{-2}\right)^{-3} \left(n^3\right)^{-3}$

$= 2^{-3} m^{(-2)(-3)} n^{3(-3)}$

$= 2^{-3} m^6 n^{-9}$

$= \dfrac{m^6}{2^3 n^9} = \dfrac{m^6}{8n^9}$

12. $2\left(3x^2 - 8x + 1\right) - 4\left(x^2 - 3x - 9\right)$

$= 6x^2 - 16x + 2 - 4x^2 + 12x + 36$

Distributive property

$= \left(6x^2 - 4x^2\right) + (-16x + 12x) + (2 + 36)$

Combine like terms

$= 2x^2 - 4x + 38$

13. $(3x + 2y)(5x - y)$

$= 3x(5x) + 3x(-y) + 2y(5x) + 2y(-y)$

FOIL

$= 15x^2 - 3xy + 10xy - 2y^2$

$= 15x^2 + 7xy - 2y^2$

14. $(x + 2y)\left(x^2 - 2xy + 4y^2\right)$

Multiply vertically.

$$
\begin{array}{r}
x^2 \ - \ 2xy \ + \ 4y^2 \\
x \ + \ 2y \\
\hline
2x^2 y \ - \ 4xy^2 \ + \ 8y^3 \\
x^3 \ - \ 2x^2 y \ + \ 4xy^2 \\
\hline
x^3 \qquad\qquad\qquad\ + \ 8y^3
\end{array}
$$

Thus,

$(x + 2y)\left(x^2 - 2xy + 4y^2\right) = x^3 + 8y^3.$

15. $\dfrac{m^3 - 3m^2 + 5m - 3}{m - 1}$

$$
\begin{array}{r}
m^2 \;\; - 2m \;\; + 3 \\
m - 1\overline{\big)\; m^3 \; - 3m^2 \; + 5m \; - 3} \\
\underline{m^3 \; - \; m^2} \\
-2m^2 \; + 5m \\
\underline{-2m^2 \; + 2m} \\
3m \; - 3 \\
\underline{3m \; - 3} \\
0
\end{array}
$$

The remainder is 0. The answer is the quotient, $m^2 - 2m + 3$.

16. $y^2 + 4yk - 12k^2$

Look for two expressions whose product is $-12k^2$ and whose sum is $4k$. These expressions are $6k$ and $-2k$. Thus,

$$y^2 + 4yk - 12k^2 = (y + 6k)(y - 2k).$$

17. $9x^4 - 25y^2$

This polynomial is the difference of two squares.

$$
\begin{aligned}
9x^4 - 25y^2 &= \left(3x^2\right)^2 - (5y)^2 \\
&= \left(3x^2 + 5y\right)\left(3x^2 - 5y\right)
\end{aligned}
$$

18. $125x^4 - 400x^3y + 195x^2y^2$

Begin by factoring out the greatest common factor, $5x^2$.

$$
\begin{aligned}
&125x^4 - 400x^3y + 195x^2y^2 \\
&= 5x^2\left(25x^2 - 80xy + 39y^2\right)
\end{aligned}
$$

Now factor $25x^2 - 80xy + 39y^2$ by trial and error.

$$
\begin{aligned}
&25x^2 - 80xy + 39y^2 \\
&= (5x - 13y)(5x - 3y)
\end{aligned}
$$

The complete factored form is

$$
\begin{aligned}
&125x^4 - 400x^3y + 195x^2y^2 \\
&= 5x^2(5x - 13y)(5x - 3y).
\end{aligned}
$$

19. $f^2 + 20f + 100$

This polynomial is a perfect square trinomial.

$$
\begin{aligned}
f^2 + 20f + 100 &= f^2 + 2(f)(10) + 10^2 \\
&= (f + 10)^2
\end{aligned}
$$

20. $100x^2 + 49$

The sum of two squares cannot be factored unless the two terms have a common factor.

This polynomial is the sum of two squares in which there is no common factor. Therefore, $100x^2 + 49$ is prime.

21. $\dfrac{3}{2x + 6} + \dfrac{2x + 3}{2x + 6}$

These rational expressions have a common denominator, so we add their numerators and keep the common denominator.

$$
\begin{aligned}
\dfrac{3}{2x + 6} + \dfrac{2x + 3}{2x + 6} \\
= \dfrac{3 + 2x + 3}{2x + 6} \\
= \dfrac{2x + 6}{2x + 6} = 1
\end{aligned}
$$

22. $\dfrac{8}{x + 1} - \dfrac{2}{x + 3}$

To add these rational expressions, we need a common denominator. The LCD is $(x + 1)(x + 3)$. Rewrite each factor with $(x + 1)(x + 3)$ as its denominator.

$$
\begin{aligned}
&\dfrac{8}{x + 1} - \dfrac{2}{x + 3} \\
&= \dfrac{8}{x + 1} \cdot \dfrac{x + 3}{x + 3} - \dfrac{2}{x + 3} \cdot \dfrac{x + 1}{x + 1} \\
&= \dfrac{8(x + 3)}{(x + 1)(x + 3)} - \dfrac{2(x + 1)}{(x + 3)(x + 1)} \\
&= \dfrac{8x + 24}{(x + 1)(x + 3)} - \dfrac{2x + 2}{(x + 1)(x + 3)} \\
&= \dfrac{(8x + 24) - (2x + 2)}{(x + 1)(x + 3)} \\
&= \dfrac{8x + 24 - 2x - 2}{(x + 1)(x + 3)} \\
&= \dfrac{6x + 22}{(x + 1)(x + 3)}
\end{aligned}
$$

or $\dfrac{2(3x + 11)}{(x + 1)(x + 3)}$

23. $\dfrac{x^2 - 25}{3x + 6} \cdot \dfrac{4x + 8}{x^2 + 10x + 25}$

$$
\begin{aligned}
&= \dfrac{(x + 5)(x - 5)}{3(x + 2)} \cdot \dfrac{4(x + 2)}{(x + 5)(x + 5)} \\
&\qquad\qquad\textit{Factor numerators and denominators} \\
&= \dfrac{(x + 5)(x - 5)(4)(x + 2)}{3(x + 2)(x + 5)(x + 5)} \quad \textit{Multiply} \\
&= \dfrac{4(x - 5)}{3(x + 5)} \qquad\qquad\quad \textit{Lowest} \\
&\qquad\qquad\qquad\qquad\qquad\;\; \textit{terms}
\end{aligned}
$$

24. $\dfrac{x^2 + 2x - 3}{x^2 - 5x + 4} \cdot \dfrac{x^2 - 3x - 4}{x^2 + 3x}$

$$
\begin{aligned}
&= \dfrac{(x - 1)(x + 3)}{(x - 1)(x - 4)} \cdot \dfrac{(x + 1)(x - 4)}{x(x + 3)} \\
&\qquad\qquad\textit{Factor numerators and denominators} \\
&= \dfrac{(x - 1)(x + 3)(x + 1)(x - 4)}{(x - 1)(x - 4)(x)(x + 3)} \quad \textit{Multiply} \\
&= \dfrac{x + 1}{x} \qquad\qquad\qquad\qquad\;\; \textit{Lowest} \\
&\qquad\qquad\qquad\qquad\qquad\qquad\; \textit{terms}
\end{aligned}
$$

25. $\dfrac{x^2 + 5x + 6}{3x} \div \dfrac{x^2 - 4}{x^2 + x - 6}$

$= \dfrac{x^2 + 5x + 6}{3x} \cdot \dfrac{x^2 + x - 6}{x^2 - 4}$ *Multiply by reciprocal of second expression*

$= \dfrac{(x+2)(x+3)}{3x} \cdot \dfrac{(x+3)(x-2)}{(x+2)(x-2)}$

Factor numerators and denominators

$= \dfrac{(x+2)(x+3)(x+3)(x-2)}{3x(x+2)(x-2)}$ *Multiply*

$= \dfrac{(x+3)^2}{3x}$

26. $\dfrac{6x^4y^3z^2}{8xyz^4} \div \dfrac{3x^2}{16y^2}$

$= \dfrac{6x^4y^3z^2}{8xyz^4} \cdot \dfrac{16y^2}{3x^2}$ *Multiply by reciprocal*

$= \dfrac{96x^4y^5z^2}{24x^3yz^4}$ *Multiply*

$= \dfrac{4xy^4}{z^2}$ *Lowest terms*

27. $\dfrac{\dfrac{12}{x+6}}{\dfrac{4}{2x+12}} = \dfrac{12}{x+6} \div \dfrac{4}{2x+12}$

$= \dfrac{12}{x+6} \cdot \dfrac{2x+12}{4}$ *Multiply by reciprocal*

$= \dfrac{12 \cdot 2(x+6)}{(x+6)4} = 6$

28. $3x + 5y = 12$

To find the x-intercept, let $y = 0$.

$$3x + 5(0) = 12$$
$$3x = 12$$
$$x = 4$$

The x-intercept is $(4, 0)$.
To find the y-intercept, let $x = 0$.

$$3(0) + 5y = 12$$
$$5y = 12$$
$$y = \dfrac{12}{5}$$

The y-intercept is $\left(0, \dfrac{12}{5}\right)$.

Plot the intercepts and draw the line through them.

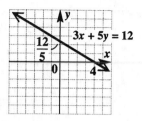

29. $A(-2, 1)$ and $B(3, -5)$
(a) The slope of line AB is

$$m = \dfrac{1 - (-5)}{-2 - 3} = \dfrac{6}{-5} = -\dfrac{6}{5}.$$

(b) The slope of a line perpendicular to line AB is the negative reciprocal of $-\dfrac{6}{5}$, which is $\dfrac{5}{6}$.

30. Through $(4, -1)$, $m = -4$

Use the point-slope form with $x_1 = 4$, $y_1 = -1$, and $m = -4$.

$$y - y_1 = m(x - x_1)$$
$$y - (-1) = -4(x - 4)$$
$$y + 1 = -4x + 16$$
$$y = -4x + 15$$

31. Through $(0, 0)$ and $(1, 4)$

Find the slope.

$$m = \dfrac{4 - 0}{1 - 0} = 4$$

Because the slope is 4 and the y-intercept is 0, the equation of the line in slope-intercept form is

$$y = 4x.$$

32. $-2x + y < -6$
Graph the line $-2x + y = -6$, which has intercepts $(3, 0)$ and $(0, -6)$, as a dashed line since the inequality involves $<$. Test $(0, 0)$, which yields $0 < -6$, a false statement. Shade the region that does not include $(0, 0)$.

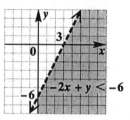

33. $f(x) = -3x + 6$
(a) The domain includes the set of all real numbers $(-\infty, \infty)$.

(b) $f(-6) = -3(-6) + 6 = 18 + 6 = 24$

34. The set $\{(\text{City}, \text{Percent})\}$ is a function. (The set $\{(\text{Percent}, \text{City})\}$ is not a function since 78, 79, and 85 would all correspond to more than one city.) The elements of the domain of the function are the names of the ten cities.

35. $2x - 7y = 14$

For any value of x, there is exactly one value of y, so the equation defines a function. Solve for y:

$$-7y = -2x + 14$$

$$y = \frac{2}{7}x - 2$$

Replace y with $f(x)$.

$$f(x) = \frac{2}{7}x - 2$$

36.

Let $x =$ the amount raised from corporate income taxes;

$x + 484.6 =$ the amount raised from individual income taxes.

Since the total is 828.2,

$$x + (x + 484.6) = 828.2$$

$$2x = 343.6$$

$$x = 171.8.$$

$$x + 484.6 = 656.4$$

The amount raised from corporate income taxes was \$171.8 billion and the amount raised from individual income taxes was \$656.4 billion.

37.   The sum of the measures of the angles of any triangle is 180°, so

$$(x + 15) + (6x + 10) + (x - 5) = 180.$$

Solve this equation.

$$8x + 20 = 180$$

$$8x = 160$$

$$x = 20$$

Substitute 20 for x to find the measures of the angles.

$$x - 5 = 20 - 5 = 15$$
$$x + 15 = 20 + 15 = 35$$
$$6x + 10 = 6(20) + 10 = 130$$

The measures of the angles of the triangle are 15°, 35°, and 130°.

38.

Let $x =$ the length of the shorter leg;

$3x + 4 =$ the length of the hypotenuse;

$2x + 10 =$ the length of the longer leg.

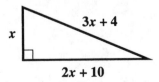

Use the Pythagorean formula with $a = x$, $b = 2x + 10$, and $c = 3x + 4$.

$$a^2 + b^2 = c^2$$

$$x^2 + (2x + 10)^2 = (3x + 4)^2$$

$$x^2 + 4x^2 + 40x + 100 = 9x^2 + 24x + 16$$

Square the binomials

$$5x^2 + 40x + 100 = 9x^2 + 24x + 16$$

Combine like terms

$$0 = 4x^2 - 16x - 84$$

Standard form

$$0 = 4(x^2 - 4x - 21)$$

Factor out 4

$$0 = 4(x - 7)(x + 3)$$

Factor

$$x - 7 = 0 \quad \text{or} \quad x + 3 = 0$$
$$x = 7 \quad \text{or} \quad x = -3$$

Discard -3 because length cannot be negative. Thus, the length of the shorter leg is 7 inches.

39.   Let $C =$ the cost of a pizza

and $r =$ the radius of the pizza.

C varies directly as r^2, so

$$C = kr^2$$

for some constant k. Since $C = 6$ when $r = 7$, substitute these values in the equation and solve for k.

$$C = kr^2$$

$$6 = k(7)^2$$

$$6 = 49k$$

$$\frac{6}{49} = k$$

So, $C = \frac{6}{49}r^2$.

When $r = 9$,

$$C = \frac{6}{49}(9)^2$$

$$= \frac{6}{49}(81) = \frac{486}{49} \approx 9.92.$$

A pizza with a 9-inch radius should cost \$9.92.

40.   Let $x =$ the number of hours it will take the man and his wife to do the job, working together.

	Rate	Time working together	Fractional part of the job done when working together
Man	$\dfrac{1}{3}$	x	$\dfrac{1}{3}x$
Wife	$\dfrac{1}{1.5}$	x	$\dfrac{1}{1.5}x$

Together, the man and his wife complete 1 whole job, so

$$\frac{1}{3}x + \frac{1}{1.5}x = 1$$

Rewrite this equation by changing the decimal to a fraction.

$$\frac{1}{3}x + \frac{1}{\frac{3}{2}}x = 1$$

$$\frac{1}{3}x + \frac{2}{3}x = 1 \qquad \textit{Reciprocal of } \tfrac{3}{2} \textit{ is } \tfrac{2}{3}$$

$$\frac{3}{3}x = 1$$

$$x = 1$$

Working together, it will take the man and his wife 1 hour to do the job.

CHAPTER 9 LINEAR SYSTEMS

Section 9.1

1. Look at the big picture for this exercise. The four given points are located in different quadrants, which should make matching the solution with the graph very easy.

   (a) $(3, 4)$ is in quadrant I—choice B is correct.

   (b) $(-2, 3)$ is in quadrant II—choice C is correct.

   (c) $(-4, -1)$ is in quadrant III—choice D is correct.

   (d) $(5, -2)$ is in quadrant IV—choice A is correct.

3. $(6, 2)$

   $$3x + y = 20$$
   $$2x + 3y = 18$$

   To decide whether $(6, 2)$ is a solution of the system, substitute 6 for x and 2 for y in each equation.

   $$3x + y = 20$$
   $$3(6) + 2 = 20 \text{ ?}$$
   $$18 + 2 = 20 \text{ ?}$$
   $$20 = 20 \quad \text{True}$$
   $$2x + 3y = 18$$
   $$2(6) + 3(2) = 18 \text{ ?}$$
   $$12 + 6 = 18 \text{ ?}$$
   $$18 = 18 \quad \text{True}$$

   Since $(6, 2)$ satisfies both equations, it is a solution of the system.

5. $(2, -3)$

   $$x + y = -1$$
   $$x + 5y = 19$$

   To decide whether $(2, -3)$ is a solution of the system, substitute 2 for x and -3 for y in each equation.

   $$x + y = -1$$
   $$2 + (-3) = -1 \text{ ?}$$
   $$-1 = -1 \quad \text{True}$$
   $$2x + 5y = 19$$
   $$2(2) + 5(-3) = 19 \text{ ?}$$
   $$4 + (-15) = 19 \text{ ?}$$
   $$-11 = 19 \quad \text{False}$$

   The ordered pair $(2, -3)$ satisfies the first equation but not the second. Because it does not satisfy *both* equations, it is not a solution of the system.

7. $(-1, -3)$

   $$3x + 5y = -18$$
   $$4x + 2y = -10$$

   Substitute -1 for x and -3 for y in each equation.

   $$3x + 5y = -18$$
   $$3(-1) + 5(-3) = -18 \text{ ?}$$
   $$-3 - 15 = -18 \text{ ?}$$
   $$-18 = -18 \quad \text{True}$$
   $$4x + 2y = -10$$
   $$4(-1) + 2(-3) = -10 \text{ ?}$$
   $$-4 - 6 = -10 \text{ ?}$$
   $$-10 = -10 \quad \text{True}$$

   Since $(-1, -3)$ satisfies both equations, it is a solution of the system.

9. $(7, -2)$

   $$4x = 26 - y$$
   $$3x = 29 + 4y$$

   Substitute 7 for x and -2 for y in each equation.

   $$4x = 26 - y$$
   $$4(7) = 26 - (-2) \text{ ?}$$
   $$28 = 26 + 2 \quad \text{?}$$
   $$28 = 28 \quad \text{True}$$

   $$3x = 29 + 4y$$
   $$3(7) = 29 + 4(-2) \text{ ?}$$
   $$21 = 29 - 8 \quad \text{?}$$
   $$21 = 21 \quad \text{True}$$

   Since $(7, -2)$ satisfies both equations, it is a solution of the system.

11. $(6, -8)$

    $$-2y = x + 10$$
    $$3y = 2x + 30$$

    Substitute 6 for x and -8 for y in each equation.

    $$-2y = x + 10$$
    $$-2(-8) = 6 + 10 \text{ ?}$$
    $$16 = 16 \quad \text{True}$$
    $$3y = 2x + 30$$
    $$3(-8) = 2(6) + 30 \text{ ?}$$
    $$-24 = 12 + 30 \quad \text{?}$$
    $$-24 = 42 \quad \text{False}$$

    The ordered pair $(6, -8)$ satisfies the first equation but not the second. Because it does not satisfy *both* equations, it is not a solution of the system.

13. From the graph, the ordered pair that is a solution of the system is in the second quadrant. Choice (a), $(-4, -4)$, is the only ordered pair given that is not in quadrant II, so it is the only valid choice.

15. $x - y = 2$
       $x + y = 6$

To graph the equations, find the intercepts.

$x - y = 2$: Let $y = 0$; then $x = 2$.
       Let $x = 0$; then $y = -2$.

Plot the intercepts, $(2, 0)$ and $(0, -2)$, and draw the line through them.

$x + y = 6$: Let $y = 0$; then $x = 6$.
       Let $x = 0$; then $y = 6$.

Plot the intercepts, $(6, 0)$ and $(0, 6)$, and draw the line through them.

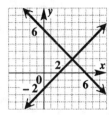

It appears that the lines intersect at the point $(4, 2)$. Check this by substituting 4 for x and 2 for y in both equations. Since $(4, 2)$ satisfies both equations, the solution set of this system is $\{(4, 2)\}$.

17. $x + y = 4$
       $y - x = 4$

To graph the equations, find the intercepts.

$x + y = 4$: Let $y = 0$; then $x = 4$.
       Let $x = 0$; then $y = 4$.

Plot the intercepts, $(0, 4)$ and $(4, 0)$, and draw the line through them.

$y - x = 4$: Let $y = 0$; then $x = -4$.
       Let $x = 0$; then $y = 4$.

Plot the intercepts, $(-4, 0)$ and $(0, 4)$, and draw the line through them.

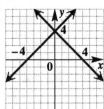

The lines intersect at their common y-intercept, $(0, 4)$, so $\{(0, 4)\}$ is the solution set of the system.

19. $x - 2y = 6$
       $x + 2y = 2$

To graph the equations, find the intercepts.

$x - 2y = 6$: Let $y = 0$; then $x = 6$.
       Let $x = 0$; then $y = -3$.

Plot the intercepts, $(6, 0)$ and $(0, -3)$, and draw the line through them.

$x + 2y = 2$: Let $y = 0$; then $x = 2$.
       Let $x = 0$; then $y = 1$.

Plot the intercepts, $(2, 0)$ and $(0, 1)$, and draw the line through them.

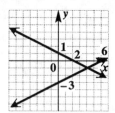

It appears that the lines intersect at the point $(4, -1)$. Since $(4, -1)$ satisfies both equations, the solution set of this system is $\{(4, -1)\}$.

21. $3x - 2y = -3$
       $-3x - y = -6$

To graph the equations, find the intercepts.

$3x - 2y = -3$: Let $y = 0$; then $x = -1$.
       Let $x = 0$; then $y = \dfrac{3}{2}$.

Plot the intercepts, $(-1, 0)$ and $\left(0, \dfrac{3}{2}\right)$, and draw the line through them.

$-3x - y = -6$: Let $y = 0$; then $x = 2$.
       Let $x = 0$; then $y = 6$.

Plot the intercepts, $(2, 0)$ and $(0, 6)$, and draw the line through them.

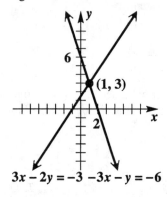

It appears that the lines intersect at the point $(1, 3)$. Since $(1, 3)$ satisfies both equations, the solution set of this system is $\{(1, 3)\}$.

23. $2x - 3y = -6$
       $y = -3x + 2$

To graph the first line, find the intercepts.

$2x - 3y = -6$: Let $y = 0$; then $x = -3$.
       Let $x = 0$; then $y = 2$.

Plot the intercepts, $(-3, 0)$ and $(0, 2)$, and draw the line through them.

To graph the second line, start by plotting the y-intercept, $(0, 2)$. From this point, go 3 units down and 1 unit to the right (because the slope is -3) to reach the point $(1, -1)$. Draw the line through $(0, 2)$ and $(1, -1)$.

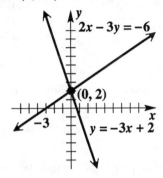

The lines intersect at their common y-intercept, $(0, 2)$, so $\{(0, 2)\}$ is the solution set of the system.

25. $2x - y = 6$
$4x - 2y = 8$

Graph the line $2x - y = 6$ through its intercepts, $(3, 0)$ and $(0, -6)$.

Graph the line $4x - 2y = 8$ through its intercepts, $(2, 0)$ and $(0, -4)$.

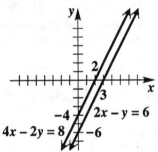

The lines each have slope 2, and hence, are parallel. Since they do not intersect, there is no solution. This is an inconsistent system and the solution set is $\emptyset$.

27. $3x = 5 - y$
$6x + 2y = 10$

$3x = 5 - y$: Let $y = 0$; then $x = \dfrac{5}{3}$.

Let $x = 0$; then $y = 5$.

Plot these intercepts, $\left(\dfrac{5}{3}, 0\right)$ and $(0, 5)$, and draw the line through them.

$6x + 2y = 10$: Let $y = 0$; then $x = \dfrac{5}{3}$.

Let $x = 0$; then $y = 5$.

Plot these intercepts, $\left(\dfrac{5}{3}, 0\right)$ and $(0, 5)$, and draw the line through them.

Since both equations have the same intercepts, they are equations of the same line.

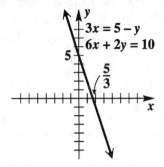

There is an infinite number of solutions. The equations are dependent equations and the solution set contains an infinite number of ordered pairs. We can write the solution set as
$\{(x, y) \mid 3x = 5 - y\}$ or as
$\{(x, y) \mid 6x + 2y = 10\}$.

29. $3x - 4y = 24$

$y = -\dfrac{3}{2}x + 3$

Graph the line $3x - 4y = 24$ through its intercepts, $(8, 0)$ and $(0, -6)$.

To graph the line $y = -\dfrac{3}{2}x + 3$, plot the y-intercept $(0, 3)$ and then go 3 units down and 2 units to the right $\left(\text{because the slope is } -\dfrac{3}{2}\right)$ to reach the point $(2, 0)$. Draw the line through $(0, 3)$ and $(2, 0)$.

It appears that the lines intersect at the point $(4, -3)$. Since $(4, -3)$ satisfies both equations, the solution set of this system is $\{(4, -3)\}$.

31. Two nonparallel lines will intersect in only one point (if they are distinct) or infinitely many points (if they are the same). They cannot intersect in exactly two points.

33. Yes, it is possible. For example, the system

$$x + y = 5$$
$$x - y = -1$$
$$\overline{2x - y = 1}$$

has the single solution $(2, 3)$. Geometrically, this is just three lines intersecting at one point.

35. From the graph, it appears that the production level for each format was about 350 million.

37. The *total* production for LPs and CDs in 1987 was about 2×100 million; or about 200 million.

39. The production of cassettes was approximately constant from 1984 to 1986 and from 1988 to 1990.

41. $y - x = -5$

$x + y = 1$

Write the equations in slope-intercept form.

$$
\begin{array}{ll}
y - x = -5 & x + y = 1 \\
\quad y = x - 5 & \quad y = -x + 1 \\
\quad m = 1 & \quad m = -1
\end{array}
$$

The lines have different slopes.

(a) The system is consistent because it has a solution. The equations are independent because they have different graphs. Therefore, the answer is "neither."

(b) The graph is a pair of intersecting lines.

(c) The system has one solution.

43. $x + 2y = 0$

$4y = -2x$

Write the equations in slope-intercept form.

$$
\begin{array}{ll}
x + 2y = 0 & 4y = -2x \\
2y = -x & \quad y = -\dfrac{1}{2}x \\
\quad y = -\dfrac{1}{2}x &
\end{array}
$$

For both lines, $m = -\dfrac{1}{2}$ and $b = 0$.

(a) Since the equations have the same slope and y-intercept, they are dependent.

(b) The graph is one line.

(c) The system has an infinite number of solutions.

45. $5x + 4y = 7$

$10x + 8y = 4$

Write the equations in slope-intercept form.

$$
\begin{array}{ll}
5x + 4y = 7 & 10x + 8y = 4 \\
4y = -5x + 7 & 8y = -10x + 4 \\
y = -\dfrac{5}{4}x + \dfrac{7}{4} & y = -\dfrac{10}{8}x + \dfrac{4}{8} \\
m = -\dfrac{5}{4},\; b = \dfrac{7}{4} & y = -\dfrac{5}{4}x + \dfrac{1}{2} \\
& m = -\dfrac{5}{4},\; b = \dfrac{1}{2}
\end{array}
$$

The lines have the same slope but different y-intercepts.

(a) The system is inconsistent because it has no solution.

(b) The graph is a pair of parallel lines.

(c) The system has no solution.

47. Supply equals demand at the point where the two lines intersect, or when $x = 40$.

49. The coordinates of the point of intersection are $(40, 30)$.

51. For $x + y = 4$, the x-intercept is $(4, 0)$ and the y-intercept is $(0, 4)$. The only screen with a graph having those intercepts is choice B. The point of intersection, $(3, 1)$ is listed at the bottom of the screen. Check $(3, 1)$ in the system of equations. Since it satisfies the equations, $\{(3, 1)\}$ is the solution set.

53. The graph of $x - y = 0$ goes through the origin. The only screen with a graph that goes through th origin is choice A.

55. $3x + y = 2$

$2x - y = -7$

First, solve the equations for y.

$$
\begin{array}{ll}
3x + y = 2 & 2x - y = -7 \\
\quad y = -3x + 2 & 2x + 7 = y
\end{array}
$$

Now enter the equations.

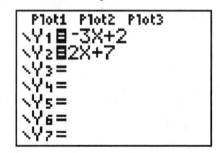

Next, graph the equations using a standard window. On the TI-82/83, just press $\boxed{\text{ZOOM}}\ \boxed{6}$.

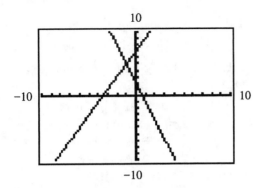

Now we'll let the calculator find the coordinates of the point of intersection of the graphs. Press $\boxed{\text{2nd}}$ $\boxed{\text{CALC}}$ $\boxed{5}$ $\boxed{\text{ENTER}}$ $\boxed{\text{ENTER}}$ to indicate the graphs for which we're trying to find the point of intersection. Now press the left cursor key, $\boxed{\triangleleft}$, four times to get close to the point of intersection. Lastly, press $\boxed{\text{ENTER}}$ to produce the following graph.

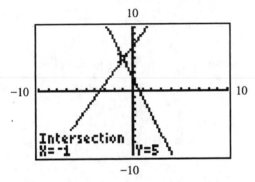

The display at the bottom of the last figure indicates that the solution set is $\{(-1, 5)\}$.

57.
$$\frac{1}{2}x + 4 = 3x - 1$$

$$2\left(\frac{1}{2}x + 4\right) = 2(3x - 1) \quad \textit{Multiply by 2}$$

$$x + 8 = 6x - 2 \quad \textit{Distributive property}$$

$$10 = 5x \quad \textit{Subtract x; add 2}$$

$$x = \frac{10}{5} = 2 \quad \textit{Divide by 5}$$

The solution set is $\{2\}$.

58.
$$\frac{1}{2}x + 4 = 3x - 1$$

$$\frac{1}{2}(2) + 4 = 3(2) - 1 \ ? \quad \textit{Let x = 2}$$

$$1 + 4 = 6 - 1 \quad ?$$

$$5 = 5 \quad \textit{True}$$

59.
$$y = \frac{1}{2}x + 4$$
$$y = 3x - 1$$

For the line $y = \frac{1}{2}x + 4$, the slope is $\frac{1}{2}$ and the y-intercept is $(0, 4)$. Plot the y-intercept and then go 1 unit up and 2 units to the right to reach the point $(2, 5)$. Draw the line through $(0, 4)$ and $(2, 5)$.

For the line $y = 3x - 1$, the slope is 3 and the y-intercept is $(0, -1)$. Plot the y-intercept and then go 3 units up and 1 unit to the right to reach the point $(1, 2)$. Draw the line through $(0, -1)$ and $(1, 2)$.

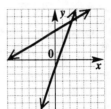

It appears that the lines intersect at the point $(2, 5)$. Since $(2, 5)$ satisfies both equations, the solution set of this system is $\{(2, 5)\}$.

60. The x-coordinate of the solution of the system, 2, is the same as the solution of the linear equation in Exercise 57.

61. The y-coordinate of the solution of the system, 5, is the same as the value obtained on both sides in the check in Exercise 58.

62. The solution of the linear equation
$$\frac{2}{3}x + 3 = -x + 8 \text{ is 3. When we substitute 3}$$
back into the equation, we get the value $\underline{5}$ on both sides, verifying that 3 is indeed the solution. Now if we graph the system
$$y = \frac{2}{3}x + 3$$
$$y = -x + 8,$$
we find that the solution set of the system is $\{(\underline{3}, \underline{5})\}$.

Section 9.2

1. No, it is not correct, because the solution set is $\{(3, 0)\}$. The y-value in the ordered pair must also be determined.

In this section, all solutions should be checked by substituting in *both* equations of the original system. Checks will not be shown here.

3.
$$x + y = 12 \quad (1)$$
$$y = 3x \quad (2)$$

Equation (2) is already solved for y. Substitute $3x$ for y in equation (1) and solve the resulting equation for x.

continued

$$x + y = 12$$
$$x + 3x = 12 \quad \textit{Let } y = 3x$$
$$4x = 12$$
$$x = 3$$

To find the y-value of the solution, substitute 3 for x in equation (2).

$$y = 3x$$
$$y = 3(3) \quad \textit{Let } x = 3$$
$$= 9$$

The solution set is $\{(3, 9)\}$.

To check this solution, substitute 3 for x and 9 for y in both equations of the given system.

5. $3x + 2y = 27$ (1)

$x = y + 4$ (2)

Equation (2) is already solved for x. Substitute $y + 4$ for x in equation (1).

$$3x + 2y = 27$$
$$3(y + 4) + 2y = 27$$
$$3y + 12 + 2y = 27$$
$$5y = 15$$
$$y = 3$$

To find x, substitute 3 for y in equation (2).

$$x = y + 4$$
$$x = 3 + 4 = 7$$

The solution set is $\{(7, 3)\}$.

7. $3x + 5y = 25$ (1)

$x - 2y = -10$ (2)

Solve equation (2) for x since its coefficient is 1.

$$x - 2y = -10$$
$$x = 2y - 10 \quad (3)$$

Substitute $2y - 10$ for x in equation (1) and solve for y.

$$3x + 5y = 25$$
$$3(2y - 10) + 5y = 25$$
$$6y - 30 + 5y = 25$$
$$11y = 55$$
$$y = 5$$

To find x, substitute 5 for y in equation (3).

$$x = 2y - 10$$
$$x = 2(5) - 10 = 0$$

The solution set is $\{(0, 5)\}$.

9. $3x + 4 = -y$ (1)

$2x + y = 0$ (2)

Solve equation (1) for y.

$$3x + 4 = -y$$
$$y = -3x - 4 \quad (3)$$

Substitute $-3x - 4$ for y in equation (2) and solve for x.

$$2x + y = 0$$
$$2x + (-3x - 4) = 0$$
$$-x - 4 = 0$$
$$-x = 4$$
$$x = -4$$

To find y, substitute -4 for x in equation (3).

$$y = -3x - 4$$
$$= -3(-4) - 4 = 8$$

The solution set is $\{(-4, 8)\}$.

11. $7x + 4y = 13$ (1)

$x + y = 1$ (2)

Solve equation (2) for y.

$$x + y = 1$$
$$y = 1 - x \quad (3)$$

Substitute $1 - x$ for y in equation (1).

$$7x + 4y = 13$$
$$7x + 4(1 - x) = 13$$
$$7x + 4 - 4x = 13$$
$$3x + 4 = 13$$
$$3x = 9$$
$$x = 3$$

To find y, substitute 3 for x in equation (3).

$$y = 1 - x$$
$$y = 1 - 3 = -2$$

The solution set is $\{(3, -2)\}$.

13. $3x - y = 5$ (1)

$y = 3x - 5$ (2)

Equation (2) is already solved for y, so we substitute $3x - 5$ for y in equation (1).

$$3x - (3x - 5) = 5$$
$$3x - 3x + 5 = 5$$
$$5 = 5 \quad \textit{True}$$

This true result means that every solution of one equation is also a solution of the other, so the system has an infinite number of solutions. We can write the solution set as $\{(x, y) \mid 3x - y = 5\}$.

15. $6x - 8y = 6$ (1)

$-3x + 2y = -2$ (2)

Solve equation (2) for y.

$$-3x + 2y = -2$$
$$2y = 3x - 2$$
$$y = \frac{3x - 2}{2} \quad (3)$$

Substitute $\frac{3x - 2}{2}$ for y in equation (1) and solve for x.

$$6x - 8y = 6$$
$$6x - 8\left(\frac{3x - 2}{2}\right) = 6$$
$$6x - 4(3x - 2) = 6$$
$$6x - 12x + 8 = 6$$
$$-6x = -2$$
$$x = \frac{-2}{-6} = \frac{1}{3}$$

To find y, let $x = \frac{1}{3}$ in equation (3).

$$y = \frac{3x - 2}{2}$$
$$y = \frac{3\left(\frac{1}{3}\right) - 2}{2}$$
$$= \frac{1 - 2}{2} = -\frac{1}{2}$$

The solution set is $\left\{\left(\frac{1}{3}, -\frac{1}{2}\right)\right\}$.

17. $2x + 8y = 3$ (1)
$x = 8 - 4y$ (2)

Equation (2) is already solved for x, so substitute $8 - 4y$ for x in equation (1).

$$2(8 - 4y) + 8y = 3$$
$$16 - 8y + 8y = 3$$
$$16 = 3 \quad False$$

This false result means that the system is inconsistent and its solution set is ∅.

19. $12x - 16y = 8$ (1)
$3x = 4y + 2$ (2)

Solve equation (2) for x.

$$3x = 4y + 2$$
$$x = \frac{4y + 2}{3} \quad (3)$$

Substitute $\frac{4y + 2}{3}$ for x in equation (1).

$$12x - 16y = 8$$
$$12\left(\frac{4y + 2}{3}\right) - 16y = 8$$
$$4(4y + 2) - 16y = 8$$
$$16y + 8 - 16y = 8$$
$$8 = 8 \quad True$$

This true result means that every solution of one equation is also a solution of the other, so the system has an infinite number of solutions. We can write the solution set as
$\{(x, y) \mid 3x = 4y + 2\}$.

21. The first student had less work to do because in solving for y in the first equation no fractions occur. However, solving for x in the second equation introduces fractions.

23. $\frac{5}{3}x + 2y = \frac{1}{3} + y$ (1)
$2x - 3 + \frac{y}{3} = -2 + x$ (2)

First, clear all fractions.

Equation (1):

$$\frac{5}{3}x + 2y = \frac{1}{3} + y$$
$$3\left(\frac{5}{3}x + 2y\right) = 3\left(\frac{1}{3} + y\right) \quad Multiply\ by\ 3$$
$$5x + 6y = 1 + 3y \quad \begin{array}{l}Distributive\\ property\end{array}$$
$$5x + 3y = 1 \quad (3)$$

Equation (2):

$$2x - 3 + \frac{y}{3} = -2 + x$$
$$3\left(2x - 3 + \frac{y}{3}\right) = 3(-2 + x) \quad Multiply\ by\ 3$$
$$6x - 9 + y = -6 + 3x \quad \begin{array}{l}Distributive\\ property\end{array}$$
$$3x + y = 3 \quad (4)$$

The system has been simplified to
$$5x + 3y = 1 \quad (3)$$
$$3x + y = 3. \quad (4)$$

Solve this system by the substitution method.

$$y = 3 - 3x \quad (5)$$
$$Solve\ (4)\ for\ y$$
$$5x + 3(3 - 3x) = 1$$
$$Substitute\ for\ y\ in\ (3)$$
$$5x + 9 - 9x = 1$$
$$-4x = -8$$
$$x = 2$$

To find y, let $x = 2$ in equation (5).

$$y = 3 - 3(2) = -3$$

The solution set is $\{(2, -3)\}$.

25. $\dfrac{x}{2} - \dfrac{y}{3} = \dfrac{5}{6}$ (1)

$\dfrac{x}{5} - \dfrac{y}{4} = \dfrac{1}{10}$ (2)

First, clear all fractions.

Equation (1):

$$\frac{x}{2} - \frac{y}{3} = \frac{5}{6}$$

$$6\left(\frac{x}{2} - \frac{y}{3}\right) = 6\left(\frac{5}{6}\right) \quad \textit{Multiply by 6}$$

$$3x - 2y = 5 \quad (3)$$

Equation (2):

$$\frac{x}{5} - \frac{y}{4} = \frac{1}{10}$$

$$20\left(\frac{x}{5} - \frac{y}{4}\right) = 20\left(\frac{1}{10}\right) \quad \textit{Multiply by 20}$$

$$4x - 5y = 2 \quad (4)$$

The system has been simplified to

$$3x - 2y = 5 \quad (3)$$
$$4x - 5y = 2. \quad (4)$$

Solve equation (3) for x.

$$3x - 2y = 5$$
$$3x = 5 + 2y$$
$$x = \frac{5 + 2y}{3} \quad (5)$$

Now substitute into equation (4).

$$4\left(\frac{5 + 2y}{3}\right) - 5y = 2$$

$$\frac{4(5 + 2y)}{3} - 5y = 2$$

Multiply by 3 to clear fractions.

$$3\left[\frac{4(5 + 2y)}{3} - 5y\right] = 3(2)$$
$$4(5 + 2y) - 15y = 6$$
$$20 + 8y - 15y = 6$$
$$-7y = -14$$
$$y = 2$$

To find x, let $y = 2$ in equation (5).

$$x = \frac{5 + 2(2)}{3} = \frac{9}{3} = 3$$

The solution set is $\{(3, 2)\}$.

27. $\dfrac{x}{5} + 2y = \dfrac{8}{5}$ (1)

$\dfrac{3x}{5} + \dfrac{y}{2} = -\dfrac{7}{10}$ (2)

First, clear all fractions.

Equation (1):

$$\frac{x}{5} + 2y = \frac{8}{5}$$

$$5\left(\frac{x}{5} + 2y\right) = 5\left(\frac{8}{5}\right) \quad \textit{Multiply by 5}$$

$$x + 10y = 8 \quad (3)$$

Equation (2):

$$\frac{3x}{5} + \frac{y}{2} = -\frac{7}{10}$$

$$10\left(\frac{3x}{5} + \frac{y}{2}\right) = 10\left(-\frac{7}{10}\right) \quad \textit{Multiply by 10}$$

$$6x + 5y = -7 \quad (4)$$

The system has been simplified to

$$x + 10y = 8 \quad (3)$$
$$6x + 5y = -7. \quad (4)$$

Solve this system by the substitution method. Solve equation (3) for x.

$$x = 8 - 10y \quad (5)$$

Substitute $8 - 10y$ for x in equation (4).

$$6(8 - 10y) + 5y = -7$$
$$48 - 60y + 5y = -7$$
$$-55y = -55$$
$$y = 1$$

To find x, let $y = 1$ in equation (5).

$$x = 8 - 10(1) = -2$$

The solution set is $\{(-2, 1)\}$.

29. $y = 2.1x + 22.8$ (1)
$y = 28.68$ (2)

Substitute 28.68 for y in equation (1).

$$28.68 = 2.1x + 22.8$$
$$5.88 = 2.1x$$
$$x = \frac{5.88}{2.1} = 2.8$$

Round this value to 3, which represents the year 1993.

31. To find the total cost, multiply the number of bicycles (x) by the cost per bicycle (400 dollars), and add the fixed cost (5000 dollars). Thus, $y_1 = 400x + 5000$ gives the total cost (in dollars).

32. Since each bicycle sells for $600, the total revenue for selling x bicycles is $600x$ (in dollars). Thus, $y_2 = 600x$ gives the total revenue.

33. $y_1 = 400x + 5000$ (1)
$y_2 = 600x$ (2)

To solve this system by the substitution method, substitute $600x$ for y_1 in equation (1).

$$600x = 400x + 5000$$
$$200x = 5000$$
$$x = 25$$

If $x = 25$, $y_2 = 600(25) = 15{,}000$.

The solution set is $\{(25,\ 15{,}000)\}$.

34. The value of x from Exercise 33 is the number of bikes it takes to break even. When <u>25</u> bikes are sold, the break-even point is reached. At that point, you have spent <u>15,000</u> dollars and taken in <u>15,000</u> dollars.

35.
$$y = 6 - x \quad (1)$$
$$y = 2x \quad\ \ (2)$$

Substitute $2x$ for y in equation (1).

$$2x = 6 - x$$
$$3x = 6$$
$$x = 2$$

Substituting 2 for x in either of the original equations gives $y = 4$.

The solution set is $\{(2, 4)\}$.

Input the equations $Y_1 = 6 - x$ and $Y_2 = 2x$ and then use the intersecting feature to obtain the following graph. See the solution to Exercise 55 in Section 9.1 for more specifics.

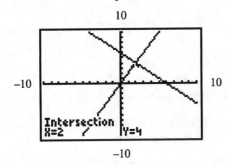

37.
$$y = -\frac{4}{3}x + \frac{19}{3} \quad (1)$$
$$y = \frac{15}{2}x - \frac{5}{2} \quad (2)$$

Substitute the expression from equation (2) into equation (1).

$$\frac{15}{2}x - \frac{5}{2} = -\frac{4}{3}x + \frac{19}{3}$$

Multiply by 6 to clear fractions.

$$6\left(\frac{15}{2}x - \frac{5}{2}\right) = 6\left(-\frac{4}{3}x + \frac{19}{3}\right)$$
$$45x - 15 = -8x + 38$$
$$53x = 53$$
$$x = 1$$

To find y, let $x = 1$ in equation (1).

$$y = -\frac{4}{3}(1) + \frac{19}{3}$$
$$= -\frac{4}{3} + \frac{19}{3} = \frac{15}{3} = 5$$

The solution set is $\{(1, 5)\}$.

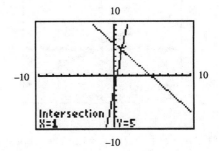

39.
$$4x + 5y = 5 \quad (1)$$
$$2x + 3y = 1 \quad (2)$$

Solve equation (2) for y.

$$2x + 3y = 1$$
$$3y = 1 - 2x$$
$$y = \frac{1 - 2x}{3} \quad (3)$$

Substitute for y in equation (1).

$$4x + 5\left(\frac{1 - 2x}{3}\right) = 5$$
$$4x + \frac{5(1 - 2x)}{3} = 5$$

Multiply by 3 to clear fractions.

$$3\left[4x + \frac{5(1 - 2x)}{3}\right] = 3(5)$$
$$12x + 5(1 - 2x) = 15$$
$$12x + 5 - 10x = 15$$
$$2x = 10$$
$$x = 5$$

To find y, let $x = 5$ in equation (3).

$$y = \frac{1 - 2(5)}{3} = \frac{-9}{3} = -3$$

The solution set is $\{(5, -3)\}$.

To graph the original system on a graphing calculator, each equation must be solved for y.

Equation (1):

$$4x + 5y = 5$$
$$5y = 5 - 4x$$
$$y = \frac{5 - 4x}{5}$$

Equation (2) was solved for y — see (3).
Thus, the equations to input are

$$Y_1 = \frac{5 - 4x}{5}$$

continued

and

$$Y_2 = \frac{1 - 2x}{3}.$$

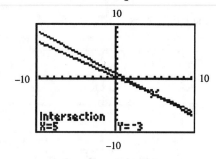

41. If the point of intersection does not appear on your screen, you will need to adjust the viewing window. Change the x- and y-minimum values or the x- and y-maximum values as necessary.

Section 9.3

1. The statement is true. Both lines would have x- and y-intercepts at $(0, 0)$. Since the point $(0, 0)$ satisfies both equations, it is a solution to the system.

3. It is impossible to have two numbers whose sum is both 1 and 2, so the given statement is true.

In Exercises 5–22, check your answers by substituting into *both* of the original equations. The check will be shown only for Exercise 5.

5. To eliminate y, add equations (1) and (2).

$$
\begin{array}{rrrrl}
x & - & y & = & -2 \quad (1) \\
x & + & y & = & 10 \quad (2) \\
\hline
2x & & & = & 8 \quad Add\ (1)\ and\ (2) \\
& & x & = & 4
\end{array}
$$

This result gives the x-value of the solution. To find the y-value of the solution, substitute 4 for x in either equation. We will use equation (2).

$$
\begin{aligned}
x + y &= 10 \\
4 + y &= 10 \quad Let\ x = 4 \\
y &= 6
\end{aligned}
$$

The solution set is $\{(4, 6)\}$.

Check by substituting 4 for x and 6 for y in both equations of the original system.

Check Equation (1)

$$
\begin{aligned}
x - y &= -2 \\
4 - 6 &= -2\ ? \quad Let\ x = 4,\ y = 6 \\
-2 &= -2 \quad True
\end{aligned}
$$

Check Equation (2)

$$
\begin{aligned}
x + y &= 10 \\
4 + 6 &= 10\ ? \quad Let\ x = 4,\ y = 6 \\
10 &= 10 \quad True
\end{aligned}
$$

7.
$$
\begin{array}{rrrrl}
2x & + & y & = & -5 \quad (1) \\
x & - & y & = & 2 \quad (2) \\
\hline
3x & & & = & -3 \quad Add\ (1)\ and\ (2) \\
& & x & = & -1
\end{array}
$$

Substitute -1 for x in equation (1) to find the y-value of the solution.

$$
\begin{aligned}
2x + y &= -5 \\
2(-1) + y &= -5 \quad Let\ x = -1 \\
-2 + y &= -5 \\
y &= -3
\end{aligned}
$$

The solution set is $\{(-1, -3)\}$.

9.
$$
\begin{array}{rrrrl}
3x & + & 2y & = & 0 \quad (1) \\
-3x & - & y & = & 3 \quad (2) \\
\hline
& & y & = & 3 \quad Add\ (1)\ and\ (2)
\end{array}
$$

Substitute 3 for y in equation (1).

$$
\begin{aligned}
3x + 2y &= 0 \\
3x + 2(3) &= 0 \\
3x + 6 &= 0 \\
3x &= -6 \\
x &= -2
\end{aligned}
$$

The solution set is $\{(-2, 3)\}$.

11.
$$
\begin{array}{rrrrl}
6x & - & 2y & = & -21 \quad (1) \\
-6x & + & 8y & = & 72 \quad (2) \\
\hline
& & 6y & = & 51 \quad Add\ (1)\ and\ (2) \\
& & y & = & \dfrac{51}{6} = \dfrac{17}{2}
\end{array}
$$

Rather than substitute $\dfrac{17}{2}$ for y in (1) or (2), we will eliminate y by multiplying equation (1) by 4 and adding that equation to equation (2).

$$
\begin{array}{rrrrl}
24x & - & 8y & = & -84 \quad (3) \\
-6x & + & 8y & = & 72 \quad (2) \\
\hline
18x & & & = & -12 \quad Add\ (3)\ and\ (2) \\
& & x & = & -\dfrac{12}{18} = -\dfrac{2}{3}
\end{array}
$$

Solving the system in this fashion reduces the chance of making an arithmetic error.

The solution set is $\left\{ \left(-\dfrac{2}{3}, \dfrac{17}{2} \right) \right\}$.

When you get a solution that has non-integer components, it is sometimes more difficult to check the problem than it was to solve it. A graphing calculator can be very helpful in this case. Just store the values for x and y in their respective memory locations, and then type the expressions as shown in the following screen. The results -21 and 72 (the right sides of the equations) indicate that we have found the correct solution.

```
-2/3→X:17/2→Y
                    8.5
6X-2Y
                    -21
-6X+8Y
                     72
```

13. $2x - y = 12$ (1)
         $3x + 2y = -3$ (2)

If we simply add the equations, we will not eliminate either variable. To eliminate y, multiply equation (1) by 2 and add the result to equation (2).

$$
\begin{array}{rcrl}
4x & - & 2y & = & 24 & (3) \\
3x & + & 2y & = & -3 & (2) \\
\hline
7x & & & = & 21 & \textit{Add (3) and (2)} \\
 & & x & = & 3
\end{array}
$$

Substitute 3 for x in equation (1).

$$
\begin{aligned}
2x - y &= 12 \\
2(3) - y &= 12 \\
-y &= 6 \\
y &= -6
\end{aligned}
$$

The solution set is $\{(3, -6)\}$.

15. $3x + 3y = 33$ (1)
         $5x - 2y = 27$ (2)

To eliminate y, multiply equation (1) by 2 and equation (2) by 3.

$$
\begin{array}{rcrl}
6x & + & 6y & = & 66 & (3) \\
15x & - & 6y & = & 81 & (4) \\
\hline
21x & & & = & 147 & \textit{Add (3) and (4)} \\
 & & x & = & 7
\end{array}
$$

Substitute 7 for x in equation (2).

$$
\begin{aligned}
5x - 2y &= 27 \\
5(7) - 2y &= 27 \\
35 - 2y &= 27 \\
-2y &= -8 \\
y &= 4
\end{aligned}
$$

The solution set is $\{(7, 4)\}$.

Note that we could have reduced equation (1) by dividing by 3 in the beginning.

17. $5x + 4y = 12$ (1)
         $3x + 5y = 15$ (2)

To eliminate x, we could multiply equation (1) by $-\frac{3}{5}$, but that would introduce fractions and make the solution more complicated. Instead, we'll work

with the least common multiple of the coefficients of x, which is 15, and choose suitable multipliers of these coefficients so that the new coefficients are opposites.

In this case, we could pick -3 times equation (1) and 5 times equation (2) *or* 3 times equation (1) and -5 times equation (2). If we wanted to eliminate y, we could multiply equation (1) by -5 and equation (2) by 4 *or* equation (1) by 5 and equation (2) by -4.

$$
\begin{array}{rcrll}
-15x & - & 12y & = & -36 & (3) -3 \times \text{Eq.(1)} \\
15x & + & 25y & = & 75 & (4) \;\; 5 \times \text{Eq.(2)} \\
\hline
 & & 13y & = & 39 & \textit{Add (3) and (4)} \\
 & & y & = & 3
\end{array}
$$

Substitute 3 for y in equation (1),

$$
\begin{aligned}
5x + 4y &= 12 \\
5x + 4(3) &= 12 \\
5x + 12 &= 12 \\
5x &= 0 \\
x &= 0
\end{aligned}
$$

The solution set is $\{(0, 3)\}$.

19. $5x - 4y = 15$ (1)
         $-3x + 6y = -9$ (2)

$$
\begin{array}{rcrll}
15x & - & 12y & = & 45 & (3) \;\; 3 \times \text{Eq.(1)} \\
-15x & + & 30y & = & -45 & (4) \;\; 5 \times \text{Eq.(2)} \\
\hline
 & & 18y & = & 0 & \textit{Add (3) and (4)} \\
 & & y & = & 0
\end{array}
$$

Substitute 0 for y in equation (1).

$$
\begin{aligned}
15x - 12y &= 45 \\
15x - 12(0) &= 45 \\
15x &= 45 \\
x &= 3
\end{aligned}
$$

The solution set is $\{(3, 0)\}$.

21. $3x - 7y = 1$ (1)
         $-5x + 4y = 4$ (2)

$$
\begin{array}{rcrll}
15x & - & 35y & = & 5 & (3) \;\; 5 \times \text{Eq.(1)} \\
-15x & + & 12y & = & 12 & (4) \;\; 3 \times \text{Eq.(2)} \\
\hline
 & & -23y & = & 17 & \textit{Add (3) and (4)} \\
 & & y & = & -\dfrac{17}{23}
\end{array}
$$

Choose multipliers to eliminate y since substituting $-\dfrac{17}{23}$ for y leads to "messy" arithmetic.

continued

$$
\begin{array}{rcl}
12x - 28y &=& 4 \quad (5) \quad 4 \times \text{Eq.}(1) \\
-35x + 28y &=& 28 \quad (6) \quad 7 \times \text{Eq.}(2) \\
\hline
-23x &=& 32 \quad Add\ (5)\ and\ (6)
\end{array}
$$

$$
x = -\frac{32}{23}
$$

The solution set is $\left\{ \left(-\dfrac{32}{23}, -\dfrac{17}{23} \right) \right\}$.

A calculator check is a good idea for this one.

```
-32/23→X: -17/23→
Y
          -.7391304348
3X-7Y
                    1
-5X+4Y
                    4
```

23. It would be easiest to solve for y in the first equation, to get $y = 5 - 2x$. The substitute $5 - 2x$ for y in the second equation to get $5x + 3(5 - 2x) = 11$. Solve this equation for x. Substitute back into either of the original equations to find the value of y. Then check the solution in both equations of the original system.

25.
$$
\begin{array}{ll}
5x - 4y - 8x - 2 = 6x + 3y - 3 & (1) \\
4x - y = -2y - 8 & (2)
\end{array}
$$

Begin by simplifying each equation.

$$
\begin{array}{ll}
-9x - 7y = -1 & (3) \\
4x + y = -8 & (4)
\end{array}
$$

Solve this system by the eliminate method. Multiply equation (4) by 7 and add the result to equation (3).

$$
\begin{array}{rcll}
-9x - 7y &=& -1 & (3) \\
28x + 7y &=& -56 & (5) \quad 7 \times \text{Eq.}(4) \\
\hline
19x &=& -57 & Add\ (3)\ and\ (5)
\end{array}
$$

$$
x = \frac{-57}{19} = -3
$$

To find y, substitute -3 for x in equation (4).

$$
\begin{array}{rl}
4(-3) + y = -8 & Let\ x = -3 \\
-12 + y = -8 & \\
y = 4 &
\end{array}
$$

The solution set is $\{(-3, 4)\}$.

27.
$$
\begin{array}{ll}
-2x + 3y = 12 + 2y & (1) \\
2x - 5y + 4 = -8 - 4y & (2)
\end{array}
$$

Simplify each equation and add the results.

$$
\begin{array}{rcl}
-2x + y &=& 12 \\
2x - y &=& -12 \\
\hline
0 &=& 0 \quad True
\end{array}
$$

The system has an infinite number of solutions. The solution set is $\{(x, y) \mid 2x - y = -12\}$.

29.
$$
\begin{array}{ll}
7x - 9 + 2y - 8 = -3y + 4x + 13 & (1) \\
4y - 8x = -8 + 9x + 32 & (2)
\end{array}
$$

Simplify each equation.

$$
\begin{array}{ll}
3x + 5y = 30 & (3) \\
-17x + 4y = 24 & (4)
\end{array}
$$

Solve this system by the elimination method. To eliminate y, multiply equation (3) by 4 and equation (4) by -5; then add the results.

$$
\begin{array}{rcll}
12x + 20y &=& 120 & (5) \quad 4 \times \text{Eq.}(3) \\
85x - 20y &=& -120 & (6) \ -5 \times \text{Eq.}(4) \\
\hline
97x &=& 0 & Add\ (5)\ and\ (6) \\
x &=& 0 &
\end{array}
$$

To find y, substitute 0 for x in equation (3).

$$
\begin{array}{rl}
3(0) + 5y = 30 & Let\ x = 0 \\
5y = 30 & \\
y = 6 &
\end{array}
$$

The solution set is $\{(0, 6)\}$.

31.
$$
\begin{array}{ll}
6x + 4y - 8 = x + 2y - 5 & (1) \\
-3x + 4y - 1 = -8x + 2y + 8 & (2)
\end{array}
$$

Simplify each equation.

$$
\begin{array}{ll}
5x + 2y = 3 & (3) \\
5x + 2y = 9 & (4)
\end{array}
$$

The left sides of (3) and (4) are equal, so they can't be equal to different numbers. There are no solutions so the solution set is $\emptyset$.

33.
$$
\begin{array}{ll}
4x - 3y = -8 & (1) \\
x + 3y = 13 & (2)
\end{array}
$$

(a) Solve the system by the elimination method.

$$
\begin{array}{rcll}
4x - 3y &=& -8 & (1) \\
x + 3y &=& 13 & (2) \\
\hline
5x &=& 5 & Add\ (1)\ and\ (2) \\
x &=& 1 &
\end{array}
$$

To find y, let $x = 1$ in equation (2).

$$
\begin{array}{rl}
x + 3y = 12 & \\
1 + 3y = 13 & \\
3y = 12 & \\
y = 4 &
\end{array}
$$

The solution set is $\{(1, 4)\}$.

(b) To solve this system by the substitution method, begin by solving equation (2) for x.

$$
\begin{array}{rl}
x + 3y = 13 & \\
x = -3y + 13 &
\end{array}
$$

Substitute $-3y + 13$ for x in equation (1).

$$4(-3y + 13) - 3y = -8$$
$$-12y + 52 - 3y = -8$$
$$-15y = -60$$
$$y = 4$$

To find x, let $y = 4$ in equation (2).

$$x + 3y = 13$$
$$x + 3(4) = 13$$
$$x + 12 = 13$$
$$x = 1$$

The solution set is $\{(1, 4)\}$.

(c) For this particular system, the elimination method is preferable because both equations are already written in the form $Ax + By = C$, and the equations can be added without multiplying either by a constant. Comparing solutions by the two methods, we see that the elimination method requires fewer steps than the substitution method for this system.

35. Yes, they should both get the same answer. However, the student who multiplied by 6 will be working with smaller numbers than the one who multiplied by 12. Working with smaller numbers is more convenient.

37. $x + \dfrac{1}{3}y = y - 2$ (1)

$\dfrac{1}{4}x + y = x + y$ (2)

Multiply each side of equation (1) by 3 to clear fractions.

$$3\left(x + \frac{1}{3}y\right) = 3(y - 2)$$
$$3x + y = 3y - 6$$
$$3x - 2y = -6 \qquad (3)$$

Multiply each side of equation (2) by 4 to clear fractions.

$$4\left(\frac{1}{4}x + y\right) = 4(x + y)$$
$$x + 4y = 4x + 4y$$
$$-3x = 0$$
$$x = 0$$

Substitute 0 for x in equation (3).

$$3x - 2y = -6$$
$$3(0) - 2y = -6$$
$$-2y = -6$$
$$y = 3$$

The solution set is $\{(0, 3)\}$.

39. $\dfrac{x}{6} + \dfrac{y}{6} = 2$ (1)

$-\dfrac{1}{2}x - \dfrac{1}{3}y = -8$ (2)

Multiply each side of equation (1) by 6 to clear fractions.

$$6\left(\frac{x}{6} + \frac{y}{6}\right) = 6(2)$$
$$6\left(\frac{x}{6}\right) + 6\left(\frac{y}{6}\right) = 6(2)$$
$$x + y = 12$$

Multiply each side of equation (2) by the LCD, 6, to clear fractions.

$$6\left(-\frac{1}{2}x - \frac{1}{3}y\right) = 6(-8)$$
$$6\left(-\frac{1}{2}x\right) + 6\left(-\frac{1}{3}y\right) = 6(-8)$$
$$-3x - 2y = -48$$

The given system of equations has been simplified as follows.

$$x + y = 12 \qquad (3)$$
$$-3x - 2y = -48 \qquad (4)$$

Multiply equation (3) by 3 and add the result to equation (4).

$$
\begin{array}{rcrcr}
3x & + & 3y & = & 36 \\
-3x & - & 2y & = & -48 \\
\hline
 & & y & = & -12
\end{array}
$$

To find x, let $y = -12$ in equation (3).

$$x + y = 12$$
$$x + (-12) = 12$$
$$x - 12 = 12$$
$$x = 24$$

The solution set is $\{(24, -12)\}$.

41. $\dfrac{x}{3} - \dfrac{3y}{4} = -\dfrac{1}{2}$ (1)

$\dfrac{x}{6} + \dfrac{y}{8} = \dfrac{3}{4}$ (2)

Multiply each side of equation (1) by 12 to clear fractions.

$$12\left(\frac{x}{3} - \frac{3y}{4}\right) = 12\left(-\frac{1}{2}\right)$$
$$4x - 9y = -6$$

Multiply each side of equation (2) by 24 to clear fractions.

$$24\left(\frac{x}{6} + \frac{y}{8}\right) = 24\left(\frac{3}{4}\right)$$
$$4x + 3y = 18$$

The given system of equations has been simplified as follows.

$$4x - 9y = -6 \qquad (3)$$
$$4x + 3y = 18 \qquad (4)$$

continued

Multiply equation (3) by -1 and add the result to equation (4).

$$\begin{array}{rcl} -4x \;+\; 9y &=& 6 \\ \underline{4x \;+\; 3y} &=& \underline{18} \\ 12y &=& 24 \\ y &=& 2 \end{array}$$

To find x, let $y = 2$ in equation (4).

$$\begin{aligned} 4x + 3y &= 18 \\ 4x + 3(2) &= 18 \\ 4x + 6 &= 18 \\ 4x &= 12 \\ x &= 3 \end{aligned}$$

The solution set is $\{(3, 2)\}$.

43.
$$y = ax + b$$
$$1141 = a(1991) + b \quad \textit{Let } x = 1991, \; y = 1141$$
$$1141 = 1991a + b$$

44. As in Exercise 43,

$$1339 = 1996a + b.$$

45.
$$\begin{aligned} 1991a + b &= 1141 \quad (1) \\ 1996a + b &= 1339 \quad (2) \end{aligned}$$

Multiply equation (1) by -1 and add the result to equation (2),

$$\begin{array}{rcl} -1991a \;-\; b &=& -1141 \\ \underline{1996a \;+\; b} &=& \underline{1339} \\ 5a &=& 198 \\ a &=& 39.6 \end{array}$$

Substitute 39.6 for a in equation (1).

$$\begin{aligned} 1991(39.6) + b &= 1141 \\ 78,843.6 + b &= 1141 \\ b &= -77,702.6 \end{aligned}$$

The solution set is $\{(39.6, -77,702.6)\}$.

46. An equation of the segment PQ is

$$y = 39.6x - 77,702.6$$

for $1991 \le x \le 1996$.

47. $y = 39.6x - 77,702.6$
$y = 39.6(1993) - 77,702.6 \quad \textit{Let } x = 1993$
$ = 78,922.8 - 77,702.6$
$ = 1220.2$ (million)

This is slightly less than the actual figure of 1244 million.

48. It is not realistic to expect the data to be in a perfectly straight line; as a result, the quantity obtained from an equation determined in this way will probably be "off" a bit. One cannot put too much faith in models such as this one, because not all data is linear in nature.

Section 9.4

1. Substitute 1 for x, 2 for y, and 3 for z in $3x + 2y - z$, which is the left side of each equation.

$$\begin{aligned} 3(1) + 2(2) - (3) &= 3 + 4 - 3 \\ &= 7 - 3 \\ &= 4 \end{aligned}$$

Choice (b) is correct since its right side is 4.

3.
$$\begin{array}{rcrl} 3x \;+\; 2y \;+\; z &=& 8 & (1) \\ 2x \;-\; 3y \;+\; 2z &=& -16 & (2) \\ x \;+\; 4y \;-\; z &=& 20 & (3) \end{array}$$

Eliminate z by adding equations (1) and (3).

$$\begin{array}{rclr} 3x \;+\; 2y \;+\; z &=& 8 & (1) \\ \underline{x \;+\; 4y \;-\; z} &=& \underline{20} & (3) \\ 4x \;+\; 6y &=& 28 & (4) \end{array}$$

To get another equation without z, multiply equation (3) by 2 and add the result to equation (2).

$$\begin{array}{rclr} 2x \;+\; 8y \;-\; 2z &=& 40 & 2 \times (3) \\ \underline{2x \;-\; 3y \;+\; 2z} &=& \underline{-16} & (2) \\ 4x \;+\; 5y &=& 24 & (5) \end{array}$$

Use equations (4) and (5) to eliminate x. Multiply equation (4) by -1 and add the result to equation (5).

$$\begin{array}{rclr} -4x \;-\; 6y &=& -28 & -1 \times (4) \\ \underline{4x \;+\; 5y} &=& \underline{24} & (5) \\ -y &=& -4 & \\ y &=& 4 & \end{array}$$

Substitute 4 for y in equation (5) to find x.

$$\begin{aligned} 4x + 5y &= 24 \quad (5) \\ 4x + 5(4) &= 24 \\ 4x + 20 &= 24 \\ 4x &= 4 \\ x &= 1 \end{aligned}$$

Substitute 1 for x and 4 for y in equation (3) to find z.

$$\begin{aligned} x + 4y - z &= 20 \quad (3) \\ 1 + 4(4) - z &= 20 \\ 1 + 16 - z &= 20 \\ 17 - z &= 20 \\ -z &= 3 \\ z &= -3 \end{aligned}$$

The solution $(1, 4, -3)$ checks in all three of the original equations.

Solution set: $\{(1, 4, -3)\}$

5.
$$2x + 5y + 2z = 0 \quad (1)$$
$$4x - 7y - 3z = 1 \quad (2)$$
$$3x - 8y - 2z = -6 \quad (3)$$

Add equations (1) and (3) to eliminate z.

$$
\begin{array}{r}
2x + 5y + 2z = 0 \quad (1) \\
3x - 8y - 2z = -6 \quad (3) \\
\hline
5x - 3y = -6 \quad (4)
\end{array}
$$

Multiply equation (1) by 3 and equation (2) by 2. Then add the results to eliminate z again.

$$
\begin{array}{rl}
6x + 15y + 6z = 0 & 3 \times (1) \\
8x - 14y - 6z = 2 & 2 \times (2) \\
\hline
14x + y = 2 \quad (5) &
\end{array}
$$

Solve the system

$$5x - 3y = -6 \quad (4)$$
$$14x + y = 2. \quad (5)$$

Multiply equation (5) by 3 then add this result to (4).

$$
\begin{array}{rl}
5x - 3y = -6 & (4) \\
42x + 3y = 6 & 3 \times (5) \\
\hline
47x = 0 &
\end{array}
$$
$$x = 0$$

To find y, substitute $x = 0$ into equation (4).

$$5x - 3y = -6 \quad (4)$$
$$5(0) - 3y = -6$$
$$y = 2$$

To find z, substitute $x = 0$ and $y = 2$ in equation (1).

$$2x + 5y + 2z = 0 \quad (1)$$
$$2(0) + 5(2) + 2z = 0$$
$$10 + 2z = 0$$
$$2z = -10$$
$$z = -5$$

Solution set: $\{(0, 2, -5)\}$

7.
$$x + y - z = -2 \quad (1)$$
$$2x - y + z = -5 \quad (2)$$
$$-x + 2y - 3z = -4 \quad (3)$$

Eliminate y and z by adding equations (1) and (2).

$$
\begin{array}{r}
x + y - z = -2 \quad (1) \\
2x - y + z = -5 \quad (2) \\
\hline
3x = -7
\end{array}
$$
$$x = -\frac{7}{3}$$

To get another equation without y, multiply equation (2) by 2 and add the result to equation (3).

$$
\begin{array}{rl}
4x - 2y + 2z = -10 & 2 \times (2) \\
-x + 2y - 3z = -4 & (3) \\
\hline
3x - z = -14 & (4)
\end{array}
$$

Substitute $-\frac{7}{3}$ for x in equation (4) to find z.

$$3x - z = -14 \quad (4)$$
$$3\left(-\frac{7}{3}\right) - z = -14$$
$$-7 - z = -14$$
$$-z = -7$$
$$z = 7$$

Substitute $-\frac{7}{3}$ for x and 7 for z in equation (1) to find y.

$$x + y - z = -2 \quad (1)$$
$$-\frac{7}{3} + y - 7 = -2$$
$$-7 + 3y - 21 = -6 \quad \textit{Multiply by 3.}$$
$$3y - 28 = -6$$
$$3y = 22$$
$$y = \frac{22}{3}$$

Solution set: $\left\{\left(-\frac{7}{3}, \frac{22}{3}, 7\right)\right\}$

A calculator check reduces the probability of making any arithmetic errors and is highly recommended. The following screen (similar to Exercise 11 solution in Section 9.3) shows the substitution of the solution for x, y, and z along with the three original equations. The evaluation of the three equations, -2, -5, and -4 (the right sides of the three equations), indicates that we have found the correct solution.

```
-7/3→X:22/3→Y:7→
Z:X+Y-Z
                    -2
2X-Y+Z
                    -5
-X+2Y-3Z
                    -4
```

9.
$$2x - 3y + 2z = -1 \quad (1)$$
$$x + 2y + z = 17 \quad (2)$$
$$2y - z = 7 \quad (3)$$

Multiply equation (2) by -2, and add the result to equation (1).

continued

$$\begin{array}{rcrcrcrl} 2x & - & 3y & + & 2z & = & -1 & (1) \\ -2x & - & 4y & - & 2z & = & -34 & \quad -2 \times (2) \\ \hline & & -7y & & & = & -35 & \\ & & & & y & = & 5 & \end{array}$$

To find z, substitute 5 for y in equation (3).

$$\begin{aligned} 2y - z &= 7 \quad (3) \\ 2(5) - z &= 7 \\ 10 - z &= 7 \\ -z &= -3 \\ z &= 3 \end{aligned}$$

To find x, substitute $y = 5$ and $z = 3$ into equation (1).

$$\begin{aligned} 2x - 3y + 2z &= -1 \quad (1) \\ 2x - 3(5) + 2(3) &= -1 \\ 2x - 9 &= -1 \\ 2x &= 8 \\ x &= 4 \end{aligned}$$

Solution set: $\{(4, 5, 3)\}$

11.
$$\begin{array}{rcrcrcrl} 4x & + & 2y & - & 3z & = & 6 & (1) \\ x & - & 4y & + & z & = & -4 & (2) \\ -x & & & + & 2z & = & 2 & (3) \end{array}$$

Equation (3) is missing y. Eliminate y again by multiplying equation (1) by 2 and adding the result to equation (2).

$$\begin{array}{rcrcrcrl} 8x & + & 4y & - & 6z & = & 12 & \quad 2 \times (1) \\ x & - & 4y & + & z & = & -4 & (2) \\ \hline 9x & & & - & 5z & = & .\,8 & (4) \end{array}$$

Use equations (3) and (4) to eliminate x. Multiply equation (3) by 9 and add the result to equation (4).

$$\begin{array}{rcrcrl} -9x & + & 18z & = & 18 & \quad 9 \times (3) \\ 9x & - & 5z & = & 8 & (4) \\ \hline & & 13z & = & 26 & \\ & & z & = & 2 & \end{array}$$

Substitute 2 for z in equation (3) to find x.

$$\begin{aligned} -x + 2z &= 2 \quad (3) \\ -x + 2(2) &= 2 \\ -x + 4 &= 2 \\ -x &= -2 \\ x &= 2 \end{aligned}$$

Substitute 2 for x and 2 for z in equation (2) to find y.

$$\begin{aligned} x - 4y + z &= -4 \quad (2) \\ 2 - 4y + 2 &= -4 \\ -4y + 4 &= -4 \\ -4y &= -8 \\ y &= 2 \end{aligned}$$

Solution set: $\{(2, 2, 2)\}$

13.
$$\begin{array}{rcrcrcrl} 2x & + & y & & & = & 6 & (1) \\ & & 3y & - & 2z & = & -4 & (2) \\ 3x & & & - & 5z & = & -7 & (3) \end{array}$$

To eliminate y, multiply equation (1) by -3 and add the result to equation (2).

$$\begin{array}{rcrcrcrl} -6x & - & 3y & & & = & -18 & \quad -3 \times (1) \\ & & 3y & - & 2z & = & -4 & (2) \\ \hline -6x & & & - & 2z & = & -22 & (4) \end{array}$$

Since equation (3) does not have a y-term, we can multiply equation (3) by 2 and add the result to equation (4) to eliminate x and solve for z.

$$\begin{array}{rcrcrl} 6x & - & 10z & = & -14 & \quad 2 \times (3) \\ -6x & - & 2z & = & -22 & (4) \\ \hline & & -12z & = & -36 & \\ & & z & = & 3 & \end{array}$$

To find x, substitute $z = 3$ into equation (3).

$$\begin{aligned} 3x - 5z &= -7 \quad (3) \\ 3x - 5(3) &= -7 \\ 3x &= 8 \\ x &= \frac{8}{3} \end{aligned}$$

To find y, substitute $z = 3$ into equation (2).

$$\begin{aligned} 3y - 2z &= -4 \quad (2) \\ 3y - 2(3) &= -4 \\ 3y &= 2 \\ y &= \frac{2}{3} \end{aligned}$$

Solution set: $\left\{ \left(\dfrac{8}{3}, \dfrac{2}{3}, 3 \right) \right\}$

15. Answers will vary.

(a) One example is two connecting walls and the ceiling (or floor) of a room, which meet at one point.

(b) One example is two opposite walls and the floor, which have no points in common. Another is the floors of three different levels of an office building.

(c) One example is the plane through the ceiling of a house, the plane through an outside wall, and the plane through one side of a slanted roof, which all meet in one line; therefore, they intersect in infinitely many points. Another example is three pages of this book, which intersect in the spine.

17.
$$2x + 2y - 6z = 5 \quad (1)$$
$$-3x + y - z = -2 \quad (2)$$
$$-x - y + 3z = 4 \quad (3)$$

Multiply equation (3) by 2 and add the result to equation (1).

$$\begin{array}{rcll}
2x + 2y - 6z &=& 5 & (1)\\
-2x - 2y + 6z &=& 8 & 2 \times (3)\\
\hline
0 &=& 13 & \textit{False}
\end{array}$$

Solution set: $\emptyset$

19.
$$-5x + 5y - 20z = -40 \quad (1)$$
$$x - y + 4z = 8 \quad (2)$$
$$3x - 3y + 12z = 24 \quad (3)$$

Dividing equation (1) by -5 gives equation (2). Dividing equation (3) by 3 also gives equation (2). The resulting equations are the same, so the three equations are dependent.
Solution set:
$$\{(x, y, z) \mid x - y + 4z = 8\}$$

21.
$$2x + y - z = 6 \quad (1)$$
$$4x + 2y - 2z = 12 \quad (2)$$
$$-x - \frac{1}{2}y + \frac{1}{2}z = -3 \quad (3)$$

Multiplying equation (1) by 2 gives equation (2). Multiplying equation (3) by -4 also gives equation (2). The resulting equations are the same, so the three equations are dependent.
Solution set:
$$\{(x, y, z) \mid 2x + y - z = 6\}$$

23.
$$x + y - 2z = 0 \quad (1)$$
$$3x - y + z = 0 \quad (2)$$
$$4x + 2y - z = 0 \quad (3)$$

Eliminate z by adding equations (2) and (3).

$$\begin{array}{rcll}
3x - y + z &=& 0 & (2)\\
4x + 2y - z &=& 0 & (3)\\
\hline
7x + y &=& 0 & (4)
\end{array}$$

To get another equation without z, multiply equation (2) by 2 and add the result to equation (1).

$$\begin{array}{rcll}
6x - 2y + 2z &=& 0 & 2 \times (2)\\
x + y - 2z &=& 0 & (1)\\
\hline
7x - y &=& 0 & (5)
\end{array}$$

Add equations (4) and (5) to find x.

$$\begin{array}{rcll}
7x + y &=& 0 & (4)\\
7x - y &=& 0 & (5)\\
\hline
14x &=& 0 &\\
x &=& 0 &
\end{array}$$

Substitute 0 for x in equation (4) to find y.

$$7x + y = 0 \quad (4)$$
$$7(0) + y = 0$$
$$0 + y = 0$$
$$y = 0$$

Substitute 0 for x and 0 for y in equation (1) to find z.

$$x + y - 2z = 0 \quad (1)$$
$$0 + 0 - 2z = 0$$
$$-2z = 0$$
$$z = 0$$

Solution set: $\{(0, 0, 0)\}$

25.
$$x + y + z - w = 5 \quad (1)$$
$$2x + y - z + w = 3 \quad (2)$$
$$x - 2y + 3z + w = 18 \quad (3)$$
$$-x - y + z + 2w = 8 \quad (4)$$

Eliminate w. Add equations (1) and (2).

$$\begin{array}{rcll}
x + y + z - w &=& 5 & (1)\\
2x + y - z + w &=& 3 & (2)\\
\hline
3x + 2y &=& 8 & (5)
\end{array}$$

Eliminate w again. Add equations (1) and (3).

$$\begin{array}{rcll}
x + y + z - w &=& 5 & (1)\\
x - 2y + 3z + w &=& 18 & (3)\\
\hline
2x - y + 4z &=& 23 & (6)
\end{array}$$

Eliminate w again. Multiply equation (2) by -2. Add the result to equation (4).

$$\begin{array}{rcll}
-4x - 2y + 2z - 2w &=& -6 & -2 \times (2)\\
-x - y + z + 2w &=& 8 & (4)\\
\hline
-5x - 3y + 3z &=& 2 & (7)
\end{array}$$

Equations (5), (6), and (7) do not contain a w-term. Since (5) does not have a z-term, we will find another equation without a z-term.

Eliminate z. Multiply equation (6) by 3 and equation (7) by -4. Then add the results.

$$\begin{array}{rcll}
6x - 3y + 12z &=& 69 & 3 \times (6)\\
20x + 12y - 12z &=& -8 & -4 \times (7)\\
\hline
26x + 9y &=& 61 & (8)
\end{array}$$

Eliminate z again. Multiply equation (5) by 9 and equation (8) by -2. Then add the results.

$$\begin{array}{rcll}
27x + 18y &=& 72 & 9 \times (5)\\
-52x - 18y &=& -122 & -2 \times (8)\\
\hline
-25x &=& -50 &\\
x &=& 2 &
\end{array}$$

To find y, substitute $x = 2$ into equation (5).

continued

$$3x + 2y = 8 \quad (5)$$
$$3(2) + 2y = 8$$
$$2y = 2$$
$$y = 1$$

To find z, substitute $x = 2$ and $y = 1$ into equation (6).

$$2x - y + 4z = 23 \quad (6)$$
$$2(2) - 1 + 4z = 23$$
$$4z = 20$$
$$z = 5$$

To find w, substitute $x = 2$, $y = 1$, and $z = 5$ into equation (1).

$$x + y + z - w = 5 \quad (1)$$
$$2 + 1 + 5 - w = 5$$
$$-w = -3$$
$$w = 3$$

Solution set: $\{(2, 1, 5, 3)\}$

For Exercises 27–36,

$$f(x) = ax^2 + bx + c \quad (a \neq 0).$$

27. $f(1) = a(1)^2 + b(1) + c$
$$= a + b + c$$
Since $f(1) = 128$, the first equation is
$$a + b + c = 128.$$

28. $f(1.5) = a(1.5)^2 + b(1.5) + c$
$$= 2.25a + 1.5b + c$$
Since $f(1.5) = 140$, the second equation is
$$2.25a + 1.5b + c = 140.$$

29. $f(3) = a(3)^2 + b(3) + c$
$$= 9a + 3b + c$$
Since $f(3) = 80$, the third equation is
$$9a + 3b + c = 80.$$

30. Using the three equations from Exercises 27–29, the system is

$$
\begin{array}{rrrrl}
a & + \quad b & + \ c & = \ 128 & (1) \\
2.25a & + \ 1.5b & + \ c & = \ 140 & (2) \\
9a & + \quad 3b & + \ c & = \ 80. & (3)
\end{array}
$$

Multiply equation (2) by -1 and add the result to equation (1).

$$
\begin{array}{rrrrl}
a & + \quad b & + \ c & = & 128 \ (1) \\
-2.25a & - \ 1.5b & - \ c & = & -140 \ (4) \\
\hline
-1.25a & - \quad .5b & & = & -12 \ (5)
\end{array}
$$

Multiply equation (3) by -1 and add the result to equation (1).

$$
\begin{array}{rrrrl}
a & + \ b & + \ c & = & 128 \ (1) \\
-9a & - \ 3b & - \ c & = & -80 \ (6) \\
\hline
-8a & - \ 2b & & = & 48 \ (7)
\end{array}
$$

Use equations (5) and (7) to eliminate b. Multiply equation (5) by -4 and add the result to equation (7).

$$
\begin{array}{rrrll}
5a & + \ 2b & = & 48 & \quad -4 \times (5) \\
-8a & - \ 2b & = & 48 & (7) \\
\hline
-3a & & = & 96 & \\
& a & = & -32 &
\end{array}
$$

To find b, substitute $a = -32$ into equation (7).

$$-8a - 2b = 48 \quad (7)$$
$$-8(-32) - 2b = 48$$
$$256 - 2b = 48$$
$$-2b = -208$$
$$b = 104$$

To find c, substitute $a = -32$ and $b = 104$ into equation (1).

$$a + b + c = 128 \quad (1)$$
$$-32 + 104 + c = 128$$
$$72 + c = 128$$
$$c = 56$$

Solution set: $\{(-32, 104, 56)\}$

31. If $(a, b, c) = (-32, 104, 56)$, then
$$f(x) = -32x^2 + 104x + 56.$$

32. $f(x) = -32x^2 + 104x + 56$
$$f(0) = -32(0)^2 + 104(0) + 56$$
$$= 56$$
The initial height is 56 ft.

33. $f(1.625) = -32(1.625)^2 + 104(1.625) + 56$
$$= -84.5 + 169 + 56$$
$$= 140.5$$
The maximum height is 140.5 ft.

34. $f(3.25) = -32(3.25)^2 + 104(3.25) + 56$
$$= -338 + 338 + 56$$
$$= 56$$
It tells us the projectile is at its original height after 3.25 seconds.

35. $f(x) = ax^2 + bx + c$
$$f(1) = a(1)^2 + b(1) + c$$
$$= a + b + c$$
So, $a + b + c = 2$.   (1)
$$f(-1) = a(-1)^2 + b(-1) + c$$
$$= a - b + c$$
So, $a - b + c = 0$.   (2)
$$f(-2) = a(-2)^2 + b(-2) + c$$
$$= 4a - 2b + c$$
So, $4a - 2b + c = 8$.   (3)

Add equations (1) and (2).

$$a + b + c = 2 \quad (1)$$
$$\underline{a - b + c = 0 \quad (2)}$$
$$2a \quad\;\; + 2c = 2 \quad (4)$$

Multiply equation (1) by 2 and add the result to equation (3).

$$2a + 2b + 2c = 4 \qquad 2 \times (1)$$
$$\underline{4a - 2b + c = 8 \quad (3)}$$
$$6a \quad\;\; + 3c = 12 \quad (5)$$

Multiply equation (4) by -3 and add the result to equation (5).

$$-6a - 6c = -6 \qquad -3 \times (4)$$
$$\underline{6a + 3c = 12 \quad (5)}$$
$$-3c = 6$$
$$c = -2$$

To find a, substitute $c = -2$ into equation (4).

$$2a + 2c = 2 \quad (4)$$
$$2a + 2(-2) = 2$$
$$2a - 4 = 2$$
$$2a = 6$$
$$a = 3$$

To find b, substitute $a = 3$ and $c = -2$ into equation (1).

$$a + b + c = 2 \quad (1)$$
$$3 + b - 2 = 2$$
$$b + 1 = 2$$
$$b = 1$$

So, $a = 3$, $b = 1$, and $c = -2$ and the equation is

$$f(x) = ax^2 + bx + c$$
$$f(x) = 3x^2 + x - 2.$$

36. $Y_1 = ax^2 + bx + c$
Using $(1, 8)$ gives
$$a(1)^2 + b(1) + c = 8$$
$$a + b + c = 8. \quad (1)$$
Using $(2, 15)$ gives
$$a(2)^2 + b(2) + c = 15$$
$$4a + 2b + c = 15. \quad (2)$$
Using $(3, 24)$ gives
$$a(3)^2 + b(3) + c = 24$$
$$9a + 3b + c = 24. \quad (3)$$
The system is

$$a + b + c = 8 \quad (1)$$
$$4a + 2b + c = 15 \quad (2)$$
$$9a + 3b + c = 24. \quad (3)$$

Multiply equation (2) by -1 and add the result to equation (1).

$$a + b + c = 8 \quad (1)$$
$$\underline{-4a - 2b - c = -15 \qquad -1 \times (2)}$$
$$-3a - b \quad\;\; = -7 \quad (4)$$

Multiply equation (3) by -1 and add the result to equation (1).

$$a + b + c = 8 \quad (1)$$
$$\underline{-9a - 3b - c = -24 \qquad -1 \times (3)}$$
$$-8a - 2b \quad\;\; = -16$$

Divide by -2.

$$4a + b = 8 \quad (5)$$

Add equations (4) and (5).

$$-3a - b = -7 \quad (4)$$
$$\underline{4a + b = 8 \quad (5)}$$
$$a \quad\;\; = 1$$

To find b, substitute $a = 1$ into equation (4).

$$-3a - b = -7 \quad (4)$$
$$-3(1) - b = -7$$
$$-3 - b = -7$$
$$-b = -4$$
$$b = 4$$

To find c, substitute $a = 1$ and $b = 4$ into equation (1).

$$a + b + c = 8 \quad (1)$$
$$1 + 4 + c = 8$$
$$c = 3$$

So, $a = 1$, $b = 4$, and $c = 3$, and the equation is

$$Y_1 = ax^2 + bx + c$$
$$Y_1 = x^2 + 4x + 3.$$

37. If one were to eliminate *different* variables in the first two steps, the result would be two equations in three variables, and it would not be possible to solve for a single variable in the next step.

Section 9.5

1. *Step 1*
   Let $x =$ the number of games won
   and $y =$ the number of games lost.

   Step 2
   Not necessary.

   Step 3
   They played 82 games, so

   $$x + y = 82. \quad (1)$$

   They won 56 more games than they lost, so

   $$x = 56 + y. \quad (2)$$

   Step 4
   Substitute $56 + y$ for x in (1).

   $$(56 + y) + y = 82$$
   $$56 + 2y = 82$$
   $$2y = 26$$
   $$y = 13$$

Substitute 13 for y in (2),

$$x = 56 + 13 = 69$$

Step 5
The Bulls win-loss record during the 1996–1997 N.B.A. regular season was 69 wins and 13 losses.

Step 6
69 is 56 more than 13 and the sum of 69 and 13 is 82. The solution is correct.

3.　*Step 1*
Let x = the number of wins,
y = the number of losses,
and z = the number of ties.

Step 2
Not necessary.

Step 3
They played 82 games, so

$$x + y + z = 82. \quad (1)$$

Their wins and losses totaled 74, so

$$x + y = 74. \quad (2)$$

They tied 18 fewer games than they lost, so

$$z = y - 18. \quad (3)$$

Step 4
Multiply (2) by -1 and add to (1).

$$\begin{array}{ll} x + y + z = 82 & (1) \\ \underline{-x - y = -74} & -1 \times (2) \\ z = 8 & \end{array}$$

Substitute 8 for z in (3).

$$8 = y - 18$$
$$26 = y$$

Substitute 26 for y in (2).

$$x + 26 = 74$$
$$x = 48$$

Step 5
The Stars won 48 games, lost 26 games, and tied 8 games.

Step 6
Adding 48, 26, and 8 gives 82 total games. The wins and losses add up to 74, and there were 18 fewer ties than losses. The solution is correct.

5.　Let W = the width of the tennis court
and L = the length.

Since the length is 42 ft more than the width,

$$L = W + 42. \quad (1)$$

The perimeter of a rectangle is given by

$$2W + 2L = P.$$

With perimeter $P = 228$ ft,

$$2W + 2L = 228. \quad (2)$$

Substitute $W + 42$ for L in equation (2).

$$2W + 2(W + 42) = 228$$
$$2W + 2W + 84 = 228$$
$$4W = 144$$
$$W = 36$$

Substitute $W = 36$ into equation (1).

$$L = W + 42 \quad (1)$$
$$L = 36 + 42$$
$$= 78$$

The length is 78 ft; the width is 36 ft.

7.　Let x = the length of the original rectangle
and y = the width of the original rectangle.

The length x is 7 ft more than the width y so,

$$x = 7 + y. \quad (1)$$

The perimeter of a rectangle is $P = 2L + 2W$. Here, $P = 32$, $L = x - 3$, and $W = y + 2$ for the new rectangle, so

$$32 = 2(x - 3) + 2(y + 2)$$
$$32 = 2x - 6 + 2y + 4$$
$$34 = 2x + 2y$$
$$\text{or} \quad x + y = 17. \quad\quad (2)$$

Solve the system. From (1), substitute $7 + y$ for x in (2).

$$(7 + y) + y = 17$$
$$7 + 2y = 17$$
$$2y = 10$$
$$y = 5$$

From (1),

$$x = 7 + y = 7 + 5 = 12.$$

The length of the rectangle is 12 ft, and the width is 5 ft.

9.　From the figure in the text, the angles marked y and $3x + 10$ are supplementary, so

$$(3x + 10) + y = 180. \quad (1)$$

Also, the angles x and y are complementary, so

$$x + y = 90. \quad (2)$$

Solve equation (2) for y to get

$$y = 90 - x. \quad (3)$$

Substitute $90 - x$ for y in equation (1).

$$(3x + 10) + (90 - x) = 180$$
$$2x + 100 = 180$$
$$2x = 80$$
$$x = 40$$

Substitute $x = 40$ into equation (3) to get

$$y = 90 - x = 90 - 40 = 50.$$

The angles measure 40° and 50°.

11. *Step 1*

Let $x =$ the number of cities
visited by Bruce
Springsteen & the
E Street Band;

$y =$ the number of cities
visited by Boyz
II Men.

Step 2
Not necessary

Step 3 The total number of cities visited was
174, so one equation is

$$x + y = 174. \quad (1)$$

Boyz II Men visited 94 cities more than Bruce
Springsteen, so another equation is

$$y = x + 94. \quad (2)$$

Step 4
Substitute $x + 94$ for y in equation (1).

$$x + y = 174$$
$$x + (x + 94) = 174$$
$$2x + 94 = 174$$
$$2x = 80$$
$$x = 40$$

Substitute 40 for x in (2) to find $y = 134$.

Step 5
Bruce Springsteen visited 40 cities and Boyz II
Men visited 134 cities.

Step 6
The sum of 40 and 134 is 174 and 134 is 94 more
than 40.

13. Let $x =$ the hockey FCI
and $y =$ the basketball FCI.

The sum is $423.12, so

$$x + y = 423.12. \quad (1)$$

The hockey FCI was $16.36 more than the
basketball FCI, so

$$x = y + 16.36. \quad (2)$$

From (2), substitute $y + 16.36$ for x in (1).

$$(y + 16.36) + y = 423.12$$
$$2y + 16.36 = 423.12$$
$$2y = 406.76$$
$$y = 203.38$$

From (2),

$$x = y + 16.36 = 203.38 + 16.36 = 219.74.$$

The hockey FCI was $219.74 and the basketball
FCI was $203.38.

15. Let $x =$ the price of a CGA monitor
and $y =$ the price of a VGA monitor.

For the first purchase, the total cost of 4 CGA
monitors and 6 VGA monitors is $4600, so

$$4x + 6y = 4600. \quad (1)$$

For the other purchase, the total cost of 6 CGA
monitors and 4 VGA monitors is $4400, so

$$6x + 4y = 4400. \quad (2)$$

To solve the system, multiply equation (1) by -2
and equation (2) by 3. Then add the results.

$$
\begin{array}{rcll}
-8x - 12y &=& -9200 & -2 \times (1) \\
18x + 12y &=& 13{,}200 & 3 \times (2) \\
\hline
10x &=& 4000 & \\
x &=& 400 &
\end{array}
$$

Since $x = 400$,

$$4x + 6y = 4600 \quad (1)$$
$$4(400) + 6y = 4600$$
$$1600 + 6y = 4600$$
$$6y = 3000$$
$$y = 500.$$

A CGA monitor costs $400, and a VGA monitor
costs $500.

17. Let $x =$ the number of units of yarn,
and $y =$ the number of units of thread.

Make a chart to organize the information in the
problem.

	Yarn	Thread	Total Hours
Hours on Machine A	1	1	8
Hours on Machine B	2	1	14

From the chart, write a system of equations.

$$x + y = 8 \quad (1)$$
$$2x + y = 14 \quad (2)$$

continued

Solve the system. Multiply equation (1) by -1 and add the result to equation (2).

$$\begin{array}{rcll} -x - y &=& -8 & -1 \times (1) \\ 2x + y &=& 14 & (2) \\ \hline x &=& 6 & \end{array}$$

Substitute $x = 6$ into equation (1) to get $y = 2$. The factory should make 6 units of yarn and 2 units of thread to keep its machines running at capacity.

19. Use the formula (rate of percent) • (base amount) = amount (percentage) of pure acid to compute parts (a) – (d).

  (a) $.10(60) = 6$ oz

  (b) $.25(60) = 15$ oz

  (c) $.40(60) = 24$ oz

  (d) $.50(60) = 30$ oz

21. The cost is the price per pound, \$.58, times the number of pounds, x, or \$.58x.

23. Let $x =$ the amount of 25% alcohol solution, and $y =$ the amount of 35% alcohol solution.

Make a table. The percent times the amount of solution gives the amount of pure alcohol in the third column.

Kind of Solution	Gallons of Solution	Amount of Pure Alcohol
.25	x	$.25x$
.35	y	$.35y$
.32	20	$.32(20) = 6.4$

The third row gives the total amounts of solution and pure alcohol. From the columns in the table, write a system of equations.

$$\begin{array}{rcll} x + y &=& 20 & (1) \\ .25x + .35y &=& 6.4 & (2) \end{array}$$

Solve the system. Multiply equation (1) by -25 and equation (2) by 100. Then add the results.

$$\begin{array}{rcll} -25x - 25y &=& -500 & -25 \times (1) \\ 25x + 35y &=& 640 & 100 \times (2) \\ \hline 10y &=& 140 & \\ y &=& 14 & \end{array}$$

Substitute $y = 14$ into equation (1).

$$\begin{array}{rcl} x + y &=& 20 \quad (1) \\ x + 14 &=& 20 \\ x &=& 6 \end{array}$$

Mix 6 gal of 25% solution and 14 gal of 35% solution.

25. Let $x =$ the amount of pure acid and $y =$ the amount of 10% acid.

Make a table.

Kind of Solution	Liters of Solution	Amount of Pure Acid
100% = 1	x	$1.00x = x$
.10	y	$.10y$
.20	27	$.20(27) = 5.4$

Solve the following system.

$$\begin{array}{rcll} x + y &=& 27 & (1) \\ x + .10y &=& 5.4 & (2) \end{array}$$

Multiply equation (2) by 10 to clear the decimals.

$$10x + y = 54 \qquad (3)$$

To eliminate y, multiply equation (1) by -1 and add the result to equation (3).

$$\begin{array}{rcll} -x - y &=& -27 & -1 \times (1) \\ 10x + y &=& 54 & (3) \\ \hline 9x &=& 27 & \\ x &=& 3 & \end{array}$$

Since $x = 3$,

$$\begin{array}{rcl} x + y &=& 27 \quad (1) \\ 3 + y &=& 27 \\ y &=& 24. \end{array}$$

Use 3 L of pure acid and 24 L of 10% acid.

27. From the "Number of Pounds" column in the text,

$$x + y = 80. \qquad (1)$$

From the "Value of Candy" column in the text,

$$3.60x + 7.20y = 4.95(80). \qquad (2)$$

Solve the system.

$$\begin{array}{rcll} -36x - 36y &=& -2880 & -36 \times (1) \\ 36x + 72y &=& 3960 & 10 \times (2) \\ \hline 36y &=& 1080 & \\ y &=& 30 & \end{array}$$

From (1), $x = 50$.
She should mix 50 lb of \$3.60/lb pecan clusters with 30 lb of \$7.20/lb chocolate truffles.

29. Let $x =$ the number of general admission tickets and $y =$ the number of student tickets.

Make a table.

Ticket	Number	Value of Tickets
General	x	$2.50x = 2.5x$
Student	y	$2.00y = 2y$
Total	184	406

Solve the system.

$$x + y = 184 \quad (1)$$
$$2.5x + 2y = 406 \quad (2)$$

Multiply equation (2) by 10 to clear the decimal.

$$25x + 20y = 4060 \quad (3)$$

To eliminate y, multiply equation (1) by -20 and add the result to equation (3).

$$
\begin{array}{rcl}
-20x - 20y &=& -3680 \\
25x + 20y &=& 4060 \quad (3) \\
\hline
5x &=& 380 \\
x &=& 76
\end{array}
$$

From (1),

$$76 + y = 184, \text{ so } y = 108.$$

76 general admission tickets and 108 student tickets were sold.

31. Let $x =$ the number of dimes and
$y =$ the number of quarters.

Make a table.

Coin	Amount	Value
Dimes ($.10)	x	$.10x$
Quarters ($.25)	y	$.25y$
Total	94	19.30

From the table, write a system of equations.

$$x + y = 94 \quad (1)$$
$$.10x + .25y = 19.30 \quad (2)$$

Solve the system. Multiply equation (1) by -10 and equation (2) by 100. Then add the results.

$$
\begin{array}{rcll}
-10x - 10y &=& -940 & -10 \times (1) \\
10x + 25y &=& 1930 & 100 \times (2) \\
\hline
15y &=& 990 & \\
y &=& 66 &
\end{array}
$$

From (1),

$$x + 66 = 94, \text{ so } x = 28.$$

She has 28 dimes and 66 quarters.

33. From the "Principal" column in the text,

$$x + y = 3000. \quad (1)$$

From the "Interest" column in the text,

$$.02x + .04y = 100. \quad (2)$$

Multiply equation (2) by 100 to clear the decimals.

$$2x + 4y = 10,000 \quad (3)$$

To eliminate x, multiply equation (1) by -2 and add the result to equation (3).

$$
\begin{array}{rcl}
-2x - 2y &=& -6000 \\
2x + 4y &=& 10,000 \quad (3) \\
\hline
2y &=& 4000 \\
y &=& 2000
\end{array}
$$

From (1), $x = 1000$.
$1000 is invested at 2%, and $2000 is invested at 4%.

35. In the formula $d = rt$, substitute 25 for r (rate or speed) and y for t (time in hours) to get

$$d = 25y \text{ mi.}$$

37. Let $x =$ speed of the freight train
and $y =$ the speed of the express train.

Make a table.

	r	t	d
Freight train	x	3	$3x$
Express train	y	3	$3y$

Since $d = rt$ and the trains are 390 km apart,

$$3x + 3y = 390. \quad (1)$$

The freight train travels 30 km/hr slower than the express train, so

$$x = y - 30. \quad (2)$$

Substitute $y - 30$ for x in equation (1) and solve for y.

$$
\begin{array}{rcl}
3x + 3y &=& 390 \quad (1) \\
3(y - 30) + 3y &=& 390 \\
3y - 90 + 3y &=& 390 \\
6y &=& 480 \\
y &=& 80
\end{array}
$$

From (2), $x = 80 - 30 = 50$.
The freight train travels at 50 km/hr, while the express train travels at 80 km/hr.

39. Let $x =$ the top speed of the snow speeder, and
$y =$ the speed of the wind.

Furthermore,

$$\text{rate into headwind} = x - y$$

and $\quad$ rate with tailwind $= x + y$.

Use these rates and the information in the problem to make a table.

	d	r	t
Into headwind	3600	$x - y$	2
With tailwind	3600	$x + y$	1.5

From the table, use the formula $d = rt$ to write a system of equations.

$$3600 = 2(x - y)$$
$$3600 = 1.5(x + y)$$

continued

Remove parentheses and move the variables to the left side.

$$2x - 2y = 3600 \quad (1)$$
$$1.5x + 1.5y = 3600 \quad (2)$$

Solve the system.

$$
\begin{array}{llll}
6x & - & 6y & = & 10,800 & (1) \times 3 \\
6x & + & 6y & = & 14,400 & (2) \times 4 \\
\hline
12x & & & = & 25,200 \\
& & x & = & 2100
\end{array}
$$

Substitute $x = 2100$ into equation (1).

$$2(2100) - 2y = 3600$$
$$4200 - 2y = 3600$$
$$-2y = -600$$
$$y = 300$$

The top speed of the snow speeder is 2100 miles per hour and the speed of the wind is 300 miles per hour.

41. Let $x =$ the number of $20 fish,
$y =$ the number of $40 fish,
and $z =$ the number of $65 fish.

The number of $40 fish is one less than twice the number of $20 fish, so

$$y = 2x - 1. \quad (1)$$

The number of fish totals 29, so

$$x + y + z = 29. \quad (2)$$

The fish are worth $1150, so

$$20x + 40y + 65z = 1150. \quad (3)$$

From (1), substitute $2x - 1$ for y in (2) and (3).

$$
\begin{array}{ll}
x + y + z = 29 & (2) \\
x + (2x - 1) + z = 29 & \\
3x + z = 30 & (4)
\end{array}
$$

$$
\begin{array}{ll}
20x + 40y + 65z = 1150 & (3) \\
20x + 40(2x - 1) + 65z = 1150 & \\
20x + 80x - 40 + 65z = 1150 & \\
100x + 65z = 1190 & (5)
\end{array}
$$

Multiply (4) by -65 and add the result to (5).

$$
\begin{array}{lll}
-195x & - & 65z & = & -1950 & \quad -65 \times (4) \\
100x & + & 65z & = & 1190 & \quad (5) \\
\hline
-95x & & & = & -760 \\
& & x & = & 8
\end{array}
$$

From (1), $y = 2(8) - 1 = 15$.
From (2), $z = 29 - x - y$
$$= 29 - 8 - 15 = 6.$$
There are 8 $20 fish, 15 $40 fish, and 6 $65 fish in the collection.

43. Let $x =$ the measure of one angle,
$y =$ the measure of another angle,
and $z =$ the measure of the last angle.

Two equations are given, so

$$z = x + 10$$
$$\text{or} \quad -x + z = 10 \quad (1)$$
$$\text{and} \quad x + y = 100. \quad (2)$$

Since the sum of the measures of the angles of a triangle is $180°$, the third equation of the system is

$$x + y + z = 180. \quad (3)$$

Equation (1) is missing y. To eliminate y again, multiply equation (2) by -1 and add the result to equation (3).

$$
\begin{array}{lllll}
-x & - & y & & = & -100 & \quad -1 \times (2) \\
x & + & y & + z & = & 180 & \quad (3) \\
\hline
& & & z & = & 80
\end{array}
$$

Since $z = 80$,

$$
\begin{array}{ll}
-x + z = 10 & (1) \\
-x + 80 = 10 & \\
-x = -70 & \\
x = 70. &
\end{array}
$$

From (2), $y = 30$.
The measures of the angles are $70°$, $30°$, and $80°$.

45. Let $x =$ the measure of the first angle,
$y =$ the measure of the second angle, and
$z =$ the measure of the third angle.

The sum of the angles in a triangle equals $180°$, so

$$x + y + z = 180. \quad (1)$$

The measure of the second angle is $10°$ more than 3 times that of the first angle, so

$$y = 3x + 10. \quad (2)$$

The third angle is equal to the sum of the other two, so

$$z = x + y. \quad (3)$$

Solve the system. Substitute z for $x + y$ in equation (1).

$$
\begin{array}{ll}
(x + y) + z = 180 & (1) \\
z + z = 180 & \\
2z = 180 & \\
z = 90 &
\end{array}
$$

Substitute $z = 90$ and $3x + 10$ for y in equation (3).

$$z = x + y \quad (3)$$
$$90 = x + (3x + 10)$$
$$80 = 4x$$
$$20 = x$$

Substitute $x = 20$ and $z = 90$ into equation (3).

$$z = x + y \quad (3)$$
$$90 = 20 + y$$
$$70 = y$$

The three angles have measures of 20°, 70°, and 90°.

47. Let $x =$ the length of the longest side,
$y =$ the length of the middle side,
and $z =$ the length of the shortest side.

Perimeter is the sum of the measures of the sides, so

$$x + y + z = 70. \quad (1)$$

The longest side is 4 cm less than the sum of the other sides, so

$$x = y + z - 4$$
$$\text{or} \quad x - y - z = -4. \quad (2)$$

Twice the shortest side is 9 cm less than the longest side, so

$$2z = x - 9$$
$$\text{or} \quad -x + 2z = -9 \quad (3)$$

Add equations (1) and (2) to eliminate y and z.

$$
\begin{array}{rcrcrcr}
x & + & y & + & z & = & 70 \quad (1) \\
x & - & y & - & z & = & -4 \quad (2) \\
\hline
2x & & & & & = & 66 \\
& & & & x & = & 33
\end{array}
$$

Since $x = 33$,

$$-x + 2z = -9 \quad (3)$$
$$-33 + 2z = -9$$
$$2z = 24$$
$$z = 12.$$

Since $x = 33$ and $z = 12$,

$$x + y + z = 70 \quad (1)$$
$$33 + y + 12 = 70$$
$$y + 45 = 70$$
$$y = 25.$$

The shortest side is 12 cm long, the middle side is 25 cm long, and the longest side is 33 cm long.

49. Let $x =$ the number of cases sent to wholesaler A,
$y =$ the number of cases sent to wholesaler B, and
$z =$ the number of cases sent to wholesaler C.

The total output is 320 cases per day, so

$$x + y + z = 320. \quad (1)$$

The number of cases to A is three times that sent to B, so

$$x = 3y. \quad (2)$$

Wholesaler C gets 160 cases less than the sum sent to A and B, so

$$z = x + y - 160. \quad (3)$$

Solve equation (3) for $x + y$ to get

$$x + y = z + 160. \quad (4)$$

Substitute $z + 160$ for $x + y$ in equation (1).

$$(x + y) + z = 320 \quad (1)$$
$$(z + 160) + z = 320$$
$$2z = 160$$
$$z = 80$$

Substitute 80 for z and $3y$ for x in equation (3).

$$z = x + y - 160 \quad (3)$$
$$80 = 3y + y - 160$$
$$240 = 4y$$
$$60 = y$$

Substitute $y = 60$ into equation (2).

$$x = 3y = 3(60) = 180$$

She should send 180 cases to Wholesaler A, 60 cases to Wholesaler B, and 80 cases to Wholesaler C.

51. Let $x =$ the amount of jelly beans,
$y =$ the amount of chocolate eggs,
and $z =$ the amount of marshmallow chicks.

The manager plans to make 15 lb of the mixture, so

$$x + y + z = 15. \quad (1)$$

She uses twice as many pounds of jelly beans as eggs and chicks, so

$$x = 2(y + z),$$
$$\text{or} \quad x - 2y - 2z = 0, \quad (2)$$

and five times as many pounds of jelly beans as eggs, so

continued

$$x = 5y. \quad (3)$$

Solve the system of equations (1), (2), and (3). Multiply equation (1) by 2 and add the result to equation (2).

$$
\begin{array}{rcll}
2x + 2y + 2z &=& 30 & 2 \times (1) \\
x - 2y - 2z &=& 0 & (2) \\
\hline
3x &=& 30 & \\
x &=& 10 &
\end{array}
$$

Since $x = 10$,

$$
\begin{array}{rcl}
x &=& 5y \quad (3) \\
10 &=& 5y \\
2 &=& y.
\end{array}
$$

Since $x = 10$ and $y = 2$,

$$
\begin{array}{rcl}
x + y + z &=& 15 \quad (1) \\
10 + 2 + z &=& 15 \\
12 + z &=& 15 \\
z &=& 3.
\end{array}
$$

She should use 10 lb of jelly beans, 2 lb of chocolate eggs, and 3 lb of marshmallow chicks.

53. $x^2 + y^2 + ax + by + c = 0$

Let $x = 2$ and $y = 1$.

$$
\begin{array}{rcl}
2^2 + 1^2 + a(2) + b(1) + c &=& 0 \\
4 + 1 + 2a + b + c &=& 0 \\
2a + b + c &=& -5
\end{array}
$$

54. $x^2 + y^2 + ax + by + c = 0$

Let $x = -1$ and $y = 0$.

$$
\begin{array}{rcl}
(-1)^2 + 0^2 + a(-1) + b(0) + c &=& 0 \\
1 - a + c &=& 0 \\
-a + c &=& -1 \\
a - c &=& 1
\end{array}
$$

55. $x^2 + y^2 + ax + by + c = 0$

Let $x = 3$ and $y = 3$.

$$
\begin{array}{rcl}
3^2 + 3^2 + a(3) + b(3) + c &=& 0 \\
9 + 9 + 3a + 3b + c &=& 0 \\
3a + 3b + c &=& -18
\end{array}
$$

56. Use the equations from Exercises 53–55 to form a system of equations.

$$
\begin{array}{rcll}
2a + b + c &=& -5 & (1) \\
a - c &=& 1 & (2) \\
3a + 3b + c &=& -18 & (3)
\end{array}
$$

Add equations (1) and (2).

$$
\begin{array}{rcll}
2a + b + c &=& -5 & (1) \\
a - c &=& 1 & (2) \\
\hline
3a + b &=& -4 & (4)
\end{array}
$$

Add equations (2) and (3).

$$
\begin{array}{rcll}
a - c &=& 1 & (2) \\
3a + 3b + c &=& -18 & (3) \\
\hline
4a + 3b &=& -17 & (5)
\end{array}
$$

Multiply equation (4) by -3 and add the result to equation (5).

$$
\begin{array}{rcll}
-9a - 3b &=& 12 & -3 \times (4) \\
4a + 3b &=& -17 & (5) \\
\hline
-5a &=& -5 & \\
a &=& 1 &
\end{array}
$$

Substitute $a = 1$ into equation (4).

$$
\begin{array}{rcl}
3a + b &=& -4 \quad (4) \\
3(1) + b &=& -4 \\
b &=& -7
\end{array}
$$

Substitute $a = 1$ into equation (2).

$$
\begin{array}{rcl}
a - c &=& 1 \quad (2) \\
1 - c &=& 1 \\
c &=& 0
\end{array}
$$

Since $a = 1$, $b = -7$, and $c = 0$, the equation of the circle is

$$
\begin{array}{rcl}
x^2 + y^2 + ax + by + c &=& 0 \\
x^2 + y^2 + x - 7y &=& 0.
\end{array}
$$

57. The graph of the circle is not the graph of a function because it fails the vertical line test.

Section 9.6

1. $\begin{bmatrix} -2 & 3 & 1 \\ 0 & 5 & -3 \\ 1 & 4 & 8 \end{bmatrix}$

(a) The elements of the second row are 0, 5, and -3.

(b) The elements of the third column are 1, -3, and 8.

(c) The matrix is square since the number of rows (three) is the same as the number of columns.

(d) The matrix obtained by interchanging the first and third rows is

$$\begin{bmatrix} 1 & 4 & 8 \\ 0 & 5 & -3 \\ -2 & 3 & 1 \end{bmatrix}.$$

(e) The matrix obtained by multiplying the first row by $-\frac{1}{2}$ is

$$\begin{bmatrix} -2\left(-\frac{1}{2}\right) & 3\left(-\frac{1}{2}\right) & 1\left(-\frac{1}{2}\right) \\ 0 & 5 & -3 \\ 1 & 4 & 8 \end{bmatrix} = \begin{bmatrix} 1 & -\frac{3}{2} & -\frac{1}{2} \\ 0 & 5 & -3 \\ 1 & 4 & 8 \end{bmatrix}.$$

(f) The matrix obtained by multiplying the third row by 3 and adding to the first row is

$$\begin{bmatrix} -2+3(1) & 3+3(4) & 1+3(8) \\ 0 & 5 & -3 \\ 1 & 4 & 8 \end{bmatrix} = \begin{bmatrix} 1 & 15 & 25 \\ 0 & 5 & -3 \\ 1 & 4 & 8 \end{bmatrix}.$$

3.
$$4x + 8y = 44$$
$$2x - y = -3$$

$$\begin{bmatrix} 4 & 8 & | & 44 \\ 2 & -1 & | & -3 \end{bmatrix}$$

$$\begin{bmatrix} 1 & \underline{2} & | & \underline{11} \\ 2 & -1 & | & -3 \end{bmatrix} \quad \tfrac{1}{4}R_1$$

$$\begin{bmatrix} 1 & 2 & | & 11 \\ 0 & \underline{-5} & | & \underline{-25} \end{bmatrix} \quad -2R_1 + R_2$$

Note: $\begin{cases} -2(2) + (-1) = -5 \\ -2(11) + (-3) = -25 \end{cases}$

$$\begin{bmatrix} 1 & 2 & | & 11 \\ 0 & 1 & | & \underline{5} \end{bmatrix} \quad -\tfrac{1}{5}R_1$$

This represents the system

$$x + 2y = 11$$
$$y = 5.$$

Substitute $y = 5$ in the first equation.

$$x + 2y = 11$$
$$x + 2(5) = 11$$
$$x + 10 = 11$$
$$x = 1$$

Solution set: $\{(1, 5)\}$

5.
$$x + y = 5$$
$$x - y = 3$$

Write the augmented matrix for this system.

$$\begin{bmatrix} 1 & 1 & | & 5 \\ 1 & -1 & | & 3 \end{bmatrix}$$

$$\begin{bmatrix} 1 & 1 & | & 5 \\ 0 & -2 & | & -2 \end{bmatrix} \quad -1R_1 + R_2$$

$$\begin{bmatrix} 1 & 1 & | & 5 \\ 0 & 1 & | & 1 \end{bmatrix} \quad -\tfrac{1}{2}R_2$$

This matrix gives the system

$$x + y = 5$$
$$y = 1.$$

Substitute $y = 1$ in the first equation.

$$x + y = 5$$
$$x + 1 = 5$$
$$x = 4$$

Solution set: $\{(4, 1)\}$

7.
$$2x + 4y = 6$$
$$3x - y = 2$$

Write the augmented matrix.

$$\begin{bmatrix} 2 & 4 & | & 6 \\ 3 & -1 & | & 2 \end{bmatrix}$$

The easiest way to get a 1 in the first row, first column position is to multiply the elements in the first row by $\tfrac{1}{2}$.

$$\begin{bmatrix} 1 & 2 & | & 3 \\ 3 & -1 & | & 2 \end{bmatrix} \quad \tfrac{1}{2}R_1$$

To get a 0 in row two, column 1, we need to subtract 3 from the 3 that is in that position. To do this we will multiply row 1 by -3 and add the result to row 2.

$$\begin{bmatrix} 1 & 2 & | & 3 \\ 0 & -7 & | & -7 \end{bmatrix} \quad -3R_1 + R_2$$

$$\begin{bmatrix} 1 & 2 & | & 3 \\ 0 & 1 & | & 1 \end{bmatrix} \quad -\tfrac{1}{7}R_2$$

This matrix gives the system

$$x + 2y = 3$$
$$y = 1.$$

Substitute $y = 1$ in the first equation.

$$x + 2y = 3$$
$$x + 2(1) = 3$$
$$x = 1$$

Solution set: $\{(1, 1)\}$

9.
$$3x + 4y = 13$$
$$2x - 3y = -14$$

Write the augmented matrix.

$$\begin{bmatrix} 3 & 4 & | & 13 \\ 2 & -3 & | & -14 \end{bmatrix}$$

$$\begin{bmatrix} 1 & 7 & | & 27 \\ 2 & -3 & | & -14 \end{bmatrix} \quad -1R_2 + R_1$$

$$\begin{bmatrix} 1 & 7 & | & 27 \\ 0 & -17 & | & -68 \end{bmatrix} \quad -2R_1 + R_2$$

$$\begin{bmatrix} 1 & 7 & | & 27 \\ 0 & 1 & | & 4 \end{bmatrix} \quad -\tfrac{1}{17}R_2$$

This matrix gives the system

$$x + 7y = 27$$
$$y = 4.$$

Substitute $y = 4$ in the first equation.

$$x + 7y = 27$$
$$x + 7(4) = 27$$
$$x + 28 = 27$$
$$x = -1$$

Solution set: $\{(-1, 4)\}$

11. $-4x + 12y = 36$
      $x - 3y = 9$

Write the augmented matrix.

$$\begin{bmatrix} -4 & 12 & | & 36 \\ 1 & -3 & | & 9 \end{bmatrix}$$

$$\begin{bmatrix} -1 & 3 & | & 9 \\ 1 & -3 & | & 9 \end{bmatrix} \quad \frac{1}{4}R_1$$

$$\begin{bmatrix} -1 & 3 & | & 9 \\ 0 & 0 & | & 18 \end{bmatrix} \quad R_1 + R_2$$

The corresponding system is

$$-x + 3y = 9$$
$$0 = 18 \quad \textit{False}$$

which is inconsistent and has no solution.

Solution set: $\emptyset$

13. Examples will vary.

(a) A matrix is a rectangular array of numbers.

(b) A horizontal arrangement of elements in a matrix is a row of a matrix.

(c) A vertical arrangement of elements in a matrix is a column of a matrix.

(d) A square matrix contains the same number of rows as columns.

(e) A matrix formed by the coefficients and constants of a linear system is an augmented matrix for the system.

(f) Row operations on a matrix allow it to be transformed into another one in which the solution of the associated system can be found more easily.

15. $x + y - z = -3$
    $2x + y + z = 4$
    $5x - y + 2z = 23$

$$\begin{bmatrix} 1 & 1 & -1 & | & -3 \\ 2 & 1 & 1 & | & 4 \\ 5 & -1 & 2 & | & 23 \end{bmatrix}$$

$$\begin{bmatrix} 1 & 1 & -1 & | & -3 \\ 0 & \underline{-1} & 3 & | & \underline{10} \\ 0 & \underline{-6} & 7 & | & \underline{38} \end{bmatrix} \quad \begin{matrix} -2R_1 + R_2 \\ -5R_1 + R_3 \end{matrix}$$

$$\begin{bmatrix} 1 & 1 & -1 & | & -3 \\ 0 & 1 & \underline{-3} & | & \underline{-10} \\ 0 & -6 & 7 & | & 38 \end{bmatrix} \quad -1R_2$$

$$\begin{bmatrix} 1 & 1 & -1 & | & -3 \\ 0 & 1 & -3 & | & -10 \\ 0 & 0 & \underline{-11} & | & \underline{-22} \end{bmatrix} \quad 6R_2 + R_3$$

$$\begin{bmatrix} 1 & 1 & -1 & | & -3 \\ 0 & 1 & -3 & | & -10 \\ 0 & 0 & 1 & | & \underline{2} \end{bmatrix} \quad -\frac{1}{11}R_3$$

This matrix gives the system

$$x + y - z = -3$$
$$y - 3z = -10$$
$$z = 2.$$

Substitute $z = 2$ in the second equation.

$$y - 3z = -10$$
$$y - 3(2) = -10$$
$$y - 6 = -10$$
$$y = -4$$

Substitute $y = -4$ and $z = 2$ in the first equation.

$$x + y - z = -3$$
$$x - 4 - 2 = -3$$
$$x - 6 = -3$$
$$x = 3$$

Solution set: $\{(3, -4, 2)\}$

17. $x + y - 3z = 1$
    $2x - y + z = 9$
    $3x + y - 4z = 8$

Write the augmented matrix.

$$\begin{bmatrix} 1 & 1 & -3 & | & 1 \\ 2 & -1 & 1 & | & 9 \\ 3 & 1 & -4 & | & 8 \end{bmatrix}$$

$$\begin{bmatrix} 1 & 1 & -3 & | & 1 \\ 0 & -3 & 7 & | & 7 \\ 0 & -2 & 5 & | & 5 \end{bmatrix} \quad \begin{matrix} -2R_1 + R_2 \\ -3R_1 + R_3 \end{matrix}$$

$$\begin{bmatrix} 1 & 1 & -3 & | & 1 \\ 0 & 1 & -\frac{7}{3} & | & -\frac{7}{3} \\ 0 & -2 & 5 & | & 5 \end{bmatrix} \quad -\frac{1}{3}R_2$$

$$\begin{bmatrix} 1 & 1 & -3 & | & 1 \\ 0 & 1 & -\frac{7}{3} & | & -\frac{7}{3} \\ 0 & 0 & \frac{1}{3} & | & \frac{1}{3} \end{bmatrix} \quad 2R_2 + R_3$$

$$\begin{bmatrix} 1 & 1 & -3 & | & 1 \\ 0 & 1 & -\frac{7}{3} & | & -\frac{7}{3} \\ 0 & 0 & 1 & | & 1 \end{bmatrix} \quad 3R_3$$

This matrix gives the system

$$x + y - 3z = 1$$
$$y - \frac{7}{3}z = -\frac{7}{3}$$
$$z = 1.$$

Substitute $z = 1$ in the second equation.

$$y - \frac{7}{3}z = -\frac{7}{3}$$
$$y - \frac{7}{3}(1) = -\frac{7}{3}$$
$$y = 0$$

Substitute $y = 0$ and $z = 1$ in the first equation.

$$x + y - 3z = 1$$
$$x + 0 - 3(1) = 1$$
$$x - 3 = 1$$
$$x = 4$$

Solution set: $\{(4, 0, 1)\}$

19.
$$x + y - z = 6$$
$$2x - y + z = -9$$
$$x - 2y + 3z = 1$$

Write the augmented matrix.

$$\begin{bmatrix} 1 & 1 & -1 & | & 6 \\ 2 & -1 & 1 & | & -9 \\ 1 & -2 & 3 & | & 1 \end{bmatrix}$$

$$\begin{bmatrix} 1 & 1 & -1 & | & 6 \\ 0 & -3 & 3 & | & -21 \\ 0 & -3 & 4 & | & -5 \end{bmatrix} \begin{matrix} \\ -2R_1 + R_2 \\ -1R_1 + R_3 \end{matrix}$$

$$\begin{bmatrix} 1 & 1 & -1 & | & 6 \\ 0 & 1 & -1 & | & 7 \\ 0 & -3 & 4 & | & -5 \end{bmatrix} \begin{matrix} \\ -\frac{1}{3}R_2 \\ \end{matrix}$$

$$\begin{bmatrix} 1 & 1 & -1 & | & 6 \\ 0 & 1 & -1 & | & 7 \\ 0 & 0 & 1 & | & 16 \end{bmatrix} \begin{matrix} \\ \\ 3R_2 + R_3 \end{matrix}$$

This matrix gives the system

$$x + y - z = 6$$
$$y - z = 7$$
$$z = 16.$$

Substitute $z = 16$ in the second equation.

$$y - 16 = 7$$
$$y = 23$$

Substitute $y = 23$ and $z = 16$ in the first equation.

$$x + 23 - 16 = 6$$
$$x + 7 = 6$$
$$x = -1$$

Solution set: $\{(-1, 23, 16)\}$

21.
$$x - y = 1$$
$$y - z = 6$$
$$x + z = -1$$

Write the augmented matrix.

$$\begin{bmatrix} 1 & -1 & 0 & | & 1 \\ 0 & 1 & -1 & | & 6 \\ 1 & 0 & 1 & | & -1 \end{bmatrix}$$

$$\begin{bmatrix} 1 & -1 & 0 & | & 1 \\ 0 & 1 & -1 & | & 6 \\ 0 & 1 & 1 & | & -2 \end{bmatrix} \begin{matrix} \\ \\ -1R_1 + R_3 \end{matrix}$$

$$\begin{bmatrix} 1 & -1 & 0 & | & 1 \\ 0 & 1 & -1 & | & 6 \\ 0 & 0 & 2 & | & -8 \end{bmatrix} \begin{matrix} \\ \\ -1R_2 + R_3 \end{matrix}$$

$$\begin{bmatrix} 1 & -1 & 0 & | & 1 \\ 0 & 1 & -1 & | & 6 \\ 0 & 0 & 1 & | & -4 \end{bmatrix} \begin{matrix} \\ \\ \frac{1}{2}R_3 \end{matrix}$$

This matrix gives the system

$$x - y = 1$$
$$y - z = 6$$
$$z = -4.$$

Substitute $z = -4$ in the second equation.

$$y - z = 6$$
$$y - (-4) = 6$$
$$y = 2$$

Substitute $y = 2$ in the first equation.

$$x - y = 1$$
$$x - 2 = 1$$
$$x = 3$$

Solution set: $\{(3, 2, -4)\}$

23.
$$x - 2y + z = 4$$
$$3x - 6y + 3z = 12$$
$$-2x + 4y - 2z = -8$$

Write the augmented matrix.

$$\begin{bmatrix} 1 & -2 & 1 & | & 4 \\ 3 & -6 & 3 & | & 12 \\ -2 & 4 & -2 & | & -8 \end{bmatrix}$$

$$\begin{bmatrix} 1 & -2 & 1 & | & 4 \\ 1 & -2 & 1 & | & 4 \\ -1 & 2 & -1 & | & -4 \end{bmatrix} \begin{matrix} \\ \frac{1}{3}R_2 \\ \frac{1}{2}R_3 \end{matrix}$$

$$\begin{bmatrix} 1 & -2 & 1 & | & 4 \\ 0 & 0 & 0 & | & 0 \\ 0 & 0 & 0 & | & 0 \end{bmatrix} \begin{matrix} \\ -1R_1 + R_2 \\ R_1 + R_3 \end{matrix}$$

This augmented matrix represents a system of dependent equations.

Solution set: $\{(x, y, z) | x - 2y + z = 4\}$

25.
$$4x + y = 5$$
$$2x + y = 3$$

Enter the augmented matrix as $[A]$.

$$\begin{bmatrix} 4 & 1 & 5 \\ 2 & 1 & 3 \end{bmatrix}$$

The TI-83 screen for A follows. (Use $\boxed{\text{MATRX}}$ EDIT.)

continued

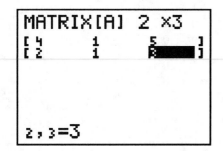

Now use the reduced row echelon form (rref) command to simplify the system. (Use MATRX MATH ALPHA B for rref and MATRX 1 for [A].)

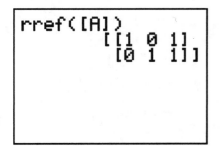

This matrix gives the system

$$1x + 0y = 1$$
$$0x + 1y = 1,$$

or simply, $x = 1$ and $y = 1$.

Solution set: $\{(1,1)\}$

27. $5x + y - 3z = -6$
        $2x + 3y + z = 5$
        $-3x - 2y + 4z = 3$

Enter the augmented matrix as [C].

$$\begin{bmatrix} 5 & 1 & -3 & -6 \\ 2 & 3 & 1 & 5 \\ -3 & -2 & 4 & 3 \end{bmatrix}$$

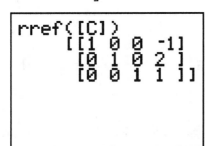

Solution set: $\{(-1, 2, 1)\}$

29. $x + z = -3$
        $y + z = 3$
        $x + y = 8$

Enter the augmented matrix as [E].

$$\begin{bmatrix} 1 & 0 & 1 & -3 \\ 0 & 1 & 1 & 3 \\ 1 & 1 & 0 & 8 \end{bmatrix}$$

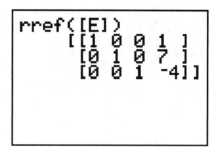

Solution set: $\{(1, 7, -4)\}$

Section 9.7

1. $\begin{vmatrix} -2 & -3 \\ 4 & -6 \end{vmatrix} = -2(-6) - (-3)(4)$

Choice (d) is correct.

3. $\begin{vmatrix} -2 & 5 \\ -1 & 4 \end{vmatrix} = -2(4) - 5(-1)$

$$= -8 + 5 = -3$$

5. $\begin{vmatrix} 1 & -2 \\ 7 & 0 \end{vmatrix} = 1(0) - (-2)(7)$

$$= 0 + 14 = 14$$

7. $\begin{vmatrix} 0 & 4 \\ 0 & 4 \end{vmatrix} = 0(4) - 4(0)$

$$= 0 - 0 = 0$$

9. $\begin{vmatrix} -1 & 2 & 4 \\ -3 & -2 & -3 \\ 2 & -1 & 5 \end{vmatrix}$ Expand by minors about the first column.

$$= -1 \begin{vmatrix} -2 & -3 \\ -1 & 5 \end{vmatrix} - (-3) \begin{vmatrix} 2 & 4 \\ -1 & 5 \end{vmatrix}$$
$$+ 2 \begin{vmatrix} 2 & 4 \\ -2 & -3 \end{vmatrix}$$
$$= -1[-2(5) - (-3)(-1)]$$
$$+ 3[2(5) - 4(-1)] + 2[2(-3) - 4(-2)]$$
$$= -1(-13) + 3(14) + 2(2)$$
$$= 13 + 42 + 4 = 59$$

11. $\begin{vmatrix} 1 & 0 & -2 \\ 0 & 2 & 3 \\ 1 & 0 & 5 \end{vmatrix}$ There are two 0's in column 2. We'll expand about that column since there is only 1 minor to evaluate.

$$= -0 \begin{vmatrix} 0 & 3 \\ 1 & 5 \end{vmatrix} + 2 \begin{vmatrix} 1 & -2 \\ 1 & 5 \end{vmatrix} - 0 \begin{vmatrix} 1 & -2 \\ 0 & 3 \end{vmatrix}$$
$$= 0 + 2[1(5) - (-2)(1)] - 0$$
$$= 2[5 + 2] = 2(7) = 14$$

13. Multiply the upper left and lower right entries. Then multiply the upper right and lower left entries. Subtract the second product from the first to obtain the determinant. For example,

$$\begin{vmatrix} 4 & 2 \\ 7 & 1 \end{vmatrix} = 4 \cdot 1 - 2 \cdot 7$$

$$= 4 - 14 = -10.$$

15. $\begin{vmatrix} 4 & 4 & 2 \\ 1 & -1 & -2 \\ 1 & 0 & 2 \end{vmatrix}$ Expand about column 2.

$$= -4 \begin{vmatrix} 1 & -2 \\ 1 & 2 \end{vmatrix} + (-1) \begin{vmatrix} 4 & 2 \\ 1 & 2 \end{vmatrix} - 0 \begin{vmatrix} 4 & 2 \\ 1 & -2 \end{vmatrix}$$

$$= -4[1(2) - (-2)(1)] - 1[4(2) - 2(1)] - 0$$

$$= -4(4) - 1(6)$$

$$= -16 - 6 = -22$$

17. $\begin{vmatrix} 3 & 5 & -2 \\ 1 & -4 & 1 \\ 3 & 1 & -2 \end{vmatrix}$ Expand about column 1.

$$= 3 \begin{vmatrix} -4 & 1 \\ 1 & -2 \end{vmatrix} - 1 \begin{vmatrix} 5 & -2 \\ 1 & -2 \end{vmatrix} + 3 \begin{vmatrix} 5 & -2 \\ -4 & 1 \end{vmatrix}$$

$$= 3[-4(-2) - 1(1)] - 1[5(-2) - (-2)(1)] + 3[5(1) - (-2)(-4)]$$

$$= 3(7) - 1(-8) + 3(-3)$$

$$= 21 + 8 - 9 = 20$$

19. $\begin{vmatrix} 3 & 0 & -2 \\ 1 & -4 & 1 \\ 3 & 1 & -2 \end{vmatrix}$ Expand about row 1.

$$= 3 \begin{vmatrix} -4 & 1 \\ 1 & -2 \end{vmatrix} - 0 \begin{vmatrix} 1 & 1 \\ 3 & -2 \end{vmatrix} + (-2) \begin{vmatrix} 1 & -4 \\ 3 & 1 \end{vmatrix}$$

$$= 3[-4(-2) - 1(1)] - 0 + (-2)[1(1) - (-4)(3)]$$

$$= 3(7) - 0 - 2(13)$$

$$= 21 - 26 = -5$$

21. By choosing that row or column to expand about, all terms will have a factor of 0, and so the sum of all these terms will be 0.

22. The slope m through the points (x_1, y_1) and (x_2, y_2) is

$$m = \frac{y_2 - y_1}{x_2 - x_1}.$$

23. $y - y_1 = m(x - x_1)$

Let $m = \dfrac{y_2 - y_1}{x_2 - x_1}$ from Exercise 22.

$$y - y_1 = \frac{y_2 - y_1}{x_2 - x_1}(x - x_1)$$

24. $y - y_1 = \dfrac{y_2 - y_1}{x_2 - x_1}(x - x_1)$

Multiply both sides by $x_2 - x_1$.

$(x_2 - x_1)(y - y_1) = (y_2 - y_1)(x - x_1)$

Subtract $(y_2 - y_1)(x - x_1)$ from both sides.

$(x_2 - x_1)(y - y_1) - (y_2 - y_1)(x - x_1) = 0$

Multiply and collect like terms.

$x_2 y - x_2 y_1 - x_1 y + x_1 y_1$
$\qquad - (xy_2 - x_1 y_2 - xy_1 + x_1 y_1) = 0$
$x_2 y - x_2 y_1 - x_1 y + x_1 y_1$
$\qquad - xy_2 + x_1 y_2 + xy_1 - x_1 y_1 = 0$
$x_2 y - x_2 y_1 - x_1 y - xy_2 + x_1 y_2 + xy_1 = 0$

25. $\begin{vmatrix} x & y & 1 \\ x_1 & y_1 & 1 \\ x_2 & y_2 & 1 \end{vmatrix} = 0$ Expand about row 1.

$$x \begin{vmatrix} y_1 & 1 \\ y_2 & 1 \end{vmatrix} - y \begin{vmatrix} x_1 & 1 \\ x_2 & 1 \end{vmatrix} + 1 \begin{vmatrix} x_1 & y_1 \\ x_2 & y_2 \end{vmatrix}$$

$$= x(y_1 - y_2) - y(x_1 - x_2) + 1(x_1 y_2 - x_2 y_1)$$

$$= xy_1 - xy_2 - x_1 y + x_2 y + x_1 y_2 - x_2 y_1$$

Thus,

$xy_1 - xy_2 - x_1 y + x_2 y + x_1 y_2 - x_2 y_1 = 0$

or

$x_2 y - x_2 y_1 - x_1 y - xy_2 + x_1 y_2 + xy_1 = 0.$

This is the same equation as that found in Exercise 24.

27. det ([B])

$$= \begin{vmatrix} -1 & 7 & 1 \\ 0 & 2 & 1 \\ -3 & 5 & 4 \end{vmatrix}$$ Expand about column 1.

$$= -1 \begin{vmatrix} 2 & 1 \\ 5 & 4 \end{vmatrix} - 0 \begin{vmatrix} 7 & 1 \\ 5 & 4 \end{vmatrix} - 3 \begin{vmatrix} 7 & 1 \\ 2 & 1 \end{vmatrix}$$

$$= -1[2(4) - 1(5)] - 0 - 3[7(1) - 1(2)]$$

$$= -1(3) - 3(5) = -3 - 15 = -18$$

29. det ([D])

$$= \begin{vmatrix} 0 & 0 & 0 \\ -2 & 3 & 7 \\ 2 & 1 & 0 \end{vmatrix}$$

$$= 0$$ since row 1 consists of only 0 elements.

31. Enter $\begin{vmatrix} \sqrt{5} & \sqrt{2} & -\sqrt{3} \\ \sqrt{7} & -\sqrt{6} & \sqrt{10} \\ -\sqrt{5} & -\sqrt{2} & \sqrt{17} \end{vmatrix}$ as [A].

Type the keys for det [A].

The result should be -22.04285452.

33.

Add the top numbers: $0 + 0 + 4 = 4$
Add the bottom numbers: $0 - 12 + 0 = -12$
Find the difference: $(4) - (-12) = 16$

35.

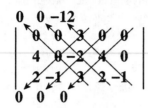

Add the top numbers: $0 + 0 - 12 = -12$
Add the bottom numbers: $0 + 0 + 0 = 0$
Find the difference: $(-12) - (0) = -12$

37.

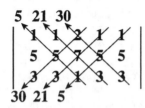

Add the top numbers: $5 + 21 + 30 = 56$
Add the bottom numbers: $30 + 21 + 5 = 56$
Find the difference: $(56) - (56) = 0$

39. $\begin{vmatrix} 5 & 3 \\ x & x \end{vmatrix} = 20$

Evaluate the determinant.

$\begin{vmatrix} 5 & 3 \\ x & x \end{vmatrix} = 5x - 3x = 2x$

Solve the equation.

$$2x = 20$$
$$x = 10$$

Solution set: $\{10\}$

41. **(a)** D is the determinant of coefficients—choice IV.

(b) D_x is the determinant with the constants replacing the x-coefficients—choice I.

(c) D_y is the determinant with the constants replacing the y-coefficients—choice III.

(d) D_z is the determinant with the constants replacing the z-coefficients—choice II.

43. $\quad 3x + 5y = -5$
$\quad -2x + 3y = 16$

$D = \begin{vmatrix} 3 & 5 \\ -2 & 3 \end{vmatrix} = 3(3) - 5(-2) = 19$

$D_x = \begin{vmatrix} -5 & 5 \\ 16 & 3 \end{vmatrix} = -5(3) - 5(16) = -95$

$D_y = \begin{vmatrix} 3 & -5 \\ -2 & 16 \end{vmatrix} = 3(16) - (-5)(-2) = 38$

$x = \dfrac{D_x}{D} = \dfrac{-95}{19} = -5; y = \dfrac{D_y}{D} = \dfrac{38}{19} = 2$

Solution set: $\{(-5, 2)\}$

45. $\quad 8x + 3y = 1$
$\quad 6x - 5y = 2$

$D = \begin{vmatrix} 8 & 3 \\ 6 & -5 \end{vmatrix} = 8(-5) - 3(6) = -58$

$D_x = \begin{vmatrix} 1 & 3 \\ 2 & -5 \end{vmatrix} = 1(-5) - 3(2) = -11$

$D_y = \begin{vmatrix} 8 & 1 \\ 6 & 2 \end{vmatrix} = 8(2) - 1(6) = 10$

$x = \dfrac{D_x}{D} = \dfrac{-11}{-58} = \dfrac{11}{58}; y = \dfrac{D_y}{D} = \dfrac{10}{-58} = -\dfrac{5}{29}$

Solution set: $\left\{ \left(\dfrac{11}{58}, -\dfrac{5}{29} \right) \right\}$

47. $\quad 2x + 3y = 4$
$\quad 5x + 6y = 7$

$D = \begin{vmatrix} 2 & 3 \\ 5 & 6 \end{vmatrix} = 2(6) - 3(5) = -3$

$D_x = \begin{vmatrix} 4 & 3 \\ 7 & 6 \end{vmatrix} = 4(6) - 3(7) = 3$

$D_y = \begin{vmatrix} 2 & 4 \\ 5 & 7 \end{vmatrix} = 2(7) - 4(5) = -6$

$x = \dfrac{D_x}{D} = \dfrac{3}{-3} = -1; y = \dfrac{D_y}{D} = \dfrac{-6}{-3} = 2$

Solution set: $\{(-1, 2)\}$

49. One example is

$$6x + 7y = 8$$
$$9x + 10y = 11.$$

$D = \begin{vmatrix} 6 & 7 \\ 9 & 10 \end{vmatrix} = 6(10) - 7(9) = -3$

$D_x = \begin{vmatrix} 8 & 7 \\ 11 & 10 \end{vmatrix} = 8(10) - 7(11) = 3$

$D_y = \begin{vmatrix} 6 & 8 \\ 9 & 11 \end{vmatrix} = 6(11) - 8(9) = -6$

$x = \dfrac{D_x}{D} = \dfrac{3}{-3} = -1; y = \dfrac{D_y}{D} = \dfrac{-6}{-3} = 2$

As in Exercises 47 and 48, the solution set is
$\{(-1, 2)\}$.

51. $\quad 2x + 3y + 2z = 15$
$\quad x - y + 2z = 5$
$\quad x + 2y - 6z = -26$

$D = \begin{vmatrix} 2 & 3 & 2 \\ 1 & -1 & 2 \\ 1 & 2 & -6 \end{vmatrix}$ Expand about row 1.

$= 2\begin{vmatrix} -1 & 2 \\ 2 & -6 \end{vmatrix} - 3\begin{vmatrix} 1 & 2 \\ 1 & -6 \end{vmatrix} + 2\begin{vmatrix} 1 & -1 \\ 1 & 2 \end{vmatrix}$

$= 2(2) - 3(-8) + 2(3)$

$= 4 + 24 + 6 = 34$

$$D_x = \begin{vmatrix} 15 & 3 & 2 \\ 5 & -1 & 2 \\ -26 & 2 & -6 \end{vmatrix} \quad \text{Expand about row 2.}$$

$$= -5 \begin{vmatrix} 3 & 2 \\ 2 & -6 \end{vmatrix} + (-1) \begin{vmatrix} 15 & 2 \\ -26 & -6 \end{vmatrix}$$

$$- 2 \begin{vmatrix} 15 & 3 \\ -26 & 2 \end{vmatrix}$$

$$= -5(-22) - 1(-38) - 2(108)$$

$$= 110 + 38 - 216 = -68$$

$$D_y = \begin{vmatrix} 2 & 15 & 2 \\ 1 & 5 & 2 \\ 1 & -26 & -6 \end{vmatrix} \quad \text{Expand about column 1.}$$

$$= 2 \begin{vmatrix} 5 & 2 \\ -26 & -6 \end{vmatrix} - 1 \begin{vmatrix} 15 & 2 \\ -26 & -6 \end{vmatrix}$$

$$+ 1 \begin{vmatrix} 15 & 2 \\ 5 & 2 \end{vmatrix}$$

$$= 2(22) - 1(-38) + 1(20)$$

$$= 44 + 38 + 20 = 102$$

$$D_z = \begin{vmatrix} 2 & 3 & 15 \\ 1 & -1 & 5 \\ 1 & 2 & -26 \end{vmatrix} \quad \text{Expand about column 1.}$$

$$= 2 \begin{vmatrix} -1 & 5 \\ 2 & -26 \end{vmatrix} - 1 \begin{vmatrix} 3 & 15 \\ 2 & -26 \end{vmatrix} + 1 \begin{vmatrix} 3 & 15 \\ -1 & 5 \end{vmatrix}$$

$$= 2(16) - 1(-108) + 1(30)$$

$$= 32 + 108 + 30 = 170$$

$$x = \frac{D_x}{D} = \frac{-68}{34} = -2; \quad y = \frac{D_y}{D} = \frac{102}{34} = 3$$

$$z = \frac{D_z}{D} = \frac{170}{34} = 5$$

Solution set: $\{(-2, 3, 5)\}$

53.
$$\begin{array}{rrrr} 2x & - 3y & + 4z & = 8 \\ 6x & - 9y & + 12z & = 24 \\ -4x & + 6y & - 8z & = -16 \end{array}$$

$$D = \begin{vmatrix} 2 & -3 & 4 \\ 6 & -9 & 12 \\ -4 & 6 & -8 \end{vmatrix} \quad \text{Expand about row 1.}$$

$$= 2 \begin{vmatrix} -9 & 12 \\ 6 & -8 \end{vmatrix} + 3 \begin{vmatrix} 6 & 12 \\ -4 & -8 \end{vmatrix} + 4 \begin{vmatrix} 6 & -9 \\ -4 & 6 \end{vmatrix}$$

$$= 2(0) + 3(0) + 4(0) = 0$$

Because $D = 0$, Cramer's rule does not apply.

55.
$$\begin{array}{rrrr} 3x & & + 5z & = 0 \\ 2x & + 3y & & = 1 \\ & -y & + 2z & = -11 \end{array}$$

$$D = \begin{vmatrix} 3 & 0 & 5 \\ 2 & 3 & 0 \\ 0 & -1 & 2 \end{vmatrix} \quad \text{Expand about row 1.}$$

$$= 3 \begin{vmatrix} 3 & 0 \\ -1 & 2 \end{vmatrix} - 0 + 5 \begin{vmatrix} 2 & 3 \\ 0 & -1 \end{vmatrix}$$

$$= 3(6) + 5(-2)$$

$$= 18 - 10 = 8$$

$$D_x = \begin{vmatrix} 0 & 0 & 5 \\ 1 & 3 & 0 \\ -11 & -1 & 2 \end{vmatrix} \quad \text{Expand about row 1.}$$

$$= 0 - 0 + 5 \begin{vmatrix} 1 & 3 \\ -11 & -1 \end{vmatrix}$$

$$= 5(32) = 160$$

$$D_y = \begin{vmatrix} 3 & 0 & 5 \\ 2 & 1 & 0 \\ 0 & -11 & 2 \end{vmatrix} \quad \text{Expand about row 1.}$$

$$= 3 \begin{vmatrix} 1 & 0 \\ -11 & 2 \end{vmatrix} - 0 + 5 \begin{vmatrix} 2 & 1 \\ 0 & -11 \end{vmatrix}$$

$$= 3(2) + 5(-22)$$

$$= 6 - 110 = -104$$

$$D_z = \begin{vmatrix} 3 & 0 & 0 \\ 2 & 3 & 1 \\ 0 & -1 & -11 \end{vmatrix} \quad \text{Expand about row 1.}$$

$$= 3 \begin{vmatrix} 3 & 1 \\ -1 & -11 \end{vmatrix} - 0 + 0$$

$$= 3(-32) = -96$$

$$x = \frac{D_x}{D} = \frac{160}{8} = 20; \quad y = \frac{D_y}{D} = \frac{-104}{8} = -13$$

$$z = \frac{D_z}{D} = \frac{-96}{8} = -12$$

Solution set: $\{(20, -13, -12)\}$

57.
$$\begin{array}{rrrr} x & - 3y & & = 13 \\ & 2y & + z & = 5 \\ -x & & + z & = -7 \end{array}$$

$$D = \begin{vmatrix} 1 & -3 & 0 \\ 0 & 2 & 1 \\ -1 & 0 & 1 \end{vmatrix} \quad \text{Expand about column 1.}$$

$$= 1 \begin{vmatrix} 2 & 1 \\ 0 & 1 \end{vmatrix} - 0 + (-1) \begin{vmatrix} -3 & 0 \\ 2 & 1 \end{vmatrix}$$

$$= 1(2) - 1(-3) = 2 + 3 = 5$$

$$D_x = \begin{vmatrix} 13 & -3 & 0 \\ 5 & 2 & 1 \\ -7 & 0 & 1 \end{vmatrix} \quad \text{Expand about column 3.}$$

$$= 0 - 1 \begin{vmatrix} 13 & -3 \\ -7 & 0 \end{vmatrix} + 1 \begin{vmatrix} 13 & -3 \\ 5 & 2 \end{vmatrix}$$

$$= -1(-21) + 1(41)$$

$$= 21 + 41 = 62$$

continued

$$D_y = \begin{vmatrix} 1 & 13 & 0 \\ 0 & 5 & 1 \\ -1 & -7 & 1 \end{vmatrix} \quad \text{Expand about column 1.}$$

$$= 1\begin{vmatrix} 5 & 1 \\ -7 & 1 \end{vmatrix} - 0 + (-1)\begin{vmatrix} 13 & 0 \\ 5 & 1 \end{vmatrix}$$

$$= 1(12) - 1(13) = 12 - 13 = -1$$

$$D_z = \begin{vmatrix} 1 & -3 & 13 \\ 0 & 2 & 5 \\ -1 & 0 & -7 \end{vmatrix} \quad \text{Expand about column 1.}$$

$$= 1\begin{vmatrix} 2 & 5 \\ 0 & -7 \end{vmatrix} - 0 - 1\begin{vmatrix} -3 & 13 \\ 2 & 5 \end{vmatrix}$$

$$= 1(-14) - 1(-41) = -14 + 41 = 27$$

$$x = \frac{D_x}{D} = \frac{62}{5}; \quad y = \frac{D_y}{D} = -\frac{1}{5}$$

$$z = \frac{D_z}{D} = \frac{27}{5}$$

Solution set: $\left\{ \left(\dfrac{62}{5}, -\dfrac{1}{5}, \dfrac{27}{5} \right) \right\}$

59.

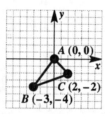

Points: $A(0,0)$, $C(2,-2)$, $B(-3,-4)$

60. Use $(0,0)$ as (x_1, y_1), $(-3,-4)$ as (x_2, y_2), and $(2,-2)$ as (x_3, y_3). Then

$$\frac{1}{2}\begin{vmatrix} x_1 & y_1 & 1 \\ x_2 & y_2 & 1 \\ x_3 & y_3 & 1 \end{vmatrix} = \frac{1}{2}\begin{vmatrix} 0 & 0 & 1 \\ -3 & -4 & 1 \\ 2 & -2 & 1 \end{vmatrix}.$$

61. $\dfrac{1}{2}\begin{vmatrix} 0 & 0 & 1 \\ -3 & -4 & 1 \\ 2 & -2 & 1 \end{vmatrix}$ Expand about row 1.

$$= \frac{1}{2}\left(0 - 0 + 1\begin{vmatrix} -3 & -4 \\ 2 & -2 \end{vmatrix} \right)$$

$$= \frac{1}{2}[1(6+8)]$$

$$= \frac{1}{2}(14) = 7$$

The area is 7 square units.

62. $\dfrac{1}{2}\begin{vmatrix} 3 & 8 & 1 \\ -1 & 4 & 1 \\ 0 & 1 & 1 \end{vmatrix}$ Expand about row 3.

$$= \frac{1}{2}\left(0 - 1\begin{vmatrix} 3 & 1 \\ -1 & 1 \end{vmatrix} + 1\begin{vmatrix} 3 & 8 \\ -1 & 4 \end{vmatrix} \right)$$

$$= \frac{1}{2}(-1[3+1] + 1[12+8])$$

$$= \frac{1}{2}(-4 + 20) = \frac{1}{2}(16) = 8$$

The area is 8 square units.

63.
$$\begin{aligned} x + 2y + z &= 10 \\ 2x - y - 3z &= -20 \\ -x + 4y + z &= 18 \end{aligned}$$

Use a calculator to evaluate each determinant.

$$D = \begin{vmatrix} 1 & 2 & 1 \\ 2 & -1 & -3 \\ -1 & 4 & 1 \end{vmatrix} = 20$$

$$D_x = \begin{vmatrix} 10 & 2 & 1 \\ -20 & -1 & -3 \\ 18 & 4 & 1 \end{vmatrix} = -20$$

$$D_y = \begin{vmatrix} 1 & 10 & 1 \\ 2 & -20 & -3 \\ -1 & 18 & 1 \end{vmatrix} = 60$$

$$D_z = \begin{vmatrix} 1 & 2 & 10 \\ 2 & -1 & -20 \\ -1 & 4 & 18 \end{vmatrix} = 100$$

$$x = \frac{D_x}{D} = \frac{-20}{20} = -1; \quad y = \frac{D_y}{D} = \frac{60}{20} = 3$$

$$z = \frac{D_z}{D} = \frac{100}{20} = 5$$

Solution set: $\{(-1, 3, 5)\}$

65.
$$\begin{aligned} -8w + 4x - 2y + z &= -28 \\ -w + x - y + z &= -10 \\ w + x + y + z &= -4 \\ 27w + 9x + 3y + z &= 2 \end{aligned}$$

Use a calculator to evaluate each determinant.

$$D = \begin{vmatrix} -8 & 4 & -2 & 1 \\ -1 & 1 & -1 & 1 \\ 1 & 1 & 1 & 1 \\ 27 & 9 & 3 & 1 \end{vmatrix} = 240$$

$$D_w = \begin{vmatrix} -28 & 4 & -2 & 1 \\ -10 & 1 & -1 & 1 \\ -4 & 1 & 1 & 1 \\ 2 & 9 & 3 & 1 \end{vmatrix} = 240$$

$$D_x = \begin{vmatrix} -8 & -28 & -2 & 1 \\ -1 & -10 & -1 & 1 \\ 1 & -4 & 1 & 1 \\ 27 & 2 & 3 & 1 \end{vmatrix} = -720$$

$$D_y = \begin{vmatrix} -8 & 4 & -28 & 1 \\ -1 & 1 & -10 & 1 \\ 1 & 1 & -4 & 1 \\ 27 & 9 & 2 & 1 \end{vmatrix} = 480$$

$$D_z = \begin{vmatrix} -8 & 4 & -2 & -28 \\ -1 & 1 & -1 & -10 \\ 1 & 1 & 1 & -4 \\ 27 & 9 & 3 & 2 \end{vmatrix} = -960$$

$$w = \frac{D_w}{D} = \frac{240}{240} = 1; \quad x = \frac{D_x}{D} = \frac{-720}{240} = -3;$$

$$y = \frac{D_y}{D} = \frac{480}{240} = 2; \quad z = \frac{D_z}{D} = \frac{-960}{240} = -4$$

Solution set: $\{(1, -3, 2, -4)\}$

Chapter 9 Review Exercises

1. From the graph, the usage rates of core competencies and total quality management both reached about 72% in 1995.

2. The graphs of core competencies and strategic alliances intersect twice. Between these times (about mid-1995 to 1997), core competencies were more popular.

3. **(a)** Since the graphs for supply and demand intersect at $(4, 300)$, supply equals demand at this point. The price is $4.

   (b) Supply equals demand for 300 half-gallons.

   (c) By reading the graphs when the price equals $2, we see that the supply is 200 half-gallons and the demand is 400 half-gallons.

4. **(a)** The system is inconsistent. Parallel lines do not intersect, so there is no ordered pair that would satisfy both equations.

   (b) The equations are equivalent and have the same line as graph. The equations are dependent.

5. $(3, 4)$

   $4x - 2y = 4$
   $5x + y = 19$

   To decide whether $(3, 4)$ is a solution of the system, substitute 3 for x and 4 for y in each equation.

   $$4x - 2y = 4$$
   $$4(3) - 2(4) = 4 \ ?$$
   $$12 - 8 = 4 \ ?$$
   $$4 = 4 \quad True$$

   $$5x + y = 19$$
   $$5(3) + 4 = 19 \ ?$$
   $$15 + 4 = 19 \ ?$$
   $$19 = 19 \quad True$$

   Since $(3, 4)$ satisfies both equations, it is a solution of the system.

6. $(-5, 2)$

   $x - 4y = -13$
   $2x + 3y = 4$

   Substitute -5 for x and 2 for y in each equation.

   $$x - 5y = -13$$
   $$-5 - 5(2) = -13 \ ?$$
   $$-5 - 10 = -13 \ ?$$
   $$-15 = -13 \quad False$$

   Since $(-5, 2)$ is not a solution of the first equation, it cannot be a solution of the system.

7. $x + y = 4$
   $2x - y = 5$

   To graph the equations, find the intercepts.

   $x + y = 4$: Let $y = 0$; then $x = 4$.
   Let $x = 0$; then $y = 4$.

   Plot the intercepts, $(4, 0)$ and $(0, 4)$, and draw the line through them.

   $2x - y = 5$: Let $y = 0$; then $x = \dfrac{5}{2}$.
   Let $x = 0$; then $y = -5$.

   Plot the intercepts, $\left(\dfrac{5}{2}, 0\right)$ and $(0, -5)$, and draw the line through them.

   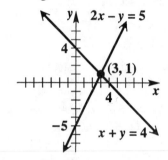

   It appears that the lines intersect at the point $(3, 1)$. Check this by substituting 3 for x and 1 for y in both equations. Since $(3, 1)$ satisfies both equations, the solution set of this system is $\{(3, 1)\}$.

8. $2x + 4 = 2y$
   $y - x = -3$

   Graph the line $2x + 4 = 24$ through its intercepts, $(-2, 0)$ and $(0, 2)$. Graph the line $y - x = -3$ through its intercepts $(3, 0)$ and $(0, -3)$.

   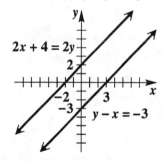

   The lines are parallel (both lines have slope 1).

   Since they do not intersect, there is no solution. This is an inconsistent system and the solution set is $\emptyset$.

9. No, this is not correct. A false statement indicates that the solution set is $\emptyset$.

10. No, two lines cannot intersect in exactly three points.

11. $3x + y = 7$ (1)
$x = 2y$ (2)

Substitute $2y$ for x in equation (1) and solve the resulting equation for y.

$$3x + y = 7$$
$$3(2y) + y = 7$$
$$6y + y = 7$$
$$7y = 7$$
$$y = 1$$

To find x, let $y = 1$ in equation (2).

$$x = 2y$$
$$x = 2(1) = 2$$

The solution set is $\{(2, 1)\}$.

12. $2x - 5y = -19$ (1)
$y = x + 2$ (2)

Substitute $x + 2$ for y in equation (1).

$$2x - 5y = -19$$
$$2x - 5(x + 2) = -19$$
$$2x - 5x - 10 = -19$$
$$-3x - 10 = -19$$
$$-3x = -9$$
$$x = 3$$

To find y, let $x = 3$ in equation (2).

$$y = x + 2$$
$$y = 3 + 2 = 5$$

The solution set is $\{(3, 5)\}$.

13. $5x + 15y = 30$ (1)
$x + 3y = 6$ (2)

Solve equation (2) for x.

$$x + 3y = 6$$
$$x = 6 - 3y \quad (3)$$

Substitute $6 - 3y$ for x in equation (1).

$$5x + 15y = 30$$
$$5(6 - 3y) + 15y = 30$$
$$30 - 15y + 15y = 30$$
$$30 = 30 \quad \textit{True}$$

This true result means that every solution of one equation is also a solution of the other, so the system has an infinite number of solutions. The solution set is $\{(x, y) \mid x + 3y = 6\}$.

14. $\dfrac{1}{3}x + \dfrac{1}{7}y = \dfrac{52}{21}$ (1)

$\dfrac{1}{2}x - \dfrac{1}{3}y = \dfrac{19}{6}$ (2)

Multiply by the LCD of the fractions in each equation to get a system with integer coefficients.

$7x + 3y = 52$ (3) *Multiply (1) by 21*
$3x - 2y = 19$ (4) *Multiply (2) by 6*

Solve equation (4) for x.

$$3x = 19 + 2y$$
$$x = \frac{19 + 2y}{3} \quad (5)$$

Substitute $\dfrac{19 + 2y}{3}$ for x in equation (3) and solve for y.

$$7\left(\frac{19 + 2y}{3}\right) + 3y = 52$$
$$7(19 + 2y) + 9y = 156 \quad \textit{Multiply by 3}$$
$$133 + 14y + 9y = 156$$
$$23y = 23$$
$$y = 1$$

Let $y = 1$ in equation (5).

$$x = \frac{19 + 2(1)}{3} = 7$$

The solution set is $\{(7, 1)\}$.

15.   His answer was incorrect since the system has infinitely many solutions (as indicated by the true statement $0 = 0$).

16.   It would be easiest to solve for x in the second equation because its coefficient is -1. No fractions would be involved.

17.   If we simply add the given equations without first multiplying one or both equations by a constant, choice (c) is the only system in which a variable will be eliminated. If we add the equations in (c) we get $3x = 17$. (The variable y was eliminated.)

18. $2x + 12y = 7$
$3x + 4y = 1$

(a) If we multiply the first equation by -3, the first term will become $-6x$. To eliminate x, we need to change the first term on the left side of the second equation from $3x$ to $6x$. In order to do this, we must multiply the second equation by 2.

(b) If we multiply the first equation by -3, the second term will become $-36y$. To eliminate y, we need to change the second term on the left side of the second equation from $4y$ to $36y$. In order to do this, we must multiply the second equation by 9.

19. $2x \quad - \quad y \quad = \quad 13$ (1)
$\underline{\quad x \quad + \quad y \quad = \quad 8} \quad (2)$
$3x \qquad\qquad = \quad 21 \quad \textit{Add (1) and (2)}$
$\qquad\qquad x \quad = \quad 7$

From (2), $y = 1$.

The solution set is $\{(7, 1)\}$.

20. $-4x + 3y = 25$ (1)
$6x - 5y = -39$ (2)

Multiply equation (1) by 3 and equation (2) by 2; then add the results.

$$\begin{array}{rcrcr} -12x & + & 9y & = & 75 \\ 12x & - & 10y & = & -78 \\ \hline & & -y & = & -3 \\ & & y & = & 3 \end{array}$$

To find x, let $y = 3$ in equation (1).

$$\begin{aligned} -4x + 3y &= 25 \\ -4x + 3(3) &= 25 \\ -4x + 9 &= 25 \\ -4x &= 16 \\ x &= -4 \end{aligned}$$

The solution set is $\{(-4, 3)\}$.

21. $3x - 4y = 9$ (1)
$6x - 8y = 18$ (2)

Multiply equation (1) by -2 and add the result to equation (2).

$$\begin{array}{rcrcr} -6x & + & 8y & = & -18 \\ 6x & - & 8y & = & 18 \\ \hline & & 0 & = & 0 \quad \textit{True} \end{array}$$

This result indicates that all solutions of equation (1) are also solutions of equation (2). The given system has an infinite number of solutions. The solution set is $\{(x, y) \mid 3x - 4y = 9\}$.

22. $2x + y = 3$ (1)
$-4x - 2y = 6$ (2)

Multiply equation (1) by 2 and add the result to equation (2).

$$\begin{array}{rcrcr} 4x & + & 2y & = & 6 \\ -4x & - & 2y & = & 6 \\ \hline & & 0 & = & 12 \quad \textit{False} \end{array}$$

This result indicates that the given system has solution set $\emptyset$.

23. $2x + 3y = -5$ (1)
$3x + 4y = -8$ (2)

Multiply equation (1) by -3 and equation (2) by 2; then add the results.

$$\begin{array}{rcrcr} -6x & - & 9y & = & 15 \\ 6x & + & 8y & = & -16 \\ \hline & & -y & = & -1 \\ & & y & = & 1 \end{array}$$

To find x, let $y = 1$ in equation (1).

$$\begin{aligned} 2x + 3y &= -5 \\ 2x + 3(1) &= -5 \\ 2x + 3 &= -5 \\ 2x &= -8 \\ x &= -4 \end{aligned}$$

The solution set is $\{(-4, 1)\}$.

24. $6x - 9y = 0$ (1)
$2x - 3y = 0$ (2)

Multiply equation (2) by -3 and add the result to equation (1).

$$\begin{array}{rcrcr} 6x & - & 9y & = & 0 \\ -6x & + & 9y & = & 0 \\ \hline & & 0 & = & 0 \quad \textit{True} \end{array}$$

This result indicates that the system has an infinite number of solutions. The solution set is $\{(x, y) \mid 2x - 3y = 0\}$.

25. $2x + y - x = 3y + 5$ (1)
$y + 2 = x - 5$ (2)

Simplify equation (1).

$$\begin{aligned} 2x + y - x &= 3y + 5 \\ x - 2y &= 5 \quad (3) \end{aligned}$$

Simplify equation (2).

$$\begin{aligned} y + 2 &= x - 5 \\ -x + y &= -7 \quad (4) \end{aligned}$$

Add equations (3) and (4) to eliminate x.

$$\begin{array}{rcrcr} x & - & 2y & = & 5 \\ -x & + & y & = & -7 \\ \hline & & -y & = & -2 \\ & & y & = & 2 \end{array}$$

Let $y = 2$ in equation (3) to find x.

$$\begin{aligned} x - 2(2) &= 5 \\ x - 4 &= 5 \\ x &= 9 \end{aligned}$$

The solution set is $\{(9, 2)\}$.

26. $\dfrac{x}{2} + \dfrac{y}{3} = 7$ (1)
$\dfrac{x}{4} + \dfrac{2y}{3} = 8$ (2)

Multiply equation (1) by 6 to clear fractions.

$$\begin{aligned} 6\left(\frac{x}{2} + \frac{y}{3}\right) &= 6(7) \\ 3x + 2y &= 42 \quad (3) \end{aligned}$$

Multiply equation (2) by 12 to clear fractions.

$$\begin{aligned} 12\left(\frac{x}{4} + \frac{2y}{3}\right) &= 12(8) \\ 3x + 8y &= 96 \quad (4) \end{aligned}$$

continued

To solve this system by the elimination method, multiply equation (3) by -1 and add the result to equation (4).

$$
\begin{array}{rcr}
-3x - 2y &=& -42 \\
3x + 8y &=& 96 \\
\hline
6y &=& 54 \\
y &=& 9
\end{array}
$$

To find x, let $y = 9$ in equation (3).

$$
\begin{aligned}
3x + 2y &= 42 \\
3x + 2(9) &= 42 \\
3x + 18 &= 42 \\
3x &= 24 \\
x &= 8
\end{aligned}
$$

The solution set is $\{(8, 9)\}$.

27. The five methods of solving a system of equations are graphing, substitution, elimination, matrix methods, and Cramer's Rule (using determinants).

Answers to the second part of this exercise may vary. The following is one possible answer.

Consider the method of graphing. One advantage of this method of solution is that it is fast and that we can easily see if the system has one solution, no solution, or infinitely many solutions. One drawback is that we cannot always read the exact coordinates of the point of intersection.

28. System B is easier to solve by the substitution method than system A because the second equation in system B is already solved for y.

Solving system A would require our solving one of the equations for one of the variables before substituting, and the expression to be substituted would involve fractions.

29.
$$
\begin{array}{rcrl}
2x + 3y - z &=& -16 & (1) \\
x + 2y + 2z &=& -3 & (2) \\
-3x + y + z &=& -5 & (3)
\end{array}
$$

To eliminate z, add equations (1) and (3).

$$
\begin{array}{rcrl}
2x + 3y - z &=& -16 & (1) \\
-3x + y + z &=& -5 & (3) \\
\hline
-x + 4y &=& -21 & (4)
\end{array}
$$

To eliminate z again, multiply equation (1) by 2 and add the result to equation (2).

$$
\begin{array}{rcrll}
4x + 6y - 2z &=& -32 & & 2 \times (1) \\
x + 2y + 2z &=& -3 & (2) \\
\hline
5x + 8y &=& -35 & (5)
\end{array}
$$

Use equations (4) and (5) to eliminate x. Multiply equation (4) by 5 and add the result to equation (5).

$$
\begin{array}{rcrll}
-5x + 20y &=& -105 & & 5 \times (4) \\
5x + 8y &=& -35 & (5) \\
\hline
28y &=& -140 \\
y &=& -5
\end{array}
$$

Substitute -5 for y in equation (4) to find x.

$$
\begin{aligned}
-x + 4y &= -21 \quad (4) \\
-x + 4(-5) &= -21 \\
-x - 20 &= -21 \\
-x &= -1 \\
x &= 1
\end{aligned}
$$

Substitute 1 for x and -5 for y in equation (2) to find z.

$$
\begin{aligned}
x + 2y + 2z &= -3 \quad (2) \\
1 + 2(-5) + 2z &= -3 \\
1 - 10 + 2z &= -3 \\
2z &= 6 \\
z &= 3
\end{aligned}
$$

The solution set is $\{(1, -5, 3)\}$.

30.
$$
\begin{array}{rcrl}
4x - y &=& 2 & (1) \\
3y + z &=& 9 & (2) \\
x + 2z &=& 7 & (3)
\end{array}
$$

To eliminate y, multiply equation (1) by 3 and add the result to equation (2).

$$
\begin{array}{rcrll}
12x - 3y &=& 6 & & 3 \times (1) \\
3y + z &=& 9 & (2) \\
\hline
12x + z &=& 15 & (4)
\end{array}
$$

To eliminate z, multiply equation (4) by -2 and add the result to equation (3).

$$
\begin{array}{rcrll}
-24x - 2z &=& -30 & & -2 \times (4) \\
x + 2z &=& 7 & (3) \\
\hline
-23x &=& -23 \\
x &=& 1
\end{array}
$$

Substitute 1 for x in equation (3) to find z.

$$
\begin{aligned}
x + 2z &= 7 \quad (3) \\
1 + 2z &= 7 \\
2z &= 6 \\
z &= 3
\end{aligned}
$$

Substitute 1 for x in equation (1) to find y.

$$
\begin{aligned}
4x - y &= 2 \quad (1) \\
4(1) - y &= 2 \\
4 - y &= 2 \\
-y &= -2 \\
y &= 2
\end{aligned}
$$

The solution set is $\{(1, 2, 3)\}$.

31.
$$
\begin{aligned}
3x - y - z &= -8 \quad (1) \\
4x + 2y + 3z &= 15 \quad (2) \\
-6x + 2y + 2z &= 10 \quad (3)
\end{aligned}
$$

To eliminate y, multiply equation (1) by 2 and add the result to equation (3).

$$
\begin{array}{rll}
6x - 2y - 2z &= -16 & \quad 2 \times (1) \\
-6x + 2y + 2z &= 10 & \quad (3) \\
\hline
0 &= -6 & \quad \textit{False}
\end{array}
$$

Since a false statement results, equations (1) and (3) have no common solution. The system is inconsistent. The solution set is $\emptyset$.

32. *Step 1*

Let x = the number of Americans in the group;

y = the number of Londoners in the group.

Step 2

Summarize the information given in the problem in a chart.

	Number of Tickets	Price per Ticket (in dollars)	Total Value
American	x	5	$5x$
Londoner	y	11	$11y$
Total	41		307

Step 3

The total number of tickets was 41, so one equation is

$$x + y = 41. \quad (1)$$

Since the total value was $307, the final column leads to

$$5x + 11y = 307. \quad (2)$$

Step 4

To eliminate x, multiply (1) by -5 and add the result to (2).

$$
\begin{array}{rll}
-5x - 5y &= -205 & \quad -5 \times (1) \\
5x + 11y &= 307 & \quad (2) \\
\hline
6y &= 102 & \\
y &= 17 &
\end{array}
$$

From (1), $y = 17$ gives $x = 24$.

Step 5

There were 24 Americans and 17 Londoners in the group.

Step 6

The sum of 24 and 17 is 41, so the number of movie-goers is correct. Since 24 Americans paid $5 each and 17 Londoners paid $11 each, the total of the admission prices is $\$5(24) + \$11(17) = \$307$, which agrees with the total amount stated in the problem.

33. Let x = the number of touchdowns

and y = the number of interceptions.

40 passes resulted in either a touchdown or an interception, so

$$x + y = 40. \quad (1)$$

The number of touchdowns was 2 less than twice the number of interceptions, so

$$x = 2y - 2. \quad (2)$$

Substitute $2y - 2$ for x in (1).

$$
\begin{aligned}
(2y - 2) + y &= 40 \\
3y &= 42 \\
y &= 14
\end{aligned}
$$

From (2), $x = 2y - 2 = 2(14) - 2 = 26$. He threw 26 touchdowns and 14 interceptions.

34. Let x = the speed of the plane

and y = the speed of the wind.

Complete the chart.

	r	t	d
With wind	$x + y$	1.75	$1.75(x + y)$
Against wind	$x - y$	2	$2(x - y)$

The distance each way is 560 miles. From the chart,

$$1.75(x + y) = 560.$$

Divide by 1.75.

$$x + y = 320 \quad (1)$$

From the chart,

$$
\begin{aligned}
2(x - y) &= 560 \\
x - y &= 280. \quad (2)
\end{aligned}
$$

Solve the system by adding equations (1) and (2) to eliminate y.

$$
\begin{array}{rll}
x + y &= 320 & \quad (1) \\
x - y &= 280 & \quad (2) \\
\hline
2x &= 600 & \\
x &= 300 &
\end{array}
$$

From (1), $x + y = 320$, $y = 20$.
The speed of the plane was 300 mph, and the speed of the wind was 20 mph.

35. Let $x =$ the amount of $2-a-pound nuts

and $y =$ the amount of $1-a-pound

chocolate candy.

Complete the chart.

	Pounds	Price per Pound (in Dollars)	Value
Nuts	x	2	$2x$
Chocolate	y	1	$1y = y$
Total	100	1.30	$1.30(100)$ $= 130$

Solve the system.

$$x + y = 100 \qquad (1)$$
$$2x + y = 130 \qquad (2)$$

Solve equation (1) for y.

$$y = 100 - x$$

Substitute $100 - x$ for y in equation (2).

$$2x + (100 - x) = 130$$
$$x = 30$$

Since $y = 100 - x$ and $x = 30$,

$$y = 100 - 30 = 70.$$

She should use 30 lb of $2-a-pound nuts and 70 lb of $1-a-pound candy.

36. Let $x =$ the measure of the largest angle,

$y =$ the measure of the middle-sized angle,

and $z =$ the measure of the smallest angle.

Since the sum of the measures of the angles of a triangle is $180°$,

$$x + y + z = 180. \qquad (1)$$

Since the largest angle measures $10°$ less than the sum of the other two,

$$x = y + z - 10$$
$$\text{or} \quad x - y - z = -10. \qquad (2)$$

Since the measure of the middle-sized angle is the average of the other two,

$$y = \frac{x + z}{2}$$
$$2y = x + z$$
$$-x + 2y - z = 0. \qquad (3)$$

Solve the system.

$$x + y + z = 180 \qquad (1)$$
$$x - y - z = -10 \qquad (2)$$
$$-x + 2y - z = 0 \qquad (3)$$

Add equations (1) and (2) to find x.

$$\begin{array}{rcl} x + y + z &=& 180 \quad (1) \\ x - y - z &=& -10 \quad (2) \\ \hline 2x &=& 170 \\ x &=& 85 \end{array}$$

Add equations (1) and (3), to find y.

$$\begin{array}{rcl} x + y + z &=& 180 \quad (1) \\ -x + 2y - z &=& 0 \quad (3) \\ \hline 3y &=& 180 \\ y &=& 60 \end{array}$$

Substitute 85 for x and 60 for y in equation (1) to find z.

$$x + y + z = 180 \quad (1)$$
$$85 + 60 + z = 180$$
$$145 + z = 180$$
$$z = 35$$

The three angles are $85°$, $60°$, and $35°$.

37. Let $x =$ the value of sales at 10%,

$y =$ the value of sales at 6%,

and $z =$ the value of sales at 5%.

Since his total sales were $280,000,

$$x + y + z = 280,000 \quad (1)$$

Since his commissions on the sales totaled $17,000,

$$.10x + .06y + .05z = 17,000.$$

Multiply by 100 to clear the decimals, so

$$10x + 6y + 5z = 1,700,000. \quad (2)$$

Since the 5% sale amounted to the sum of the other two sales,

$$z = x + y. \quad (3)$$

Solve the system.

$$\begin{array}{rcl} x + y + z &=& 280,000 \quad (1) \\ 10x + 6y + 5z &=& 1,700,000 \quad (2) \\ z &=& x + y \quad (3) \end{array}$$

Since equation (3) is given in terms of z, substitute $x + y$ for z in equations (1) and (2).

$$x + y + z = 280,000 \quad (1)$$
$$x + y + (x + y) = 280,000$$
$$2x + 2y = 280,000$$
$$x + y = 140,000 \quad (4)$$

$$10x + 6y + 5z = 1,700,000 \quad (2)$$
$$10x + 6y + 5(x + y) = 1,700,000$$
$$10x + 6y + 5x + 5y = 1,700,000$$
$$15x + 11y = 1,700,000 \quad (5)$$

To eliminate x, multiply equation (4) by -11 and add the result to equation (5).

$$-11x - 11y = -1,540,000 \qquad -11 \times (4)$$
$$\underline{15x + 11y = 1,700,000 \quad (5)}$$
$$4x \qquad = 160,000$$
$$x = 40,000$$

From (4), $y = 100,000$.
From (3), $z = 40,000 + 100,000 = 140,000$.
He sold $\$40,000$ at 10%, $\$100,000$ at 6%, and $\$140,000$ at 5%.

38. Let $x =$ the number of liters of 8% solution,

$\qquad y =$ the number of liters of 10% solution,

and $z =$ the number of liters of 20% solution.

Since the amount of the mixture will be 8 L,

$$x + y + z = 8. \quad (1)$$

Since the final solution will be 12.5% hydrogen peroxide,

$$.08x + .10y + .20z = .125(8).$$

Multiply by 100 to clear the decimals.

$$8x + 10y + 20z = 100 \quad (2)$$

Since the amount of 8% solution used must be 2 L more than the amount of 20% solution,

$$x = z + 2. \quad (3)$$

Solve the system.

$$
\begin{aligned}
x + y + z &= 8 &(1)\\
8x + 10y + 20z &= 100 &(2)\\
x &= z + 2 &(3)
\end{aligned}
$$

Since equation (3) is given in terms of x, substitute $z + 2$ for x in equations (1) and (2).

$$
\begin{aligned}
x + y + z &= 8 \quad (1)\\
(z + 2) + y + z &= 8\\
y + 2z &= 6 \quad (4)
\end{aligned}
$$

$$
\begin{aligned}
8x + 10y + 20z &= 100 \quad (2)\\
8(z + 2) + 10y + 20z &= 100\\
8z + 16 + 10y + 20z &= 100\\
10y + 28z &= 84 \quad (5)
\end{aligned}
$$

To eliminate y, multiply equation (4) by -10 and add the result to equation (5).

$$-10y - 20z = -60 \qquad -10 \times (4)$$
$$\underline{10y + 28z = 84 \quad (5)}$$
$$8z = 24$$
$$z = 3$$

From (3), $x = z + 2 = 3 + 2 = 5$.
From (4), $y = 6 - 2z = 6 - 2(3) = 0$.
Mix 5 L of 8% solution, none of 10% solution, and 3 L of 20% solution.

39. Let $x =$ the number of home runs hit by Mantle,

$\qquad y =$ the number of home runs hit by Maris,

and $z =$ the number of home runs

$\qquad$ hit by Blanchard.

They combined for 136 home runs, so

$$x + y + z = 136. \quad (1)$$

Mantle hit 7 fewer than Maris, so

$$x = y - 7. \quad (2)$$

Maris hit 40 more than Blanchard, so

$$y = z + 40 \quad \text{or} \quad z = y - 40. \quad (3)$$

Substitute $y - 7$ for x and $y - 40$ for z in (1).

$$
\begin{aligned}
x + y + z &= 136 \quad (1)\\
(y - 7) + y + (y - 40) &= 136\\
3y - 47 &= 136\\
3y &= 183\\
y &= 61
\end{aligned}
$$

From (2), $x = y - 7 = 61 - 7 = 54$.
From (3), $z = y - 40 = 61 - 40 = 21$.
Mantle hit 54 home runs, Maris hit 61 home runs, and Blanchard hit 21 home runs.

40. $\quad 2x + 5y = -4$
$\qquad 4x - y = 14$

Write the augmented matrix.

$$\begin{bmatrix} 2 & 5 & | & -4 \\ 4 & -1 & | & 14 \end{bmatrix}$$

$$\begin{bmatrix} 2 & 5 & | & -4 \\ 0 & -11 & | & 22 \end{bmatrix} \quad -2R_1 + R_2$$

$$\begin{bmatrix} 2 & 5 & | & -4 \\ 0 & 1 & | & -2 \end{bmatrix} \quad -\frac{1}{11}R_2$$

This matrix gives the system

$$
\begin{aligned}
2x + 5y &= -4\\
y &= -2.
\end{aligned}
$$

Substitute $y = -2$ in the first equation.

$$
\begin{aligned}
2x + 5y &= -4\\
2x + 5(-2) &= -4\\
2x - 10 &= -4\\
2x &= 6\\
x &= 3
\end{aligned}
$$

Solution set: $\{(3, -2)\}$

41.
$$6x + 3y = 9$$
$$-7x + 2y = 17$$

Write the augmented matrix.

$$\begin{bmatrix} 6 & 3 & | & 9 \\ -7 & 2 & | & 17 \end{bmatrix}$$

$$\begin{bmatrix} 1 & \frac{1}{2} & | & \frac{3}{2} \\ -7 & 2 & | & 17 \end{bmatrix} \quad \frac{1}{6}R_1$$

$$\begin{bmatrix} 1 & \frac{1}{2} & | & \frac{3}{2} \\ 0 & \frac{11}{2} & | & \frac{55}{2} \end{bmatrix} \quad 7R_1 + R_2$$

$$\begin{bmatrix} 1 & \frac{1}{2} & | & \frac{3}{2} \\ 0 & 1 & | & 5 \end{bmatrix} \quad \frac{2}{11}R_2$$

This matrix gives the system

$$x + \frac{1}{2}y = \frac{3}{2}$$
$$y = 5.$$

Substitute $y = 5$ in the first equation.

$$x + \frac{1}{2}y = \frac{3}{2}$$
$$x + \frac{1}{2}(5) = \frac{3}{2}$$
$$x + \frac{5}{2} = \frac{3}{2}$$
$$x = -1$$

Solution set: $\{(-1, 5)\}$

42.
$$x + 2y - z = 1$$
$$3x + 4y + 2z = -2$$
$$-2x - y + z = -1$$

$$\begin{bmatrix} 1 & 2 & -1 & | & 1 \\ 3 & 4 & 2 & | & -2 \\ -2 & -1 & 1 & | & -1 \end{bmatrix}$$

$$\begin{bmatrix} 1 & 2 & -1 & | & 1 \\ 0 & -2 & 5 & | & -5 \\ 0 & 3 & -1 & | & 1 \end{bmatrix} \quad \begin{matrix} -3R_1 + R_2 \\ 2R_1 + R_3 \end{matrix}$$

$$\begin{bmatrix} 1 & 2 & -1 & | & 1 \\ 0 & 1 & 4 & | & -4 \\ 0 & 3 & -1 & | & 1 \end{bmatrix} \quad R_3 + R_2$$

$$\begin{bmatrix} 1 & 2 & -1 & | & 1 \\ 0 & 1 & 4 & | & -4 \\ 0 & 0 & -13 & | & 13 \end{bmatrix} \quad -3R_2 + R_3$$

$$\begin{bmatrix} 1 & 2 & -1 & | & 1 \\ 0 & 1 & 4 & | & -4 \\ 0 & 0 & 1 & | & -1 \end{bmatrix} \quad -\frac{1}{13}R_3$$

This matrix gives the system

$$x + 2y - z = 1$$
$$y + 4z = -4$$
$$z = -1.$$

Substitute $z = -1$ in the second equation.

$$y + 4z = -4$$
$$y + 4(-1) = -4$$
$$y = 0$$

Substitute $y = 0$ and $z = -1$ in the first equation.

$$x + 2y - z = 1$$
$$x + 2(0) - (-1) = 1$$
$$x + 1 = 1$$
$$x = 0$$

Solution set: $\{(0, 0, -1)\}$

43.
$$x + 3y \qquad = 7$$
$$3x \qquad + z = 2$$
$$y - 2z = 4$$

$$\begin{bmatrix} 1 & 3 & 0 & | & 7 \\ 3 & 0 & 1 & | & 2 \\ 0 & 1 & -2 & | & 4 \end{bmatrix}$$

$$\begin{bmatrix} 1 & 3 & 0 & | & 7 \\ 0 & -9 & 1 & | & -19 \\ 0 & 1 & -2 & | & 4 \end{bmatrix} \quad -3R_1 + R_2$$

$$\begin{bmatrix} 1 & 3 & 0 & | & 7 \\ 0 & 1 & -2 & | & 4 \\ 0 & -9 & 1 & | & -19 \end{bmatrix} \quad R_2 \leftrightarrow R_3$$

We use $\leftrightarrow$ to represent the interchanging of 2 rows.

$$\begin{bmatrix} 1 & 3 & 0 & | & 7 \\ 0 & 1 & -2 & | & 4 \\ 0 & 0 & -17 & | & 17 \end{bmatrix} \quad 9R_2 + R_3$$

$$\begin{bmatrix} 1 & 3 & 0 & | & 7 \\ 0 & 1 & -2 & | & 4 \\ 0 & 0 & 1 & | & -1 \end{bmatrix} \quad -\frac{1}{17}R_3$$

This matrix gives the system

$$x + 3y = 7$$
$$y - 2z = 4$$
$$z = -1.$$

Substitute $z = -1$ in the second equation.

$$y - 2z = 4$$
$$y - 2(-1) = 4$$
$$y + 2 = 4$$
$$y = 2$$

Substitute $y = 2$ in the first equation.

$$x + 3y = 7$$
$$x + 3(2) = 7$$
$$x + 6 = 7$$
$$x = 1$$

Solution set: $\{(1, 2, -1)\}$

44. (a) $\begin{vmatrix} 3 & 2 \\ 2 & 3 \end{vmatrix} = 3(3) - 2(2)$

$= 9 - 4 = 5$

(b) $\begin{vmatrix} 4 & 2 \\ -3 & 2 \end{vmatrix} = 4(2) - 2(-3)$

$= 8 + 6 = 14$

(c) $\begin{vmatrix} -1 & 1 \\ 8 & 8 \end{vmatrix} = -1(8) - 1(8)$

$= -8 - 8 = -16$

(d) $\begin{vmatrix} 1 & 2 \\ 6 & 12 \end{vmatrix} = 1(12) - 2(6)$

$= 12 - 12 = 0$

The answer is (d).

45. $\begin{vmatrix} 2 & -9 \\ 8 & 4 \end{vmatrix} = 2(4) - (-9)(8)$

$= 8 + 72 = 80$

46. $\begin{vmatrix} 7 & 0 \\ 5 & -3 \end{vmatrix} = 7(-3) - 0(5)$

$= -21$

47. $\begin{vmatrix} 2 & 10 & 4 \\ 0 & 1 & 3 \\ 0 & 6 & -1 \end{vmatrix}$ Expand about column 1.

$= 2 \begin{vmatrix} 1 & 3 \\ 6 & -1 \end{vmatrix} - 0 + 0$

$= 2[1(-1) - 3(6)]$

$= 2(-19) = -38$

48. $\begin{vmatrix} -1 & 7 & 2 \\ 3 & 0 & 5 \\ -1 & 2 & 6 \end{vmatrix}$ Expand about row 2.

$-3 \begin{vmatrix} 7 & 2 \\ 2 & 6 \end{vmatrix} + 0 - 5 \begin{vmatrix} -1 & 7 \\ -1 & 2 \end{vmatrix}$

$= -3(42 - 4) - 5(-2 + 7)$

$= -114 - 25$

$= -139$

49. If $D = 0$, Cramer's rule does not apply.

50. For three unknowns, three equations are needed.

51. $3x - 4y = -32$
$2x + y = -3$

$D = \begin{vmatrix} 3 & -4 \\ 2 & 1 \end{vmatrix} = 3(1) - (-4)(2) = 11$

$D_x = \begin{vmatrix} -32 & -4 \\ -3 & 1 \end{vmatrix} = -32(1) - (-4)(-3)$

$= -32 - 12 = -44$

$D_y = \begin{vmatrix} 3 & -32 \\ 2 & -3 \end{vmatrix} = 3(-3) - (-32)(2)$

$= -9 + 64 = 55$

$x = \dfrac{D_x}{D} = \dfrac{-44}{11} = -4; \; y = \dfrac{D_y}{D} = \dfrac{55}{11} = 5$

Solution set: $\{(-4, 5)\}$

52. $-4x + 3y = -12$
$2x + 6y = 15$

$D = \begin{vmatrix} -4 & 3 \\ 2 & 6 \end{vmatrix} = -4(6) - 3(2)$

$= -24 - 6 = -30$

$D_x = \begin{vmatrix} -12 & 3 \\ 15 & 6 \end{vmatrix} = -12(6) - 3(15)$

$= -72 - 45 = -117$

$D_y = \begin{vmatrix} -4 & -12 \\ 2 & 15 \end{vmatrix} = -4(15) - (-12)(2)$

$= -60 + 24 = -36$

$x = \dfrac{D_x}{D} = \dfrac{-117}{-30} = \dfrac{39}{10}; \; y = \dfrac{D_y}{D} = \dfrac{-36}{-30} = \dfrac{6}{5}$

Solution set: $\left\{ \left(\dfrac{39}{10}, \dfrac{6}{5} \right) \right\}$

53. $4x + y + z = 11$
$x - y - z = 4$
$\phantom{4x + {}}y + 2z = 0$

Expand about row 3 to find D, D_x, D_y, and D_z.

$D = \begin{vmatrix} 4 & 1 & 1 \\ 1 & -1 & -1 \\ 0 & 1 & 2 \end{vmatrix}$

$= 0 - 1 \begin{vmatrix} 4 & 1 \\ 1 & -1 \end{vmatrix} + 2 \begin{vmatrix} 4 & 1 \\ 1 & -1 \end{vmatrix}$

$= -1(-5) + 2(-5)$

$= 5 - 10 = -5$

$D_x = \begin{vmatrix} 11 & 1 & 1 \\ 4 & -1 & -1 \\ 0 & 1 & 2 \end{vmatrix}$

$= 0 - 1 \begin{vmatrix} 11 & 1 \\ 4 & -1 \end{vmatrix} + 2 \begin{vmatrix} 11 & 1 \\ 4 & -1 \end{vmatrix}$

$= -1(-15) + 2(-15)$

$= 15 - 30 = -15$

$D_y = \begin{vmatrix} 4 & 11 & 1 \\ 1 & 4 & -1 \\ 0 & 0 & 2 \end{vmatrix} = 0 - 0 + 2 \begin{vmatrix} 4 & 11 \\ 1 & 4 \end{vmatrix}$

$= 2(5) = 10$

$D_z = \begin{vmatrix} 4 & 1 & 11 \\ 1 & -1 & 4 \\ 0 & 1 & 0 \end{vmatrix} = 0 - 1 \begin{vmatrix} 4 & 11 \\ 1 & 4 \end{vmatrix} + 0$

$= -1(5) = -5$

$x = \dfrac{D_x}{D} = \dfrac{-15}{-5} = 3; \; y = \dfrac{D_y}{D} = \dfrac{10}{-5} = -2$

$z = \dfrac{D_z}{D} = \dfrac{-5}{-5} = 1$

Solution set: $\{(3, -2, 1)\}$

54.
$$-x + 3y - 4z = 4$$
$$2x + 4y + z = -14$$
$$3x - y + 2z = -8$$

$$D = \begin{vmatrix} -1 & 3 & -4 \\ 2 & 4 & 1 \\ 3 & -1 & 2 \end{vmatrix} \quad \text{Expand about column 1.}$$

$$= -1\begin{vmatrix} 4 & 1 \\ -1 & 2 \end{vmatrix} - 2\begin{vmatrix} 3 & -4 \\ -1 & 2 \end{vmatrix} + 3\begin{vmatrix} 3 & -4 \\ 4 & 1 \end{vmatrix}$$

$$= -1(9) - 2(2) + 3(19)$$

$$= -9 - 4 + 57 = 44$$

$$D_x = \begin{vmatrix} 4 & 3 & -4 \\ -14 & 4 & 1 \\ -8 & -1 & 2 \end{vmatrix} \quad \text{Expand about column 3.}$$

$$= -4\begin{vmatrix} -14 & 4 \\ -8 & -1 \end{vmatrix} - 1\begin{vmatrix} 4 & 3 \\ -8 & -1 \end{vmatrix}$$

$$+ 2\begin{vmatrix} 4 & 3 \\ -14 & 4 \end{vmatrix}$$

$$= -4(46) - 1(20) + 2(58)$$

$$= -184 - 20 + 116 = -88$$

$$D_y = \begin{vmatrix} -1 & 4 & -4 \\ 2 & -14 & 1 \\ 3 & -8 & 2 \end{vmatrix} \quad \text{Expand about column 1.}$$

$$= -1\begin{vmatrix} -14 & 1 \\ -8 & 2 \end{vmatrix} - 2\begin{vmatrix} 4 & -4 \\ -8 & 2 \end{vmatrix}$$

$$+ 3\begin{vmatrix} 4 & -4 \\ -14 & 1 \end{vmatrix}$$

$$= -1(-20) - 2(-24) + 3(-52)$$

$$= 20 + 48 - 156 = -88$$

$$D_z = \begin{vmatrix} -1 & 3 & 4 \\ 2 & 4 & -14 \\ 3 & -1 & -8 \end{vmatrix} \quad \text{Expand about column 1.}$$

$$= -1\begin{vmatrix} 4 & -14 \\ -1 & -8 \end{vmatrix} - 2\begin{vmatrix} 3 & 4 \\ -1 & -8 \end{vmatrix}$$

$$+ 3\begin{vmatrix} 3 & 4 \\ 4 & -14 \end{vmatrix}$$

$$= -1(-46) - 2(-20) + 3(-58)$$

$$= 46 + 40 - 174 = -88$$

Since $D_x = D_y = D_z$,

$$x = y = z = \frac{-88}{44} = -2.$$

Solution set: $\{(-2, -2, -2)\}$

55.
$$2x + y + 3z = 1 \quad (1)$$
$$x - 2y + z = -3 \quad (2)$$
$$-3x + y - 2z = -4 \quad (3)$$

Eliminate x from equations (1) and (2) by multiplying equation (2) by -2. Then add the results.

$$\begin{array}{l} 2x + y + 3z = 1 \quad (1) \\ \underline{-2x + 4y - 2z = 6} \quad -2 \times (2) \\ 5y + z = 7 \quad (4) \end{array}$$

Eliminate x from equations (2) and (3) by multiplying equation (2) by 3. Then add the results.

$$\begin{array}{l} 3x - 6y + 3z = -9 \quad 3 \times (2) \\ \underline{-3x + y - 2z = -4} \quad (3) \\ -5y + z = -13 \quad (5) \end{array}$$

The resulting system is

$$5y + z = 7 \quad (4)$$
$$-5y + z = -13. \quad (5)$$

56.
$$\begin{array}{l} 5y + z = 7 \quad (4) \\ \underline{-5y + z = -13} \quad (5) \\ 2z = -6 \quad \text{Add.} \\ z = -3 \end{array}$$

Substitute $z = -3$ into equation (4).

$$5y + z = 7 \quad (4)$$
$$5y - 3 = 7$$
$$5y = 10$$
$$y = 2$$

Solution set: $\{(y, z) = (2, -3)\}$

57. Substitute $y = 2$ and $z = -3$ into equation (2).

$$x - 2y + z = -3 \quad (2)$$
$$x - 2(2) + (-3) = -3$$
$$x - 7 = -3$$
$$x = 4$$

Solution set: $\{(4, 2, -3)\}$

58. The augmented matrix is

$$\begin{bmatrix} 2 & 1 & 3 & | & 1 \\ 1 & -2 & 1 & | & -3 \\ -3 & 1 & -2 & | & -4 \end{bmatrix}.$$

$$\begin{bmatrix} 1 & -2 & 1 & | & -3 \\ 2 & 1 & 3 & | & 1 \\ -3 & 1 & -2 & | & -4 \end{bmatrix} \quad R_1 \leftrightarrow R_2$$

$$\begin{bmatrix} 1 & -2 & 1 & | & -3 \\ 0 & 5 & 1 & | & 7 \\ 0 & -5 & 1 & | & -13 \end{bmatrix} \quad \begin{array}{l} -2R_1 + R_2 \\ 3R_1 + R_3 \end{array}$$

$$\begin{bmatrix} 1 & -2 & 1 & | & -3 \\ 0 & 1 & \frac{1}{5} & | & \frac{7}{5} \\ 0 & -5 & 1 & | & -13 \end{bmatrix} \quad \frac{1}{5}R_2$$

$$\begin{bmatrix} 1 & -2 & 1 & | & -3 \\ 0 & 1 & \frac{1}{5} & | & \frac{7}{5} \\ 0 & 0 & 2 & | & -6 \end{bmatrix} \quad 5R_2 + R_3$$

$$\begin{bmatrix} 1 & -2 & 1 & | & -3 \\ 0 & 1 & \frac{1}{5} & | & \frac{7}{5} \\ 0 & 0 & 1 & | & -3 \end{bmatrix} \quad \frac{1}{2}R_3$$

This matrix gives the system

$$x - 2y + z = -3$$
$$y + \frac{1}{5}z = \frac{7}{5}$$
$$z = -3.$$

Substitute $z = -3$ in the second equation.

$$y + \frac{1}{5}z = \frac{7}{5}$$
$$y + \frac{1}{5}(-3) = \frac{7}{5}$$
$$y - \frac{3}{5} = \frac{7}{5}$$
$$y = 2$$

Substitute $y = 2$ and $z = -3$ in the first equation.

$$x - 2y + z = -3$$
$$x - 2(2) + (-3) = -3$$
$$x - 7 = -3$$
$$x = 4$$

The solution set is $\{(4, 2, -3)\}$, the same as that found in Exercise 57.

59. (a)

$$D = \begin{vmatrix} 2 & 1 & 3 \\ 1 & -2 & 1 \\ -3 & 1 & -2 \end{vmatrix} \quad \text{Expand about row 1.}$$

$$= 2\begin{vmatrix} -2 & 1 \\ 1 & -2 \end{vmatrix} - 1\begin{vmatrix} 1 & 1 \\ -3 & -2 \end{vmatrix} + 3\begin{vmatrix} 1 & -2 \\ -3 & 1 \end{vmatrix}$$
$$= 2(4 - 1) - 1(-2 + 3) + 3(1 - 6)$$
$$= 2(3) - 1(1) + 3(-5)$$
$$= 6 - 1 - 15 = -10$$

(b)

$$D_x = \begin{vmatrix} 1 & 1 & 3 \\ -3 & -2 & 1 \\ -4 & 1 & -2 \end{vmatrix} \quad \text{Expand about row 1.}$$

$$= 1\begin{vmatrix} -2 & 1 \\ 1 & -2 \end{vmatrix} - 1\begin{vmatrix} -3 & 1 \\ -4 & -2 \end{vmatrix} + 3\begin{vmatrix} -3 & -2 \\ -4 & 1 \end{vmatrix}$$
$$= 1(4 - 1) - 1(6 + 4) + 3(-3 - 8)$$
$$= 1(3) - 1(10) + 3(-11)$$
$$= 3 - 10 - 33 = -40$$

(c)

$$D_y = \begin{vmatrix} 2 & 1 & 3 \\ 1 & -3 & 1 \\ -3 & -4 & -2 \end{vmatrix} \quad \text{Expand about row 1.}$$

$$= 2\begin{vmatrix} -3 & 1 \\ -4 & -2 \end{vmatrix} - 1\begin{vmatrix} 1 & 1 \\ -3 & -2 \end{vmatrix} + 3\begin{vmatrix} 1 & -3 \\ -3 & -4 \end{vmatrix}$$
$$= 2(6 + 4) - 1(-2 + 3) + 3(-4 - 9)$$
$$= 2(10) - 1(1) + 3(-13)$$
$$= 20 - 1 - 39 = -20$$

(d)

$$D_z = \begin{vmatrix} 2 & 1 & 1 \\ 1 & -2 & -3 \\ -3 & 1 & -4 \end{vmatrix} \quad \text{Expand about row 1.}$$

$$= 2\begin{vmatrix} -2 & -3 \\ 1 & -4 \end{vmatrix} - 1\begin{vmatrix} 1 & -3 \\ -3 & -4 \end{vmatrix} + 1\begin{vmatrix} 1 & -2 \\ -3 & 1 \end{vmatrix}$$
$$= 2(8 + 3) - 1(-4 - 9) + 1(1 - 6)$$
$$= 2(11) - 1(-13) + 1(-5)$$
$$= 22 + 13 - 5 = 30$$

60.
$$x = \frac{D_x}{D} = \frac{-40}{-10} = 4$$
$$y = \frac{D_y}{D} = \frac{-20}{-10} = 2$$
$$z = \frac{D_z}{D} = \frac{30}{-10} = -3$$

The solution set, $\{(4, 2, -3)\}$, is the same as before.

61.
$$\frac{2}{3}x + \frac{1}{6}y = \frac{19}{2} \quad (1)$$
$$\frac{1}{3}x - \frac{2}{9}y = 2 \quad (2)$$

Multiply equation (1) by 6 and equation (2) by 9 to clear the fractions.

$$4x + y = 57 \quad (3) \ 6 \times (1)$$
$$3x - 2y = 18 \quad (4) \ 9 \times (2)$$

To eliminate y, multiply equation (3) by 2 and add the result to equation (4).

$$\begin{array}{rlll} 8x & + 2y & = 114 & 2 \times (3) \\ 3x & - 2y & = 18 & (4) \\ \hline 11x & & = 132 & \\ x & & = 12 & \end{array}$$

Substitute 12 for x in equation (3) to find y.

$$4x + y = 57 \quad (3)$$
$$4(12) + y = 57$$
$$48 + y = 57$$
$$y = 9$$

Solution set: $\{(12, 9)\}$

62.
$$\begin{array}{rcl} 2x + 5y - z &=& 12 \quad (1) \\ -x + y - 4z &=& -10 \quad (2) \\ -8x - 20y + 4z &=& 31 \quad (3) \end{array}$$

Multiply equation (1) by 4 and add the result to equation (3).

$$\begin{array}{rll} 8x + 20y - 4z = 48 & 4 \times (1) \\ -8x - 20y + 4z = 31 & (3) \\ \hline 0 = 79 & \textit{False} \end{array}$$

Since a false statement results, the system is inconsistent. The solution set is $\emptyset$.

63. $x = 7y + 10$ (1)
$2x + 3y = 3$ (2)

Since equation (1) is given in terms of x, substitute $7y + 10$ for x in equation (2) and solve for y.

$$2(7y + 10) + 3y = 3$$
$$14y + 20 + 3y = 3$$
$$17y = -17$$
$$y = -1$$

Since $x = 7y + 10$ and $y = -1$,

$$x = 7(-1) + 10 = -7 + 10 = 3.$$

Solution set: $\{(3, -1)\}$

64. $x + 4y = 17$ (1)
$-3x + 2y = -9$ (2)

To eliminate x, multiply equation (1) by 3 and add the result to equation (2).

$$
\begin{array}{rl}
3x + 12y = 51 & 3 \times (1) \\
-3x + 2y = -9 & (2) \\
\hline
14y = 42 & \\
y = 3 &
\end{array}
$$

Substitute 3 for y in equation (1) to find x.

$$x + 4y = 17 \quad (1)$$
$$x + 4(3) = 17$$
$$x + 12 = 17$$
$$x = 5$$

Solution set: $\{(5, 3)\}$

65. $-7x + 3y = 12$ (1)
$5x + 2y = 8$ (2)

To eliminate y, multiply equation (1) by 2 and equation (2) by -3. Then add the results.

$$
\begin{array}{rl}
-14x + 6y = 24 & 2 \times (1) \\
-15x - 6y = -24 & -3 \times (2) \\
\hline
-29x = 0 & \\
x = 0 &
\end{array}
$$

Substitute 0 for x in equation (1) to find y.

$$-7x + 3y = 12 \quad (1)$$
$$-7(0) + 3y = 12$$
$$3y = 12$$
$$y = 4$$

Solution set: $\{(0, 4)\}$

66. $2x - 5y = 8$ (1)
$3x + 4y = 10$ (2)

To eliminate y, multiply equation (1) by 4 and equation (2) by 5 and add the results.

$$
\begin{array}{rl}
8x - 20y = 32 & 4 \times (1) \\
15x + 20y = 50 & 5 \times (2) \\
\hline
23x = 82 & \\
x = \dfrac{82}{23} &
\end{array}
$$

Instead of substituting to find y, we'll choose different multipliers and eliminate x from the original system.

$$
\begin{array}{rl}
6x - 15y = 24 & 3 \times (1) \\
-6x - 8y = -20 & -2 \times (2) \\
\hline
-23y = 4 & \\
y = -\dfrac{4}{23} &
\end{array}
$$

Solution set: $\left\{\left(\dfrac{82}{23}, -\dfrac{4}{23}\right)\right\}$

67. *Step 1*
Let $x =$ the number of
    Godzilla movies;
$y =$ the number of
    Aerosmith CDs.

Step 2

Type of Gift	Number Bought	Cost of each (in dollars)	Total Value
movie	x	14.95	$14.95x$
CD	y	16.88	$16.88y$
Totals	7	—	114.30

Step 3
From the second and fourth columns of the table, we obtain the system

$$x + y = 7 \qquad (1)$$
$$14.95x + 16.88y = 114.30 \qquad (2)$$

Step 4
Multiply equation (1) by -14.95 and add the result to equation (2).

$$
\begin{array}{rcl}
-14.95x - 14.95y & = & -104.65 \\
14.95x + 16.88y & = & 114.30 \\
\hline
1.93y & = & 9.65 \\
y & = & 5
\end{array}
$$

From (1), $x = 2$.

Step 5
Cheryl bought 2 movies and 5 CDs.

Step 6
Two \$14.95 movies and five \$16.88 CDs give us 7 gifts worth \$114.30.

68. Let $x = $ the number of gold medals won by Canada,

$y = $ the number of silver medals won by Canada,

and $z = $ the number of bronze medals won by Canada.

They won a total of 22 medals, so

$$x + y + z = 22. \quad (1)$$

There were 5 fewer gold medals than bronze, so

$$x = z - 5. \quad (2)$$

There were 3 fewer bronze than silver, so

$$z = y - 3 \quad \text{or} \quad y = z + 3. \quad (3)$$

Substitute $z - 5$ for x and $z + 3$ for y in (1).

$$x + y + z = 22 \quad (1)$$
$$(z - 5) + (z + 3) + z = 22$$
$$3z - 2 = 22$$
$$3z = 24$$
$$z = 8$$

From (2), $x = z - 5 = 8 - 5 = 3$.
From (3), $y = z + 3 = 8 + 3 = 11$.
Canada won 3 gold medals, 11 silver medals, and 8 bronze medals.

Chapter 9 Test

1. No; The graph for Babe Ruth lies completely below the graph for Aaron, indicating that Ruth's total was always lower than Aaron's.

2. Aaron had the most and Ruth had the fewest.

3. When each equation of the system

$$x + y = 7$$
$$x - y = 5$$

is graphed, the point of intersection appears to be $(6, 1)$. To check, substitute 6 for x and 1 for y in each of the equations. Since $(6, 1)$ makes both equations true, the solution set of the system is

$$\{(6, 1)\}.$$

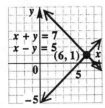

4.
$$2x - 3y = 24 \quad (1)$$
$$y = -\frac{2}{3}x \quad (2)$$

Since equation (2) is solved for y, substitute $-\frac{2}{3}x$ for y in equation (1) and solve for x.

$$2x - 3y = 24 \quad (1)$$
$$2x - 3\left(-\frac{2}{3}x\right) = 24$$
$$2x + 2x = 24$$
$$4x = 24$$
$$x = 6$$

Since $y = -\frac{2}{3}x$ and $x = 6$,

$$y = -\frac{2}{3}(6) = -4.$$

Solution set: $\{(6, -4)\}$

5.
$$12x - 5y = 8 \quad (1)$$
$$3x = \frac{5}{4}y + 2$$
or
$$x = \frac{5}{12}y + \frac{2}{3} \quad (2)$$

Substitute $\frac{5}{12}y + \frac{2}{3}$ for x in equation (1) and solve for y.

$$12x - 5y = 8 \quad (1)$$
$$12\left(\frac{5}{12}y + \frac{2}{3}\right) - 5y = 8$$
$$5y + 8 - 5y = 8$$
$$8 = 8 \quad \textit{True}$$

Equations (1) and (2) are dependent.
Solution set: $\{(x, y) \mid 12x - 5y = 8\}$

6.
$$3x + y = 12 \quad (1)$$
$$2x - y = 3 \quad (2)$$

To eliminate y, add equations (1) and (2).

$$\begin{array}{rcl} 3x + y &=& 12 \quad (1) \\ 2x - y &=& 3 \quad (2) \\ \hline 5x &=& 15 \\ x &=& 3 \end{array}$$

Substitute 3 for x in equation (1) to find y.

$$3x + y = 12 \quad (1)$$
$$3(3) + y = 12$$
$$9 + y = 12$$
$$y = 3$$

Solution set: $\{(3, 3)\}$

7. $-5x + 2y = -4$ (1)
          $6x + 3y = -6$ (2)

To eliminate x, multiply equation (1) by 6 and equation (2) by 5. Then add the results.

$$\begin{array}{rl} -30x + 12y = -24 & 6 \times (1) \\ \underline{30x + 15y = -30} & 5 \times (2) \\ 27y = -54 & \\ y = -2 & \end{array}$$

Substitute -2 for y in equation (1) to find x.

$$\begin{array}{r} -5x + 2y = -4 \quad (1) \\ -5x + 2(-2) = -4 \\ -5x - 4 = -4 \\ -5x = 0 \\ x = 0 \end{array}$$

Solution set: $\{(0, -2)\}$

8. $3x + 4y = 8$ (1)
          $8y = 7 - 6x$

or $6x + 8y = 7$ (2)

Multiply equation (1) by -2 and add the result to equation (2).

$$\begin{array}{rl} -6x - 8y = -16 & -2 \times (1) \\ \underline{6x + 8y = 7} & \\ 0 = -9 & \textit{False} \end{array}$$

Since a false statement results, the system is inconsistent. The solution set is $\emptyset$.

9. $3x + 5y + 3z = 2$ (1)
          $6x + 5y + z = 0$ (2)
          $3x + 10y - 2z = 6$ (3)

To eliminate x, multiply equation (1) by -1 and add the result to equation (3).

$$\begin{array}{rl} -3x - 5y - 3z = -2 & -1 \times (1) \\ \underline{3x + 10y - 2z = 6} & (3) \\ 5y - 5z = 4 & (4) \end{array}$$

To eliminate x again, multiply equation (1) by -2 and add the result to equation (2).

$$\begin{array}{rl} -6x - 10y - 6z = -4 & -2 \times (1) \\ \underline{6x + 5y + z = 0} & (2) \\ -5y - 5z = -4 & (5) \end{array}$$

To eliminate y, add equations (4) and (5).

$$\begin{array}{rl} 5y - 5z = 4 & (4) \\ \underline{-5y - 5z = -4} & (5) \\ -10z = 0 & \\ z = 0 & \end{array}$$

Substitute 0 for z in equation (4) to find y.

$$\begin{array}{r} 5y - 5z = 4 \quad (4) \\ 5y - 5(0) = 4 \\ 5y - 0 = 4 \\ 5y = 4 \\ y = \frac{4}{5} \end{array}$$

Substitute $\frac{4}{5}$ for y and 0 for z in equation (1) to find x.

$$\begin{array}{r} 3x + 5y + 3z = 2 \quad (1) \\ 3x + 5\left(\frac{4}{5}\right) + 3(0) = 2 \\ 3x + 4 + 0 = 2 \\ 3x = -2 \\ x = -\frac{2}{3} \end{array}$$

Solution set: $\left\{\left(-\frac{2}{3}, \frac{4}{5}, 0\right)\right\}$

10.   Let $x =$ the speed of the fast car and $y =$ the speed of the slow car.

Make a table.

	r	t	d
Fast car	x	3.5	$3.5x$
Slow car	y	3.5	$3.5y$

Since the fast car travels 30 mph faster than the slow car,

$$x - y = 30. \quad (1)$$

Since the cars travel a total of 420 miles,

$$3.5x + 3.5y = 420.$$

Multiply by 10 to clear the decimals.

$$35x + 35y = 4200 \quad (2)$$

To eliminate y, multiply equation (1) by 35 and add the result to equation (2).

$$\begin{array}{rl} 35x - 35y = 1050 & 35 \times (1) \\ \underline{35x + 35y = 4200} & (2) \\ 70x = 5250 & \\ x = 75 & \end{array}$$

Substitute 75 for x in equation (1) to find y.

$$\begin{array}{r} x - y = 30 \quad (1) \\ 75 - y = 30 \\ -y = -45 \\ y = 45 \end{array}$$

The fast car is traveling at 75 mph, and the slow car is traveling at 45 mph.

11. Let $x =$ the number of liters of 20% solution
and $y =$ the number of liters of 50% solution.

Make a table.

Kind of Solution	Liters of Solution	Liters of Pure Alcohol
.20	x	$.20x$
.50	y	$.50y$
.40	12	$.40(12) = 4.8$

Since 12 L of the mixture are needed,

$$x + y = 12. \quad (1)$$

Since the amount of pure alcohol in the 20% solution plus the amount of pure alcohol in the 50% solution must equal the amount of alcohol in the mixture,

$$.2x + .5y = 4.8.$$

Multiply by 10 to clear the decimals.

$$2x + 5y = 48 \quad (2)$$

Multiply equation (1) by -2 and add the result to equation (2).

$$
\begin{array}{rl}
-2x - 2y = -24 & -2 \times (1) \\
\underline{2x + 5y = 48} & (2) \\
3y = 24 & \\
y = 8 &
\end{array}
$$

From (1), $x + y = 12$, $x = 4$.
4 L of 20% solution and 8 L of 50% solution are needed.

12. *Step 1*
Let $x =$ the number of visitors
           to the Magic Kingdom
           (in millions);
    $y =$ the number of visitors
           to Disneyland
           (in millions).

Step 2
Not necessary

Step 3
Disneyland had 1.2 million more visitors than the Magic Kingdom, so

$$y = x + 1.2. \quad (1)$$

Together they had 28.8 million visitors, so

$$x + y = 28.8. \quad (2)$$

Step 4
From (1), substitute $x + 1.2$ for y in (2).

$$
\begin{aligned}
x + (x + 1.2) &= 28.8 \\
2x + 1.2 &= 28.8 \\
2x &= 27.6 \\
x &= 13.8
\end{aligned}
$$

From (1), $y = 13.8 + 1.2 = 15.0$.

Step 5
In 1996, the Magic Kingdom had 13.8 million visitors and Disneyland had 15.0 million visitors.

Step 6
15.0 is 1.2 more than 13.8 and the sum of 15.0 and 13.8 is 28.8, as required.

13. Let $x =$ the cost of a sheet of colored paper
and $y =$ the cost of a marker pen.

For the first purchase $8x$ represents the cost of the paper and $3y$ the cost of the pens. The total cost was $6.50, so

$$8x + 3y = 6.50. \quad (1)$$

For the second purchase,

$$2x + 2y = 3.00. \quad (2)$$

To eliminate x, multiply equation (2) by -4 and add the result to equation (1).

$$
\begin{array}{rl}
8x + 3y = 6.50 & (1) \\
\underline{-8x - 8y = -12.00} & -4 \times (2) \\
-5y = -5.50 & \\
y = 1.10 &
\end{array}
$$

Substitute $y = 1.10$ in equation (2).

$$
\begin{aligned}
2x + 2y &= 3.00 \quad (2) \\
2x + 2(1.10) &= 3.00 \\
2x + 2.20 &= 3.00 \\
2x &= .80 \\
x &= .40
\end{aligned}
$$

She paid $1.10 for each marker pen and $.40 for each sheet of colored paper.

14.
$$
\begin{aligned}
3x + 2y &= 4 \\
5x + 5y &= 9
\end{aligned}
$$

Write the augmented matrix.

$$\begin{bmatrix} 3 & 2 & | & 4 \\ 5 & 5 & | & 9 \end{bmatrix}$$

$$\begin{bmatrix} 1 & \frac{2}{3} & | & \frac{4}{3} \\ 5 & 5 & | & 9 \end{bmatrix} \quad \frac{1}{3}R_1$$

$$\begin{bmatrix} 1 & \frac{2}{3} & | & \frac{4}{3} \\ 0 & \frac{5}{3} & | & \frac{7}{3} \end{bmatrix} \quad -5R_1 + R_2$$

$$\begin{bmatrix} 1 & \frac{2}{3} & | & \frac{4}{3} \\ 0 & 1 & | & \frac{7}{5} \end{bmatrix} \quad \frac{3}{5}R_2$$

continued

This matrix gives the system

$$x + \frac{2}{3}y = \frac{4}{3}$$
$$y = \frac{7}{5}.$$

Substitute $y = \frac{7}{5}$ in the first equation.

$$x + \frac{2}{3}\left(\frac{7}{5}\right) = \frac{4}{3}$$
$$x + \frac{14}{15} = \frac{4}{3}$$
$$x = \frac{20}{15} - \frac{14}{15} = \frac{6}{15} = \frac{2}{5}$$

Solution set: $\left\{\left(\frac{2}{5}, \frac{7}{5}\right)\right\}$

15. $\begin{aligned} x + 3y + 2z &= 11 \\ 3x + 7y + 4z &= 23 \\ 5x + 3y - 5z &= -14 \end{aligned}$

Write the augmented matrix.

$$\begin{bmatrix} 1 & 3 & 2 & | & 11 \\ 3 & 7 & 4 & | & 23 \\ 5 & 3 & -5 & | & -14 \end{bmatrix}$$

$$\begin{bmatrix} 1 & 3 & 2 & | & 11 \\ 0 & -2 & -2 & | & -10 \\ 0 & -12 & -15 & | & -69 \end{bmatrix} \quad \begin{aligned} -3R_1 + R_2 \\ -5R_1 + R_3 \end{aligned}$$

$$\begin{bmatrix} 1 & 3 & 2 & | & 11 \\ 0 & 1 & 1 & | & 5 \\ 0 & -12 & -15 & | & -69 \end{bmatrix} \quad -\tfrac{1}{2}R_2$$

$$\begin{bmatrix} 1 & 3 & 2 & | & 11 \\ 0 & 1 & 1 & | & 5 \\ 0 & 0 & -3 & | & -9 \end{bmatrix} \quad 12R_2 + R_3$$

$$\begin{bmatrix} 1 & 3 & 2 & | & 11 \\ 0 & 1 & 1 & | & 5 \\ 0 & 0 & 1 & | & 3 \end{bmatrix} \quad -\tfrac{1}{3}R_3$$

This matrix gives the system

$$\begin{aligned} x + 3y + 2z &= 11 \\ y + z &= 5 \\ z &= 3. \end{aligned}$$

Substitute $z = 3$ in the second equation.

$$\begin{aligned} y + z &= 5 \\ y + 3 &= 5 \\ y &= 2 \end{aligned}$$

Substitute $y = 2$ and $z = 3$ in the first equation.

$$\begin{aligned} x + 3y + 2z &= 11 \\ x + 3(2) + 2(3) &= 11 \\ x + 6 + 6 &= 11 \\ x &= -1 \end{aligned}$$

Solution set: $\{(-1, 2, 3)\}$

16. $\begin{aligned} 4x - 2y \quad\;\; &= -8 \\ 3y - 5z &= 14 \\ 2x \quad\;\; + z &= -10 \end{aligned}$

Write the augmented matrix.

$$\begin{bmatrix} 4 & -2 & 0 & | & -8 \\ 0 & 3 & -5 & | & 14 \\ 2 & 0 & 1 & | & -10 \end{bmatrix}$$

$$\begin{bmatrix} 1 & -\tfrac{1}{2} & 0 & | & -2 \\ 0 & 3 & -5 & | & 14 \\ 2 & 0 & 1 & | & -10 \end{bmatrix} \quad \tfrac{1}{4}R_1$$

$$\begin{bmatrix} 1 & -\tfrac{1}{2} & 0 & | & -2 \\ 0 & 3 & -5 & | & 14 \\ 0 & 1 & 1 & | & -6 \end{bmatrix} \quad -2R_1 + R_3$$

$$\begin{bmatrix} 1 & -\tfrac{1}{2} & 0 & | & -2 \\ 0 & 1 & -\tfrac{5}{3} & | & \tfrac{14}{3} \\ 0 & 1 & 1 & | & -6 \end{bmatrix} \quad \tfrac{1}{3}R_2$$

$$\begin{bmatrix} 1 & -\tfrac{1}{2} & 0 & | & -2 \\ 0 & 1 & -\tfrac{5}{3} & | & \tfrac{14}{3} \\ 0 & 0 & \tfrac{8}{3} & | & -\tfrac{32}{3} \end{bmatrix} \quad -1R_2 + R_3$$

$$\begin{bmatrix} 1 & -\tfrac{1}{2} & 0 & | & -2 \\ 0 & 1 & -\tfrac{5}{3} & | & \tfrac{14}{3} \\ 0 & 0 & 1 & | & -4 \end{bmatrix} \quad \tfrac{3}{8}R_3$$

This matrix gives the system

$$\begin{aligned} x - \frac{1}{2}y &= -2 \\ y - \frac{5}{3}z &= \frac{14}{3} \\ z &= -4. \end{aligned}$$

Substitute $z = -4$ in the second equation.

$$\begin{aligned} y - \frac{5}{3}z &= \frac{14}{3} \\ y - \frac{5}{3}(-4) &= \frac{14}{3} \\ y + \frac{20}{3} &= \frac{14}{3} \\ y &= -2 \end{aligned}$$

Substitute $y = -2$ in the first equation.

$$\begin{aligned} x - \frac{1}{2}y &= -2 \\ x - \frac{1}{2}(-2) &= -2 \\ x + 1 &= -2 \\ x &= -3 \end{aligned}$$

Solution set: $\{(-3, -2, -4)\}$

17. $\begin{vmatrix} 6 & -3 \\ 5 & -2 \end{vmatrix} = 6(-2) - (-3)(5)$
$$= -12 + 15 = 3$$

18. $\begin{vmatrix} 4 & 1 & 0 \\ -2 & 7 & 3 \\ 0 & 5 & 2 \end{vmatrix}$ Expand about row 1.

$$= 4\begin{vmatrix} 7 & 3 \\ 5 & 2 \end{vmatrix} - 1\begin{vmatrix} -2 & 3 \\ 0 & 2 \end{vmatrix} + 0$$

$$= 4[7(2) - 3(5)] - 1[-2(2) - 3(0)]$$

$$= 4(-1) - 1(-4)$$

$$= -4 + 4 = 0$$

19. $\begin{array}{rcl} 3x & - & y & = & -8 \\ 2x & + & 6y & = & 3 \end{array}$

$$D = \begin{vmatrix} 3 & -1 \\ 2 & 6 \end{vmatrix} = 18 - (-2) = 20$$

$$D_x = \begin{vmatrix} -8 & -1 \\ 3 & 6 \end{vmatrix} = -48 - (-3) = -45$$

$$D_y = \begin{vmatrix} 3 & -8 \\ 2 & 3 \end{vmatrix} = 9 - (-16) = 25$$

$$x = \frac{D_x}{D} = \frac{-45}{20} = -\frac{9}{4}; \; y = \frac{D_y}{D} = \frac{25}{20} = \frac{5}{4}$$

Solution set: $\left\{ \left(-\dfrac{9}{4}, \dfrac{5}{4} \right) \right\}$

20. $\begin{array}{rcl} x & + & y & + & z & = & 6 \\ 2x & - & 2y & + & z & = & 5 \\ -x & + & 3y & + & z & = & 0 \end{array}$

$$D = \begin{vmatrix} 1 & 1 & 1 \\ 2 & -2 & 1 \\ -1 & 3 & 1 \end{vmatrix} \; \text{Expand about row 1.}$$

$$= 1\begin{vmatrix} -2 & 1 \\ 3 & 1 \end{vmatrix} - 1\begin{vmatrix} 2 & 1 \\ -1 & 1 \end{vmatrix} + 1\begin{vmatrix} 2 & -2 \\ -1 & 3 \end{vmatrix}$$

$$= 1(-2 - 3) - 1(2 + 1) + 1(6 - 2)$$

$$= -5 - 3 + 4 = -4$$

$$D_x = \begin{vmatrix} 6 & 1 & 1 \\ 5 & -2 & 1 \\ 0 & 3 & 1 \end{vmatrix} \; \text{Expand about row 3.}$$

$$= 0 - 3\begin{vmatrix} 6 & 1 \\ 5 & 1 \end{vmatrix} + 1\begin{vmatrix} 6 & 1 \\ 5 & -2 \end{vmatrix}$$

$$= -3(6 - 5) + 1(-12 - 5)$$

$$= -3 - 17 = -20$$

$$D_y = \begin{vmatrix} 1 & 6 & 1 \\ 2 & 5 & 1 \\ -1 & 0 & 1 \end{vmatrix} \; \text{Expand about row 3.}$$

$$= -1\begin{vmatrix} 6 & 1 \\ 5 & 1 \end{vmatrix} - 0 + 1\begin{vmatrix} 1 & 6 \\ 2 & 5 \end{vmatrix}$$

$$= -1(6 - 5) + 1(5 - 12)$$

$$= -1 - 7 = -8$$

$$D_z = \begin{vmatrix} 1 & 1 & 6 \\ 2 & -2 & 5 \\ -1 & 3 & 0 \end{vmatrix} \; \text{Expand about row 3.}$$

$$= -1\begin{vmatrix} 1 & 6 \\ -2 & 5 \end{vmatrix} - 3\begin{vmatrix} 1 & 6 \\ 2 & 5 \end{vmatrix} + 0$$

$$= -1(5 + 12) - 3(5 - 12)$$

$$= -17 + 21 = 4$$

$$x = \frac{D_x}{D} = \frac{-20}{-4} = 5; \; y = \frac{D_y}{D} = \frac{-8}{-4} = 2$$

$$z = \frac{D_z}{D} = \frac{4}{-4} = -1$$

Solution set: $\{(5, 2, -1)\}$

Cumulative Review Exercises Chapters 1–9

1. The integer factors of 40 are $-1, 1, -2, 2, -4, 4,$ $-5, 5, -8, 8, -10, 10, -20, 20, -40,$ and $40.$

2. $\dfrac{3x^2 + 2y^2}{10y + 3} = \dfrac{3 \cdot 1^2 + 2 \cdot 5^2}{10(5) + 3}$ *Let x = 1, y = 5*

$$= \frac{3 \cdot 1 + 2 \cdot 25}{50 + 3}$$

$$= \frac{3 + 50}{50 + 3} = \frac{53}{53} = 1$$

3. The commutative property says that $3 \cdot 6 = 6 \cdot 3,$ so that is the property that justifies the given statement.

4. $\quad 7(2x + 3) - 4(2x + 1) = 2(x + 1)$

$$14x + 21 - 8x - 4 = 2x + 2$$

$$6x + 17 = 2x + 2$$

$$4x = -15$$

$$x = -\frac{15}{4}$$

Solution set: $\left\{ -\dfrac{15}{4} \right\}$

5. $|6x - 8| = 4$

$\quad 6x - 8 = 4 \quad$ or $\quad 6x - 8 = -4$

$\qquad 6x = 12 \qquad\qquad\quad 6x = 4$

$\qquad\; x = 2 \quad$ or $\qquad x = \dfrac{4}{6} = \dfrac{2}{3}$

Solution set: $\left\{ \dfrac{2}{3}, 2 \right\}$

6. $\qquad\qquad ax + by = cx + d$

To solve for x, get all terms with x alone on one side of the equals sign.

$$ax - cx = d - by$$

$$x(a - c) = d - by$$

$$x = \frac{d - by}{a - c}$$

or $\qquad x = \dfrac{by - d}{c - a}$

if the x-terms were put on the right side of the equals sign.

7. $.04x + .06(x - 1) = 1.04$

Multiply both sides by 100 to clear the decimals.

$$4x + 6(x - 1) = 104$$

$$4x + 6x - 6 = 104$$

continued

$$10x - 6 = 104$$
$$10x = 110$$
$$x = 11$$

Solution set: $\{11\}$

8.
$$\frac{2}{3}y + \frac{5}{12}y \leq 20$$

Multiply both sides by 12.

$$12\left(\frac{2}{3}y + \frac{5}{12}y\right) \leq 12(20)$$
$$8y + 5y \leq 240$$
$$13y \leq 240$$
$$y \leq \frac{240}{13}$$

Solution set: $\left(-\infty, \dfrac{240}{13}\right]$

9. $|3x + 2| \leq 4$
$$-4 \leq 3x + 2 \leq 4$$
$$-6 \leq 3x \leq 2 \qquad \textit{Subtract 2.}$$
$$-2 \leq x \leq \frac{2}{3} \qquad \textit{Divide by 3.}$$

Solution set: $\left[-2, \dfrac{2}{3}\right]$

10. $|12t + 7| \geq 0$
The solution set is $(-\infty, \infty)$ since the absolute value of any number is greater than or equal to 0.

11. Let $x =$ the number of "guilty" votes
and $x + 10 =$ the number of "not guilty" votes.
The total number of votes was 200, so

$$x + (x + 10) = 200$$
$$2x + 10 = 200$$
$$2x = 190$$
$$x = 95,$$
$$\text{and} \quad x + 10 = 105.$$

There were 105 "not guilty" votes and 95 "guilty" votes.

12. Let $l =$ the length of the book;
$l - 2.5 =$ the width of the book.
$$P = 2l + 2w$$
$$38 = 2l + 2(l - 2.5)$$
$$38 = 2l + 2l - 5$$
$$43 = 4l$$
$$l = \frac{43}{4} \text{ or } 10\frac{3}{4}$$
$$l - 2.5 = \frac{43}{4} - \frac{10}{4} = \frac{33}{4} \text{ or } 8\frac{1}{4}$$

The width is $8\frac{1}{4}$ inches and the length is $10\frac{3}{4}$ inches.

13. Let $x =$ the measure of the equal angles and
$2x - 4 =$ the measure of the third angle.

The sum of the measures of the angles in a triangle is 180, so

$$x + x + 2x - 4 = 180$$
$$4x - 4 = 180$$
$$4x = 184$$
$$x = 46.$$

So, $2x - 4 = 2(46) - 4 = 92 - 4 = 88.$
The measures of the angles are 46°, 46°, and 88°.

14. $-3\left(-5x^2 + 3x - 10\right) - \left(x^2 - 4x + 7\right)$
$$= 15x^2 - 9x + 30 - x^2 + 4x - 7$$
$$= 14x^2 - 5x + 23$$

15. $(3x - 7)(2y + 4)$

$$\qquad\quad \textbf{F} \qquad\quad \textbf{O} \qquad\quad \textbf{I} \qquad\quad \textbf{L}$$
$$= (3x)(2y) + (3x)(4) + (-7)(2y) + (-7)(4)$$
$$= 6xy + 12x - 14y - 28$$

16. $\dfrac{3k^2 + 17k^2 - 27k + 7}{k + 7}$

$$
\begin{array}{r}
3k^2 \quad - 4k \quad + 1 \\
k + 7 \overline{\smash{\big)}\, 3k^3 \quad + 17k^2 \quad - 27k \quad + 7} \\
\underline{3k^3 \quad + 21k^2 \qquad\qquad\qquad} \\
- 4k^2 \quad - 27k \\
\underline{- 4k^2 \quad - 28k \qquad} \\
k \quad + 7 \\
\underline{k \quad + 7} \\
0
\end{array}
$$

The remainder is 0, so the answer is the quotient, $3k^2 - 4k + 1$.

17. $36,500,000,000$

Move the decimal point to the right of the first nonzero digit, which is 3. Count the number of places the decimal point was moved, which is 10.

Because moving the decimal point made the number smaller, the exponent on 10 will be positive. Thus,

$$36,500,000,000 = 3.65 \times 10^{10}.$$

18. $\left(\dfrac{x^{-4}y^3}{x^2y^4}\right)^{-1} = \left(\dfrac{x^2y^4}{x^{-4}y^3}\right)^1$
$$= \frac{x^2 \cdot x^4 \cdot y^4}{y^3}$$
$$= x^6y$$

19. Factor $10m^2 + 7mp - 12p^2$ by trial and error.
$$10m^2 + 7mp - 12p^2$$
$$= (5m - 4p)(2m + 3p)$$

20. $64t^2 - 48t + 9$ is a perfect square trinomial.

$$64t^2 - 48t + 9$$
$$= (8t)^2 - 2(8t)(3) + (3)^2$$
$$= (8t - 3)^2$$

21.
$$6x^2 - 7x - 3 = 0$$
$$(3x + 1)(2x - 3) = 0$$

$$3x + 1 = 0 \quad \text{or} \quad 2x - 3 = 0$$
$$x = -\frac{1}{3} \quad \text{or} \quad x = \frac{3}{2}$$

The solution set is $\left\{ -\dfrac{1}{3}, \dfrac{3}{2} \right\}$.

22.
$$r^2 - 121 = 0$$
$$(r + 11)(r - 11) = 0$$
$$r = -11 \quad \text{or} \quad r = 11$$

The solution set is $\{-11, 11\}$.

23.
$$\frac{-3x + 6}{2x + 4} - \frac{-3x - 8}{2x + 4}$$
$$= \frac{(-3x + 6) - (-3x - 8)}{2x + 4}$$
$$= \frac{-3x + 6 + 3x + 8}{2x + 4}$$
$$= \frac{14}{2x + 4}$$
$$= \frac{2 \cdot 7}{2(x + 2)} = \frac{7}{x + 2}$$

24.
$$\frac{16k^2 - 9}{8k + 6} \div \frac{16k^2 - 24k + 9}{6}$$
$$= \frac{16k^2 - 9}{8k + 6} \cdot \frac{6}{16k^2 - 24k + 9}$$
$$= \frac{(4k + 3)(4k - 3)}{2(4k + 3)} \cdot \frac{6}{(4k - 3)(4k - 3)}$$
$$= \frac{2 \cdot 3}{2(4k - 3)} = \frac{3}{4k - 3}$$

25.
$$\frac{4}{x + 1} + \frac{3}{x - 2} = 4$$

Multiply each side by the LCD, $(x + 1)(x - 2)$.

$$(x + 1)(x - 2)\left(\frac{4}{x + 1} + \frac{3}{x - 2} \right)$$
$$= (x + 1)(x - 2)(4)$$
$$4(x - 2) + 3(x + 1) = 4(x^2 - x - 2)$$
$$4x - 8 + 3x + 3 = 4x^2 - 4x - 8$$
$$0 = 4x^2 - 11x - 3$$
$$(4x + 1)(x - 3) = 0$$
$$x = -\frac{1}{4} \quad \text{or} \quad x = 3$$

Check $x = -\dfrac{1}{4}$: $\dfrac{16}{3} - \dfrac{4}{3} = 4$ *True*

Check $x = 3$: $1 + 3 = 4$ *True*

Thus, the solution set is $\left\{ -\dfrac{1}{4}, 3 \right\}$.

26. A horizontal line through the point (x, k) has equation $y = k$. Since point A has coordinates $(-2, 6)$, $k = 6$. The equation of the horizontal line through A is $y = 6$.

27. A vertical line through the point (k, y) has equation $x = k$. Since point B has coordinates $(4, -2)$, $k = 4$. The equation of the vertical line through B is $x = 4$.

28. Let $(x_1, y_1) = (-2, 6)$ and $(x_2, y_2) = (4, -2)$. Then,

$$m = \frac{y_2 - y_1}{x_2 - x_1} = \frac{-2 - 6}{4 - (-2)} = \frac{-8}{6} = -\frac{4}{3}.$$

The slope is $-\dfrac{4}{3}$.

29. Perpendicular lines have slopes that are negative reciprocals of each other. The slope of line AB is $-\dfrac{4}{3}$ (from Exercise 25). The negative reciprocal of $-\dfrac{4}{3}$ is $\dfrac{3}{4}$, so the slope of a line perpendicular to line AB is $\dfrac{3}{4}$.

30. Let $m = -\dfrac{4}{3}$ and $(x_1, y_1) = (4, -2)$ in the point-slope form.

$$y - y_1 = m(x - x_1)$$
$$y - (-2) = -\frac{4}{3}(x - 4)$$
$$y + 2 = -\frac{4}{3}x + \frac{16}{3}$$

Multiply by 3 to clear the fractions, and then write the equation in standard form, $Ax + By = C$.

$$3y + 6 = -4x + 16$$
$$4x + 3y = 10$$

31. First locate the point $(-1, -3)$ on a graph. Then use the definition of slope to find a second point on the line.

$$m = \frac{\text{change in } y}{\text{change in } x} = \frac{2}{3}$$

From $(-1, -3)$, move 2 units up and 3 units to the right. The line through $(-1, -3)$ and the new point, $(2, -1)$, is the graph.

continued

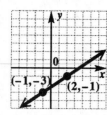

32. Use the points $(0, 3502)$ and $(4, 3915)$ to find the slope of the line.

$$m = \frac{3915 - 3502}{4 - 0} = \frac{413}{4} = 103.25$$

The value of y when x is 0 is 3502, so the slope-intercept form of the line is

$$y = 103.25x + 3502.$$

The slope represents the average yearly increase in health benefit cost during the period.

33. $-3x - 2y \le 6$

Graph the line $-3x - 2y = 6$ as a solid line through its intercepts, $(-2, 0)$ and $(0, -3)$, since the inequality involves $\le$.
To determine the region that belongs to the graph, test $(0, 0)$.

$$-3x - 2y \le 6$$
$$-3(0) - 2(0) \le 6$$
$$0 \le 6 \quad True$$

Since the result is true, shade the region that includes $(0, 0)$.

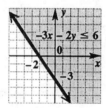

34. $f(x) = x^2 + 3x - 6$

(a) $f(-3) = (-3)^2 + 3(-3) - 6$
$$= 9 - 9 - 6 = -6$$

(b) $f(a) = (a)^2 + 3(a) - 6$
$$= a^2 + 3a - 6$$

35.
$$-2x + 3y = -15 \quad (1)$$
$$4x - y = 15 \quad (2)$$

To eliminate x, multiply equation (1) by 2 and add the result to equation (2).

$$
\begin{array}{rl}
-4x + 6y = -30 & 2 \times (1) \\
4x - y = 15 & (2) \\
\hline
5y = -15 & \\
y = -3 &
\end{array}
$$

Substitute -3 for y in equation (2) to find x.

$$4x - y = 15 \quad (2)$$
$$4x - (-3) = 15$$
$$4x + 3 = 15$$
$$4x = 12$$
$$x = 3$$

Solution set: $\{(3, -3)\}$

36.
$$
\begin{array}{rrrrl}
x & + y & + z & = & 10 \quad (1) \\
x & - y & - z & = & 0 \quad (2) \\
-x & + y & - z & = & -4 \quad (3)
\end{array}
$$

Add equations (1) and (2) to eliminate y and z. The result is

$$2x = 10$$
$$x = 5.$$

Add equations (2) and (3) to eliminate x and y. The result is

$$-2z = -4$$
$$z = 2.$$

Substitute 5 for x and 2 for z in equation (1) to find y.

$$x + y + z = 10 \quad (1)$$
$$5 + y + 2 = 10$$
$$y + 7 = 10$$
$$y = 3$$

Solution set: $\{(5, 3, 2)\}$

37. *Step 1*
Let $x =$ the number of adults' tickets sold;
$\quad y =$ the number of children's tickets sold.

Step 2

Kind of Ticket	Number Sold	Cost of Each (in dollars)	Total Value (in dollars)
Adult	x	6	$6x$
Child	y	2	$2y$
Total	454	—	2528

Step 3
The total number of tickets sold was 454, so

$$x + y = 454. \quad (1)$$

Since the total value was \$2528, the final column leads to

$$6x + 2y = 2528. \quad (2)$$

Step 4
Multiply both sides of equation (1) by -2 and add this result to equation (2).

$$
\begin{array}{rcl}
-2x \;-\; 2y &=& -908 \\
6x \;+\; 2y &=& 2528 \\
\hline
4x &=& 1620 \\
x &=& 405
\end{array}
$$

From (1), $y = 49$.

Step 5
There were 405 adults and 49 children at the game.

Step 6
The total number of tickets sold was $405 + 49 = 454$. Since 405 adults paid $6 each and 49 children paid $2 each, the value of tickets sold should be $405(6) + 49(2) = 2528$, or 2528. The result agrees with the given information.

38. Let $x =$ the average retail price of Elmo and $y =$ the average retail price of Kid.

Elmo cost $8.63 less than Kid, so

$$x = y - 8.63. \quad (1)$$

Together they cost $63.89, so

$$x + y = 63.89. \quad (2)$$

From (1), substitute $y - 8.63$ for x in equation (2).

$$
\begin{aligned}
(y - 8.63) + y &= 63.89 \\
2y - 8.63 &= 63.89 \\
2y &= 72.52 \\
y &= 36.26
\end{aligned}
$$

From (1), $x = y - 8.63 = 36.26 - 8.63 = 27.63$. The average retail price of Elmo was $27.63 and of Kid was $36.26.

39. Let $x =$ the cost of a pound of peanuts and $y =$ the cost of a pound of cashews.

The cost of 6 lb of peanuts and 12 lb of cashews is $60, so

$$6x + 12y = 60. \quad (1)$$

The cost of 3 lb of peanuts and 4 lb of cashews is $22, so

$$3x + 4y = 22. \quad (2)$$

To eliminate x, multiply equation (2) by -2 and add the result to equation (1).

$$
\begin{array}{rcll}
6x \;+\; 12y &=& 60 & (1) \\
-6x \;-\; 8y &=& -44 & -2 \times (2) \\
\hline
4y &=& 16 & \\
y &=& 4 &
\end{array}
$$

Substitute 4 for y in equation (2) to find x.

$$
\begin{aligned}
3x + 4y &= 22 \quad (2) \\
3x + 4(4) &= 22 \\
3x + 16 &= 22 \\
3x &= 6 \\
x &= 2
\end{aligned}
$$

Peanuts cost $2/lb, and cashews cost $4/lb.

40. **(a)** The lines intersect at $(8, 3000)$, so the cost equals the revenue at $x = 8$ (which is 800 items). The revenue is $3000.

(b) On the sale of 1100 parts $(x = 11)$, the revenue is about $4100 and the cost is about $3600.

$$
\begin{aligned}
\text{Profit} &= \text{Revenue} - \text{Cost} \\
&\approx 4100 - 3600 \\
&= 500
\end{aligned}
$$

The profit is about $500.

CHAPTER 10 ROOTS AND RADICALS

Section 10.1

1. Every nonnegative number has two square roots. This statement is *false* since zero is a nonnegative number that has only one square root, namely 0.

3. Every positive number has two real square roots. This statement is *true*. One of the real square roots is a positive number and the other is its opposite.

5. The cube root of every real number has the same sign as the number itself. This statement is *true*. The cube root of a positive real number is positive and the cube root of a negative real number is negative.

7. The square roots of 16 are -4 and 4 because $(-4)(-4) = 16$ and $4 \cdot 4 = 16$.

9. The square roots of 144 are -12 and 12 because $(-12)(-12) = 144$ and
$$12 \cdot 12 = 144.$$

11. The square roots of $\dfrac{25}{196}$ are $-\dfrac{5}{14}$ and $\dfrac{5}{14}$ because
$$\left(-\frac{5}{14}\right)\left(-\frac{5}{14}\right) = \frac{25}{196}$$
and
$$\frac{5}{14} \cdot \frac{5}{14} = \frac{25}{196}.$$

13. The square roots of 900 are -30 and 30 because
$$(-30)(-30) = 900 \quad \text{and} \quad 30 \cdot 30 = 900.$$

15. $\sqrt{49}$ represents the positive square root of 49. Since $7 \cdot 7 = 49$,
$$\sqrt{49} = 7.$$

17. $-\sqrt{121}$ represents the negative square root of 121. Since $11 \cdot 11 = 121$,
$$-\sqrt{121} = -11.$$

19. $-\sqrt{\dfrac{144}{121}}$ represents the negative square root of $\dfrac{144}{121}$. Since $\dfrac{12}{11} \cdot \dfrac{12}{11} = \dfrac{144}{121}$,
$$-\sqrt{\frac{144}{121}} = -\frac{12}{11}.$$

21. $\sqrt{-121}$ is not a real number because there is no real number whose square is -121.

23. The square of $\sqrt{100}$ is
$$\left(\sqrt{100}\right)^2 = 100,$$
by the definition of square root.

25. The square of $-\sqrt{19}$ is
$$\left(-\sqrt{19}\right)^2 = 19,$$
since the square of a negative number is positive.

27. The square of $\sqrt{3x^2 + 4}$ is
$$\left(\sqrt{3x^2 + 4}\right)^2 = 3x^2 + 4.$$

29. For the statement "$\sqrt{a}$ represents a positive number" to be true, a must be positive because the square root of a negative number is not a real number and $\sqrt{0} = 0$.

31. For the statement "$\sqrt{a}$ is not a real number" to be true, a must be negative.

33. $\sqrt{25}$

The number 25 is a perfect square, 5^2, so $\sqrt{25}$ is a *rational* number.
$$\sqrt{25} = 5$$

35. $\sqrt{29}$

Because 29 is not a perfect square, $\sqrt{29}$ is *irrational*. Using a calculator, we obtain
$$\sqrt{29} \approx 5.385.$$

37. $-\sqrt{64}$

The number 64 is a perfect square, 8^2, so $-\sqrt{64}$ is *rational*.
$$-\sqrt{64} = -8$$

39. $\sqrt{-29}$

There is no real number whose square is -29. Therefore, $\sqrt{-29}$ is *not a real number*.

41. $-\sqrt{121}$ and $\sqrt{-121}$ are different because $-\sqrt{121}$ is the negative square root of a positive number, while $\sqrt{-121}$ is the square root of a negative number (which is not a real number).

43. $\sqrt{571} \approx 23.896$

45. $\sqrt{798} \approx 28.249$

47. $\sqrt{3.94} \approx 1.985$

49. $\sqrt{3^2 + 4^2} = \sqrt{9 + 16} = \sqrt{25} = 5$

51. $\sqrt{8^2 + 15^2} = \sqrt{64 + 225} = \sqrt{289} = 17$

53. $\sqrt{2^2 + 3^2} = \sqrt{4 + 9} = \sqrt{13} \approx 3.606$

55. $\sqrt[3]{12} \approx 2.289$

57. $\sqrt[3]{130.6} \approx 5.074$

59. $\sqrt[3]{-87} = -\sqrt[3]{87} \approx -4.431$

61. $a = 8, b = 15$

Substitute the given values in the Pythagorean formula and then solve for c^2.

$$c^2 = a^2 + b^2$$
$$c^2 = 8^2 + 15^2$$
$$= 64 + 225$$
$$= 289$$

Now find the positive square root of 289 to obtain the length of the hypotenuse, c.

$$c = \sqrt{289} = 17$$

63. $a = 6, c = 10$

Substitute the given values in the Pythagorean formula and then solve for b^2.

$$c^2 = a^2 + b^2$$
$$10^2 = 6^2 + b^2$$
$$100 = 36 + b^2$$
$$64 = b^2$$

Now find the positive square root of 64 to obtain the length of the leg b.

$$b = \sqrt{64} = 8$$

65. $a = 11, b = 4$

$$c^2 = a^2 + b^2$$
$$c^2 = 11^2 + 4^2$$
$$= 121 + 16$$
$$= 137$$
$$c = \sqrt{137} \approx 11.705$$

67. The given information involves a right triangle with hypotenuse 25 centimeters and a leg of length 7 centimeters. Let a represent the length of the other leg, and use the Pythagorean formula.

$$c^2 = a^2 + b^2$$
$$25^2 = a^2 + 7^2$$
$$625 = a^2 + 49$$
$$576 = a^2$$
$$a = \sqrt{576} = 24$$

The length of the rectangle is 24 centimeters.

69. *Step 1*   Let x represent the vertical distance of the kite above Margaret's hand.

Step 2   The kite string forms the hypotenuse of a right triangle.

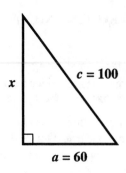

Step 3   Use the Pythagorean formula.

$$a^2 + x^2 = c^2$$
$$60^2 + x^2 = 100^2$$

Step 4 $3600 + x^2 = 10,000$
$$x^2 = 6400$$
$$x = \sqrt{6400} = 80$$

Step 5   The kite is 80 feet above her hand.

Step 6   From the figure, we see that we must have

$$60^2 + 80^2 = 100^2 \ ?$$
$$3600 + 6400 = 10,000. \quad True$$

71. *Step 1*   Let x represent the distance that the two cars are apart after 3 hours.

Step 2   Form a right triangle. The distance between the two cars after three hours, which is unknown, is the hypotenuse of the right triangle. The length of the northward leg is $a = 3(25) = 75$, and the length of the westward leg is $b = 3(60) = 180$.

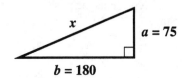

Step 3   Use the Pythagorean formula.

$$x^2 = a^2 + b^2$$
$$x^2 = 75^2 + 180^2$$
$$= 5625 + 32,400$$

Step 4 $x^2 = 38,025$
$$x = \sqrt{38,025} = 195$$

Step 5   The cars are 195 miles apart after 3 hours.

Step 6   From the figure, we see that we must have

$$75^2 + 180^2 = 195^2 \ ?$$
$$5625 + 32,400 = 38,025. \quad True$$

73. Use the Pythagorean formula with $a = 5$, $b = 8$, and $c = x$.

$$c^2 = a^2 + b^2$$
$$x^2 = 5^2 + 8^2$$
$$= 25 + 64$$
$$= 89$$
$$x = \sqrt{89} \approx 9.434$$

75. Answers will vary.

For example, if we choose $a = 2$ and $b = 7$,

$$\sqrt{a^2 + b^2} = \sqrt{2^2 + 7^2} = \sqrt{53},$$

while

$$a + b = 2 + 7 = 9.$$

$\sqrt{53} \neq 9$, so we have

$$\sqrt{a^2 + b^2} \neq a + b.$$

If $a = 0$ and $b = 1$,

$$\sqrt{a^2 + b^2} = \sqrt{0^2 + 1^2} = \sqrt{0 + 1}$$
$$= \sqrt{1} = 1,$$

and

$$a + b = 0 + 1 = 1.$$

This is a pair of numbers for which $\sqrt{a^2 + b^2}$ is equal to $a + b$ (which is not true in general).

77. Let $(x_1, y_1) = (5, -3)$ and $(x_2, y_2) = (7, 2)$.

Use the distance formula.

$$d = \sqrt{(x_2 - x_1)^2 + (y_2 - y_1)^2}$$
$$= \sqrt{(7 - 5)^2 + [2 - (-3)]^2}$$
$$= \sqrt{2^2 + 5^2}$$
$$= \sqrt{4 + 25}$$
$$= \sqrt{29}$$

79. $(x_1, y_1) = \left(-\dfrac{1}{4}, \dfrac{2}{3}\right)$, $(x_2, y_2) = \left(\dfrac{3}{4}, -\dfrac{1}{3}\right)$

$$d = \sqrt{(x_2 - x_1)^2 + (y_2 - y_1)^2}$$
$$= \sqrt{\left[\dfrac{3}{4} - \left(-\dfrac{1}{4}\right)\right]^2 + \left(-\dfrac{1}{3} - \dfrac{2}{3}\right)^2}$$
$$= \sqrt{1^2 + (-1)^2}$$
$$= \sqrt{1 + 1} = \sqrt{2}$$

81. $\sqrt[3]{1000} = 10$ because $10^3 = 1000$.

83. $\sqrt[3]{-27} = -3$ because $(-3)^3 = -27$.

85. $\sqrt[4]{625} = 5$ because 5 is positive and $5^4 = 625$.

87. $\sqrt[4]{-1}$ is not a real number because the fourth power of a real number cannot be negative.

89. $\sqrt[3]{(-27)} = -3$ because $(-3)^3 = -27$.

91. $\sqrt[5]{243} = 3$ because $3^5 = 243$.

93. The length of each side of the large square is c, so the area of the square on the left is c^2.

94. The length of each side of the small square in the middle of the figure on the left is $b - a$, so the area of this square is $(b - a)^2$.

95. Each of these rectangles has a width of a and a length of b, so the area of each one is ab. The sum of the areas is $2ab$.

96. The length of a side of the small square in the figure on the right is $b - a$ (the height on the far right minus the height on the far left), so the area of this square is

$$(b - a)^2 = b^2 - 2ab + a^2$$
$$= a^2 - 2ab + b^2.$$

97. The area of the figure on the left is c^2. The area of the figure on the right is the sum of the areas from Exercises 95 and 96, which is $2ab + (a^2 - 2ab + b^2)$. The two figures must have the same area, so

$$c^2 = 2ab + (a^2 - 2ab + b^2).$$

98. $c^2 = 2ab + (a^2 - 2ab + b^2)$
$$= 2ab + a^2 - 2ab + b^2$$
$$c^2 = a^2 + b^2$$

The final result is the Pythagorean formula.

Section 10.2

1. $\sqrt{4} = +2$ or -2

This statement is *false* since $\sqrt{4}$ represents only the principal (positive) square root.

3. $\sqrt{-6} \cdot \sqrt{6} = -6$

This statement is *false* since $\sqrt{-6}$ is not a real number.

5. $\sqrt[3]{3} \cdot \sqrt[3]{2} = \sqrt[3]{6}$

By *Properties of Radicals* (before Example 8 in the text),

$$\sqrt[3]{3} \cdot \sqrt[3]{2} = \sqrt[3]{3 \cdot 2} = \sqrt[3]{6},$$

so the given statement is *true*.

7. $\sqrt{4} \cdot \sqrt{9} = \sqrt{4 \cdot 9}$

This statement is *true* by the product rule for radicals.

9. $\sqrt{3} \cdot \sqrt{27} = \sqrt{3 \cdot 27} = \sqrt{81} = 9$

11. $\sqrt{6} \cdot \sqrt{15} = \sqrt{6 \cdot 15} = \sqrt{90} = \sqrt{9 \cdot 10}$
$$= \sqrt{9} \cdot \sqrt{10} = 3\sqrt{10}$$

13. $\sqrt{13} \cdot \sqrt{13} = \sqrt{13 \cdot 13} = 13$

15. $\sqrt{13} \cdot \sqrt{r} \quad (r \ge 0) = \sqrt{13r}$

17. **(a)** $\sqrt{47}$ is in simplified form since 47 has no perfect square factor (other than 1).

The other three choices could be simplified as follows.

$$\sqrt{45} = \sqrt{9 \cdot 5} = 3\sqrt{5}$$
$$\sqrt{48} = \sqrt{16 \cdot 3} = 4\sqrt{3}$$
$$\sqrt{44} = \sqrt{4 \cdot 11} = 2\sqrt{11}$$

19. $\sqrt{45} = \sqrt{9 \cdot 5} = \sqrt{9} \cdot \sqrt{5} = 3\sqrt{5}$

21. $\sqrt{75} = \sqrt{25 \cdot 3} = \sqrt{25} \cdot \sqrt{3} = 5\sqrt{3}$

23. $\sqrt{125} = \sqrt{25 \cdot 5} = \sqrt{25} \cdot \sqrt{5} = 5\sqrt{5}$

25. $-\sqrt{700} = -\sqrt{100 \cdot 7}$
$$= -\sqrt{100} \cdot \sqrt{7} = -10\sqrt{7}$$

27. $3\sqrt{27} = 3\sqrt{9 \cdot 3}$
$$= 3 \cdot \sqrt{9} \cdot \sqrt{3}$$
$$= 3 \cdot 3\sqrt{3}$$
$$= 9\sqrt{3}$$

29. $\sqrt{3} \cdot \sqrt{18} = \sqrt{3 \cdot 18} = \sqrt{54} = \sqrt{9 \cdot 6}$
$$= \sqrt{9} \cdot \sqrt{6} = 3\sqrt{6}$$

31. $\sqrt{12} \cdot \sqrt{48} = \sqrt{12 \cdot 48}$
$$= \sqrt{12 \cdot 12 \cdot 4}$$
$$= \sqrt{12 \cdot 12} \cdot \sqrt{4}$$
$$= 12 \cdot 2$$
$$= 24$$

33. $\sqrt{12} \cdot \sqrt{30} = \sqrt{12 \cdot 30} = \sqrt{360}$
$$= \sqrt{36 \cdot 10} = \sqrt{36} \cdot \sqrt{10} = 6\sqrt{10}$$

35. The product rule says that the product of two radicals is the radical whose radicand is the product of their radicands.

Similarly, by the quotient rule, the quotient of two radicals is the radical whose radicand is the quotient of their radicands.

37. $\sqrt{\dfrac{16}{225}} = \dfrac{\sqrt{16}}{\sqrt{225}} = \dfrac{4}{15}$

39. $\sqrt{\dfrac{7}{16}} = \dfrac{\sqrt{7}}{\sqrt{16}} = \dfrac{\sqrt{7}}{4}$

41. $\sqrt{\dfrac{5}{7}} \cdot \sqrt{35} = \dfrac{\sqrt{5}}{\sqrt{7}} \cdot \dfrac{\sqrt{5} \cdot \sqrt{7}}{1}$
$$= \sqrt{5} \cdot \sqrt{5} = 5$$

43. $\sqrt{\dfrac{5}{2}} \cdot \sqrt{\dfrac{125}{8}} = \sqrt{\dfrac{5}{2} \cdot \dfrac{125}{8}}$
$$= \sqrt{\dfrac{625}{16}}$$
$$= \dfrac{\sqrt{625}}{\sqrt{16}} = \dfrac{25}{4}$$

45. $\dfrac{30\sqrt{10}}{5\sqrt{2}} = \dfrac{30}{5}\sqrt{\dfrac{10}{2}} = 6\sqrt{5}$

47. $\sqrt{m^2} = m \quad (m \ge 0)$

49. $\sqrt{y^4} = \sqrt{(y^2)^2} = y^2$

51. $\sqrt{36z^2} = \sqrt{36} \cdot \sqrt{z^2} = 6z$

53. $\sqrt{400x^6} = \sqrt{20 \cdot 20 \cdot x^3 \cdot x^3} = 20x^3$

55. $\sqrt{z^5} = \sqrt{z^4 \cdot z}$
$$= \sqrt{z^4} \cdot \sqrt{z}$$
$$= z^2\sqrt{z}$$

57. $\sqrt{x^6 y^{12}} = \sqrt{(x^3)^2 \cdot (y^6)^2} = x^3 y^6$

59. $\sqrt[3]{40}$

8 is a perfect cube that is a factor of 40.

$$\sqrt[3]{40} = \sqrt[3]{8 \cdot 5}$$
$$= \sqrt[3]{8} \cdot \sqrt[3]{5} = 2\sqrt[3]{5}$$

61. $\sqrt[3]{54}$

27 is a perfect cube that is a factor of 54.

$$\sqrt[3]{54} = \sqrt[3]{27 \cdot 2}$$
$$= \sqrt[3]{27} \cdot \sqrt[3]{2} = 3\sqrt[3]{2}$$

63. $\sqrt[4]{80}$

16 is a perfect fourth power that is a factor of 80.

$$\sqrt[4]{80} = \sqrt[4]{16 \cdot 5}$$
$$= \sqrt[4]{16} \cdot \sqrt[4]{5} = 2\sqrt[4]{5}$$

65. $\sqrt[3]{\dfrac{8}{27}}$

8 and 27 are both perfect cubes.

$$\sqrt[3]{\dfrac{8}{27}} = \dfrac{\sqrt[3]{8}}{\sqrt[3]{27}} = \dfrac{2}{3}$$

67. $\sqrt[3]{-\dfrac{216}{125}} = -\sqrt[3]{\dfrac{216}{125}} = -\dfrac{\sqrt[3]{216}}{\sqrt[3]{125}} = -\dfrac{6}{5}$

69. $\sqrt[3]{5} \cdot \sqrt[3]{25} = \sqrt[3]{125} = 5$

71. $\sqrt[4]{4} \cdot \sqrt[4]{3} = \sqrt[4]{4 \cdot 3} = \sqrt[4]{12}$

Since 12 has no factor, other than 1, that is a perfect fourth power, this expression cannot be simplified further.

73. $\sqrt[3]{4x} \cdot \sqrt[3]{8x^2}$

$\begin{aligned} &= \sqrt[3]{4x} \cdot \sqrt[3]{8} \cdot \sqrt[3]{x^2} & \textit{Product rule} \\ &= \sqrt[3]{4x} \cdot 2 \cdot \sqrt[3]{x^2} & \sqrt[3]{8} = 2 \\ &= 2\sqrt[3]{4x^3} & \textit{Product rule} \\ &= 2 \cdot \sqrt[3]{4} \cdot \sqrt[3]{x^3} & \textit{Product rule} \\ &= 2x\sqrt[3]{4} & \sqrt[3]{x^3} = x \end{aligned}$

75. (a) $\sqrt{20} \approx 4.472135955$

(b) $\sqrt{5} \approx 2.236067977$

Multiply both sides of this equation by 2 to obtain

$$2\sqrt{5} \approx 4.472135955.$$

(c) These approximations suggest that $\sqrt{20}$ is equal to $2\sqrt{5}$, but this cannot be considered a proof of their equality since there is no guarantee that the two decimal approximations have the same digit in all of their corresponding decimal places. We have shown that the first nine digits after the decimal point agree in these two approximations , but what about the tenth, eleventh, and twelfth decimal places, and so on?

77. Use the formula for the volume of a cube.

$\begin{aligned} V &= s^3 \\ 216 &= s^3 & \textit{Let V = 216} \\ \sqrt[3]{216} &= s \\ 6 &= s \end{aligned}$

The depth of the container is 6 centimeters.

79. Use the formula for the volume of a sphere.

$$V = \frac{4}{3}\pi r^3$$

Let $V = 288\pi$ and solve for r.

$\begin{aligned} \frac{4}{3}\pi r^3 &= 288\pi \\ \frac{3}{4}\left(\frac{4}{3}\pi r^3\right) &= \frac{3}{4}(288\pi) \\ \pi r^3 &= 216\pi \\ r^3 &= 216 \\ r &= \sqrt[3]{216} = 6 \end{aligned}$

The radius is 6 inches.

81. When we have variables under radical signs, such as $\sqrt{a}$ and $\sqrt{b}$, it is important to know that both a and b are nonnegative before using the product

rule. This is because $\sqrt{a}$ and $\sqrt{b}$ would not be real numbers if a and b are negative.

83. In the first triangle, the length of each leg is 1. Using the Pythagorean formula, $1^2 + 1^2 = 2$, so the hypotenuse has length $\sqrt{2}$.

The hypotenuse from the first triangle is a leg in the second triangle. Using the Pythagorean formula, $\left(\sqrt{2}\right)^2 + 1^2 = 2 + 1 = 3$, so the hypotenuse of the second triangle has length $\sqrt{3}$.

By continuing this pattern, we obtain the lengths in the figure.

84. The lengths are equal to whole numbers when the radicands are perfect squares. In the figure, these lengths are $\sqrt{4}$ and $\sqrt{9}$. The next two whole number lengths would be $\sqrt{16}$ and $\sqrt{25}$.

85. The consecutive differences between the radicands in Exercise 84 are:

$$9 - 4 = 5$$
$$16 - 9 = 7$$
$$25 - 16 = 9$$

86. The differences between radicands of consecutive whole number lengths increase by 2 each time $(5, 7, 9, \ldots)$, so we can predict that the next difference will be

$$x - 25 = 11.$$

so $x = 36$ and the next whole number length is $\sqrt{36} = 6$.

The next difference will be

$$x - 36 = 13.$$

Thus, $x = 49$ and the next whole number length is $\sqrt{49} = 7$.

Section 10.3

1. $5\sqrt{2} + 6\sqrt{2} = (5 + 6)\sqrt{2}$
$$= 11\sqrt{2}$$

is an example of the _distributive_ property.

3. $\sqrt{2} + 2\sqrt{3}$ cannot be simplified because the _radicands_ are different.

5. $14\sqrt{7} - 19\sqrt{7} = (14 - 19)\sqrt{7}$
$$\textit{Distributive property}$$
$$= -5\sqrt{7}$$

7. $\sqrt{17} + 4\sqrt{17} = 1\sqrt{17} + 4\sqrt{17}$
$$= (1 + 4)\sqrt{17}$$
$$= 5\sqrt{17}$$

9. $6\sqrt{7} - \sqrt{7} = 6\sqrt{7} - 1\sqrt{7}$

$= (6-1)\sqrt{7} = 5\sqrt{7}$

11. $\sqrt{45} + 4\sqrt{20} = \sqrt{9 \cdot 5} + 4\sqrt{4 \cdot 5}$

$= \sqrt{9} \cdot \sqrt{5} + 4\left(\sqrt{4} \cdot \sqrt{5}\right)$

$= 3\sqrt{5} + 4\left(2\sqrt{5}\right)$

$= 3\sqrt{5} + 8\sqrt{5} = 11\sqrt{5}$

13. $5\sqrt{72} - 3\sqrt{50}$

$= 5\sqrt{36 \cdot 2} - 3\sqrt{25 \cdot 2}$

$= 5 \cdot \sqrt{36} \cdot \sqrt{2} - 3 \cdot \sqrt{25} \cdot \sqrt{2}$

$= 5 \cdot 6 \cdot \sqrt{2} - 3 \cdot 5 \cdot \sqrt{2}$

$= 30\sqrt{2} - 15\sqrt{2}$

$= 15\sqrt{2}$

15. $-5\sqrt{32} + 2\sqrt{45}$

$= -5\left(\sqrt{16} \cdot \sqrt{2}\right) + 2\left(\sqrt{9} \cdot \sqrt{5}\right)$

$= -5\left(4\sqrt{2}\right) + 2\left(3\sqrt{5}\right)$

$= -20\sqrt{2} + 6\sqrt{5}$

17. $5\sqrt{7} - 3\sqrt{28} + 6\sqrt{63}$

$= 5\sqrt{7} - 3\sqrt{4 \cdot 7} + 6\sqrt{9 \cdot 7}$

$= 5\sqrt{7} - 3 \cdot \sqrt{4} \cdot \sqrt{7} + 6 \cdot \sqrt{9} \cdot \sqrt{7}$

$= 5\sqrt{7} - 3 \cdot 2 \cdot \sqrt{7} + 6 \cdot 3 \cdot \sqrt{7}$

$= 5\sqrt{7} - 6\sqrt{7} + 18\sqrt{7}$

$= (5 - 6 + 18)\sqrt{7}$

$= 17\sqrt{7}$

19. $2\sqrt{8} - 5\sqrt{32} - 2\sqrt{48}$

$= 2\left(\sqrt{4} \cdot \sqrt{2}\right) - 5\left(\sqrt{16} \cdot \sqrt{2}\right)$

$\quad - 2\left(\sqrt{16} \cdot \sqrt{3}\right)$

$= 2\left(2\sqrt{2}\right) - 5\left(4\sqrt{2}\right) - 2\left(4\sqrt{3}\right)$

$= 4\sqrt{2} - 20\sqrt{2} - 8\sqrt{3}$

$= -16\sqrt{2} - 8\sqrt{3}$

(Because $\sqrt{2}$ and $\sqrt{3}$ are unlike radicals, this difference cannot be simplified further.)

21. $4\sqrt{50} + 3\sqrt{12} - 5\sqrt{45}$

$= 4\sqrt{25 \cdot 2} + 3\sqrt{4 \cdot 3} - 5\sqrt{9 \cdot 5}$

$= 4 \cdot \sqrt{25} \cdot \sqrt{2} + 3 \cdot \sqrt{4} \cdot \sqrt{3} - 5 \cdot \sqrt{9} \cdot \sqrt{5}$

$= 4 \cdot 5 \cdot \sqrt{2} + 3 \cdot 2 \cdot \sqrt{3} - 5 \cdot 3 \cdot \sqrt{5}$

$= 20\sqrt{2} + 6\sqrt{3} - 15\sqrt{5}$

23. $\dfrac{1}{4}\sqrt{288} + \dfrac{1}{6}\sqrt{72}$

$= \dfrac{1}{4}\left(\sqrt{144} \cdot \sqrt{2}\right) + \dfrac{1}{6}\left(\sqrt{36} \cdot \sqrt{2}\right)$

$= \dfrac{1}{4}\left(12\sqrt{2}\right) + \dfrac{1}{6}\left(6\sqrt{2}\right)$

$= 3\sqrt{2} + 1\sqrt{2} = 4\sqrt{2}$

25. $2\sqrt{3} + 4\sqrt{3} = (2+4)\sqrt{3} = 6\sqrt{3}$

The distributive property is used in the first step, where $a = \sqrt{3}$, $b = 2$, and $c = 4$.

27. $\sqrt{6} \cdot \sqrt{2} + 9\sqrt{3} = \sqrt{6 \cdot 2} + 9\sqrt{3}$

$= \sqrt{12} + 9\sqrt{3}$

$= \sqrt{4 \cdot 3} + 9\sqrt{3}$

$= 2\sqrt{3} + 9\sqrt{3} = 11\sqrt{3}$

29. $\sqrt{9x} + \sqrt{49x} - \sqrt{25x}$

$= \sqrt{9} \cdot \sqrt{x} + \sqrt{49} \cdot \sqrt{x} - \sqrt{25} \cdot \sqrt{x}$

$= 3\sqrt{x} + 7\sqrt{x} - 5\sqrt{x}$

$= (3 + 7 - 5)\sqrt{x}$

$= 5\sqrt{x}$

31. $\sqrt{6x^2} + x\sqrt{24}$

$= \sqrt{x^2 \cdot 6} + x\sqrt{4 \cdot 6}$

$= x\sqrt{6} + x \cdot 2\sqrt{6}$

$= (x + 2x)\sqrt{6} = 3x\sqrt{6}$

33. $3\sqrt{8x^2} - 4x\sqrt{2} - x\sqrt{8}$

$= 3\sqrt{4x^2 \cdot 2} - 4x\sqrt{2} - x\sqrt{4 \cdot 2}$

$= 3 \cdot \sqrt{4x^2} \cdot \sqrt{2} - 4x\sqrt{2} - x \cdot \sqrt{4} \cdot \sqrt{2}$

$= 3 \cdot 2x \cdot \sqrt{2} - 4x\sqrt{2} - x \cdot 2 \cdot \sqrt{2}$

$= 6x\sqrt{2} - 4x\sqrt{2} - 2x\sqrt{2}$

$= (6x - 4x - 2x)\sqrt{2}$

$= 0 \cdot \sqrt{2} = 0$

35. $-8\sqrt{32k} + 6\sqrt{8k}$

$= -8\left(\sqrt{16 \cdot 2k}\right) + 6\left(\sqrt{4 \cdot 2k}\right)$

$= -8\left(4\sqrt{2k}\right) + 6\left(2\sqrt{2k}\right)$

$= -32\sqrt{2k} + 12\sqrt{2k}$

$= (-32 + 12)\sqrt{2k}$

$= -20\sqrt{2k}$

37. $2\sqrt{125x^2z} + 8x\sqrt{80z}$

$\quad = 2\sqrt{25x^2 \cdot 5z} + 8x\sqrt{16 \cdot 5z}$

$\quad = 2\sqrt{25x^2}\sqrt{5z} + 8x\sqrt{16}\sqrt{5z}$

$\quad = 2 \cdot 5x\sqrt{5z} + 8x \cdot 4\sqrt{5z}$

$\quad = 10x\sqrt{5z} + 32x\sqrt{5z}$

$\quad = (10x + 32x)\sqrt{5z}$

$\quad = 42x\sqrt{5z}$

39. $4\sqrt[3]{16} - 3\sqrt[3]{54}$

Recall that 8 and 27 are perfect cubes.

$\quad 4\sqrt[3]{16} - 3\sqrt[3]{54}$

$\quad = 4\left(\sqrt[3]{8 \cdot 2}\right) - 3\left(\sqrt[3]{27 \cdot 2}\right)$

$\quad = 4\left(\sqrt[3]{8} \cdot \sqrt[3]{2}\right) - 3\left(\sqrt[3]{27} \cdot \sqrt[3]{2}\right)$

$\quad = 4\left(2\sqrt[3]{2}\right) - 3\left(3\sqrt[3]{2}\right)$

$\quad = 8\sqrt[3]{2} - 9\sqrt[3]{2}$

$\quad = (8 - 9)\sqrt[3]{2} = -1\sqrt[3]{2} = -\sqrt[3]{2}$

41. $6\sqrt[3]{8p^2} - 2\sqrt[3]{27p^2}$

$\quad = 6 \cdot \sqrt[3]{8} \cdot \sqrt[3]{p^2} - 2 \cdot \sqrt[3]{27} \cdot \sqrt[3]{p^2}$

$\quad = 6 \cdot 2 \cdot \sqrt[3]{p^2} - 2 \cdot 3 \cdot \sqrt[3]{p^2}$

$\quad = 12\sqrt[3]{p^2} - 6\sqrt[3]{p^2}$

$\quad = 6\sqrt[3]{p^2}$

43. $5\sqrt[4]{m^3} + 8\sqrt[4]{16m^3}$

$\quad = 5\sqrt[4]{m^3} + 8\left(\sqrt[4]{16} \cdot \sqrt[4]{m^3}\right)$

$\quad = 5\sqrt[4]{m^3} + 8\left(2\sqrt[4]{m^3}\right)$

$\quad = 5\sqrt[4]{m^3} + 16\sqrt[4]{m^3} = 21\sqrt[4]{m^3}$

45. Only like radicals can be added or subtracted. Thus, we can add $2\sqrt{3}$ and $5\sqrt{3}$ using the distributive property:

$\quad 2\sqrt{3} + 5\sqrt{3} = (2 + 5)\sqrt{3} = 7\sqrt{3}.$

We cannot subtract in the expression $2\sqrt{3} - 2\sqrt[3]{3}$, because $\sqrt{3}$ and $\sqrt[3]{3}$ are unlike radicals.

47. $5x^2y + 3x^2y - 14x^2y$

$\quad = (5 + 3 - 14)x^2y$

$\quad = -6x^2y$

48. $5(p - 2q)^2(a + b) + 3(p - 2q)^2(a + b)$

$\quad - 14(p - 2q)^2(a + b)$

$\quad = (5 + 3 - 14)(p - 2q)^2(a + b)$

$\quad = -6(p - 2q)^2(a + b)$

49. $5a^2\sqrt{xy} + 3a^2\sqrt{xy} - 14a^2\sqrt{xy}$

$\quad = (5 + 3 - 14)a^2\sqrt{xy}$

$\quad = -6a^2\sqrt{xy}$

50. In Exercises 47–49, each problem is of the form

$\quad 5A + 3A - 14A = (5 + 3 - 14)A = -6A.$

They are different in that A stands for x^2y in Exercise 47, $(p - 2q)^2 \cdot (a + b)$ in Exercise 48, and $a^2\sqrt{xy}$ in Exercise 49. Also, the first variable factor is raised to the second factor, and the second variable factor is raised to the first power. The answers are different because the variables are different: x and y, then $p - 2q$ and $a + b$, and then a and $\sqrt{xy}$.

51. $\sqrt{(-3 - 6)^2 + (2 - 4)^2}$

$\quad = \sqrt{(-9)^2 + (-2)^2}$

$\quad = \sqrt{81 + 4}$

$\quad = \sqrt{85} \approx 9.220$

53. $\sqrt{(2 - (-2))^2 + (-1 - 2)^2}$

$\quad = \sqrt{(4)^2 + (-3)^2}$

$\quad = \sqrt{16 + 9}$

$\quad = \sqrt{25} = 5$

55. Use the formula for the perimeter of a rectangle.

$\quad P = 2L + 2W$

$\quad = 2\left(7\sqrt{2}\right) + 2\left(4\sqrt{2}\right)$

$\quad = 14\sqrt{2} + 8\sqrt{2}$

$\quad = 22\sqrt{2}$

57. $y = 1.4\sqrt{x - 2.5} + 87.5$

(a) $y = 1.4\sqrt{10 - 2.5} + 87.5$ *Let x = 10*

$\quad = 1.4\sqrt{7.5} + 87.5$

$\quad \approx 3.83 + 87.5$

$\quad = 91.33$

So U.S. exports of electronics were $10 billion in 1991.

(b) $y = 1.4\sqrt{50 - 2.5} + 87.5$ *Let x = 50*

$\quad = 1.4\sqrt{47.5} + 87.5$

$\quad \approx 9.65 + 87.5$

$\quad = 97.15$

So U.S. exports of electronics were $50 billion in 1997.

Section 10.4

1. Rationalizing the denominator means to change the denominator from a <u>radical</u> to a rational number.

3. The expression $\sqrt{\dfrac{3m}{2}}$, where $m \geq 0$, is not simplified because the radical contains a <u>fraction</u>.

5. $\dfrac{8}{\sqrt{2}}$

To rationalize the denominator, multiply the numerator and denominator by $\sqrt{2}$.

$$\dfrac{8}{\sqrt{2}} = \dfrac{8 \cdot \sqrt{2}}{\sqrt{2} \cdot \sqrt{2}} = \dfrac{8\sqrt{2}}{2} = 4\sqrt{2}$$

7. $\dfrac{-\sqrt{11}}{\sqrt{3}} = \dfrac{-\sqrt{11} \cdot \sqrt{3}}{\sqrt{3} \cdot \sqrt{3}} = \dfrac{-\sqrt{33}}{3}$

9. $\dfrac{7\sqrt{3}}{\sqrt{5}} = \dfrac{7\sqrt{3} \cdot \sqrt{5}}{\sqrt{5} \cdot \sqrt{5}} = \dfrac{7\sqrt{15}}{5}$

11. $\dfrac{24\sqrt{10}}{16\sqrt{3}} = \dfrac{3 \cdot 8 \cdot \sqrt{10}}{2 \cdot 8 \cdot \sqrt{3}}$

$$= \dfrac{3\sqrt{10} \cdot \sqrt{3}}{2\sqrt{3} \cdot \sqrt{3}}$$

$$= \dfrac{3\sqrt{30}}{2 \cdot 3} = \dfrac{\sqrt{30}}{2}$$

13. $\dfrac{16}{\sqrt{27}} = \dfrac{16}{\sqrt{9 \cdot 3}} = \dfrac{16}{\sqrt{9} \cdot \sqrt{3}} = \dfrac{16}{3\sqrt{3}}$

$$= \dfrac{16 \cdot \sqrt{3}}{3\sqrt{3} \cdot \sqrt{3}} = \dfrac{16\sqrt{3}}{9}$$

15. $\dfrac{-3}{\sqrt{50}} = \dfrac{-3}{\sqrt{25 \cdot 2}}$

$$= \dfrac{-3}{5\sqrt{2}}$$

$$= \dfrac{-3 \cdot \sqrt{2}}{5\sqrt{2} \cdot \sqrt{2}}$$

$$= \dfrac{-3\sqrt{2}}{5 \cdot 2} = \dfrac{-3\sqrt{2}}{10}$$

17. $\dfrac{63}{\sqrt{45}} = \dfrac{63}{\sqrt{9} \cdot \sqrt{5}} = \dfrac{63}{3\sqrt{5}} = \dfrac{21}{\sqrt{5}}$

$$= \dfrac{21 \cdot \sqrt{5}}{\sqrt{5} \cdot \sqrt{5}} = \dfrac{21\sqrt{5}}{5}$$

19. $\dfrac{\sqrt{24}}{\sqrt{8}}$

When rationalizing the denominator, there are often several ways to approach the problem. Three ways to simplify this expression are shown here.

$$\dfrac{\sqrt{24}}{\sqrt{8}} = \dfrac{\sqrt{4} \cdot \sqrt{6}}{\sqrt{4} \cdot \sqrt{2}} = \dfrac{\sqrt{6}}{\sqrt{2}}$$

$$= \dfrac{\sqrt{6} \cdot \sqrt{2}}{\sqrt{2} \cdot \sqrt{2}} = \dfrac{\sqrt{12}}{2}$$

$$= \dfrac{\sqrt{4} \cdot \sqrt{3}}{2} = \dfrac{2\sqrt{3}}{2} = \sqrt{3}$$

or

$$\dfrac{\sqrt{24}}{\sqrt{8}} = \dfrac{\sqrt{24} \cdot \sqrt{2}}{\sqrt{8} \cdot \sqrt{2}} = \dfrac{\sqrt{48}}{\sqrt{16}}$$

$$= \dfrac{\sqrt{16} \cdot \sqrt{3}}{4} = \dfrac{4\sqrt{3}}{4} = \sqrt{3}$$

or

$$\dfrac{\sqrt{24}}{\sqrt{8}} = \sqrt{\dfrac{24}{8}} \quad \textit{Quotient rule}$$

$$= \sqrt{3}$$

21. $\sqrt{\dfrac{1}{2}} = \dfrac{\sqrt{1}}{\sqrt{2}} \quad \textit{Quotient rule}$

$$= \dfrac{1 \cdot \sqrt{2}}{\sqrt{2} \cdot \sqrt{2}} = \dfrac{\sqrt{2}}{2}$$

23. $\sqrt{\dfrac{13}{5}} = \dfrac{\sqrt{13}}{\sqrt{5}} = \dfrac{\sqrt{13} \cdot \sqrt{5}}{\sqrt{5} \cdot \sqrt{5}} = \dfrac{\sqrt{65}}{5}$

25. The given expression is being multiplied by $\dfrac{\sqrt{3}}{\sqrt{3}}$, which is 1. According to the identity property for multiplication, multiplying an expression by 1 does not change the value of the expression.

27. $\sqrt{\dfrac{7}{13}} \cdot \sqrt{\dfrac{13}{3}} = \sqrt{\dfrac{7}{13} \cdot \dfrac{13}{3}} \quad \textit{Product rule}$

$$= \sqrt{\dfrac{7}{3}} = \dfrac{\sqrt{7}}{\sqrt{3}}$$

$$= \dfrac{\sqrt{7} \cdot \sqrt{3}}{\sqrt{3} \cdot \sqrt{3}} = \dfrac{\sqrt{21}}{3}$$

29. $\sqrt{\dfrac{21}{7}} \cdot \sqrt{\dfrac{21}{8}} = \dfrac{\sqrt{21}}{\sqrt{7}} \cdot \dfrac{\sqrt{21}}{\sqrt{8}} = \dfrac{21}{\sqrt{7 \cdot 2 \cdot 4}}$

$$= \dfrac{21}{2\sqrt{14}} = \dfrac{21 \cdot \sqrt{14}}{2 \cdot \sqrt{14} \cdot \sqrt{14}}$$

$$= \dfrac{21\sqrt{14}}{2 \cdot 14} = \dfrac{3\sqrt{14}}{4}$$

31. $\sqrt{\dfrac{1}{12}} \cdot \sqrt{\dfrac{1}{3}} = \sqrt{\dfrac{1}{12} \cdot \dfrac{1}{3}}$

$$= \sqrt{\dfrac{1}{36}} = \dfrac{\sqrt{1}}{\sqrt{36}} = \dfrac{1}{6}$$

33. $\sqrt{\dfrac{2}{9}} \cdot \sqrt{\dfrac{9}{2}} = \sqrt{\dfrac{2}{9} \cdot \dfrac{9}{2}} = \sqrt{1} = 1$

35. $\sqrt{\dfrac{7}{x}} = \dfrac{\sqrt{7}}{\sqrt{x}} = \dfrac{\sqrt{7} \cdot \sqrt{x}}{\sqrt{x} \cdot \sqrt{x}} = \dfrac{\sqrt{7x}}{x}$

37. $\sqrt{\dfrac{4x^3}{y}} = \dfrac{\sqrt{4x^3}}{\sqrt{y}} = \dfrac{\sqrt{4x^2} \cdot \sqrt{x}}{\sqrt{y}} = \dfrac{2x\sqrt{x}}{\sqrt{y}}$

$$= \dfrac{2x\sqrt{x} \cdot \sqrt{y}}{\sqrt{y} \cdot \sqrt{y}} = \dfrac{2x\sqrt{xy}}{y}$$

39.
$$\sqrt{\frac{18x^3}{6y}} = \sqrt{\frac{3x^3}{y}} = \frac{\sqrt{3x^3}}{\sqrt{y}}$$
$$= \frac{\sqrt{x^2} \cdot \sqrt{3x}}{\sqrt{y}} = \frac{x\sqrt{3x}}{\sqrt{y}}$$
$$= \frac{x\sqrt{3x} \cdot \sqrt{y}}{\sqrt{y} \cdot \sqrt{y}} = \frac{x\sqrt{3xy}}{y}$$

41.
$$\sqrt{\frac{9a^2r^5}{7t}} = \frac{\sqrt{9a^2r^5}}{\sqrt{7t}} = \frac{\sqrt{9a^2r^4 \cdot r}}{\sqrt{7t}}$$
$$= \frac{\sqrt{9a^2r^4} \cdot \sqrt{r}}{\sqrt{7t}} = \frac{3ar^2\sqrt{r}}{\sqrt{7t}}$$
$$= \frac{3ar^2\sqrt{r} \cdot \sqrt{7t}}{\sqrt{7t} \cdot \sqrt{7t}} = \frac{3ar^2\sqrt{7rt}}{7t}$$

43. We need to multiply the numerator and

denominator of $\dfrac{\sqrt[3]{2}}{\sqrt[3]{5}}$ by enough factors of 5 to

make the radicand in the denominator a perfect cube. In this case we have one factor of 5, so we need to multiply by two more factors of 5 to make three factors of 5. Thus, the correct choice for a rationalizing factor in this problem is

$\sqrt[3]{5^2} = \sqrt[3]{25}$, which corresponds to choice (b).

45. $\sqrt[3]{\dfrac{3}{2}}$

Multiply the numerator and the denominator by enough factors of 2 to make the radicand in the denominator a perfect cube. This will eliminate the radical in the denominator. Here, we multiply by $\sqrt[3]{2^2}$ or $\sqrt[3]{4}$.

$$\sqrt[3]{\frac{3}{2}} = \frac{\sqrt[3]{3}}{\sqrt[3]{2}} = \frac{\sqrt[3]{3} \cdot \sqrt[3]{2^2}}{\sqrt[3]{2} \cdot \sqrt[3]{2^2}} = \frac{\sqrt[3]{3 \cdot 2^2}}{\sqrt[3]{2 \cdot 2^2}} = \frac{\sqrt[3]{12}}{\sqrt[3]{2^3}}$$
$$= \frac{\sqrt[3]{12}}{2}$$

47. $\dfrac{\sqrt[3]{4}}{\sqrt[3]{7}} = \dfrac{\sqrt[3]{4} \cdot \sqrt[3]{7^2}}{\sqrt[3]{7} \cdot \sqrt[3]{7^2}}$
$$= \frac{\sqrt[3]{4} \cdot \sqrt[3]{49}}{\sqrt[3]{7^3}} = \frac{\sqrt[3]{196}}{7}$$

49. To make the radicand in the denominator, $4y^2$, into a perfect cube, we must multiply 4 by 2 to get the perfect cube 8 and y^2 by y to get the perfect cube y^3. So we multiply the numerator and denominator by $\sqrt[3]{2y}$.

$$\sqrt[3]{\frac{3}{4y^2}} = \frac{\sqrt[3]{3}}{\sqrt[3]{4y^2}} = \frac{\sqrt[3]{3} \cdot \sqrt[3]{2y}}{\sqrt[3]{4y^2} \cdot \sqrt[3]{2y}}$$
$$= \frac{\sqrt[3]{6y}}{\sqrt[3]{8y^3}} = \frac{\sqrt[3]{6y}}{2y}$$

51. $\dfrac{\sqrt[3]{7m}}{\sqrt[3]{36n}} = \dfrac{\sqrt[3]{7m}}{\sqrt[3]{6^2n}}$
$$= \frac{\sqrt[3]{7m} \cdot \sqrt[3]{6n^2}}{\sqrt[3]{6^2n} \cdot \sqrt[3]{6n^2}}$$
$$= \frac{\sqrt[3]{42mn^2}}{\sqrt[3]{6^3n^3}} = \frac{\sqrt[3]{42mn^2}}{6n}$$

53. **(a)** $p = k \cdot \sqrt{\dfrac{L}{g}}$

$$p = 6 \cdot \sqrt{\frac{9}{32}} \qquad Let\ k = 6,\ L = 9,\ g = 32$$
$$= \frac{6\sqrt{9}}{\sqrt{32}}$$
$$= \frac{6 \cdot 3}{\sqrt{16 \cdot 2}}$$
$$= \frac{18}{4\sqrt{2}}$$
$$= \frac{9}{2\sqrt{2}}$$
$$= \frac{9 \cdot \sqrt{2}}{2\sqrt{2} \cdot \sqrt{2}} \qquad Rationalize$$
$$\qquad\qquad the\ denominator$$
$$= \frac{9\sqrt{2}}{4}$$

The period of the pendulum is $\dfrac{9\sqrt{2}}{4}$ seconds.

(b) Using a calculator, we obtain

$$\frac{9\sqrt{2}}{4} \approx 3.182 \text{ seconds.}$$

Section 10.5

1. $\sqrt{49} + \sqrt{36} = 13$
$$\left(\sqrt{49} + \sqrt{36} = 7 + 6\right)$$

3. $\sqrt{2} \cdot \sqrt{8} = 4$
$$\left(\sqrt{2} \cdot \sqrt{8} = \sqrt{16}\right)$$

5. $\sqrt{2}\left(\sqrt{32} - \sqrt{8}\right) = 4$
$$\left(\sqrt{64} - \sqrt{16} = 8 - 4\right)$$

7. $\sqrt[3]{8} + \sqrt[3]{27} = 5$
$$\left(\sqrt[3]{8} + \sqrt[3]{27} = 2 + 3\right)$$

9. $3\sqrt{5} + 2\sqrt{45} = 3\sqrt{5} + 2\sqrt{9 \cdot 5}$

$\qquad = 3\sqrt{5} + 2 \cdot \sqrt{9} \cdot \sqrt{5}$

$\qquad = 3\sqrt{5} + 2 \cdot 3 \cdot \sqrt{5}$

$\qquad = 3\sqrt{5} + 6\sqrt{5}$

$\qquad = 9\sqrt{5}$

11. $8\sqrt{50} - 4\sqrt{72}$

$\qquad = 8\left(\sqrt{25} \cdot \sqrt{2}\right) - 4\left(\sqrt{36} \cdot \sqrt{2}\right)$

$\qquad = 8\left(5\sqrt{2}\right) - 4\left(6\sqrt{2}\right)$

$\qquad = 40\sqrt{2} - 24\sqrt{2}$

$\qquad = 16\sqrt{2}$

13. $\sqrt{5}\left(\sqrt{3} - \sqrt{7}\right) = \sqrt{5} \cdot \sqrt{3} - \sqrt{5} \cdot \sqrt{7}$

$\qquad = \sqrt{15} - \sqrt{35}$

15. $2\sqrt{5}\left(\sqrt{2} + 3\sqrt{5}\right)$

$\qquad = 2\sqrt{5} \cdot \sqrt{2} + 2\sqrt{5} \cdot 3\sqrt{5}$

$\qquad = 2\sqrt{10} + 2 \cdot 3 \cdot \sqrt{5} \cdot \sqrt{5}$

$\qquad = 2\sqrt{10} + 6 \cdot 5$

$\qquad = 2\sqrt{10} + 30$

17. $3\sqrt{14} \cdot \sqrt{2} - \sqrt{28} = 3\sqrt{14 \cdot 2} - \sqrt{28}$

$\qquad = 3\sqrt{28} - \sqrt{28}$

$\qquad = 2\sqrt{28}$

$\qquad = 2\sqrt{4 \cdot 7}$

$\qquad = 2 \cdot \sqrt{4} \cdot \sqrt{7}$

$\qquad = 2 \cdot 2 \cdot \sqrt{7}$

$\qquad = 4\sqrt{7}$

19. $\left(2\sqrt{6} + 3\right)\left(3\sqrt{6} + 7\right)$

$\qquad = 2\sqrt{6} \cdot 3\sqrt{6} + 7 \cdot 2\sqrt{6} + 3 \cdot 3\sqrt{6}$

$\qquad + 3 \cdot 7 \quad FOIL$

$\qquad = 2 \cdot 3 \cdot \sqrt{6} \cdot \sqrt{6} + 14\sqrt{6} + 9\sqrt{6} + 21$

$\qquad = 6 \cdot 6 + 23\sqrt{6} + 21$

$\qquad = 36 + 23\sqrt{6} + 21$

$\qquad = 57 + 23\sqrt{6}$

21. $\left(5\sqrt{7} - 2\sqrt{3}\right)\left(3\sqrt{7} + 4\sqrt{3}\right)$

$\qquad = 5\sqrt{7}\left(3\sqrt{7}\right) + 5\sqrt{7}\left(4\sqrt{3}\right)$

$\qquad - 2\sqrt{3}\left(3\sqrt{7}\right) - 2\sqrt{3}\left(4\sqrt{3}\right)$

$\qquad = 15 \cdot 7 + 20\sqrt{21} - 6\sqrt{21} - 8 \cdot 3$

$\qquad = 105 + 14\sqrt{21} - 24$

$\qquad = 81 + 14\sqrt{21}$

23. $\left(2\sqrt{7} + 3\right)^2$

$\qquad = \left(2\sqrt{7}\right)^2 + 2\left(2\sqrt{7}\right)(3) + (3)^2$

$\qquad\qquad Square\ of\ a\ binomial$

$\qquad = 4 \cdot 7 + 12\sqrt{7} + 9$

$\qquad = 28 + 12\sqrt{7} + 9$

$\qquad = 37 + 12\sqrt{7}$

25. $\left(5 - \sqrt{2}\right)\left(5 + \sqrt{2}\right) = (5)^2 - \left(\sqrt{2}\right)^2$

$\qquad\qquad Product\ of\ the\ sum\ and$

$\qquad\qquad difference\ of\ two\ terms$

$\qquad\qquad = 25 - 2 = 23$

27. $\left(\sqrt{8} - \sqrt{7}\right)\left(\sqrt{8} + \sqrt{7}\right)$

$\qquad = \left(\sqrt{8}\right)^2 - \left(\sqrt{7}\right)^2$

$\qquad\qquad Product\ of\ the\ sum\ and$

$\qquad\qquad difference\ of\ two\ terms$

$\qquad\qquad (Difference\ of\ two\ squares)$

$\qquad = 8 - 7 = 1$

29. $\left(\sqrt{2} + \sqrt{3}\right)\left(\sqrt{6} - \sqrt{2}\right)$

$\qquad = \sqrt{2}\left(\sqrt{6}\right) - \sqrt{2}\left(\sqrt{2}\right) + \sqrt{3}\left(\sqrt{6}\right)$

$\qquad - \sqrt{3}\left(\sqrt{2}\right) \qquad FOIL$

$\qquad = \sqrt{12} - 2 + \sqrt{18} - \sqrt{6} \quad Product\ rule$

$\qquad = \sqrt{4} \cdot \sqrt{3} - 2 + \sqrt{9} \cdot \sqrt{2} - \sqrt{6}$

$\qquad = 2\sqrt{3} - 2 + 3\sqrt{2} - \sqrt{6}$

31. $\left(\sqrt{10} - \sqrt{5}\right)\left(\sqrt{5} + \sqrt{20}\right)$

$\qquad = \sqrt{10} \cdot \sqrt{5} + \sqrt{10} \cdot \sqrt{20} - \sqrt{5} \cdot \sqrt{5}$

$\qquad - \sqrt{5} \cdot \sqrt{20} \qquad FOIL$

$\qquad = \sqrt{50} + \sqrt{200} - 5 - \sqrt{100}$

$\qquad = \sqrt{25 \cdot 2} + \sqrt{100 \cdot 2} - 5 - 10$

$\qquad = 5\sqrt{2} + 10\sqrt{2} - 15$

$\qquad = 15\sqrt{2} - 15$

33. $\left(5\sqrt{7} - 2\sqrt{3}\right)\left(3\sqrt{7} + 3\sqrt{3}\right)$

$\qquad = 5\sqrt{7}\left(3\sqrt{7}\right) + 5\sqrt{7}\left(3\sqrt{3}\right)$

$\qquad - 2\sqrt{3}\left(3\sqrt{7}\right) - 2\sqrt{3}\left(3\sqrt{3}\right)$

$\qquad = (15)(7) + 15\sqrt{21} - 6\sqrt{21} - 6(3)$

$\qquad = 105 + 9\sqrt{21} - 18$

$\qquad = 87 + 9\sqrt{21}$

35. "-2" is the coefficient of $\sqrt{15}$ and multiplication must be performed before addition. Since $-2\sqrt{15}$ cannot be simplified, the expression cannot be written in a simpler form.

37. $\dfrac{1}{3+\sqrt{2}} = \dfrac{1\left(3-\sqrt{2}\right)}{\left(3+\sqrt{2}\right)\left(3-\sqrt{2}\right)}$

Multiply numerator and denominator by the conjugate of the denominator

$= \dfrac{3-\sqrt{2}}{3^2 - \left(\sqrt{2}\right)^2}$

$(a+b)(a-b) = a^2 - b^2$

$= \dfrac{3-\sqrt{2}}{9-2} = \dfrac{3-\sqrt{2}}{7}$

39. $\dfrac{14}{2-\sqrt{11}} = \dfrac{14\left(2+\sqrt{11}\right)}{\left(2-\sqrt{11}\right)\left(2+\sqrt{11}\right)}$

Multiply numerator and denominator by the conjugate of the denominator

$= \dfrac{14\left(2+\sqrt{11}\right)}{(2)^2 - \left(\sqrt{11}\right)^2}$

$(a+b)(a-b) = a^2 - b^2$

$= \dfrac{14\left(2+\sqrt{11}\right)}{4-11}$

$= \dfrac{14\left(2+\sqrt{11}\right)}{-7}$

$= -2\left(2+\sqrt{11}\right)$

$= -4 - 2\sqrt{11}$

41. $\dfrac{\sqrt{2}}{2-\sqrt{2}} = \dfrac{\sqrt{2}\left(2+\sqrt{2}\right)}{\left(2-\sqrt{2}\right)\left(2+\sqrt{2}\right)}$ *Multiply by the conjugate*

$= \dfrac{2\sqrt{2}+2}{2^2 - \left(\sqrt{2}\right)^2}$

$= \dfrac{2\sqrt{2}+2}{4-2} = \dfrac{2\sqrt{2}+2}{2}$

$= \dfrac{2\left(\sqrt{2}+1\right)}{2}$ *Factor out 2*

$= \sqrt{2}+1 \text{ or } 1+\sqrt{2}$

43. $\dfrac{\sqrt{5}}{\sqrt{2}+\sqrt{3}} = \dfrac{\sqrt{5}\left(\sqrt{2}-\sqrt{3}\right)}{\left(\sqrt{2}+\sqrt{3}\right)\left(\sqrt{2}-\sqrt{3}\right)}$

Multiply by the conjugate

$= \dfrac{\sqrt{5}\cdot\sqrt{2} - \sqrt{5}\cdot\sqrt{3}}{\left(\sqrt{2}\right)^2 - \left(\sqrt{3}\right)^2}$

$= \dfrac{\sqrt{10}-\sqrt{15}}{2-3}$

$= \dfrac{\sqrt{10}-\sqrt{15}}{-1} = -\sqrt{10}+\sqrt{15}$

45. $\dfrac{\sqrt{12}}{\sqrt{3}+1} = \dfrac{\sqrt{12}\left(\sqrt{3}-1\right)}{\left(\sqrt{3}+1\right)\left(\sqrt{3}-1\right)}$

$= \dfrac{\sqrt{36}-\sqrt{12}}{\left(\sqrt{3}\right)^2 - 1^2} = \dfrac{6-\sqrt{12}}{3-1}$

$= \dfrac{6-\sqrt{12}}{2} = \dfrac{6-\sqrt{4\cdot3}}{2}$

$= \dfrac{6-\sqrt{4}\cdot\sqrt{3}}{2} = \dfrac{6-2\sqrt{3}}{2}$

$= \dfrac{2\left(3-\sqrt{3}\right)}{2} = 3-\sqrt{3}$

47. $\dfrac{\sqrt{5}+2}{2-\sqrt{3}} = \dfrac{\left(\sqrt{5}+2\right)\left(2+\sqrt{3}\right)}{\left(2-\sqrt{3}\right)\left(2+\sqrt{3}\right)}$

$= \dfrac{2\sqrt{5}+\sqrt{5}\cdot\sqrt{3}+4+2\sqrt{3}}{(2)^2 - \left(\sqrt{3}\right)^2}$

$= \dfrac{2\sqrt{5}+\sqrt{15}+4+2\sqrt{3}}{4-3}$

$= \dfrac{2\sqrt{5}+\sqrt{15}+4+2\sqrt{3}}{1}$

$= 2\sqrt{5}+\sqrt{15}+4+2\sqrt{3}$

49. $\dfrac{6\sqrt{11}-12}{6} = \dfrac{6\left(\sqrt{11}-2\right)}{6}$ *Factor numerator*

$= \sqrt{11}-2$ *Lowest terms*

51. $\dfrac{2\sqrt{3}+10}{16} = \dfrac{2\left(\sqrt{3}+5\right)}{2\cdot8}$ *Factor numerator and denominator*

$= \dfrac{\sqrt{3}+5}{8}$ *Lowest terms*

53. $\dfrac{12-\sqrt{40}}{4} = \dfrac{12-\sqrt{4}\cdot\sqrt{10}}{4}$

$= \dfrac{12-2\sqrt{10}}{4}$

$= \dfrac{2\left(6-\sqrt{10}\right)}{2\cdot2} = \dfrac{6-\sqrt{10}}{2}$

55. $\left(\sqrt{5x}+\sqrt{30}\right)\left(\sqrt{6x}+\sqrt{3}\right)$

$= \sqrt{5x}\cdot\sqrt{6x}+\sqrt{5x}\cdot\sqrt{3}$

$\quad + \sqrt{30}\cdot\sqrt{6x}+\sqrt{30}\cdot\sqrt{3}$ *FOIL*

$= \sqrt{30x^2}+\sqrt{15x}+\sqrt{180x}+\sqrt{90}$

$= \sqrt{30\cdot x^2}+\sqrt{15x}+\sqrt{36\cdot5x}+\sqrt{9\cdot10}$

$= x\sqrt{30}+\sqrt{15x}+6\sqrt{5x}+3\sqrt{10}$

57. $\left(3\sqrt{t}+\sqrt{7}\right)\left(2\sqrt{t}-\sqrt{14}\right)$

$= 3\sqrt{t}\cdot 2\sqrt{t}-3\sqrt{t}\cdot\sqrt{14}$

$\quad + 2\sqrt{t}\cdot\sqrt{7}-\sqrt{7}\cdot\sqrt{14}$ *FOIL*

$= 3\cdot 2\cdot\sqrt{t}\cdot\sqrt{t}-3\sqrt{14t}+2\sqrt{7t}-\sqrt{98}$

$= 6t-3\sqrt{14t}+2\sqrt{7t}-\sqrt{49}\cdot\sqrt{2}$

$= 6t-3\sqrt{14t}+2\sqrt{7t}-7\sqrt{2}$

59. $\left(\sqrt{3m}+\sqrt{2n}\right)\left(\sqrt{5m}-\sqrt{5n}\right)$

$= \sqrt{3m}\cdot\sqrt{5m}+\sqrt{3m}\left(-\sqrt{5n}\right)$

$\quad + \sqrt{2n}\cdot\sqrt{5m}+\sqrt{2n}\left(-\sqrt{5n}\right)$ *FOIL*

$= \sqrt{15m^2}-\sqrt{15mn}+\sqrt{10mn}-\sqrt{10n^2}$

$= m\sqrt{15}-\sqrt{15mn}+\sqrt{10mn}-n\sqrt{10}$

61. $\sqrt[3]{4}\left(\sqrt[3]{2}-3\right)$

$= \sqrt[3]{4}\left(\sqrt[3]{2}\right)+\sqrt[3]{4}(-3)$ *Distributive property*

$= \sqrt[3]{8}-3\sqrt[3]{4}$ *Product rule*

$= 2-3\sqrt[3]{4}$ $\sqrt[3]{8}=2$

63. $2\sqrt[4]{2}\left(3\sqrt[4]{8}+5\sqrt[4]{4}\right)$

$= 2\cdot 3\cdot\sqrt[4]{2}\cdot\sqrt[4]{8}+2\cdot 5\cdot\sqrt[4]{2}\cdot\sqrt[4]{4}$

$\qquad$ *Distributive property*

$= 6\sqrt[4]{16}+10\sqrt[4]{8}$ *Product rule*

$= 6\cdot 2+10\sqrt[4]{8}$ $\sqrt[4]{16}=2$

$= 12+10\sqrt[4]{8}$

65. $\left(\sqrt[3]{2}-1\right)\left(\sqrt[3]{4}+3\right)$

$= \sqrt[3]{8}+3\sqrt[3]{2}-\sqrt[3]{4}-3$ *FOIL*

$= 2+3\sqrt[3]{2}-\sqrt[3]{4}-3$

$= -1+3\sqrt[3]{2}-\sqrt[3]{4}$

67. $\left(\sqrt[3]{5}-\sqrt[3]{4}\right)\left(\sqrt[3]{25}+\sqrt[3]{20}+\sqrt[3]{16}\right)$

$= \sqrt[3]{5}\left(\sqrt[3]{25}+\sqrt[3]{20}+\sqrt[3]{16}\right)$

$\quad - \sqrt[3]{4}\left(\sqrt[3]{25}+\sqrt[3]{20}+\sqrt[3]{16}\right)$

$\qquad$ *Distributive property*

$= \sqrt[3]{5}\cdot\sqrt[3]{25}+\sqrt[3]{5}\cdot\sqrt[3]{20}+\sqrt[3]{5}\cdot\sqrt[3]{16}$

$\quad - \sqrt[3]{4}\cdot\sqrt[3]{25}-\sqrt[3]{4}\cdot\sqrt[3]{20}-\sqrt[3]{4}\cdot\sqrt[3]{16}$

$\qquad$ *Distributive property*

$= \sqrt[3]{125}+\sqrt[3]{100}+\sqrt[3]{80}-\sqrt[3]{100}$

$\quad - \sqrt[3]{80}-\sqrt[3]{64}$ *Product rule*

$= \sqrt[3]{125}-\sqrt[3]{64}$

$= 5-4=1$

69. $6(5+3x) = (6)(5)+(6)(3x)$

$\qquad\qquad = 30+18x$

70. 30 and $18x$ cannot be combined because they are not like terms.

71. $\left(2\sqrt{10}+5\sqrt{2}\right)\left(3\sqrt{10}-3\sqrt{2}\right)$

$= 2\sqrt{10}\left(3\sqrt{10}\right)+2\sqrt{10}\left(-3\sqrt{2}\right)$

$\quad + 5\sqrt{2}\left(3\sqrt{10}\right)+5\sqrt{2}\left(-3\sqrt{2}\right)$ *FOIL*

$= 6\cdot 10-6\sqrt{20}+15\sqrt{20}-15\cdot 2$

$= 60+9\sqrt{20}-30$

$= 30+9\sqrt{4}\cdot\sqrt{5}$

$= 30+9\left(2\sqrt{5}\right)$

$= 30+18\sqrt{5}$

72. 30 and $18\sqrt{5}$ cannot be combined because they are not like radicals.

73. In the expression $30+18x$, make the first term $30x$, so that

$$30x+18x = 48x.$$

In the expression $30+18\sqrt{5}$, make the first term $30\sqrt{5}$, so that

$$30\sqrt{5}+18\sqrt{5} = 48\sqrt{5}.$$

74. One similarity between combining like terms and combining like radicals is that you combine them by adding or subtracting their coefficients. In both situations, the expression that is common to all terms is not changed in the result. When we combine like terms or like radicals, we are using the distributive property, although we do not usually write the step which shows this.

75. $r = \dfrac{-h+\sqrt{h^2+.64S}}{2}$

Substitute 12 for h and 400 for S.

$r = \dfrac{-12+\sqrt{12^2+.64(400)}}{2}$

$= \dfrac{-12+\sqrt{144+256}}{2}$

$= \dfrac{-12+\sqrt{400}}{2}$

$= \dfrac{-12+20}{2}$

$= \dfrac{8}{2}=4$

The radius should be 4 inches.

Section 10.6

1. $\sqrt{x}=7$

Use the *squaring property of equality* to square each side of the equation.

$$\left(\sqrt{x}\right)^2 = 7^2$$

$$x = 49$$

Now check this proposed solution in the original equation.

$$\text{Check } x = 49$$
$$\sqrt{x} = 7$$
$$\sqrt{49} = 7 \text{ ? } \quad \text{Let } x = 49$$
$$7 = 7 \quad \text{True}$$

Since this statement is true, the solution set of the original equation is $\{49\}$.

3.
$$\sqrt{y+2} = 3$$
$$\left(\sqrt{y+2}\right)^2 = 3^2 \quad \text{Square each side}$$
$$y + 2 = 9$$
$$y = 7$$

$$\text{Check } y = 7$$
$$\sqrt{y+2} = 3$$
$$\sqrt{7+2} = 3 \text{ ? } \quad \text{Let } y = 7$$
$$\sqrt{9} = 3 \text{ ?}$$
$$3 = 3 \quad \text{True}$$

Since this statement is true, the solution set of the original equation is $\{7\}$.

5.
$$\sqrt{r-4} = 9$$
$$\left(\sqrt{r-4}\right)^2 = 9^2 \quad \text{Square each side}$$
$$r - 4 = 81$$
$$r = 85$$

$$\text{Check } r = 85$$
$$\sqrt{r-4} = 9$$
$$\sqrt{85-4} = 9 \text{ ? } \quad \text{Let } r = 85$$
$$\sqrt{81} = 9 \text{ ?}$$
$$9 = 9 \quad \text{True}$$

Since this statement is true, the solution set of the original equation is $\{85\}$.

7.
$$\sqrt{4-t} = 7$$
$$\left(\sqrt{4-t}\right)^2 = 7^2 \quad \text{Square each side}$$
$$4 - t = 49$$
$$-t = 45$$
$$t = -45$$

$$\text{Check } t = -45$$
$$\sqrt{4-t} = 7$$
$$\sqrt{4-(-45)} = 7 \text{ ? } \quad \text{Let } t = -45$$
$$\sqrt{49} = 7 \text{ ?}$$
$$7 = 7 \quad \text{True}$$

Since this statement is true, the solution set of the original equation is $\{-45\}$.

9.
$$\sqrt{2t+3} = 0$$
$$\left(\sqrt{2t+3}\right)^2 = 0^2 \quad \text{Square each side}$$
$$2t + 3 = 0$$
$$2t = -3$$
$$t = -\frac{3}{2}$$

$$\text{Check } t = -\frac{3}{2}$$
$$\sqrt{2t+3} = 0$$
$$\sqrt{2\left(-\frac{3}{2}\right)+3} = 0 \text{ ? } \quad \text{Let } t = -\frac{3}{2}$$
$$\sqrt{-3+3} = 0 \text{ ?}$$
$$\sqrt{0} = 0 \text{ ?}$$
$$0 = 0 \quad \text{True}$$

Since this statement is true, the solution set of the original equation is $\left\{-\frac{3}{2}\right\}$.

11.
$$\sqrt{3x-8} = -2$$
$$\left(\sqrt{3x-8}\right)^2 = (-2)^2 \quad \text{Square each side}$$
$$3x - 8 = 4$$
$$3x = 12$$
$$x = 4$$

$$\text{Check } x = 4$$
$$\sqrt{3x-8} = -2$$
$$\sqrt{3(4)-8} = -2 \text{ ? } \quad \text{Let } x = 4$$
$$\sqrt{12-8} = -2 \text{ ?}$$
$$\sqrt{4} = -2 \text{ ?}$$
$$2 = -2 \quad \text{False}$$

Since this statement is false, the solution set of the original equation is $\emptyset$.

13.
$$\sqrt{m} - 4 = 7$$

Add 4 to both sides of the equation before squaring.

$$\sqrt{m} = 11$$
$$\left(\sqrt{m}\right)^2 = (11)^2$$
$$m = 121$$

$$\text{Check } m = 121$$
$$\sqrt{m} - 4 = 7$$
$$\sqrt{121} - 4 = 7 \text{ ? } \quad \text{Let } m = 121$$
$$11 - 4 = 7 \text{ ?}$$
$$7 = 7 \quad \text{True}$$

Since this statement is true, the solution set of the original equation is $\{121\}$.

15. $\sqrt{10x - 8} = 3\sqrt{x}$

$\left(\sqrt{10x - 8}\right)^2 = \left(3\sqrt{x}\right)^2$ *Square sides*

$10x - 8 = (3)^2\left(\sqrt{x}\right)^2$ *$(ab)^2 = a^2b^2$*

$10x - 8 = 9x$

$x = 8$

Check x = 8

$\sqrt{10x - 8} = 3\sqrt{x}$

$\sqrt{10(8) - 8} = 3\sqrt{8}$?   *Let x = 8*

$\sqrt{72} = 3\sqrt{8}$?

$\sqrt{36 \cdot 2} = 3 \cdot 2\sqrt{2}$?

$6\sqrt{2} = 6\sqrt{2}$ *True*

Since this statement is true, the solution set of the original equation is $\{8\}$.

17. $5\sqrt{x} = \sqrt{10x + 15}$

$\left(5\sqrt{x}\right)^2 = \left(\sqrt{10x + 15}\right)^2$

$25x = 10x + 15$

$15x = 15$

$x = 1$

Check x = 1

$5\sqrt{x} = \sqrt{10x + 15}$

$5\sqrt{1} = \sqrt{10 \cdot 1 + 15}$?   *Let x = 1*

$5 \cdot 1 = \sqrt{25}$?

$5 = 5$ *True*

Since this statement is true, the solution set of the original equation is $\{1\}$.

19. $\sqrt{3x - 5} = \sqrt{2x + 1}$

$\left(\sqrt{3x - 5}\right)^2 = \left(\sqrt{2x + 1}\right)^2$

$3x - 5 = 2x + 1$

$x = 6$

Check x = 6

$\sqrt{3x - 5} = \sqrt{2x + 1}$

$\sqrt{3(6) - 5} = \sqrt{2(6) + 1}$? *Let x = 6*

$\sqrt{13} = \sqrt{13}$ *True*

Since this statement is true, the solution set of the original equation is $\{6\}$.

21. $k = \sqrt{k^2 - 5k - 15}$

$(k)^2 = \left(\sqrt{k^2 - 5k - 15}\right)^2$

$k^2 = k^2 - 5k - 15$

$0 = -5k - 15$

$5k = -15$

$k = -3$

Check k = -3

$k = \sqrt{k^2 - 5k - 15}$

$-3 = \sqrt{(-3)^2 - 5(-3) - 15}$?

Let k = -3

$-3 = \sqrt{9 + 15 - 15}$?

$-3 = \sqrt{9}$?

$-3 = 3$ *False*

Since this statement is false, the solution set of the original equation is $\emptyset$.

23. $7x = \sqrt{49x^2 + 2x - 10}$

$(7x)^2 = \left(\sqrt{49x^2 + 2x - 10}\right)^2$

$49x^2 = 49x^2 + 2x - 10$

$0 = 2x - 10$

$10 = 2x$

$5 = x$

Check x = 5

$7x = \sqrt{49x^2 + 2x - 10}$

$7(5) = \sqrt{49(5)^2 + 2(5) - 10}$?

Let x = 5

$35 = \sqrt{1225 + 10 - 10}$?

$35 = \sqrt{1225}$?

$35 = 35$ *True*

Since this statement is true, the solution set of the original equation is $\{5\}$.

25. Since $\sqrt{x}$ must be greater than or equal to zero for any replacement for x, it cannot equal -8, a negative number.

27. Arrange the equation so that one radical is alone on one side of the equation; then square both sides. Combine like terms. If there is still a radical term, repeat these steps. Solve the equation, and check all potential solutions in the original equation.

29. $\sqrt{2x+1} = x - 7$

$\left(\sqrt{2x+1}\right)^2 = (x-7)^2$ *Square each side*

$2x + 1 = x^2 - 14x + 49$ *Square binomial on the right*

$0 = x^2 - 16x + 48$ *Set equal to 0*

$0 = (x-4)(x-12)$ *Factor*

$x - 4 = 0$ or $x - 12 = 0$

Set each factor equal to 0

$x = 4$ or $x = 12$ *Solve*

Both of the potential solutions must be checked in the original equation.

Check $x = 4$

$\sqrt{2x+1} = x - 7$

$\sqrt{2(4)+1} = 4 - 7$? *Let x = 4*

$\sqrt{9} = -3$?

$3 = -3$ *False*

Check $x = 12$

$\sqrt{2x+1} = x - 7$

$\sqrt{2(12)+1} = 12 - 7$? *Let x = 12*

$\sqrt{25} = 5$?

$5 = 5$ *True*

Of the two potential solutions, 12 checks in the original equation, but 4 does not. Thus, the solution set is $\{12\}$.

31. $\sqrt{3k+10} + 5 = 2k$

$\sqrt{3k+10} = 2k - 5$ *Isolate the radical*

$\left(\sqrt{3k+10}\right)^2 = (2k-5)^2$ *Square each side*

$3k + 10 = 4k^2 - 20k + 25$

$0 = 4k^2 - 23k + 15$

$0 = (k-5)(4k-3)$

$k = 5$ or $k = \dfrac{3}{4}$

Check $k = 5$

$\sqrt{3k+10} + 5 = 2k$

$\sqrt{3 \cdot 5 + 10} + 5 = 2 \cdot 5$? *Let k = 5*

$\sqrt{25} + 5 = 10$?

$5 + 5 = 10$?

$10 = 10$ *True*

Check $k = \dfrac{3}{4}$

$\sqrt{3k+10} + 5 = 2k$

$\sqrt{3\left(\dfrac{3}{4}\right) + 10} + 5 = 2\left(\dfrac{3}{4}\right)$? *Let k = $\dfrac{3}{4}$*

$\sqrt{\dfrac{9}{4} + 10} + 5 = \dfrac{3}{2}$?

$\sqrt{\dfrac{49}{4}} + 5 = \dfrac{3}{2}$?

$\dfrac{7}{2} + 5 = \dfrac{3}{2}$?

$\dfrac{17}{2} = \dfrac{3}{2}$ *False*

Of the two potential solutions, 5 checks in the original equation, but $\dfrac{3}{4}$ does not. Thus, the solution set is $\{5\}$.

33. $\sqrt{5x+1} - 1 = x$

$\sqrt{5x+1} = x + 1$ *Isolate the radical*

$\left(\sqrt{5x+1}\right)^2 = (x+1)^2$

$5x + 1 = x^2 + 2x + 1$

$0 = x^2 - 3x$

$0 = x(x-3)$

$x = 0$ or $x = 3$

Check $x = 0$

$\sqrt{5x+1} - 1 = x$

$\sqrt{5(0)+1} - 1 = 0$? *Let x = 0*

$\sqrt{1} - 1 = 0$?

$1 - 1 = 0$?

$0 = 0$ *True*

Check $x = 3$

$\sqrt{5x+1} - 1 = x$

$\sqrt{5(3)+1} - 1 = 3$? *Let x = 3*

$\sqrt{16} - 1 = 3$?

$4 - 1 = 3$?

$3 = 3$ *True*

Both potential solutions check, so the solution set is $\{0, 3\}$.

35. $\sqrt{6t+7}+3=t+5$

$\sqrt{6t+7}=t+2$ *Isolate the radical*

$\left(\sqrt{6t+7}\right)^2=(t+2)^2$

$6t+7=t^2+4t+4$

$0=t^2-2t-3$

$0=(t-3)(t+1)$

$t=3$ or $t=-1$

Check $t=3$

$\sqrt{6t+7}+3=t+5$

$\sqrt{6\cdot3+7}+3=3+5$? *Let t = 3*

$\sqrt{25}+3=8$?

$5+3=8$?

$8=8$ *True*

Check $t=-1$

$\sqrt{6t+7}+3=t+5$

$\sqrt{6(-1)+7}+3=-1+5$? *Let t = –1*

$\sqrt{1}+3=4$?

$1+3=4$?

$4=4$ *True*

Both potential solutions check, so the solution set is $\{-1,3\}$.

37. $x-4-\sqrt{2x}=0$

$x-4=\sqrt{2x}$ *Isolate the radical*

$(x-4)^2=\left(\sqrt{2x}\right)^2$

$x^2-8x+16=2x$

$x^2-10x+16=0$

$(x-8)(x-2)=0$

$x=8$ or $x=2$

Check $x=8$

$x-4-\sqrt{2x}=0$

$8-4-\sqrt{2(8)}=0$?   *Let x = 8*

$4-\sqrt{16}=0$?

$4-4=0$?

$0=0$ *True*

Check $x=2$

$x-4-\sqrt{2x}=0$

$2-4-\sqrt{2(2)}=0$?   *Let x = 2*

$-2-\sqrt{4}=0$?

$-2-2=0$?

$-4=0$ *False*

Of the two potential solutions, 8 checks in the original equation, but 2 does not. Thus, the solution set is $\{8\}$.

39. $\sqrt{x}+6=2x$

$\sqrt{x}=2x-6$

$\left(\sqrt{x}\right)^2=(2x-6)^2$

$x=4x^2-24x+36$

$0=4x^2-25x+36$

$0=(4x-9)(x-4)$

$x=\dfrac{9}{4}$ or $x=4$

Check $x=\dfrac{9}{4}$

$\sqrt{x}+6=2x$

$\sqrt{\dfrac{9}{4}}+6=2\cdot\dfrac{9}{4}$? *Let x = $\dfrac{9}{4}$*

$\dfrac{3}{2}+6=\dfrac{9}{2}$?

$\dfrac{15}{2}=\dfrac{9}{2}$ *False*

Check $x=4$

$\sqrt{x}+6=2x$

$\sqrt{4}+6=2\cdot4$?   *Let x = 4*

$2+6=8$?

$8=8$ *True*

Of the two potential solutions, 4 checks in the original equation, but $\dfrac{9}{4}$ does not. Thus, the solution set is $\{4\}$.

41. $\sqrt{x+1}-\sqrt{x-4}=1$

Rewrite the equation so that there is one radical on each side.

$\sqrt{x+1}=1+\sqrt{x-4}$

Square both sides. On the right-hand side, use the formula for the square of a binomial.

$\left(\sqrt{x+1}\right)^2=\left(1+\sqrt{x-4}\right)^2$

$x+1=1^2+2\cdot1\cdot\sqrt{x-4}+\left(\sqrt{x-4}\right)^2$

$x+1=1+2\sqrt{x-4}+x-4$

$x+1=2\sqrt{x-4}+x-3$

$4=2\sqrt{x-4}$

$2=\sqrt{x-4}$

We still have a radical on the right, so we must square both sides again.

$$(2)^2 = \left(\sqrt{x-4}\right)^2$$

$$4 = x - 4$$

$$8 = x$$

Check $x = 8$

$$\sqrt{x+1} - \sqrt{x-4} = 1$$

$$\sqrt{8+1} - \sqrt{8-4} = 1 ? \quad \textit{Let } x = 8$$

$$\sqrt{9} - \sqrt{4} = 1 ?$$

$$3 - 2 = 1 ?$$

$$1 = 1 \quad \textit{True}$$

The solution set is $\{8\}$.

43. $\quad \sqrt{x} = \sqrt{x-5} + 1$

$$\left(\sqrt{x}\right)^2 = \left(\sqrt{x-5}+1\right)^2 \quad \begin{array}{l}\textit{Square}\\\textit{both sides}\end{array}$$

$$x = x - 5 + 2\sqrt{x-5} + 1$$

$(a+b)^2 = a^2 + 2ab + b^2$

$$4 = 2\sqrt{x-5} \qquad \textit{Combine terms}$$

$$2 = \sqrt{x-5} \qquad \textit{Divide by 2}$$

Since we still have a radical, square both sides a second time.

$$(2)^2 = \left(\sqrt{x-5}\right)^2$$

$$4 = x - 5$$

$$9 = x$$

Check $x = 9$

$$\sqrt{x} = \sqrt{x-5} + 1$$

$$\sqrt{9} = \sqrt{9-5} + 1 ? \quad \textit{Let } x = 9$$

$$3 = \sqrt{4} + 1 \quad ?$$

$$3 = 2 + 1 \quad ?$$

$$3 = 3 \qquad \textit{True}$$

The solution set is $\{9\}$.

45. The error occurs in the first step. We cannot square term by term. The left side must be squared as a binomial, using the formula

$$(a+b)^2 = a^2 + 2ab + b^2.$$

47. Refer to the right triangle shown in the figure in the textbook. Note that the given distances are the lengths of the hypotenuse (193.0 feet) and one of the legs (110.0 feet) of the triangle. Use the Pythagorean formula with $a = 110.0$, $c = 193.0$, and $b =$ the height of the building.

$$a^2 + b^2 = c^2$$

$$(110.0)^2 + b^2 = (193.0)^2$$

$$12,100 + b^2 = 37,249$$

$$b^2 = 25,149$$

$$b \approx 158.6$$

The height of the building, to the nearest tenth, is 158.6 feet.

49. $\quad s = 30\sqrt{\dfrac{a}{p}}$

Use a calculator and round answers to the nearest tenth.

(a) $s = 30\sqrt{\dfrac{862}{156}} \quad \begin{array}{l}\textit{Let } a = 862\\\textit{and } p = 156\end{array}$

$$\approx 70.5 \text{ miles per hour}$$

(b) $s = 30\sqrt{\dfrac{382}{96}} \quad \begin{array}{l}\textit{Let } a = 382\\\textit{and } p = 96\end{array}$

$$\approx 59.8 \text{ miles per hour}$$

(c) $s = 30\sqrt{\dfrac{84}{26}} \quad \begin{array}{l}\textit{Let } a = 84\\\textit{and } p = 26\end{array}$

$$\approx 53.9 \text{ miles per hour}$$

51. $\quad y = \dfrac{1.76 + \sqrt{.0176 + .044x}}{.022}$

(a) Let $x = 1$ to determine the year when the number of calls reached 1 million.

$$y = \dfrac{1.76 + \sqrt{.0176 + .044(1)}}{.022}$$

$$= \dfrac{1.76 + \sqrt{.0616}}{.022}$$

$$\approx 91.28,$$

which represents 1991.

(b) If $y = f(x)$, then

$$f(3) = \dfrac{1.76 + \sqrt{.0176 + .044(3)}}{.022}$$

$$= \dfrac{1.76 + \sqrt{.1496}}{.022}$$

$$\approx 97.58,$$

which represents 1997.

Thus, 1997 is the year when about 3 million calls were made.

(c) To approximate the number of calls made in 1995, let $y = 95$ and solve for x.

$$95 = \dfrac{1.76 + \sqrt{.0176 + .044x}}{.022}$$

$$95(.022) = 1.76 + \sqrt{.0176 + .044x}$$

$$2.09 - 1.76 = \sqrt{.0176 + .044x}$$

$$(.33)^2 = \left(\sqrt{.0176 + .044x}\right)^2$$

$$.1089 = .0176 + .044x$$

$$.0913 = .044x$$

$$x = \dfrac{.0913}{.044} = 2.075$$

continued

Thus, in 1995 about 2.1 million calls were made (according to the equation).

53. From the graph, the solution of the equation is 4. *Check x = 4.*

$$\sqrt{5x - 4} = 4$$
$$\sqrt{5(4) - 4} = 4 \ ? \quad \textit{Let x = 4}$$
$$\sqrt{16} = 4 \ ?$$
$$4 = 4 \quad \textit{True}$$

55. From the graph, the solution of the equation is -2. *Check x = -2.*

$$\sqrt{1 - 3x} = \sqrt{5 - x}$$
$$\sqrt{1 - 3(-2)} = \sqrt{5 - (-2)} \ ? \quad \textit{Let x = -2}$$
$$\sqrt{7} = \sqrt{7} \quad \textit{True}$$

Note that $\sqrt{7} \approx 2.6457513$, the y-value shown in the screen.

57. The second point of intersection occurs where $x = -1$. *Check x = -1*

$$\sqrt{2x^2 + 4x + 3} = -x$$
$$\sqrt{2(-1)^2 + 4(-1) + 3} = -(-1) \ ? \quad \textit{Let x = -1}$$
$$\sqrt{1} = 1 \quad ?$$
$$1 = 1 \quad \textit{True}$$

Section 10.7

1. $49^{1/2}$ can be written as $49^{.5}$ or $\sqrt{49}$, which is 7 and not -7. Thus, all of the choices are equal to 7 except (a) -7, which is the answer.

3. $-64^{1/3}$ can be written as $-\sqrt[3]{64}$, which is -4. Also, $-\sqrt{16} = -4$. Thus, all of the choices are equal to -4 except (c) 4, which is the answer.

5. $25^{1/2} = \sqrt{25} = \sqrt{5^2} = 5$

7. $64^{1/3} = \sqrt[3]{64} = \sqrt[3]{4^3} = 4$

9. $16^{1/4} = \sqrt[4]{16} = \sqrt[4]{2^4} = 2$

11. $32^{1/5} = \sqrt[5]{32} = \sqrt[5]{2^5} = 2$

13. $4^{3/2} = \left(4^{1/2}\right)^3$
$$= \left(\sqrt{4}\right)^3 = 2^3 = 8$$

15. $27^{2/3} = \left(27^{1/3}\right)^2$
$$= \left(\sqrt[3]{27}\right)^2 = 3^2 = 9$$

17. $16^{3/4} = \left(16^{1/4}\right)^3$
$$= \left(\sqrt[4]{16}\right)^3 = 2^3 = 8$$

19. $32^{2/5} = \left(32^{1/5}\right)^2$
$$= \left(\sqrt[5]{32}\right)^2 = 2^2 = 4$$

21. $-8^{2/3} = -\left(8^{1/3}\right)^2$
$$= -\left(\sqrt[3]{8}\right)^2 = -2^2 = -4$$

23. $-64^{1/3} = -\sqrt[3]{64} = -4$

25. $49^{-3/2} = \dfrac{1}{49^{3/2}} = \dfrac{1}{\left(49^{1/2}\right)^3} = \dfrac{1}{\left(\sqrt{49}\right)^3}$
$$= \dfrac{1}{7^3} = \dfrac{1}{343}$$

27. $216^{-2/3} = \dfrac{1}{216^{2/3}} = \dfrac{1}{\left(216^{1/3}\right)^2} = \dfrac{1}{\left(\sqrt[3]{216}\right)^2}$
$$= \dfrac{1}{6^2} = \dfrac{1}{36}$$

29. $-16^{-5/4} = -\dfrac{1}{16^{5/4}} = -\dfrac{1}{\left(16^{1/4}\right)^5}$
$$= -\dfrac{1}{\left(\sqrt[4]{16}\right)^5} = -\dfrac{1}{2^5} = -\dfrac{1}{32}$$

31. $2^{1/2} \cdot 2^{5/2} = 2^{1/2 + 5/2} \quad \textit{Product rule}$
$$= 2^{6/2}$$
$$= 2^3$$

33. $6^{1/4} \cdot 6^{-3/4} = 6^{1/4 + (-3/4)} \quad \textit{Product rule}$
$$= 6^{-2/4}$$
$$= 6^{-1/2} = \dfrac{1}{6^{1/2}}$$

35. $\dfrac{15^{3/4}}{15^{5/4}} = 15^{3/4 - 5/4} \quad \textit{Quotient rule}$
$$= 15^{-2/4}$$
$$= 15^{-1/2}$$
$$= \dfrac{1}{15^{1/2}}$$

37. $\dfrac{11^{-2/7}}{11^{-3/7}} = 11^{-2/7 - (-3/7)} \quad \textit{Quotient rule}$
$$= 11^{1/7}$$

39. $\left(8^{3/2}\right)^2 = 8^{(3/2)(2)} \quad \textit{Power rule}$
$$= 8^3$$

41. $\left(6^{1/3}\right)^{3/2} = 6^{(1/3)(3/2)} \quad \textit{Power rule}$
$$= 6^{1/2}$$

43. $\left(\dfrac{25}{4}\right)^{3/2} = \dfrac{25^{3/2}}{4^{3/2}}$
$$= \dfrac{\left(25^{1/2}\right)^3}{\left(4^{1/2}\right)^3}$$
$$= \dfrac{\left(\sqrt{25}\right)^3}{\left(\sqrt{4}\right)^3}$$
$$= \dfrac{5^3}{2^3}$$

45. $\dfrac{2^{2/5} \cdot 2^{-3/5}}{2^{7/5}}$

$= \dfrac{2^{2/5}}{2^{3/5} \cdot 2^{7/5}}$

$= \dfrac{2^{2/5}}{2^{10/5}}$

$= \dfrac{1}{2^{8/5}}$

47. $\dfrac{6^{-2/9}}{6^{1/9} \cdot 6^{-5/9}}$

$= \dfrac{6^{5/9}}{6^{1/9} \cdot 6^{2/9}}$

$= \dfrac{6^{5/9}}{6^{3/9}}$

$= 6^{2/9}$

49. $\dfrac{z^{2/3}}{z^{-1/3}} = z^{2/3} \cdot z^{1/3}$

$= z^{2/3+1/3}$

$= z^{3/3} = z$

51. $\left(m^3 n^{1/4}\right)^{2/3} = \left(m^3\right)^{2/3}\left(n^{1/4}\right)^{2/3}$

$= m^{3 \cdot (2/3)} n^{(1/4) \cdot (2/3)}$

$= m^2 n^{1/6}$

53. $\left(\dfrac{a^{1/2}}{b^{1/3}}\right)^{4/3} = \dfrac{\left(a^{1/2}\right)^{4/3}}{\left(b^{1/3}\right)^{4/3}}$

$= \dfrac{a^{(1/2) \cdot (4/3)}}{b^{(1/3) \cdot (4/3)}}$

$= \dfrac{a^{2/3}}{b^{4/9}}$

55. $\sqrt[6]{4^3} = 4^{3/6} = 4^{1/2} = \sqrt{4} = 2$

57. $\sqrt[8]{16^2} = 16^{2/8} = 16^{1/4} = \sqrt[4]{16} = 2$

59. $\sqrt[4]{a^2} = a^{2/4} = a^{1/2} = \sqrt{a}$

61. $\sqrt[6]{k^4} = k^{4/6} = k^{2/3} = \sqrt[3]{k^2}$

63. $\sqrt{2} = 2^{1/2}$

$\sqrt[3]{2} = 2^{1/3}$

64. $\sqrt{2} \cdot \sqrt[3]{2} = 2^{1/2} \cdot 2^{1/3}$

65. The least common denominator of $\dfrac{1}{2}$ and $\dfrac{1}{3}$ is 6.

66. $\sqrt{2} \cdot \sqrt[3]{2} = 2^{3/6} \cdot 2^{2/6}$

67. $2^{3/6} \cdot 2^{2/6} = 2^{3/6+2/6} = 2^{5/6}$

68. $2^{5/6} = \sqrt[6]{2^5}$ or $\sqrt[6]{32}$

69. $\sqrt[6]{64} = 64^{1/6} = 2$

71. $\sqrt[7]{84} = 84^{1/7} \approx 1.883$

73. $\sqrt[5]{987} = 987^{1/5} \approx 3.971$

75. $\sqrt[4]{19^3} = 19^{3/4} \approx 9.100$

77. $\sqrt{(6)} \approx 2.449489743$

79. $1/((4\ ^{x}\sqrt{16})\char94 3) = .125$

81. (a) Using a fractional exponent, $d = 1.22\sqrt{x}$ can be written as

$$d = 1.22x^{1/2}.$$

(b) Let $x = 30,000$.

$$d = 1.22\sqrt{30,000} \approx 211.31$$

The distance is approximately 211.31 miles.

83. We want the laws of exponents to apply, so that

$$7^{1/2} \cdot 7^{1/2} = 7^1 = 7.$$

By definition of square root, we also know that

$$\sqrt{7} \cdot \sqrt{7} = 7.$$

Thus, it makes sense to define $7^{1/2}$ as $\sqrt{7}$.

Chapter 10 Review Exercises

1. The square roots of 49 are -7 and 7 because $(-7)^2 = 49$ and $7^2 = 49$.

2. The square roots of 81 are -9 and 9 because $(-9)^2 = 81$ and $9^2 = 81$.

3. The square roots of 196 are -14 and 14 because $(-14)^2 = 196$ and $14^2 = 196$.

4. The square roots of 121 are -11 and 11 because $(-11)^2 = 121$ and $11^2 = 121$.

5. The square roots of 225 are -15 and 15 because $(-15)^2 = 225$ and $15^2 = 225$.

6. The square roots of 729 are -27 and 27 because $(-27)^2 = 729$ and $27^2 = 729$.

7. $\sqrt{16} = 4$ because $4^2 = 16$.

8. $-\sqrt{36}$ represents the negative square root of 36. Since $6 \cdot 6 = 36$,

$$-\sqrt{36} = -6.$$

9. $\sqrt[3]{1000} = 10$ because $10^3 = 1000$.

10. $\sqrt[4]{81} = 3$ because 3 is positive and $3^4 = 81$.

11. $\sqrt{-8100}$ is not a real number.

12. $-\sqrt{4225}$ represents the negative square root of 4225. Since $65 \cdot 65 = 4225$,

$$-\sqrt{4225} = -65.$$

13. $\sqrt{\dfrac{49}{36}} = \dfrac{\sqrt{49}}{\sqrt{36}} = \dfrac{7}{6}$

14. $\sqrt{\dfrac{100}{81}} = \dfrac{\sqrt{100}}{\sqrt{81}} = \dfrac{10}{9}$

15. If $\sqrt{a}$ is not a real number, then a must be a negative number.

16. Use the Pythagorean formula with $a = 15$, $b = x$, and $c = 17$.

$$c^2 = a^2 + b^2$$
$$17^2 = 15^2 + x^2$$
$$289 = 225 + x^2$$
$$64 = x^2$$
$$x = \sqrt{64} = 8$$

17. $\sqrt{23}$

This number is *irrational* because 23 is not a perfect square.

$$\sqrt{23} \approx 4.796$$

18. $\sqrt{169}$

This number is *rational* because 169 is a perfect square.

$$\sqrt{169} = 13$$

19. $-\sqrt{25}$

This number is *rational* because 25 is a perfect square.

$$-\sqrt{25} = -5$$

20. $\sqrt{-4}$

This is not a real number.

21. $\sqrt{5} \cdot \sqrt{15} = \sqrt{5} \cdot \sqrt{5} \cdot \sqrt{3}$
$\qquad = \sqrt{25} \cdot \sqrt{3} = 5\sqrt{3}$

22. $-\sqrt{27} = -\sqrt{9 \cdot 3} = -\sqrt{9} \cdot \sqrt{3} = -3\sqrt{3}$

23. $\sqrt{160} = \sqrt{16 \cdot 10} = \sqrt{16} \cdot \sqrt{10} = 4\sqrt{10}$

24. $\sqrt[3]{-125} = -5$ because $(-5)^3 = -125$.

25. $\sqrt[3]{1728} = 12$ because $12^3 = 1728$.

26. $\sqrt{12} \cdot \sqrt{27} = \sqrt{4 \cdot 3} \cdot \sqrt{9 \cdot 3}$
$\qquad = 2\sqrt{3} \cdot 3\sqrt{3}$
$\qquad = 2 \cdot 3 \cdot \left(\sqrt{3}\right)^2$
$\qquad = 2 \cdot 3 \cdot 3 = 18$

27. $\sqrt{32} \cdot \sqrt{48} = \sqrt{16 \cdot 2} \cdot \sqrt{16 \cdot 3}$
$\qquad = 4\sqrt{2} \cdot 4\sqrt{3}$
$\qquad = 4 \cdot 4 \cdot \sqrt{2 \cdot 3}$
$\qquad = 16\sqrt{6}$

28. $\sqrt{50} \cdot \sqrt{125} = \sqrt{25 \cdot 2} \cdot \sqrt{25 \cdot 5}$
$\qquad = 5\sqrt{2} \cdot 5\sqrt{5}$
$\qquad = 5 \cdot 5 \cdot \sqrt{2 \cdot 5}$
$\qquad = 25\sqrt{10}$

29. $-\sqrt{\dfrac{121}{400}} = -\dfrac{\sqrt{121}}{\sqrt{400}} = -\dfrac{11}{20}$

30. $\sqrt{\dfrac{3}{49}} = \dfrac{\sqrt{3}}{\sqrt{49}} = \dfrac{\sqrt{3}}{7}$

31. $\sqrt{\dfrac{7}{169}} = \dfrac{\sqrt{7}}{\sqrt{169}} = \dfrac{\sqrt{7}}{13}$

32. $\sqrt{\dfrac{1}{6}} \cdot \sqrt{\dfrac{5}{6}} = \sqrt{\dfrac{1}{6} \cdot \dfrac{5}{6}}$
$\qquad = \sqrt{\dfrac{5}{36}}$
$\qquad = \dfrac{\sqrt{5}}{\sqrt{36}} = \dfrac{\sqrt{5}}{6}$

33. $\sqrt{\dfrac{2}{5}} \cdot \sqrt{\dfrac{2}{45}} = \sqrt{\dfrac{2}{5} \cdot \dfrac{2}{45}}$
$\qquad = \sqrt{\dfrac{4}{225}}$
$\qquad = \dfrac{\sqrt{4}}{\sqrt{225}} = \dfrac{2}{15}$

34. $\dfrac{3\sqrt{10}}{\sqrt{5}} = \dfrac{3 \cdot \sqrt{5}\sqrt{2}}{\sqrt{5}}$
$\qquad = 3\sqrt{2}$

35. $\dfrac{24\sqrt{12}}{6\sqrt{3}} = \dfrac{24 \cdot \sqrt{4} \cdot \sqrt{3}}{6\sqrt{3}}$
$\qquad = 4\sqrt{4} = 4 \cdot 2 = 8$

36. $\dfrac{8\sqrt{150}}{4\sqrt{75}} = \dfrac{8 \cdot \sqrt{75} \cdot \sqrt{2}}{4\sqrt{75}}$
$\qquad = 2\sqrt{2}$

37. $\sqrt{p} \cdot \sqrt{p} = p$

38. $\sqrt{k} \cdot \sqrt{m} = \sqrt{km}$

39. $\sqrt{r^{18}} = r^9$ because $\left(r^9\right)^2 = r^{18}$.

40. $\sqrt{x^{10}y^{16}} = x^5 y^8$ because $\left(x^5 y^8\right)^2 = x^{10}y^{16}$.

41. $\sqrt{a^{15}b^{21}} = \sqrt{a^{14}b^{20} \cdot ab}$
$\qquad = \sqrt{a^{14}b^{20}} \cdot \sqrt{ab}$
$\qquad = a^7 b^{10} \sqrt{ab}$

42. $\sqrt{121x^6 y^{10}} = 11x^3 y^5$ because
$$\left(11x^3 y^5\right)^2 = 121x^6 y^{10}.$$

43. Using a calculator,

$$\sqrt{.5} \approx .7071067812$$

and

$$\frac{\sqrt{2}}{2} \approx \frac{1.414213562}{2}$$

$$= .7071067812$$

It looks like these two expressions represent the same number. In fact, they do represent the same number because

$$\sqrt{.5} = \sqrt{\frac{5}{10}} = \sqrt{\frac{1}{2}} = \frac{\sqrt{1}}{\sqrt{2}}$$

$$= \frac{1 \cdot \sqrt{2}}{\sqrt{2} \cdot \sqrt{2}} = \frac{\sqrt{2}}{2}.$$

44. $3\sqrt{2} + 6\sqrt{2} = (3 + 6)\sqrt{2} = 9\sqrt{2}$

45. $3\sqrt{75} + 2\sqrt{27}$

$$= 3\left(\sqrt{25} \cdot \sqrt{3}\right) + 2\left(\sqrt{9} \cdot \sqrt{3}\right)$$

$$= 3\left(5\sqrt{3}\right) + 2\left(3\sqrt{3}\right)$$

$$= 15\sqrt{3} + 6\sqrt{3} = 21\sqrt{3}$$

46. $4\sqrt{12} + \sqrt{48}$

$$= 4\left(\sqrt{4} \cdot \sqrt{3}\right) + \sqrt{16} \cdot \sqrt{3}$$

$$= 4\left(2\sqrt{3}\right) + 4\sqrt{3}$$

$$= 8\sqrt{3} + 4\sqrt{3} = 12\sqrt{3}$$

47. $4\sqrt{24} - 3\sqrt{54} + \sqrt{6}$

$$= 4\left(\sqrt{4} \cdot \sqrt{6}\right) - 3\left(\sqrt{9} \cdot \sqrt{6}\right) + \sqrt{6}$$

$$= 4\left(2\sqrt{6}\right) - 3\left(3\sqrt{6}\right) + \sqrt{6}$$

$$= 8\sqrt{6} - 9\sqrt{6} + 1\sqrt{6}$$

$$= 0\sqrt{6} = 0$$

48. $2\sqrt{7} - 4\sqrt{28} + 3\sqrt{63}$

$$= 2\sqrt{7} - 4\left(\sqrt{4} \cdot \sqrt{7}\right) + 3\left(\sqrt{9} \cdot \sqrt{7}\right)$$

$$= 2\sqrt{7} - 4\left(2\sqrt{7}\right) + 3\left(3\sqrt{7}\right)$$

$$= 2\sqrt{7} - 8\sqrt{7} + 9\sqrt{7} = 3\sqrt{7}$$

49. $\frac{2}{5}\sqrt{75} + \frac{3}{4}\sqrt{160}$

$$= \frac{2}{5}\left(\sqrt{25} \cdot \sqrt{3}\right) + \frac{3}{4}\left(\sqrt{16} \cdot \sqrt{10}\right)$$

$$= \frac{2}{5}\left(5\sqrt{3}\right) + \frac{3}{4}\left(4\sqrt{10}\right)$$

$$= 2\sqrt{3} + 3\sqrt{10}$$

50. $\frac{1}{3}\sqrt{18} + \frac{1}{4}\sqrt{32}$

$$= \frac{1}{3}\left(\sqrt{9} \cdot \sqrt{2}\right) + \frac{1}{4}\left(\sqrt{16} \cdot \sqrt{2}\right)$$

$$= \frac{1}{3}\left(3\sqrt{2}\right) + \frac{1}{4}\left(4\sqrt{2}\right)$$

$$= 1\sqrt{2} + 1\sqrt{2} = 2\sqrt{2}$$

51. $\sqrt{15} \cdot \sqrt{2} + 5\sqrt{30} = \sqrt{30} + 5\sqrt{30}$

$$= 1\sqrt{30} + 5\sqrt{30}$$

$$= 6\sqrt{30}$$

52. $\sqrt{4x} + \sqrt{36x} - \sqrt{9x}$

$$= \sqrt{4}\sqrt{x} + \sqrt{36}\sqrt{x} - \sqrt{9}\sqrt{x}$$

$$= 2\sqrt{x} + 6\sqrt{x} - 3\sqrt{x} = 5\sqrt{x}$$

53. $\sqrt{16p} + 3\sqrt{p} - \sqrt{49p}$

$$= \sqrt{16}\sqrt{p} + 3\sqrt{p} - \sqrt{49}\sqrt{p}$$

$$= 4\sqrt{p} + 3\sqrt{p} - 7\sqrt{p}$$

$$= 0\sqrt{p} = 0$$

54. $\sqrt{20m^2} - m\sqrt{45}$

$$= \sqrt{4m^2 \cdot 5} - m\left(\sqrt{9} \cdot \sqrt{5}\right)$$

$$= \sqrt{4m^2} \cdot \sqrt{5} - m\left(3\sqrt{5}\right)$$

$$= 2m\sqrt{5} - 3m\sqrt{5} = -m\sqrt{5}$$

55. $3k\sqrt{8k^2 n} + 5k^2\sqrt{2n}$

$$= 3k\left(\sqrt{4k^2} \cdot \sqrt{2n}\right) + 5k^2\sqrt{2n}$$

$$= 3k\left(2k\sqrt{2n}\right) + 5k^2\sqrt{2n}$$

$$= 6k^2\sqrt{2n} + 5k^2\sqrt{2n}$$

$$= \left(6k^2 + 5k^2\right)\sqrt{2n}$$

$$= 11k^2\sqrt{2n}$$

56. $\frac{8\sqrt{2}}{\sqrt{5}} = \frac{8\sqrt{2} \cdot \sqrt{5}}{\sqrt{5} \cdot \sqrt{5}} = \frac{8\sqrt{10}}{5}$

57. $\frac{5}{\sqrt{5}} = \frac{5 \cdot \sqrt{5}}{\sqrt{5} \cdot \sqrt{5}} = \frac{5\sqrt{5}}{5} = \sqrt{5}$

58. $\frac{12}{\sqrt{24}} = \frac{12}{\sqrt{4 \cdot 6}} = \frac{12}{2\sqrt{6}}$

$$= \frac{12 \cdot \sqrt{6}}{2\sqrt{6} \cdot \sqrt{6}} = \frac{12\sqrt{6}}{2 \cdot 6}$$

$$= \frac{12\sqrt{6}}{12} = \sqrt{6}$$

59. $\frac{\sqrt{2}}{\sqrt{15}} = \frac{\sqrt{2} \cdot \sqrt{15}}{\sqrt{15} \cdot \sqrt{15}} = \frac{\sqrt{30}}{15}$

60. $\sqrt{\frac{2}{5}} = \frac{\sqrt{2}}{\sqrt{5}} = \frac{\sqrt{2} \cdot \sqrt{5}}{\sqrt{5} \cdot \sqrt{5}} = \frac{\sqrt{10}}{5}$

61. $\sqrt{\dfrac{5}{14}} \cdot \sqrt{28} = \sqrt{\dfrac{5}{14} \cdot 28}$

$= \sqrt{5 \cdot 2} = \sqrt{10}$

62. $\sqrt{\dfrac{2}{7}} \cdot \sqrt{\dfrac{1}{3}} = \sqrt{\dfrac{2}{7} \cdot \dfrac{1}{3}}$

$= \sqrt{\dfrac{2}{21}} = \dfrac{\sqrt{2}}{\sqrt{21}}$

$= \dfrac{\sqrt{2} \cdot \sqrt{21}}{\sqrt{21} \cdot \sqrt{21}} = \dfrac{\sqrt{42}}{21}$

63. $\sqrt{\dfrac{r^2}{16x}} = \dfrac{\sqrt{r^2}}{\sqrt{16x}}$

$= \dfrac{r \cdot \sqrt{x}}{\sqrt{16x} \cdot \sqrt{x}}$

$= \dfrac{r\sqrt{x}}{\sqrt{16x^2}} = \dfrac{r\sqrt{x}}{4x}$

64. $\sqrt[3]{\dfrac{1}{3}} = \dfrac{\sqrt[3]{1}}{\sqrt[3]{3}} = \dfrac{1 \cdot \sqrt[3]{3^2}}{\sqrt[3]{3} \cdot \sqrt[3]{3^2}}$

$= \dfrac{\sqrt[3]{3^2}}{\sqrt[3]{3^3}} = \dfrac{\sqrt[3]{9}}{3}$

65. $\sqrt[3]{\dfrac{2}{7}} = \dfrac{\sqrt[3]{2}}{\sqrt[3]{7}} = \dfrac{\sqrt[3]{2} \cdot \sqrt[3]{7^2}}{\sqrt[3]{7} \cdot \sqrt[3]{7^2}}$

$= \dfrac{\sqrt[3]{2 \cdot 7^2}}{\sqrt[3]{7^3}} = \dfrac{\sqrt[3]{98}}{7}$

66. $\dfrac{\sqrt{6}}{4}$ and $\sqrt{\dfrac{48}{128}}$

To show that these expressions represent the same number, work with the second expression in the following way: First simplify the fraction inside the radical, then use the quotient rule to split the expression into two radicals, rationalize the denominator, and simplify the result.

$\sqrt{\dfrac{48}{128}} = \sqrt{\dfrac{3}{8}} = \dfrac{\sqrt{3}}{\sqrt{8}} = \dfrac{\sqrt{3} \cdot \sqrt{2}}{\sqrt{8} \cdot \sqrt{2}}$

$= \dfrac{\sqrt{6}}{\sqrt{16}} = \dfrac{\sqrt{6}}{4}$

67. $-\sqrt{3}\left(\sqrt{5} + \sqrt{27}\right)$

$= -\sqrt{3}\left(\sqrt{5}\right) + \left(-\sqrt{3}\right)\left(\sqrt{27}\right)$

$= -\sqrt{3 \cdot 5} - \sqrt{3 \cdot 27}$

$= -\sqrt{15} - \sqrt{81}$

$= -\sqrt{15} - 9$

68. $3\sqrt{2}\left(\sqrt{3} + 2\sqrt{2}\right)$

$= 3\sqrt{2}\left(\sqrt{3}\right) + 3\sqrt{2}\left(2\sqrt{2}\right)$

$= 3\sqrt{6} + 6 \cdot 2$

$= 3\sqrt{6} + 12$

69. $\left(2\sqrt{3} - 4\right)\left(5\sqrt{3} + 2\right)$

$= 2\sqrt{3}\left(5\sqrt{3}\right) + \left(2\sqrt{3}\right)(2) - 4\left(5\sqrt{3}\right)$

$\quad - 4(2)$ *FOIL*

$= 10 \cdot 3 + 4\sqrt{3} - 20\sqrt{3} - 8$

$= 30 - 16\sqrt{3} - 8$

$= 22 - 16\sqrt{3}$

70. $\left(5\sqrt{7} + 2\right)^2$

$= \left(5\sqrt{7}\right)^2 + 2\left(5\sqrt{7}\right)(2) + 2^2$

Square of a binomial

$= 25 \cdot 7 + 20\sqrt{7} + 4$

$= 175 + 20\sqrt{7} + 4$

$= 179 + 20\sqrt{7}$

71. $\left(\sqrt{5} - \sqrt{7}\right)\left(\sqrt{5} + \sqrt{7}\right)$

$= \left(\sqrt{5}\right)^2 - \left(\sqrt{7}\right)^2$

$= 5 - 7 = -2$

72. $\left(2\sqrt{3} + 5\right)\left(2\sqrt{3} - 5\right)$

$= \left(2\sqrt{3}\right)^2 - (5)^2$

$= 4 \cdot 3 - 25$

$= 12 - 25 = -13$

73. $\dfrac{1}{2 + \sqrt{5}}$

$= \dfrac{1\left(2 - \sqrt{5}\right)}{\left(2 + \sqrt{5}\right)\left(2 - \sqrt{5}\right)}$ *Multiply by the conjugate*

$= \dfrac{2 - \sqrt{5}}{(2)^2 - \left(\sqrt{5}\right)^2}$

$= \dfrac{2 - \sqrt{5}}{4 - 5}$

$= \dfrac{2 - \sqrt{5}}{-1} = -2 + \sqrt{5}$

74. $\dfrac{2}{\sqrt{2}-3} = \dfrac{2\left(\sqrt{2}+3\right)}{\left(\sqrt{2}-3\right)\left(\sqrt{2}+3\right)}$

$= \dfrac{2\sqrt{2}+6}{\left(\sqrt{2}\right)^2-(3)^2}$

$= \dfrac{2\sqrt{2}+6}{2-9}$

$= \dfrac{2\sqrt{2}+6}{-7}$

$= \dfrac{\left(2\sqrt{2}+6\right)(-1)}{-7(-1)}$

$= \dfrac{-2\sqrt{2}-6}{7}$

75. $\dfrac{\sqrt{8}}{\sqrt{2}+6}$

$= \dfrac{\sqrt{8}\left(\sqrt{2}-6\right)}{\left(\sqrt{2}+6\right)\left(\sqrt{2}-6\right)}$ *Multiply by the conjugate*

$= \dfrac{\sqrt{16}-6\sqrt{8}}{2-36}$

$= \dfrac{4-6\cdot\sqrt{4\cdot2}}{-34}$

$= \dfrac{4-6\cdot2\sqrt{2}}{-34}$

$= \dfrac{4-12\sqrt{2}}{-34}$

$= \dfrac{-2\left(-2+6\sqrt{2}\right)}{-2(17)}$ *Factor numerator and denominator*

$= \dfrac{-2+6\sqrt{2}}{17}$ *Lowest terms*

76. $\dfrac{\sqrt{3}}{1+\sqrt{3}} = \dfrac{\sqrt{3}\left(1-\sqrt{3}\right)}{\left(1+\sqrt{3}\right)\left(1-\sqrt{3}\right)}$

$= \dfrac{\sqrt{3}-\sqrt{9}}{1-3}$

$= \dfrac{\sqrt{3}-3}{-2}$

$= \dfrac{\left(\sqrt{3}-3\right)(-1)}{-2(-1)}$

$= \dfrac{-\sqrt{3}+3}{2}$

77. $\dfrac{\sqrt{5}-1}{\sqrt{2}+3} = \dfrac{\left(\sqrt{5}-1\right)\left(\sqrt{2}-3\right)}{\left(\sqrt{2}+3\right)\left(\sqrt{2}-3\right)}$

$= \dfrac{\sqrt{10}-3\sqrt{5}-\sqrt{2}+3}{2-9}$

$= \dfrac{\sqrt{10}-3\sqrt{5}-\sqrt{2}+3}{-7}$

$= \dfrac{-\sqrt{10}+3\sqrt{5}+\sqrt{2}-3}{7}$

78. $\dfrac{2+\sqrt{6}}{\sqrt{3}-1} = \dfrac{\left(2+\sqrt{6}\right)\left(\sqrt{3}+1\right)}{\left(\sqrt{3}-1\right)\left(\sqrt{3}+1\right)}$

$= \dfrac{2\sqrt{3}+2+\sqrt{18}+\sqrt{6}}{3-1}$

$= \dfrac{2\sqrt{3}+2+\sqrt{9\cdot2}+\sqrt{6}}{2}$

$= \dfrac{2\sqrt{3}+2+3\sqrt{2}+\sqrt{6}}{2}$

79. $\dfrac{15+10\sqrt{6}}{15} = \dfrac{5\left(3+2\sqrt{6}\right)}{5(3)}$ *Factor*

$= \dfrac{3+2\sqrt{6}}{3}$ *Lowest terms*

80. $\dfrac{3+9\sqrt{7}}{12} = \dfrac{3\left(1+3\sqrt{7}\right)}{3(4)}$ *Factor*

$= \dfrac{1+3\sqrt{7}}{4}$ *Lowest terms*

81. $\dfrac{6+\sqrt{192}}{2} = \dfrac{6+\sqrt{64\cdot3}}{2}$

$= \dfrac{6+8\sqrt{3}}{2}$

$= \dfrac{2\left(3+4\sqrt{3}\right)}{2}$

$= 3+4\sqrt{3}$

82. $\sqrt{m}-5 = 0$

$\sqrt{m} = 5$ *Isolate the radical*

$\left(\sqrt{m}\right)^2 = 5^2$ *Square both sides*

$m = 25$

Check $m=25$: $5-5=0$ *True*

The solution set is $\{25\}$.

83. $\sqrt{p}+4 = 0$

$\sqrt{p} = -4$

Since a square root cannot equal a negative number, there is no solution and the solution set is $\emptyset$.

84.
$$\sqrt{k+1} = 7$$
$$\left(\sqrt{k+1}\right)^2 = 7^2$$
$$k + 1 = 49$$
$$k = 48$$

Check $k = 48$: $\sqrt{49} = 7$ *True*

The solution set is $\{48\}$.

85.
$$\sqrt{5m+4} = 3\sqrt{m}$$
$$\left(\sqrt{5m+4}\right)^2 = \left(3\sqrt{m}\right)^2$$
$$5m + 4 = 9m$$
$$4 = 4m$$
$$1 = m$$

Check $m = 1$: $\sqrt{9} = 3\sqrt{1}$ *True*

The solution set is $\{1\}$.

86.
$$\sqrt{2p+3} = \sqrt{5p-3}$$
$$\left(\sqrt{2p+3}\right)^2 = \left(\sqrt{5p-3}\right)^2$$
$$2p + 3 = 5p - 3$$
$$6 = 3p$$
$$2 = p$$

Check $p = 2$: $\sqrt{7} = \sqrt{7}$ *True*

The solution set is $\{2\}$.

87.
$$\sqrt{4y+1} = y - 1$$
$$\left(\sqrt{4y+1}\right)^2 = (y-1)^2$$
$$4y + 1 = y^2 - 2y + 1$$
$$0 = y^2 - 6y$$
$$0 = y(y - 6)$$
$$y = 0 \text{ or } y = 6$$

Check $y = 0$: $\quad 1 = -1$ *False*
Check $y = 6$: $\sqrt{25} = 5$ *True*

Of the two potential solutions, 6 checks in the original equation, but 0 does not. Thus, the solution set is $\{6\}$.

88.
$$\sqrt{-2k-4} = k + 2$$
$$\left(\sqrt{-2k-4}\right)^2 = (k+2)^2$$
$$-2k - 4 = k^2 + 4k + 4$$
$$0 = k^2 + 6k + 8$$
$$0 = (k+2)(k+4)$$
$$k = -2 \text{ or } k = -4$$

Check $k = -2$: $\sqrt{0} = 0$ *True*
Check $k = -4$: $\sqrt{4} = -2$ *False*

Of the two potential solutions, -2 checks in the original equation, but -4 does not. Thus, the solution set is $\{-2\}$.

89.
$$\sqrt{2-x} + 3 = x + 7$$
$$\sqrt{2-x} = x + 4 \quad \textit{Isolate the radical}$$
$$\left(\sqrt{2-x}\right)^2 = (x+4)^2$$
$$2 - x = x^2 + 8x + 16$$
$$0 = x^2 + 9x + 14$$
$$0 = (x+2)(x+7)$$
$$x = -2 \text{ or } x = -7$$

Check $x = -2$: $\quad 2 + 3 = 5$ *True*
Check $x = -7$: $\quad 3 + 3 = 0$ *False*

Of the two potential solutions, -2 checks in the original equation, but -7 does not. Thus, the solution set is $\{-2\}$.

90.
$$\sqrt{x} - x + 2 = 0$$
$$\sqrt{x} = x - 2 \quad \textit{Isolate the radical}$$
$$\left(\sqrt{x}\right)^2 = (x-2)^2$$
$$x = x^2 - 4x + 4$$
$$0 = x^2 - 5x + 4$$
$$0 = (x-4)(x-1)$$
$$x = 4 \text{ or } x = 1$$

Check $x = 4$: $\quad 0 = 0$ *True*
Check $x = 1$: $\quad 2 = 0$ *False*

Of the two potential solutions, 4 checks in the original equation, but 1 does not. Thus, the solution set is $\{4\}$.

91. **(a)** The domain elements are billions of dollars in exports of electronics.

(b) The range elements are years, represented by the last two digits of the year.

(c)
$$y = 1.4\sqrt{x - 2.5} + 87.5$$
$$y = 1.4\sqrt{13 - 2.5} + 87.5 \quad \textit{Let x = 13}$$
$$= 1.4\sqrt{10.5} + 87.5$$
$$\approx 92.04, \text{ which represents the year 1992.}$$

(d) The answer to part (c) seems reasonable because \$13 billion is between \$7.5 billion and \$19.6 billion, so the year should be between 1990 and 1993.

(e)
$$y = 1.4\sqrt{x - 2.5} + 87.5$$
$$95 = 1.4\sqrt{x - 2.5} + 87.5 \quad \textit{Let y = 95}$$
$$7.5 = 1.4\sqrt{x - 2.5}$$
$$\frac{7.5}{1.4} = \sqrt{x - 2.5}$$
$$\left(\frac{7.5}{1.4}\right)^2 = \left(\sqrt{x - 2.5}\right)^2$$
$$\frac{56.25}{1.96} = x - 2.5$$
$$x = \frac{56.25}{1.96} + 2.5 \approx 31.2$$

This answer is reasonable since it is between $25.8 billion and $36.4 billion.

92.
$$x = \frac{9 + \sqrt{81 + 20(187 - y)}}{10}$$
$$10x = 9 + \sqrt{81 + 20(187 - y)}$$
$$10x - 9 = \sqrt{81 + 20(187 - y)}$$
$$(10x - 9)^2 = \left(\sqrt{81 + 20(187 - y)}\right)^2$$
$$100x^2 - 180x + 81 = 81 + 20(187 - y)$$
$$100x^2 - 180x = 3740 - 20y$$
$$20y = -100x^2 + 180x + 3740$$
$$y = \frac{-100x^2 + 180x + 3740}{20}$$
$$f(x) = \frac{-100x^2 + 180x + 3740}{20}$$

93. $81^{1/2} = \sqrt{81} = 9$

94. $-125^{1/3} = -\sqrt[3]{125} = -5$

95. $7^{2/3} \cdot 7^{7/3} = 7^{2/3+7/3}$
$$= 7^{9/3} = 7^3 \text{ or } 343$$

96. $\dfrac{13^{4/5}}{13^{-3/5}} = 13^{4/5-(-3/5)} = 13^{7/5}$

97. $\dfrac{x^{1/4} \cdot x^{5/4}}{x^{3/4}} = \dfrac{x^{1/4+5/4}}{x^{3/4}} = \dfrac{x^{6/4}}{x^{3/4}}$
$$= x^{6/4-3/4} = x^{3/4}$$

98. $\sqrt[8]{49^4} = 49^{4/8} = 49^{1/2} = \sqrt{49} = 7$

99. $64^{2/3} = \left(\sqrt[3]{64}\right)^2 = 4^2 = 16$

100. $2\sqrt{27} + 3\sqrt{75} - \sqrt{300}$
$$= 2\sqrt{9 \cdot 3} + 3\sqrt{25 \cdot 3} - \sqrt{100 \cdot 3}$$
$$= 2 \cdot 3\sqrt{3} + 3 \cdot 5\sqrt{3} - 10\sqrt{3}$$
$$= 6\sqrt{3} + 15\sqrt{3} - 10\sqrt{3}$$
$$= 11\sqrt{3}$$

101. $\dfrac{1}{5 + \sqrt{2}} = \dfrac{1\left(5 - \sqrt{2}\right)}{\left(5 + \sqrt{2}\right)\left(5 - \sqrt{2}\right)}$
$$= \dfrac{5 - \sqrt{2}}{(5)^2 - \left(\sqrt{2}\right)^2}$$
$$= \dfrac{5 - \sqrt{2}}{25 - 2}$$
$$= \dfrac{5 - \sqrt{2}}{23}$$

102. $\sqrt{\dfrac{1}{3}} \cdot \sqrt{\dfrac{24}{5}} = \sqrt{\dfrac{1}{3} \cdot \dfrac{24}{5}} = \sqrt{\dfrac{8}{5}} = \dfrac{\sqrt{8}}{\sqrt{5}}$
$$= \dfrac{\sqrt{8} \cdot \sqrt{5}}{\sqrt{5} \cdot \sqrt{5}} = \dfrac{\sqrt{40}}{5}$$
$$= \dfrac{\sqrt{4 \cdot 10}}{5} = \dfrac{2\sqrt{10}}{5}$$

103. $\sqrt{50y^2} = \sqrt{25y^2 \cdot 2}$
$$= \sqrt{25y^2} \cdot \sqrt{2}$$
$$= 5y\sqrt{2}$$

104. $\sqrt[3]{-125} = -5$ because $(-5)^3 = -125$.

105. $-\sqrt{5}\left(\sqrt{2} + \sqrt{75}\right)$
$$= -\sqrt{5}\left(\sqrt{2}\right) + \left(-\sqrt{5}\right)\left(\sqrt{75}\right)$$
$$= -\sqrt{10} - \sqrt{375}$$
$$= -\sqrt{10} - \sqrt{25 \cdot 15}$$
$$= -\sqrt{10} - 5\sqrt{15}$$

106. $\sqrt{\dfrac{16r^3}{3s}} = \dfrac{\sqrt{16r^3}}{\sqrt{3s}} = \dfrac{\sqrt{16r^2} \cdot \sqrt{r}}{\sqrt{3s}}$
$$= \dfrac{4r\sqrt{r}}{\sqrt{3s}} = \dfrac{4r\sqrt{r} \cdot \sqrt{3s}}{\sqrt{3s} \cdot \sqrt{3s}}$$
$$= \dfrac{4r\sqrt{3rs}}{3s}$$

107. $\dfrac{12 + 6\sqrt{13}}{12} = \dfrac{6\left(2 + \sqrt{13}\right)}{6(2)}$
$$= \dfrac{2 + \sqrt{13}}{2}$$

108. $-\sqrt{162} + \sqrt{8} = -\sqrt{81 \cdot 2} + \sqrt{4 \cdot 2}$
$$= -9\sqrt{2} + 2\sqrt{2}$$
$$= -7\sqrt{2}$$

109. $\left(\sqrt{5} - \sqrt{2}\right)^2$
$$= \left(\sqrt{5}\right)^2 - 2\sqrt{5}\sqrt{2} + \left(\sqrt{2}\right)^2$$
Square of a binomial
$$= 5 - 2\sqrt{10} + 2$$
$$= 7 - 2\sqrt{10}$$

110. $\left(6\sqrt{7} + 2\right)\left(4\sqrt{7} - 1\right)$
$$= 6\sqrt{7}\left(4\sqrt{7}\right) - 1\left(6\sqrt{7}\right)$$
$$\quad + 2\left(4\sqrt{7}\right) + 2(-1)$$
$$= 24 \cdot 7 - 6\sqrt{7} + 8\sqrt{7} - 2$$
$$= 168 - 2 + 2\sqrt{7}$$
$$= 166 + 2\sqrt{7}$$

111. $-\sqrt{121} = -11$

112. $\dfrac{x^{8/3}}{x^{2/3}} = x^{8/3 - 2/3} = x^{6/3} = x^2$

113.
$$\sqrt{x+2} = x - 4$$
$$\left(\sqrt{x+2}\right)^2 = (x-4)^2$$
$$x + 2 = x^2 - 8x + 16$$
$$0 = x^2 - 9x + 14$$
$$0 = (x-2)(x-7)$$
$$x = 2 \quad \text{or} \quad x = 7$$

Check $x = 2$: $\sqrt{4} = -2$ *False*

Check $x = 7$: $\sqrt{9} = 3$ *True*

The solution set is $\{7\}$.

114. $\sqrt{k} + 3 = 0$
$$\sqrt{k} = -3$$

Since a square root cannot equal a negative number, there is no solution and the solution set is $\emptyset$.

115. $\sqrt{1 + 3t} - t = -3$
$$\sqrt{1 + 3t} = t - 3$$
$$\left(\sqrt{1 + 3t}\right)^2 = (t-3)^2$$
$$1 + 3t = t^2 - 6t + 9$$
$$0 = t^2 - 9t + 8$$
$$0 = (t-1)(t-8)$$
$$t = 1 \text{ or } t = 8$$

Check $t = 1$: $2 - 1 = -3$ *False*

Check $t = 8$: $5 - 8 = -3$ *True*

The solution set is $\{8\}$.

In Exercises 116–120, consider the points
$$A\left(2\sqrt{14}, 5\sqrt{7}\right) \text{ and } B\left(-3\sqrt{14}, 10\sqrt{7}\right).$$

116. $m = \dfrac{y_2 - y_1}{x_2 - x_1} = \dfrac{10\sqrt{7} - 5\sqrt{7}}{-3\sqrt{14} - 2\sqrt{14}}$

An equivalent expression for the slope, obtained by using the two points in the reverse order is
$$\frac{5\sqrt{7} - 10\sqrt{7}}{2\sqrt{14} - \left(-3\sqrt{14}\right)} = \frac{5\sqrt{7} - 10\sqrt{7}}{2\sqrt{14} + 3\sqrt{14}}.$$

117. $\dfrac{10\sqrt{7} - 5\sqrt{7}}{-3\sqrt{14} - 2\sqrt{14}} = \dfrac{5\sqrt{7}}{-5\sqrt{14}}$ or $-\dfrac{\sqrt{7}}{\sqrt{14}}$

118. $-\dfrac{\sqrt{7}}{\sqrt{14}} = -\sqrt{\dfrac{7}{14}} = -\sqrt{\dfrac{1}{2}}$

119. $-\sqrt{\dfrac{1}{2}} = -\dfrac{\sqrt{1}}{\sqrt{2}} = -\dfrac{1}{\sqrt{2}}$
$$= -\frac{1}{\sqrt{2}} \cdot \frac{\sqrt{2}}{\sqrt{2}} = -\frac{\sqrt{2}}{2}$$

120. The slope is negative, so the line AB falls from left to right.

Chapter 10 Test

1. The square roots of 196 are -14 and 14 because $(-14)^2 = 196$ and $14^2 = 196$.

2. (a) $\sqrt{142}$ is *irrational* because 142 is not a perfect square.

(b) $\sqrt{142} \approx 11.916$

3.
$$\sqrt[3]{216} = \sqrt[3]{8 \cdot 27}$$
$$= \sqrt[3]{8} \cdot \sqrt[3]{27}$$
$$= 2 \cdot 3 = 6$$

4. $-\sqrt{27} = -\sqrt{9 \cdot 3} = -\sqrt{9} \cdot \sqrt{3} = -3\sqrt{3}$

5. $\sqrt{\dfrac{128}{25}} = \dfrac{\sqrt{128}}{\sqrt{25}} = \dfrac{\sqrt{64 \cdot 2}}{5} = \dfrac{8\sqrt{2}}{5}$

6. $\sqrt[3]{32} = \sqrt[3]{8 \cdot 4} = \sqrt[3]{8} \cdot \sqrt[3]{4} = 2\sqrt[3]{4}$

7.
$$\frac{20\sqrt{18}}{5\sqrt{3}} = \frac{4\sqrt{9 \cdot 2}}{\sqrt{3}}$$
$$= \frac{4 \cdot 3\sqrt{2}}{\sqrt{3}}$$
$$= \frac{12\sqrt{2} \cdot \sqrt{3}}{\sqrt{3} \cdot \sqrt{3}}$$
$$= \frac{12\sqrt{6}}{3} = 4\sqrt{6}$$

8.
$$3\sqrt{28} + \sqrt{63} = 3\left(\sqrt{4 \cdot 7}\right) + \sqrt{9 \cdot 7}$$
$$= 3\left(2\sqrt{7}\right) + 3\sqrt{7}$$
$$= 6\sqrt{7} + 3\sqrt{7} = 9\sqrt{7}$$

9.
$$3\sqrt{27x} - 4\sqrt{48x} + 2\sqrt{3x}$$
$$= 3\left(\sqrt{9 \cdot 3x}\right) - 4\left(\sqrt{16 \cdot 3x}\right) + 2\sqrt{3x}$$
$$= 3\left(3\sqrt{3x}\right) - 4\left(4\sqrt{3x}\right) + 2\sqrt{3x}$$
$$= 9\sqrt{3x} - 16\sqrt{3x} + 2\sqrt{3x} = -5\sqrt{3x}$$

10.
$$\sqrt[3]{32x^2y^3} = \sqrt[3]{8y^3 \cdot 4x^2}$$
$$= \sqrt[3]{8y^3} \cdot \sqrt[3]{4x^2}$$
$$= 2y\sqrt[3]{4x^2}$$

11. $\left(6 - \sqrt{5}\right)\left(6 + \sqrt{5}\right)$

$= (6)^2 - \left(\sqrt{5}\right)^2$

$= 36 - 5 = 31$

12. $\left(2 - \sqrt{7}\right)\left(3\sqrt{2} + 1\right)$

$= 2\left(3\sqrt{2}\right) + 2(1) - \sqrt{7}\left(3\sqrt{2}\right) - \sqrt{7}(1)$

$= 6\sqrt{2} + 2 - 3\sqrt{14} - \sqrt{7}$

13. $\left(\sqrt{5} + \sqrt{6}\right)^2$

$= \left(\sqrt{5}\right)^2 + 2\left(\sqrt{5}\right)\left(\sqrt{6}\right) + \left(\sqrt{6}\right)^2$

$= 5 + 2\sqrt{30} + 6$

$= 11 + 2\sqrt{30}$

14. Use the Pythagorean formula with $c = 9$ and $b = 3$.

$$c^2 = a^2 + b^2$$
$$9^2 = a^2 + 3^2$$
$$81 = a^2 + 9$$
$$72 = a^2$$
$$\sqrt{72} = a$$

(a) $a = \sqrt{72} = \sqrt{36 \cdot 2} = 6\sqrt{2}$ inches

(b) $a = \sqrt{72} \approx 8.485$ inches

15. $\dfrac{5\sqrt{2}}{\sqrt{7}} = \dfrac{5\sqrt{2} \cdot \sqrt{7}}{\sqrt{7} \cdot \sqrt{7}} = \dfrac{5\sqrt{14}}{7}$

16. $\sqrt{\dfrac{2}{3x}} = \dfrac{\sqrt{2}}{\sqrt{3x}} = \dfrac{\sqrt{2} \cdot \sqrt{3x}}{\sqrt{3x} \cdot \sqrt{3x}} = \dfrac{\sqrt{6x}}{3x}$

17. $\dfrac{-2}{\sqrt[3]{4}} = \dfrac{-2 \cdot \sqrt[3]{2}}{\sqrt[3]{4} \cdot \sqrt[3]{2}} = \dfrac{-2\sqrt[3]{2}}{\sqrt[3]{8}}$

$= \dfrac{-2\sqrt[3]{2}}{2} = -\sqrt[3]{2}$

18. $\dfrac{-3}{4 - \sqrt{3}} = \dfrac{-3\left(4 + \sqrt{3}\right)}{\left(4 - \sqrt{3}\right)\left(4 + \sqrt{3}\right)}$

$= \dfrac{-12 - 3\sqrt{3}}{(4)^2 - \left(\sqrt{3}\right)^2}$

$= \dfrac{-12 - 3\sqrt{3}}{16 - 3}$

$= \dfrac{-12 - 3\sqrt{3}}{13}$

19. $\sqrt{x + 1} = 5 - x$

$\left(\sqrt{x + 1}\right)^2 = (5 - x)^2$

$x + 1 = 25 - 10x + x^2$

$0 = x^2 - 11x + 24$

$0 = (x - 3)(x - 8)$

$x = 3$ or $x = 8$

Check $x = 3$: $\sqrt{4} = 2$ *True*

Check $x = 8$: $\sqrt{9} = -3$ *False*

The solution set is $\{3\}$.

20. $3\sqrt{x} - 1 = 2x$

$3\sqrt{x} = 2x + 1$ *Isolate the radical*

$\left(3\sqrt{x}\right)^2 = (2x + 1)^2$

$9x = 4x^2 + 4x + 1$

$0 = 4x^2 - 5x + 1$

$0 = (4x - 1)(x - 1)$

$x = \dfrac{1}{4}$ or $x = 1$

Check $x = \dfrac{1}{4}$: $\dfrac{1}{2} = \dfrac{1}{2}$ *True*

Check $x = 1$: $2 = 2$ *True*

The solution set is $\left\{\dfrac{1}{4}, 1\right\}$.

21. $8^{4/3} = \left(\sqrt[3]{8}\right)^4 = 2^4 = 16$

22. $-125^{2/3} = -\left(\sqrt[3]{125}\right)^2 = -(5)^2 = -25$

23. $5^{3/4} \cdot 5^{1/4} = 5^{3/4 + 1/4} = 5^{4/4} = 5^1 = 5$

24. $\dfrac{\left(3^{1/4}\right)^3}{3^{7/4}} = \dfrac{3^{(1/4) \cdot 3}}{3^{7/4}} = \dfrac{3^{3/4}}{3^{7/4}} = 3^{3/4 - 7/4}$

$= 3^{-4/4} = 3^{-1} = \dfrac{1}{3}$

25. Nothing is wrong with the steps taken so far, but the potential solution must be checked.

Let $x = 12$ in the original equation.

$$\sqrt{2x + 1} + 5 = 0$$
$$\sqrt{2(12) + 1} + 5 = 0 \ ? \quad Let\ x = 12$$
$$\sqrt{25} + 5 = 0 \ ?$$
$$5 + 5 = 0 \ ?$$
$$10 = 0 \quad False$$

12 is not a solution because it does not satisfy the original equation. The equation has no solution, so the solution set is $\emptyset$.

Cumulative Review Exercises Chapters 1–10

1. $3(6+7)+6 \cdot 4 - 3^2$

    $= 3(13) + 24 - 9$

    $= 39 + 24 - 9$

    $= 63 - 9 = 54$

2. $\dfrac{3(6+7)+3}{2(4)-1} = \dfrac{3(13)+3}{8-1}$

    $= \dfrac{39+3}{7}$

    $= \dfrac{42}{7} = 6$

3. $|-6| - |-3| = 6 - 3 = 3$

4. $-9 + 14 + 11 + (-3 + 5)$

    $= -9 + 14 + 11 + 2$

    $= 5 + 11 + 2$

    $= 16 + 2 = 18$

5. $13 - [-4 - (-2)]$

    $= 13 - (-4 + 2)$

    $= 13 - (-2)$

    $= 13 + 2 = 15$

6. $-2.523 + 8.674 - 1.928$

    $= 6.151 - 1.928$

    $= 4.223$

7. $5(k-4) - k = k - 11$

    $5k - 20 - k = k - 11$

    $4k - 20 = k - 11$

    $3k = 9$

    $k = 3$

    The solution set is $\{3\}$.

8. $-\dfrac{3}{4}y \le 12$

    $-\dfrac{4}{3}\left(-\dfrac{3}{4}y\right) \ge -\dfrac{4}{3}(12)$

    $y \ge -16$

    The solution set is $[-16, \infty)$.

9. $5z + 3 - 4 > 2z + 9 + z$

    $5z - 1 > 3z + 9$

    $2z > 10$

    $z > 5$

    The solution set is $(5, \infty)$.

10. $V = LWH$

    $= (11.5)(9)(2)$

    $= 207$ cubic inches

11. $-4x + 5y = -20$

    Find the intercepts.

    If $y = 0$, $x = 5$, so the x-intercept is $(5, 0)$.

If $x = 0$, $y = -4$, so the y-intercept is $(0, -4)$.

Draw the line that passes through the points $(5, 0)$ and $(0, -4)$.

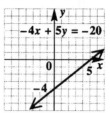

12. $x = 2$

For any value of y, the value of x is 2, so this is a vertical line through $(2, 0)$.

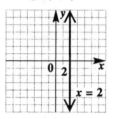

13. $(9, -2), (-3, 8)$

The slope m of the line through these points is

$$m = \frac{y_2 - y_1}{x_2 - x_1} = \frac{8 - (-2)}{-3 - 9}$$

$$= \frac{10}{-12} = -\frac{5}{6}.$$

14. $\left(3x^6\right)\left(2x^2 y\right)^2$

    $= \left(3x^6\right)(2)^2 \left(x^2\right)^2 (y)^2$

    $= \left(3x^6\right) \cdot 4x^4 y^2$

    $= 12x^{10} y^2$

15. $\left(\dfrac{3^2 y^{-2}}{2^{-1} y^3}\right)^{-3} = \left(\dfrac{2^{-1} y^3}{3^2 y^{-2}}\right)^3$

    $= \left(\dfrac{y^3 \cdot y^2}{2^1 \cdot 3^2}\right)^3$

    $= \dfrac{\left(y^5\right)^3}{(18)^3} = \dfrac{y^{15}}{5832}$

16. $\left(10x^3 + 3x^2 - 9\right) - \left(7x^3 - 8x^2 + 4\right)$

    $= 10x^3 + 3x^2 - 9 - 7x^3 + 8x^2 - 4$

    $= 3x^3 + 11x^2 - 13$

17.
$$
\begin{array}{r}
4t^2 \;-\; 8t \;+\; 5 \\
2t+3 \overline{\smash{\big)}\ 8t^3 \;-\; 4t^2 \;-\; 14t \;+\; 15} \\
\underline{8t^3 \;+\; 12t^2} \\
-16t^2 \;-\; 14t \\
\underline{-16t^2 \;-\; 24t} \\
10t \;+\; 15 \\
\underline{10t \;+\; 15} \\
0
\end{array}
$$

The remainder is 0, so the answer is the quotient, $4t^2 - 8t + 5$.

18. $m^2 + 12m + 32 = (m+8)(m+4)$

19. $25t^4 - 36 = \left(5t^2\right)^2 - (6)^2$
$$= \left(5t^2 + 6\right)\left(5t^2 - 6\right)$$

20. $12a^2 + 4ab - 5b^2 = (6a+5b)(2a-b)$

21. $81z^2 + 72z + 16$
$$= (9z)^2 + 2(9z)(4) + 4^2$$
$$= (9z+4)^2$$

22.
$$x^2 - 7x = -12$$
$$x^2 - 7x + 12 = 0$$
$$(x-3)(x-4) = 0$$
$$x = 3 \text{ or } x = 4$$

The solution set is $\{3, 4\}$.

23. $(x+4)(x-1) = -6$
$$x^2 + 3x - 4 = -6$$
$$x^2 + 3x + 2 = 0$$
$$(x+2)(x+1) = 0$$
$$x = -2 \text{ or } x = -1$$

The solution set is $\{-2, -1\}$.

24. $\dfrac{3}{x^2 + 5x - 14} = \dfrac{3}{(x+7)(x-2)}$

The expression is undefined when x is -7 or 2, because those values make the denominator equal zero.

25. $\dfrac{x^2 - 3x - 4}{x^2 + 3x} \cdot \dfrac{x^2 + 2x - 3}{x^2 - 5x + 4}$
$$= \dfrac{(x-4)(x+1)}{x(x+3)} \cdot \dfrac{(x-1)(x+3)}{(x-4)(x-1)} \quad \textit{Factor}$$
$$= \dfrac{x+1}{x} \qquad\qquad\qquad \textit{Lowest terms}$$

26. $\dfrac{t^2 + 4t - 5}{t + 5} \div \dfrac{t - 1}{t^2 + 8t + 15}$
$$= \dfrac{t^2 + 4t - 5}{t + 5} \cdot \dfrac{t^2 + 8t + 15}{t - 1}$$
$$\textit{Multiply by the reciprocal}$$
$$= \dfrac{(t+5)(t-1)}{t+5} \cdot \dfrac{(t+5)(t+3)}{t-1} \quad \textit{Factor}$$
$$= (t+5)(t+3) \qquad\qquad \textit{Lowest terms}$$

27. $\dfrac{y}{y^2 - 1} + \dfrac{y}{y + 1}$
$$= \dfrac{y}{(y+1)(y-1)} + \dfrac{y(y-1)}{(y+1)(y-1)}$$
$$= \dfrac{y + y(y-1)}{(y+1)(y-1)}$$
$$= \dfrac{y + y^2 - y}{(y+1)(y-1)} = \dfrac{y^2}{(y+1)(y-1)}$$

28. $\dfrac{2}{x+3} - \dfrac{4}{x-1}$
$$= \dfrac{2(x-1)}{(x+3)(x-1)} - \dfrac{4(x+3)}{(x-1)(x+3)}$$
$$= \dfrac{2(x-1) - 4(x+3)}{(x+3)(x-1)}$$
$$= \dfrac{2x - 2 - 4x - 12}{(x+3)(x-1)}$$
$$= \dfrac{-2x - 14}{(x+3)(x-1)}$$

29. $\dfrac{\dfrac{2}{3} + \dfrac{1}{2}}{\dfrac{1}{9} - \dfrac{1}{6}} = \dfrac{\dfrac{4}{6} + \dfrac{3}{6}}{\dfrac{2}{18} - \dfrac{3}{18}}$
$$= \dfrac{\dfrac{7}{6}}{\dfrac{-1}{18}} = \dfrac{7}{6} \div \dfrac{-1}{18}$$
$$= \dfrac{7}{6} \cdot \dfrac{18}{-1} = 7 \cdot (-3) = -21$$

30. $2x - 5y > 10$

The boundary, $2x - 5y = 10$, is the line that passes through $(5, 0)$ and $(0, -2)$; draw it as a dashed line because of the $>$ symbol. Use $(0, 0)$ as a test point. Because

$$2(0) - 5(0) > 10$$

is a false statement, shade the side of the dashed boundary that does not include the origin, $(0, 0)$.

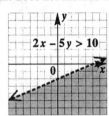

31. $4x - y = 19 \quad (1)$
$3x + 2y = -5 \quad (2)$

We will solve this system by the elimination method. Multiply both sides of equation (1) by 2, and then add the result to equation (2).

$$
\begin{array}{rrcr}
8x & - \ 2y & = & 38 \\
3x & + \ 2y & = & -5 \\
\hline
11x & & = & 33 \\
& x & = & 3
\end{array}
$$

Let $x = 3$ in equation (1).

continued

$$4(3) - y = 19$$
$$12 - y = 19$$
$$-y = 7$$
$$y = -7$$

The solution set is $\{(3, -7)\}$.

32. $2x - y = 6 \quad (1)$
$3y = 6x - 18 \quad (2)$

We will solve this system by the substitution method. Solve equation (2) for y by dividing both sides by 3.

$$y = 2x - 6$$

Substitute $2x - 6$ for y in (1).

$$2x - (2x - 6) = 6$$
$$2x - 2x + 6 = 6$$
$$6 = 6$$

This true statement indicates that the two original equations both describe the same line. This system has an infinite number of solutions.

33. Let $x =$ average speed of the slower car
(departing from Des Moines);
$x + 7 =$ average speed of the faster car
(departing from Chicago).

In 3 hours, the slower car travels $3x$ miles and the faster car travels $3(x + 7)$ miles. The total distance traveled is 321 miles, so

$$3x + 3(x + 7) = 321$$
$$3x + 3x + 21 = 321$$
$$6x = 300$$
$$x = 50.$$

The car departing from Des Moines averaged 50 miles per hour and traveled $3(50) = 150$ miles. The car departing from Chicago averaged 57 miles per hour and traveled $3(57) = 171$ miles.

34. Let $x =$ the number of subscribers to CNN (in millions);
$x + .1 =$ the number of subscribers to ESPN (in millions).

Together, the two networks had 135.7 million subscribers, so

$$x + (x + .1) = 135.7$$
$$2x + .1 = 135.7$$
$$2x = 135.6$$
$$x = 67.8$$

So in 1995, CNN had 67.8 million subscribers and ESPN had 67.9 million subscribers.

35. $\sqrt{27} - 2\sqrt{12} + 6\sqrt{75}$
$= \sqrt{9} \cdot \sqrt{3} - 2\sqrt{4} \cdot \sqrt{3} + 6\sqrt{25} \cdot \sqrt{3}$
$= 3\sqrt{3} - 2\left(2\sqrt{3}\right) + 6\left(5\sqrt{3}\right)$
$= 3\sqrt{3} - 4\sqrt{3} + 30\sqrt{3} = 29\sqrt{3}$

36. $\dfrac{2}{\sqrt{3} + \sqrt{5}} = \dfrac{2\left(\sqrt{3} - \sqrt{5}\right)}{\left(\sqrt{3} + \sqrt{5}\right)\left(\sqrt{3} - \sqrt{5}\right)}$

$= \dfrac{2\left(\sqrt{3} - \sqrt{5}\right)}{3 - 5}$

$= \dfrac{2\left(\sqrt{3} - \sqrt{5}\right)}{-2}$

$= \dfrac{\sqrt{3} - \sqrt{5}}{-1} = -\sqrt{3} + \sqrt{5}$

37. $\sqrt{200x^2y^5} = \sqrt{100x^2y^4 \cdot 2y}$
$= \sqrt{100x^2y^4} \cdot \sqrt{2y}$
$= 10xy^2\sqrt{2y}$

38. $16^{5/4} = \left(\sqrt[4]{16}\right)^5 = 2^5 = 32$

39. $\left(3\sqrt{2} + 1\right)\left(4\sqrt{2} - 3\right)$
$= 3\sqrt{2}\left(4\sqrt{2}\right) - 3\sqrt{2}(3) + 1\left(4\sqrt{2}\right)$
$\quad + 1(-3)$
$= 12 \cdot 2 - 9\sqrt{2} + 4\sqrt{2} - 3$
$= 24 - 3 - 5\sqrt{2}$
$= 21 - 5\sqrt{2}$

40. $\sqrt{x} + 2 = x - 10$
$\sqrt{x} = x - 12$
$\left(\sqrt{x}\right)^2 = (x - 12)^2$
$x = x^2 - 24x + 144$
$0 = x^2 - 25x + 144$
$0 = (x - 16)(x - 9)$

$x = 16 \quad$ or $\quad x = 9$

Check $x = 16$: $\ 6 = 6 \quad$ *True*

Check $x = 9$: $\ 5 = -1 \quad$ *False*

The solution set is $\{16\}$.

CHAPTER 11 QUADRATIC EQUATIONS, INEQUALITIES, AND GRAPHS

Section 11.1

1. $x^2 = 0$ has one real solution, 0. The correct choice is C.

3. $x^2 = -4$ has no real number solutions since no real number squared is negative. The correct choice is A.

5. $x^2 = 9$ has two integer solutions, ± 3. The correct choice is B.

7. If k is a positive perfect square, then $x^2 = k$ has two rational solutions.

   If k is a positive perfect square, $\sqrt{k}$ and $-\sqrt{k}$ will be two different rational numbers, so this statement is *true*.

9. If $-10 < k < 0$, then $x^2 = k$ has no real solutions.

   There is no real number whose square is negative, so this statement is *true*.

11. It is not correct to say that the solution set of $x^2 = 81$ is $\{9\}$, because -9 also satisfies the equation.

    When we solve an equation, we want to find *all* values of the variable that satisfy the equation. The completely correct answer is that the solution set of $x^2 = 81$ is $\{\pm 9\}$.

13. $x^2 = 81$

    Use the square root property to get

    $$x = \sqrt{81} = 9 \quad \text{or} \quad x = -\sqrt{81} = -9.$$

    The solution set is $\{\pm 9\}$.

15. $k^2 = 14$

    Use the square root property to get

    $$k = \sqrt{14} \quad \text{or} \quad k = -\sqrt{14}.$$

    The solution set is $\left\{\pm\sqrt{14}\right\}$.

17. $t^2 = 48$

    $$t = \sqrt{48} \quad \text{or} \quad t = -\sqrt{48}$$

    Write $\sqrt{48}$ in simplest form.

    $$\sqrt{48} = \sqrt{16} \cdot \sqrt{3} = 4\sqrt{3}$$

    The solution set is $\left\{\pm 4\sqrt{3}\right\}$.

19. $y^2 = -100$

    This equation has no real solution because the square of a real number cannot be negative. The square root property cannot be used because it requires that k be positive. The solution set is $\emptyset$.

21. $z^2 = 2.25$

    $$z = \sqrt{2.25} \quad \text{or} \quad z = -\sqrt{2.25}$$
    $$z = 1.5 \quad \text{or} \quad z = -1.5$$

    The solution set is $\{\pm 1.5\}$.

23. $3x^2 - 8 = 64$

    $$3x^2 = 72$$
    $$x^2 = 24$$

    Now use the square root property.

    $$x = \pm\sqrt{24} = \pm\sqrt{4 \cdot 6} = \pm 2\sqrt{6}$$

    The solution set is $\left\{\pm 2\sqrt{6}\right\}$.

25. $(x - 3)^2 = 25$

    Use the square root property.

    $$x - 3 = \sqrt{25} \quad \text{or} \quad x - 3 = -\sqrt{25}$$
    $$x - 3 = 5 \quad \text{or} \quad x - 3 = -5$$
    $$x = 8 \quad \text{or} \quad x = -2$$

    The solution set is $\{-2, 8\}$.

27. $(z + 5)^2 = -13$

    The square root of -13 is not a real number, so there is no real solution for this equation. The solution set is $\emptyset$.

29. $(x - 8)^2 = 27$

    Begin by using the square root property.

    $$x - 8 = \sqrt{27} \quad \text{or} \quad x - 8 = -\sqrt{27}$$

    Now simplify the radical.

    $$\sqrt{27} = \sqrt{9} \cdot \sqrt{3} = 3\sqrt{3}$$

    $$x - 8 = 3\sqrt{3} \quad \text{or} \quad x - 8 = -3\sqrt{3}$$
    $$x = 8 + 3\sqrt{3} \quad \text{or} \quad x = 8 - 3\sqrt{3}$$

    The solution set is $\left\{8 \pm 3\sqrt{3}\right\}$.

31. $(3k + 2)^2 = 49$

    $$3k + 2 = \sqrt{49} \quad \text{or} \quad 3k + 2 = -\sqrt{49}$$
    $$3k + 2 = 7 \quad \text{or} \quad 3k + 2 = -7$$
    $$3k = 5 \quad \text{or} \quad 3k = -9$$
    $$k = \frac{5}{3} \quad \text{or} \quad k = -3$$

    The solution set is $\left\{-3, \dfrac{5}{3}\right\}$.

33. $(4x - 3)^2 = 9$

$$4x - 3 = \sqrt{9} \quad \text{or} \quad 4x - 3 = -\sqrt{9}$$
$$4x - 3 = 3 \quad \text{or} \quad 4x - 3 = -3$$
$$4x = 6 \quad \text{or} \quad 4x = 0$$
$$x = \frac{6}{4} = \frac{3}{2} \quad \text{or} \quad x = 0$$

The solution set is $\left\{ 0, \dfrac{3}{2} \right\}$.

35. $(5 - 2x)^2 = 30$

$$5 - 2x = \sqrt{30} \quad \text{or} \quad 5 - 2x = -\sqrt{30}$$
$$-2x = -5 + \sqrt{30} \quad \text{or} \quad -2x = -5 - \sqrt{30}$$
$$x = \frac{-5 + \sqrt{30}}{-2} \quad \text{or} \quad x = \frac{-5 - \sqrt{30}}{-2}$$
$$x = \frac{-5 + \sqrt{30}}{-2} \cdot \frac{-1}{-1} \quad \text{or} \quad x = \frac{-5 - \sqrt{30}}{-2} \cdot \frac{-1}{-1}$$
$$x = \frac{5 - \sqrt{30}}{2} \quad \text{or} \quad x = \frac{5 + \sqrt{30}}{2}$$

The solution set is $\left\{ \dfrac{5 \pm \sqrt{30}}{2} \right\}$.

37. $(3k + 1)^2 = 18$

$$3k + 1 = \sqrt{18} \quad \text{or} \quad 3k + 1 = -\sqrt{18}$$
$$3k = -1 + 3\sqrt{2} \quad \text{or} \quad 3k = -1 - 3\sqrt{2}$$
$$\sqrt{18} = \sqrt{9 \cdot 2} = 3\sqrt{2}$$
$$k = \frac{-1 + 3\sqrt{2}}{3} \quad \text{or} \quad k = \frac{-1 - 3\sqrt{2}}{3}$$

The solution set is $\left\{ \dfrac{-1 \pm 3\sqrt{2}}{3} \right\}$.

39. $\left(\dfrac{1}{2}x + 5 \right)^2 = 12$

$$\frac{1}{2}x + 5 = \sqrt{12} \quad \text{or} \quad \frac{1}{2}x + 5 = -\sqrt{12}$$
$$\frac{1}{2}x = -5 + 2\sqrt{3} \quad \text{or} \quad \frac{1}{2}x = -5 - 2\sqrt{3}$$
$$\sqrt{12} = \sqrt{4 \cdot 3} = 2\sqrt{3}$$
$$x = 2\left(-5 + 2\sqrt{3} \right) \quad \text{or} \quad x = 2\left(-5 - 2\sqrt{3} \right)$$
$$x = -10 + 4\sqrt{3} \quad \text{or} \quad x = -10 - 4\sqrt{3}$$

The solution set is $\left\{ -10 \pm 4\sqrt{3} \right\}$.

41. $(4k - 1)^2 - 48 = 0$

$$(4k - 1)^2 = 48$$
$$4k - 1 = \sqrt{48} \quad \text{or} \quad 4k - 1 = -\sqrt{48}$$
$$4k - 1 = 4\sqrt{3} \quad \text{or} \quad 4k - 1 = -4\sqrt{3}$$
$$4k = 1 + 4\sqrt{3} \quad \text{or} \quad 4k = 1 - 4\sqrt{3}$$
$$k = \frac{1 + 4\sqrt{3}}{4} \quad \text{or} \quad k = \frac{1 - 4\sqrt{3}}{4}$$

The solution set is $\left\{ \dfrac{1 \pm 4\sqrt{3}}{4} \right\}$.

43. Johnny's first solution, $\dfrac{5 + \sqrt{30}}{2}$, is equivalent to Linda's second solution, $\dfrac{-5 - \sqrt{30}}{-2}$. This can be verified by multiplying $\dfrac{5 + \sqrt{30}}{2}$ by 1 in the form $\dfrac{-1}{-1}$. Similarly, Johnny's second solution is equivalent to Linda's first one.

45. $(k + 2.14)^2 = 5.46$

$$k + 2.14 = \sqrt{5.46} \quad \text{or} \quad k + 2.14 = -\sqrt{5.46}$$
$$k = -2.14 + \sqrt{5.46} \quad \text{or} \quad k = -2.14 - \sqrt{5.46}$$
$$k \approx .20 \quad \text{or} \quad k \approx -4.48$$

To the nearest hundredth, the solution set is $\{-4.48, .20\}$.

47. $(2.11p + 3.42)^2 = 9.58$

$$2.11p + 3.42 = \pm \sqrt{9.58}$$

Remember that this represents two equations

$$2.11p = -3.42 \pm \sqrt{9.58}$$
$$p = \frac{-3.42 \pm \sqrt{9.58}}{2.11}$$
$$\approx -3.09 \text{ or } -.15$$

To the nearest hundredth, the solution set is $\{-3.09, -.15\}$.

49. $x^2 + 6x + 9 = 100$

$$(x + 3)^2 = 100$$

50. $(x + 3)^2 = 100$

$$x + 3 = \pm \sqrt{100}$$
$$x + 3 = \pm 10$$
$$x + 3 = 10 \quad \text{or} \quad x + 3 = -10$$

51. $x + 3 = -10 \quad \text{or} \quad x + 3 = 10$

$$x = -13 \quad \text{or} \quad x = 7$$

The solution sets are $\{-13\}$ and $\{7\}$.

52. The solution set of the original equation is $\{-13, 7\}$.

53. $x^2 + 4x + 4 = 25$

$$(x+2)^2 = 25$$
$$x + 2 = \pm\sqrt{25}$$
$$x + 2 = \pm 5$$
$$x + 2 = 5 \quad \text{or} \quad x + 2 = -5$$
$$x = 3 \quad \text{or} \qquad x = -7$$

The solution set is $\{-7, 3\}$.

54. $4k^2 - 12k + 9 = 81$

$$(2k - 3)^2 = 81$$
$$2k - 3 = \pm\sqrt{81}$$
$$2k - 3 = \pm 9$$

$$2k - 3 = 9 \qquad \text{or} \qquad 2k - 3 = -9$$
$$2k = 12 \qquad \text{or} \qquad 2k = -6$$
$$k = 6 \qquad \text{or} \qquad k = -3$$

The solution set is $\{-3, 6\}$.

55. $y = -.017x^2 + x$

$$y = -.017(1)^2 + 1 \quad \textit{Let x = 1}$$
$$= -.017 + 1$$
$$= .983$$

After 1 second, the height of the object is .983 foot.

57. $y = -.003x^2 + 1.727x$

$$y = -.003(2)^2 + 1.727(2) \quad \textit{Let x = 2}$$
$$= -.012 + 3.454$$
$$= 3.442$$

After 2 seconds, the height of the object is 3.442 feet.

59. $d = 16t^2$

$$4 = 16t^2 \quad \textit{Let d = 4}$$
$$t^2 = \frac{4}{16} = \frac{1}{4}$$
$$t = \pm\sqrt{\frac{1}{4}} = \pm\frac{1}{2}$$

Reject $-\frac{1}{2}$ as a solution, since negative time does not make sense. About $\frac{1}{2}$ second elapses between the dropping of the coin and the shot.

61. $A = \pi r^2$

$$81\pi = \pi r^2 \quad \textit{Let A = 81}\pi$$
$$81 = r^2 \quad \textit{Divide by }\pi$$
$$r = 9 \quad \text{or} \quad r = -9$$

Discard -9 since the radius cannot be negative. The radius is 9 inches.

63. $A = P(1 + r)^2$

If $A = 110.25$ and $P = 100$,

$$110.25 = 100(1 + r)^2$$
$$1.1025 = (1 + r)^2$$
$$(1 + r)^2 = 1.1025$$
$$1 + r = \pm\sqrt{1.1025}$$
$$1 + r = \pm 1.05$$
$$r = -1 \pm 1.05.$$
$$-1 + 1.05 = .05$$
$$-1 - 1.05 = -2.05$$

Reject the solution -2.05. The rate is $r = .05$ or 5%.

Section 11.2

1. To solve the equation $x^2 - 8x = 4$ by completing the square, the first step is to add _16_ to both sides of the equation. The 16 comes from one-half of the coefficient of x squared; that is,

$$\left[\frac{1}{2}(-8)\right]^2 = (-4)^2 = 16.$$

3. To solve $(t + 2)(t - 5) = 18$ by completing the square, we should start by _multiplying_ $(t + 2)(t - 5)$ to get $t^2 - 3t - 10$. This will enable us to get all the terms with variables on one side of the equals sign.

5. $2x^2 - 4x = 9$

Before completing the square, the coefficient of x^2 must be 1. Dividing each side of the equation by 2 is the correct way to begin solving the equation, and this corresponds to choice (d).

7. $y^2 + 14y$

Take half of the coefficient of y and square it.

$$\frac{1}{2}(14) = 7, \text{ and } 7^2 = 49.$$

Adding 49 to the expression $y^2 + 14y$ will make it a perfect square.

9. $k^2 - 5k$

Take half of the coefficient of k and square it.

$$\frac{1}{2}(-5) = -\frac{5}{2}, \text{ and } \left(-\frac{5}{2}\right)^2 = \frac{25}{4}.$$

Adding $\frac{25}{4}$ to the expression $k^2 - 5k$ will make it a perfect square.

11. $r^2 + \dfrac{1}{2}r$

Take half of the coefficient of r and square it.

$$\frac{1}{2}\left(\frac{1}{2}\right) = \frac{1}{4}, \text{ and } \left(\frac{1}{4}\right)^2 = \frac{1}{16}.$$

Adding $\dfrac{1}{16}$ to the expression $r^2 + \dfrac{1}{2}r$ will make it a perfect square.

13. $x^2 - 4x = -3$

Take half of the coefficient of x and square it. Half of -4 is -2, and $(-2)^2 = 4$. Add 4 to each side of the equation, and write the left side as a perfect square.

$$x^2 - 4x + 4 = -3 + 4$$
$$(x - 2)^2 = 1$$

Use the square root property.

$$x - 2 = \sqrt{1} \quad \text{or} \quad x - 2 = -\sqrt{1}$$
$$x - 2 = 1 \quad \text{or} \quad x - 2 = -1$$
$$x = 3 \quad \text{or} \quad x = 1$$

A check verifies that the solution set is $\{1, 3\}$.

15. $x^2 + 2x - 5 = 0$

Add 5 to each side.

$$x^2 + 2x = 5$$

Take half the coefficient of x and square it.

$$\frac{1}{2}(2) = 1, \text{ and } 1^2 = 1.$$

Add 1 to each side of the equation, and write the left side as a perfect square.

$$x^2 + 2x + 1 = 5 + 1$$
$$(x + 1)^2 = 6$$

Use the square root property.

$$x + 1 = \sqrt{6} \quad \text{or} \quad x + 1 = -\sqrt{6}$$
$$x = -1 + \sqrt{6} \quad \text{or} \quad x = -1 - \sqrt{6}$$

A check verifies that the solution set is $\left\{-1 \pm \sqrt{6}\right\}$. Using a calculator for your check is highly recommended.

17. $z^2 + 6z + 9 = 0$

The left-hand side of this equation is already a perfect square.

$$(z + 3)^2 = 0$$
$$z + 3 = 0$$
$$z = -3$$

A check verifies that the solution set is $\{-3\}$.

19. $4y^2 + 4y = 3$

Divide each side by 4 so that the coefficient of y^2 is 1.

$$y^2 + y = \frac{3}{4}$$

The coefficient of y is 1. Take half of 1, square the result, and add this square to each side.

$$\frac{1}{2}(1) = \frac{1}{2} \text{ and } \left(\frac{1}{2}\right)^2 = \frac{1}{4}$$

$$y^2 + y + \frac{1}{4} = \frac{3}{4} + \frac{1}{4}$$

The left-hand side can then be written as a perfect square.

$$\left(y + \frac{1}{2}\right)^2 = 1$$

Use the square root property.

$$y + \frac{1}{2} = 1 \quad \text{or} \quad y + \frac{1}{2} = -1$$
$$y = -\frac{1}{2} + 1 \quad \text{or} \quad y = -\frac{1}{2} - 1$$
$$y = \frac{1}{2} \quad \text{or} \quad y = -\frac{3}{2}$$

A check verifies that the solution set is

$$\left\{-\frac{3}{2}, \frac{1}{2}\right\}.$$

21. $2p^2 - 2p + 3 = 0$

Divide each side by 2.

$$p^2 - p + \frac{3}{2} = 0$$

Subtract $\dfrac{3}{2}$ from both sides.

$$p^2 - p = -\frac{3}{2}$$

Take half the coefficient of p and square it.

$$\frac{1}{2}(-1) = -\frac{1}{2}, \text{ and } \left(-\frac{1}{2}\right)^2 = \frac{1}{4}.$$

Add $\dfrac{1}{4}$ to each side of the equation.

$$p^2 - p + \frac{1}{4} = -\frac{3}{2} + \frac{1}{4}$$

Factor on the left side and add on the right.

$$\left(p - \frac{1}{2}\right)^2 = -\frac{5}{4}$$

The square root of $-\dfrac{5}{4}$ is not a real number, so the solution set is $\emptyset$.

23. $3k^2 + 7k = 4$

Divide each side by 3.

$$k^2 + \frac{7}{3}k = \frac{4}{3}$$

Take half of the coefficient of k and square it.

$$\frac{1}{2}\left(\frac{7}{3}\right) = \frac{7}{6} \quad \text{and} \quad \left(\frac{7}{6}\right)^2 = \frac{49}{36}.$$

Add $\frac{49}{36}$ to each side of the equation.

$$k^2 + \frac{7}{3}k + \frac{49}{36} = \frac{4}{3} + \frac{49}{36}$$

$$\left(k + \frac{7}{6}\right)^2 = \frac{97}{36}$$

Use the square root property.

$$k + \frac{7}{6} = \sqrt{\frac{97}{36}} \qquad \text{or} \qquad k + \frac{7}{6} = -\sqrt{\frac{97}{36}}$$

$$k + \frac{7}{6} = \frac{\sqrt{97}}{6} \qquad \text{or} \qquad k + \frac{7}{6} = -\frac{\sqrt{97}}{6}$$

$$k = -\frac{7}{6} + \frac{\sqrt{97}}{6} \qquad \text{or} \qquad k = -\frac{7}{6} - \frac{\sqrt{97}}{6}$$

$$k = \frac{-7 + \sqrt{97}}{6} \qquad \text{or} \qquad k = \frac{-7 - \sqrt{97}}{6}$$

A check verifies that the solution set is

$$\left\{\frac{-7 \pm \sqrt{97}}{6}\right\}.$$

25. $(x + 3)(x - 1) = 5$

$$x^2 + 2x - 3 = 5$$

$$x^2 + 2x = 8$$

$$x^2 + 2x + 1 = 8 + 1$$

$$(x + 1)^2 = 9$$

$$x + 1 = 3 \quad \text{or} \quad x + 1 = -3$$

$$x = 2 \quad \text{or} \quad x = -4$$

A check verifies that the solution set is $\{-4, 2\}$.

27. $-x^2 + 2x = -5$

Divide each side by -1.

$$x^2 - 2x = 5$$

Take half of the coefficient of x and square it. Half of -2 is -1, and $(-1)^2 = 1$. Add 1 to each side of the equation, and write the left side as a perfect square.

$$x^2 - 2x + 1 = 5 + 1$$

$$(x - 1)^2 = 6$$

Use the square root property.

$$x - 1 = \sqrt{6} \qquad \text{or} \qquad x - 1 = -\sqrt{6}$$

$$x = 1 + \sqrt{6} \qquad \text{or} \qquad x = 1 - \sqrt{6}$$

A check verifies that the solution set is

$$\left\{1 \pm \sqrt{6}\right\}.$$

29. $3r^2 - 2 = 6r + 3$

$$3r^2 - 6r = 5$$

$$r^2 - 2r = \frac{5}{3}$$

$$r^2 - 2r + 1 = \frac{5}{3} + 1$$

$$(r - 1)^2 = \frac{8}{3}$$

$$r - 1 = \pm\sqrt{\frac{8}{3}}$$

Simplify the radical.

$$\sqrt{\frac{8}{3}} = \frac{\sqrt{8}}{\sqrt{3}} = \frac{2\sqrt{2}}{\sqrt{3}} \cdot \frac{\sqrt{3}}{\sqrt{3}} = \frac{2\sqrt{6}}{3}$$

$$r = 1 \pm \frac{2\sqrt{6}}{3}$$

$$r = \frac{3}{3} \pm \frac{2\sqrt{6}}{3} = \frac{3 \pm 2\sqrt{6}}{3}$$

(a) The solution set with exact values is

$$\left\{\frac{3 \pm 2\sqrt{6}}{3}\right\}.$$

(b) $\dfrac{3 + 2\sqrt{6}}{3} \approx 2.633$

$\dfrac{3 - 2\sqrt{6}}{3} \approx -.633$

The solution set with approximate values is

$\{-.633, 2.633\}$.

31. $(x + 1)(x + 3) = 2$

$$x^2 + 3x + x + 3 = 2$$

$$x^2 + 4x = -1$$

$$x^2 + 4x + 4 = -1 + 4$$

$$(x + 2)^2 = 3$$

$$x + 2 = \pm\sqrt{3}$$

$$x = -2 \pm \sqrt{3}$$

(a) The solution set with exact values is

$$\left\{-2 \pm \sqrt{3}\right\}.$$

(b) $-2 + \sqrt{3} \approx -.268$

$-2 - \sqrt{3} \approx -3.732$

The solution set with approximate values is

$\{-3.732, -.268\}$.

33. The student should have divided both sides of the equation by 2 as his first step. This gives

$$x^2 - 5x = -4.$$

Now add the square of half the coefficient of x; that is, $\left[\frac{1}{2}(-5)\right]^2 = \left(-\frac{5}{2}\right)^2 = \frac{25}{4}$.

$$x^2 - 5x + \frac{25}{4} = -4 + \frac{25}{4}$$

Factor the left side and simplify the right side.

$$\left(x - \frac{5}{2}\right)^2 = \frac{9}{4}$$

Use the square root property.

$$x - \frac{5}{2} = \pm\sqrt{\frac{9}{4}} = \pm\frac{3}{2}$$

$$x = \frac{5}{2} \pm \frac{3}{2}$$

$$= 4 \text{ or } 1$$

The correct solution set is $\{1, 4\}$.

35.   Let $x =$ the width of the pen;
$175 - x =$ the length of the pen.

Use the formula for the area of a rectangle, $A = LW$.

$$7500 = (175 - x)x$$
$$7500 = 175x - x^2$$
$$x^2 - 175x + 7500 = 0$$

Solve this quadratic equation by completing the square.

$$x^2 - 175x = -7500$$
$$x^2 - 175x + \frac{30,625}{4} = -\frac{30,000}{4} + \frac{30,625}{4}$$
$$\left(\frac{175}{2}\right)^2 = \frac{30,625}{4}$$
$$\left(x - \frac{175}{2}\right)^2 = \frac{625}{4}$$
$$x - \frac{175}{2} = \pm\sqrt{\frac{625}{4}} = \pm\frac{25}{2}$$
$$x = \frac{175}{2} \pm \frac{25}{2}$$
$$x = \frac{175}{2} + \frac{25}{2} \quad \text{or} \quad x = \frac{175}{2} - \frac{25}{2}$$
$$x = \frac{200}{2} \quad \text{or} \quad x = \frac{150}{2}$$
$$x = 100 \quad \text{or} \quad x = 75$$

If $x = 100$,

$$175 - x = 175 - 100 = 75.$$

If $x = 75$,

$$175 - x = 175 - 75 = 100.$$

The dimensions of the pen are 75 feet by 100 feet.

37. $s = -16t^2 + 96t$

Find the value of t when $s = 80$.

$$80 = -16t^2 + 96t \quad \textit{Let s = 80}$$
$$-16t^2 + 96t = 80$$
$$t^2 - 6t = -5 \qquad \textit{Divide by -16}$$
$$t^2 - 6t + 9 = -5 + 9 \qquad \textit{Add 9}$$
$$(t - 3)^2 = 4 \qquad \textit{Factor; add}$$
$$t - 3 = \pm\sqrt{4} = \pm 2$$
$$t = 3 + 2 = 5 \quad \text{or} \quad t = 3 - 2 = 1$$

The object will reach a height of 80 feet at 1 second (on the way up) and at 5 seconds (on the way down).

39.
$$s = -13t^2 + 104t$$
$$195 = -13t^2 + 104t \quad \textit{Let s = 195}$$
$$-15 = t^2 - 8t \qquad \textit{Divide by -13}$$
$$t^2 - 8t + 16 = -15 + 16$$
$$\textit{Add }\left[\frac{1}{2}(-8)\right]^2 = 16$$
$$(t - 4)^2 = 1$$
$$t - 4 = \pm\sqrt{1} = \pm 1$$
$$t = 4 \pm 1$$
$$= 3 \text{ or } 5$$

The object will be at a height of 195 feet at 3 and 5 seconds.

41.   Let $x =$ the distance the slower
car traveled;
$x + 7 =$ the distance the faster
car traveled.

Since the cars traveled at right angles, a right triangle is formed with hypotenuse of length 17. Use the Pythagorean formula with $a = x$, $b = x + 7$, and $c = 17$.

$$a^2 + b^2 = c^2$$
$$x^2 + (x + 7)^2 = 17^2$$
$$x^2 + (x^2 + 14x + 49) = 289$$
$$2x^2 + 14x = 240$$
$$x^2 + 7x = 120$$
$$x^2 + 7x + \frac{49}{4} = 120 + \frac{49}{4}$$
$$\textit{Add }\left[\frac{1}{2}(7)\right]^2 = \frac{49}{4}$$

$$\left(x + \frac{7}{2}\right)^2 = \frac{529}{4}$$

$$x + \frac{7}{2} = \pm\sqrt{\frac{529}{4}} = \pm\frac{23}{2}$$

$$x = -\frac{7}{2} \pm \frac{23}{2}$$

$$= 8 \text{ or } -15$$

Discard -15 since distance cannot be negative.
The slower car traveled 8 miles.

Section 11.3

1. For the quadratic equation $4x^2 + 5x - 9 = 0$, the values of a, b, and c are respectively $\underline{4}$, $\underline{5}$, and $\underline{-9}$. The equation is already in the standard form $4x^2 + 5x - 9 = 0$.

3. When using the quadratic formula, if the discriminant $b^2 - 4ac$ is positive, the equation has $\underline{2}$ real solutions. These are the two values given in the quadratic formula,

$$\frac{-b \pm \sqrt{b^2 - 4ac}}{2a}.$$

5. $3x^2 = 4x + 2$

First write the equation in standard form, $ax^2 + bx + c = 0$.

$$3x^2 - 4x - 2 = 0$$

Now identify $a = 3$, $b = -4$, and $c = -2$.

7. $3x^2 = -7x$

Write the equation in standard form.

$$3x^2 + 7x = 0$$

Now identify $a = 3$, $b = 7$, and $c = 0$.

9. $(x - 3)(x + 4) = 0$
$$x^2 + x - 12 = 0 \quad \textit{FOIL}$$

$$a = 1, b = 1, \text{ and } c = -12$$

11. $9(x - 1)(x + 2) = 8$
$$9(x^2 + x - 2) = 8 \quad \textit{FOIL}$$
$$9x^2 + 9x - 18 = 8 \quad \begin{array}{l}\textit{Distributive}\\ \textit{property}\end{array}$$
$$9x^2 + 9x - 26 = 0 \quad \textit{Standard form}$$
$$a = 9, b = 9, \text{ and } c = -26$$

13. If $a = 0$, then the equation

$$ax^2 + bx + c = 0$$

would become $bx + c = 0$, which is a linear equation, not a quadratic equation. Therefore, the restriction $a \neq 0$ is necessary in the definition of a quadratic equation.

15. No, because $2a$ should be the denominator for $-b$ as well. The correct formula is

$$x = \frac{-b \pm \sqrt{b^2 - 4ac}}{2a}.$$

17. $k^2 = -12k + 13$

Write the equation in standard form.

$$k^2 + 12k - 13 = 0$$

Here, $a = 1$, $b = 12$, and $c = -13$.

Substitute these values into the quadratic formula.

$$k = \frac{-b \pm \sqrt{b^2 - 4ac}}{2a}$$

$$k = \frac{-12 \pm \sqrt{12^2 - 4(1)(-13)}}{2(1)}$$

$$= \frac{-12 \pm \sqrt{144 + 52}}{2}$$

$$= \frac{-12 \pm \sqrt{196}}{2}$$

$$= \frac{-12 \pm 14}{2}$$

$$k = \frac{-12 + 14}{2} = \frac{2}{2} = 1$$

$$\text{or } k = \frac{-12 - 14}{2} = \frac{-26}{2} = -13$$

The solution set is $\{-13, 1\}$.

19. $p^2 - 4p + 4 = 0$

Substitute $a = 1$, $b = -4$, and $c = 4$ into the quadratic formula.

$$p = \frac{-b \pm \sqrt{b^2 - 4ac}}{2a}$$

$$p = \frac{-(-4) \pm \sqrt{(-4)^2 - 4(1)(4)}}{2(1)}$$

$$= \frac{4 \pm \sqrt{16 - 16}}{2}$$

$$= \frac{4 \pm 0}{2} = \frac{4}{2} = 2.$$

The solution set is $\{2\}$. Note that the discriminant is 0.

21. $2x^2 + 12x = -5$

Write the equation in standard form.

$$2x^2 + 12x + 5 = 0$$

Substitute $a = 2$, $b = 12$, and $c = 5$ into the quadratic formula.

$$x = \frac{-b \pm \sqrt{b^2 - 4ac}}{2a}$$

$$x = \frac{-12 \pm \sqrt{12^2 - 4(2)(5)}}{2(2)}$$

$$= \frac{-12 \pm \sqrt{144 - 40}}{4}$$

$$= \frac{-12 \pm \sqrt{104}}{4} = \frac{-12 \pm \sqrt{4} \cdot \sqrt{26}}{4}$$

$$= \frac{-12 \pm 2\sqrt{26}}{4} = \frac{2\left(-6 \pm \sqrt{26}\right)}{2 \cdot 2}$$

$$= \frac{-6 \pm \sqrt{26}}{2}$$

The solution set is $\left\{ \dfrac{-6 \pm \sqrt{26}}{2} \right\}$.

23. $2y^2 = 5 + 3y$

Write the equation in standard form.

$$2y^2 - 3y - 5 = 0$$

Substitute $a = 2$, $b = -3$, and $c = -5$ into the quadratic formula.

$$y = \frac{-b \pm \sqrt{b^2 - 4ac}}{2a}$$

$$y = \frac{-(-3) \pm \sqrt{(-3)^2 - 4(2)(-5)}}{2(2)}$$

$$= \frac{3 \pm \sqrt{9 + 40}}{4}$$

$$= \frac{3 \pm \sqrt{49}}{4} = \frac{3 \pm 7}{4}$$

$$y = \frac{3 + 7}{4} = \frac{10}{4} = \frac{5}{2}$$

$$\text{or} \quad y = \frac{3 - 7}{4} = \frac{-4}{4} = -1$$

The solution set is $\left\{ -1, \dfrac{5}{2} \right\}$.

25. $6x^2 + 6x = 0$

Substitute $a = 6$, $b = 6$, and $c = 0$ into the quadratic formula.

$$x = \frac{-b \pm \sqrt{b^2 - 4ac}}{2a}$$

$$x = \frac{-6 \pm \sqrt{6^2 - 4(6)(0)}}{2(6)}$$

$$= \frac{-6 \pm \sqrt{36 - 0}}{12}$$

$$= \frac{-6 \pm 6}{12}$$

$$x = \frac{-6 + 6}{12} = \frac{0}{12} = 0$$

$$\text{or} \quad x = \frac{-6 - 6}{12} = \frac{-12}{12} = -1$$

The solution set is $\{-1, 0\}$.

27. $7x^2 = 12x$

Write the equation in standard form.

$$7x^2 - 12x = 0$$

Substitute $a = 7$, $b = -12$, and $c = 0$ into the quadratic formula.

$$x = \frac{-b \pm \sqrt{b^2 - 4ac}}{2a}$$

$$x = \frac{-(-12) \pm \sqrt{(-12)^2 - 4(7)(0)}}{2(7)}$$

$$= \frac{12 \pm \sqrt{144 - 0}}{14}$$

$$= \frac{12 \pm 12}{14}$$

$$x = \frac{12 + 12}{14} = \frac{24}{14} = \frac{12}{7}$$

$$\text{or} \quad x = \frac{12 - 12}{14} = \frac{0}{14} = 0$$

The solution set is $\left\{ 0, \dfrac{12}{7} \right\}$.

29. $x^2 - 24 = 0$

Substitute $a = 1$, $b = 0$, and $c = -24$ into the quadratic formula.

$$x = \frac{-b \pm \sqrt{b^2 - 4ac}}{2a}$$

$$x = \frac{-0 \pm \sqrt{0^2 - 4(1)(-24)}}{2(1)}$$

$$= \frac{\pm \sqrt{96}}{2} = \frac{\pm \sqrt{16} \cdot \sqrt{6}}{2}$$

$$= \frac{\pm 4\sqrt{6}}{2} = \pm 2\sqrt{6}$$

The solution set is $\left\{ \pm 2\sqrt{6} \right\}$.

31. $25x^2 - 4 = 0$

Substitute $a = 25$, $b = 0$, and $c = -4$ into the quadratic formula.

$$x = \frac{-b \pm \sqrt{b^2 - 4ac}}{2a}$$

$$x = \frac{-0 \pm \sqrt{0^2 - 4(25)(-4)}}{2(25)}$$

$$= \frac{\pm\sqrt{400}}{50}$$

$$= \frac{\pm 20}{50} = \pm\frac{2}{5}$$

The solution set is $\left\{ \pm\frac{2}{5} \right\}$.

33. $3x^2 - 2x + 5 = 10x + 1$

Write the equation in standard form.

$$3x^2 - 12x + 4 = 0$$

Substitute $a = 3$, $b = -12$, and $c = 4$ into the quadratic formula.

$$x = \frac{-b \pm \sqrt{b^2 - 4ac}}{2a}$$

$$x = \frac{-(-12) \pm \sqrt{(-12)^2 - 4(3)(4)}}{2(3)}$$

$$= \frac{12 \pm \sqrt{144 - 48}}{6}$$

$$= \frac{12 \pm \sqrt{96}}{6} = \frac{12 \pm \sqrt{16} \cdot \sqrt{6}}{6}$$

$$= \frac{12 \pm 4\sqrt{6}}{6} = \frac{2\left(6 \pm 2\sqrt{6}\right)}{2 \cdot 3}$$

$$= \frac{6 \pm 2\sqrt{6}}{3}.$$

The solution set is $\left\{ \dfrac{6 \pm 2\sqrt{6}}{3} \right\}$.

35. $-2x^2 = -3x + 2$

Write the equation in standard form.

$$-2x^2 + 3x - 2 = 0$$

Substitute $a = -2$, $b = 3$, and $c = -2$ into the quadratic formula.

$$x = \frac{-b \pm \sqrt{b^2 - 4ac}}{2a}$$

$$x = \frac{-3 \pm \sqrt{3^2 - 4(-2)(-2)}}{2(-2)}$$

$$= \frac{-3 \pm \sqrt{9 - 16}}{-4}$$

$$= \frac{-3 \pm \sqrt{-7}}{-4}$$

Because $\sqrt{-7}$ does not represent a real number, the solution set is $\emptyset$.

37. $2x^2 + x + 5 = 0$

Substitute $a = 2$, $b = 1$, and $c = 5$ into the quadratic formula.

$$x = \frac{-1 \pm \sqrt{1^2 - 4(2)(5)}}{2(2)}$$

$$= \frac{-1 \pm \sqrt{1 - 40}}{4}$$

$$= \frac{-1 \pm \sqrt{-39}}{4}$$

Because $\sqrt{-39}$ does not represent a real number, the solution set is $\emptyset$.

39. $(x + 3)(x + 2) = 15$

$$x^2 + 5x + 6 = 15$$

$$x^2 + 5x - 9 = 0$$

$$a = 1, b = 5, \text{ and } c = -9$$

$$x = \frac{-5 \pm \sqrt{5^2 - 4(1)(-9)}}{2(1)}$$

$$= \frac{-5 \pm \sqrt{25 + 36}}{2}$$

$$= \frac{-5 \pm \sqrt{61}}{2}$$

The solution set is $\left\{ \dfrac{-5 \pm \sqrt{61}}{2} \right\}$.

41. $2x^2 + 2x = 5$

$$2x^2 + 2x - 5 = 0$$

$$a = 2, b = 2, \text{ and } c = -5$$

$$x = \frac{-2 \pm \sqrt{2^2 - 4(2)(-5)}}{2(2)}$$

$$= \frac{-2 \pm \sqrt{4 + 40}}{4}$$

$$= \frac{-2 \pm \sqrt{44}}{4}$$

continued

$$= \frac{-2 \pm 2\sqrt{11}}{4}$$

$$= \frac{2\left(-1 \pm \sqrt{11}\right)}{2 \cdot 2}$$

$$= \frac{-1 \pm \sqrt{11}}{2}$$

(a) The solution set with exact values is

$$\left\{ \frac{-1 \pm \sqrt{11}}{2} \right\}.$$

(b) The solution set with approximate values (to the nearest thousandth) is $\{-2.158, 1.158\}$.

43.
$$x^2 = 1 + x$$
$$x^2 - x - 1 = 0$$
$$a = 1, b = -1, \text{ and } c = -1$$

$$x = \frac{-(-1) \pm \sqrt{(-1)^2 - 4(1)(-1)}}{2(1)}$$

$$= \frac{1 \pm \sqrt{1 + 4}}{2}$$

$$= \frac{1 \pm \sqrt{5}}{2}$$

(a) The solution set with exact values is

$$\left\{ \frac{1 \pm \sqrt{5}}{2} \right\}.$$

(b) The solution set with approximate values (to the nearest thousandth) is $\{-.618, 1.618\}$.

45. $\frac{3}{2}k^2 - k - \frac{4}{3} = 0$

Eliminate the denominators by multiplying each side by the least common denominator, 6.

$$9k^2 - 6k - 8 = 0$$

Substitute $a = 9$, $b = -6$, and $c = -8$ into the quadratic formula.

$$k = \frac{-(-6) \pm \sqrt{(-6)^2 - 4(9)(-8)}}{2(9)}$$

$$= \frac{6 \pm \sqrt{36 + 288}}{18}$$

$$= \frac{6 \pm \sqrt{324}}{18}$$

$$= \frac{6 \pm 18}{18}$$

$$k = \frac{6 + 18}{18} = \frac{24}{18} = \frac{4}{3}$$

$$\text{or } k = \frac{6 - 18}{18} = \frac{-12}{18} = -\frac{2}{3}$$

The solution set is $\left\{ -\frac{2}{3}, \frac{4}{3} \right\}$.

47. $\frac{1}{2}x^2 + \frac{1}{6}x = 1$

Eliminate the denominators by multiplying each side by the least common denominator, 6.

$$3x^2 + x = 6$$
$$3x^2 + x - 6 = 0$$

$a = 3$, $b = 1$, and $c = -6$

$$x = \frac{-1 \pm \sqrt{1^2 - 4(3)(-6)}}{2(3)}$$

$$= \frac{-1 \pm \sqrt{1 + 72}}{6}$$

$$= \frac{-1 \pm \sqrt{73}}{6}$$

The solution set is $\left\{ \frac{-1 \pm \sqrt{73}}{6} \right\}$.

49. $.5x^2 = x + .5$

To eliminate the decimals, multiply each side by 10.

$$5x^2 = 10x + 5$$
$$5x^2 - 10x - 5 = 0$$

Use the quadratic formula with

$a = 5$, $b = -10$, and $c = -5$.

$$x = \frac{-(-10) \pm \sqrt{(-10)^2 - 4(5)(-5)}}{2(5)}$$

$$= \frac{10 \pm \sqrt{100 + 100}}{10}$$

$$= \frac{10 \pm \sqrt{200}}{10} = \frac{10 \pm \sqrt{100} \cdot \sqrt{2}}{10}$$

$$= \frac{10 \pm 10\sqrt{2}}{10} = \frac{10\left(1 \pm \sqrt{2}\right)}{10}$$

$$= 1 \pm \sqrt{2}$$

The solution set is $\left\{ 1 \pm \sqrt{2} \right\}$.

51. $\frac{3}{8}x^2 - x + \frac{17}{24} = 0$

Multiply each side by the least common denominator, 24.

$$9x^2 - 24x + 17 = 0$$

Use the quadratic formula with

$a = 9$, $b = -24$, and $c = 17$.

$$x = \frac{-(-24) \pm \sqrt{(-24)^2 - 4(9)(17)}}{2(9)}$$

$$= \frac{24 \pm \sqrt{576 - 612}}{18}$$

$$= \frac{24 \pm \sqrt{-36}}{18}$$

Because $\sqrt{-36}$ does not represent a real number, the solution set is $\emptyset$.

53. If an applied problem leads to a quadratic equation, the solution(s) must make sense in the original problem. For example, a length cannot be negative.

55. $S = 2\pi rh + \pi r^2$ for r

Write this equation in the standard form of a quadratic equation, treating r as the variable and S, π, and h as constants.

$$\pi r^2 + (2\pi h)r - S = 0$$

Use $a = \pi$, $b = 2\pi h$, and $c = -S$ in the quadratic formula.

$$r = \frac{-b \pm \sqrt{b^2 - 4ac}}{2a}$$

$$r = \frac{-2\pi h \pm \sqrt{(2\pi h)^2 - 4(\pi)(-S)}}{2(\pi)}$$

$$= \frac{-2\pi h \pm \sqrt{4\pi^2 h^2 + 4\pi S}}{2\pi}$$

$$= \frac{-2\pi h \pm \sqrt{4(\pi^2 h^2 + \pi S)}}{2\pi}$$

$$= \frac{-2\pi h \pm 2\sqrt{\pi^2 h^2 + \pi S}}{2\pi}$$

$$r = \frac{-\pi h \pm \sqrt{\pi^2 h^2 + \pi S}}{\pi}$$

57. From Example 6, $P^2 = a^3$.

$$P^2 = (.387)^3 \quad \text{Let } a = .387$$

$$\approx .05796$$

Use the square root property.

$$P \approx .241 \text{ year}$$

Multiplying by 365 gives us about 88 days.

59. $P^2 = a^3$

$$P^2 = (1.524)^3$$

$$\approx 3.5396$$

$$P \approx 1.881 \text{ years}$$

61. $P^2 = a^3$

$$P^2 = (19.2)^3$$

$$= 7077.888$$

$$P \approx 84.130 \text{ years}$$

63. $P^2 = a^3$

$$1^2 = 1^3 \quad \text{Let } P = 1, a = 1$$

$$1 = 1,$$

which is a true statement.

65. The distance traveled is 14 feet.

Use $d = 2t^2 - 5t + 2$ with $d = 14$.

$$14 = 2t^2 - 5t + 2$$

$$0 = 2t^2 - 5t - 12 \quad \text{Standard form}$$

$$a = 2, b = -5, c = -12$$

$$t = \frac{-b \pm \sqrt{b^2 - 4ac}}{2a}$$

$$t = \frac{-(-5) \pm \sqrt{(-5)^2 - 4(2)(-12)}}{2(2)}$$

$$= \frac{5 \pm \sqrt{25 + 96}}{4}$$

$$= \frac{5 \pm \sqrt{121}}{4}$$

$$= \frac{5 \pm 11}{4}$$

$$t = \frac{5 + 11}{4} = \frac{16}{4} = 4$$

$$\text{or } t = \frac{5 - 11}{4} = \frac{-6}{4} = -\frac{3}{2}$$

Since t represents the time in seconds, t must be a positive number.

It will take the projectile 4 seconds to travel 14 feet.

67. $h = -2.7x^2 + 30x + 6.5$

$$12 = -2.7x^2 + 30x + 6.5 \quad \text{Let } h = 12$$

$$2.7x^2 - 30x + 5.5 = 0$$

$$27x^2 - 300x + 55 = 0 \quad \text{Multiply by 10}$$

$$a = 27, b = -300, \text{ and } c = 55$$

$$x = \frac{-(-300) \pm \sqrt{(-300)^2 - 4(27)(55)}}{2(27)}$$

$$= \frac{300 \pm \sqrt{90,000 - 5940}}{54}$$

$$= \frac{300 \pm \sqrt{84,060}}{54}$$

$$\approx .2 \text{ or } 10.9$$

The ball is 12 feet above the moon's surface twice, at approximately .2 second and 10.9 seconds.

69. Let $x =$ the length of the side of the square.

If the area minus the length of a side is 870, then

$$x^2 - x = 870$$
$$x^2 - x - 870 = 0.$$

$a = 1, b = -1$, and $c = -870$

$$x = \frac{-(-1) \pm \sqrt{(-1)^2 - 4(1)(-870)}}{2(1)}$$

$$= \frac{1 \pm \sqrt{1 + 3480}}{2}$$

$$= \frac{1 \pm \sqrt{3481}}{2}$$

$$= \frac{1 \pm 59}{2}$$

$$x = \frac{1 + 59}{2} = \frac{60}{2} = 30$$

or $x = \frac{1 - 59}{2} = \frac{-58}{2} = -29$

We reject the solution -29.

The side of the square is 30 units.

71. $4x^2 - 4x + 3 = 0$
$a = 4, b = -4, c = 3$

$$b^2 - 4ac = (-4)^2 - 4(4)(3)$$
$$= 16 - 48 = -32$$

Since the discriminant is negative, the solutions are (d) two different imaginary numbers.

73. $7x^2 - 32x - 15 = 0$
$a = 7, b = -32, c = -15$

$$b^2 - 4ac = (-32)^2 - 4(7)(-15)$$
$$= 1024 + 420 = 1444$$

Since $38^2 = 1444$, the discriminant is positive and the square of an integer. The solutions are (a) two different rational numbers.

75. $$4y^2 + 36y = -81$$
$$4y^2 + 36y + 81 = 0$$
$a = 4, b = 36, c = 81$

$$b^2 - 4ac = 36^2 - 4(4)(81)$$
$$= 1296 - 1296 = 0$$

Since the discriminant is 0, the solution is (b) exactly one rational number.

77. $$25k^2 = -20k + 2$$
$$25k^2 + 20k - 2 = 0$$
$a = 25, b = 20, c = -2$

$$b^2 - 4ac = 20^2 - 4(25)(-2)$$
$$= 400 + 200 = 600$$

Since the discriminant is positive but not the square of an integer, the solutions are (c) two different irrational numbers.

78. The discriminant, $b^2 - 4ac$, is the radicand in the quadratic formula.

79. **(a)** $18x^2 - 9x - 2$

$a = 18, b = -9$, and $c = -2$

$$b^2 - 4ac = (-9)^2 - 4(18)(-2)$$
$$= 81 + 144 = 225$$

(b) $5x^2 + 7x - 6$

$a = 5, b = 7$, and $c = -6$

$$b^2 - 4ac = 7^2 - 4(5)(-6)$$
$$= 49 + 120 = 169$$

(c) $48x^2 + 14x + 1$

$a = 48, b = 14$, and $c = 1$

$$b^2 - 4ac = 14^2 - 4(48)(1)$$
$$= 196 - 192 = 4$$

(d) $x^2 - 5x - 24$

$a = 1, b = -5$, and $c = -24$

$$b^2 - 4ac = (-5)^2 - 4(1)(-24)$$
$$= 25 + 96 = 121$$

80. Each discriminant is a perfect square:

$$225 = 15^2$$
$$169 = 13^2$$
$$4 = 2^2$$
$$121 = 11^2.$$

81. **(a)** $18x^2 - 9x - 2 = (3x - 2)(6x + 1)$

(b) $5x^2 + 7x - 6 = (5x - 3)(x + 2)$

(c) $48x^2 + 14x + 1 = (8x + 1)(6x + 1)$

(d) $x^2 - 5x - 24 = (x - 8)(x + 3)$

82. **(a)** $2x^2 + x - 5$

$a = 2, b = 1$, and $c = -5$

$$b^2 - 4ac = 1^2 - 4(2)(-5)$$
$$= 1 + 40 = 41$$

(b) $2x^2 + x + 5$

$a = 2, b = 1$, and $c = 5$

$$b^2 - 4ac = 1^2 - 4(2)(5)$$
$$= 1 - 40 = -39$$

(c) $x^2 + 6x + 6$

$a = 1, b = 6$, and $c = 6$

$$b^2 - 4ac = 6^2 - 4(1)(6)$$
$$= 36 - 24 = 12$$

(d) $3x^2 + 2x - 9$

$a = 3$, $b = 2$, and $c = -9$

$$b^2 - 4ac = 2^2 - 4(3)(-9)$$
$$= 4 + 108 = 112$$

83. None of the discriminants in Exercise 82 is a perfect square.

84. The trinomial $ax^2 + bx + c$ is factorable if the discriminant $b^2 - 4ac$ is a perfect square.

(a) $42x^2 + 117x + 66$

$a = 42$, $b = 117$, and $c = 66$

$$b^2 - 4ac = (117)^2 - 4(42)(66)$$
$$= 13,689 - 11,088$$
$$= 2601 = 51^2$$

Since 2601 is a perfect square, the trinomial is factorable.

(b) $99x^2 + 186x - 24$

$a = 99$, $b = 186$, and $c = -24$

$$b^2 - 4ac = (186)^2 - 4(99)(-24)$$
$$= 34,596 + 9504$$
$$= 44,100 = 210^2$$

Since $44,100$ is a perfect square, the trinomial is factorable.

(c) $58x^2 + 184x + 27$

$a = 58$, $b = 184$, and $c = 27$

$$b^2 - 4ac = (184)^2 - 4(58)(27)$$
$$= 33,856 - 6264$$
$$= 27,592$$

Since $27,592$ is not a perfect square, the trinomial is not factorable.

Section 11.4

1. $\sqrt{-9} = \sqrt{-1 \cdot 9} = \sqrt{-1} \cdot \sqrt{9}$
$$= i \cdot 3 = 3i$$

3. $\sqrt{-20} = i\sqrt{20}$
$$= i\sqrt{4 \cdot 5}$$
$$= i \cdot 2 \cdot \sqrt{5}$$
$$= 2i\sqrt{5}$$

5. $\sqrt{-18} = i\sqrt{18}$
$$= i\sqrt{9 \cdot 2}$$
$$= i \cdot 3 \cdot \sqrt{2}$$
$$= 3i\sqrt{2}$$

7. $\sqrt{-125} = i\sqrt{125}$
$$= i\sqrt{25 \cdot 5}$$
$$= i \cdot 5 \cdot \sqrt{5}$$
$$= 5i\sqrt{5}$$

9. $(2 + 8i) + (3 - 5i)$
$$= (2 + 3) + (8 - 5)i$$
Add real parts; add imaginary parts
$$= 5 + 3i$$

11. $(8 - 3i) - (2 + 6i)$
$$= (8 - 3i) + (-2 - 6i)$$
Change $2 + 6i$ to its negative; then add
$$= (8 - 2) + (-3 - 6)i$$
$$= 6 - 9i$$

13. $(3 - 4i) + (6 - i) - (3 + 2i)$
$$= (9 - 5i) - (3 + 2i)$$
Add the first two complex numbers
$$= (9 - 5i) + (-3 - 2i)$$
Use the definition of subtraction
$$= (9 - 3) + (-5 - 2)i$$
$$= 6 - 7i$$

15. $(3 + 2i)(4 - i)$
$$= 3(4) + 3(-i) + (2i)(4) + (2i)(-i) \; FOIL$$
$$= 12 - 3i + 8i - 2i^2$$
$$= 12 + 5i - 2(-1) \qquad\qquad i^2 = -1$$
$$= 12 + 5i + 2$$
$$= 14 + 5i$$

17. $(5 - 4i)(3 - 2i)$
$$= 15 - 10i - 12i + 8i^2 \qquad FOIL$$
$$= 15 - 22i + 8(-1) \qquad i^2 = -1$$
$$= 15 - 22i - 8$$
$$= 7 - 22i$$

19. $(3 + 6i)(3 - 6i)$
$$= 3(3) + 3(-6i) + (6i)(3)$$
$$\qquad + (6i)(-6i) \qquad FOIL$$
$$= 9 - 18i + 18i - 36i^2$$
$$= 9 + 0 - 36(-1) \qquad i^2 = -1$$
$$= 9 + 36 = 45$$

This product can also be found by using the formula for the product of the sum and difference of two terms, $(a + b)(a - b) = a^2 - b^2$.

$(3 + 6i)(3 - 6i)$
$$= 3^2 - (6i)^2$$
$$= 9 - 36i^2$$
$$= 9 - 36(-1) \quad i^2 = -1$$
$$= 9 + 36 = 45$$

continued

The quickest way to find the product is to notice that $3 + 6i$ and $3 - 6i$ are conjugates and use the rule for the product of conjugates.

$$(a + bi)(a - bi) = a^2 + b^2$$
$$(3 + 6i)(3 - 6i) = 3^2 + 6^2$$

Let a = 3, b = 6
$$= 9 + 36 = 45$$

21. $\sqrt{(-169)} + \sqrt{(-25)}$

$$= i\sqrt{169} + i\sqrt{25}$$

$$= i \cdot 13 + i \cdot 5 = 18i$$

23. $(5 - i) + (-2 + 3i) - (7 + 2i)$

$$= (3 + 2i) + (-7 - 2i)$$

$$= -4 + 0i = -4$$

25. $\text{conj}(6 - 4i) = 6 + 4i$

27. $(15 - 5i)/(3 + i)$

$$= \frac{15 - 5i}{3 + i} \cdot \frac{3 - i}{3 - i} \quad \text{\textit{Multiply numerator and denominator by conjugate of denominator}}$$

$$= \frac{45 - 15i - 15i + 5i^2}{3^2 + 1^2} \quad \text{\textit{FOIL; Product of conjugates}}$$

$$= \frac{45 - 5 - 30i}{9 + 1} \quad i^2 = -1$$

$$= \frac{40 - 30i}{10} = \frac{10(4 - 3i)}{10}$$

$$= 4 - 3i$$

29. $\dfrac{17 + i}{5 + 2i}$

$$= \frac{17 + i}{5 + 2i} \cdot \frac{5 - 2i}{5 - 2i} \quad \text{\textit{Multiply numerator and denominator by conjugate of denominator}}$$

$$= \frac{85 - 34i + 5i - 2i^2}{5^2 + 2^2} \quad \text{\textit{FOIL; Product of conjugates}}$$

$$= \frac{85 - 29i - 2(-1)}{25 + 4} \quad i^2 = -1$$

$$= \frac{87 - 29i}{29} = \frac{29(3 - i)}{29}$$

$$= 3 - i \quad \text{\textit{Standard form}}$$

31. $\dfrac{40}{2 + 6i} = \dfrac{2(20)}{2(1 + 3i)}$

$$= \frac{20}{1 + 3i}$$

$$= \frac{20}{1 + 3i} \cdot \frac{1 - 3i}{1 - 3i}$$

$$= \frac{20(1 - 3i)}{(1 + 3i)(1 - 3i)}$$

$$= \frac{20(1 - 3i)}{1 - 9i^2}$$

$$= \frac{20(1 - 3i)}{1 - 9(-1)}$$

$$= \frac{20(1 - 3i)}{10}$$

$$= 2(1 - 3i)$$

$$= 2 - 6i \quad \text{\textit{Standard form}}$$

33. $\dfrac{i}{4 - 3i} = \dfrac{i}{4 - 3i} \cdot \dfrac{4 + 3i}{4 + 3i}$

$$= \frac{i(4 + 3i)}{(4 - 3i)(4 + 3i)}$$

$$= \frac{4i + 3i^2}{16 - 9i^2}$$

$$= \frac{4i + 3(-1)}{16 - 9(-1)}$$

$$= \frac{-3 + 4i}{25}$$

$$= -\frac{3}{25} + \frac{4}{25}i \quad \text{\textit{Standard form}}$$

35. $\dfrac{-29 - 3i}{2 + 9i} = \dfrac{-29 - 3i}{2 + 9i} \cdot \dfrac{2 - 9i}{2 - 9i}$

$$= \frac{-58 + 261i - 6i + 27i^2}{4 - 81i^2}$$

$$= \frac{-58 + 255i + 27(-1)}{4 - 81(-1)}$$

$$= \frac{-85 + 255i}{85}$$

$$= -\frac{85}{85} + \frac{255}{85}i$$

$$= -1 + 3i \quad \text{\textit{Standard form}}$$

36. $(-1 + 3i)(2 + 9i)$

$$= -2 - 9i + 6i + 27i^2$$

$$= -2 - 3i + 27(-1)$$

$$= -29 - 3i$$

The product of $-1 + 3i$ and $2 + 9i$ is $-29 - 3i$, which *is* the original dividend.

37. $\dfrac{4 - 3i}{i} = \dfrac{4 - 3i}{0 + i} \cdot \dfrac{0 - i}{0 - i}$

$$= \frac{(4 - 3i)(-i)}{(i)(-i)}$$

$$= \frac{-4i + 3i^2}{-i^2}$$

$$= \frac{-4i + 3(-1)}{-(-1)}$$

$$= \frac{-3 - 4i}{1}$$

$$= -3 - 4i$$

38. $(-3 - 4i)i = -3i - 4i^2$

$$= -3i - 4(-1)$$

$$= 4 - 3i$$

The product of $-3 - 4i$ and i is $4 - 3i$, which *is* the original dividend.

39. $(3 - 2i)(4 + i) = 12 + 3i - 8i - 2i^2$

$$= 12 - 5i - 2(-1)$$
$$= 14 - 5i$$

Because the product is the dividend, the given statement is true.

40. The answer to a division problem involving complex numbers is correct if the product of the quotient and divisor is equal to the dividend.

41. $(a + 1)^2 = -4$

$a + 1 = \sqrt{-4}$ or $a + 1 = -\sqrt{-4}$

Square root property

$a + 1 = 2i$ or $a + 1 = -2i$

$a = -1 + 2i$ or $a = -1 - 2i$

The solution set is $\{-1 \pm 2i\}$.

43. $(k - 3)^2 = -5$

$k - 3 = \sqrt{-5}$ or $k - 3 = -\sqrt{-5}$

$k - 3 = i\sqrt{5}$ or $k - 3 = -i\sqrt{5}$

$k = 3 + i\sqrt{5}$ or $k = 3 - i\sqrt{5}$

The solution set is $\left\{3 \pm i\sqrt{5}\right\}$.

45. $(3x + 2)^2 = -18$

$$3x + 2 = \pm\sqrt{-18}$$
$$3x + 2 = \pm i\sqrt{18}$$
$$3x + 2 = \pm 3i\sqrt{2}$$
$$3x = -2 \pm 3i\sqrt{2}$$
$$x = \frac{-2 \pm 3i\sqrt{2}}{3}$$
$$x = -\frac{2}{3} \pm i\sqrt{2} \quad \textit{Standard form}$$

The solution set is $\left\{-\frac{2}{3} \pm i\sqrt{2}\right\}$.

47. $m^2 - 2m + 2 = 0$

$a = 1, b = -2,$ and $c = 2$

$$m = \frac{-b \pm \sqrt{b^2 - 4ac}}{2a}$$

$$m = \frac{-(-2) \pm \sqrt{(-2)^2 - 4(1)(2)}}{2(1)}$$

$$= \frac{2 \pm \sqrt{4 - 8}}{2} = \frac{2 \pm \sqrt{-4}}{2}$$

$$= \frac{2 \pm 2i}{2} = \frac{2(1 \pm i)}{2}$$

$$= 1 \pm i \qquad\qquad \textit{Lowest terms}$$

The solution set is $\{1 \pm i\}$.

49. $2r^2 + 3r + 5 = 0$

$a = 2, b = 3,$ and $c = 5$

$$r = \frac{-b \pm \sqrt{b^2 - 4ac}}{2a}$$

$$r = \frac{-3 \pm \sqrt{3^2 - 4(2)(5)}}{2(2)}$$

$$= \frac{-3 \pm \sqrt{9 - 40}}{4} = \frac{-3 \pm \sqrt{-31}}{4}$$

$$= \frac{-3 \pm i\sqrt{31}}{4} = -\frac{3}{4} \pm \frac{\sqrt{31}}{4}i$$

The solution set is $\left\{-\frac{3}{4} \pm \frac{\sqrt{31}}{4}i\right\}$.

51. $p^2 - 3p + 4 = 0$

$a = 1, b = -3,$ and $c = 4$

$$p = \frac{-(-3) \pm \sqrt{(-3)^2 - 4(1)(4)}}{2(1)}$$

$$p = \frac{3 \pm \sqrt{9 - 16}}{2} = \frac{3 \pm \sqrt{-7}}{2}$$

$$= \frac{3 \pm i\sqrt{7}}{2}$$

$$= \frac{3}{2} \pm \frac{\sqrt{7}}{2}i \qquad \textit{Standard form}$$

The solution set is $\left\{\frac{3}{2} \pm \frac{\sqrt{7}}{2}i\right\}$.

53. $\qquad 5x^2 + 3 = 2x$

$5x^2 - 2x + 3 = 0 \qquad \textit{Standard form}$

$a = 5, b = -2,$ and $c = 3$

$$x = \frac{-b \pm \sqrt{b^2 - 4ac}}{2a}$$

$$x = \frac{-(-2) \pm \sqrt{(-2)^2 - 4(5)(3)}}{2(5)}$$

$$= \frac{2 \pm \sqrt{4 - 60}}{10} = \frac{2 \pm \sqrt{-56}}{10}$$

$$= \frac{2 \pm i\sqrt{4 \cdot 14}}{10} = \frac{2 \pm 2i\sqrt{14}}{10}$$

$$= \frac{2\left(1 \pm i\sqrt{14}\right)}{2 \cdot 5} \qquad \textit{Factor}$$

$$= \frac{1 \pm i\sqrt{14}}{5} \qquad \textit{Lowest terms}$$

continued

$$= \frac{1}{5} \pm \frac{\sqrt{14}}{5}i \qquad \text{Standard form}$$

The solution set is $\left\{ \frac{1}{5} \pm \frac{\sqrt{14}}{5}i \right\}$.

55.
$$2m^2 + 7 = -2m$$
$$2m^2 + 2m + 7 = 0 \qquad \text{Standard form}$$
$$a = 2, b = 2, \text{ and } c = 7$$

$$m = \frac{-b \pm \sqrt{b^2 - 4ac}}{2a}$$

$$m = \frac{-2 \pm \sqrt{2^2 - 4(2)(7)}}{2(2)}$$

$$= \frac{-2 \pm \sqrt{4 - 56}}{4} = \frac{-2 \pm \sqrt{-52}}{4}$$

$$= \frac{-2 \pm i\sqrt{52}}{4} = \frac{-2 \pm 2i\sqrt{13}}{4}$$

$$= \frac{2\left(-1 \pm i\sqrt{13}\right)}{4} \qquad \text{Factor}$$

$$= \frac{-1 \pm i\sqrt{13}}{2} \qquad \text{Lowest terms}$$

$$= -\frac{1}{2} \pm \frac{\sqrt{13}}{2}i \qquad \text{Standard form}$$

The solution set is $\left\{ -\frac{1}{2} \pm \frac{\sqrt{13}}{2}i \right\}$.

57.
$$r^2 + 3 = r$$
$$r^2 - r + 3 = 0 \qquad \text{Standard form}$$
$$a = 1, b = -1, \text{ and } c = 3$$

$$r = \frac{-b \pm \sqrt{b^2 - 4ac}}{2a}$$

$$r = \frac{-(-1) \pm \sqrt{(-1)^2 - 4(1)(3)}}{2(1)}$$

$$= \frac{1 \pm \sqrt{1 - 12}}{2} = \frac{1 \pm \sqrt{-11}}{2}$$

$$= \frac{1 \pm i\sqrt{11}}{2} = \frac{1}{2} \pm \frac{\sqrt{11}}{2}i$$

The solution set is $\left\{ \frac{1}{2} \pm \frac{\sqrt{11}}{2}i \right\}$.

59. If the discriminant $b^2 - 4ac$ is negative, then the quadratic equation will have solutions that are not real numbers.

61. "Every real number is a complex number" is *true*. A real number m can be represented as $m + 0i$.

63. "Every complex number is a real number" is *false*. Examples of complex numbers which are not real numbers are $3i, -i\sqrt{2}, 4 + 6i$, and $\sqrt{3} - 2i$.

65. To add complex numbers, add the real parts to find the real part of the sum and add the imaginary parts to find the imaginary part of the sum. For example,
$$(5 + 2i) + (6 + 3i) = (5 + 6) + (2 + 3)i$$
$$= 11 + 5i.$$

Subtraction is done similarly, but we subtract the real and imaginary parts instead:
$$(5 + 2i) - (6 + 3i) = (5 - 6) + (2 - 3)i$$
$$= -1 - i.$$

To multiply complex numbers, use the FOIL method and replace i^2 with -1. Then combine terms. For example,
$$(5 + 2i)(6 + 3i) = 30 + 15i + 12i + 6i^2$$
$$= 30 + 15i + 12i + 6(-1) = 24 + 27i.$$

To divide complex numbers, multiply both the numerator and the denominator by the conjugate of the denominator. For example,
$$\frac{3 + i}{2 - i} = \frac{3 + i}{2 - i} \cdot \frac{2 + i}{2 + i} = \frac{6 + 5i + i^2}{4 - i^2}$$
$$= \frac{5 + 5i}{5} = 1 + i.$$

Section 11.5

1. $\frac{14}{x} = x - 5$

This is a rational equation, so multiply both sides by the LCD, x.

3. $\left(r^2 + r\right)^2 - 8\left(r^2 + r\right) + 12 = 0$

This is quadratic in form, so substitute a variable for $r^2 + r$.

5. Many answers are possible. See the chart in the text for possible reasons for agreeing or not.

7. $$\frac{14}{x} = x - 5$$

To clear the fraction, multiply each term by the LCD, x.
$$x\left(\frac{14}{x}\right) = x(x) - 5x$$
$$14 = x^2 - 5x$$
$$-x^2 + 5x + 14 = 0$$
$$x^2 - 5x - 14 = 0$$
$$(x + 2)(x - 7) = 0$$

$x + 2 = 0$ or $x - 7 = 0$

$x = -2$ or $x = 7$

Check $x = -2$: $-7 \overset{?}{=} -7$ *True*

Check $x = 7$: $2 \overset{?}{=} 2$ *True*

Solution set: $\{-2, 7\}$

9.
$$1 - \frac{3}{x} - \frac{28}{x^2} = 0$$

Multiply by the LCD, x^2.

$$x^2(1) - x^2\left(\frac{3}{x}\right) - x^2\left(\frac{28}{x^2}\right) = x^2 \cdot 0$$

$$x^2 - 3x - 28 = 0$$

$$(x + 4)(x - 7) = 0$$

$x + 4 = 0$ or $x - 7 = 0$

$x = -4$ or $x = 7$

Check $x = -4$: $1 + \frac{3}{4} - \frac{7}{4} \overset{?}{=} 0$ *True*

Check $x = 7$: $1 - \frac{3}{7} - \frac{4}{7} \overset{?}{=} 0$ *True*

Solution set: $\{-4, 7\}$

11.
$$\frac{1}{x} + \frac{2}{x+2} = \frac{17}{35}$$

Multiply by the LCD, $35x(x + 2)$.

$$35x(x+2)\left(\frac{1}{x}\right) + 35x(x+2)\left(\frac{2}{x+2}\right)$$
$$= 35x(x+2)\left(\frac{17}{35}\right)$$

$$35(x + 2) + 35x(2) = 17x(x + 2)$$

$$35x + 70 + 70x = 17x^2 + 34x$$

$$70 + 105x = 17x^2 + 34x$$

$$0 = 17x^2 - 71x - 70$$

$$0 = (17x + 14)(x - 5)$$

$17x + 14 = 0$ or $x - 5 = 0$

$x = -\dfrac{14}{17}$ or $x = 5$

Check $x = -\dfrac{14}{17}$: $-\dfrac{17}{14} + \dfrac{17}{10} \overset{?}{=} \dfrac{17}{35}$ *True*

Check $x = 5$: $\dfrac{1}{5} + \dfrac{2}{7} \overset{?}{=} \dfrac{17}{35}$ *True*

Solution set: $\left\{-\dfrac{14}{17}, 5\right\}$

13.
$$\frac{2}{x+1} + \frac{3}{x+2} = \frac{7}{2}$$

Multiply by the LCD, $2(x + 1)(x + 2)$.

$$2(x+1)(x+2)\left(\frac{2}{x+1} + \frac{3}{x+2}\right)$$
$$= 2(x+1)(x+2)\left(\frac{7}{2}\right)$$

$$2(x+2)(2) + 2(x+1)(3)$$
$$= (x+1)(x+2)(7)$$

$$4x + 8 + 6x + 6 = (x^2 + 3x + 2)(7)$$

$$10x + 14 = 7x^2 + 21x + 14$$

$$7x^2 + 11x = 0$$

$$x(7x + 11) = 0$$

$x = 0$ or $7x + 11 = 0$

$$x = -\frac{11}{7}$$

Check $x = -\dfrac{11}{7}$: $-\dfrac{7}{2} + 7 \overset{?}{=} \dfrac{7}{2}$ *True*

Check $x = 0$: $2 + \dfrac{3}{2} \overset{?}{=} \dfrac{7}{2}$ *True*

Solution set: $\left\{-\dfrac{11}{7}, 0\right\}$

15.
$$\frac{3}{2x} - \frac{1}{2(x+2)} = 1$$

Multiply by the LCD, $2x(x + 2)$.

$$2x(x+2)\left(\frac{3}{2x} - \frac{1}{2(x+2)}\right)$$
$$= 2x(x+2) \cdot 1$$

$$3(x + 2) - x(1) = 2x(x + 2)$$

$$3x + 6 - x = 2x^2 + 4x$$

$$0 = 2x^2 + 2x - 6$$

$$0 = x^2 + x - 3$$

Use $a = 1$, $b = 1$, $c = -3$ in the quadratic formula.

$$x = \frac{-b \pm \sqrt{b^2 - 4ac}}{2a}$$

$$x = \frac{-1 \pm \sqrt{1^2 - 4(1)(-3)}}{2(1)}$$

$$= \frac{-1 \pm \sqrt{1 + 12}}{2} = \frac{-1 \pm \sqrt{13}}{2}$$

Use a calculator to check both potential solutions. Both solutions check.

Solution set: $\left\{\dfrac{-1 + \sqrt{13}}{2}, \dfrac{-1 - \sqrt{13}}{2}\right\}$

17. The grader's rate $= \dfrac{1 \text{ job}}{\text{time to finish}}$

$$= \frac{1}{m} \text{ job per hour.}$$

19. Let $x =$ rate of the boat in still water.
With the speed of the current at 15 mph, then
$x - 15 =$ rate going upstream and
$x + 15 =$ rate going downstream.
Complete a table using the information in the problem, the rates given above, and the formula $d = rt$ or $t = \dfrac{d}{r}$.

	d	r	t
Upstream	4	$x - 15$	$\dfrac{4}{x - 15}$
Downstream	16	$x + 15$	$\dfrac{16}{x + 15}$

The time, 48 min, is written as $\dfrac{48}{60} = \dfrac{4}{5}$ hr. The time upstream plus the time downstream equals $\dfrac{4}{5}$. So, from the table, the equation is written as
$$\frac{4}{x - 15} + \frac{16}{x + 15} = \frac{4}{5}.$$
Multiply by the LCD, $5(x - 15)(x + 15)$.
$$5(x - 15)(x + 15)\left(\frac{4}{x - 15} + \frac{16}{x + 15}\right)$$
$$= 5(x - 15)(x + 15) \cdot \frac{4}{5}$$
$$20(x + 15) + 80(x - 15)$$
$$= 4(x - 15)(x + 15)$$
$$20x + 300 + 80x - 1200$$
$$= 4\left(x^2 - 225\right)$$
$$100x - 900 = 4x^2 - 900$$
$$0 = 4x^2 - 100x$$
$$0 = 4x(x - 25)$$

$4x = 0$ or $x - 25 = 0$
$x = 0$ or $x = 25$

Reject $x = 0$ mph as a possible boat speed.
Yoshiaki's boat had a top speed of 25 mph.

21. Let $x =$ Harry's average speed.
$x - 20 =$ Karen's average speed.

	d	r	t
Harry	300	x	$\dfrac{300}{x}$
Karen	300	$x - 20$	$\dfrac{300}{x - 20}$

It takes Harry $1\dfrac{1}{4}$ or $\dfrac{5}{4}$ hours less time than Karen.
$$\frac{300}{x} = \frac{300}{x - 20} - \frac{5}{4}$$
Multiply by the LCD, $4x(x - 20)$.

$$4x(x - 20)\left(\frac{300}{x}\right) = 4x(x - 20)\left(\frac{300}{x - 20} - \frac{5}{4}\right)$$
$$1200(x - 20) = 4x(300) - x(x - 20) \cdot 5$$
$$1200x - 24,000 = 1200x - 5x^2 + 100x$$
$$5x^2 - 100x - 24,000 = 0$$
$$x^2 - 20x - 4800 = 0$$
$$(x - 80)(x + 60) = 0$$

$x - 80 = 0$ or $x + 60 = 0$
$x = 80$ or $x = -60$

Reject $x = -60$. Harry's average speed is 80 km/hr.

23. Let $x =$ the number of minutes it takes for the cold water tap alone to fill the washer.
$x + 9 =$ the number of minutes it takes for the hot water tap alone to fill the washer.

Working together, both taps can fill the washer in 6 minutes. Complete a chart using the above information.

Tap	Rate	Time	Fractional Part of Washer Filled
Cold	$\dfrac{1}{x}$	6	$\dfrac{1}{x}(6)$
Hot	$\dfrac{1}{x + 9}$	6	$\dfrac{1}{x + 9}(6)$

Since together the hot and cold taps fill one washer, the sum of their fractional parts is 1; that is,
$$\frac{6}{x} + \frac{6}{x + 9} = 1.$$
Multiply by the LCD, $x(x + 9)$.
$$x(x + 9)\left(\frac{6}{x} + \frac{6}{x + 9}\right) = x(x + 9) \cdot 1$$
$$6(x + 9) + 6x = x(x + 9)$$
$$6x + 54 + 6x = x^2 + 9x$$
$$0 = x^2 - 3x - 54$$
$$0 = (x - 9)(x + 6)$$

$x - 9 = 0$ or $x + 6 = 0$
$x = 9$ or $x = -6$

Reject -6 as a possible time. The cold water tap can fill the washer in 9 minutes.

25.
$$2x = \sqrt{11x + 3}$$
$$(2x)^2 = \left(\sqrt{11x + 3}\right)^2$$
$$4x^2 = 11x + 3$$
$$4x^2 - 11x - 3 = 0$$
$$(4x + 1)(x - 3) = 0$$

$$4x + 1 = 0 \quad \text{or} \quad x - 3 = 0$$
$$x = -\frac{1}{4} \quad \text{or} \quad x = 3$$

Check $x = -\frac{1}{4}$: $\quad -\frac{1}{2} \overset{?}{=} \sqrt{\frac{1}{4}}$ $\quad$ *False*

Check $x = 3$: $\quad 6 \overset{?}{=} \sqrt{36}$ $\quad$ *True*

Solution set: $\{3\}$

27.
$$3y = (16 - 10y)^{1/2}$$
$$(3y)^2 = \left[(16 - 10y)^{1/2}\right]^2$$
$$9y^2 = 16 - 10y$$
$$9y^2 + 10y - 16 = 0$$
$$(9y - 8)(y + 2) = 0$$

$$9y - 8 = 0 \quad \text{or} \quad y + 2 = 0$$
$$y = \frac{8}{9} \quad \text{or} \quad y = -2$$

Check $y = \frac{8}{9}$: $\quad \frac{8}{3} \overset{?}{=} \sqrt{\frac{64}{9}}$ $\quad$ *True*

Check $y = -2$: $-6 \overset{?}{=} \sqrt{36}$ $\quad$ *False*

Solution set: $\left\{ \frac{8}{9} \right\}$

29.
$$p - 2\sqrt{p} = 8$$
$$p - 8 = 2\sqrt{p}$$
$$(p - 8)^2 = \left(2\sqrt{p}\right)^2$$
$$p^2 - 16p + 64 = 4p$$
$$p^2 - 20p + 64 = 0$$
$$(p - 4)(p - 16) = 0$$

$$p - 4 = 0 \quad \text{or} \quad p - 16 = 0$$
$$p = 4 \quad \text{or} \quad p = 16$$

Check $p = 4$: $\quad 4 - 4 \overset{?}{=} 8$ $\quad$ *False*

Check $p = 16$: $16 - 8 \overset{?}{=} 8$ $\quad$ *True*

Solution set: $\{16\}$

31.
$$m = \sqrt{\frac{6 - 13m}{5}}$$
$$m^2 = \frac{6 - 13m}{5}$$
$$5m^2 = 6 - 13m$$
$$5m^2 + 13m - 6 = 0$$
$$(5m - 2)(m + 3) = 0$$

$$5m - 2 = 0 \quad \text{or} \quad m + 3 = 0$$
$$m = \frac{2}{5} \quad \text{or} \quad m = -3$$

Check $m = \frac{2}{5}$: $\quad \frac{2}{5} \overset{?}{=} \sqrt{\frac{4}{25}}$ $\quad$ *True*

Check $m = -3$: $-3 \overset{?}{=} \sqrt{9}$ $\quad$ *False*

Solution set: $\left\{ \frac{2}{5} \right\}$

33.
$$.126\pi r = \sqrt{4.7r}$$
$$(.126\pi r)^2 = \left(\sqrt{4.7r}\right)^2$$
$$(.126\pi)^2 r^2 = 4.7r$$
$$(.126\pi)^2 r^2 - 4.7r = 0$$
$$r\left[(.126\pi)^2 r - 4.7\right] = 0$$

$$r = 0 \quad \text{or} \quad r = \frac{4.7}{(.126\pi)^2} \approx 30$$

Reject $r = 0$, the distance is about 30 meters.

35. $t^4 - 18t^2 + 81 = 0$
Let $u = t^2$, so $u^2 = t^4$. The equation becomes
$$u^2 - 18u + 81 = 0.$$
$$(u - 9)^2 = 0$$
$$u - 9 = 0$$
$$u = 9$$

To find t, substitute t^2 for u.
$$t^2 = 9$$
$$t = 3 \quad \text{or} \quad t = -3$$

Check $t = \pm 3$: $81 - 162 + 81 \overset{?}{=} 0$ $\quad$ *True*

Solution set: $\{-3, 3\}$

37. $4k^4 - 13k^2 + 9 = 0$
Let $u = k^2$ and $u^2 = k^4$ to get
$$4u^2 - 13u + 9 = 0$$
$$(4u - 9)(u - 1) = 0$$
$$4u - 9 = 0 \quad \text{or} \quad u - 1 = 0$$
$$u = \frac{9}{4} \quad \text{or} \quad u = 1.$$

To find k, substitute k^2 for u.
$$k^2 = \frac{9}{4} \quad \text{or} \quad k^2 = 1$$
$$k = \pm\frac{3}{2} \quad \text{or} \quad k = \pm 1$$

Check $k = \pm\frac{3}{2}$: $\frac{81}{4} - \frac{117}{4} + 9 \overset{?}{=} 0$ $\quad$ *True*

Check $k = \pm 1$: $\quad 4 - 13 + 9 \overset{?}{=} 0$ $\quad$ *True*

Solution set: $\left\{ -\frac{3}{2}, -1, 1, \frac{3}{2} \right\}$

39. $(x + 3)^2 + 5(x + 3) + 6 = 0$

Let $u = x + 3$, so $u^2 = (x + 3)^2$.

$$u^2 + 5u + 6 = 0$$
$$(u + 3)(u + 2) = 0$$
$$u + 3 = 0 \quad \text{or} \quad u + 2 = 0$$
$$u = -3 \quad \text{or} \quad u = -2$$

To find x, substitute $x + 3$ for u.

$$x + 3 = -3 \quad \text{or} \quad x + 3 = -2$$
$$x = -6 \quad \text{or} \quad x = -5$$

Check $x = -6$: $9 - 15 + 6 \overset{?}{=} 0$ *True*

Check $x = -5$: $4 - 10 + 6 \overset{?}{=} 0$ *True*

Solution set: $\{-6, -5\}$

41. $(t + 5)^2 + 6 = 7(t + 5)$

Let $u = t + 5$, so $u^2 = (t + 5)^2$.

$$u^2 + 6 = 7u$$
$$u^2 - 7u + 6 = 0$$
$$(u - 6)(u - 1) = 0$$
$$u - 6 = 0 \quad \text{or} \quad u - 1 = 0$$
$$u = 6 \quad\quad u = 1$$

To find t, substitute $t + 5$ for u.

$$t + 5 = 6 \quad \text{or} \quad t + 5 = 1$$
$$t = 1 \quad\quad\quad t = -4$$

Check $t = 1$: $36 + 6 \overset{?}{=} 42$ *True*

Check $t = -4$: $1 + 6 \overset{?}{=} 7$ *True*

Solution set: $\{-4, 1\}$

43. $2 + \dfrac{5}{3k - 1} = \dfrac{-2}{(3k - 1)^2}$

Let $u = 3k - 1$, so $u^2 = (3k - 1)^2$.

$$2 + \frac{5}{u} = -\frac{2}{u^2}$$

Multiply by the LCD, u^2.

$$u^2\left(2 + \frac{5}{u}\right) = u^2\left(-\frac{2}{u^2}\right)$$
$$2u^2 + 5u = -2$$
$$2u^2 + 5u + 2 = 0$$
$$(2u + 1)(u + 2) = 0$$
$$2u + 1 = 0 \quad \text{or} \quad u + 2 = 0$$
$$u = -\frac{1}{2} \quad\quad\quad u = -2$$

To find k, substitute $3k - 1$ for u.

$$3k - 1 = -\frac{1}{2} \quad \text{or} \quad 3k - 1 = -2$$
$$3k = \frac{1}{2} \quad\quad\quad\quad 3k = -1$$
$$k = \frac{1}{6} \quad \text{or} \quad k = -\frac{1}{3}$$

Check $k = \dfrac{1}{6}$: $2 - 10 \overset{?}{=} -8$ *True*

Check $k = -\dfrac{1}{3}$: $2 - \dfrac{5}{2} \overset{?}{=} -\dfrac{1}{2}$ *True*

Solution set: $\left\{-\dfrac{1}{3}, \dfrac{1}{6}\right\}$

45. $2 - 6(m - 1)^{-2} = (m - 1)^{-1}$

Let $u = m - 1$ to get

$$2 - 6u^{-2} = u^{-1}$$

or

$$2 - \frac{6}{u^2} = \frac{1}{u}.$$

Multiply by the LCD, u^2.

$$2u^2 - 6 = u$$
$$2u^2 - u - 6 = 0$$
$$(2u + 3)(u - 2) = 0$$
$$2u + 3 = 0 \quad \text{or} \quad u - 2 = 0$$
$$u = -\frac{3}{2} \quad \text{or} \quad u = 2$$

To find m, substitute $m - 1$ for u.

$$m - 1 = -\frac{3}{2} \quad \text{or} \quad m - 1 = 2$$
$$m = -\frac{1}{2} \quad \text{or} \quad m = 3$$

Check $m = -\dfrac{1}{2}$: $2 - \dfrac{8}{3} \overset{?}{=} -\dfrac{2}{3}$ *True*

Check $m = 3$: $2 - \dfrac{3}{2} \overset{?}{=} \dfrac{1}{2}$ *True*

Solution set: $\left\{-\dfrac{1}{2}, 3\right\}$

47. $4k^{4/3} - 13k^{2/3} + 9 = 0$

Let $x = k^{2/3}$, so $x^2 = k^{4/3}$.

$$4x^2 - 13x + 9 = 0$$
$$(4x - 9)(x - 1) = 0$$
$$4x - 9 = 0 \quad \text{or} \quad x - 1 = 0$$
$$x = \frac{9}{4} \quad \text{or} \quad x = 1$$
$$k^{2/3} = \frac{9}{4} \quad \text{or} \quad k^{2/3} = 1$$
$$\left(k^{2/3}\right)^3 = \left(\frac{9}{4}\right)^3 \quad \text{or} \quad \left(k^{2/3}\right)^3 = 1^3$$
$$k^2 = \frac{729}{64} \quad \text{or} \quad k^2 = 1$$
$$k = \pm\sqrt{\frac{729}{64}} \quad \text{or} \quad k = \pm\sqrt{1}$$
$$k = \pm\frac{27}{8} \quad \text{or} \quad k = \pm 1$$

Check $k = \pm\dfrac{27}{8}$: $\dfrac{81}{4} - \dfrac{117}{4} + 9 \overset{?}{=} 0$ *True*

Check $k = \pm 1$: $4 - 13 + 9 \overset{?}{=} 0$ *True*

Solution set: $\left\{-\dfrac{27}{8}, -1, 1, \dfrac{27}{8}\right\}$

49. $x^{2/3} + x^{1/3} - 2 = 0$

Let $u = x^{1/3}$, so $u^2 = x^{2/3}$.

$u^2 + u - 2 = 0$

$(u + 2)(u - 1) = 0$

$u + 2 = 0$ or $u - 1 = 0$

$u = -2$ or $u = 1$

To find x, substitute $x^{1/3}$ for u.

$x^{1/3} = -2$ or $x^{1/3} = 1$

Cube both sides of each equation.

$\left(x^{1/3}\right)^3 = (-2)^3$ $\left(x^{1/3}\right)^3 = 1^3$

$x = -8$ or $x = 1$

Check $x = -8$: $4 - 2 - 2 \overset{?}{=} 0$ *True*

Check $x = 1$: $1 + 1 - 2 \overset{?}{=} 0$ *True*

Solution set: $\{-8, 1\}$

51. $2\left(1 + \sqrt{y}\right)^2 = 13\left(1 + \sqrt{y}\right) - 6$

Let $u = 1 + \sqrt{y}$.

$2u^2 = 13u - 6$

$2u^2 - 13u + 6 = 0$

$(2u - 1)(u - 6) = 0$

$2u - 1 = 0$ or $u - 6 = 0$

$u = \dfrac{1}{2}$ or $u = 6$

Replace u with $1 + \sqrt{y}$.

$1 + \sqrt{y} = \dfrac{1}{2}$ or $1 + \sqrt{y} = 6$

$\sqrt{y} = -\dfrac{1}{2}$ $\sqrt{y} = 5$

Not possible, $y = 25$

since $\sqrt{y} \ge 0$.

Check $y = 25$: $72 \overset{?}{=} 78 - 6$ *True*

Solution set: $\{25\}$

53. $2x^4 + x^2 - 3 = 0$

Let $m = x^2$, so $m^2 = x^4$.

$2m^2 + m - 3 = 0$

$(2m + 3)(m - 1) = 0$

$2m + 3 = 0$ or $m - 1 = 0$

$m = -\dfrac{3}{2}$ or $m = 1$

To find x, substitute x^2 for m.

$x^2 = -\dfrac{3}{2}$ or $x^2 = 1$

$x = \pm\sqrt{-\dfrac{3}{2}}$ $x = \pm\sqrt{1}$

$x = \pm\dfrac{\sqrt{3}}{\sqrt{2}} \cdot \dfrac{\sqrt{2}}{\sqrt{2}}i$ $x = \pm 1$

$x = \pm\dfrac{\sqrt{6}}{2}i$

Check $x = \pm\dfrac{\sqrt{6}}{2}i$: $\dfrac{9}{2} - \dfrac{3}{2} - 3 \overset{?}{=} 0$ *True*

Check $x = \pm 1$: $2 + 1 - 3 \overset{?}{=} 0$ *True*

Solution set: $\left\{-1, 1, -\dfrac{\sqrt{6}}{2}i, \dfrac{\sqrt{6}}{2}i\right\}$

55.   The solutions stated are the values of u, $u = \dfrac{1}{2}$ or $u = 1$. We must give the values of the original variable, m. Substitute $m - 1$ for u.

$u = \dfrac{1}{2}$ or $u = 1$

$m - 1 = \dfrac{1}{2}$ $m - 1 = 1$

$m = \dfrac{3}{2}$ or $m = 2$

Solution set: $\left\{\dfrac{3}{2}, 2\right\}$

57. $x^4 - 16x^2 + 48 = 0$

Let $m = x^2$, so $m^2 = x^4$.

$m^2 - 16m + 48 = 0$

$(m - 4)(m - 12) = 0$

$m - 4 = 0$ or $m - 12 = 0$

$m = 4$ or $m = 12$

To find x, substitute x^2 for m.

$x^2 = 4$ or $x^2 = 12$

$x = \pm\sqrt{4}$ $x = \pm\sqrt{12}$

$x = \pm 2$ or $x = \pm 2\sqrt{3}$

Check $x = \pm 2$: $16 - 64 + 48 \overset{?}{=} 0$ *True*

Check $x = \pm 2\sqrt{3}$: $144 - 192 + 48 \overset{?}{=} 0$ *True*

Solution set: $\left\{-2\sqrt{3}, -2, 2, 2\sqrt{3}\right\}$

59.
$$\sqrt{2x+3} = 2 + \sqrt{x-2}$$
Square both sides.
$$\left(\sqrt{2x+3}\right)^2 = \left(2 + \sqrt{x-2}\right)^2$$
$$2x + 3 = 4 + 4\sqrt{x-2} + (x-2)$$
$$2x + 3 = x + 2 + 4\sqrt{x-2}$$
Isolate the radical term on one side.
$$x + 1 = 4\sqrt{x-2}$$
Square both sides again.
$$(x+1)^2 = \left(4\sqrt{x-2}\right)^2$$
$$x^2 + 2x + 1 = 16(x-2)$$
$$x^2 + 2x + 1 = 16x - 32$$
$$x^2 - 14x + 33 = 0$$
$$(x-11)(x-3) = 0$$
$$x - 11 = 0 \quad \text{or} \quad x - 3 = 0$$
$$x = 11 \quad \text{or} \quad x = 3$$
Check $x = 11$: $\sqrt{25} \overset{?}{=} 2 + \sqrt{9}$ *True*

Check $x = 3$: $\sqrt{9} \overset{?}{=} 2 + \sqrt{1}$ *True*

Solution set: $\{3, 11\}$

61. $2m^6 + 11m^3 + 5 = 0$
Let $y = m^3$, so $y^2 = m^6$.
$$2y^2 + 11y + 5 = 0$$
$$(2y+1)(y+5) = 0$$
$$2y + 1 = 0 \quad \text{or} \quad y + 5 = 0$$
$$y = -\frac{1}{2} \quad \text{or} \quad y = -5$$
To find m, substitute m^3 for y.
$$m^3 = -\frac{1}{2} \quad \text{or} \quad m^3 = -5$$
Take the cube root of both sides of each equation.
$$m = \sqrt[3]{-\frac{1}{2}} \quad \text{or} \quad m = \sqrt[3]{-5}$$
$$m = -\sqrt[3]{\frac{1}{2}} \qquad m = -\sqrt[3]{5}$$
$$= -\frac{\sqrt[3]{1}}{\sqrt[3]{2}} \cdot \frac{\sqrt[3]{2^2}}{\sqrt[3]{2^2}}$$
$$= -\frac{\sqrt[3]{4}}{2}$$
Check $m = -\frac{\sqrt[3]{4}}{2}$: $\frac{1}{2} - \frac{11}{2} + 5 \overset{?}{=} 0$ *True*

Check $m = -\sqrt[3]{5}$: $50 - 55 + 5 \overset{?}{=} 0$ *True*

Solution set: $\left\{-\sqrt[3]{5}, -\frac{\sqrt[3]{4}}{2}\right\}$

63. $2 - (y-1)^{-1} = 6(y-1)^{-2}$
Let $m = (y-1)^{-1}$, so $m^2 = (y-1)^{-2}$.
$$2 - m = 6m^2$$
$$0 = 6m^2 + m - 2$$
$$0 = (3m+2)(2m-1)$$
$$3m + 2 = 0 \quad \text{or} \quad 2m - 1 = 0$$
$$m = -\frac{2}{3} \quad \text{or} \quad m = \frac{1}{2}$$
To find y, substitute $(y-1)^{-1}$ for m.
$$(y-1)^{-1} = -\frac{2}{3} \quad \text{or} \quad (y-1)^{-1} = \frac{1}{2}$$
$$\frac{1}{y-1} = -\frac{2}{3} \qquad \frac{1}{y-1} = \frac{1}{2}$$
$$3 = -2y + 2 \qquad 2 = y - 1$$
$$2y = -1 \qquad y = 3$$
$$y = -\frac{1}{2}$$
Check $y = -\frac{1}{2}$: $2 + \frac{2}{3} \overset{?}{=} \frac{8}{3}$ *True*

Check $y = 3$: $2 - \frac{1}{2} \overset{?}{=} \frac{3}{2}$ *True*

Solution set: $\left\{-\frac{1}{2}, 3\right\}$

For Exercises 65–69, use the equation
$$\frac{x^2}{(x-3)^2} + \frac{3x}{x-3} - 4 = 0.$$

65. Substituting 3 for x would cause both denominators to be 0, and division by 0 is undefined.

66. $(x-3)^2\left(\frac{x^2}{(x-3)^2}\right) + (x-3)^2\left(\frac{3x}{x-3}\right)$
$$\qquad - (x-3)^2 \cdot 4 = (x-3)^2 \cdot 0$$
$$x^2 + 3x(x-3) - 4\left(x^2 - 6x + 9\right) = 0$$
$$x^2 + 3x^2 - 9x - 4x^2 + 24x - 36 = 0$$
$$15x - 36 = 0$$
$$15x = 36$$
$$x = \frac{36}{15} = \frac{12}{5}$$
The solution is $\frac{12}{5}$.

67.
$$\frac{x^2}{(x-3)^2} + \frac{3x}{x-3} - 4 = 0$$
$$\left(\frac{x}{x-3}\right)^2 + 3\left(\frac{x}{x-3}\right) - 4 = 0$$

68. If a fraction is equal to 1, then the numerator must be equal to the denominator. But the numerator can never equal the denominator, since the denominator is 3 less than the numerator.

69. From Exercise 67,

$$\left(\frac{x}{x-3}\right)^2 + 3\left(\frac{x}{x-3}\right) - 4 = 0.$$

Let $t = \dfrac{x}{x-3}$, so $t^2 = \left(\dfrac{x}{x-3}\right)^2$.

$$t^2 + 3t - 4 = 0$$
$$(t-1)(t+4) = 0$$
$$t - 1 = 0 \quad \text{or} \quad t + 4 = 0$$
$$t = 1 \quad \text{or} \quad t = -4$$

To find x, substitute $\dfrac{x}{x-3}$ for t.

$$\frac{x}{x-3} = 1 \quad \text{or} \quad \frac{x}{x-3} = -4$$

The equation $\dfrac{x}{x-3} = 1$ has no solution since there is no value of x for which $x = x - 3$. (See Exercise 68.) Therefore, $t = 1$ is impossible.

$$\frac{x}{x-3} = -4$$
$$x = -4(x-3)$$
$$x = -4x + 12$$
$$5x = 12$$
$$x = \frac{12}{5}$$

Solution set: $\left\{\dfrac{12}{5}\right\}$

70. $x^2(x-3)^{-2} + 3x(x-3)^{-1} - 4 = 0$

Let $s = (x-3)^{-1}$, so $s^2 = (x-3)^{-2}$.

$$x^2 s^2 + 3xs - 4 = 0$$
$$(xs+4)(xs-1) = 0$$
$$xs + 4 = 0 \quad \text{or} \quad xs - 1 = 0$$
$$s = -\frac{4}{x} \quad \text{or} \quad s = \frac{1}{x}$$

To find x, substitute $\dfrac{1}{x-3}$ for s.

$$\frac{1}{x-3} = -\frac{4}{x} \quad \text{or} \quad \frac{1}{x-3} = \frac{1}{x}$$
$$x = -4(x-3) \qquad\qquad$$
$$x = -4x + 12 \qquad x = x - 3$$
$$5x = 12 \qquad\qquad 0 = -3 \;\text{False}$$
$$x = \frac{12}{5} \qquad \text{Thus, } s = \frac{1}{x} \text{ is}$$
$$\text{impossible.}$$

Solution set: $\left\{\dfrac{12}{5}\right\}$

Section 11.6

1. The first step in solving a formula that has the specified variable in the denominator is to multiply both sides by the LCD to clear the equation of fractions.

3. We must recognize that a formula like

$$gw^2 = kw + 24$$

is quadratic in w. So the first step is to write the formula in standard form (with 0 on one side in decreasing powers of w). This allows us to apply the quadratic formula to solve for w.

5. No. There is only one time after the rock starts moving when it hits the ground.

7. Since the triangle is a right triangle, use the Pythagorean formula with legs m and n and hypotenuse p.

$$m^2 + n^2 = p^2$$
$$m^2 = p^2 - n^2$$
$$m = \sqrt{p^2 - n^2}$$

Only the positive square root is given since m represents the side of a triangle.

9. Solve $d = kt^2$ for t.

$$kt^2 = d$$
$$t^2 = \frac{d}{k} \qquad \textit{Divide by k.}$$
$$t = \pm\sqrt{\frac{d}{k}} \qquad \textit{Use square root property.}$$
$$= \frac{\pm\sqrt{d}}{\sqrt{k}} \cdot \frac{\sqrt{k}}{\sqrt{k}} \qquad \textit{Rationalize denominator.}$$
$$t = \frac{\pm\sqrt{dk}}{k} \qquad \textit{Simplify}$$

11. Solve $F = \dfrac{kA}{v^2}$ for v.

$$v^2 F = kA \qquad \textit{Multiply by } v^2.$$
$$v^2 = \frac{kA}{F} \qquad \textit{Divide by F.}$$
$$v = \pm\sqrt{\frac{kA}{F}} \qquad \textit{Use square root property.}$$
$$= \frac{\pm\sqrt{kA}}{\sqrt{F}} \cdot \frac{\sqrt{F}}{\sqrt{F}} \qquad \textit{Rationalize denominator.}$$
$$v = \frac{\pm\sqrt{kAF}}{F} \qquad \textit{Simplify}$$

13. Solve $V = \dfrac{1}{3}\pi r^2 h$ for r.

$$3V = \pi r^2 h \qquad \textit{Multiply by 3.}$$
$$\frac{3V}{\pi h} = r^2 \qquad \textit{Divide by } \pi h.$$
$$r = \pm\sqrt{\frac{3V}{\pi h}} \qquad \textit{Use square root property.}$$

continued

$$= \frac{\pm \sqrt{3V} \cdot \sqrt{\pi h}}{\sqrt{\pi h} \cdot \sqrt{\pi h}} \quad \textit{Rationalize denominator.}$$

$$r = \frac{\pm \sqrt{3\pi V h}}{\pi h} \quad \textit{Simplify}$$

15. Solve $At^2 + Bt = -C$ for t.

$$At^2 + Bt + C = 0$$

Use the quadratic formula.

$$t = \frac{-B \pm \sqrt{B^2 - 4AC}}{2A}$$

17. Solve $D = \sqrt{kh}$ for h.

$$D^2 = kh \qquad \textit{Square both sides.}$$

$$\frac{D^2}{k} = h \qquad \textit{Divide by k.}$$

19. Solve $p = \sqrt{\dfrac{kl}{g}}$ for l.

$$p^2 = \frac{kl}{g} \qquad \textit{Square both sides.}$$

$$p^2 g = kl \qquad \textit{Multiply by g.}$$

$$\frac{p^2 g}{k} = l \qquad \textit{Divide by k.}$$

21. If g is positive, the only way to have a real value for p is to have kl positive, since the quotient of two positive numbers is positive. If k and l have different signs, their product is negative, leading to a negative radicand.

23.
$$s = -16t^2 + 45t + 400$$

$$200 = -16t^2 + 45t + 400 \quad \textit{Let s=200.}$$

$$0 = -16t^2 + 45t + 200$$

$$0 = 16t^2 - 45t - 200 \quad \textit{Multiply by -1.}$$

$$a = 16, b = -45, c = -200$$

$$t = \frac{-b \pm \sqrt{b^2 - 4ac}}{2a}$$

$$t = \frac{-(-45) \pm \sqrt{(-45)^2 - 4(16)(-200)}}{2(16)}$$

$$= \frac{45 \pm \sqrt{2025 + 12,800}}{32}$$

$$= \frac{45 \pm \sqrt{14,825}}{32}$$

$$t = \frac{45 + \sqrt{14,825}}{32} \approx 5.2 \text{ or}$$

$$t = \frac{45 - \sqrt{14,825}}{32} \approx -2.4$$

Reject the negative solution since time cannot be negative. The ball will reach a height of 200 feet above the ground after 5.2 seconds.

25. The height of the building could be determined by finding s when $t = 0$.

27. $s = 144t - 16t^2$

(a) Substitute 128 for s.

$$128 = 144t - 16t^2$$

$$0 = -16t^2 + 144t - 128$$

Divide by -16.

$$0 = t^2 - 9t + 8$$

$$0 = (t - 8)(t - 1)$$

$$t - 8 = 0 \quad \text{or} \quad t - 1 = 0$$

$$t = 8 \quad \text{or} \qquad t = 1$$

The object will be 128 feet above the ground at two times, going up and coming down, or at 1 second and at 8 seconds.

(b) Substitute 0 for s since the ground is at 0 height.

$$0 = 144t - 16t^2$$

Divide by 16.

$$0 = 9t - t^2$$

$$0 = t(9 - t)$$

$$t = 0 \quad \text{or} \quad 9 - t = 0$$

$$9 = t$$

The object will be on the ground at the starting point ($t = 0$). Then, it will hit the ground again at 9 seconds.

29. (a) From Example 4,

$$y = x - \frac{g}{1922}x^2.$$

For the moon, $g = 3320$, so

$$y = x - \frac{3320}{1922}x^2 \approx x - 1.727x^2.$$

(b) $y = .25 - 1.727(.25)^2 \quad \textit{Let x=.25}$

$$\approx .1421 \text{ foot}$$

$$y = .5 - 1.727(.5)^2 \qquad \textit{Let x=.5}$$

$$\approx .06825 \text{ foot}$$

(c) $0 = x - 1.727x^2 \qquad \textit{Let y=0.}$

$$0 = (1 - 1.727x)x$$

$$1 - 1.727x = 0 \quad \text{or} \quad x = 0$$

$$x = \frac{1}{1.727} \approx .5790$$

The vehicle is on the surface of the moon at 0 seconds and .5790 second.

(d) One time represents the time before the vehicle leaves the surface. The other time represents the time it returns to the surface.

31. Apply the Pythagorean formula.

$$(5m)^2 = (2m)^2 + (2m + 3)^2$$
$$25m^2 = 4m^2 + 4m^2 + 12m + 9$$
$$17m^2 - 12m - 9 = 0$$
$$a = 17, b = -12, c = -9$$
$$m = \frac{-(-12) \pm \sqrt{(-12)^2 - 4(17)(-9)}}{2(17)}$$
$$= \frac{12 \pm \sqrt{144 + 612}}{34} = \frac{12 \pm \sqrt{756}}{34}$$
$$m = \frac{12 + \sqrt{756}}{34} \approx 1.16 \text{ or}$$
$$m = \frac{12 - \sqrt{756}}{34} \approx -.46$$

Reject the negative solution.
If $m = 1.16$, then

$$5m = 5(1.16) = 5.80,$$
$$2m = 2(1.16) = 2.32, \text{ and}$$
$$2m + 3 = 2(1.16) + 3 = 5.32.$$

The lengths of the sides of the triangle are approximately 2.3, 5.3, and 5.8.

33. Let x = the length of the wire.
Use the Pythagorean formula.

$$x^2 = 100^2 + 400^2 \quad \textit{Height=400}$$
$$= 10,000 + 160,000$$
$$= 170,000$$
$$x = \pm\sqrt{170,000} \approx \pm 412.3$$

Reject the negative solution. The length of the wire is about 412.3 feet.

35. Let x = the distance traveled by the eastbound ship and
$x + 70$ = the distance traveled by the southbound ship.

Since the ships are traveling at right angles to one another, the distance d between them can be found using the Pythagorean formula.

$$c^2 = a^2 + b^2$$
$$d^2 = x^2 + (x + 70)^2$$

Let $d = 170$, and solve for x.

$$170^2 = x^2 + (x + 70)^2$$
$$28,900 = x^2 + x^2 + 140x + 4900$$
$$0 = 2x^2 + 140x - 24,000$$
$$0 = x^2 + 70x - 12,000$$
$$0 = (x + 150)(x - 80)$$

$$x + 150 = 0 \quad \text{or} \quad x - 80 = 0$$
$$x = -150 \quad \text{or} \quad x = 80$$

Distance cannot be negative, so reject -150. If $x = 80$, then $x + 70 = 150$. The eastbound ship

traveled 80 miles, and the southbound ship traveled 150 miles.

37. Let x = length of the shorter leg;
$2x + 2$ = length of the longer leg;
$2x + 2 + 1$ = length of the hypotenuse.

Use the Pythagorean formula.

$$x^2 + (2x + 2)^2 = (2x + 3)^2$$
$$x^2 + 4x^2 + 8x + 4 = 4x^2 + 12x + 9$$
$$5x^2 + 8x + 4 = 4x^2 + 12x + 9$$
$$x^2 - 4x - 5 = 0$$
$$(x - 5)(x + 1) = 0$$

$$x - 5 = 0 \quad \text{or} \quad x + 1 = 0$$
$$x = 5 \qquad\qquad x = -1$$

Since x represents length, discard -1 as a solution. If $x = 5$, then

$$2x + 2 = 2(5) + 2 = 12 \text{ and}$$
$$2x + 2 + 1 = 2(5) + 3 = 13.$$

The lengths should be 5 cm, 12 cm, and 13 cm.

39. Let x = the width of the rectangle;
$2x - 1$ = the length of the rectangle.

Use the Pythagorean formula.

$$x^2 + (2x - 1)^2 = (2.5)^2$$
$$x^2 + 4x^2 - 4x + 1 = 6.25$$
$$5x^2 - 4x - 5.25 = 0$$

Multiply by 4.
$$20x^2 - 16x - 21 = 0$$
$$(2x - 3)(10x + 7) = 0$$

$$2x - 3 = 0 \quad \text{or} \quad 10x + 7 = 0$$
$$x = \frac{3}{2} \quad \text{or} \qquad x = -\frac{7}{10}$$

Discard the negative solution.
If $x = \frac{3}{2}$, then

$$2x - 1 = 2\left(\frac{3}{2}\right) - 1 = 2.$$

The width of the rectangle is 1.5 cm, and the length is 2 cm.

41. Let x = the width of the uncovered strip of flooring.

From the problem,
(length of the rug) • (width of the rug) = 234.

The rug is centered in the room a distance x from the walls (width of the strip x), so the length of the rug

continued

= length of the room $- 2 \cdot$ (width of the strip)

= $20 - 2x$,

and the width of the rug

= width of the room $- 2 \cdot$ (width of the strip)

= $15 - 2x$.

The equation (length of rug) $\cdot$ (width of rug) $= 234$ becomes

$$(20 - 2x)(15 - 2x) = 234.$$
$$300 - 70x + 4x^2 = 234$$
$$4x^2 - 70x + 66 = 0$$

Divide by 2.
$$2x^2 - 35x + 33 = 0$$
$$(2x - 33)(x - 1) = 0$$
$$2x - 33 = 0 \quad \text{or} \quad x - 1 = 0$$
$$x = \frac{33}{2} \quad \text{or} \quad x = 1$$

Reject $\dfrac{33}{2} = 16\dfrac{1}{2}$ since $16\dfrac{1}{2}$ is wider than the room itself. The width of the uncovered strip is 1 foot.

43. Let x be the width of grass.

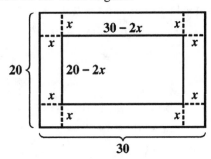

The area of the flower garden is $(20 - 2x)(30 - 2x)$.

Area of back yard	less	area of flower garden	is	area of grass.

$$20 \cdot 30 \quad - \quad (20 - 2x)(30 - 2x) = \quad 184$$
$$600 - \left(600 - 100x + 4x^2\right) = 184$$
$$600 - 600 + 100x - 4x^2 = 184$$
$$100x - 4x^2 = 184$$
$$4x^2 - 100x + 184 = 0$$

Divide by 4.
$$x^2 - 25x + 46 = 0$$
$$(x - 23)(x - 2) = 0$$
$$x - 23 = 0 \quad \text{or} \quad x - 2 = 0$$
$$x = 23 \quad \text{or} \quad x = 2$$

Discard $x = 23$ since the width of the garden cannot be

$$20 - 2x = 20 - 2(23) = -26.$$

Thus, if $x = 2$, then

$$30 - 2x = 30 - 2(2) = 26,$$

and

$$20 - 2x = 20 - 2(2) = 16.$$

The flower garden is 26 m by 16 m.

45. Let $r =$ the interest rate.

$$A = P(1 + r)^2$$

Let $A = 2142.25$ and $P = 2000$. Solve for r.

$$2142.25 = 2000(1 + r)^2$$
$$1.071125 = (1 + r)^2$$

Use the square root property.

$$1 + r \approx 1.035 \quad \text{or} \quad 1 + r \approx -1.035$$
$$r \approx .035 \quad \text{or} \quad r \approx -2.035$$

Since the interest rate cannot be negative, reject -2.035. The interest rate is about 3.5%.

47. Supply and demand are equal when

$$3p - 200 = \frac{3200}{p}.$$

Solve for p.
$$3p^2 - 200p = 3200$$
$$3p^2 - 200p - 3200 = 0$$

Use the quadratic formula with $a = 3$, $b = -200$, and $c = -3200$.

$$p = \frac{-(-200) \pm \sqrt{(-200)^2 - 4(3)(-3200)}}{2(3)}$$
$$= \frac{200 \pm \sqrt{40,000 + 38,400}}{6}$$
$$= \frac{200 \pm \sqrt{78,400}}{6} = \frac{200 \pm 280}{6}$$
$$p = \frac{480}{6} = 80 \quad \text{or} \quad p = \frac{-80}{6} = -\frac{40}{3}$$

Discard the negative solution. The supply and demand are equal when the price is 80 cents or $.80.

49. Let F denote the Froude number. Solve

$$F = \frac{v^2}{gl}$$

for v.
$$v^2 = Fgl$$
$$v = \pm\sqrt{Fgl}$$

v is positive, so
$$V = \sqrt{Fgl}.$$

For the rhinoceros, $l = 1.2$ and $F = 2.57$.
$$V = \sqrt{(2.57)(9.8)(1.2)} \approx 5.5$$

or 5.5 meters per second.

51. Write a proportion.

$$\frac{x-4}{3x-19} = \frac{4}{x-3}$$

Multiply by the LCD, $(3x-19)(x-3)$.

$$(3x-19)(x-3)\left(\frac{x-4}{3x-19}\right)$$

$$= (3x-19)(x-3)\left(\frac{4}{x-3}\right)$$

$$(x-3)(x-4) = (3x-19)4$$
$$x^2 - 7x + 12 = 12x - 76$$
$$x^2 - 19x + 88 = 0$$
$$(x-8)(x-11) = 0$$

$$x - 8 = 0 \quad \text{or} \quad x - 11 = 0$$
$$x = 8 \quad \text{or} \quad x = 11$$

If $x = 8$, then

$$3x - 19 = 3(8) - 19 = 5.$$

If $x = 11$, then

$$3x - 19 = 3(11) - 19 = 14.$$

Thus, AC = 5 or AC = 14.

53. Solve $p = \dfrac{E^2 R}{(r+R)^2}$ $(E > 0)$ for R.

$$p(r+R)^2 = E^2 R$$
$$p(r^2 + 2rR + R^2) = E^2 R$$
$$pr^2 + 2prR + pR^2 = E^2 R$$
$$pR^2 + 2prR - E^2 R + pr^2 = 0$$
$$pR^2 + (2pr - E^2)R + pr^2 = 0$$

$$a = p, \, b = 2pr - E^2, \, c = pr^2$$

$$R = \frac{-(2pr - E^2) \pm \sqrt{(2pr - E^2)^2 - 4p \cdot pr^2}}{2p}$$

$$= \frac{E^2 - 2pr \pm \sqrt{4p^2r^2 - 4prE^2 + E^4 - 4p^2r^2}}{2p}$$

$$= \frac{E^2 - 2pr \pm \sqrt{E^4 - 4prE^2}}{2p}$$

$$= \frac{E^2 - 2pr \pm \sqrt{E^2(E^2 - 4pr)}}{2p}$$

$$R = \frac{E^2 - 2pr \pm E\sqrt{E^2 - 4pr}}{2p}$$

55. Solve $10p^2 c^2 + 7pcr = 12r^2$ for r.

$$0 = 12r^2 - 7pcr - 10p^2 c^2$$

$$a = 12, \, b = -7pc, \, c = -10p^2 c^2$$

$$r = \frac{-(-7pc) \pm \sqrt{(-7pc)^2 - 4(12)(-10p^2 c^2)}}{2(12)}$$

$$= \frac{7pc \pm \sqrt{49p^2 c^2 + 480p^2 c^2}}{24}$$

$$= \frac{7pc \pm \sqrt{529p^2 c^2}}{24} = \frac{7pc \pm 23pc}{24}$$

$$r = \frac{7pc + 23pc}{24} = \frac{30pc}{24} = \frac{5pc}{4} \quad \text{or}$$

$$r = \frac{7pc - 23pc}{24} = \frac{-16pc}{24} = -\frac{2pc}{3}$$

57. Solve $LI^2 + RI + \dfrac{1}{c} = 0$ for I.

$$cLI^2 + cRI + 1 = 0 \quad \textit{Multiply by } c.$$

$$a = cL, \, b = cR, \, c = 1$$

$$I = \frac{-cR \pm \sqrt{(cR)^2 - 4(cL)(1)}}{2(cL)}$$

$$= \frac{-cR \pm \sqrt{c^2 R^2 - 4cL}}{2cL}$$

Section 11.7

1. $g(x) = x^2 - 5$ written in the form
$$g(x) = a(x-h)^2 + k \text{ is}$$
$$g(x) = 1(x-0)^2 + (-5).$$

Here, $h = 0$ and $k = -5$, so the vertex (h, k) is $(0, -5)$. The graph is shifted 5 units down from the graph of $f(x) = x^2$. Since $a = 1 > 0$, the graph opens upward. The correct figure is F.

3. $F(x) = (x-1)^2$ written in the form
$$F(x) = a(x-h)^2 + k \text{ is}$$
$$F(x) = 1(x-1)^2 + 0.$$

Here, $h = 1$ and $k = 0$, so the vertex (h, k) is $(1, 0)$. The graph is shifted 1 unit to the right of the graph of $f(x) = x^2$. Since $a = 1 > 0$, the graph opens upward. The correct figure is C.

5. $H(x) = (x-1)^2 + 1$ is written in the form
$$H(x) = a(x-h)^2 + k.$$

Here, $h = 1$ and $k = 1$, so the vertex (h, k) is $(1, 1)$. The graph is shifted 1 unit to the right and 1 unit up from the graph of $f(x) = x^2$. Since $a = 1 > 0$, the graph opens upward. The correct figure is E.

7. **(a)** The vertex is the point that contains the smallest or largest y-value of the parabola.

(b) The axis of the parabola is a line through the vertex about which the parabola is symmetric.

For Exercises 9–16, we write $f(x)$ in the form $f(x) = a(x-h)^2 + k$ and then list the vertex (h, k).

9. $f(x) = -3x^2 = -3(x-0)^2 + 0$
The vertex (h, k) is $(0, 0)$.

11. $f(x) = x^2 + 4 = 1(x-0)^2 + 4$
The vertex (h, k) is $(0, 4)$.

13. $f(x) = (x-1)^2 = 1(x-1)^2 + 0$
The vertex (h, k) is $(1, 0)$.

15. $f(x) = (x + 3)^2 - 4 = 1[x - (-3)]^2 - 4$
The vertex (h, k) is $(-3, -4)$.

17. In Exercise 15, the parabola is shifted 3 units to the left and 4 units down. The parabola in Exercise 16 is shifted 5 units to the right and 8 units down.

19. $f(x) = -3x^2 + 1$
$f(x) = -3(x - 0)^2 + 1$

Since $a = -3 < 0$, the graph opens downward. Since $|a| = |-3| = 3 > 1$, the graph is narrower than the graph of $f(x) = x^2$.

21. $f(x) = \dfrac{2}{3}x^2 - 4$
$f(x) = \dfrac{2}{3}(x - 0)^2 - 4$

Since $a = \dfrac{2}{3} > 0$, the graph opens upward. Since $|a| = \left|\dfrac{2}{3}\right| = \dfrac{2}{3} < 1$, the graph is wider than the graph of $f(x) = x^2$.

23. Consider $f(x) = a(x - h)^2 + k$.

(a) If $h > 0$ and $k > 0$ in $f(x) = a(x - h)^2 + k$, the shift is to the right and upward, so the vertex is in quadrant I.

(b) If $h > 0$ and $k < 0$, the shift is to the right and downward, so the vertex is in quadrant IV.

(c) If $h < 0$ and $k > 0$, the shift is to the left and upward, so the vertex is in quadrant II.

(d) If $h < 0$ and $k < 0$, the shift is to the left and downward, so the vertex is in quadrant III.

25. $f(x) = -2x^2$ written in the form
$f(x) = a(x - h)^2 + k$ is
$f(x) = -2(x - 0)^2 + 0$.

Here, $h = 0$ and $k = 0$, so the vertex (h, k) is $(0, 0)$. Since $a = -2 < 0$, the graph opens downward. Since $|a| = |-2| = 2 > 1$, the graph is narrower than the graph of $f(x) = x^2$. By evaluating the function with $x = 2$ and $x = -2$, we see that the points $(2, -8)$ and $(-2, -8)$ are on the graph.

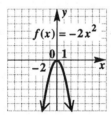

27. $f(x) = x^2 - 1$ written in the form
$f(x) = a(x - h)^2 + k$ is
$f(x) = 1(x - 0)^2 + (-1)$.

Here, $h = 0$ and $k = -1$, so the vertex is $(0, -1)$. The graph opens upward and has the same shape as $f(x) = x^2$ because $a = 1$. Two other points on the graph are $(-2, 3)$ and $(2, 3)$.

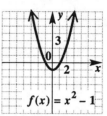

29. $f(x) = -x^2 + 2$ written in the form
$f(x) = a(x - h)^2 + k$ is
$f(x) = -1(x - 0)^2 + 2$.

Here, $h = 0$ and $k = 2$, so the vertex (h, k) is $(0, 2)$. Since $a = -1 < 0$, the graph opens downward. Since $|a| = |-1| = 1$, the graph has the same shape as $f(x) = x^2$. The points $(2, -2)$ and $(-2, -2)$ are on the graph.

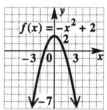

31. $f(x) = .5(x - 4)^2$ written in the form
$f(x) = a(x - h)^2 + k$ is
$f(x) = .5(x - 4)^2 + 0$.

Here, $h = 4$ and $k = 0$, so the vertex (h, k) is $(4, 0)$. The graph opens upward since a is positive and is wider than $f(x) = x^2$ because $|a| = |.5| < 1$. Two other points on the graph are $(2, 2)$ and $(6, 2)$.

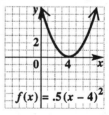

33. $f(x) = (x + 2)^2 - 1$ written in the form
$f(x) = a(x - h)^2 + k$ is
$f(x) = 1[x - (-2)]^2 + (-1)$.

Since $h = -2$ and $k = -1$, the vertex (h, k) is $(-2, -1)$. Here, $a = 1$, so the graph opens upward and has the same shape as $f(x) = x^2$. The points $(-1, 0)$ and $(-3, 0)$ are on the graph.

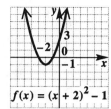

$$f(x) = (x + 2)^2 - 1$$

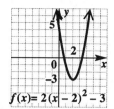

$$f(x) = 2(x - 2)^2 - 3$$

35. $f(x) = 2(x - 2)^2 - 4$ written in the form
$f(x) = a(x - h)^2 + k$ is
$f(x) = 2(x - 2)^2 + (-4)$.

Here, $h = 2$ and $k = -4$, so the vertex (h, k) is
$(2, -4)$. The graph opens upward and is narrower
than $f(x) = x^2$ because $|a| = |2| > 1$. Two other
points on the graph are $(0, 4)$ and $(4, 4)$.
We can substitute any value for x, so the domain
is $(-\infty, \infty)$. The value of y is greater than or
equal to -4, so the range is $[-4, \infty)$.

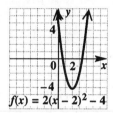

$$f(x) = 2(x - 2)^2 - 4$$

37. $f(x) = -.5(x + 1)^2 + 2$ written in the form
$f(x) = a(x - h)^2 + k$ is
$f(x) = -.5[x - (-1)]^2 + 2$.

Since $h = -1$ and $k = 2$, the vertex (h, k) is
$(-1, 2)$. Here, $a = -.5 < 0$, so the graph opens
downward. Also, $|a| = |-.5| = .5 < 1$, so the
graph is wider than the graph of $f(x) = x^2$. The
points $(1, 0)$ and $(-3, 0)$ are on the graph.
We can substitute any value for x, so the domain
is $(-\infty, \infty)$. The value of y is less than or equal to
2, so the range is $(-\infty, 2]$.

$$f(x) = -.5(x + 1)^2 + 2$$

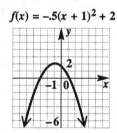

39. $f(x) = 2(x - 2)^2 - 3$ written in the form
$f(x) = a(x - h)^2 + k$ is
$f(x) = 2(x - 2)^2 + (-3)$.

Here, $h = 2$ and $k = -3$, so the vertex (h, k) is
$(2, -3)$. The graph opens upward and is narrower
than $f(x) = x^2$ because $|a| = |2| > 1$. Two other
points on the graph are $(3, -1)$ and $(1, -1)$.
We can substitute any value for x, so the domain
is $(-\infty, \infty)$. The value of y is greater than or
equal to -3, so the range is $[-3, \infty)$.

41. The graph of $y = x^2 + 6$ would be shifted 6 units
upward from the graph of $y = x^2$.

42. To graph $y = x + 6$, plot the intercepts $(-6, 0)$
and $(0, 6)$, and draw the line through them.

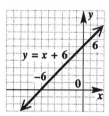

$$y = x + 6$$

43. When considering the graph of $y = x + 6$, the
y-intercept is 6. The graph of $y = x$ has
y-intercept 0. Therefore, the graph of $y = x + 6$ is
shifted 6 units upward compared to the graph of
$y = x$.

44. The graph of $y = (x - 6)^2$ is shifted 6 units to the
right compared to the graph of $y = x^2$.

45. To graph $y = x - 6$, plot the intercepts $(6, 0)$ and
$(0, -6)$, and draw the line through them.

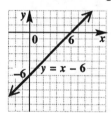

$$y = x - 6$$

46. When considering the graph of $y = x - 6$, its
x-intercept is 6 as compared to the graph of $y = x$
with x-intercept 0. The graph of $y = x - 6$ is
shifted 6 units to the right compared to the graph
of $y = x$.

47. **(a)** $|x - (-p)| = |x + p|$

(b) The focus should have coordinates $(p, 0)$
because the distance from the focus to the origin
should equal the distance from the directrix to the
origin.

(c) The distance from (x, y) to $(p, 0)$ is

$$\sqrt{(x - p)^2 + (y - 0)^2} = \sqrt{(x - p)^2 + y^2}.$$

(d) Using the results from parts (a) and (c), these distances should be equal.

$$\sqrt{(x-p)^2 + y^2} = |x + p|$$

Square both sides.

$$(x-p)^2 + y^2 = (x+p)^2$$
$$x^2 - 2px + p^2 + y^2 = x^2 + 2px + p^2$$
$$y^2 = 4px$$

49. (a)

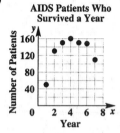

AIDS Patients Who Survived a Year

(b) Since the parabola would be opening downward, the value of a would be *negative*.

(c) Use $ax^2 + bx + c = y$ with $(2, 130)$, $(3, 155)$, and $(7, 115)$.

$$\begin{aligned} 4a + 2b + c &= 130 \quad (1) \\ 9a + 3b + c &= 155 \quad (2) \\ 49a + 7b + c &= 115 \quad (3) \end{aligned}$$

Eliminate c twice.

$$\begin{array}{rll} -4a - 2b - c = & -130 & -1 \times (1) \\ 49a + 7b + c = & 115 & (3) \\ \hline 45a + 5b = & -15 & \\ 9a + b = & -3 & (4) \end{array}$$

$$\begin{array}{rll} -9a - 3b - c = & -155 & -1 \times (2) \\ 49a + 7b + c = & 115 & (3) \\ \hline 40a + 4b = & -40 & \\ 10a + b = & -10 & (5) \end{array}$$

Now eliminate b from (4) and (5).

$$\begin{array}{rll} -9a - b = & 3 & -1 \times (4) \\ 10a + b = & -10 & (5) \\ \hline a = & -7 & \end{array}$$

Use (5) to find b.

$$\begin{aligned} 10(-7) + b &= -10 \\ -70 + b &= -10 \\ b &= 60 \end{aligned}$$

Use (1) to find c.

$$\begin{aligned} 4a + 2b + c &= 130 \quad (1) \\ 4(-7) + 2(60) + c &= 130 \\ -28 + 120 + c &= 130 \\ c &= 38 \end{aligned}$$

The quadratic function is

$$y = f(x) = -7x^2 + 60x + 38.$$

(d) Many answers are possible. Since the data gives one y-value for each year, $\{1, 2, 3, 4, 5, 6, 7\}$ may be the most appropriate.

51. $x^2 - x - 20 = 0$

From the screens, we see that the x-values of the x-intercepts are -4 and 5, so the solution set is $\{-4, 5\}$.

53. $-2x^2 + 5x + 3 = 0$

From the screens, we see that the x-values of the x-intercepts are $-.5$ and 3, so the solution set is $\{-.5, 3\}$.

55. The only possible choice for the solution set for the equation $f(x) = 0$ would be choice (a) $\{-4, 1\}$. This is because the graph crosses the x-axis on each side of the y-axis, indicating that there is one negative solution and one positive solution. Choice (d) also has a negative and a positive value, but in this case, the negative value must have the greater absolute value of the two numbers.

Section 11.8

1. If there is an x^2-term in the equation, the axis is vertical. If there is a y^2-term, the axis is horizontal.

3. **(a)** Use the discriminant, $b^2 - 4ac$, of the function. If it is positive, there are two x-intercepts. If it is zero, there is one x-intercept (at the vertex), and if it is negative, there is no x-intercept.

(b) If the vertex is at $(1, -3)$ and the graph opens downward, the vertex is the highest point of the graph. Because the y-coordinate of the vertex is negative, the parabola lies below the x-axis so the graph has no x-intercepts.

5.
$$y = 2x^2 + 4x + 5$$

Complete the square on x to find the vertex.

$$\frac{y}{2} = x^2 + 2x + \frac{5}{2} \qquad \textit{Divide by 2.}$$

$$\frac{y}{2} - \frac{5}{2} = x^2 + 2x \qquad \textit{Get the constant term on the left.}$$

$$\frac{y}{2} - \frac{5}{2} + 1 = x^2 + 2x + 1 \qquad \textit{Half of 2 is 1;} \; (1)^2 = 1. \textit{ Add 1 to both sides.}$$

$$\frac{y}{2} - \frac{3}{2} = (x+1)^2 \qquad \textit{Combine terms on the left and factor on the right.}$$

$$\frac{y}{2} = (x+1)^2 + \frac{3}{2} \qquad \textit{Add } \frac{3}{2}.$$

$$y = 2(x+1)^2 + 3 \qquad \textit{Multiply by 2.}$$

The vertex is $(-1, 3)$.
Because $a = 2$, the graph opens upward and is narrower than the graph of $y = x^2$.

For $y = 2x^2 + 4x + 5$, $a = 2$, $b = 4$, and $c = 5$. The discriminant is

$$b^2 - 4ac = 4^2 - 4(2)(5)$$
$$= 16 - 40 = -24.$$

The discriminant is negative, so the parabola has no x-intercepts.

7. $y = -x^2 + 5x + 3$
Use the vertex formula.
$a = -1, b = 5, c = 3$

$$\frac{-b}{2a} = \frac{-5}{2(-1)} = \frac{5}{2}$$

$y = f(x)$, so

$$f\left(\frac{-b}{2a}\right) = f\left(\frac{5}{2}\right)$$

$$= -\left(\frac{5}{2}\right)^2 + 5\left(\frac{5}{2}\right) + 3$$

$$= -\frac{25}{4} + \frac{25}{2} + 3$$

$$= \frac{-25 + 50 + 12}{4} = \frac{37}{4}.$$

The vertex is

$$\left(\frac{-b}{2a}, f\left(\frac{-b}{2a}\right)\right) = \left(\frac{5}{2}, \frac{37}{4}\right).$$

Because $a = -1$, the parabola opens downward and has the same shape as the graph of $y = x^2$.

$$b^2 - 4ac = 5^2 - 4(-1)(3)$$
$$= 25 + 12 = 37$$

The discriminant is positive, so the parabola has two x-intercepts.

9. $$x = \frac{1}{3}y^2 + 6y + 24$$

Complete the square on the y-terms to find the vertex.

$$3x = y^2 + 18y + 72$$
$$3x - 72 = y^2 + 18y$$
$$3x - 72 + 81 = y^2 + 18y + 81$$
$$3x + 9 = (y + 9)^2$$
$$3x = (y + 9)^2 - 9$$
$$x = \frac{1}{3}(y + 9)^2 - 3$$

The vertex is at $(-3, -9)$.
The graph is a horizontal parabola. Because of the coefficient, $\frac{1}{3}$, the parabola opens to the right and it is wider than the graph of $y = x^2$.

11. $y = f(x) = x^2 + 8x + 10$
To find the y-intercept, let $x = 0$.
$f(0) = 10$, so the y-intercept is $(0, 10)$.
To find the x-intercepts, let $y = 0$.

$$0 = x^2 + 8x + 10$$
$$x = \frac{-8 \pm \sqrt{64 - 40}}{2} = \frac{-8 \pm \sqrt{24}}{2}$$
$$= \frac{-8 \pm 2\sqrt{6}}{2} = -4 \pm \sqrt{6}$$

Complete the square to find the vertex.

$$y - 10 = x^2 + 8x$$
$$\frac{1}{2}(8) = 4; 4^2 = 16$$
$$y - 10 + 16 = x^2 + 8x + 16$$
$$y + 6 = (x + 4)^2$$
$$y = (x + 4)^2 - 6$$

The vertex is $(-4, -6)$. Since $a = 1$, the graph opens upward and is the same shape as the graph of $y = x^2$. For an additional point on the graph, let $x = -2$ (two units to the right of the axis) to get $f(-2) = -2$. So the point $(-2, -2)$ is on the graph. By symmetry, the point $(-6, -2)$ (two units to the left of the axis) is on the graph. The x-intercepts are approximately $(-6.45, 0)$ and $(-1.55, 0)$.

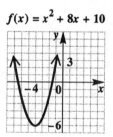

$f(x) = x^2 + 8x + 10$

From the graph, we see that the domain is $(-\infty, \infty)$ and the range is $[-6, \infty)$.

13. $y = -2x^2 + 4x - 5$
If $x = 0$, $y = -5$, so the y-intercept is $(0, -5)$.
To find the x-intercepts, let $y = 0$.

$$0 = -2x^2 + 4x - 5$$
$$x = \frac{-4 \pm \sqrt{16 - 40}}{2(-2)}$$

The discriminant is negative, so there are no x-intercepts.
Use the formula to find the x-value of the vertex.

$$\frac{-b}{2a} = \frac{-4}{2(-2)} = 1$$

If $x = 1$, $y = -3$, so the vertex is $(1, -3)$. By symmetry, $(2, -5)$ is also on the graph. Since $a = -2$, the graph opens downward and is narrower than the graph of $y = x^2$.

continued

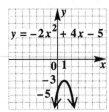

From the graph, we see that the domain is $(-\infty, \infty)$ and the range is $(-\infty, -3]$.

15. $x = -\dfrac{1}{5}y^2 + 2y - 4$

The roles of x and y are reversed, so this is a horizontal parabola.

To find the x-intercept, let $y = 0$.
If $y = 0$, $x = -4$, so the x-intercept is $(-4, 0)$.
To find the y-intercepts, let $x = 0$.

$$0 = -\frac{1}{5}y^2 + 2y - 4$$
$$0 = y^2 - 10y + 20$$
$$y = \frac{10 \pm \sqrt{100 - 80}}{2} = \frac{10 \pm \sqrt{20}}{2}$$
$$= \frac{10 \pm 2\sqrt{5}}{2} = 5 \pm \sqrt{5}$$

The y-intercepts are approximately $(0, 7.2)$ and $(0, 2.8)$.
Complete the square to find the vertex.

$$x = -\frac{1}{5}y^2 + 2y - 4$$
$$-5x = y^2 - 10y + 20$$
$$-5x - 20 = y^2 - 10y$$
$$-5x - 20 + 25 = y^2 - 10y + 25$$
$$-5x + 5 = (y - 5)^2$$
$$-5x = (y - 5)^2 - 5$$
$$x = -\frac{1}{5}(y - 5)^2 + 1$$

When $y = 5$, $x = 1$, so the vertex is $(1, 5)$. Since $a = -\dfrac{1}{5}$, the graph opens to the left and is wider than the graph of $y = x^2$. For an additional point on the graph, let $y = 7$ (two units above the axis) to get $x = \dfrac{1}{5}$. So the point $\left(\dfrac{1}{5}, 7\right)$ is on the graph. By symmetry, the point $\left(\dfrac{1}{5}, 3\right)$ (two units below the axis) is on the graph. The graph does not pass the vertical line test, so it does not represent a function.

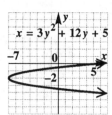

17. $x = 3y^2 + 12y + 5$
If $y = 0$, $x = 5$, so the x-intercept is $(5, 0)$.
To find the y-intercepts, let $x = 0$.

$$0 = 3y^2 + 12y + 5$$
$$y = \frac{-12 \pm \sqrt{144 - 60}}{6} = \frac{-12 \pm \sqrt{84}}{6}$$
$$= \frac{-12 \pm 2\sqrt{21}}{6} = -2 \pm \frac{1}{3}\sqrt{21}$$

The y-intercepts are approximately $(0, -.5)$ and $(0, -3.5)$. Use the formula to find the y-value of the vertex.

$$\frac{-b}{2a} = \frac{-12}{2(3)} = -2$$

If $y = -2$, $x = -7$, so the vertex is $(-7, -2)$. By symmetry, $(5, -4)$ is also on the graph. Since $a = 3$, the graph opens to the right and is narrower than the graph of $y = x^2$. The graph does not pass the vertical line test, so it does not represent a function.

19. The graph of $y = 2x^2 + 4x - 3$ is a vertical parabola opening upward, so choice F is correct.
(F)

21. The graph of $y = -\dfrac{1}{2}x^2 - x + 1$ is a vertical parabola opening downward, so choices A and C are possibilities. The graph in C is wider than the graph in A, so it must correspond to $a = -\dfrac{1}{2}$ while the graph in A must correspond to $a = -1$.
(C)

23. The graph of $x = -y^2 - 2y + 4$ is a horizontal parabola opening to the left, so choice D is correct.
(D)

25. Let $x = $ the width of the rectangle, and
$\qquad y = $ the length of the rectangle.

The sum of the sides is 100 m, so

$$x + x + y + y = 100$$
$$2x + 2y = 100$$
$$2y = 100 - 2x.$$
$$y = 50 - x \qquad \textit{Divide by 2.}$$

Since Area = Width • Length,

$$A = xy.$$

Substitute $50 - x$ for y.
$$A = x(50 - x)$$
$$= 50x - x^2 \text{ or } -x^2 + 50x$$

The width x of the rectangle with maximum area will occur at the vertex of $A = -x^2 + 50x$. Use the vertex formula with $a = -1$ and $b = 50$ to get

$$x = \frac{-b}{2a} = \frac{-50}{2(-1)} = 25.$$

A width of 25 m will produce the maximum area. Note that the length is also 25 m, so the rectangle is a square.

27. $h = 32t - 16t^2$ or $h = -16t^2 + 32t$

Here, $a = -16 < 0$, so the parabola opens downward. The time it takes to reach the maximum height and the maximum height are given by the vertex of the parabola. Use the vertex formula to find that

$$t = \frac{-b}{2a} = \frac{-32}{2(-16)} = \frac{-32}{-32} = 1,$$
and
$$h = -16(1)^2 + 32(1)$$
$$= -16 + 32 = 16.$$

The vertex is $(1, 16)$, so the maximum height is 16 feet which occurs when the time is 1 second. The object hits the ground when $h = 0$.

$$0 = -16t^2 + 32t$$
$$0 = -16t(t - 2)$$

$$-16t = 0 \quad \text{or} \quad t - 2 = 0$$
$$t = 0 \quad \text{or} \qquad t = 2$$

It takes 2 seconds for the object to hit the ground.

29. The graph of the height of the projectile,

$$s(t) = -4.9t^2 + 40t,$$

is a parabola that opens downward since $a = -4.9 < 0$. The time is the t-coordinate of the highest point, or vertex; the maximum height is the s-coordinate of the vertex.
To find the vertex, use the vertex formula with $a = -4.9$ and $b = 40$.

$$t = \frac{-b}{2a} = \frac{-40}{2(-4.9)} = \frac{40}{9.8} \approx 4.1$$

Therefore, the projectile will reach its maximum height after about 4.1 seconds.
The maximum height is given by

$$s(4.1) = -4.9(4.1)^2 + 40(4.1)$$
$$= -4.9(16.81) + 164$$
$$\approx 81.6.$$

The maximum height is about 81.6 m.

31. $f(x) = -20.57x^2 + 758.9x - 3140$

(a) The coefficient of x^2 is negative because the parabola opens downward.

(b) Use the vertex formula.

$$x = \frac{-b}{2a} = \frac{-758.9}{2(-20.57)} \approx 18.45$$
$$f(18.45) \approx 3860$$

The vertex is approximately $(18.45, 3860)$.

(c) 18 corresponds to 2018, so in 2018 social security assets will reach their maximum value of $3860 billion.

33. The number of people on the plane is $100 - x$ since x is the number of unsold seats. The price per seat is $200 + 4x$.

(a) The total revenue received for the flight is found by multiplying the number of seats by the price per seat. Thus, the revenue is

$$R(x) = (100 - x)(200 + 4x)$$
$$= 20{,}000 + 200x - 4x^2.$$

(b) Use the formula for the vertex.

$$x = \frac{-b}{2a} = \frac{-200}{2(-4)} = 25$$

$R(25) = 22{,}500$, so the vertex is $(25, 22{,}500)$. $R(0) = 20{,}000$, so the R-intercept is $(0, 20{,}000)$. From the factored form for R, we see that the positive x-intercept is $(100, 0)$. (The factor $200 + 4x$ leads to a negative x-intercept, meaningless in this problem.)

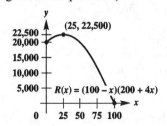

(c) The number of unsold seats x that produce the maximum revenue is 25, the x-value of the vertex.

(d) The maximum revenue is $22{,}500$, the y-value of the vertex.

35. $f(x) = x^2 - 8x + 18$

$$\frac{-b}{2a} = \frac{-(-8)}{2(1)} = 4$$

$f(4) = 2$, so the vertex is $(4, 2)$, which matches choice B.

37. $f(x) = x^2 - 8x + 14$

$$\frac{-b}{2a} = \frac{-(-8)}{2(1)} = 4$$

$f(4) = -2$, so the vertex is $(4, -2)$, which matches choice A.

Section 11.9

1. $\qquad 2x^2 + x - 3 = 0$
$(2x + 3)(x - 1) = 0$

$2x + 3 = 0 \qquad$ or $\qquad x - 1 = 0$

$x = -\dfrac{3}{2} \qquad$ or $\qquad x = 1$

Solution set: $\left\{ -\dfrac{3}{2}, 1 \right\}$

The numbers -1 and 5 divide the number line into three regions: A, B, and C.

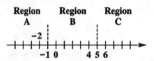

Test a number from each region in the original inequality.

Region A: Let $x = -2$.
$\qquad (x + 1)(x - 5) > 0$
$(-2 + 1)(-2 - 5) > 0 \qquad$?
$\qquad -1(-7) > 0 \qquad$?
$\qquad\qquad 7 > 0 \qquad\qquad$ *True*

Region B: Let $x = 0$.
$(0 + 1)(0 - 5) > 0 \qquad$?
$\qquad\qquad -5 > 0 \qquad$ *False*

Region C: Let $x = 6$.
$(6 + 1)(6 - 5) > 0 \qquad$?
$\qquad\qquad 7 > 0 \qquad\qquad$ *True*

The solution set includes the numbers in Regions A and C, excluding -1 and 5 because of $>$.
Solution set: $(-\infty, -1) \cup (5, \infty)$

3. $\qquad 2x^2 + x - 3 \le 0$

Region A: Let $x = -2$.
$2(-2)^2 + (-2) - 3 \le 0 \qquad$?
$\qquad\qquad 3 \le 0 \qquad\qquad$ *False*

Region B: Let $x = 0$.
$\qquad\qquad -3 \le 0 \qquad\qquad$ *True*

Region C: Let $x = 2$.
$2(2)^2 + (2) - 3 \le 0 \qquad$?
$\qquad\qquad 7 \le 0 \qquad\qquad$ *False*

The open interval that satisfies the strict inequality corresponds to Region B; that is, the interval
$\left(-\dfrac{3}{2}, 1 \right)$.

5. $\qquad (x + 1)(x - 5) > 0$
Solve the equation
$\qquad (x + 1)(x - 5) = 0$.

$x + 1 = 0 \qquad$ or $\qquad x - 5 = 0$
$\qquad x = -1 \quad$ or $\qquad\qquad x = 5$

7. $\qquad (r + 4)(r - 6) < 0$
Solve the equation
$\qquad (r + 4)(r - 6) = 0$.

$r + 4 = 0 \qquad$ or $\qquad r - 6 = 0$
$\qquad r = -4 \quad$ or $\qquad\qquad r = 6$

These numbers divide the number line into three regions: A, B, and C.

Test a number from each region in the original inequality.

Region A: Let $r = -5$.
$\qquad (r + 4)(r - 6) < 0$
$(-5 + 4)(-5 - 6) < 0 \qquad$?
$\qquad -1(-11) < 0 \qquad$?
$\qquad\qquad 11 < 0 \qquad\qquad$ *False*

Region B: Let $r = 0$.
$\qquad\qquad 4(-6) < 0 \qquad$?
$\qquad\qquad -24 < 0 \qquad\qquad$ *True*

Region C: Let $r = 7$.
$\qquad (7 + 4)(7 - 6) < 0 \qquad$?
$\qquad\qquad 11(1) < 0 \qquad$?
$\qquad\qquad 11 < 0 \qquad\qquad$ *False*

The solution set includes Region B, where the expression is negative.

Solution set: $(-4, 6)$

9. $x^2 - 4x + 3 \geq 0$

Solve the equation

$x^2 - 4x + 3 = 0.$

$(x - 1)(x - 3) = 0$

$x - 1 = 0$ or $x - 3 = 0$

$x = 1$ or $x = 3$

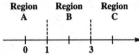

Region A: Let $x = 0$.

$\qquad 3 \geq 0 \qquad$ *True*

Region B: Let $x = 2$.

$2^2 - 4(2) + 3 \geq 0 \quad$?

$\qquad -1 \geq 0 \qquad$ *False*

Region C: Let $x = 4$.

$4^2 - 4(4) + 3 \geq 0 \quad$?

$\qquad 3 \geq 0 \qquad$ *True*

The solution set includes the numbers in Regions A and C, including 1 and 3 because of $\geq$.

Solution set: $(-\infty, 1] \cup [3, \infty)$

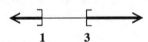

11. $10a^2 + 9a \geq 9$

$10a^2 + 9a - 9 \geq 0$

Solve the equation

$10a^2 + 9a - 9 = 0.$

$(2a + 3)(5a - 3) = 0$

$2a + 3 = 0$ or $5a - 3 = 0$

$a = -\dfrac{3}{2}$ or $a = \dfrac{3}{5}$

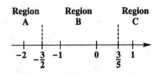

Test a number from each region in the original inequality.

Region A: Let $a = -2$.

$10(-2)^2 + 9(-2) \geq 9 \quad$?

$40 - 18 \geq 9 \quad$?

$22 \geq 9 \qquad$ *True*

Region B: Let $a = 0$.

$0 \geq 9 \qquad$ *False*

Region C: Let $a = 1$.

$10(1)^2 + 9(1) \geq 9 \quad$?

$10 + 9 \geq 9 \quad$?

$19 \geq 9 \qquad$ *True*

The solution set includes the numbers in Region A and C, including $-\dfrac{3}{2}$ and $\dfrac{3}{5}$ because of $\geq$.

Solution set: $\left(-\infty, -\dfrac{3}{2}\right] \cup \left[\dfrac{3}{5}, \infty\right)$

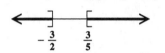

13. $9p^2 + 3p < 2$

Solve the equation

$9p^2 + 3p - 2 = 0.$

$(3p - 1)(3p + 2) = 0$

$3p - 1 = 0$ or $3p + 2 = 0$

$p = \dfrac{1}{3}$ or $p = -\dfrac{2}{3}$

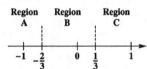

Test a number from each region in the inequality

$9p^2 + 3p < 2.$

Region A: Let $p = -1$.

$9(-1)^2 + 3(-1) < 2 \quad$?

$6 < 2 \qquad$ *False*

Region B: Let $p = 0$.

$0 < 2 \qquad$ *True*

Region C: Let $p = 1$.

$9(1)^2 + 3(1) < 2 \quad$?

$12 < 2 \qquad$ *False*

The solution set includes Region B, but not the endpoints.

Solution set: $\left(-\dfrac{2}{3}, \dfrac{1}{3}\right)$

15.
$$6x^2 + x \geq 1$$
$$6x^2 + x - 1 \geq 0$$

Solve the equation
$$6x^2 + x - 1 = 0.$$
$$(2x + 1)(3x - 1) = 0$$

$$2x + 1 = 0 \quad \text{or} \quad 3x - 1 = 0$$

$$x = -\frac{1}{2} \quad \text{or} \quad x = \frac{1}{3}$$

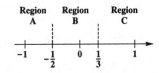

Test a number from each region in the inequality.
$$6x^2 + x \geq 1.$$

Region A: Let $x = -1$.
$$6(-1)^2 + (-1) \geq 1 \quad ?$$
$$5 \geq 1 \qquad \text{True}$$

Region B: Let $x = 0$.
$$0 \geq 1 \qquad \text{False}$$

Region C: Let $x = 1$.
$$6(1)^2 + 1 \geq 1 \quad ?$$
$$7 \geq 1 \qquad \text{True}$$

The solution set includes the numbers in Region A and C, including $-\frac{1}{2}$ and $\frac{1}{3}$ because of $\geq$.

Solution set: $\left(-\infty, -\frac{1}{2}\right] \cup \left[\frac{1}{3}, \infty\right)$

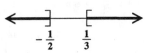

17. $y^2 - 6y + 6 \geq 0$
Solve the equation
$$y^2 - 6y + 6 = 0.$$

Since $y^2 - 6y + 6$ does not factor, let $a = 1$, $b = -6$, and $c = 6$ in the quadratic formula.

$$y = \frac{-(-6) \pm \sqrt{(-6)^2 - 4(1)(6)}}{2(1)}$$

$$= \frac{6 \pm \sqrt{12}}{2} = \frac{6 \pm 2\sqrt{3}}{2}$$

$$= \frac{2\left(3 \pm \sqrt{3}\right)}{2} = 3 \pm \sqrt{3}$$

$$y = 3 + \sqrt{3} \approx 4.7 \text{ or}$$
$$y = 3 - \sqrt{3} \approx 1.3$$

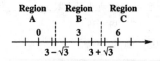

Test a number from each region in the inequality
$$y^2 - 6y + 6 \geq 0.$$

Region A: Let $y = 0$.
$$6 \geq 0 \qquad \text{True}$$

Region B: Let $y = 3$.
$$3^2 - 6(3) + 6 \geq 0 \quad ?$$
$$-3 \geq 0 \qquad \text{False}$$

Region C: Let $y = 5$.
$$5^2 - 6(5) + 6 \geq 0 \quad ?$$
$$1 \geq 0 \qquad \text{True}$$

The solution set includes the numbers in Regions A and C, including $3 - \sqrt{3}$ and $3 + \sqrt{3}$ because of $\geq$.

Solution set: $\left(-\infty, 3 - \sqrt{3}\right] \cup \left[3 + \sqrt{3}, \infty\right)$

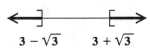

19. Include the endpoints if the inequality is $\leq$ or $\geq$. Exclude the endpoints if the inequality is $<$ or $>$.

21. $(4 - 3x)^2 \geq -2$

Since $(4 - 3x)^2$ is either 0 or positive, $(4 - 3x)^2$ will always be greater than -2. Therefore, the solution set is $(-\infty, \infty)$.

23. $(3x + 5)^2 \leq -4$

Since $(3x + 5)^2$ is never negative, $(3x + 5)^2$ will never be less than or equal to a negative number. Therefore, the solution set is $\emptyset$.

25. Change the inequality to an equation and solve it. Use the solutions of the equation to determine the regions on the number line where the inequality is true. Use test points to do this. Determine whether or not the endpoints are included and indicate them on the number line. Use interval notation to write the solution set. Examples will vary.

27. $(2r + 1)(3r - 2)(4r + 7) < 0$

The numbers $-\frac{1}{2}, \frac{2}{3}$, and $-\frac{7}{4}$ are solutions of the cubic equation
$$(2r + 1)(3r - 2)(4r + 7) = 0.$$

These numbers divide the number line into 4 regions.

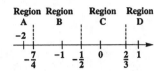

Region A: Let $r = -2$.
$$-3(-8)(-1) < 0 \quad ?$$
$$-24 < 0 \qquad True$$

Region B: Let $r = -1$.
$$-1(-5)(3) < 0 \quad ?$$
$$15 < 0 \qquad False$$

Region C: Let $r = 0$.
$$1(-2)(7) < 0 \quad ?$$
$$-14 < 0 \qquad True$$

Region D: Let $r = 1$.
$$3(1)(11) < 0 \quad ?$$
$$33 < 0 \qquad False$$

The solution set includes numbers in Regions A and C, excluding endpoints.

Solution set: $\left(-\infty, -\dfrac{7}{4}\right) \cup \left(-\dfrac{1}{2}, \dfrac{2}{3}\right)$

29. $(z+2)(4z-3)(2z+7) \geq 0$

The numbers -2, $\dfrac{3}{4}$, and $-\dfrac{7}{2}$ are solutions of the cubic equation

$$(z+2)(4z-3)(2z+7) = 0.$$

These numbers divide the number line into 4 regions.

Region A: Let $z = -4$.
$$-2(-19)(-1) \geq 0 \quad ?$$
$$-38 \geq 0 \qquad False$$

Region B: Let $z = -3$.
$$-1(-15)(1) \geq 0 \quad ?$$
$$15 \geq 0 \qquad True$$

Region C: Let $z = 0$.
$$2(-3)(7) \geq 0 \quad ?$$
$$-42 \geq 0 \qquad False$$

Region D: Let $z = 1$.
$$3(1)(9) \geq 0 \quad ?$$
$$27 \geq 0 \qquad True$$

The solution set includes numbers in Regions B and D, including the endpoints.

Solution set: $\left[-\dfrac{7}{2}, -2\right] \cup \left[\dfrac{3}{4}, \infty\right)$

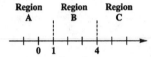

31. $\dfrac{x-1}{x-4} > 0$

Solve the equation

$$\dfrac{x-1}{x-4} = 0.$$

Multiply by the LCD, $x - 4$.

$$x - 1 = 0$$
$$x = 1$$

Find the number that makes the denominator 0.

$$x - 4 = 0$$
$$x = 4$$

The numbers 1 and 4 divide the number line into three regions.

Test a number from each region in the inequality

$$\dfrac{x-1}{x-4} > 0.$$

Region A: Let $x = 0$.
$$\dfrac{0-1}{0-4} > 0 \quad ?$$
$$\dfrac{1}{4} > 0 \qquad True$$

Region B: Let $x = 2$.
$$\dfrac{2-1}{2-4} > 0 \quad ?$$
$$\dfrac{1}{-2} > 0 \qquad False$$

Region C: Let $x = 5$.
$$\dfrac{5-1}{5-4} > 0 \quad ?$$
$$4 > 0 \qquad True$$

The solution set includes numbers in Regions A and C, excluding endpoints.

Solution set: $(-\infty, 1) \cup (4, \infty)$

33. $\dfrac{2y+3}{y-5} \le 0$

Solve the equation

$$\dfrac{2y+3}{y-5} = 0.$$
$$2y+3 = 0$$
$$y = -\dfrac{3}{2}$$

Find the number that makes the denominator 0.

$$y-5 = 0$$
$$y = 5$$

The numbers $-\dfrac{3}{2}$ and 5 divide the number line into three regions.

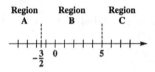

Test a number from each region in the inequality

$$\dfrac{2y+3}{y-5} \le 0.$$

Region A: Let $y = -2$.
$$\dfrac{2(-2)+3}{(-2)-5} \le 0 \qquad ?$$
$$\dfrac{1}{7} \le 0 \qquad \qquad False$$

Region B: Let $y = 0$.
$$\dfrac{2(0)+3}{0-5} \le 0 \qquad ?$$
$$-\dfrac{3}{5} \le 0 \qquad \qquad True$$

Region C: Let $y = 6$.
$$\dfrac{2(6)+3}{6-5} \le 0 \qquad ?$$
$$15 \le 0 \qquad \qquad False$$

The solution set includes the points in Region B. The endpoint 5 is not included since it makes the left side undefined. The endpoint $-\dfrac{3}{2}$ is included because it makes the left side equal to 0.

Solution set: $\left[-\dfrac{3}{2}, 5\right)$

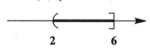

35. $\dfrac{8}{x-2} \ge 2$

Solve the equation

$$\dfrac{8}{x-2} = 2.$$
$$8 = 2(x-2)$$
$$8 = 2x - 4$$
$$12 = 2x$$
$$6 = x$$

Find the number that makes the denominator 0.

$$x-2 = 0$$
$$x = 2$$

The numbers 2 and 6 divide the number line into three regions.

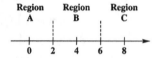

Test a number from each region in the inequality

$$\dfrac{8}{x-2} \ge 2.$$

Region A: Let $x = 0$.
$$\dfrac{8}{0-2} \ge 2 \qquad ?$$
$$-4 \ge 2 \qquad \qquad False$$

Region B: Let $x = 3$.
$$\dfrac{8}{3-2} \ge 2 \qquad ?$$
$$8 \ge 2 \qquad \qquad True$$

Region C: Let $x = 7$.
$$\dfrac{8}{7-2} \ge 2 \qquad ?$$
$$\dfrac{8}{5} \ge 2 \qquad \qquad False$$

The solution set includes numbers in Region B, including 6 but excluding 2, which makes the fraction undefined.

Solution set: $(2, 6]$

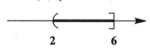

37. $\dfrac{3}{2t-1} < 2$

Solve the equation

$$\dfrac{3}{2t-1} = 2.$$
$$3 = 2(2t-1)$$
$$3 = 4t - 2$$
$$5 = 4t$$
$$\dfrac{5}{4} = t$$

Find the number that makes the denominator 0.

$2t - 1 = 0$

$t = \dfrac{1}{2}$

The numbers $\dfrac{1}{2}$ and $\dfrac{5}{4}$ divide the number line into three regions.

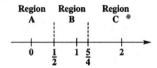

Test a number from each region in the inequality

$$\dfrac{3}{2t - 1} < 2.$$

Region A: Let $t = 0$.

$$\dfrac{3}{2(0) - 1} < 2 \quad ?$$

$$-3 < 2 \qquad True$$

Region B: Let $t = 1$.

$$\dfrac{3}{2(1) - 1} < 2 \quad ?$$

$$3 < 2 \qquad False$$

Region C: Let $t = 2$.

$$\dfrac{3}{2(2) - 1} < 2 \quad ?$$

$$1 < 2 \qquad True$$

The solution set includes numbers in Regions A and C, excluding endpoints.

Solution set: $\left(-\infty, \dfrac{1}{2}\right) \cup \left(\dfrac{5}{4}, \infty\right)$

39. $\dfrac{a}{a + 2} \geq 2$

Solve the equation

$\dfrac{a}{a + 2} = 2.$

$a = 2(a + 2)$

$a = 2a + 4$

$-a = 4$

$a = -4$

Find the number that makes the denominator 0.

$a + 2 = 0$

$a = -2$

The numbers -4 and -2 divide the number line into three regions.

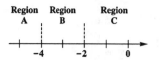

Test a number from each region in the inequality

$$\dfrac{a}{a + 2} \geq 2.$$

Region A: Let $a = -5$.

$$\dfrac{-5}{-5 + 2} \geq 2 \quad ?$$

$$\dfrac{5}{3} \geq 2 \qquad False$$

Region B: Let $a = -3$.

$$\dfrac{-3}{-3 + 2} \geq 2 \quad ?$$

$$3 \geq 2 \qquad True$$

Region C: Let $a = 0$.

$$\dfrac{0}{0 + 2} \geq 2 \quad ?$$

$$0 \geq 2 \qquad False$$

The solution set includes numbers in Region B, including -4 but excluding -2, which makes the fraction undefined.

Solution set: $[-4, -2)$

41. $\dfrac{x}{x - 4} < 3$

Solve the equation

$\dfrac{x}{x - 4} = 3.$

$x = 3(x - 4)$

$x = 3x - 12$

$12 = 2x$

$6 = x$

Find the number that makes the denominator 0.

$x - 4 = 0$

$x = 4$

The numbers 4 and 6 divide the number line into three regions.

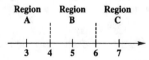

Test a number from each region in the inequality

$$\dfrac{x}{x - 4} < 3.$$

continued

Region A: Let $x = 0$.

$$\frac{0}{0 - 4} < 3 \quad ?$$

$$0 < 3 \qquad True$$

Region B: Let $x = 5$.

$$\frac{5}{5 - 4} < 3 \quad ?$$

$$5 < 3 \qquad False$$

Region C: Let $x = 7$.

$$\frac{7}{7 - 4} < 3 \quad ?$$

$$\frac{7}{3} < 3 \qquad True$$

The solution set includes numbers in Regions A and C, excluding endpoints.

Solution set: $(-\infty, 4) \cup (6, \infty)$

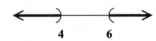

43. $\dfrac{4k}{2k - 1} < k$

Solve the equation

$$\frac{4k}{2k - 1} = k.$$

$$4k = k(2k - 1)$$

$$4k = 2k^2 - k$$

$$0 = 2k^2 - 5k$$

$$0 = k(2k - 5)$$

$$k = 0 \quad \text{or} \quad 2k - 5 = 0$$

$$k = \frac{5}{2}$$

Find the number that makes the denominator 0.

$$2k - 1 = 0$$

$$k = \frac{1}{2}$$

The numbers 0, $\dfrac{1}{2}$, and $\dfrac{5}{2}$ divide the number line into four regions.

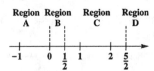

Test a number from each region in the inequality

$$\frac{4k}{2k - 1} < k.$$

Region A: Let $k = -1$.

$$\frac{4(-1)}{2(-1) - 1} < -1 \quad ?$$

$$\frac{4}{3} < -1 \qquad False$$

Region B: Let $k = \dfrac{1}{4}$.

$$\frac{4 \cdot \left(\dfrac{1}{4}\right)}{2\left(\dfrac{1}{4}\right) - 1} < \frac{1}{4} \quad ?$$

$$-2 < \frac{1}{4} \qquad True$$

Region C: Let $k = 1$.

$$\frac{4(1)}{2(1) - 1} < 1 \quad ?$$

$$4 < 1 \qquad False$$

Region D: Let $k = 3$.

$$\frac{4(3)}{2(3) - 1} < 3 \quad ?$$

$$\frac{12}{5} < 3 \qquad True$$

The solution set includes numbers in Regions B and D. None of the endpoints are included.

Solution set: $\left(0, \dfrac{1}{2}\right) \cup \left(\dfrac{5}{2}, \infty\right)$

45. $\dfrac{2x - 3}{x^2 + 1} \geq 0$

The denominator is positive for all real numbers x, so it has no effect on the solution set for the inequality.

$$2x - 3 \geq 0.$$

$$2x \geq 3$$

$$x \geq \frac{3}{2}$$

Solution set: $\left[\dfrac{3}{2}, \infty\right)$

47. $\dfrac{(3x - 5)^2}{x + 2} > 0$

The numerator is positive for all real numbers x except $x = \dfrac{5}{3}$, which makes it equal to 0. If we solve the inequality $x + 2 > 0$, then we only have to be sure to exclude $\dfrac{5}{3}$ from that solution set to determine the solution set of the original inequality.

$$x + 2 > 0$$
$$x > -2$$

Solution set: $\left(-2, \dfrac{5}{3}\right) \cup \left(\dfrac{5}{3}, \infty\right)$

49.
$$\dfrac{-1}{p-3} \geq 1$$
$$\dfrac{-1}{p-3} - 1 \geq 0$$

50.
$$\dfrac{-1}{p-3} - \dfrac{p-3}{p-3} \geq 0$$

51.
$$\dfrac{-1 - (p-3)}{p-3} \geq 0$$
$$\dfrac{-1 - p + 3}{p-3} \geq 0$$
$$\dfrac{-p+2}{p-3} \geq 0$$

52. $-p + 2 = 0$
$$2 = p$$

2 makes the numerator 0.
$$p - 3 = 0$$
$$p = 3$$

3 makes the denominator 0.
The solution (see Example 4) gives us the solution set $[2, 3)$.

53. "The height is greater than .140 foot" is described by the inequality

$$y > .140.$$
$$x - 1.727x^2 > .140$$
$$0 > 1.727x^2 - x + .140$$

This is equivalent to

$$1.727x^2 - x + .140 < 0.$$

To solve the corresponding equation,

$$1.727x^2 - x + .140 = 0,$$

use the quadratic formula.

$$x = \dfrac{-(-1) \pm \sqrt{(-1)^2 - 4(1.727)(.140)}}{2(1.727)}$$
$$= \dfrac{1 \pm \sqrt{.03288}}{3.454}$$
$$\approx .342 \quad \text{or} \quad .237$$

For this problem, our common sense tells us that $y > .140$ between approximately .237 and .342 second. We can check a time in this interval, say .3, to verify this.

$$1.727(.3)^2 - (.3) + .140 < 0 \quad ?$$
$$-.00457 < 0 \qquad True$$

55. (a) The x-intercepts determine the solutions of the equation $x^2 - 4x + 3 = 0$. From the graph, the solution set is $\{1, 3\}$.

(b) The x-values of the points on the graph that are *above* the x-axis form the solution set of the inequality $x^2 - 4x + 3 > 0$. From the graph, the solution set is $(-\infty, 1) \cup (3, \infty)$.

(c) The x-values of the points on the graph that are *below* the x-axis form the solution set of the inequality $x^2 - 4x + 3 < 0$. From the graph, the solution set is $(1, 3)$.

56. (a) The x-intercepts determine the solutions of the equation $3x^2 + 10x - 8 = 0$. From the graph, the solution set is $\left\{-4, \dfrac{2}{3}\right\}$.

(b) The x-values of the points on the graph that are *above* the x-axis form the solution set of the inequality $3x^2 + 10x - 8 > 0$. From the graph, the solution set for $3x^2 + 10x - 8 \geq 0$ is $(-\infty, -4] \cup \left[\dfrac{2}{3}, \infty\right)$.

(c) The x-values of the points on the graph that are *below* the x-axis form the solution set of the inequality $3x^2 + 10x - 8 < 0$. From the graph, the solution set is $\left(-4, \dfrac{2}{3}\right)$.

57. (a) The x-intercepts determine the solutions of the equation $-x^2 + 3x + 10 = 0$. From the graph, the solution set is $\{-2, 5\}$.

(b) The x-values of the points on the graph that are *above* the x-axis form the solution set of the inequality $-x^2 + 3x + 10 > 0$. From the graph, the solution set for $-x^2 + 3x + 10 \geq 0$ is $[-2, 5]$.

(c) The x-values of the points on the graph that are *below* the x-axis form the solution set of the inequality $-x^2 + 3x + 10 < 0$. From the graph, the solution set for $-x^2 + 3x + 10 \leq 0$ is $(-\infty, -2] \cup [5, \infty)$.

58. (a) The x-intercepts determine the solutions of the equation $-2x^2 - x + 15 = 0$. From the graph, the solution set is $\left\{-3, \dfrac{5}{2}\right\}$.

(b) The x-values of the points on the graph that are *above* the x-axis form the solution set of the inequality $-2x^2 - x + 15 > 0$. From the graph, the solution set for $-2x^2 - x + 15 \geq 0$ is

$$\left[-3, \frac{5}{2}\right].$$

(c) The x-values of the points on the graph that are *below* the x-axis form the solution set of the inequality $-2x^2 - x + 15 < 0$. From the graph, the solution set for $-2x^2 - x + 15 \leq 0$ is

$$(-\infty, -3] \cup \left[\frac{5}{2}, \infty\right).$$

59. The height is less than 10 feet is described by the inequality

$$f(x) < 10.$$
$$x - .0066x^2 < 10$$
$$0 < .0066x^2 - x + 10$$

This is equivalent to

$$.0066x^2 - x + 10 > 0.$$

To solve the corresponding equation,

$$.0066x^2 - x + 10 = 0,$$

use the quadratic formula.

$$x = \frac{-(-1) \pm \sqrt{(-1)^2 - 4(.0066)(10)}}{2(.0066)}$$

$$= \frac{1 \pm \sqrt{.736}}{.0132}$$

$$\approx 140.8 \text{ or } 10.8$$

To determine how long the rocket is in the air, let $f(x) = 0$ and solve for x.

$$0 = x - .0066x^2$$
$$0 = x(1 - .0066x)$$
$$x = 0 \quad \text{or} \quad 1 - .0066x = 0$$
$$x = \frac{1}{.0066}$$
$$\approx 151.5$$

Thus, our common sense tells us that $f(x) < 10$ between 0 and 10.8 seconds and between 140.8 and 151.5 seconds.

Chapter 11 Review Exercises

1. $y^2 = 144$

$$y = \sqrt{144} \quad \text{or} \quad y = -\sqrt{144}$$
$$y = 12 \quad \text{or} \quad y = -12$$

The solution set is $\{\pm 12\}$.

2. $x^2 = 37$

$$x = \sqrt{37} \quad \text{or} \quad x = -\sqrt{37}$$

The solution set is $\left\{\pm\sqrt{37}\right\}$.

3. $m^2 = 128$

$$m = \pm\sqrt{128}$$
$$= \pm\sqrt{64 \cdot 2}$$
$$= \pm 8\sqrt{2}$$

The solution set is $\left\{\pm 8\sqrt{2}\right\}$.

4. $(k + 2)^2 = 25$

$$k + 2 = \sqrt{25} \quad \text{or} \quad k + 2 = -\sqrt{25}$$
$$k + 2 = 5 \quad \text{or} \quad k + 2 = -5$$
$$k = 3 \quad \text{or} \quad k = -7$$

The solution set is $\{-7, 3\}$.

5. $(r - 3)^2 = 10$

$$r - 3 = \sqrt{10} \quad \text{or} \quad r - 3 = -\sqrt{10}$$
$$r = 3 + \sqrt{10} \quad \text{or} \quad r = 3 - \sqrt{10}$$

The solution set is $\left\{3 \pm \sqrt{10}\right\}$.

6. $(2p + 1)^2 = 14$

$$2p + 1 = \sqrt{14} \quad \text{or} \quad 2p + 1 = -\sqrt{14}$$
$$2p = -1 + \sqrt{14} \quad \text{or} \quad 2p = -1 - \sqrt{14}$$
$$p = \frac{-1 + \sqrt{14}}{2} \quad \text{or} \quad p = \frac{-1 - \sqrt{14}}{2}$$

The solution set is $\left\{\dfrac{-1 \pm \sqrt{14}}{2}\right\}$.

7. $(3k + 2)^2 = -3$

$$3k + 2 = \sqrt{-3} \quad \text{or} \quad 3k + 2 = -\sqrt{-3}$$

Because $\sqrt{-3}$ does not represent a real number, the solution set is $\emptyset$.

8. **(a)** $x^2 = 0$ has 1 real solution, namely 0, and **(b)** $x^2 = -4$ and **(c)** $(x + 5)^2 = -16$ have no real solutions because the square of a real number cannot be negative. The equation that has two real solutions is **(d)** $(x + 6)^2 = 25$.

9. $m^2 + 6m + 5 = 0$

Rewrite the equation with the variable terms on one side and the constant on the other side.

$$m^2 + 6m = -5$$

Take half the coefficient of m and square it.

$$\frac{1}{2}(6) = 3, \text{ and } (3)^2 = 9.$$

Add 9 to each side of the equation.

$$m^2 + 6m + 9 = -5 + 9$$
$$m^2 + 6m + 9 = 4$$
$$(m + 3)^2 = 4 \quad Factor$$

$$m + 3 = \sqrt{4} \quad \text{or} \quad m + 3 = -\sqrt{4}$$
$$m + 3 = 2 \quad \text{or} \quad m + 3 = -2$$
$$m = -1 \quad \text{or} \quad m = -5$$

The solution set is $\{-5, -1\}$.

10. $p^2 + 4p = 7$

Take half the coefficient of p and square it.

$$\frac{1}{2}(4) = 2, \text{ and } (2)^2 = 4.$$

Add 4 to each side of the equation.

$$p^2 + 4p + 4 = 7 + 4$$
$$(p + 2)^2 = 11$$

$$p + 2 = \sqrt{11} \quad \text{or} \quad p + 2 = -\sqrt{11}$$
$$p = -2 + \sqrt{11} \quad \text{or} \quad p = -2 - \sqrt{11}$$

The solution set is $\left\{-2 \pm \sqrt{11}\right\}$.

11. $-x^2 + 5 = 2x$

Divide each side of the equation by -1 to make the coefficient of the squared term equal to 1.

$$x^2 - 5 = -2x$$

Rewrite the equation with the variable terms on one side and the constant on the other side.

$$x^2 + 2x = 5$$

Take half the coefficient of x and square it.

$$\frac{1}{2}(2) = 1, \text{ and } 1^2 = 1.$$

Add 1 to both sides of the equation.

$$x^2 + 2x + 1 = 5 + 1$$
$$(x + 1)^2 = 6$$

$$x + 1 = \sqrt{6} \quad \text{or} \quad x + 1 = -\sqrt{6}$$
$$x = -1 + \sqrt{6} \quad \text{or} \quad x = -1 - \sqrt{6}$$

The solution set is $\left\{-1 \pm \sqrt{6}\right\}$.

12. $2y^2 - 3 = -8y$

Divide both sides by 2 to get the y^2 coefficient equal to 1.

$$y^2 - \frac{3}{2} = -4y$$

Rewrite the equation with the variable terms on one side and the constant on the other side.

$$y^2 + 4y = \frac{3}{2}$$

Take half the coefficient of y and square it.

$$\frac{1}{2}(4) = 2, \text{ and } 2^2 = 4.$$

Add 4 to both sides of the equation.

$$y^2 + 4y + 4 = \frac{3}{2} + 4$$
$$(y + 2)^2 = \frac{11}{2}$$

$$y + 2 = \pm\sqrt{\frac{11}{2}}$$
$$y + 2 = \pm\frac{\sqrt{11}}{\sqrt{2}} \cdot \frac{\sqrt{2}}{\sqrt{2}}$$
$$y + 2 = \pm\frac{\sqrt{22}}{2}$$
$$y = -2 \pm \frac{\sqrt{22}}{2}$$
$$y = \frac{-4}{2} \pm \frac{\sqrt{22}}{2}$$
$$y = \frac{-4 \pm \sqrt{22}}{2}$$

The solution set is $\left\{\dfrac{-4 \pm \sqrt{22}}{2}\right\}$.

13. $5k^2 - 3k - 2 = 0$

Divide both sides by 5 to get the k^2 coefficient equal to 1.

$$k^2 - \frac{3}{5}k - \frac{2}{5} = 0$$

Rewrite the equation with the variable terms on one side and the constant on the other side.

$$k^2 - \frac{3}{5}k = \frac{2}{5}$$

Take half the coefficient of k and square it.

$$\frac{1}{2}\left(-\frac{3}{5}\right) = -\frac{3}{10}, \text{ and } \left(-\frac{3}{10}\right)^2 = \frac{9}{100}$$

$$k^2 - \frac{3}{5}k + \frac{9}{100} = \frac{2}{5} + \frac{9}{100}$$
$$\left(k - \frac{3}{10}\right)^2 = \frac{40}{100} + \frac{9}{100}$$
$$\left(k - \frac{3}{10}\right)^2 = \frac{49}{100}$$

continued

$$k - \frac{3}{10} = \sqrt{\frac{49}{100}} \quad \text{or} \quad k - \frac{3}{10} = -\sqrt{\frac{49}{100}}$$

$$k - \frac{3}{10} = \frac{7}{10} \quad \text{or} \quad k - \frac{3}{10} = -\frac{7}{10}$$

$$k = \frac{10}{10} \quad \text{or} \quad k = -\frac{4}{10}$$

$$k = 1 \quad \text{or} \quad k = -\frac{2}{5}$$

The solution set is $\left\{ -\frac{2}{5}, 1 \right\}$.

14. $(4a + 1)(a - 1) = -7$

Multiply on the left side and then simplify. Get all variable terms on one side and the constant on the other side.

$$4a^2 - 4a + a - 1 = -7$$
$$4a^2 - 3a = -6$$

Divide both sides by 4 so that the coefficient of a^2 will be 1.

$$a^2 - \frac{3}{4}a = -\frac{6}{4} = -\frac{3}{2}$$

Square half the coefficient of a and add it to both sides.

$$a^2 - \frac{3}{4}a + \frac{9}{64} = -\frac{3}{2} + \frac{9}{64}$$

$$\left(a - \frac{3}{8}\right)^2 = -\frac{96}{64} + \frac{9}{64}$$

$$\left(a - \frac{3}{8}\right)^2 = -\frac{87}{64}$$

The square root of $-\frac{87}{64}$ is not a real number, so the solution set is $\emptyset$.

15. $h = -16t^2 + 32t + 50$

Let $h = 30$ and solve for t (which must have a positive value since it represents a number of seconds).

$$30 = -16t^2 + 32t + 50$$
$$16t^2 - 32t - 20 = 0$$

Divide both sides by 16.

$$t^2 - 2t - \frac{20}{16} = 0$$

$$t^2 - 2t = \frac{5}{4}$$

Half of -2 is -1, and $(-1)^2 = 1$.

Add 1 to both sides of the equation.

$$t^2 - 2t + 1 = \frac{5}{4} + 1$$

$$(t - 1)^2 = \frac{9}{4}$$

$$t - 1 = \sqrt{\frac{9}{4}} \quad \text{or} \quad t - 1 = -\sqrt{\frac{9}{4}}$$

$$t - 1 = \frac{3}{2} \quad \text{or} \quad t - 1 = -\frac{3}{2}$$

$$t = 1 + \frac{3}{2} \quad \text{or} \quad t = 1 - \frac{3}{2}$$

$$t = \frac{5}{2} = 2\frac{1}{2} \quad \text{or} \quad t = -\frac{1}{2}$$

Reject the negative value of t. The object will reach a height of 30 feet after $2\frac{1}{2}$ seconds.

16. Use the Pythagorean formula with legs x and $x + 2$ and hypotenuse $x + 4$.

$$a^2 + b^2 = c^2$$
$$(x)^2 + (x + 2)^2 = (x + 4)^2$$
$$x^2 + x^2 + 4x + 4 = x^2 + 8x + 16$$
$$x^2 - 4x - 12 = 0$$
$$(x - 6)(x + 2) = 0$$

$$x - 6 = 0 \quad \text{or} \quad x + 2 = 0$$
$$x = 6 \quad \text{or} \quad x = -2$$

Reject the negative value because x represents a length. The value of x is 6. The lengths of the three sides are 6, 8, and 10.

17. Take half the coefficient of x and square the result.

$$\frac{1}{2} \cdot k = \frac{k}{2}$$

$$\left(\frac{k}{2}\right)^2 = \frac{k^2}{4}$$

Add $\left(\frac{k}{2}\right)^2$ or $\frac{k^2}{4}$ to $x^2 + kx$ to get the perfect square $x^2 + kx + \frac{k^2}{4}$.

18. $x^2 - 9 = 0$

(a) $(x + 3)(x - 3) = 0$

$$x + 3 = 0 \quad \text{or} \quad x - 3 = 0$$
$$x = -3 \quad \text{or} \quad x = 3$$

The solution set is $\{\pm 3\}$.

(b) $x^2 = 9$

$$x = \pm\sqrt{9} = \pm 3$$

The solution set is $\{\pm 3\}$.

(c) $a = 1$, $b = 0$, and $c = -9$

$$x = \frac{-0 \pm \sqrt{0^2 - 4(1)(-9)}}{2(1)}$$

$$= \frac{\pm \sqrt{36}}{2}$$

$$= \frac{\pm 6}{2} = \pm 3$$

The solution set is $\{\pm 3\}$.

(d) We will always get the same results, no matter which method of solution is used. This is because there is only one solution set, and the solutions will always match.

19. $x^2 - 2x - 4 = 0$

This equation is in standard form with $a = 1$, $b = -2$, and $c = -4$. Substitute these values into the quadratic formula.

$$x = \frac{-b \pm \sqrt{b^2 - 4ac}}{2a}$$

$$x = \frac{-(-2) \pm \sqrt{(-2)^2 - 4(1)(-4)}}{2(1)}$$

$$= \frac{2 \pm \sqrt{4 + 16}}{2}$$

$$= \frac{2 \pm \sqrt{20}}{2} = \frac{2 \pm 2\sqrt{5}}{2}$$

$$= \frac{2\left(1 \pm \sqrt{5}\right)}{2} = 1 \pm \sqrt{5}$$

The solution set is $\left\{1 \pm \sqrt{5}\right\}$.

20. $3k^2 + 2k = -3$

$3k^2 + 2k + 3 = 0$

$a = 3$, $b = 2$, and $c = 3$

$$k = \frac{-2 \pm \sqrt{2^2 - 4(3)(3)}}{2(3)}$$

$$= \frac{-2 \pm \sqrt{4 - 36}}{6}$$

$$= \frac{-2 \pm \sqrt{-32}}{6}$$

Because $\sqrt{-32}$ does not represent a real number, the solution set is $\emptyset$.

21. $2p^2 + 8 = 4p + 11$

$2p^2 - 4p - 3 = 0$

Use the quadratic formula with

$a = 2$, $b = -4$, and $c = -3$.

$$p = \frac{-(-4) \pm \sqrt{(-4)^2 - 4(2)(-3)}}{2(2)}$$

$$= \frac{4 \pm \sqrt{16 + 24}}{4} = \frac{4 \pm \sqrt{40}}{4}$$

$$= \frac{4 \pm \sqrt{4 \cdot 10}}{4} = \frac{4 \pm 2\sqrt{10}}{4}$$

$$= \frac{2\left(2 \pm \sqrt{10}\right)}{2(2)} = \frac{2 \pm \sqrt{10}}{2}$$

The solution set is $\left\{\dfrac{2 \pm \sqrt{10}}{2}\right\}$.

22. $-4x^2 + 7 = 2x$

$-4x^2 - 2x + 7 = 0$

$a = -4$, $b = -2$, and $c = 7$

$$x = \frac{-(-2) \pm \sqrt{(-2)^2 - 4(-4)(7)}}{2(-4)}$$

$$= \frac{2 \pm \sqrt{4 + 112}}{-8} = \frac{2 \pm \sqrt{116}}{-8}$$

$$= \frac{2 \pm 2\sqrt{29}}{-8} = \frac{2\left(1 \pm \sqrt{29}\right)}{2(-4)}$$

$$= \frac{1 \pm \sqrt{29}}{-4}$$

The solution set is $\left\{\dfrac{1 \pm \sqrt{29}}{-4}\right\}$.

23. $\dfrac{1}{4}p^2 = 2 - \dfrac{3}{4}p$

$\dfrac{1}{4}p^2 + \dfrac{3}{4}p - 2 = 0$

Multiply both sides by the least common denominator, 4.

$$4\left(\frac{1}{4}p^2 + \frac{3}{4}p - 2\right) = 4(0)$$

$$p^2 + 3p - 8 = 0$$

Use the quadratic formula with

$a = 1$, $b = 3$, and $c = -8$.

$$p = \frac{-3 \pm \sqrt{3^2 - 4(1)(-8)}}{2(1)}$$

$$= \frac{-3 \pm \sqrt{9 + 32}}{2}$$

$$= \frac{-3 \pm \sqrt{41}}{2}$$

The solution set is $\left\{\dfrac{-3 \pm \sqrt{41}}{2}\right\}$

24. (a) $a^2 + 5a + 2 = 0$

$$a = 1, b = 5, c = 2$$
$$b^2 - 4ac = 5^2 - 4(1)(2)$$
$$= 25 - 8 = 17$$

Since the discriminant is positive, but not a perfect square, there are two different irrational number solutions. The answer is (c).

(b) $$4x^2 = 3 - 4x$$
$$4x^2 + 4x - 3 = 0$$
$$a = 4, b = 4, c = -3$$
$$b^2 - 4ac = 4^2 - 4(4)(-3)$$
$$= 16 + 48$$
$$= 64 \text{ or } 8^2$$

Since the discriminant is positive, and a perfect square, there are two different rational number solutions. The answer is (a).

(c) $$4x^2 = 6x - 8$$
$$4x^2 - 6x + 8 = 0$$
$$a = 4, b = -6, c = 8$$
$$b^2 - 4ac = (-6)^2 - 4(4)(8)$$
$$= 36 - 128 = -92$$

Since the discriminant is negative, there are two different imaginary number solutions. The answer is (d).

(d) $9z^2 + 30z + 25 = 0$

$$a = 9, b = 30, c = 25$$
$$b^2 - 4ac = 30^2 - 4(9)(25)$$
$$= 900 - 900 = 0$$

Since the discriminant is zero, there is exactly one rational number solution. The answer is (b).

25. To divide complex numbers, we multiply numerator and denominator by the conjugate of the denominator (the divisor).

26. $i^2 = -1$

(a) Since $i^3 = i^2 \cdot i$,
$$i^3 = (-1)(i) = -i.$$

(b) Since $i^4 = i^3 \cdot i$,
$$i^4 = (-i)(i) = -i^2$$
$$= -(-1) = 1.$$

(c) Since $i^{48} = \left(i^4\right)^{12}$,
$$i^{48} = (1)^{12} = 1.$$

27. $(3 + 5i) + (2 - 6i)$
$$= (3 + 2) + (5 - 6)i$$
$$= 5 - i$$

28. $(-2 - 8i) - (4 - 3i)$
$$= (-2 - 8i) + (-4 + 3i)$$
$$= (-2 - 4) + (-8 + 3)i$$
$$= -6 - 5i$$

29. $(6 - 2i)(3 + i)$
$$= 18 + 6i - 6i - 2i^2 \quad \textit{FOIL}$$
$$= 18 - 2(-1) \qquad \textit{i}^2 = -1$$
$$= 18 + 2 = 20$$

30. $(2 + 3i)(2 - 3i)$
$$= 2^2 + 3^2 \qquad \textit{Product of conjugates}$$
$$= 4 + 9 = 13$$

31. $\dfrac{1 + i}{1 - i} = \dfrac{1 + i}{1 - i} \cdot \dfrac{1 + i}{1 + i}$

Multiply by the conjugate of the denominator

$$= \frac{1 + 2i + i^2}{1 - i^2}$$
$$= \frac{1 + 2i + (-1)}{1 - (-1)} \qquad \textit{i}^2 = -1$$
$$= \frac{2i}{2} = i$$

32. $\dfrac{5 + 6i}{2 + 3i} = \dfrac{5 + 6i}{2 + 3i} \cdot \dfrac{2 - 3i}{2 - 3i}$

$$= \frac{10 - 3i - 18i^2}{4 - 9i^2}$$
$$= \frac{10 - 3i - 18(-1)}{4 - 9(-1)}$$
$$= \frac{10 + 18 - 3i}{4 + 9}$$
$$= \frac{28 - 3i}{13}$$
$$= \frac{28}{13} - \frac{3}{13}i \qquad \textit{Standard form}$$

33. $\dfrac{1}{7 - i} = \dfrac{1}{7 - i} \cdot \dfrac{7 + i}{7 + i}$

$$= \frac{7 + i}{(7 - i)(7 + i)}$$
$$= \frac{7 + i}{49 - i^2}$$
$$= \frac{7 + i}{49 - (-1)}$$
$$= \frac{7 + i}{50}$$
$$= \frac{7}{50} + \frac{1}{50}i \qquad \textit{Standard form}$$

34. The real number a can be written as $a + 0i$. The conjugate of a is $a - 0i$ or a. Thus, the conjugate of a real number is the real number itself.

35. The product of a complex number and its conjugate is always a real number.

$$(a + bi)(a - bi) = a^2 - b^2i^2$$
$$= a^2 - b^2(-1)$$
$$= a^2 + b^2$$

Therefore, the product of a complex number and its conjugate is never an imaginary number.

36. $(m + 2)^2 = -3$

Use the square root property.

$$m + 2 = \pm\sqrt{-3}$$
$$m + 2 = \pm i\sqrt{3}$$
$$m = -2 \pm i\sqrt{3}$$

The solution set is $\left\{-2 \pm i\sqrt{3}\right\}$.

37. $(3p - 2)^2 = -8$

Use the square root property.

$$3p - 2 = \pm\sqrt{-8}$$
$$3p - 2 = \pm 2i\sqrt{2}$$
$$3p = 2 \pm 2i\sqrt{2}$$
$$p = \frac{2 \pm 2i\sqrt{2}}{3} = \frac{2}{3} \pm \frac{2\sqrt{2}}{3}i$$

The solution set is $\left\{\dfrac{2}{3} \pm \dfrac{2\sqrt{2}}{3}i\right\}$.

38. $3k^2 = 2k - 1$

Rewrite the equation in standard form.

$$3k^2 - 2k + 1 = 0$$

Use the quadratic formula with

$a = 3$, $b = -2$, and $c = 1$.

$$k = \frac{-b \pm \sqrt{b^2 - 4ac}}{2a}$$

$$k = \frac{-(-2) \pm \sqrt{(-2)^2 - 4(3)(1)}}{2(3)}$$

$$= \frac{2 \pm \sqrt{4 - 12}}{6}$$

$$= \frac{2 \pm \sqrt{-8}}{6} = \frac{2 \pm 2i\sqrt{2}}{2 \cdot 3}$$

$$= \frac{2\left(1 \pm i\sqrt{2}\right)}{2 \cdot 3} = \frac{1 \pm i\sqrt{2}}{3}$$

$$= \frac{1}{3} \pm \frac{\sqrt{2}}{3}i \qquad \textit{Standard form}$$

The solution set is $\left\{\dfrac{1}{3} \pm \dfrac{\sqrt{2}}{3}i\right\}$.

39. $h^2 + 3h = -8$

Rewrite the equation in standard form.

$$h^2 + 3h + 8 = 0$$

Use the quadratic formula with

$a = 1$, $b = 3$, and $c = 8$.

$$h = \frac{-b \pm \sqrt{b^2 - 4ac}}{2a}$$

$$h = \frac{-3 \pm \sqrt{3^2 - 4(1)(8)}}{2(1)}$$

$$= \frac{-3 \pm \sqrt{9 - 32}}{2} = \frac{-3 \pm \sqrt{-23}}{2}$$

$$= -\frac{3}{2} \pm \frac{\sqrt{23}}{2}i \qquad \textit{Standard form}$$

The solution set is $\left\{-\dfrac{3}{2} \pm \dfrac{\sqrt{23}}{2}i\right\}$.

40. $4q^2 + 2 = 3q$

$$4q^2 - 3q + 2 = 0 \quad \textit{Standard form}$$

Use the quadratic formula with

$a = 4$, $b = -3$, and $c = 2$.

$$q = \frac{-(-3) \pm \sqrt{(-3)^2 - 4(4)(2)}}{2 \cdot 4}$$

$$= \frac{3 \pm \sqrt{9 - 32}}{8}$$

$$= \frac{3 \pm \sqrt{-23}}{8}$$

$$= \frac{3 \pm i\sqrt{23}}{8}$$

$$= \frac{3}{8} \pm \frac{\sqrt{23}}{8}i \qquad \textit{Standard form}$$

The solution set is $\left\{\dfrac{3}{8} \pm \dfrac{\sqrt{23}}{8}i\right\}$.

41. $9z^2 + 2z + 1 = 0$

Use the quadratic formula with

$a = 9$, $b = 2$, and $c = 1$.

$$z = \frac{-b \pm \sqrt{b^2 - 4ac}}{2a}$$

$$z = \frac{-2 \pm \sqrt{2^2 - 4(9)(1)}}{2(9)}$$

$$= \frac{-2 \pm \sqrt{4 - 36}}{18}$$

$$= \frac{-2 \pm \sqrt{-32}}{18}$$

continued

$$= \frac{-2 \pm i\sqrt{32}}{18} = \frac{-2 \pm i\sqrt{16 \cdot 2}}{18}$$

$$= \frac{-2 \pm 4i\sqrt{2}}{18} = \frac{2\left(-1 \pm 2i\sqrt{2}\right)}{2 \cdot 9}$$

$$= \frac{-1 \pm 2i\sqrt{2}}{9}$$

$$= -\frac{1}{9} \pm \frac{2\sqrt{2}}{9}i \qquad \textit{Standard form}$$

The solution set is $\left\{ -\dfrac{1}{9} \pm \dfrac{2\sqrt{2}}{9}i \right\}$.

42. $\dfrac{1}{y} + \dfrac{2}{y+1} = 2$

$$y(y+1)\left(\frac{1}{y} + \frac{2}{y+1}\right) \qquad \begin{array}{l}\textit{Multiply by} \\ \textit{the LCD, } y(y+1).\end{array}$$

$$= y(y+1) \cdot 2$$

$$(y+1) + 2y = 2y^2 + 2y$$

$$0 = 2y^2 - y - 1$$

$$0 = (2y+1)(y-1)$$

$$2y + 1 = 0 \qquad \text{or} \quad y - 1 = 0$$

$$y = -\frac{1}{2} \quad \text{or} \qquad y = 1$$

Check $y = -\dfrac{1}{2}$: $-2 + 4 \overset{?}{=} 2$ *True*

Check $y = 1$: $1 + 1 \overset{?}{=} 2$ *True*

Solution set: $\left\{ -\dfrac{1}{2}, 1 \right\}$

43. $8(3x+5)^2 + 2(3x+5) - 1 = 0$

Let $u = 3x + 5$. The equation becomes

$$8u^2 + 2u - 1 = 0$$

$$(2u+1)(4u-1) = 0$$

$$2u + 1 = 0 \qquad \text{or} \quad 4u - 1 = 0$$

$$u = -\frac{1}{2} \quad \text{or} \qquad u = \frac{1}{4}$$

To find x, substitute $3x + 5$ for u.

$$3x + 5 = -\frac{1}{2} \qquad \text{or} \quad 3x + 5 = \frac{1}{4}$$

$$3x = -\frac{11}{2} \qquad\qquad 3x = -\frac{19}{4}$$

$$x = -\frac{11}{6} \quad \text{or} \qquad x = -\frac{19}{12}$$

Check $x = -\dfrac{11}{6}$: $2 - 1 - 1 \overset{?}{=} 0$ *True*

Check $x = -\dfrac{19}{12}$: $.5 + .5 - 1 \overset{?}{=} 0$ *True*

Solution set: $\left\{ -\dfrac{11}{6}, -\dfrac{19}{12} \right\}$

44. $$-2r = \sqrt{\frac{48 - 20r}{2}}$$

Square both sides.

$$(-2r)^2 = \left(\sqrt{\frac{48 - 20r}{2}}\right)^2$$

$$4r^2 = \frac{48 - 20r}{2}$$

$$4r^2 = 24 - 10r$$

$$4r^2 + 10r - 24 = 0$$

$$2r^2 + 5r - 12 = 0$$

$$(r+4)(2r-3) = 0$$

$$r + 4 = 0 \qquad \text{or} \quad 2r - 3 = 0$$

$$r = -4 \quad \text{or} \qquad r = \frac{3}{2}$$

Check $r = -4$: $8 \overset{?}{=} \sqrt{64}$ *True*

Check $r = \dfrac{3}{2}$: $-3 \overset{?}{=} \sqrt{9}$ *False*

Solution set: $\{-4\}$

45. $2x^{2/3} - x^{1/3} - 28 = 0$

Let $u = x^{1/3}$, so $u^2 = \left(x^{1/3}\right)^2 = x^{2/3}$.

The equation becomes

$$2u^2 - u - 28 = 0.$$

$$(2u+7)(u-4) = 0$$

$$2u + 7 = 0 \qquad \text{or} \quad u - 4 = 0$$

$$u = -\frac{7}{2} \quad \text{or} \qquad u = 4$$

To find x, substitute $x^{1/3}$ for u.

$$x^{1/3} = -\frac{7}{2} \qquad \text{or} \qquad x^{1/3} = 4$$

$$\left(x^{1/3}\right)^3 = \left(-\frac{7}{2}\right)^3 \qquad \left(x^{1/3}\right)^3 = 4^3$$

$$x = -\frac{343}{8} \qquad \text{or} \qquad x = 64$$

Check $x = -\dfrac{343}{8}$: $24.5 + 3.5 - 28 \overset{?}{=} 0$ *True*

Check $x = 64$: $32 - 4 - 28 \overset{?}{=} 0$ *True*

Solution set: $\left\{ -\dfrac{343}{8}, 64 \right\}$

46. $5x^4 - 2x^2 - 3 = 0$

Let $y = x^2$, so $y^2 = x^4$.

$$5y^2 - 2y - 3 = 0$$

$$(5y+3)(y-1) = 0$$

$$5y + 3 = 0 \qquad \text{or} \quad y - 1 = 0$$

$$y = -\frac{3}{5} \quad \text{or} \qquad y = 1$$

To find x, substitute x^2 for y.

$$x^2 = -\frac{3}{5} \quad \text{or} \quad x^2 = 1$$

$$x = \pm i\sqrt{\frac{3}{5}} \quad \text{or} \quad x = \pm 1$$

$$x = \frac{\pm\sqrt{15}}{5}i$$

Check $x = \pm \dfrac{\sqrt{15}}{5}i$: $\dfrac{9}{5} + \dfrac{6}{5} - 3 \overset{?}{=} 0$ *True*

Check $x = \pm 1$: $5 - 2 - 3 \overset{?}{=} 0$ *True*

Solution set: $\left\{-1, 1, -\dfrac{\sqrt{15}}{5}i, \dfrac{\sqrt{15}}{5}i\right\}$

47. Let x = Lisa's speed on the trip to pick up Laurie.

Make a chart. Use $d = rt$, or $t = \dfrac{d}{r}$.

	Distance	Rate	Time
To Laurie	8	x	$\dfrac{8}{x}$
To the mall	11	$x + 15$	$\dfrac{11}{x + 15}$

$$\begin{array}{ccccc}
\text{Time to pick} & & \text{time to} & & 24 \text{ min} \\
\text{up Laurie} & + & \text{mall} & = & \text{(or .4 hr).} \\
\dfrac{8}{x} & + & \dfrac{11}{x + 15} & = & .4
\end{array}$$

Multiply by the LCD, $x(x + 15)$.

$$x(x+15)\left(\frac{8}{x} + \frac{11}{x+15}\right) = x(x+15)(.4)$$

$$8(x+15) + 11x = .4x(x+15)$$

$$8x + 120 + 11x = .4x^2 + 6x$$

$$0 = .4x^2 - 13x - 120$$

Multiply by 5 to clear the decimal.

$$0 = 2x^2 - 65x - 600$$

$$0 = (x - 40)(2x + 15)$$

$$x - 40 = 0 \quad \text{or} \quad 2x + 15 = 0$$

$$x = 40 \quad \text{or} \qquad x = -\frac{15}{2}$$

Speed cannot be negative, so $-\dfrac{15}{2}$ is not a solution. Lisa's speed on the trip to pick up Laurie was 40 mph.

48. Let x = the time for Laketa alone and $x - 2$ = the time for Ed alone.

Worker	Rate	Time Together	Part of Job Done
Laketa	$\dfrac{1}{x}$	3	$\dfrac{3}{x}$
Ed	$\dfrac{1}{x-2}$	3	$\dfrac{3}{x-2}$

$$\begin{array}{ccccc}
\text{Part by} & & \text{part} & & 1 \text{ whole} \\
\text{Lisa} & + & \text{by Ed} & = & \text{job.} \\
\dfrac{3}{x} & + & \dfrac{3}{x-2} & = & 1
\end{array}$$

Multiply by the LCD, $x(x - 2)$.

$$x(x-2)\left(\frac{3}{x} + \frac{3}{x-2}\right) = x(x-2)\cdot 1$$

$$3x - 6 + 3x = x^2 - 2x$$

$$0 = x^2 - 8x + 6$$

Use the quadratic formula.

$$x = \frac{-b \pm \sqrt{b^2 - 4ac}}{2a}$$

$$x = \frac{8 \pm \sqrt{(-8)^2 - 4(1)(6)}}{2}$$

$$= \frac{8 \pm \sqrt{64 - 24}}{2} = \frac{8 \pm \sqrt{40}}{2}$$

$$= \frac{8 \pm 2\sqrt{10}}{2} = \frac{2\left(4 \pm \sqrt{10}\right)}{2} = 4 \pm \sqrt{10}$$

$$x = 4 + \sqrt{10} \approx 7.2 \text{ or}$$

$$x = 4 - \sqrt{10} \approx .8$$

Reject .8 as Lisa's time, because that would make Ed's time negative. So, Lisa's time alone is about 7.2 hours and Ed's time alone is about $x - 2 = 5.2$ hours.

49. The equation $x = \sqrt{2x + 4}$ can't have a negative solution, because the square root can't be negative.

50. Solve $S = \dfrac{Id^2}{k}$ for d.

Multiply both sides by k, then divide by I.

$$\frac{Sk}{I} = d^2$$

$$d = \pm\sqrt{\frac{Sk}{I}} = \frac{\pm\sqrt{Sk}}{\sqrt{I}}$$

$$= \frac{\pm\sqrt{Sk}}{\sqrt{I}}\cdot\frac{\sqrt{I}}{\sqrt{I}}$$

$$d = \frac{\pm\sqrt{SkI}}{I}$$

51. Solve $k = \dfrac{rF}{wv^2}$ for v.

Multiply both sides by v^2, then divide by k.

$$v^2 = \frac{rF}{kw}$$

$$v = \pm\sqrt{\frac{rF}{kw}} = \frac{\pm\sqrt{rF}}{\sqrt{kw}}$$

$$= \frac{\pm\sqrt{rF}}{\sqrt{kw}}\cdot\frac{\sqrt{kw}}{\sqrt{kw}}$$

$$v = \frac{\pm\sqrt{rFkw}}{kw}$$

52. Solve $2\pi R^2 + 2\pi RH - S = 0$ for R.

$$(2\pi)R^2 + (2\pi H)R - S = 0$$

Use the quadratic formula with

$$a = 2\pi, \ b = 2\pi H, \text{ and } c = -S.$$

$$R = \frac{-b \pm \sqrt{b^2 - 4ac}}{2a}$$

$$R = \frac{-2\pi H \pm \sqrt{(2\pi H)^2 - 4(2\pi)(-S)}}{2(2\pi)}$$

$$= \frac{-2\pi H \pm \sqrt{4\pi^2 H^2 + 8\pi S}}{4\pi}$$

$$= \frac{-2\pi H \pm 2\sqrt{\pi^2 H^2 + 2\pi S}}{4\pi}$$

$$= \frac{2\left(-\pi H \pm \sqrt{\pi^2 H^2 + 2\pi S}\right)}{4\pi}$$

$$R = \frac{-\pi H \pm \sqrt{\pi^2 H^2 + 2\pi S}}{2\pi}$$

53. $s = 16t^2 + 15t$

$$25 = 16t^2 + 15t \qquad \textit{Let } s = 25.$$

$$0 = 16t^2 + 15t - 25$$

$$a = 16, \ b = 15, \ c = -25$$

$$t = \frac{-15 \pm \sqrt{15^2 - 4(16)(-25)}}{2(16)}$$

$$= \frac{-15 \pm \sqrt{1825}}{32}$$

$$t = \frac{-15 + \sqrt{1825}}{32} \approx .87 \text{ or}$$

$$t = \frac{-15 - \sqrt{1825}}{32} \approx -1.80$$

Reject the negative solution since time cannot be negative. The object has fallen 25 feet in approximately .87 second.

54. Let $\ x = $ the length of the longer leg;

$\dfrac{3}{4}x = $ the length of the shorter leg;

$2x - 9 = $ the length of the hypotenuse.

Use the Pythagorean formula.

$$c^2 = a^2 + b^2$$

$$(2x - 9)^2 = x^2 + \left(\frac{3}{4}x\right)^2$$

$$4x^2 - 36x + 81 = x^2 + \frac{9}{16}x^2$$

$$16\left(4x^2 - 36x + 81\right) = 16\left(x^2 + \frac{9}{16}x^2\right)$$

$$64x^2 - 576x + 1296 = 16x^2 + 9x^2$$

$$39x^2 - 576x + 1296 = 0$$

Divide by 3.

$$13x^2 - 192x + 432 = 0$$

$$(13x - 36)(x - 12) = 0$$

$$13x - 36 = 0 \qquad \text{or} \quad x - 12 = 0$$

$$x = \frac{36}{13} \quad \text{or} \qquad x = 12$$

Reject $x = \dfrac{36}{13}$ since $2\left(\dfrac{36}{13}\right) - 9$ is negative.

If $x = 12$, then

$$\frac{3}{4}x = \frac{3}{4}(12) = 9$$

and

$$2x - 9 = 2(12) - 9 = 15.$$

The lengths of the three sides are 9 feet, 12 feet, and 15 feet.

55. Let $\ x = $ the width of the original rectangle and

$x + 2 = $ the length of the original rectangle.

Adding 1 m to the width and subtracting 1 m from the length of the rectangle gives $x + 1$ in each case, which is the length of a side of the square.

$$(\text{length of one side})^2 = \text{Area of square}$$

$$(x + 1)^2 = 121$$

$$x + 1 = 11 \quad \text{or} \quad x + 1 = -11$$

$$x = 10 \quad \text{or} \qquad x = -12$$

Since width cannot be negative, the original rectangle had a width of 10 m and a length $x + 2 = 12$ m.

56. Let $x = $ the width of the border.

$$\text{Area of mat} = \text{length} \cdot \text{width}$$

$$352 = (2x + 20)(2x + 14)$$

$$352 = 4x^2 + 68x + 280$$

$$0 = 4x^2 + 68x - 72$$

$$0 = x^2 + 17x - 18$$

$$0 = (x + 18)(x - 1)$$

$$x + 18 = 0 \qquad \text{or} \quad x - 1 = 0$$

$$x = -18 \quad \text{or} \qquad x = 1$$

Reject the negative answer for length. The mat is 1 in wide.

57. Let $\quad x = $ the length of the middle side,

$2x - 110 = $ the length of the longest side

$$(\text{the hypotenuse}),$$

and $\quad 50 = $ the length of the shortest side.

Use the Pythagorean formula.

$$a^2 + b^2 = c^2$$

$$x^2 + 50^2 = (2x - 110)^2$$

$$x^2 + 2500 = 4x^2 - 440x + 12{,}100$$

$$0 = 3x^2 - 440x + 9600$$

$$0 = (x - 120)(3x - 80)$$

$$x - 120 = 0 \quad \text{or} \quad 3x - 80 = 0$$
$$x = 120 \quad \text{or} \qquad x = \frac{80}{3}$$

Reject $\frac{80}{3}$ as the length of the middle side, because that is smaller than the length of the shortest side. So, the middle side is 120 m long.

58. $y = 3x^2 - 2$
Write in $y = a(x - h)^2 + k$ form as
$y = 3(x - 0)^2 - 2$. The vertex (h, k) is $(0, -2)$.

59. $y = 6 - 2x^2$
Write in $y = a(x - h)^2 + k$ form as
$y = -2(x - 0)^2 + 6$. The vertex (h, k) is $(0, 6)$.

60. $f(x) = -(x - 1)^2$
Write in $y = a(x - h)^2 + k$ form as
$y = -1(x - 1)^2 + 0$. The vertex (h, k) is $(1, 0)$.

61. $f(x) = (x + 2)^2$
Write in $y = a(x - h)^2 + k$ form as
$y = 1[x - (-2)]^2 + 0$. The vertex (h, k) is $(-2, 0)$.

62. $y = (x - 3)^2 + 7$
The equation is in the form $y = a(x - h)^2 + k$, so the vertex (h, k) is $(3, 7)$.

63.
$$y = -3x^2 + 4x - 2$$
$$\frac{y}{-3} = x^2 - \frac{4}{3}x + \frac{2}{3}$$
$$-\frac{y}{3} - \frac{2}{3} = x^2 - \frac{4}{3}x$$
$$-\frac{y}{3} - \frac{2}{3} + \frac{4}{9} = x^2 - \frac{4}{3}x + \frac{4}{9}$$
$$-\frac{y}{3} - \frac{2}{9} = \left(x - \frac{2}{3}\right)^2$$
$$-\frac{y}{3} = \left(x - \frac{2}{3}\right)^2 + \frac{2}{9}$$
$$y = -3\left(x - \frac{2}{3}\right)^2 - \frac{2}{3}$$

The equation is now in the form
$$y = a(x - h)^2 + k,$$
so the vertex (h, k) is $\left(\frac{2}{3}, -\frac{2}{3}\right)$.

64.
$$y = 4x^2 + 4x - 2$$
Complete the square to find the vertex.
$$\frac{y}{4} = x^2 + x - \frac{1}{2}$$
$$\frac{y}{4} + \frac{1}{2} = x^2 + x$$
$$\frac{y}{4} + \frac{1}{2} + \frac{1}{4} = x^2 + x + \frac{1}{4}$$

$$\frac{y}{4} + \frac{3}{4} = \left(x + \frac{1}{2}\right)^2$$
$$\frac{y}{4} = \left(x + \frac{1}{2}\right)^2 - \frac{3}{4}$$
$$y = 4\left(x + \frac{1}{2}\right)^2 - 3$$

The equation is now in the form
$y = a(x - h)^2 + k$, so the vertex (h, k) is $\left(-\frac{1}{2}, -3\right)$. Since $a = 4 > 0$, the parabola opens upward. Also, $|a| = |4| = 4 > 1$, so the graph is narrower than the graph of $y = x^2$. The points $(-2, 6)$, $(0, -2)$, and $(1, 6)$ are on the graph.

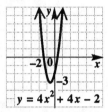
$y = 4x^2 + 4x - 2$

65. $x = 2y^2 + 8y + 3$
Complete the square to find the vertex. Since the roles of x and y are reversed, this is a horizontal parabola.

$$\frac{x}{2} = y^2 + 4y + \frac{3}{2}$$
$$\frac{x}{2} - \frac{3}{2} = y^2 + 4y$$
$$\frac{x}{2} - \frac{3}{2} + 4 = y^2 + 4y + 4$$
$$\frac{x}{2} + \frac{5}{2} = (y + 2)^2$$
$$\frac{x}{2} = (y + 2)^2 - \frac{5}{2}$$
$$x = 2(y + 2)^2 - 5$$

The equation is in the form $x = a(y - k)^2 + h$, so the vertex (h, k) is $(-5, -2)$. Here, $a = 2 > 0$, so the parabola opens to the right. Also, $|a| = |2| = 2 > 1$, so the graph is narrower than the graph of $y = x^2$. The points $(3, 0)$ and $(3, -4)$ are on the graph.

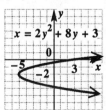
$x = 2y^2 + 8y + 3$

66. If the discriminant is negative, there are no x-intercepts.

67. $f(x) = a(x - h)^2 + k$

Since $a < 0$, the parabola opens downward. Since $h > 0$ and $k < 0$, the x-coordinate is positive and the y-coordinate is negative. Therefore, the vertex is in quadrant IV. The correct graph is (a).

68. The missile reaches its maximum height at the vertex of the parabola with equation $f(x) = -.017x^2 + x$. Use the vertex formula.

$$x = \frac{-b}{2a} = \frac{-1}{2(-.017)} \approx 29.41 \text{ seconds}$$

To find the maximum height, evaluate $f(29.41)$, which is about 14.71 feet.

69. **(a)** Use $ax^2 + bx + c = y$ with $(0, 12.39)$, $(4, 15.78)$, and $(7, 22.71)$.

$$c = 12.39 \quad (1)$$
$$16a + 4b + c = 15.78 \quad (2)$$
$$49a + 7b + c = 22.71 \quad (3)$$

(b) Rewrite equations (2) and (3) with $c = 12.39$.

$$16a + 4b + 12.39 = 15.78$$
$$49a + 7b + 12.39 = 22.71$$

$$16a + 4b = 3.39 \quad (4)$$
$$49a + 7b = 10.32 \quad (5)$$

Now eliminate b.

$$
\begin{array}{rl}
-112a - 28b = -23.73 & -7 \times (4) \\
\underline{196a + 28b = 41.28} & 4 \times (5) \\
84a = 17.55 &
\end{array}
$$
$$a \approx .2089$$

Use (4) to approximate b.

$$16\left(\frac{17.55}{84}\right) + 4b = 3.39$$

$$4b \approx .0471$$
$$b \approx .0118$$

Thus, $f(x) = .2089x^2 + .0118x + 12.39$.

70. Let x and $40 - x$ denote the two numbers. The product P is given by

$$P = x(40 - x)$$
$$= -x^2 + 40x.$$

The maximum of P can be found by locating the vertex of the graph of P.

$$x = \frac{-b}{2a} = \frac{-40}{2(-1)} = 20$$

So x is 20 and $40 - x$ is also 20.

71. **(a)** The solution set is $\{-.5, 4\}$ since that is where the graph of

$$y = 2x^2 - 7x - 4$$

crosses the x-axis.

(b) The solution set of $2x^2 - 7x - 4 > 0$ can be found by determining the values of x such that y is greater than zero (above the x-axis). The solution set is $(-\infty, -.5) \cup (4, \infty)$.

(c) The solution set of $2x^2 - 7x - 4 \leq 0$ can be found by determining the values of x such that y is less than or equal to zero (on or below the x-axis). The solution set is $[-.5, 4]$.

72. $(x - 4)(2x + 3) > 0$

Solve the equation

$$(x - 4)(2x + 3) = 0.$$
$$x - 4 = 0 \quad \text{or} \quad 2x + 3 = 0$$
$$x = 4 \quad \text{or} \quad x = -\frac{3}{2}$$

The numbers $-\frac{3}{2}$ and 4 divide the number line into three regions.

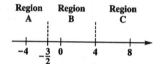

Test a number from each region in the inequality

$$(x - 4)(2x + 3) > 0.$$

Region A:	Let $x = -2$.	
	$-6(-1) > 0$	?
	$6 > 0$	*True*
Region B:	Let $x = 0$.	
	$-4(3) > 0$	?
	$-12 > 0$	*False*
Region C:	Let $x = 5$.	
	$1(13) > 0$	?
	$13 > 0$	*True*

The solution set includes numbers in Regions A and C, excluding endpoints.

Solution set: $\left(-\infty, -\frac{3}{2}\right) \cup (4, \infty)$.

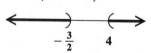

73. $$x^2 + x \leq 12$$

Solve the equation

$$x^2 + x = 12.$$
$$x^2 + x - 12 = 0$$
$$(x + 4)(x - 3) = 0$$
$$x + 4 = 0 \quad \text{or} \quad x - 3 = 0$$
$$x = -4 \quad \text{or} \quad x = 3$$

The numbers -4 and 3 divide the number line into three regions.

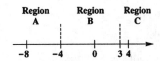

Test a number from each region in the inequality

$$x^2 + x \le 12.$$

Region A: Let $x = -5$.

$$25 - 5 \le 12 \quad ?$$
$$20 \le 12 \qquad \textit{False}$$

Region B: Let $x = 0$.

$$0 \le 12 \qquad \textit{True}$$

Region C: Let $x = 4$.

$$16 + 4 \le 12 \quad ?$$
$$20 \le 12 \qquad \textit{False}$$

The numbers in Region B, including -4 and 3, are solutions.

Solution set: $[-4, 3]$

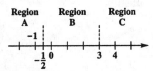

74. $\qquad\qquad 2k^2 > 5k + 3$

Solve the equation

$$2k^2 = 5k + 3.$$
$$2k^2 - 5k - 3 = 0$$
$$(2k + 1)(k - 3) = 0$$
$$2k + 1 = 0 \quad \text{or} \quad k - 3 = 0$$
$$k = -\frac{1}{2} \quad \text{or} \quad k = 3$$

The numbers $-\dfrac{1}{2}$ and 3 divide the number line into three regions.

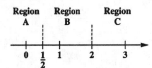

Test a number from each region in the inequality

$$2k^2 > 5k + 3.$$

Region A: Let $k = -1$.

$$2 > -5 + 3 \quad ?$$
$$2 > -2 \qquad \textit{True}$$

Region B: Let $k = 0$.

$$0 > 3 \qquad \textit{False}$$

Region C: Let $k = 4$.

$$32 > 20 + 3 \quad ?$$
$$32 > 23 \qquad \textit{True}$$

The numbers in Regions A and C, not including $-\dfrac{1}{2}$ or 3, are solutions.

Solution set: $\left(-\infty, -\dfrac{1}{2}\right) \cup (3, \infty)$

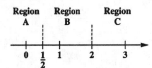

75. $(4m + 3)^2 \le -4$

The square of a real number is never negative. So, the solution set of this inequality is $\emptyset$.

76. $\dfrac{6}{2z - 1} < 2$

Solve the equation

$$\frac{6}{2z - 1} = 2.$$
$$6 = 2(2z - 1)$$
$$6 = 4z - 2$$
$$8 = 4z$$
$$2 = z$$

Find the number that makes the denominator 0.

$$2z - 1 = 0$$
$$z = \frac{1}{2}$$

The numbers $\dfrac{1}{2}$ and 2 divide the number line into three regions.

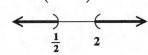

Test a number from each region in the inequality

$$\frac{6}{2z - 1} < 2.$$

Region A: Let $z = 0$.

$$-6 < 2 \qquad \textit{True}$$

Region B: Let $z = 1$.

$$6 < 2 \qquad \textit{False}$$

Region C: Let $z = 3$.

$$\frac{6}{5} < 2 \qquad \textit{True}$$

The solution set includes numbers in Regions A and C, excluding endpoints.

Solution set: $\left(-\infty, \dfrac{1}{2}\right) \cup (2, \infty)$

77. $\dfrac{3y + 4}{y - 2} \leq 1$

Solve the equation

$\dfrac{3y + 4}{y - 2} = 1.$

$3y + 4 = y - 2$

$2y = -6$

$y = -3$

Find the number that makes the denominator 0.

$y - 2 = 0$

$y = 2$

The numbers -3 and 2 divide the number line into three regions.

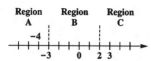

Test a number from each region in the inequality

$\dfrac{3y + 4}{y - 2} \leq 1.$

Region A: Let $y = -4$.

$\dfrac{-8}{-6} \leq 1$?

$\dfrac{4}{3} \leq 1$ *False*

Region B: Let $y = 0$.

$\dfrac{4}{-2} \leq 1$?

$-2 \leq 1$ *True*

Region C: Let $y = 3$.

$\dfrac{13}{1} \leq 1$?

$13 \leq 1$ *False*

The numbers in Region B, including -3 but not 2, are solutions.

Solution set: $[-3, 2)$

78. Solve $V = r^2 + R^2 h$ for R.

$V - r^2 = R^2 h$

$R^2 h = V - r^2$

$R^2 = \dfrac{V - r^2}{h}$

$R = \pm \sqrt{\dfrac{V - r^2}{h}} = \dfrac{\pm \sqrt{V - r^2}}{\sqrt{h}}$

$= \dfrac{\pm \sqrt{V - r^2}}{\sqrt{h}} \cdot \dfrac{\sqrt{h}}{\sqrt{h}}$

$= \dfrac{\pm \sqrt{Vh - r^2 h}}{h}$

79. $3t^2 - 6t = -4$

$3t^2 - 6t + 4 = 0$

Use the quadratic formula.

$t = \dfrac{-b \pm \sqrt{b^2 - 4ac}}{2a}$

$t = \dfrac{-(-6) \pm \sqrt{(-6)^2 - 4(3)(4)}}{2(3)}$

$= \dfrac{6 \pm \sqrt{-12}}{6} = \dfrac{6 \pm 2i\sqrt{3}}{6}$

$= \dfrac{2\left(3 \pm i\sqrt{3}\right)}{6} = \dfrac{3 \pm i\sqrt{3}}{3}$

Solution set: $\left\{ \dfrac{3 + i\sqrt{3}}{3}, \dfrac{3 - i\sqrt{3}}{3} \right\}$

80. $x^4 - 1 = 0$

$\left(x^2 + 1\right)\left(x^2 - 1\right) = 0$

$x^2 + 1 = 0$ or $x^2 - 1 = 0$

$x^2 = -1$ $(x + 1)(x - 1) = 0$

$x = \pm \sqrt{-1}$ $x + 1 = 0$ or $x - 1 = 0$

$x = \pm i$ or $x = -1$ or $x = 1$

Solution set: $\{-i, i, -1, 1\}$

81. $\left(b^2 - 2b\right)^2 = 11\left(b^2 - 2b\right) - 24$

Let $u = b^2 - 2b$. The equation becomes

$u^2 = 11u - 24.$

$u^2 - 11u + 24 = 0$

$(u - 8)(u - 3) = 0$

$u - 8 = 0$ or $u - 3 = 0$

$u = 8$ or $u = 3$

To find b, substitute $b^2 - 2b$ for u.

$b^2 - 2b = 8$ or $b^2 - 2b = 3$

$b^2 - 2b - 8 = 0$ $b^2 - 2b - 3 = 0$

$(b - 4)(b + 2) = 0$ $(b - 3)(b + 1) = 0$

$b - 4 = 0$ or $b + 2 = 0$ $b - 3 = 0$ or $b + 1 = 0$

$b = 4$ or $b = -2$ or $b = 3$ or $b = -1$

The potential solutions all check.

Solution set: $\{-2, -1, 3, 4\}$

82. $(r - 1)(2r + 3)(r + 6) < 0$

Solve the equation

$(r - 1)(2r + 3)(r + 6) = 0.$

$r - 1 = 0$ or $2r + 3 = 0$ or $r + 6 = 0$

$r = 1$ or $r = -\dfrac{3}{2}$ or $r = -6$

The numbers -6, $-\dfrac{3}{2}$, and 1 divide the number line into four regions.

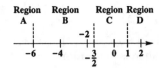

Test a number from each region in the inequality

$$(r-1)(2r+3)(r+6) < 0.$$

Region A: Let $r = -7$.
$$-8(-11)(-1) < 0 \qquad ?$$
$$-88 < 0 \qquad \textit{True}$$

Region B: Let $r = -2$.
$$-3(-1)(4) < 0 \qquad ?$$
$$12 < 0 \qquad \textit{False}$$

Region C: Let $r = 0$.
$$-1(3)(6) < 0 \qquad ?$$
$$-18 < 0 \qquad \textit{True}$$

Region D: Let $r = 2$.
$$1(7)(8) < 0 \qquad ?$$
$$56 < 0 \qquad \textit{False}$$

The numbers in Regions A and C, not including $-6, -\frac{3}{2}$, or 1, are solutions.

Solution set: $(-\infty, -6) \cup \left(-\frac{3}{2}, 1\right)$

83.
$$\frac{2}{x-4} + \frac{1}{x} = \frac{11}{5}$$
Multiply by the LCD, $5x(x-4)$.
$$5x(x-4)\left(\frac{2}{x-4} + \frac{1}{x}\right) = 5x(x-4)\left(\frac{11}{5}\right)$$
$$10x + 5(x-4) = 11x(x-4)$$
$$10x + 5x - 20 = 11x^2 - 44x$$
$$0 = 11x^2 - 59x + 20$$
$$0 = (x-5)(11x-4)$$

$$11x - 4 = 0 \quad \text{or} \quad x - 5 = 0$$
$$x = \frac{4}{11} \quad \text{or} \quad x = 5$$

Check $x = \frac{4}{11}$: $-\frac{11}{20} + \frac{11}{4} \overset{?}{=} \frac{11}{5}$ *True*

Check $x = 5$: $2 + \frac{1}{5} \overset{?}{=} \frac{11}{5}$ *True*

Solution set: $\left\{\frac{4}{11}, 5\right\}$

84. $(3k + 11)^2 = 7$

$$3k + 11 = \sqrt{7} \qquad \text{or } 3k + 11 = -\sqrt{7}$$
$$3k = -11 + \sqrt{7} \qquad\qquad 3k = -11 - \sqrt{7}$$
$$k = \frac{-11 + \sqrt{7}}{3} \quad \text{or} \quad k = \frac{-11 - \sqrt{7}}{3}$$

Solution set: $\left\{\dfrac{-11 + \sqrt{7}}{3}, \dfrac{-11 - \sqrt{7}}{3}\right\}$

85. Solve $p = \sqrt{\dfrac{yz}{6}}$ for y.

$$p^2 = \left(\sqrt{\frac{yz}{6}}\right)^2 \qquad \textit{Square}$$
$$p^2 = \frac{yz}{6}$$
$$\frac{6p^2}{z} = y \text{ or } y = \frac{6p^2}{z}$$

86. $(8k - 7)^2 \geq -1$
The square of any real number is always greater than or equal to 0, so any real number satisfies this inequality. The solution set is $(-\infty, \infty)$.

87.
$$-5x^2 = -8x + 3$$
$$-5x^2 + 8x - 3 = 0$$
$$5x^2 - 8x + 3 = 0$$
$$(5x - 3)(x - 1) = 0$$

$$5x - 3 = 0 \quad \text{or} \quad x - 1 = 0$$
$$x = \frac{3}{5} \quad \text{or} \quad x = 1$$

Solution set: $\left\{\dfrac{3}{5}, 1\right\}$

88.
$$6 + \frac{15}{s^2} = -\frac{19}{s}$$
Multiply by the LCD, s^2.
$$s^2\left(6 + \frac{15}{s^2}\right) = s^2\left(-\frac{19}{s}\right)$$
$$6s^2 + 15 = -19s$$
$$6s^2 + 19s + 15 = 0$$
$$(3s + 5)(2s + 3) = 0$$

$$3s + 5 = 0 \quad \text{or} \quad 2s + 3 = 0$$
$$s = -\frac{5}{3} \quad \text{or} \quad s = -\frac{3}{2}$$

Check $s = -\frac{5}{3}$: $6 + \frac{27}{5} \overset{?}{=} \frac{57}{5}$ *True*

Check $s = -\frac{3}{2}$: $6 + \frac{20}{3} \overset{?}{=} \frac{38}{3}$ *True*

Solution set: $\left\{-\dfrac{5}{3}, -\dfrac{3}{2}\right\}$

89.
$$\frac{-2}{x+5} \leq -5$$
Solve the equation
$$\frac{-2}{x+5} = -5.$$
$$-2 = -5(x+5)$$
$$-2 = -5x - 25$$
$$5x = -23$$
$$x = -\frac{23}{5}$$

continued

Find the number that makes the denominator 0.

$$x + 5 = 0$$
$$x = -5$$

The numbers -5 and $-\dfrac{23}{5}$ divide the number line into three regions.

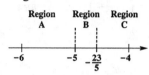

Test a number from each region in the inequality

$$\dfrac{-2}{x+5} \le -5.$$

Region A: Let $x = -6$.

$$\dfrac{-2}{-1} \le -5 \quad ?$$
$$2 \le -5 \qquad \textit{False}$$

Region B: Let $x = -\dfrac{24}{5}$.

$$-\dfrac{2}{\frac{1}{5}} \le -5 \quad ?$$
$$-10 \le -5 \qquad \textit{True}$$

Region C: Let $x = 0$.

$$\dfrac{-2}{5} \le -5 \qquad \textit{False}$$

The numbers in Region B, including $-\dfrac{23}{5}$ but not -5, are solutions.

Solution set: $\left(-5, -\dfrac{23}{5}\right]$

90. $y = \dfrac{4}{3}(x-2)^2 + 1$

The equation is in the form $y = a(x-h)^2 + k$, so the vertex (h, k) is $(2, 1)$. $a = \dfrac{4}{3}$, so the parabola opens upward and is narrower than the graph of $y = x^2$. Two other points on the graph are $\left(0, \dfrac{19}{3}\right)$ and $\left(4, \dfrac{19}{3}\right)$.

$$y = \tfrac{4}{3}(x-2)^2 + 1$$

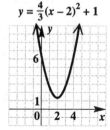

91. $x = 2(y+3)^2 - 4$

$$x = 2[y - (-3)]^2 + (-4)$$

The equation is in the form

$$x = a(y-k)^2 + h,$$

so the vertex (h, k) is $(-4, -3)$. $a = 2$, so the parabola opens to the right and is narrower than the graph of $y = x^2$. Two other points on the graph are $(4, -1)$ and $(4, -5)$.

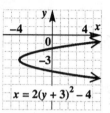

$$x = 2(y+3)^2 - 4$$

92. $f(x) = -2x^2 + 8x - 5$

Replace $f(x)$ with y, and then complete the square to find the vertex.

$$y = -2x^2 + 8x - 5$$
$$\dfrac{y}{-2} = x^2 - 4x + \dfrac{5}{2}$$
$$-\dfrac{y}{2} - \dfrac{5}{2} = x^2 - 4x$$
$$-\dfrac{y}{2} - \dfrac{5}{2} + 4 = x^2 - 4x + 4$$
$$-\dfrac{y}{2} + \dfrac{3}{2} = (x-2)^2$$
$$-\dfrac{y}{2} = (x-2)^2 - \dfrac{3}{2}$$
$$y = -2(x-2)^2 + 3$$

The equation is in the form $y = a(x-h)^2 + k$, so the vertex (h, k) is $(2, 3)$. Here, $a = -2 < 0$, so the parabola opens downward.

Also, $|a| = |-2| = 2 > 1$, so the graph is narrower than the graph of $y = x^2$. The points $(0, -5)$, $(1, 1)$, and $(3, 1)$ are on the graph.

$$f(x) = -2x^2 + 8x - 5$$

93. $x = -\dfrac{1}{2}y^2 + 6y - 14$

Complete the square to find the vertex. Since the roles of x and y are reversed, this is a horizontal parabola.

$$x = -\frac{1}{2}y^2 + 6y - 14$$
$$-2x = y^2 - 12y + 28$$
$$-2x - 28 = y^2 - 12y$$
$$-2x - 28 + 36 = y^2 - 12y + 36$$
$$-2x + 8 = (y - 6)^2$$
$$-2x = (y - 6)^2 - 8$$
$$x = -\frac{1}{2}(y - 6)^2 + 4$$

The equation is in the form $x = a(y - k)^2 + h$, so the vertex (h, k) is $(4, 6)$. Here, $a = -\frac{1}{2} < 0$, so the parabola opens to the left.

Also, $|a| = \left|-\frac{1}{2}\right| = \frac{1}{2} < 1$, so the graph is wider than the graph of $y = x^2$. The points $(-14, 0)$, $(2, 4)$, and $(2, 8)$ are on the graph.

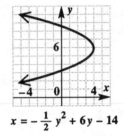

$$x = -\frac{1}{2}y^2 + 6y - 14$$

94. The student improperly used the square root property. If $b^2 = 12$, then $b = \pm\sqrt{12} = \pm 2\sqrt{3}$. The student forgot the " $\pm$ " sign.

95. To write the equation, work backwards. First we know that

$$x = -5 \quad \text{or} \quad x = 6.$$

Then,

$$x + 5 = 0 \quad \text{or} \quad x - 6 = 0,$$

and

$$(x + 5)(x - 6) = 0.$$

Multiply the factors. In standard form the equation is,

$$x^2 - x - 30 = 0.$$

96. $s(t) = -16t^2 + 160t$
The equation represents a parabola. Since $a = -16 < 0$, the parabola opens downward. The time and maximum height occur at the vertex (h, k) of the parabola, given by

$$\left(\frac{-b}{2a}, s\left(\frac{-b}{2a}\right)\right).$$

Using the standard form of the equation, $a = -16$ and $b = 160$, so

$$h = \frac{-b}{2a} = \frac{-160}{2(-16)} = 5,$$

and $k = s(h) = -16(5)^2 + 160(5)$
$$= -400 + 800 = 400.$$

The vertex is $(5, 400)$.

(a) The time at which the maximum height is reached is 5 seconds.

(b) The maximum height is 400 feet.

97. Let $L = $ the length of the rectangle and
$W = $ the width.

The perimeter of the rectangle is 600 m, so

$$2L + 2W = 600$$
$$2W = 600 - 2L$$
$$W = 300 - L.$$

Since the area is length times width, substitute $300 - L$ for W.

$$A = LW$$
$$= L(300 - L)$$
$$= 300L - L^2 \text{ or } -L^2 + 300L$$

Use the vertex formula.

$$L = \frac{-b}{2a} = \frac{-300}{2(-1)} = 150$$

So $L = 150$ meters and
$W = 300 - L = 300 - 150 = 150$ meters.

98. The quadratic formula could be used to solve for x^2 (not x) in the equation $x^4 - 5x^2 + 6 = 0$. After you solve for x^2, you would have to remember to use the square root property to solve for x.

99. (a) $3x - (4x + 2) = 0$
$$3x - 4x - 2 = 0$$
$$-x - 2 = 0$$
$$x = -2$$

Solution set: $\{-2\}$

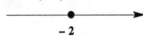

(b) $3x - (4x + 2) > 0$
$$3x - 4x - 2 > 0$$
$$-x - 2 > 0$$
$$-x > 2$$
$$x < -2$$

Solution set: $(-\infty, -2)$

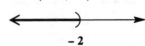

(c) $3x - (4x + 2) < 0$

$$3x - 4x - 2 < 0$$

$$-x - 2 < 0$$

$$-x < 2$$

$$x > -2$$

Solution set: $(-2, \infty)$

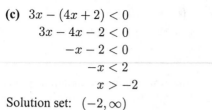

100. (a) $x^2 - 6x + 5 = 0$

$$(x - 1)(x - 5) = 0$$

$$x - 1 = 0 \quad \text{or} \quad x - 5 = 0$$

$$x = 1 \quad \text{or} \quad x = 5$$

Solution set: $\{1, 5\}$

(b) $x^2 - 6x + 5 > 0$

From the equation

$$x^2 - 6x + 5 = 0$$

in part (a), the solutions are $x = 1$ or $x = 5$. These numbers divide the number line into three regions.

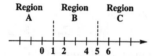

Test a point from each region in the inequality

$$x^2 - 6x + 5 > 0.$$

Region A: Let $x = 0$.

$$5 > 0 \qquad \text{True}$$

Region B: Let $x = 2$.

$$4 - 12 + 5 > 0 \quad ?$$

$$-3 > 0 \qquad \text{False}$$

Region C: Let $x = 6$.

$$36 - 36 + 5 > 0 \quad ?$$

$$5 > 0 \qquad \text{True}$$

The numbers in Regions A and C, not including the endpoints, are solutions.

Solution set: $(-\infty, 1) \cup (5, \infty)$

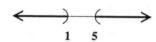

(c) $x^2 - 6x + 5 < 0$

The regions are the same as in part (b). Solutions to this inequality would be the numbers in Region B, again not including the endpoints.

Solution set: $(1, 5)$

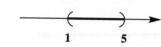

101. (a)

$$\frac{-5x + 20}{x - 2} = 0$$

$$(x - 2)\left(\frac{-5x + 20}{x - 2}\right) = (x - 2) \cdot 0$$

$$-5x + 20 = 0$$

$$-5x = -20$$

$$x = 4$$

Solution set: $\{4\}$

(b) $\dfrac{-5x + 20}{x - 2} > 0$

From part (a), the solution of the corresponding equation is $x = 4$.

The number that makes the denominator zero is $x = 2$.

These numbers divide the number line into three regions.

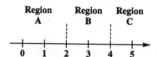

Test a number from each region in the inequality

$$\frac{-5x + 20}{x - 2} > 0.$$

Region A: Let $x = 0$.

$$\frac{20}{-2} > 0 \quad ?$$

$$-10 > 0 \qquad \text{False}$$

Region B: Let $x = 3$.

$$\frac{5}{1} > 0 \quad ?$$

$$5 > 0 \qquad \text{True}$$

Region C: Let $x = 5$.

$$\frac{-5}{3} > 0 \quad ?$$

$$-\frac{5}{3} > 0 \qquad \text{False}$$

The numbers in Region B, not including the endpoints, are solutions.

Solution set: $(2, 4)$

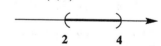

(c) $\dfrac{-5x + 20}{x - 2} < 0$

The regions are the same as in part (b). Solutions to this inequality would be the numbers in Regions A and C, again not including the endpoints.

Solution set: $(-\infty, 2) \cup (4, \infty)$

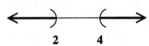

102. $\{-2\} \cup (-\infty, -2) \cup (-2, \infty) = (-\infty, \infty)$

103. $\{1, 5\} \cup (-\infty, 1) \cup (5, \infty) \cup (1, 5) = (-\infty, \infty)$

104. The number 2 cannot possibly be part of any of the solution sets since it makes the denominator zero.
$\{4\} \cup (2, 4) \cup (-\infty, 2) \cup (4, \infty)$
$= (-\infty, 2) \cup (2, \infty)$

105. If we solve a linear, quadratic, or rational equation and the two inequalities associated with it, the union of the three sets will be $(-\infty, \infty)$; the only exception will be in the case of the rational equation and inequalities, where the number or numbers that cause the _denominator_ to be zero will be deleted.

106. If set S is the solution of the other inequality, then
$S \cup \{-5, 3\} \cup (-\infty, -5) \cup (3, \infty) = (-\infty, \infty)$.
Therefore, $S = (-5, 3)$.

Chapter 11 Test

1. $x^2 = 39$

$x = \sqrt{39}$ or $x = -\sqrt{39}$

The solution set is $\left\{ \pm\sqrt{39} \right\}$.

2. $(y + 3)^2 = 64$

$\begin{aligned}
y + 3 &= \sqrt{64} \quad &\text{or} \quad y + 3 &= -\sqrt{64} \\
y + 3 &= 8 \quad &\text{or} \quad y + 3 &= -8 \\
y &= 5 \quad &\text{or} \quad y &= -11
\end{aligned}$

The solution set is $\{-11, 5\}$.

3. $x^2 - 4x = 6$

$x^2 - 4x + 4 = 6 + 4$

$\frac{1}{2}(-4) = -2, \text{ and } (-2)^2 = 4$

$(x - 2)^2 = 10$

$\begin{aligned}
x - 2 &= \sqrt{10} \quad &\text{or} \quad x - 2 &= -\sqrt{10} \\
x &= 2 + \sqrt{10} \quad &\text{or} \quad x &= 2 - \sqrt{10}
\end{aligned}$

The solution set is $\left\{ 2 \pm \sqrt{10} \right\}$.

4. $2x^2 + 12x - 3 = 0$

$x^2 + 6x - \dfrac{3}{2} = 0$

$x^2 + 6x = \dfrac{3}{2}$

$x^2 + 6x + 9 = \dfrac{3}{2} + 9$

$\frac{1}{2}(6) = 3, \text{ and } 3^2 = 9$

$(x + 3)^2 = \dfrac{21}{2}$

$x + 3 = \sqrt{\dfrac{21}{2}}$ or $x + 3 = -\sqrt{\dfrac{21}{2}}$

Note that

$$\sqrt{\dfrac{21}{2}} = \dfrac{\sqrt{21}}{\sqrt{2}} = \dfrac{\sqrt{21} \cdot \sqrt{2}}{\sqrt{2} \cdot \sqrt{2}} = \dfrac{\sqrt{42}}{2}.$$

$x + 3 = \dfrac{\sqrt{42}}{2}$ or $x + 3 = -\dfrac{\sqrt{42}}{2}$

$x = -3 + \dfrac{\sqrt{42}}{2}$ or $x = -3 - \dfrac{\sqrt{42}}{2}$

$x = \dfrac{-6 + \sqrt{42}}{2}$ or $x = \dfrac{-6 - \sqrt{42}}{2}$

The solution set is $\left\{ \dfrac{-6 \pm \sqrt{42}}{2} \right\}$.

5. $3w^2 + 2 = 6w$

$3w^2 - 6w + 2 = 0$

Use $a = 3$, $b = -6$, and $c = 2$.

$w = \dfrac{-(-6) \pm \sqrt{(-6)^2 - 4(3)(2)}}{2(3)}$

$= \dfrac{6 \pm \sqrt{36 - 24}}{6}$

$= \dfrac{6 \pm \sqrt{12}}{6} = \dfrac{6 \pm 2\sqrt{3}}{6}$

$= \dfrac{2\left(3 \pm \sqrt{3}\right)}{2(3)} = \dfrac{3 \pm \sqrt{3}}{3}$

The solution set is $\left\{ \dfrac{3 \pm \sqrt{3}}{3} \right\}$.

6. $4x^2 + 8x + 11 = 0$

Use $a = 4$, $b = 8$, and $c = 11$.

$x = \dfrac{-8 \pm \sqrt{8^2 - 4(4)(11)}}{2(4)}$

$= \dfrac{-8 \pm \sqrt{64 - 176}}{8} = \dfrac{-8 \pm \sqrt{-112}}{8}$

continued

$$= \frac{-8 \pm i\sqrt{112}}{8} = \frac{-8 \pm i\sqrt{16 \cdot 7}}{8}$$

$$= \frac{-8 \pm 4i\sqrt{7}}{8} = \frac{4\left(-2 \pm i\sqrt{7}\right)}{4 \cdot 2}$$

$$= \frac{-2 \pm i\sqrt{7}}{2} = -1 \pm \frac{\sqrt{7}}{2}i$$

The solution set is $\left\{-1 \pm \frac{\sqrt{7}}{2}i\right\}$.

7. $p^2 - 2p - 1 = 0$

Solve by completing the square.

$$p^2 - 2p = 1$$
$$p^2 - 2p + 1 = 1 + 1$$
$$(p - 1)^2 = 2$$

$$p - 1 = \sqrt{2} \qquad \text{or} \qquad p - 1 = -\sqrt{2}$$
$$p = 1 + \sqrt{2} \qquad \text{or} \qquad p = 1 - \sqrt{2}$$

The solution set is $\left\{1 \pm \sqrt{2}\right\}$.

8. $(2x + 1)^2 = 18$

Use the square root property.

$$2x + 1 = \pm\sqrt{18}$$
$$2x + 1 = \pm 3\sqrt{2}$$
$$2x = -1 \pm 3\sqrt{2}$$
$$x = \frac{-1 \pm 3\sqrt{2}}{2}$$

The solution set is $\left\{\frac{-1 \pm 3\sqrt{2}}{2}\right\}$.

9. $(x - 5)(2x - 1) = 1$
$$2x^2 - 11x + 5 = 1$$
$$2x^2 - 11x + 4 = 0$$

Use $a = 2$, $b = -11$, and $c = 4$ in the quadratic formula.

$$x = \frac{-(-11) \pm \sqrt{(-11)^2 - 4(2)(4)}}{2(2)}$$

$$= \frac{11 \pm \sqrt{121 - 32}}{4} = \frac{11 \pm \sqrt{89}}{4}$$

The solution set is $\left\{\frac{11 \pm \sqrt{89}}{4}\right\}$.

10.
$$t^2 + 25 = 10t$$
$$t^2 - 10t + 25 = 0$$
$$(t - 5)^2 = 0$$
$$t - 5 = 0$$
$$t = 5$$

The solution set is $\{5\}$.

11. $s = -16t^2 + 64t$

Let $s = 64$ and solve for t.

$$64 = -16t^2 + 64t$$
$$16t^2 - 64t + 64 = 0$$
$$t^2 - 4t + 4 = 0$$
$$(t - 2)^2 = 0$$
$$t - 2 = 0$$
$$t = 2$$

The object will reach a height of 64 feet after 2 seconds.

12. $P^2 = a^3$
$$P^2 = (39.4)^3$$
$$= 61,162.984$$
$$P \approx 247.3 \text{ years}$$

13. $(3 + i) + (-2 + 3i) - (6 - i)$
$$= (3 + i) + (-2 + 3i) + (-6 + i)$$
$$= (3 - 2 - 6) + (1 + 3 + 1)i$$
$$= -5 + 5i$$

14. $(6 + 5i)(-2 + i)$
$$= -12 + 6i - 10i + 5i^2$$
$$= -12 - 4i + 5(-1)$$
$$= -12 - 4i - 5$$
$$= -17 - 4i$$

15. $(3 - 8i)(3 + 8i)$
$$= 3^2 - (8i)^2$$
$$= 9 - 64i^2$$
$$= 9 - 64(-1)$$
$$= 9 + 64 = 73$$

16. $\dfrac{15 - 5i}{7 + i} = \dfrac{15 - 5i}{7 + i} \cdot \dfrac{7 - i}{7 - i}$

$$= \frac{(15 - 5i)(7 - i)}{(7 + i)(7 - i)}$$

$$= \frac{105 - 15i - 35i + 5i^2}{49 - i^2}$$

$$= \frac{105 - 50i + 5(-1)}{49 - (-1)}$$

$$= \frac{100 - 50i}{50}$$

$$= \frac{100}{50} - \frac{50}{50}i$$

$$= 2 - i \qquad \textit{Standard form}$$

17. $2x^2 - 8x - 3 = 0$
$$b^2 - 4ac = (-8)^2 - 4(2)(-3)$$
$$= 64 + 24 = 88$$
The discriminant is positive but not a perfect square, so there will be two distinct irrational number solutions.

18.
$$9x^4 + 4 = 37x^2$$
$$9x^4 - 37x^2 + 4 = 0$$
Let $u = x^2$, so $u^2 = (x^2)^2 = x^4$.
The equation becomes
$$9u^2 - 37u + 4 = 0.$$
$$(9u - 1)(u - 4) = 0$$

$9u - 1 = 0$ or $u - 4 = 0$
$u = \dfrac{1}{9}$ or $u = 4$

To find x, substitute x^2 for u.

$x^2 = \dfrac{1}{9}$ or $x^2 = 4$

$x = \dfrac{1}{3}$ or $x = -\dfrac{1}{3}$ or $x = 2$ or $x = -2$

Check $x = \pm\dfrac{1}{3}$: $\dfrac{1}{9} + 4 \overset{?}{=} \dfrac{37}{9}$ *True*

Check $x = \pm 2$: $144 + 4 \overset{?}{=} 37(4)$ *True*

Solution set: $\left\{-2, -\dfrac{1}{3}, \dfrac{1}{3}, 2\right\}$

19. $12 = (2d + 1)^2 + (2d + 1)$
Let $u = 2d + 1$. The equation becomes
$12 = u^2 + u.$
$0 = u^2 + u - 12$
$0 = (u + 4)(u - 3)$

$u + 4 = 0$ or $u - 3 = 0$
$u = -4$ or $u = 3$

To find d, substitute $2d + 1$ for u.

$2d + 1 = -4$ or $2d + 1 = 3$
$2d = -5$ $\qquad$ $2d = 2$
$d = -\dfrac{5}{2}$ or $d = 1$

Check $d = -\dfrac{5}{2}$: $12 \overset{?}{=} 16 - 4$ *True*

Check $d = 1$: $12 \overset{?}{=} 9 + 3$ *True*

Solution set: $\left\{-\dfrac{5}{2}, 1\right\}$

20. Solve $S = 4\pi r^2$ for r.
$$\dfrac{S}{4\pi} = r^2$$
$$r = \pm\sqrt{\dfrac{S}{4\pi}} = \dfrac{\pm\sqrt{S}}{2\sqrt{\pi}}$$
$$= \dfrac{\pm\sqrt{S}}{2\sqrt{\pi}} \cdot \dfrac{\sqrt{\pi}}{\sqrt{\pi}}$$
$$r = \dfrac{\pm\sqrt{\pi S}}{2\pi}$$

21. Let $x =$ the width of the walk.
The area of the walk is equal to the area of the outer figure minus the area of the pool.
$$152 = (10 + 2x)(24 + 2x) - (24)(10)$$
$$152 = 240 + 68x + 4x^2 - 240$$
$$0 = 4x^2 + 68x - 152$$
$$0 = x^2 + 17x - 38$$
$$0 = (x + 19)(x - 2)$$

$x + 19 = 0$ or $x - 2 = 0$
$x = -19$ or $x = 2$

Reject -19 since width can't be negative. The walk is 2 feet wide.

22. Let $x =$ the height of the tower, and
$2x + 2 =$ the distance from the point to the top.

The distance from the base to the point is 30 m. These three segments form a right triangle, so the Pythagorean formula applies.
$$a^2 + b^2 = c^2$$
$$x^2 + 30^2 = (2x + 2)^2$$
$$x^2 + 900 = 4x^2 + 8x + 4$$
$$0 = 3x^2 + 8x - 896$$
$$0 = (x - 16)(3x + 56)$$

$x - 16 = 0$ or $3x + 56 = 0$
$x = 16$ or $x = -\dfrac{56}{3}$

Reject $-\dfrac{56}{3}$ since height can't be negative. The tower is 16 m high.

23. $f(x) = \dfrac{1}{2}x^2 - 2$
$$f(x) = \dfrac{1}{2}(x - 0)^2 - 2$$
The graph is a parabola in $f(x) = a(x - h)^2 + k$ form with vertex (h, k) at $(0, -2)$. Since $a = \dfrac{1}{2} > 0$, the parabola opens upward. Also,
$$|a| = \left|\dfrac{1}{2}\right| = \dfrac{1}{2} < 1,$$ so the graph of the parabola

continued

is wider than the graph of $f(x) = x^2$. The points $(2, 0)$ and $(-2, 0)$ are on the graph.

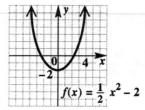

$$f(x) = \frac{1}{2}x^2 - 2$$

From the graph, we see that the x-values can be any real number, so the domain is $(-\infty, \infty)$. The y-values are greater than or equal to -2, so the range is $[-2, \infty)$.

24. $f(x) = -x^2 + 4x - 1$

Replace $f(x)$ with y and complete the square to find the vertex.

$$y = -x^2 + 4x - 1$$
$$-y = x^2 - 4x + 1$$
$$-y - 1 = x^2 - 4x$$
$$-y - 1 + 4 = x^2 - 4x + 4$$
$$-y + 3 = (x - 2)^2$$
$$-y = (x - 2)^2 - 3$$
$$y = -(x - 2)^2 + 3$$

The graph is a parabola with vertex (h, k) at $(2, 3)$. Since $a = -1 < 0$, the parabola opens downward.

Also, $|a| = |-1| = 1$, so the graph has the same shape as the graph of $f(x) = x^2$. The points $(0, -1)$ and $(4, -1)$ are on the graph.

$$f(x) = -x^2 + 4x - 1$$

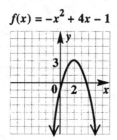

25. $x = -(y - 2)^2 + 2$

The equation is in $x = a(y - k)^2 + h$ form. The graph is a horizontal parabola with vertex (h, k) at $(2, 2)$. Since $a = -1 < 0$, the graph opens to the left. Also, $|a| = |-1| = 1$, so the graph has the same shape as the graph of $y = x^2$. The points $(-2, 0)$ and $(-2, 4)$ are on the same graph.

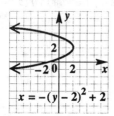

$$x = -(y - 2)^2 + 2$$

26. $f(x) = 22.56x^2 - 129.8x + 611.8$

(a) 1992 corresponds to $x = 2$.

$$f(2) = 442.44 \approx 442$$

Based on the model, about 442 million salmon were caught in 1992.

(b) Find the vertex.

$$x = \frac{-b}{2a} = \frac{-(-129.8)}{2(22.56)} \approx 2.877$$
$$f(2.88) \approx 425$$

Rounding 2.88 to 3 gives us a minimum of 425 million salmon in 1993.

27.
$$2x^2 + 7x > 15$$
$$2x^2 + 7x - 15 > 0$$

Solve the equation

$$2x^2 + 7x - 15 = 0.$$
$$(2x - 3)(x + 5) = 0$$

$$2x - 3 = 0 \quad \text{or} \quad x + 5 = 0$$
$$x = \frac{3}{2} \quad \text{or} \quad x = -5$$

The numbers -5 and $\frac{3}{2}$ divide the number line into three regions.

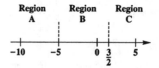

Test a number from each region in the inequality

$$2x^2 + 7x > 15.$$

Region A: Let $x = -6$.

$$72 - 42 > 15 \quad ?$$
$$30 > 15 \qquad True$$

Region B: Let $x = 0$.

$$0 > 15 \qquad False$$

Region C: Let $x = 2$.

$$8 + 14 > 15 \quad ?$$
$$22 > 15 \qquad True$$

The numbers in Regions A and C, not including -5 and $\frac{3}{2}$, are solutions.

Solution set: $(-\infty, -5) \cup \left(\frac{3}{2}, \infty\right)$

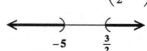

28. $\dfrac{5}{t-4} \le 1$

Solve the equation

$$\dfrac{5}{t-4} = 1.$$
$$5 = t - 4$$
$$t = 9$$

Find the number that makes the denominator 0.

$$t - 4 = 0$$
$$t = 4$$

The numbers 4 and 9 divide the number line into three regions.

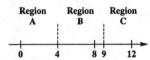

Test a number from each region in the inequality

$$\dfrac{5}{t-4} \le 1.$$

Region A: Let $t = 0$.

$$\dfrac{5}{-4} \le 1 \qquad \textit{True}$$

Region B: Let $t = 7$.

$$\dfrac{5}{3} \le 1 \qquad \textit{False}$$

Region C: Let $t = 10$.

$$\dfrac{5}{6} \le 1 \qquad \textit{True}$$

The numbers in Regions A and C, including 9 but not 4, are solutions.

Solution set: $(-\infty, 4) \cup [9, \infty)$

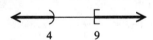

Cumulative Review Exercises Chapters 1–11

1. $S = \left\{ -\dfrac{7}{3}, -2, -\sqrt{3}, 0, .7, \sqrt{12}, \sqrt{-8}, 7, \dfrac{32}{3} \right\}$

(a) The elements of S that are integers are $-2, 0,$ and 7.

(b) The elements of S that are rational numbers are $-\dfrac{7}{3}, -2, 0, .7, 7,$ and $\dfrac{32}{3}$.

2. $2(-3)^2 + (-8)(-5) + (-17)$
$$= 2(9) + 40 - 17$$
$$= 18 + 23 = 41$$

3. $7 - (4 + 3t) + 2t = -6(t - 2) - 5$
$$7 - 4 - 3t + 2t = -6t + 12 - 5$$
$$3 - t = -6t + 7$$
$$5t = 4$$
$$t = \dfrac{4}{5}$$

Check $t = \dfrac{4}{5}$: $\dfrac{11}{5} = \dfrac{11}{5}$

Solution set: $\left\{ \dfrac{4}{5} \right\}$

4. $|6x - 9| = |-4x + 2|$

$6x - 9 = -4x + 2$ or $6x - 9 = -(-4x + 2)$
$\quad 10x = 11 \qquad\qquad\qquad 6x - 9 = 4x - 2$
$\qquad\qquad\qquad\qquad\qquad\qquad\quad 2x = 7$

$$x = \dfrac{11}{10} \qquad \text{or} \qquad x = \dfrac{7}{2}$$

Check $x = \dfrac{11}{10}$: $\left| -\dfrac{24}{10} \right| \overset{?}{=} \left| -\dfrac{24}{10} \right|$ *True*

Check $x = \dfrac{7}{2}$: $|12| \overset{?}{=} |-12|$ *True*

Solution set: $\left\{ \dfrac{11}{10}, \dfrac{7}{2} \right\}$

5. $$2x = \sqrt{\dfrac{5x + 2}{3}}$$

Square both sides.

$$(2x)^2 = \left(\sqrt{\dfrac{5x + 2}{3}} \right)^2$$
$$4x^2 = \dfrac{5x + 2}{3}$$
$$12x^2 = 5x + 2$$
$$12x^2 - 5x - 2 = 0$$
$$(3x - 2)(4x + 1) = 0$$

$3x - 2 = 0$ or $4x + 1 = 0$

$$x = \dfrac{2}{3} \quad \text{or} \qquad x = -\dfrac{1}{4}$$

Check $x = \dfrac{2}{3}$: $\dfrac{4}{3} \overset{?}{=} \sqrt{\dfrac{16}{9}}$ *True*

Check $x = -\dfrac{1}{4}$: $-\dfrac{1}{2} \overset{?}{=} \sqrt{\dfrac{1}{4}}$ *False*

Solution set: $\left\{ \dfrac{2}{3} \right\}$

6.

$$\dfrac{3}{x-3} - \dfrac{2}{x-2} = \dfrac{3}{x^2 - 5x + 6}$$
$$\dfrac{3}{x-3} - \dfrac{2}{x-2} = \dfrac{3}{(x-3)(x-2)}$$

Multiply by the LCD, $(x - 3)(x - 2)$.

$$(x-3)(x-2)\left(\dfrac{3}{x-3} - \dfrac{2}{x-2} \right)$$
$$= (x-3)(x-2)\left[\dfrac{3}{(x-3)(x-2)} \right]$$
$$3(x-2) - 2(x-3) = 3$$
$$3x - 6 - 2x + 6 = 3$$
$$x = 3$$

The number 3 is not allowed as a solution since it makes the denominator 0. The solution set is $\emptyset$.

7. $(r - 5)(2r + 3) = 1$

$2r^2 - 7r - 15 = 1$

$2r^2 - 7r - 16 = 0$

Use the quadratic formula.

$$t = \frac{-b \pm \sqrt{b^2 - 4ac}}{2a}$$

$$t = \frac{-(-7) \pm \sqrt{(-7)^2 - 4(2)(-16)}}{2(2)}$$

$$= \frac{7 \pm \sqrt{49 + 128}}{4} = \frac{7 \pm \sqrt{177}}{4}$$

Solution set: $\left\{ \dfrac{7 + \sqrt{177}}{4}, \dfrac{7 - \sqrt{177}}{4} \right\}$

8. $b^4 - 5b^2 + 4 = 0$

Let $u = b^2$, so $u^2 = \left(b^2\right)^2 = b^4$.

The equation becomes

$u^2 - 5u + 4 = 0.$

$(u - 4)(u - 1) = 0$

$u - 4 = 0 \quad$ or $\quad u - 1 = 0$

$u = 4 \quad$ or $\qquad u = 1$

To find b, substitute b^2 for u.

$\quad b^2 = 4 \qquad$ or $\qquad b^2 = 1$

$b = 2$ or $b = -2 \quad$ or $\quad b = 1$ or $b = -1$

The potential solutions check.

Solution set: $\{-2, -1, 1, 2\}$

9. $-2x + 4 \le -x + 3$

$-x \le -1$

Multiply by -1, and reverse the direction
of the inequality.

$x \ge 1$

Solution set: $[1, \infty)$

10. $|3y - 7| \le 1$

$-1 \le 3y - 7 \le 1$

$6 \le 3y \le 8$

$2 \le y \le \dfrac{8}{3}$

Solution set: $\left[2, \dfrac{8}{3}\right]$

11. $x^2 - 4x + 3 < 0$

Solve the equation

$x^2 - 4x + 3 = 0.$

$(x - 3)(x - 1) = 0$

$x - 3 = 0 \quad$ or $\quad x - 1 = 0$

$x = 3 \quad$ or $\qquad x = 1$

The numbers 1 and 3 divide the number line into
three regions.

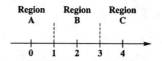

Test a number from each region in the inequality

$$x^2 - 4x + 3 < 0.$$

Region A: Let $x = 0$.

$3 < 0 \qquad$ *False*

Region B: Let $x = 2$.

$4 - 8 + 3 < 0 \quad$?

$-1 < 0 \qquad$ *True*

Region C: Let $x = 4$.

$16 - 16 + 3 < 0 \quad$?

$3 < 0 \qquad$ *False*

The numbers from Region B, not including 1 or 3,
are solutions.

Solution set: $(1, 3)$

12. $\dfrac{3}{y + 2} > 1$

Solve the equation

$\dfrac{3}{y + 2} = 1.$

$3 = y + 2$

$y = 1$

Find the number that makes the denominator 0.

$y + 2 = 0$

$y = -2$

The numbers -2 and 1 divide the number line into
three regions.

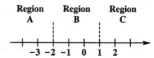

Test a number from each region in the inequality

$$\frac{3}{y + 2} > 1.$$

Region A: Let $y = -4$.

$\dfrac{3}{-2} > 1 \qquad$ *False*

Region B: Let $y = 0$.

$\dfrac{3}{2} > 1 \qquad$ *True*

Region C: Let $y = 2$.

$\dfrac{3}{4} > 1 \qquad$ *False*

The numbers from Region B, not including -2 or
1, are solutions.

Solution set: $(-2, 1)$

13. $-2x + 7y = 16$
Solve the equation for y.

$$7y = 2x + 16$$

$$y = \frac{2}{7}x + \frac{16}{7}$$

So the slope is $\frac{2}{7}$ and the y-intercept is $\left(0, \frac{16}{7}\right)$.

Let $y = 0$ in $-2x + 7y = 16$ to find the x-intercept.

$$-2x + 7(0) = 16$$
$$-2x = 16$$
$$x = -8$$

The x-intercept is $(-8, 0)$.

14. Solve $5x + 2y = 6$ for y.

$$2y = -5x + 6$$
$$y = -\frac{5}{2}x + 3$$

So the slope of the given line is $-\frac{5}{2}$ and the required form is

$$y = -\frac{5}{2}x + b.$$

$$-3 = -\frac{5}{2}(2) + b \quad \textit{Let x=2, y=-3.}$$
$$-3 = -5 + b$$
$$2 = b$$

The equation is $y = -\frac{5}{2}x + 2$.

15. The negative reciprocal of the slope in Exercise 14 is

$$-\frac{1}{-\dfrac{5}{2}} = \frac{2}{5}.$$

So $y = \frac{2}{5}x + b$.

$$1 = \frac{2}{5}(-4) + b \quad \textit{Let x= -4, y=1.}$$
$$1 = -\frac{8}{5} + b$$
$$\frac{13}{5} = b$$

The equation is $y = \frac{2}{5}x + \frac{13}{5}$.

16. $(7x + 4)(2x - 3)$
$$= 14x^2 - 21x + 8x - 12$$
$$= 14x^2 - 13x - 12$$

17. $\left(\frac{2}{3}t + 9\right)^2 = \left(\frac{2}{3}t\right)^2 + 2\left(\frac{2}{3}t\right)(9) + 9^2$
$$= \frac{4}{9}t^2 + 12t + 81$$

18. $(3t^3 + 5t^2 - 8t + 7) - (6t^3 + 4t - 8)$
$$= 3t^3 + 5t^2 - 8t + 7 - 6t^3 - 4t + 8$$
$$= -3t^3 + 5t^2 - 12t + 15$$

19. Divide $4x^3 + 2x^2 - x + 26$ by $x + 2$.
Use synthetic division.

$$
\begin{array}{r|rrrr}
-2 & 4 & 2 & -1 & 26 \\
   &   & -8 & 12 & -22 \\
\hline
   & 4 & -6 & 11 & 4
\end{array}
$$

The answer is

$$4x^2 - 6x + 11 + \frac{4}{x + 2}.$$

20. $16x - x^3 = x(16 - x^2)$
$$= x(4 + x)(4 - x)$$

21. $(3x + 2)^2 - 4(3x + 2) - 5$
Let $m = 3x + 2$.
$$= m^2 - 4m - 5$$
$$= (m - 5)(m + 1)$$
Substitute $3x + 2$ for m.
$$= [(3x + 2) - 5][(3x + 2) + 1]$$
$$= (3x - 3)(3x + 3)$$
$$= 3(x - 1) \cdot 3(x + 1)$$
$$= 9(x - 1)(x + 1)$$

22. $8x^3 + 27y^3$
Use the sum of two cubes formula,

$$a^3 + b^3 = (a + b)(a^2 - ab + b^2),$$

with $a = 2x$ and $b = 3y$.

$$8x^3 + 27y^3$$
$$= (2x + 3y)[(2x)^2 - (2x)(3y) + (3y)^2]$$
$$= (2x + 3y)(4x^2 - 6xy + 9y^2)$$

23. $9x^2 - 30xy + 25y^2$
Use the perfect square formula,

$$a^2 - 2ab + b^2 = (a - b)^2,$$

with $a = 3x$ and $b = 5y$.

$$9x^2 - 30xy + 25y^2$$
$$= [(3x)^2 - 2(3x)(5y) + (5y)^2]$$
$$= (3x - 5y)^2$$

24. $\dfrac{x^2 - 3x - 10}{x^2 + 3x + 2} \cdot \dfrac{x^2 - 2x - 3}{x^2 + 2x - 15}$
$$= \frac{(x - 5)(x + 2)}{(x + 2)(x + 1)} \cdot \frac{(x - 3)(x + 1)}{(x + 5)(x - 3)}$$
$$= \frac{x - 5}{x + 5}$$

25. $\dfrac{3}{2-k} - \dfrac{5}{k} + \dfrac{6}{k^2 - 2k}$

$= \dfrac{3}{2-k} - \dfrac{5}{k} + \dfrac{6}{k(k-2)}$

$= \dfrac{-3}{k-2} - \dfrac{5}{k} + \dfrac{6}{k(k-2)}$

The LCD is $k(k-2)$.

$= \dfrac{-3k}{(k-2)k} - \dfrac{5(k-2)}{k(k-2)} + \dfrac{6}{k(k-2)}$

$= \dfrac{-3k - 5(k-2) + 6}{k(k-2)}$

$= \dfrac{-3k - 5k + 10 + 6}{k(k-2)}$

$= \dfrac{-8k + 16}{k(k-2)}$

$= \dfrac{-8(k-2)}{k(k-2)} = -\dfrac{8}{k}$

26. $\dfrac{\dfrac{r}{s} - \dfrac{s}{r}}{\dfrac{r}{s} + 1}$

Multiply the numerator and denominator by the LCD of all the fractions, rs.

$= \dfrac{\left(\dfrac{r}{s} - \dfrac{s}{r}\right)rs}{\left(\dfrac{r}{s} + 1\right)rs} = \dfrac{r^2 - s^2}{r^2 + rs}$

$= \dfrac{(r-s)(r+s)}{r(r+s)} = \dfrac{r-s}{r}$

27. $4x - 5y = 15$

Draw the line through its intercepts, $\left(\dfrac{15}{4}, 0\right)$ and

$(0, -3)$. The graph passes the vertical line test, so the relation is a function. As with any line that is not horizontal or vertical, the domain and range are both $(-\infty, \infty)$.

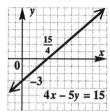

28. $4x - 5y < 15$

Draw a dashed line through the points $\left(\dfrac{15}{4}, 0\right)$

and $(0, -3)$. Check the origin:

$4(0) - 5(0) < 15$?

$\qquad\qquad 0 < 15 \qquad$ *True*

Shade the region that contains the origin.

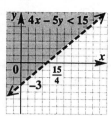

29. $f(x) = -2(x-1)^2 + 3$

The equation is in $f(x) = a(x-h)^2 + k$ form, so the graph is a parabola with vertex (h, k) at $(1, 3)$. Since $a = -2 < 0$, the parabola opens downward. Also $|a| = |-2| = 2 > 1$, so the graph is narrower than the graph of $f(x) = x^2$. The points $(0, 1)$ and $(2, 1)$ are on the graph.

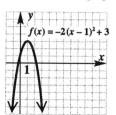

30. **(a)** The points are $(0, 116.26)$, $(10, 132.02)$, $(20, 141.85)$, $(30, 155.38)$, and $(33, 156.36)$.

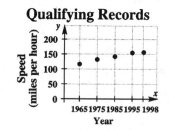

(b) The set of ordered pairs is a function since each year corresponds to a unique speed.

(c) The slope m of the line passing through $(0, 116.26)$ and $(20, 141.85)$ is

$$m = \dfrac{141.85 - 116.26}{20 - 0} = \dfrac{25.59}{20} = 1.2795.$$

Note that $x = 0$ represents 1965, so the point $(0, 116.26)$ can be considered to be the y-intercept.

The slope-intercept form of the linear equation that models these data is

$$y = 1.2795x + 116.26.$$

(d) 1998 corresponds to $x = 33$.

$$y = 1.2795(33) + 116.26$$
$$= 158.4835,$$

which is an overestimate of the actual 156.36.

31. The relation $x = 5$ does not define a function because its graph is a vertical line, which is not the graph of a function by the vertical line test.

32. $f(x) = 2(x-1)^2 - 5$

    (a) $f(-2) = 2(-2-1)^2 - 5$

                  $= 2(-3)^2 - 5$

                  $= 2(9) - 5 = 13$

    (b) Any value can be substituted for x, so the domain is $(-\infty, \infty)$. The graph of f is a parabola that opens upward with vertex $(1, -5)$. The vertex is a minimum point so the y-values are all greater than or equal to -5. Thus, the range is $[-5, \infty)$.

33. $\begin{aligned} 2x - 4y &= 10 \quad (1) \\ 9x + 3y &= 3 \quad (2) \end{aligned}$

Simplify the equations.

$$x - 2y = 5 \quad (3) \quad \tfrac{1}{2} \times (1)$$

$$3x + y = 1 \quad (4) \quad \tfrac{1}{3} \times (2)$$

To eliminate y, multiply (4) by 2 and add the result to (3).

$$\begin{aligned} x - 2y &= 5 \quad (3) \\ 6x + 2y &= 2 \quad\ \ 2 \times (4) \\ \hline 7x\ \ \ \ \ &= 7 \\ x &= 1 \end{aligned}$$

Substitute $x = 1$ into (4).

$$\begin{aligned} 3x + y &= 1 \quad (4) \\ 3(1) + y &= 1 \\ y &= -2 \end{aligned}$$

Solution set: $\{(1, -2)\}$

34. $\begin{aligned} x + y + 2z &= 3 \quad (1) \\ -x + y + z &= -5 \quad (2) \\ 2x + 3y - z &= -8 \quad (3) \end{aligned}$

Eliminate z by adding (2) and (3).

$$\begin{aligned} -x + y + z &= -5 \quad (2) \\ 2x + 3y - z &= -8 \quad (3) \\ \hline x + 4y\ \ \ \ \ &= -13 \quad (4) \end{aligned}$$

To get another equation without z, multiply equation (3) by 2 and add the result to equation (1).

$$\begin{aligned} x + y + 2z &= 3 \quad (1) \\ 4x + 6y - 2z &= -16 \quad\ \ 2 \times (3) \\ \hline 5x + 7y\ \ \ \ \ &= -13 \quad (5) \end{aligned}$$

To eliminate x, multiply (4) by -5 and add the result to (5).

$$\begin{aligned} -5x - 20y &= 65 \quad\ \ -5 \times (4) \\ 5x + 7y &= -13 \quad (5) \\ \hline -13y &= 52 \\ y &= -4 \end{aligned}$$

Use (4) to find x.

$$\begin{aligned} x + 4y &= -13 \quad (4) \\ x + 4(-4) &= -13 \\ x - 16 &= -13 \\ x &= 3 \end{aligned}$$

Use (2) to find z.

$$\begin{aligned} -x + y + z &= -5 \quad (2) \\ -3 - 4 + z &= -5 \\ -7 + z &= -5 \\ z &= 2 \end{aligned}$$

Solution set: $\{(3, -4, 2)\}$

35. Let $x =$ the speed of the boat in still water and $y =$ the speed of the current.

Use the facts that the rate upriver is $x - y$ and that the rate downriver is $x + y$ along with the given information to make a chart.

	d	r	t
Upriver	20	$x - y$	1
Downriver	20	$x + y$	.5

Use $d = rt$ to write a system of equations.

$$\begin{aligned} 20 &= (x - y)(1) \quad (1) \\ 20 &= (x + y)(.5) \quad (2) \end{aligned}$$

Multiply (2) by 2 and add the result to (1).

$$\begin{aligned} 20 &= x - y \quad (1) \\ 40 &= x + y \quad\ \ 2 \times (2) \\ \hline 60 &= 2x \\ 30 &= x \end{aligned}$$

From (1), $20 = 30 - y$, so $y = 10$.
The speed of the boat is 30 mph and the speed of the current is 10 mph.

36. $\sqrt[3]{\dfrac{27}{16}} = \dfrac{\sqrt[3]{27}}{\sqrt[3]{16}} = \dfrac{\sqrt[3]{3^3}}{\sqrt[3]{8 \cdot 2}} = \dfrac{3}{2\sqrt[3]{2}}$

$\phantom{\sqrt[3]{\dfrac{27}{16}}} = \dfrac{3 \cdot \sqrt[3]{4}}{2\sqrt[3]{2} \cdot \sqrt[3]{4}} = \dfrac{3\sqrt[3]{4}}{2\sqrt[3]{8}}$

$\phantom{\sqrt[3]{\dfrac{27}{16}}} = \dfrac{3\sqrt[3]{4}}{2 \cdot 2} = \dfrac{3\sqrt[3]{4}}{4}$

37. $\dfrac{2}{\sqrt{7} - \sqrt{5}} = \dfrac{2(\sqrt{7} + \sqrt{5})}{(\sqrt{7} - \sqrt{5})(\sqrt{7} + \sqrt{5})}$

$\phantom{\dfrac{2}{\sqrt{7} - \sqrt{5}}} = \dfrac{2(\sqrt{7} + \sqrt{5})}{7 - 5}$

$\phantom{\dfrac{2}{\sqrt{7} - \sqrt{5}}} = \dfrac{2(\sqrt{7} + \sqrt{5})}{2} = \sqrt{7} + \sqrt{5}$

38. Let x = the width of the rectangle.
The perimeter of the rectangle is 20 inches. Since the formula for perimeter is $P = 2L + 2W$,

$$20 = 2L + 2x.$$

Solve the equation for L.

$$20 - 2x = 2L$$
$$10 - x = L$$

The area of the rectangle is 21 in^2.
Since the formula for area is LW,

$$21 = (10 - x)x$$
$$21 = 10x - x^2$$
$$x^2 - 10x + 21 = 0$$
$$(x - 7)(x - 3) = 0$$

$$x - 7 = 0 \quad \text{or} \quad x - 3 = 0$$
$$x = 7 \quad \text{or} \quad x = 3.$$

Reject 7 for the width, since width is not longer than length. Thus, the width is 3 inches, and the length is $10 - x = 7$ inches.

39. Let x = Tri's rate on the bicycle and $x - 10$ = Tri's rate while walking.

Make a chart. Use $d = rt$, or $t = \dfrac{d}{r}$.

	Distance	Rate	Time
Bicycle	12	x	$\dfrac{12}{x}$
Walking	8	$x - 10$	$\dfrac{8}{x - 10}$

$$\begin{array}{c} \text{Tri's time on} \\ \text{the bicycle} \end{array} + \begin{array}{c} \text{Tri's time} \\ \text{walking} \end{array} = \begin{array}{c} 5 \\ \text{hours.} \end{array}$$

$$\frac{12}{x} + \frac{8}{x - 10} = 5$$

Multiply by the LCD, $x(x - 10)$.

$$x(x - 10)\left(\frac{12}{x} + \frac{8}{x - 10}\right) = x(x - 10) \cdot 5$$
$$12(x - 10) + 8x = 5x(x - 10)$$
$$12x - 120 + 8x = 5x^2 - 50x$$
$$0 = 5x^2 - 70x + 120$$
$$0 = x^2 - 14x + 24$$
$$0 = (x - 12)(x - 2)$$

$$x - 12 = 0 \quad \text{or} \quad x - 2 = 0$$
$$x = 12 \quad \text{or} \quad x = 2$$

Reject 2 for Tri's bicycle speed, since it would yield a negative walking speed. Thus, his bicycle speed was 12 mph, and his walking speed was $x - 10 = 2$ mph.

40. Let x = the distance traveled by the southbound car and
$2x - 38$ = the distance traveled by the eastbound car.

Since the cars are traveling at right angles with one another, the Pythagorean formula can be applied.

$$a^2 + b^2 = c^2$$
$$x^2 + (2x - 38)^2 = 95^2$$
$$x^2 + 4x^2 - 152x + 1444 = 9025$$
$$5x^2 - 152x - 7581 = 0$$

$$x = \frac{-(-152) \pm \sqrt{(-152)^2 - 4(5)(-7581)}}{2(5)}$$
$$= \frac{152 \pm \sqrt{174,724}}{10} = \frac{152 \pm 418}{10}$$

So $x = \dfrac{152 \pm 418}{10} = 57$ (the other value is negative). The southbound car traveled 57 miles, and the eastbound car traveled
$2x - 38 = 2(57) - 38 = 76$ miles.

CHAPTER 12   INVERSE, EXPONENTIAL, AND LOGARITHMIC FUNCTIONS

Section 12.1

1.   This function is not one-to-one because both Illinois and Wisconsin are paired with the same range element, 40.

3.   The function in the table that pairs a city with a distance is a one-to-one function because for each city there is one distance and each distance has only one city to which it is paired.

   If the distance from Indianapolis to Denver had 1 mile added to it, it would be $1058 + 1 = 1059$ mi, the same as the distance from Los Angeles to Denver. In this case, one distance would have two cities to which it is paired, and the function would not be one-to-one.

5.   If a function is made up of ordered pairs in such a way that the same y-value appears in a correspondence with two different x-values, then the function is not one-to-one.

7.   All of the graphs pass the vertical line test, so they all represent functions. The graph in choice (a) is the only one that passes the horizontal line test, so it is the one-to-one function.

9. $\{(3, 6), (2, 10), (5, 12)\}$ is a one-to-one function, since each x-value corresponds to only one y-value and each y-value corresponds to only one x-value. To find the inverse, interchange x and y in each ordered pair. The inverse is

   $$\{(6, 3), (10, 2), (12, 5)\}.$$

11. $\{(-1, 3), (2, 7), (4, 3), (5, 8)\}$ is not a one-to-one function. The ordered pairs $(-1, 3)$ and $(4, 3)$ have the same y-values for two different x-values.

13.   The graph of $f(x) = 2x + 4$ is a nonvertical, nonhorizontal line. By the horizontal line test, $f(x)$ is a one-to-one function. To find the inverse, replace $f(x)$ with y.

   $$y = 2x + 4$$
   Interchange x and y.
   $$x = 2y + 4$$
   Solve for y.
   $$2y = x - 4$$
   $$y = \frac{x - 4}{2}$$
   Replace y with $f^{-1}(x)$.
   $$f^{-1}(x) = \frac{x - 4}{2}$$

15.   Write $g(x) = \sqrt{x - 3}$ as $y = \sqrt{x - 3}$. Since $x \geq 3$, $y \geq 0$. The graph of g is half of a horizontal parabola that opens to the right. The graph passes the horizontal line test, so g is one-to-one. To find the inverse, interchange x and y to get

   $$x = \sqrt{y - 3}.$$

   Note that now $y \geq 3$, so $x \geq 0$.
   Solve for y by squaring both sides.

   $$x^2 = y - 3$$
   $$x^2 + 3 = y$$
   Replace y with $g^{-1}(x)$.
   $$g^{-1}(x) = x^2 + 3, \ x \geq 0$$

17. $f(x) = 3x^2 + 2$ is not a one-to-one function because two x-values, such as 1 and -1, both have the same y-value, in this case 5. The graph of this function is a vertical parabola which does not pass the horizontal line test.

19.   The graph of $f(x) = x^3 - 4$ is the graph of $g(x) = x^3$ shifted down 4 units. (Recall that $g(x) = x^3$ is the elongated S-shaped curve.) The graph of f passes the horizontal line test, so f is one-to-one.

   Replace $f(x)$ with y.
   $$y = x^3 - 4$$
   Interchange x and y.
   $$x = y^3 - 4$$
   Solve for y.
   $$x + 4 = y^3$$
   Take the cube root of each side.
   $$\sqrt[3]{x + 4} = y$$
   Replace y with $f^{-1}(x)$.
   $$f^{-1}(x) = \sqrt[3]{x + 4}$$

21.   (a) $f(x) = 2^x$

   To find $f(3)$, substitute 3 for x.
   $$f(3) = 2^3 = 8$$

   (b) Since f is one-to-one and $f(3) = 8$, it follows that $f^{-1}(8) = 3$.

23.   (a) $f(x) = 2^x$

   To find $f(0)$, substitute 0 for x.
   $$f(0) = 2^0 = 1$$

   (b) Since f is one-to-one and $f(0) = 1$, it follows that $f^{-1}(1) = 0$.

25.   (a) The function is one-to-one since any horizontal line intersects the graph at most once.

(b) In the graph, the two points marked on the line are $(-1, 5)$ and $(2, -1)$. Interchange x and y in each ordered pair to get $(5, -1)$ and $(-1, 2)$. Plot these points, then draw a dashed line through them to obtain the graph of the inverse function.

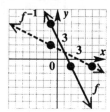

27. **(a)** The function is not one-to-one since there are horizontal lines that intersect the graph more than once. For example, the line $y = 1$ intersects the graph twice.

29. **(a)** The function is one-to-one since any horizontal line intersects the graph at most once.

(b) In the graph, the four points marked on the curve are $(-4, 2)$, $(-1, 1)$, $(1, -1)$, and $(4, -2)$. Interchange x and y in each ordered pair to get $(2, -4)$, $(1, -1)$, $(-1, 1)$, and $(-2, 4)$. Plot these points, then draw a dashed curve (symmetric to the original graph about the line $y = x$) through them to obtain the graph of the inverse.

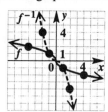

31. $f(x) = 2x - 1$ or $y = 2x - 1$
The graph is a line through $(-2, -5)$, $(0, -1)$, and $(3, 5)$. Plot these points and draw the solid line through them. Then the inverse will be a line through $(-5, -2)$, $(-1, 0)$, and $(5, 3)$. Plot these points and draw the dashed line through them.

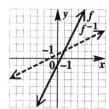

33. $g(x) = -4x$ or $y = -4x$
The graph is a line through $(0, 0)$ and $(1, -4)$. For the inverse, interchange x and y in each ordered pair to get the points $(0, 0)$ and $(-4, 1)$. Draw a dashed line through these points to obtain the graph of the inverse function.

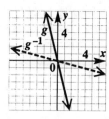

35. $f(x) = \sqrt{x}$, $x \geq 0$
Complete the table of values.

x	$f(x)$
0	0
1	1
4	2

Plot these points and connect them with a solid smooth curve.
Since $f(x)$ is one-to-one, make a table of values for $f^{-1}(x)$ by interchanging x and y.

x	$f^{-1}(x)$
0	0
1	1
2	4

Plot these points and connect them with a dashed smooth curve.

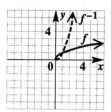

37. $f(x) = x^3 - 2$
Complete the table of values.

x	$f(x)$
-1	-3
0	-2
1	-1
2	6

Plot these points and connect them with a solid smooth curve.
Make a table of values for f^{-1}.

x	$f^{-1}(x)$
-3	-1
-2	0
-1	1
6	2

Plot these points and connect them with a dashed smooth curve.

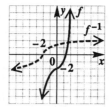

39.
$$f(x) = 4x - 5$$

Replace $f(x)$ with y.
$$y = 4x - 5$$

Interchange x and y.
$$x = 4y - 5$$

Solve for y.
$$x + 5 = 4y$$
$$\frac{x + 5}{4} = y$$

Replace y with $f^{-1}(x)$.
$$\frac{x + 5}{4} = f^{-1}(x)$$

40. Insert each number in the inverse function found in Exercise 39,

$$f^{-1}(x) = \frac{x + 5}{4}.$$

$$f^{-1}(47) = \frac{47 + 5}{4} = \frac{52}{4} = 13 = M,$$

$$f^{-1}(95) = \frac{95 + 5}{4} = \frac{100}{4} = 25 = Y,$$

and so on.

The decoded message is as follows:
My graphing calculator is the greatest thing since sliced bread.

41. A one-to-one code is essential to this process because if the code is not one-to-one, an encoded number would refer to two different letters.

42. Answers will vary according to the student's name. For example, Jane Doe is encoded as follows:

1004   5   2748   129   68   3379   129.

43. $Y_1 = f(x) = 2x - 7$
Replace $f(x)$ with y.
$$y = 2x - 7$$

Interchange x and y.
$$x = 2y - 7$$

Solve for y.
$$x + 7 = 2y$$
$$\frac{x + 7}{2} = y$$

Replace y with $f^{-1}(x)$.
$$\frac{x + 7}{2} = f^{-1}(x) = Y_2$$

Now graph Y_1 and Y_2.

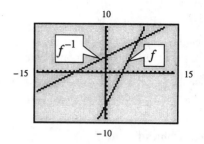

45. $Y_1 = f(x) = x^3 + 5$
Replace $f(x)$ with y.
$$y = x^3 + 5$$

Interchange x and y.
$$x = y^3 + 5$$

Solve for y.
$$x - 5 = y^3$$

Take the cube root of each side.
$$\sqrt[3]{x - 5} = y$$

Replace y with $f^{-1}(x)$.
$$\sqrt[3]{x - 5} = f^{-1}(x) = Y_2$$

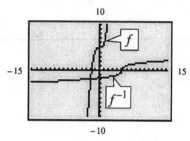

47. $Y_1 = x^2 + 3x + 4$

Graph Y_1 and its inverse in the same square window on a graphing calculator. On a TI-83, graph Y_1 and then enter

DrawInv Y_1

on the home screen. DrawInv is choice 8 under the DRAW menu. Y_1 is choice 1 under VARS, Y-VARS, Function.

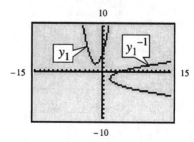

49. The "inverse" of the function in Exercise 47 does not actually satisfy the definition of inverse because the original function is required to be one-to-one. The original function $y = x^2 + 3x + 4$ is not one-to-one as shown by the fact that $(0, 4)$ and $(-3, 4)$ lie on the graph.

Section 12.2

1. Since the graph of $F(x) = a^x$ always contains the point $(0, 1)$, the correct response is (c).

3. Since the graph of $F(x) = a^x$ always approaches the x-axis, the correct response is (a).

5. $f(x) = 3^x$
Make a table of values.

$$f(-2) = 3^{-2} = \frac{1}{3^2} = \frac{1}{9},$$

$$f(-1) = 3^{-1} = \frac{1}{3^1} = \frac{1}{3}, \text{ and so on.}$$

x	-2	-1	0	1	2
$f(x)$	$\frac{1}{9}$	$\frac{1}{3}$	1	3	9

Plot the points from the table and draw a smooth curve through them.

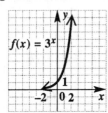

7. $g(x) = \left(\frac{1}{3}\right)^x$
Make a table of values.

$$g(-2) = \left(\frac{1}{3}\right)^{-2} = \left(\frac{3}{1}\right)^2 = 9,$$

$$g(-1) = \left(\frac{1}{3}\right)^{-1} = \left(\frac{3}{1}\right)^1 = 3, \text{ and so on.}$$

x	-2	-1	0	1	2
$g(x)$	9	3	1	$\frac{1}{3}$	$\frac{1}{9}$

Plot the points from the table and draw a smooth curve through them.

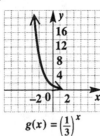

9. $y = 4^{-x}$
This equation can be rewritten as

$$y = \left(4^{-1}\right)^x = \left(\frac{1}{4}\right)^x,$$

which shows that it is *falling* from left to right. Make a table of values.

x	-2	-1	0	1	2
y	16	4	1	$\frac{1}{4}$	$\frac{1}{16}$

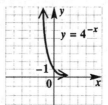

11. $y = 2^{2x-2}$
Make a table of values. It will help to find values for $2x - 2$ before you find y.

x	-2	-1	0	1	2	3
$2x - 2$	-6	-4	-2	0	2	4
y	$\frac{1}{64}$	$\frac{1}{16}$	$\frac{1}{4}$	1	4	16

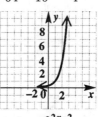

$y = 2^{2x-2}$

13. **(a)** For an exponential function defined by $f(x) = a^x$, if $a > 1$, the graph *rises* from left to right. (See Example 1, $f(x) = 2^x$, in your text.) If $0 < a < 1$, the graph *falls* from left to right. (See Example 2, $g(x) = \left(\frac{1}{2}\right)^x = 2^{-x}$, in your text.)

(b) An exponential function defined by $f(x) = a^x$ is one-to-one and has an inverse, since each value of $f(x)$ corresponds to one and only one value of x.

15. $$100^x = 1000$$
Write each side as a power of 10.
$$\left(10^2\right)^x = 10^3$$
$$10^{2x} = 10^3$$
For $a > 0$ and $a \neq 1$, if $a^x = a^y$, then $x = y$. Set the exponents equal to each other.

$$2x = 3$$
$$x = \frac{3}{2}$$

Check $x = \frac{3}{2}$: $100^{3/2} \overset{?}{=} 1000$ *True*

Solution set: $\left\{\frac{3}{2}\right\}$

17. $16^{2x+1} = 64^{x+3}$

Write each side as a power of 4.

$$\left(4^2\right)^{2x+1} = \left(4^3\right)^{x+3}$$
$$4^{4x+2} = 4^{3x+9}$$

Set the exponents equal.

$$4x + 2 = 3x + 9$$
$$x = 7$$

Check $x = 7$: $16^{15} \overset{?}{=} 64^{10}$ *True*

Solution set: $\{7\}$

19. $5^x = \dfrac{1}{125}$

$$5^x = \left(\dfrac{1}{5}\right)^3$$

Write each side as a power of 5.

$$5^x = 5^{-3}$$

Set the exponents equal.

$$x = -3$$

Check $x = -3$: $5^{-3} \overset{?}{=} \dfrac{1}{125}$ *True*

Solution set: $\{-3\}$

21. $5^x = .2$

$$5^x = \dfrac{2}{10} = \dfrac{1}{5}$$

Write each side as a power of 5.

$$5^x = 5^{-1}$$

Set the exponents equal.

$$x = -1$$

Check $x = -1$: $5^{-1} \overset{?}{=} .2$ *True*

Solution set: $\{-1\}$

23. $\left(\dfrac{3}{2}\right)^x = \dfrac{8}{27}$

$$\left(\dfrac{3}{2}\right)^x = \left(\dfrac{2}{3}\right)^3$$

Write each side as a power of $\dfrac{3}{2}$.

$$\left(\dfrac{3}{2}\right)^x = \left(\dfrac{3}{2}\right)^{-3}$$

Set the exponents equal.

$$x = -3$$

Check $x = -3$: $\left(\dfrac{3}{2}\right)^{-3} \overset{?}{=} \dfrac{8}{27}$ *True*

Solution set: $\{-3\}$

25. $12^{2.6} \approx 639.545$

27. $.5^{3.921} \approx .066$

29. $2.718^{2.5} \approx 12.179$

31.   The reason many scientific calculators cannot calculate something like $(-2)^4$ is that in an exponential function, the base must be positive. This is necessary so the function can be defined for all real numbers. For example, $f(x) = (-4)^x$ would not be defined for $x = \dfrac{1}{2}$ since it is impossible to take the square root of a negative number.

33.   **(a)** The increase for the exponential-type curve in the year 2000 is about .5°C.

     (b) The increase for the linear graph in the year 2000 is about .35°C.

35.   **(a)** The increase for the exponential-type curve in the year 2020 is about 1.6°C.

     (b) The increase for the linear graph in the year 2020 is about .5°C.

37. $f(x) = 7147(1.0366)^x$

     (a) 1950 corresponds to $x = 0$.

$$f(0) = 7147(1.0366)^0$$
$$= 7147(1) = 7147$$

The answer has units in millions of short tons. In case you were wondering, a short ton is 2000 pounds and a long ton is 2240 pounds.

     (b) 1985 corresponds to $x = 35$.

$$f(35) = 7147(1.0366)^{35} \approx 25,149$$

     (c) 1990 corresponds to $x = 40$.

$$f(40) = 7147(1.0366)^{40} \approx 30,100$$

The actual amount, $25,010$ million short tons, is less than the $30,100$ million short tons that the model provides.

39. $A(t) = 100(3.2)^{-.5t}$

     (a) The initial measurement is when $t = 0$.

$$A(0) = 100(3.2)^{-.5(0)}$$
$$= 100(3.2)^0 = 100(1) = 100$$

The initial measurement was 100 g.

     (b) The measurement after 2 months is when $t = 2$.

$$A(2) = 100(3.2)^{-.5(2)}$$
$$= 100(3.2)^{-1} = 31.25$$

The measurement after 2 months was 31.25 g.

     (c) The measurement after 10 months is when $t = 10$.

$$A(10) = 100(3.2)^{-.5(10)}$$
$$= 100(3.2)^{-5} \approx .30$$

The measurement after 10 months was about .30 g.

(d) Use the results of parts (a) – (c) to make a table of values.

t	0	2	10
$A(t)$	100	31.25	.3

Plot the points from the table and draw a smooth curve through them.

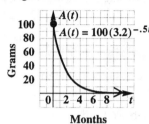

41. $V(t) = 5000(2)^{-.15t}$

$2500 = 5000(2)^{-.15t}$ *Let* $V(t) = 2500$.

$\dfrac{1}{2} = (2)^{-.15t}$ *Divide by 5000.*

$2^{-1} = 2^{-.15t}$

$-1 = -.15t$ *Equate exponents*

$t = \dfrac{-1}{-.15} \approx 6.67$

The value of the machine will be $2500 in approximately 6.67 years after it was purchased.

43. $S(x) = 74,741(1.17)^x$

Since the year 1976 corresponds to $x = 0$, the year 1986 corresponds to $x = 10$ $(1986 - 1976 = 10)$.

$$S(10) = 74,741(1.17)^{10}$$
$$\approx 359,267$$

The average salary in 1986 was about $360,000.

45. $16^{3/4} = \left(\sqrt[4]{16}\right)^3 = (2)^3 = 8$

46. $16^{3/4} = \sqrt[4]{16^3} = \sqrt[4]{4096} = 8$

47. $\sqrt{16^3} = 64$ and $\sqrt{64} = 8$.

48. In Exercise 47, we are finding $\sqrt{\sqrt{16^3}}$, that is, taking the square root twice.

$$\sqrt{\sqrt{16^3}} = \sqrt{(16^3)^{1/2}} = \sqrt{16^{3/2}}$$
$$= (16^{3/2})^{1/2} = 16^{3/4}$$

49. Since $16^{.5} = \sqrt{16} = 4$ and $16^1 = 16$, a reasonable prediction for $16^{.75}$ must be between 4 and 16. From a calculator,

$$16^{.75} = 8.$$

50. $\sqrt[100]{16^{75}} = 16^{75/100} = 16^{3/4}$

$$= \left(\sqrt[4]{16}\right)^3 = (2)^3 = 8$$

51. The display indicates that for the year 1965, the model gives a value of 102.287 million tons,

which is slightly less than the actual value of 103.4 million tons.

Section 12.3

1. **(a)** $\log_4 16$ is equal to 2, because 2 is the exponent to which 4 must be raised in order to obtain 16. **(C)**

(b) $\log_3 81$ is equal to 4, because 4 is the exponent to which 3 must be raised in order to obtain 81. **(F)**

(c) $\log_3\left(\dfrac{1}{3}\right)$ is equal to -1, because -1 is the exponent to which 3 must be raised in order to obtain $\dfrac{1}{3}$. **(B)**

(d) $\log_{10} .01$ is equal to -2, because -2 is the exponent to which 10 must be raised in order to obtain .01. **(A)**

(e) $\log_5 \sqrt{5}$ is equal to $\dfrac{1}{2}$, because $\dfrac{1}{2}$ is the exponent to which 5 must be raised in order to obtain $\sqrt{5}$. **(E)**

(f) $\log_{13} 1$ is equal to 0, because 0 is the exponent to which 13 must be raised in order to obtain 1. **(D)**

3. The base is 4, the exponent (logarithm) is 5, and the number is 1024, so $4^5 = 1024$ becomes $\log_4 1024 = 5$ in logarithmic form.

5. $\dfrac{1}{2}$ is the base and -3 is the exponent, so $\left(\dfrac{1}{2}\right)^{-3} = 8$ becomes $\log_{1/2} 8 = -3$ in logarithmic form.

7. The base is 10, the exponent (logarithm) is -3, and the number is .001, so $10^{-3} = .001$ becomes $\log_{10} .001 = -3$ in logarithmic form.

9. In $\log_4 64 = 3$, 4 is the base and 3 is the logarithm (exponent), so $\log_4 64 = 3$ becomes $4^3 = 64$ in exponential form.

11. The base is 10, logarithm (exponent) is -4, and the number is $\dfrac{1}{10,000}$, so $\log_{10} \dfrac{1}{10,000} = -4$ becomes $10^{-4} = \dfrac{1}{10,000}$ in exponential form.

13. In $\log_6 1 = 0$, 6 is the base and 0 is the logarithm (exponent), so $\log_6 1 = 0$ becomes $6^0 = 1$ in exponential form.

15. To evaluate $\log_9 3$, one has to ask "9 raised to what power gives you a result of 3?" We know that the square root (a radical) of 9 is 3. Therefore, the teacher's hint was to see what root of 9 equals 3. The answer is the reciprocal of the root index.

17. $x = \log_{27} 3$

Write in exponential form.

$$27^x = 3$$

Write each side as a power of 3.

$$(3^3)^x = 3$$
$$3^{3x} = 3^1$$

Set the exponents equal.

$$3x = 1$$
$$x = \frac{1}{3}$$

Solution set: $\left\{\dfrac{1}{3}\right\}$

19. $\log_x 9 = \dfrac{1}{2}$

Change to exponential form.

$$x^{1/2} = 9$$
$$\left(x^{1/2}\right)^2 = 9^2 \qquad \textit{Square}$$
$$x^1 = 81$$
$$x = 81$$

Solution set: $\{81\}$

21. $\log_x 125 = -3$

Write in exponential form.

$$x^{-3} = 125$$
$$\frac{1}{x^3} = 125$$
$$1 = 125\left(x^3\right)$$
$$\frac{1}{125} = x^3$$

Take the cube root of both sides.

$$\sqrt[3]{\frac{1}{125}} = \sqrt[3]{x^3}$$
$$x = \sqrt[3]{\frac{1}{5^3}} = \frac{1}{5}$$

Solution set: $\left\{\dfrac{1}{5}\right\}$

23. $\log_{12} x = 0$

Write in exponential form.

$$12^0 = x$$
$$1 = x$$

Solution set: $\{1\}$

25. $\log_x x = 1$

Write in exponential form.

$$x^1 = x$$

This equation is true for all the numbers x that are allowed as the base of a logarithm; that is, all positive numbers x, $x \neq 1$.

Solution set: $\{x \mid x > 0,\ x \neq 1\}$

27. $\log_x \dfrac{1}{25} = -2$

Write in exponential form.

$$x^{-2} = \frac{1}{25}$$
$$\frac{1}{x^2} = \frac{1}{25}$$
$$x^2 = 25 \qquad \textit{Denominators must be equal.}$$
$$x = \pm 5$$

Reject $x = -5$ since the base of a logarithm must be positive.

Solution set: $\{5\}$

29. $\log_8 32 = x$

$$8^x = 32 \qquad \textit{Exponential form}$$

Write each side as a power of 2.

$$(2^3)^x = 2^5$$
$$2^{3x} = 2^5$$
$$3x = 5 \qquad \textit{Equate exponents}$$
$$x = \frac{5}{3}$$

Solution set: $\left\{\dfrac{5}{3}\right\}$

31. $\log_\pi \pi^4 = x$

$$\pi^x = \pi^4 \qquad \textit{Exponential form}$$
$$x = 4 \qquad \textit{Equate exponents}$$

Solution set: $\{4\}$

33. $\log_6 \sqrt{216} = x$

$$\log_6 216^{1/2} = x \qquad \textit{Equivalent form}$$
$$6^x = 216^{1/2} \qquad \textit{Exponential form}$$
$$6^x = \left(6^3\right)^{1/2} \qquad \textit{Same base}$$
$$6^x = 6^{3/2}$$
$$x = \frac{3}{2} \qquad \textit{Equate exponents}$$

Solution set: $\left\{\dfrac{3}{2}\right\}$

35. $y = \log_3 x$

Change to exponential form.

$$3^y = x$$

Refer to Section 12.2, Exercise 5, for the graph of $f(x) = 3^x$. Since $y = \log_3 x$ $\left(\text{or } 3^y = x\right)$ is the inverse of $f(x) = y = 3^x$, its graph is symmetric about the line $y = x$ to the graph of $f(x) = 3^x$. The graph can be plotted by reversing the ordered pairs in the table of values belonging to $f(x) = 3^x$.

continued

x	$\dfrac{1}{9}$	$\dfrac{1}{3}$	1	3	9
y	-2	-1	0	1	2

Plot the points, and draw a smooth curve through them.

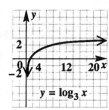

$y = \log_3 x$

37. $y = \log_{1/3} x$

Change to exponential form.

$$\left(\frac{1}{3}\right)^y = x$$

Refer to Section 12.2, Exercise 7, for the graph of $g(x) = \left(\dfrac{1}{3}\right)^x$. Since $y = \log_{1/3} x$ $\left(\text{or } \left(\dfrac{1}{3}\right)^y = x\right)$ is the inverse of $y = \left(\dfrac{1}{3}\right)^x$, its graph is symmetric about the line $y = x$ to the graph of $y = \left(\dfrac{1}{3}\right)^x$. The graph can be plotted by reversing the ordered pairs in the table of values belonging to $g(x) = \left(\dfrac{1}{3}\right)^x$.

x	9	3	1	$\dfrac{1}{3}$	$\dfrac{1}{9}$
y	-2	-1	0	1	2

Plot the points, and draw a smooth curve through them.

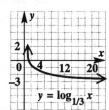

$y = \log_{1/3} x$

39. The number 1 is not used as a base for a logarithmic function since the function would look like $x = 1^y$ in exponential form. Then, for any real value of y, the statement $1 = 1$ would always be the result since every power of 1 is equal to 1.

41. The range of $F(x) = a^x$ is the domain of $G(x) = \log_a x$, that is, $(0, \infty)$.
The domain of $F(x) = a^x$ is the range of $G(x) = \log_a x$, that is, $(-\infty, \infty)$.

43. The values of t are on the horizontal axis, and the values of $f(t)$ are on the vertical axis. Read the value of $f(t)$ from the graph for the given value of t. At $t = 0$, $f(0) = 8$.

45. To find $f(60)$, find 60 on the t-axis, then go up to the graph and across to the $f(t)$ axis to read the value of $f(60)$. At $t = 60$, $f(60) = 24$.

47. $f(x) = 11.34 + 317.01 \log_2 x$

(a) 1984 corresponds to $x = 4$.

$$f(4) = 11.34 + 317.01 \log_2 4$$
$$= 11.34 + 317.01(2)$$
$$= 11.34 + 634.02 = 645.36$$

The model gives 645 sites for 1984.

(b) 1988 corresponds to $x = 8$.

$$f(8) = 11.34 + 317.01 \log_2 8$$
$$= 11.34 + 317.01(3)$$
$$= 11.34 + 951.03 = 962.37$$

The model gives 962 sites for 1988.

49. $M(t) = 6 \log_4 (2t + 4)$

(a) $t = 0$ corresponds to January 1998.

$$M(0) = 6 \log_4 (2 \cdot 0 + 4)$$
$$= 6 \log_4 4$$
$$= 6(1) = 6$$

There were 6 mice in the house in January 1998.

(b) $t = 1$ corresponds to February 1998, so $t = 6$ corresponds to July 1998.

$$M(6) = 6 \log_4 (2 \cdot 6 + 4)$$
$$= 6 \log_4 16$$
$$= 6(2) = 12$$

There were 12 mice in the house in July 1998.

(c) $t = 30$ corresponds to July 2000.

$$M(30) = 6 \log_4 (2 \cdot 30 + 4)$$
$$= 6 \log_4 64$$
$$= 6(3) = 18$$

There were 18 mice in the house in July 2000.

(d) Make a table for M using parts (a) – (c).

t	0	6	30
$M(t)$	6	12	18

Plot the points, and draw a smooth logarithmic curve through them.

$$M(t) = 6 \log_4 (2t + 4)$$

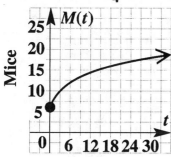

Months since January 1998

51. $g(x) = \log_3 x$

On a TI-83, assign 3^x to Y_1. Then enter

DrawInv Y_1

on the home screen to obtain the figure that follows. (See Exercise 47 in Section 12.1 for TI-83 specifics.)

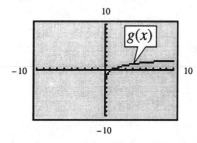

53. $g(x) = \log_{1/3} x$

Assign (1/3)^x to Y_1 and enter DrawInv Y_1.

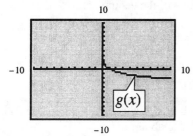

Section 12.4

1. By the product rule,

$$\log_{10} (3 \cdot 4) = \log_{10} 3 + \log_{10} 4.$$

3. By the power rule,

$$\log_{10} 3^4 = 4 \log_{10} 3.$$

5. By a special property (see page 779 in the text),

$$\log_3 3^4 = 4.$$

7. Use the quotient rule for logarithms.

$$\log_7 \frac{4}{5} = \log_7 4 - \log_7 5$$

9. $\log_2 8^{1/4}$

$$= \frac{1}{4} \log_2 8 \qquad \textit{Power rule}$$

$$= \frac{1}{4} \log_2 2^3$$

$$= \frac{1}{4}(3) = \frac{3}{4}$$

11. $\log_4 \dfrac{3\sqrt{x}}{y}$

$$= \log_4 \frac{3 \cdot x^{1/2}}{y}$$

Use the quotient rule for logarithms.

$$= \log_4 \left(3 \cdot x^{1/2}\right) - \log_4 y$$

Use the product and power rules for logarithms.

$$= \log_4 3 + \frac{1}{2} \log_4 x - \log_4 y$$

13. $\log_3 \dfrac{\sqrt[3]{4}}{x^2 y} = \log_3 \dfrac{4^{1/3}}{x^2 y}$

Use the quotient rule for logarithms.

$$= \log_3 4^{1/3} - \log_3 \left(x^2 y\right)$$

Use the product rule for logarithms.

$$= \log_3 4^{1/3} - \left(\log_3 x^2 + \log_3 y\right)$$

$$= \log_3 4^{1/3} - \log_3 x^2 - \log_3 y$$

Use the power rule for logarithms.

$$= \frac{1}{3} \log_3 4 - 2 \log_3 x - \log_3 y$$

15. $\log_3 \sqrt{\dfrac{xy}{5}}$

$$= \log_3 \left(\frac{xy}{5}\right)^{1/2}$$

$$= \frac{1}{2} \log_3 \left(\frac{xy}{5}\right) \qquad \textit{Power rule}$$

$$= \frac{1}{2} \left[\log_3 (xy) - \log_3 5\right] \qquad \textit{Quotient rule}$$

$$= \frac{1}{2} \left(\log_3 x + \log_3 y - \log_3 5\right) \qquad \textit{Product rule}$$

$$= \frac{1}{2} \log_3 x + \frac{1}{2} \log_3 y - \frac{1}{2} \log_3 5$$

17. $\log_2 \dfrac{\sqrt[3]{x} \cdot \sqrt[5]{y}}{r^2}$

$$= \log_2 \frac{x^{1/3} y^{1/5}}{r^2}$$

$$= \log_2 \left(x^{1/3} y^{1/5}\right) - \log_2 r^2 \qquad \textit{Quotient rule}$$

$$= \log_2 x^{1/3} + \log_2 y^{1/5} - \log_2 r^2 \quad \textit{Product rule}$$

$$= \frac{1}{3} \log_2 x + \frac{1}{5} \log_2 y - 2 \log_2 r \quad \textit{Power rule}$$

19. The distributive property tells us that the *product* $a(x + y)$ equals the sum $ax + ay$. In the notation $\log_a (x + y)$, the parentheses do not indicate multiplication. They indicate that $x + y$ is the result of raising a to some power.

21. By the product rule for logarithms,

$$\log_b x + \log_b y = \log_b xy.$$

23. $3 \log_a m - \log_a n$

$$= \log_a m^3 - \log_a n \quad \textit{Power rule}$$

$$= \log_a \frac{m^3}{n} \quad \textit{Quotient rule}$$

25. $\left(\log_a r - \log_a s \right) + 3 \log_a t$

Use the quotient and power rules for logarithms.

$$= \log_a \frac{r}{s} + \log_a t^3$$

$$= \log_a \frac{rt^3}{s} \quad \textit{Product rule}$$

27. $3 \log_a 5 - 4 \log_a 3$

$$= \log_a 5^3 - \log_a 3^4 \quad \textit{Power rule}$$

$$= \log_a \frac{5^3}{3^4} \quad \textit{Quotient rule}$$

$$= \log_a \frac{125}{81}$$

29. $\log_{10} (x + 3) + \log_{10} (x - 3)$

$$= \log_{10} (x + 3)(x - 3) \quad \textit{Product rule}$$

$$= \log_{10} \left(x^2 - 9 \right)$$

31. By the power rule for logarithms,

$$3 \log_p x + \frac{1}{2} \log_p y - \frac{3}{2} \log_p z - 3 \log_p a$$

$$= \log_p x^3 + \log_p y^{1/2} - \log_p z^{3/2} - \log_p a^3$$

Group the terms into sums.

$$= \left(\log_p x^3 + \log_p y^{1/2} \right)$$

$$\quad - \left(\log_p z^{3/2} + \log_p a^3 \right)$$

$$= \log_p x^3 y^{1/2} - \log_p z^{3/2} a^3 \quad \textit{Product rule}$$

$$= \log_p \frac{x^3 y^{1/2}}{z^{3/2} a^3} \quad \textit{Quotient rule}$$

In Exercises 33–40, $\log_{10} 2 \approx .3010$ and $\log_{10} 9 \approx .9542$.

33. By the product rule for logarithms,

$$\log_{10} 18 = \log_{10} (2 \cdot 9)$$

$$= \log_{10} 2 + \log_{10} 9$$

$$\approx .3010 + .9542$$

$$= 1.2552.$$

35. By the quotient rule for logarithms,

$$\log_{10} \frac{2}{9} = \log_{10} 2 - \log_{10} 9$$

$$\approx .3010 - .9542$$

$$= -.6532$$

37. By the product and power rules for logarithms,

$$\log_{10} 36 = \log_{10} 2^2 \cdot 9$$

$$= 2 \log_{10} 2 + \log_{10} 9$$

$$\approx 2(.3010) + .9542$$

$$= 1.5562.$$

39. By the power rule for logarithms,

$$\log_{10} 3 = \log_{10} 9^{1/2}$$

$$= \frac{1}{2} \log_{10} 9$$

$$\approx \frac{1}{2} (.9542)$$

$$= .4771.$$

41. $\log_6 60 - \log_6 10 = \log_6 \frac{60}{10}$

$$= \log_6 6 = 1$$

The statement is true.

43. $\dfrac{\log_{10} 7}{\log_{10} 14} \overset{?}{=} \dfrac{1}{2}$

$$2 \log_{10} 7 \overset{?}{=} 1 \log_{10} (7 \cdot 2)$$

$$\textit{Cross products are equal}$$

$$2 \log_{10} 7 \overset{?}{=} \log_{10} 7 + \log_{10} 2$$

$$\log_{10} 7 \overset{?}{=} \log_{10} 2 \quad \textit{Subtract } log_{10} \, 7$$

The statement is false.

45. The exponent of a quotient is the difference between the exponent of the numerator and the exponent of the denominator.

47. $\log_2 8 - \log_2 4 = \log_2 \dfrac{8}{4}$

$$= \log_2 2 = 1$$

49. $\log_3 81 = \log_3 3^4 = 4$

50. $\log_3 81$ is the exponent to which 3 must be raised in order to obtain 81.

51. Using the result from Exercise 49,

$$3^{\log_3 81} = 3^4 = 81.$$

52. $\log_2 19$ is the exponent to which 2 must be raised in order to obtain 19.

53. Keeping in mind the result from Exercise 51,

$$2^{\log_2 19} = 19.$$

54. To find $k^{\log_k m}$, first assume $\log_k m = y$. This means, changing to an exponential equation, $k^y = m$. Therefore,

$$k^{\log_k m} = k^y = m.$$

Section 12.5

1. Since $\log x = \log_{10} x$, the base is 10. The correct response is (c).

3. $10^0 = 1$ and $10^1 = 10$, so $\log 1 = 0$ and $\log 10 = 1$. Thus, the value of $\log 5.6$ must lie between 0 and 1. The correct response is (c).

5. $\log 10^{19.2} = \log_{10} 10^{19.2} = 19.2$ by the special property, $\log_b b^x = x$.

7. To four decimal places,
$$\log 43 \approx 1.6335.$$

9. $\log 328.4 \approx 2.5164$

11. $\log .0326 \approx -1.4868$

13. $\log \left(4.76 \times 10^9\right) \approx 9.6776$
On a TI-83, enter

$$\boxed{\text{LOG}}\ 4.76\ \boxed{\text{2nd}}\ \boxed{\text{EE}}\ 9\ ).$$

15. $\ln 7.84 \approx 2.0592$

17. $\ln .0556 \approx -2.8896$

19. $\ln 388.1 \approx 5.9613$

21. $\ln \left(8.59 \times e^2\right) \approx 4.1506$
On a TI-83, enter

$$\boxed{\text{LN}}\ 8.59\ \boxed{\text{X}}\ \boxed{\text{2nd}}\ \boxed{e^x}\ 2\ )\ ).$$

23. $\ln 10 \approx 2.3026$

25. (a) $\log 356.8 \approx 2.552424846$

   (b) $\log 35.68 \approx 1.552424846$

   (c) $\log 3.568 \approx 0.552424846$

   (d) The whole number part of the answers (2, 1, or 0) varies, whereas the decimal part (.552424846) remains the same, indicating that the whole number part corresponds to the placement of the decimal point and the decimal part corresponds to the digits 3, 5, 6, and 8.

27. When you try to find $\log (-1)$ on a calculator, an error message is displayed. This is because the domain of the logarithmic function is $(0, \infty)$; -1 is not in the domain.

29. $pH = -\log \left[H_3O^+\right]$
$$= -\log \left(2.5 \times 10^{-2}\right)$$
$$\approx 1.6$$
Since the pH is less than 3.0, the wetland is classified as a *bog*.

31. Ammonia has a hydronium ion concentration of 2.5×10^{-12}.
$$pH = -\log \left[H_3O^+\right]$$
$$pH = -\log \left(2.5 \times 10^{-12}\right) \approx 11.6$$

33. Grapes have a hydronium ion concentration of 5.0×10^{-5}.
$$pH = -\log \left[H_3O^+\right]$$
$$pH = -\log \left(5.0 \times 10^{-5}\right) \approx 4.3$$

35. Human blood plasma has a pH of 7.4.
$$pH = -\log \left[H_3O^+\right]$$
$$7.4 = -\log \left[H_3O^+\right]$$
$$\log_{10} \left[H_3O^+\right] = -7.4$$
$$\left[H_3O^+\right] = 10^{-7.4} \approx 4.0 \times 10^{-8}$$

37. Spinach has a pH value of 5.4.
$$pH = -\log \left[H_3O^+\right]$$
$$5.4 = -\log \left[H_3O^+\right]$$
$$\log_{10} \left[H_3O^+\right] = -5.4$$
$$\left[H_3O^+\right] = 10^{-5.4} \approx 4.0 \times 10^{-6}$$

39. $P(x) = 70,967e^{.0526x}$

   (a) 1987 corresponds to $x = 2$.
$$P(2) = 70,967e^{.0526(2)}$$
$$\approx 78,839.6$$

   The approximate expenditures for 1987 is $78,840$ million dollars.

   (b) 1990 corresponds to $x = 5$.
$$P(5) \approx 92,316$$

   (c) 1993 corresponds to $x = 8$.
$$P(8) \approx 108,095$$

   (d) 1985 corresponds to $x = 0$.
$$P(0) = 70,967 \quad e^0 = 1$$

41. $B(x) = 8768e^{.072x}$
1998 corresponds to $x = 18$.
$$B(18) = 8768e^{.072(18)}$$
$$\approx 32,044$$
The approximate consumer expenditures for 1998 is $32,044$ million dollars.

43. $A(t) = 2.00e^{-.053t}$

   (a) Let $t = 4$.
$$A(4) = 2.00e^{-.053(4)}$$
$$= 2.00e^{-.212}$$
$$\approx 1.62$$

   About 1.62 grams would be present.

   (b) $A(10) = 2.00e^{-.053(10)}$
$$\approx 1.18$$

   About 1.18 grams would be present.

   (c) $A(20) = 2.00e^{-.053(20)}$
$$\approx .69$$

   About .69 grams would be present.

(d) The initial amount is the amount $A(t)$ present at time $t = 0$.

$$A(0) = 2.00e^{-.053(0)}$$
$$= 2.00e^0 = 2.00(1) = 2.00$$

Initially, 2.00 grams were present.

45. $N(r) = -5000 \ln r$

$N(.9) = -5000 \ln .9$
$\approx -5000(-.1054) = 527$

About 527 years have elapsed.

47. $N(r) = -5000 \ln r$

$N(.5) = -5000 \ln .5$
$\approx -5000(-.6931) \approx 3466$

About 3466 years have elapsed.

49. The change-of-base rule is

$$\log_a x = \frac{\log_b x}{\log_b a}.$$

Use common logarithms ($b = 10$).

$$\log_6 12 = \frac{\log_{10} 12}{\log_{10} 6} = \frac{\log 12}{\log 6} \approx 1.3869$$

51. Use natural logarithms ($b = e$).

$$\log_{12} 6 = \frac{\log_e 6}{\log_e 12} = \frac{\ln 6}{\ln 12} \approx .7211$$

53. $\log_{\sqrt{2}} \pi = \dfrac{\ln \pi}{\ln \sqrt{2}} \approx 3.3030$

55. Let $m =$ the number of letters in your first name and $n =$ the number of letters in your last name.

Answers will vary, but suppose the name is Paul Bunyan, with $m = 4$ and $n = 6$.

(a) $\log_m n = \log_4 6$ is the exponent to which 4 must be raised in order to obtain 6.

(b) Use the change-of-base rule.

$$\log_4 6 = \frac{\log 6}{\log 4}$$
$$\approx 1.29248125$$

(c) Here, $m = 4$. Use the power key ($y^x, x^y, \wedge$) on your calculator.

$$4^{1.29248125} \approx 6$$

The result is 6, the value of n.

57. $x = 3$ corresponds to 1988. So in 1988, total expenditures for pollution abatement and control were approximately $83,097.53$ million dollars.

59. **(a)** The expression $\dfrac{1}{x}$ in the base cannot be evaluated, since division by 0 is not defined.

(b) When $x = 100,000$, $Y_1 \approx 2.7183$. The decimal approximation appears to be close to the decimal approximation for e, $2.71828\ldots$.

(c) $\left(1 + \dfrac{1}{1,000,000}\right)^{1,000,000} \approx 2.718280469$

$$e = e^1 \approx 2.718281828$$

The two values differ in the sixth decimal place.

(d) As the values of x approach infinity, the value of $\left(1 + \dfrac{1}{x}\right)^x$ approaches $\underline{e}$.

61. To graph $g(x) = \log_3 x$, assign either $\dfrac{\log x}{\log 3}$ or $\dfrac{\ln x}{\ln 3}$ to Y_1.

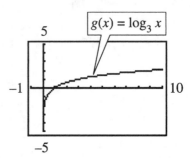

63. To graph $g(x) = \log_{1/3} x$, assign either $\dfrac{\log x}{\log \frac{1}{3}}$ or $\dfrac{\ln x}{\ln \frac{1}{3}}$ to Y_1.

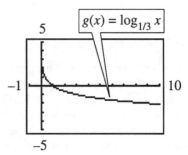

Section 12.6

1. $5^x = 125$
$\log 5^x = \log 125$

2. $x \log 5 = \log 125$

3. $\dfrac{x \log 5}{\log 5} = \dfrac{\log 125}{\log 5}$

$$x = \frac{\log 125}{\log 5}$$

4. $\dfrac{\log 125}{\log 5} = 3$ (calculator)

Solution set: $\{3\}$

5. $$7^x = 5$$

Take the logarithm of each side.

$$\log 7^x = \log 5$$

Use the power rule for logarithms.

$$x \log 7 = \log 5$$

$$x = \frac{\log 5}{\log 7} \approx .827$$

Solution set: $\{.827\}$

7. $$9^{-x+2} = 13$$

$$\log 9^{-x+2} = \log 13$$

$$(-x+2)\log 9 = \log 13 \; (*)$$

$$-x \log 9 + 2 \log 9 = \log 13$$

$$-x \log 9 = \log 13 - 2 \log 9$$

$$x \log 9 = 2 \log 9 - \log 13$$

$$x = \frac{2 \log 9 - \overset{\cdot}{\log} 13}{\log 9}$$

$$\approx .833$$

$(*)$ Alternative solutions steps:

$$(-x+2)\log 9 = \log 13$$

$$-x + 2 = \frac{\log 13}{\log 9}$$

$$2 - \frac{\log 13}{\log 9} = x$$

Solution set: $\{.833\}$

9. $$2^{y+3} = 5^y$$

$$\log 2^{y+3} = \log 5^y$$

$$(y+3)\log 2 = y \log 5$$

$$y \log 2 + 3 \log 2 = y \log 5$$

Get y-terms on one side.

$$y \log 2 - y \log 5 = -3 \log 2$$

$$y(\log 2 - \log 5) = -3 \log 2 \quad \textit{Factor out y.}$$

$$y = \frac{-3 \log 2}{\log 2 - \log 5}$$

$$\approx 2.269$$

Solution set: $\{2.269\}$

11. $$e^{.006x} = 30$$

Take base e logarithms on both sides.

$$\ln e^{.006x} = \ln 30$$

$$.006x \ln e = \ln 30$$

$$.006x = \ln 30 \quad \ln e = 1$$

$$x = \frac{\ln 30}{.006} \approx 566.866$$

Solution set: $\{566.866\}$

13. $$e^{-.103x} = 7$$

$$\ln e^{-.103x} = \ln 7$$

$$-.103x \ln e = \ln 7$$

$$-.103x = \ln 7 \quad \ln e = 1$$

$$x = \frac{\ln 7}{-.103} \approx -18.892$$

Solution set: $\{-18.892\}$

15. $$100 e^{.045x} = 300$$

$$e^{.045x} = \frac{300}{100} = 3$$

$$\ln e^{.045x} = \ln 3$$

$$.045x \left(\ln e\right) = \ln 3$$

$$.045x = \ln 3$$

$$x = \frac{\ln 3}{.045} \approx 24.414$$

Solution set: $\{24.414\}$

17. Let's try Exercise 11.

$$e^{.006x} = 30$$

$$\log e^{.006x} = \log 30$$

$$.006x \left(\log e\right) = \log 30$$

$$x = \frac{\log 30}{.006 \log e} \approx 566.866$$

The natural logarithm is easier because $\ln e = 1$, whereas $\log e$ needs to be calculated.

19. $\log_3 (6x + 5) = 2$

$$6x + 5 = 3^2 \qquad \textit{Exponential form}$$

$$6x + 5 = 9$$

$$6x = 4$$

$$x = \frac{4}{6} = \frac{2}{3}$$

Solution set: $\left\{\dfrac{2}{3}\right\}$

21. $\log_7 (x + 1)^3 = 2$

$$(x+1)^3 = 7^2 \qquad \textit{Exponential form}$$

$$x + 1 = \sqrt[3]{49} \qquad \textit{Cube root}$$

$$x = -1 + \sqrt[3]{49}$$

Solution set: $\left\{-1 + \sqrt[3]{49}\right\}$

23. The apparent solution, 2, causes the expression $\log_4 (x - 3)$ to be $\log_4 (-1)$, and we cannot take the logarithm of a negative number.

25. $\log(6x + 1) = \log 3$

$\quad\quad 6x + 1 = 3 \quad\quad$ *Property 4*

$\quad\quad\quad 6x = 2$

$\quad\quad\quad\quad x = \dfrac{2}{6} = \dfrac{1}{3}$

Solution set: $\left\{\dfrac{1}{3}\right\}$

27. $\log_5(3t + 2) - \log_5 t = \log_5 4$

$\quad\quad \log_5 \dfrac{3t + 2}{t} = \log_5 4$

$\quad\quad\quad \dfrac{3t + 2}{t} = 4$

$\quad\quad\quad 3t + 2 = 4t$

$\quad\quad\quad\quad 2 = t$

Solution set: $\{2\}$

29. $\log 4x - \log(x - 3) = \log 2$

$\quad\quad \log \dfrac{4x}{x - 3} = \log 2$

$\quad\quad\quad \dfrac{4x}{x - 3} = 2$

$\quad\quad\quad 4x = 2(x - 3)$

$\quad\quad\quad 4x = 2x - 6$

$\quad\quad\quad 2x = -6$

$\quad\quad\quad\quad x = -3$

Reject $x = -3$, because $4x = -12$, which yields an equation in which the logarithm of a negative number must be found.

Solution set: $\emptyset$

31. $\log_2 x + \log_2(x - 7) = 3$

$\quad\quad \log_2[x(x - 7)] = 3$

$\quad\quad\quad x(x - 7) = 2^3 \quad$ *Exponential form*

$\quad\quad\quad x^2 - 7x = 8$

$\quad\quad\quad x^2 - 7x - 8 = 0$

$\quad\quad\quad (x - 8)(x + 1) = 0$

$x - 8 = 0 \quad$ or $\quad x + 1 = 0$

$x = 8 \quad$ or $\quad\quad x = -1$

Reject $x = -1$, because it yields an equation in which the logarithm of a negative number must be found.

Solution set: $\{8\}$

33. $\log 5x - \log(2x - 1) = \log 4$

$\quad\quad \log \dfrac{5x}{2x - 1} = \log 4$

$\quad\quad\quad \dfrac{5x}{2x - 1} = 4$

$\quad\quad\quad 5x = 8x - 4$

$\quad\quad\quad 4 = 3x$

$\quad\quad\quad \dfrac{4}{3} = x$

Solution set: $\left\{\dfrac{4}{3}\right\}$

35. $\log_2 x + \log_2(x - 6) = 4$

$\quad\quad \log_2[x(x - 6)] = 4$

$\quad\quad\quad x(x - 6) = 2^4 \quad$ *Exponential form*

$\quad\quad\quad x^2 - 6x = 16$

$\quad\quad\quad x^2 - 6x - 16 = 0$

$\quad\quad\quad (x - 8)(x + 2) = 0$

$x - 8 = 0 \quad$ or $\quad x + 2 = 0$

$x = 8 \quad$ or $\quad\quad x = -2$

Reject $x = -2$, because it yields an equation in which the logarithm of a negative number must be found.

Solution set: $\{8\}$

37. Use the formula $A = P\left(1 + \dfrac{r}{n}\right)^{nt}$ with $P = 2000$, $r = .04$, $n = 4$, and $t = 6$.

$$A = 2000\left(1 + \dfrac{.04}{4}\right)^{4 \cdot 6}$$

$$= 2000(1.01)^{24} \approx 2539.47$$

The account will contain $2539.47.

39. Find $A(t) = 400e^{-.032t}$ when $t = 25$.

$$A(25) = 400e^{-.032(25)}$$

$$= 400e^{-.8} \approx 179.73$$

About 180 grams of lead will be left.

41. Use $A = P\left(1 + \dfrac{r}{n}\right)^{nt}$ with $P = 5000$, $r = .07$, and $t = 12$.

(a) If the interest is compounded annually, $n = 1$.

$$A = 5000\left(1 + \dfrac{.07}{1}\right)^{1 \cdot 12}$$

$$= 5000(1.07)^{12} \approx 11,260.96$$

There will be $11,260.96 in the account.

(b) If the interest is compounded semiannually, $n = 2$.

$$A = 5000\left(1 + \dfrac{.07}{2}\right)^{2 \cdot 12}$$

$$= 5000(1.035)^{24} \approx 11,416.64$$

There will be $11,416.64 in the account.

(c) If the interest is compounded quarterly, $n = 4$.

$$A = 5000\left(1 + \dfrac{.07}{4}\right)^{4 \cdot 12}$$

$$= 5000(1.0175)^{48} \approx 11,497.99$$

There will be $11,497.99 in the account.

(d) If the interest is compounded daily, $n = 365$.

$$A = 5000\left(1 + \frac{.07}{365}\right)^{365 \cdot 12}$$
$$\approx 11,580.90$$

There will be $11,580.90 in the account.

(e) Use the continuous compound interest formula.

$$A = Pe^{rt}$$
$$A = 5000e^{.07(12)}$$
$$= 5000e^{.84} \approx 11,581.83$$

There will be $11,581.83 in the account.

43. In the continuous compound interest formula, let $A = 1850$, $r = .065$, and $t = 40$.

$$A = Pe^{rt}$$
$$1850 = Pe^{.065(40)}$$
$$1850 = Pe^{2.6}$$
$$P = \frac{1850}{e^{2.6}} \approx 137.41$$

Deposit $137.41 today.

45.
$$P(x) = 70,967e^{.0526x}$$
$$133,500 = 70,967e^{.0526x}$$
$$e^{.0526x} = \frac{133,500}{70,967}$$
$$\ln e^{.0526x} = \ln \frac{133,500}{70,967}$$
$$.0526x\,(\ln e) = \ln \frac{133,500}{70,967}$$
$$.0526x = \ln \frac{133,500}{70,967}$$
$$x = \frac{\ln \dfrac{133,500}{70,967}}{.0526} \approx 12.0$$

Since $x = 0$ corresponds to 1985, $x = 12$ corresponds to 1997.

47. $y = 2e^{-.125t}$
When $t = 0$, $y = 2e^0 = 2(1) = 2$. Thus, the original value is 2. Half the original value is 1.

$$1 = 2e^{-.125t}$$
$$.5 = e^{-.125t}$$
$$\ln .5 = \ln e^{-.125t}$$
$$\ln .5 = -.125t\,(\ln e)$$
$$-.125t = \ln .5$$
$$t = \frac{\ln .5}{-.125} \approx 5.55$$

It will take about 5.55 hours.

49. $y = y_0 e^{-.0239t}$

(a) Let $y_0 = 5$ and $t = 20$.

$$y = 5e^{-.0239(20)}$$
$$= 5e^{-.478} \approx 3.10$$

About 3.10 grams will be present after 20 years.

(b) Let $y_0 = 5$ and $t = 60$.

$$y = 5e^{-.0239(60)}$$
$$= 5e^{-1.434} \approx 1.19$$

About 1.19 grams will be present after 60 years.

(c) To find the half-life of radioactive strontium, determine how long it takes until 2.5 grams of the original 5 grams remain. Let $y = 2.5$ and $y_0 = 5$.

$$2.5 = 5e^{-.0239t}$$
$$.5 = e^{-.0239t}$$
$$\ln .5 = \ln e^{-.0239t}$$
$$\ln .5 = -.0239t\,(\ln e)$$
$$-.0239t = \ln .5$$
$$t = \frac{\ln .5}{-.0239} \approx 29.002$$

The half-life is about 29 years.

51.
$$7^x = 5$$
$$7^x - 5 = 0$$
Graph $Y_1 = 7^x - 5$ and find the x-intercept.

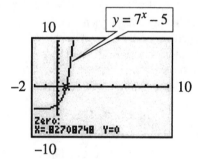

Solution set: $\{.827\}$

53.
$$\log(6x + 1) = \log 3$$
$$\log(6x + 1) - \log 3 = 0$$
Graph $Y_1 = \log(6x + 1) - \log 3$ and find the x-intercept.

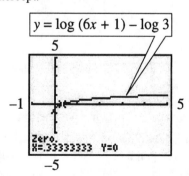

Solution set: $\{.333\}$

Chapter 12 Review Exercises

1. Since a horizontal line intersects the graph in two points, the function is not one-to-one.

2. Since every horizontal line intersects the graph in no more than one point, the function is one-to-one.

3. Each element of the domain corresponds to only one element of the range, and vice versa. Thus, the function is one-to-one.

4. The function $f(x) = -3x + 7$ is a linear function. By the horizontal line test, it is a one-to-one function. To find the inverse, replace $f(x)$ with y.
    $$y = -3x + 7$$
    Interchange x and y.
    $$x = -3y + 7$$
    Solve for y.
    $$x - 7 = -3y$$
    $$\frac{x - 7}{-3} = y$$
    or $\frac{7 - x}{3} = y$
    Replace y with $f^{-1}(x)$.
    $$f^{-1}(x) = \frac{x - 7}{-3} \text{ or } \frac{7 - x}{3}$$

5. $f(x) = \sqrt[3]{6x - 4}$
    The cube root causes each value of x to be matched with only one value of $f(x)$. The function is one-to-one.
    To find the inverse, replace $f(x)$ with y.
    $$y = \sqrt[3]{6x - 4}$$
    Cube both sides.
    $$y^3 = 6x - 4$$
    Interchange x and y.
    $$x^3 = 6y - 4$$
    Solve for y.
    $$x^3 + 4 = 6y$$
    $$\frac{x^3 + 4}{6} = y$$
    Replace y with $f^{-1}(x)$
    $$\frac{x^3 + 4}{6} = f^{-1}(x)$$

6. $f(x) = -x^2 + 3$
    This is an equation of a vertical parabola which opens downward.
    Since a horizontal line will intersect the graph in two points, the function is not one-to-one.

7. The graph is a linear function through $(0, 1)$ and $(3, 0)$. The graph of $f^{-1}(x)$ will include the points $(1, 0)$ and $(0, 3)$, found by interchanging x and y.

Plot these points, and draw a straight line through them.

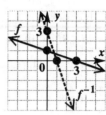

8. The graph is a curve through $(1, 2)$, $(0, 1)$, and $\left(-1, \frac{1}{2}\right)$. Interchange x and y to get $(2, 1)$, $(1, 0)$, and $\left(\frac{1}{2}, -1\right)$, which are on the graph of $f^{-1}(x)$. Plot these points, and draw a smooth curve through them.

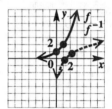

9. $f(x) = 3^x$
    Make a table of values.

x	-2	-1	0	1	2
$f(x)$	$\frac{1}{9}$	$\frac{1}{3}$	1	3	9

    Plot the points from the table and draw a smooth curve through them.

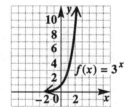

10. $f(x) = \left(\frac{1}{3}\right)^x$
    Make a table of values.

x	-2	-1	0	1	2
$f(x)$	9	3	1	$\frac{1}{3}$	$\frac{1}{9}$

    Plot the points from the table and draw a smooth curve through them.

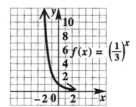

11. $y = 3^{x+1}$

Make a table of values.

x	-3	-2	-1	0	1
y	$\frac{1}{9}$	$\frac{1}{3}$	1	3	9

Plot the points from the table and draw a smooth curve through them.

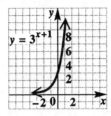

12. $y = 2^{2x+3}$

Make a table of values.

x	-2	$-\frac{3}{2}$	-1	0	$\frac{1}{2}$
y	$\frac{1}{2}$	1	2	8	16

Plot the points from the table and draw a smooth curve through them.

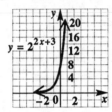

13. $4^{3x} = 8^{x+4}$

Write each side as a power of 2.

$$\left(2^2\right)^{3x} = \left(2^3\right)^{(x+4)}$$
$$2^{6x} = 2^{(3x+12)}$$
$$6x = 3x + 12 \quad \textit{Equate exponents}$$
$$3x = 12$$
$$x = 4$$

Solution set: $\{4\}$

14. $\left(\dfrac{1}{27}\right)^{x-1} = 9^{2x}$

$$\left[\left(\dfrac{1}{3}\right)^3\right]^{x-1} = \left(3^2\right)^{2x}$$

Write each side as a power of 3.

$$\left(3^{-3}\right)^{x-1} = \left(3^2\right)^{2x}$$
$$3^{(-3x+3)} = 3^{4x}$$
$$-3x + 3 = 4x \quad \textit{Equate exponents}$$
$$3 = 7x$$
$$\dfrac{3}{7} = x$$

Solution set: $\left\{\dfrac{3}{7}\right\}$

15. $W(x) = .67(1.123)^x$

(a) 1965 corresponds to $x = 5$.

$$W(5) = .67(1.123)^5 \approx 1.2,$$

which is less than the actual value of 1.4 million tons.

(b) 1975 corresponds to $x = 15$.

$$W(15) = .67(1.123)^{15} \approx 3.8,$$

which is less than the actual value of 4.5 million tons.

(c) 1990 corresponds to $x = 30$.

$$W(30) = .67(1.123)^{30} \approx 21.8,$$

which is more than the actual value of 16.2 million tons.

16. $g(x) = \log_3 x$

Replace $g(x)$ with y, and write in exponential form.

$$y = \log_3 x$$
$$3^y = x$$

Make a table of values. Since $x = 3^y$ is the inverse of $f(x) = y = 3^x$ in Exercise 9, simply reverse the ordered pairs in the table of values belonging to $f(x) = 3^x$.

x	$\frac{1}{9}$	$\frac{1}{3}$	1	3	9
y	-2	-1	0	1	2

Plot the points from the table and draw a smooth curve through them.

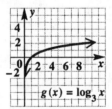

17. $g(x) = \log_{1/3} x$

Replace $g(x)$ with y, and write in exponential form.

$$y = \log_{1/3} x$$
$$\left(\dfrac{1}{3}\right)^y = x$$

Make a table of values. Since $x = \left(\dfrac{1}{3}\right)^y$ is the inverse of $f(x) = y = \left(\dfrac{1}{3}\right)^x$ in Exercise 10, simply reverse the ordered pairs in the table of values belonging to $f(x) = \left(\dfrac{1}{3}\right)^x$.

continued

x	9	3	1	$\frac{1}{3}$	$\frac{1}{9}$
y	-2	-1	0	1	2

Plot the points from the table and draw a smooth curve through them.

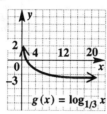

18. $\log_8 64 = x$

$8^x = 64$ *Exponential form*

Write each side as a power of 8.

$8^x = 8^2$

$x = 2$ *Equate exponents*

Solution set: $\{2\}$

19. $\log_2 \sqrt{8} = x$

$2^x = \sqrt{8}$ *Exponential form*

$2^x = 8^{1/2}$

Write each side as a power of 2.

$2^x = \left(2^3\right)^{1/2}$

$2^x = 2^{3/2}$

$x = \frac{3}{2}$ *Equate exponents*

Solution set: $\left\{\dfrac{3}{2}\right\}$

20. $\log_7 \dfrac{1}{49} = x$

$7^x = \dfrac{1}{49}$ *Exponential form*

Write each side as a power of 7.

$7^x = 7^{-2}$

$x = -2$ *Equate exponents*

Solution set: $\{-2\}$

21. $\log_4 x = \dfrac{3}{2}$

$x = 4^{3/2}$ *Exponential form*

$x = \left(\sqrt{4}\right)^3 = 2^3 = 8$

Solution set: $\{8\}$

22. $\log_k 4 = 1$

$k^1 = 4$ *Exponential form*

$k = 4$

Solution set: $\{4\}$

23. $\log_b b^2 = 2$

$b^2 = b^2$ *Exponential form*

This is an identity. Thus, b can be any real

number, $b > 0$ and $b \neq 1$.

Solution set: $\{b \mid b > 0,\ b \neq 1\}$

24. $\log_b a$ is the exponent to which b must be raised in order to obtain a.

25. From Exercise 24,

$$b^{\log_b a} = a.$$

26. $S(x) = 100 \log_2 (x + 2)$

When $x = 6$,

$$S(x) = 100 \log_2 (x + 2)$$
$$= 100(3) = 300.$$

After 6 weeks the sales were 300 thousand dollars or $300,000$.

To graph the function, make a table of values that includes the ordered pair from above.

x	0	2	6
$S(x)$	100	200	300

Plot the ordered pairs and draw the graph through them.

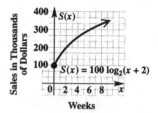

27. $\log_2 3xy^2$

$= \log_2 3 + \log_2 x + \log_2 y^2$ *Product rule*

$= \log_2 3 + \log_2 x + 2\log_2 y$ *Power rule*

28. $\log_4 \dfrac{\sqrt{x} \cdot w^2}{z}$

$= \log_4 \left(\sqrt{x} \cdot w^2\right) - \log_4 z$ *Quotient rule*

$= \log_4 x^{1/2} + \log_4 w^2 - \log_4 z$ *Product rule*

$= \dfrac{1}{2} \log_4 x + 2\log_4 w - \log_4 z$ *Power rule*

29. $\log_b 3 + \log_b x - 2\log_b y$

Use the product and power rules for logarithms.

$= \log_b (3 \cdot x) - \log_b y^2$

$= \log_b \dfrac{3x}{y^2}$ *Quotient rule*

30. $\log_3 (x + 7) - \log_3 (4x + 6)$

$= \log_3 \dfrac{x + 7}{4x + 6}$ *Quotient rule*

31. $\log 28.9 \approx 1.4609$

32. $\log .257 \approx -.5901$

33. $\ln 28.9 \approx 3.3638$

34. $\ln .257 \approx -1.3587$

35. $\log_{16} 13 = \dfrac{\log 13}{\log 16} \approx .9251$

36. $\log_4 12 = \dfrac{\log 12}{\log 4} \approx 1.7925$

37. Milk has a hydronium ion concentration of 4.0×10^{-7}.

$$pH = -\log\left[H_3O^+\right]$$
$$pH = -\log\left(4.0 \times 10^{-7}\right) \approx 6.4$$

38. Crackers have a hydronium ion concentration of 3.8×10^{-9}.

$$pH = -\log\left[H_3O^+\right]$$
$$pH = -\log\left(3.8 \times 10^{-9}\right) \approx 8.4$$

39. Orange juice has a pH of 4.6.
$$pH = -\log\left[H_3O^+\right]$$
$$4.6 = -\log\left[H_3O^+\right]$$
$$\log_{10}\left[H_3O^+\right] = -4.6$$
$$\left[H_3O^+\right] = 10^{-4.6} \approx 2.5 \times 10^{-5}$$

40. $Q(t) = 500e^{-.05t}$

(a) Let $t = 0$.

$$Q(0) = 500e^{-.05(0)}$$
$$= 500e^0 = 500(1) = 500$$

There are 500 grams.

(b) Let $t = 4$.

$$Q(4) = 500e^{-.05(4)}$$
$$= 500e^{-.2} \approx 409.4$$

There will be about 409 grams in 4 days.

41. $3^x = 9.42$

$\log 3^x = \log 9.42$

$x \log 3 = \log 9.42$

$x = \dfrac{\log 9.42}{\log 3} \approx 2.042$

Solution set: $\{2.042\}$

42. $e^{.06x} = 3$

Take base e logarithms on both sides.

$\ln e^{.06x} = \ln 3$

$.06x \ln e = \ln 3$

$.06x = \ln 3 \qquad\qquad \ln e = 1$

$x = \dfrac{\ln 3}{.06} \approx 18.310$

Solution set: $\{18.310\}$

43. **(a)** Solve $7^x = 23$ by using property 3 with common logarithms.

$$7^x = 23$$
$$\log 7^x = \log 23$$
$$x \log 7 = \log 23$$
$$x = \dfrac{\log 23}{\log 7}$$

(b) Solve $7^x = 23$ by using property 3 with natural logarithms.

$$7^x = 23$$
$$\ln 7^x = \ln 23$$
$$x \ln 7 = \ln 23$$
$$x = \dfrac{\ln 23}{\ln 7}$$

(c) Use the change-of-base rule with the solution in part (a).

$$x = \dfrac{\log 23}{\log 7} = \log_7 23$$

(d) $x = \dfrac{\log 23}{\log 7} \neq \log_{23} 7$

The answer is (d).

44. $\log_3(9x + 8) = 2$

$9x + 8 = 3^2 \quad$ *Exponential form*

$9x + 8 = 9$

$9x = 1$

$x = \dfrac{1}{9}$

Solution set: $\left\{\dfrac{1}{9}\right\}$

45. $\log_3(p + 2) - \log_3 p = \log_3 2$

$\log_3 \dfrac{p+2}{p} = \log_3 2 \quad$ *Quotient rule*

$\dfrac{p+2}{p} = 2 \qquad\qquad$ *Property 4*

$p + 2 = 2p$

$2 = p$

Solution set: $\{2\}$

46. $\log(2x + 3) = \log 3x + 2$

$\log(2x + 3) - \log 3x = 2$

$\log_{10} \dfrac{2x+3}{3x} = 2 \quad$ *Quotient rule*

$\dfrac{2x+3}{3x} = 10^2 \quad$ *Exponential form*

$\dfrac{2x+3}{3x} = 100$

$2x + 3 = 300x$

$3 = 298x$

$x = \dfrac{3}{298}$

Solution set: $\left\{\dfrac{3}{298}\right\}$

47. $\log_4 x + \log_4 (8 - x) = 2$

$\qquad \log_4 [x(8 - x)] = 2 \qquad$ *Product rule*

$\qquad\quad x(8 - x) = 4^2 \qquad$ *Exponential form*

$\qquad\qquad 8x - x^2 = 16$

$\qquad\quad x^2 - 8x + 16 = 0$

$\qquad (x - 4)(x - 4) = 0$

$\qquad\qquad\quad x - 4 = 0$

$\qquad\qquad\qquad\quad x = 4$

Solution set: $\{4\}$

48. $\log_2 x + \log_2 (x + 15) = 4$

$\qquad \log_2 [x(x + 15)] = 4 \quad$ *Product rule*

$\qquad\quad x(x + 15) = 2^4 \quad$ *Exponential form*

$\qquad\qquad x^2 + 15x = 16$

$\qquad\quad x^2 + 15x - 16 = 0$

$\qquad (x + 16)(x - 1) = 0$

$x + 16 = 0 \qquad$ or $\qquad x - 1 = 0$

$\quad x = -16 \quad$ or $\qquad\quad x = 1$

Reject $x = -16$, because it yields an equation in which the logarithm of a negative number must be found.

Solution set: $\{1\}$

49. (a)

$$\log (2x + 3) = \log x + 1$$

$\log (2x + 3) - \log x = 1$

$\qquad \log_{10} \dfrac{2x + 3}{x} = 1 \qquad$ *Quotient rule*

$\qquad\quad \dfrac{2x + 3}{x} = 10^1 \quad$ *Exponential form*

$\qquad\quad 2x + 3 = 10x$

$\qquad\qquad\quad 3 = 8x$

$\qquad\qquad \dfrac{3}{8} = x$

Solution set: $\left\{\dfrac{3}{8}\right\}$

(b) From the graph, the x-value of the x-intercept is .375, the decimal equivalent of $\dfrac{3}{8}$. Note that the solutions of $Y_1 - Y_2 = 0$ are the same as the solutions of $Y_1 = Y_2$.

50. When the power rule was applied in the second step, the domain was changed from $\{x \mid x \neq 0\}$ to $\{x \mid x > 0\}$. Instead of using the power rule for logarithms, we can change the original equation to the exponential form $x^2 = 10^2$ and get $x = \pm 10$. As you can see in the erroneous solution, the valid solution -10 was "lost." The solution set is $\{\pm 10\}$.

51. $A = P\left(1 + \dfrac{r}{n}\right)^{nt}$

Let $P = 20,000$, $r = .07$, and $t = 5$. For $n = 4$ (quarterly compounding),

$$A = 20,000\left(1 + \dfrac{.07}{4}\right)^{4 \cdot 5} \approx 28,295.56.$$

There will be $\$28,295.56$ in the account after 5 years.

52. In the continuous compounding formula, let $P = 10,000$, $r = .06$, and $t = 3$.

$$A = Pe^{rt}$$
$$A = 10,000e^{.06(3)} \approx 11,972.17$$

There will be $\$11,972.17$ in the account after 3 years.

53. Use $A = P\left(1 + \dfrac{r}{n}\right)^{nt}$.

Plan A: Let $P = 1000$, $r = .04$, $n = 4$, and $t = 3$.

$$A = 1000\left(1 + \dfrac{.04}{4}\right)^{4 \cdot 3} \approx 1126.83$$

Plan B: Let $P = 1000$, $r = .039$, $n = 12$, and $t = 3$.

$$A = 1000\left(1 + \dfrac{.039}{12}\right)^{12 \cdot 3} \approx 1123.91$$

Plan A is the better plan by $\$2.92$.

54. Let $Q(t) = \dfrac{1}{2}(500) = 250$ to find the half-life of the radioactive substance.

$$Q(t) = 500e^{-.05t}$$
$$250 = 500e^{-.05t}$$
$$.5 = e^{-.05t}$$
$$\ln .5 = \ln e^{-.05t}$$
$$\ln .5 = -.05t \,(\ln e)$$
$$-.05t = \ln .5$$
$$t = \dfrac{\ln .5}{-.05} \approx 13.9$$

The half-life is about 13.9 days.

55. $R(x) = .0597e^{.0553x}$

1990 corresponds to $x = 20$.

$R(20) = .0597e^{.0553(20)}$

$\qquad\quad \approx .1804$

In 1990, about 18.04% of municipal solid waste was recovered.

56. $N(r) = -5000 \ln r$

Replace $N(r)$ by 2000, and solve for r.

$$2000 = -5000 \ln r$$

$$\ln r = -\frac{2000}{5000}$$

$$\log_e r = -.4$$

Change to exponential form and approximate.

$$r = e^{-.4} \approx .67$$

About 67% of the words are common to both of the evolving languages.

57. $S = C(1 - r)^n$

Let $C = 30,000$, $r = .15$, and $n = 12$.

$$S = 30,000(1 - .15)^{12}$$

$$= 30,000(.85)^{12} \approx 4267$$

The scrap value is about \$4267.

58. Let $S = \frac{1}{2}C$ and $n = 6$.

$$S = C(1 - r)^n$$

$$\frac{1}{2}C = C(1 - r)^6$$

$$.5 = (1 - r)^6 \ (*)$$

$$\ln .5 = \ln (1 - r)^6$$

$$\ln .5 = 6 \ln (1 - r)$$

$$\ln (1 - r) = \frac{\ln .5}{6}$$

$$\ln (1 - r) \approx -.1155$$

$$1 - r = e^{-.1155}$$

$$1 - r \approx .89$$

$$r = .11$$

The rate is approximately 11%.

$(*)$ Alternative solution steps without logarithms:

$$.5 = (1 - r)^6$$

$$\sqrt[6]{.5} = 1 - r$$

$$r = 1 - \sqrt[6]{.5} \approx .11$$

59. $f(x) = 2^x$

x	$f(x)$
-2	$2^{-2} = \frac{1}{4}$
-1	$2^{-1} = \frac{1}{2}$
0	$2^0 = 1$
1	$2^1 = 2$
2	$2^2 = 4$
3	$2^3 = 8$

Plot the ordered pairs from the table, and draw a smooth curve through them.

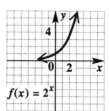

$f(x) = 2^x$

60. $g(x) = \log_2 x$

x	$g(x)$
$\frac{1}{4}$	$\log_2 \frac{1}{4} = \log_2 2^{-2} = -2$
$\frac{1}{2}$	$\log_2 \frac{1}{2} = \log_2 2^{-1} = -1$
1	$\log_2 1 = \log_2 2^0 = 0$
2	$\log_2 2 = \log_2 2^1 = 1$
4	$\log_2 4 = \log_2 2^2 = 2$
8	$\log_2 8 = \log_2 2^3 = 3$

Plot the ordered pairs from the table, and draw a smooth curve through them.

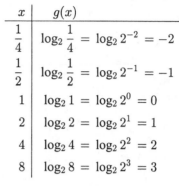

$g(x) = \log_2 x$

61. The ordered pairs in Exercises 59 and 60 are reverses of each other. In other words, if (x, y) is an ordered pair in Exercise 59, (y, x) is an ordered pair in Exercise 60. Functions f and g are inverses of each other.

62. The graph of f in Exercise 59 has a _horizontal_ asymptote, $y = 0$, while the graph of g in Exercise 60 has a _vertical_ asymptote, $x = 0$.

63. $2^2 \cdot 2^3 = 2^5$ because _2_ + _3_ = _5_.

64. It is a fact that $32 = 4 \cdot 8$. Therefore, using properties of logarithms,

$$\log_2 32 = \log_2 \underline{4} + \log_2 \underline{8}.$$

65. $\log_2 13 = \dfrac{\log 13}{\log 2} = 3.700439718$

66. Based on the result of Exercise 65, $\log_2 13$ means that $2^{3.700439718} = 13$. $\log_2 13$ is the exponent to which 2 must be raised in order to obtain 13.

67. $2^{\log_2 13} = 13$ since (using the results of Exercises 65 and 66)

$$2^{\log_2 13} = 2^{3.700439718} = 13.$$

68. Using a calculator,

$$2^{3.700439718} = 13.$$

The number in Exercise 65 is the exponent to which 2 must be raised in order to obtain 13.

69. Based on the result in Exercise 65, the point $(13, \underline{3.700439718})$ lies on the graph of $g(x) = \log_2 x$.

70. $2^{x+1} = 8^{2x+3}$

$$2^{x+1} = \left(2^3\right)^{2x+3}$$

$$2^{x+1} = 2^{6x+9}$$

$$x + 1 = 6x + 9$$

$$-8 = 5x$$

$$-\frac{8}{5} = x$$

Solution set: $\left\{ -\dfrac{8}{5} \right\}$

71. $\log_3 (x + 9) = 4$

$$x + 9 = 3^4 \quad \textit{Exponential form}$$

$$x + 9 = 81$$

$$x = 72$$

Solution set: $\{72\}$

72. $\log_2 32 = x$

$$2^x = 32 \quad \textit{Exponential form}$$

Write each side as a power of 2.

$$2^x = 2^5$$

$$x = 5 \quad \textit{Equate exponents}$$

Solution set: $\{5\}$

73. $\log_x \dfrac{1}{81} = 2$

$$x^2 = \frac{1}{81} \quad \textit{Exponential form}$$

$$x^2 = \left(\frac{1}{9}\right)^2$$

$$x = \pm \frac{1}{9} \quad \textit{Square root property}$$

The base x cannot be negative.

Solution set: $\left\{ \dfrac{1}{9} \right\}$

74. $27^x = 81$

Write each side as a power of 3.

$$\left(3^3\right)^x = 3^4$$

$$3^{3x} = 3^4$$

$$3x = 4 \quad \textit{Equate exponents}$$

$$x = \frac{4}{3}$$

Solution set: $\left\{ \dfrac{4}{3} \right\}$

75. $2^{2x-3} = 8$

Write each side as a power of 2.

$$2^{2x-3} = 2^3$$

$$2x - 3 = 3 \quad \textit{Equate exponents}$$

$$2x = 6$$

$$x = 3$$

Solution set: $\{3\}$

76. $\log_3 (x + 1) - \log_3 x = 2$

$$\log_3 \frac{x+1}{x} = 2 \quad \textit{Quotient rule}$$

$$\frac{x+1}{x} = 3^2 \quad \textit{Exponential form}$$

$$9x = x + 1$$

$$8x = 1$$

$$x = \frac{1}{8}$$

Solution set: $\left\{ \dfrac{1}{8} \right\}$

77. $\log (3x - 1) = \log 10$

$$3x - 1 = 10$$

$$3x = 11$$

$$x = \frac{11}{3}$$

Solution set: $\left\{ \dfrac{11}{3} \right\}$

78. $f(t) = 5000(2)^{-.15t}$

(a) The original value is found when $t = 0$.

$$f(0) = 5000(2)^{-.15(0)}$$

$$= 5000(2)^0 = 5000(1) = 5000$$

The original value is $5000.

(b) The value after 5 years is found when $t = 5$.

$$f(5) = 5000(2)^{-.15(5)} \approx 2973.018$$

The value after 5 years is about $2973.

(c) The value after 10 years is found when $t = 10$.

$$f(10) = 5000(2)^{-.15(10)} \approx 1767.767$$

The value after 10 years is about $1768.

Chapter 12 Test

1. **(a)** $f(x) = x^2 + 9$

This function is not one-to-one. The graph of $f(x)$ is a vertical parabola. A horizontal line will intersect the graph more than once.

(b) This function is one-to-one. A horizontal line will not intersect the graph in more than one point.

2. $f(x) = \sqrt[3]{x+7}$

Replace $f(x)$ with y.
$$y = \sqrt[3]{x+7}$$
Interchange x and y.
$$x = \sqrt[3]{y+7}$$
Solve for y.
$$x^3 = y + 7$$
$$x^3 - 7 = y$$
Replace y with $f^{-1}(x)$.
$$f^{-1}(x) = x^3 - 7$$

3. By the horizontal line test, $f(x)$ is a one-to-one function and has an inverse. Choose some points on the graph of $f(x)$, such as $(4, 0)$, $(3, -1)$, and $(0, -2)$. To graph the inverse, interchange the x- and y-values to get $(0, 4)$, $(-1, 3)$, and $(-2, 0)$. Plot these points and draw a smooth curve through them.

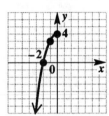

4. $f(x) = 6^x$
Make a table of values.

x	-2	-1	0	1
$f(x)$	$\frac{1}{36}$	$\frac{1}{6}$	1	6

Plot these points and draw a smooth exponential curve through them.

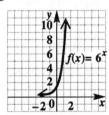

5. $g(x) = \log_6 x$
Make a table of values.

Powers of 6	6^{-2}	6^{-1}	6^0	6^1
x	$\frac{1}{36}$	$\frac{1}{6}$	1	6
$g(x)$	-2	-1	0	1

Plot these points and draw a smooth logarithmic curve through them.

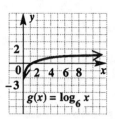

6. $y = 6^x$ and $y = \log_6 x$ are inverse functions. To use the graph from Exercise 4 to obtain the graph of the function in Exercise 5, interchange the x- and y-coordinates of the ordered pairs

$$\left(-2, \frac{1}{36}\right), \left(-1, \frac{1}{6}\right), (0, 1), \text{ and } (1, 6) \text{ to get}$$

$$\left(\frac{1}{36}, -2\right), \left(\frac{1}{6}, -1\right), (1, 0), \text{ and } (6, 1). \text{ Plot}$$

these points and draw a smooth logarithmic curve through them.

7. $5^x = \dfrac{1}{625}$

$$5^x = \left(\frac{1}{5}\right)^4$$

Write each side as a power of 5.
$$5^x = 5^{-4}$$

$\quad x = -4 \quad$ *Equate exponents*

Solution set: $\{-4\}$

8. $2^{3x-7} = 8^{2x+2}$

Write each side as a power of 2.
$$2^{3x-7} = 2^{3(2x+2)}$$

$3x - 7 = 3(2x + 2) \quad$ *Equate exponents*
$$3x - 7 = 6x + 6$$
$$-13 = 3x$$
$$-\frac{13}{3} = x$$

Solution set: $\left\{-\dfrac{13}{3}\right\}$

9. $R(x) = 2821(.9195)^x$

(a) 1990 corresponds to $x = 0$.

$$R(0) = 2821(.9195)^0$$
$$= 2821(1) = 2821$$

In 1990, the toxic release inventory was 2821 million pounds.

(b) 1992 corresponds to $x = 2$.

$$R(2) = 2821(.9195)^2$$
$$\approx 2385 \text{ million pounds}$$

(c) 1993 corresponds to $x = 3$.

$$R(3) = 2821(.9195)^3$$
$$\approx 2193 \text{ million pounds}$$

The actual amount of 2157.4 million pounds is less than the amount provided by the model.

10. The base is 4, the exponent (logarithm) is -2, and the number is .0625, so $4^{-2} = .0625$ becomes $\log_4 .0625 = -2$ in logarithmic form.

11. The base is 7, the logarithm (exponent) is 2, and the number is 49, so $\log_7 49 = 2$ becomes $7^2 = 49$ in exponential form.

12. $\log_{1/2} x = -5$

$$x = \left(\frac{1}{2}\right)^{-5} \quad \textit{Exponential form}$$

$$x = \left(\frac{2}{1}\right)^5 = 32$$

Solution set: $\{32\}$

13. $x = \log_9 3$

$9^x = 3$ *Exponential form*

Write each side as a power of 3.

$$\left(3^2\right)^x = 3$$

$$3^{2x} = 3^1$$

$2x = 1$ *Equate exponents*

$$x = \frac{1}{2}$$

Solution set: $\left\{\dfrac{1}{2}\right\}$

14. $\log_x 16 = 4$

$x^4 = 16$ *Exponential form*

$x^2 = \pm 4$ *Square root property*

Reject -4 since $x^2 \geq 0$.

$x^2 = 4$

$x = \pm 2$ *Square root property*

Reject -2 since the base cannot be negative.

Solution set: $\{2\}$

15. The value of $\log_2 32$ is _5_. This means that if we raise _2_ to the _fifth_ power, we result is _32_.

16. $\log_3 x^2 y$

$= \log_3 x^2 + \log_3 y$ *Product rule*

$= 2 \log_3 x + \log_3 y$ *Power rule*

17. $\log_5 \left(\dfrac{\sqrt{x}}{yz}\right)$

$= \log_5 \sqrt{x} - \log_5 yz$ *Quotient rule*

$= \log_5 x^{1/2} - \left(\log_5 y + \log_5 z\right)$ *Product rule*

$= \dfrac{1}{2} \log_5 x - \log_5 y - \log_5 z$ *Power rule*

18. $3 \log_b s - \log_b t$

$= \log_b s^3 - \log_b t$ *Power rule*

$= \log_b \dfrac{s^3}{t}$ *Quotient rule*

19. $\dfrac{1}{4} \log_b r + 2 \log_b s - \dfrac{2}{3} \log_b t$

Use the power rule for logarithms.

$= \log_b r^{1/4} + \log_b s^2 - \log_b t^{2/3}$

Use the product and quotient rules for logarithms.

$= \log_b \dfrac{r^{1/4} s^2}{t^{2/3}}$

20. **(a)** $\log 23.1 \approx 1.3636$

    (b) $\ln .82 \approx -.1985$

21. **(a)** $\log_3 19 = \dfrac{\log_{10} 19}{\log_{10} 3} = \dfrac{\log 19}{\log 3}$

    (b) $\log_3 19 = \dfrac{\log_e 19}{\log_e 3} = \dfrac{\ln 19}{\ln 3}$

    (c) The four-decimal-place approximation of either fraction is 2.6801.

22. $3^x = 78$

$\ln 3^x = \ln 78$

$x \ln 3 = \ln 78$ *Power rule*

$$x = \frac{\ln 78}{\ln 3} \approx 3.9656$$

Solution set: $\{3.9656\}$

23. $\log_8 (x + 5) + \log_8 (x - 2) = 1$

Use the product rule for logarithms.

$$\log_8 [(x + 5)(x - 2)] = 1$$
$$(x + 5)(x - 2) = 8^1$$
$$x^2 + 3x - 10 = 8$$
$$x^2 + 3x - 18 = 0$$
$$(x + 6)(x - 3) = 0$$

$x + 6 = 0$ or $x - 3 = 0$

$x = -6$ or $x = 3$

Reject $x = -6$, because $x + 5 = -1$, which yields an equation in which the logarithm of a negative number must be found.

Solution set: $\{3\}$

24. $A = P\left(1 + \dfrac{r}{n}\right)^{nt}$

$$A = 10{,}000\left(1 + \frac{.045}{4}\right)^{4 \cdot 5} \approx 12{,}507.51$$

\$10,000 invested at 4.5% annual interest, compounded quarterly, will increase to \$12,507.51 in 5 years.

25. $A = Pe^{rt}$

(a) $A = 15{,}000e^{.05(5)} \approx 19{,}260.38$

There will be \$19,260.38 in the account.

(b) Let $A = 2(15{,}000) = 30{,}000$ and solve for t.

$$30,000 = 15,000e^{.05t}$$
$$2 = e^{.05t}$$
$$\ln 2 = \ln e^{.05t}$$
$$\ln 2 = .05t\,(\ln e)$$
$$.05t = \ln 2$$
$$t = \frac{\ln 2}{.05} \approx 13.9$$

The principal will double in about 13.9 years.

Cumulative Review Exercises Chapters 1–12

For Exercises 1–4,
$$S = \left\{-\frac{9}{4}, -2, -\sqrt{2}, 0, .6, \sqrt{11}, \sqrt{-8}, 6, \frac{30}{3}\right\}.$$

1. The integers are $-2, 0, 6$, and $\frac{30}{3}$ (or 10).

2. The rational numbers are $-\frac{9}{4}, -2, 0, .6, 6$, and $\frac{30}{3}$ (or 10).

Each can be expressed as a quotient of two integers.

3. The irrational numbers are $-\sqrt{2}$ and $\sqrt{11}$.

4. All are real numbers except $\sqrt{-8}$.

5. $|-8| + 6 - |-2| - (-6 + 2)$
$= 8 + 6 - 2 - (-4)$
$= 14 - 2 + 4 = 16$

6. $-12 - |-3| - 7 - |-5|$
$= -12 - 3 - 7 - 5 = -27$

7. $2(-5) + (-8)(4) - (-3)$
$= -10 - 32 + 3 = -39$

8. $7 - (3 + 4a) + 2a = -5(a - 1) - 3$
$7 - 3 - 4a + 2a = -5a + 5 - 3$
$4 - 2a = -5a + 2$
$3a = -2$
$a = -\frac{2}{3}$

Solution set: $\left\{-\frac{2}{3}\right\}$

9. $2m + 2 \le 5m - 1$
$-3m \le -3$

Divide by -3; reverse the inequality.
$m \ge 1$
Solution set: $[1, \infty)$

10. $|2x - 5| = 9$

$2x - 5 = 9$ or $2x - 5 = -9$
$2x = 14$ $2x = -4$
$x = 7$ or $x = -2$

Solution set: $\{-2, 7\}$

11. $|3p| - 4 = 12$
$|3p| = 16$

$3p = 16$ or $3p = -16$
$p = \frac{16}{3}$ or $p = -\frac{16}{3}$

Solution set: $\left\{\pm\frac{16}{3}\right\}$

12. $|3k - 8| \le 1$
$-1 \le 3k - 8 \le 1$
$7 \le 3k \le 9$
$\frac{7}{3} \le k \le 3$

Solution set: $\left[\frac{7}{3}, 3\right]$

13. $|4m + 2| > 10$

$4m + 2 > 10$ or $4m + 2 < -10$
$4m > 8$ $4m < -12$
$m > 2$ or $m < -3$
Solution set: $(-\infty, -3) \cup (2, \infty)$

14. $5x + 2y = 10$

Find the x- and y-intercepts. To find the x-intercept, let $y = 0$.

$$5x + 2(0) = 10$$
$$5x = 10$$
$$x = 2$$

The x-intercept is $(2, 0)$.
To find the y-intercept, let $x = 0$.

$$5(0) + 2y = 10$$
$$2y = 10$$
$$y = 5$$

The y-intercept is $(0, 5)$.
Plot the intercepts and draw the line through them.

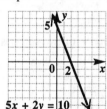

15. $-4x + y \le 5$

Graph the line $-4x + y = 5$, which has intercepts $(0, 5)$ and $\left(-\frac{5}{4}, 0\right)$, as a solid line because the inequality involves $\le$. Test $(0, 0)$, which yields $0 \le 5$, a true statement. Shade the region that includes $(0, 0)$.

continued

$-4x + y \leq 5$

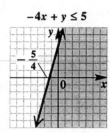

16. $(2p + 3)(3p - 1) = 6p^2 - 2p + 9p - 3$
$$= 6p^2 + 7p - 3$$

17. $(4k - 3)^2 = (4k)^2 - 2(4k)(3) + 3^2$
$$= 16k^2 - 24k + 9$$

18. $(3m^3 + 2m^2 - 5m) - (8m^3 + 2m - 4)$
$$= 3m^3 + 2m^2 - 5m - 8m^3 - 2m + 4$$
$$= 3m^3 - 8m^3 + 2m^2 - 5m - 2m + 4$$
$$= -5m^3 + 2m^2 - 7m + 4$$

19.
$$\begin{array}{r}
2t^3 + 5t^2 - 3t + 4 \\
3t + 1 \overline{\smash{\big)}\, 6t^4 + 17t^3 - 4t^2 + 9t + 4} \\
\underline{6t^4 + 2t^3} \\
15t^3 - 4t^2 \\
\underline{15t^3 + 5t^2} \\
-9t^2 + 9t \\
\underline{-9t^2 - 3t} \\
12t + 4 \\
\underline{12t + 4} \\
0
\end{array}$$

The quotient is
$$2t^3 + 5t^2 - 3t + 4.$$

20. $8x + x^3 = x(8 + x^2)$

21. $24y^2 - 7y - 6 = (8y + 3)(3y - 2)$

22. $5z^3 - 19z^2 - 4z = z(5z^2 - 19z - 4)$
$$= z(5z + 1)(z - 4)$$

23. $16a^2 - 25b^4$

Use the difference of two squares formula,
$$x^2 - y^2 = (x + y)(x - y),$$
where $x = 4a$ and $y = 5b^2$.
$$16a^2 - 25b^4 = (4a + 5b^2)(4a - 5b^2)$$

24. $8c^3 + d^3$

Use the sum of two cubes formula,
$$x^3 + y^3 = (x + y)(x^2 - xy + y^2),$$
where $x = 2c$ and $y = d$.
$$8c^3 + d^3 = (2c + d)(4c^2 - 2cd + d^2)$$

25. $16r^2 + 56rq + 49q^2$
$$= (4r)^2 + 2(4r)(7q) + (7q)^2$$

Use the perfect square formula,

$$x^2 + 2xy + y^2 = (x + y)^2,$$
where $x = 4r$ and $y = 7q$.
$$16r^2 + 56rq + 49q^2 = (4r + 7q)^2$$

26. $\dfrac{(5p^3)^4(-3p^7)}{2p^2(4p^4)} = \dfrac{(5^4 p^{12})(-3p^7)}{8p^6}$
$$= \dfrac{(625)(-3)p^{19}}{8p^6}$$
$$= -\dfrac{1875p^{13}}{8}$$

27. $\dfrac{x^2 - 9}{x^2 + 7x + 12} \div \dfrac{x - 3}{x + 5}$

Multiply by the reciprocal.
$$= \dfrac{x^2 - 9}{x^2 + 7x + 12} \cdot \dfrac{x + 5}{x - 3}$$
$$= \dfrac{(x + 3)(x - 3)}{(x + 3)(x + 4)} \cdot \dfrac{(x + 5)}{(x - 3)} \quad \text{Factor}$$
$$= \dfrac{x + 5}{x + 4}$$

28. $\dfrac{2}{k + 3} - \dfrac{5}{k - 2}$

The LCD is $(k + 3)(k - 2)$.
$$= \dfrac{2(k - 2)}{(k + 3)(k - 2)} - \dfrac{5(k + 3)}{(k - 2)(k + 3)}$$
$$= \dfrac{2k - 4 - 5k - 15}{(k + 3)(k - 2)}$$
$$= \dfrac{-3k - 19}{(k + 3)(k - 2)}$$

29. $\dfrac{3}{p^2 - 4p} - \dfrac{4}{p^2 + 2p}$
$$= \dfrac{3}{p(p - 4)} - \dfrac{4}{p(p + 2)}$$

The LCD is $p(p - 4)(p + 2)$.
$$= \dfrac{3(p + 2)}{p(p - 4)(p + 2)} - \dfrac{4(p - 4)}{p(p + 2)(p - 4)}$$
$$= \dfrac{3p + 6 - 4p + 16}{p(p - 4)(p + 2)}$$
$$= \dfrac{22 - p}{p(p - 4)(p + 2)}$$

30. The points are $(1986, 70,000,000)$ and $(1994, 25,300,000)$.
$$m = \dfrac{70,000,000 - 25,300,000}{1986 - 1994}$$
$$= \dfrac{44,700,000}{-8}$$
$$= -5,587,500$$

The slope is $-5,587,500$ (in $/year).

31. Through $(5, -1)$; parallel to $3x - 4y = 12$

Find the slope of

$$3x - 4y = 12$$
$$-4y = -3x + 12$$
$$y = \frac{3}{4}x - 3.$$

The slope is $\dfrac{3}{4}$, so a line parallel to it also has

slope $\dfrac{3}{4}$.

Let $m = \dfrac{3}{4}$ and $(x_1, y_1) = (5, -1)$ in the point-slope form.

$$y - y_1 = m(x - x_1)$$
$$y - (-1) = \frac{3}{4}(x - 5)$$
$$y + 1 = \frac{3}{4}(x - 5)$$

Multiply by 4 to clear the fraction.
$$4(y + 1) = 3(x - 5)$$
$$4y + 4 = 3x - 15$$
$$19 = 3x - 4y$$

Write in standard form.
$$3x - 4y = 19$$

32. $5x - 3y = 14$ (1)

$2x + 5y = 18$ (2)

Multiply equation (1) by 5 and equation (2) by 3. Then add the results.

$$
\begin{array}{rll}
25x - 15y = & 70 & 5 \times (1) \\
6x + 15y = & 54 & 3 \times (2) \\
\hline
31x = & 124 & Add \\
x = & 4 &
\end{array}
$$

Substitute 4 for x in equation (1) to find y.

$$5x - 3y = 14 \quad (1)$$
$$5(4) - 3y = 14$$
$$20 - 3y = 14$$
$$-3y = -6$$
$$y = 2$$

Solution set: $\{(4, 2)\}$

33. $x + 2y + 3z = 11$ (1)

$3x - y + z = 8$ (2)

$2x + 2y - 3z = -12$ (3)

To eliminate z, add equations (1) and (3).

$$
\begin{array}{rll}
x + 2y + 3z = & 11 & (1) \\
2x + 2y - 3z = & -12 & (3) \\
\hline
3x + 4y \phantom{{}- 3z} = & -1 & (4)
\end{array}
$$

To eliminate z again, multiply equation (2) by 3 and add the result to equation (3).

$$
\begin{array}{rll}
9x - 3y + 3z = & 24 & 3 \times (2) \\
2x + 2y - 3z = & -12 & (3) \\
\hline
11x - y \phantom{{}+ 3z} = & 12 & (5)
\end{array}
$$

Multiply equation (5) by 4 and add the result to equation (4).

$$
\begin{array}{rll}
44x - 4y = & 48 & 4 \times (5) \\
3x + 4y = & -1 & (4) \\
\hline
47x \phantom{{}- 4y} = & 47 & \\
x = & 1 &
\end{array}
$$

Substitute 1 for x in equation (5) to find y.

$$11x - y = 12 \quad (5)$$
$$11(1) - y = 12$$
$$11 - y = 12$$
$$-y = 1$$
$$y = -1$$

Substitute 1 for x and -1 for y in equation (2) to find z.

$$3x - y + z = 8 \quad (2)$$
$$3(1) - (-1) + z = 8$$
$$3 + 1 + z = 8$$
$$4 + z = 8$$
$$z = 4$$

Solution set: $\{(1, -1, 4)\}$

34. $\begin{vmatrix} -2 & -1 \\ 5 & 3 \end{vmatrix} = (-2)(3) - (-1)(5)$

$$= -6 + 5 = -1$$

35. Let $x =$ the amount of candy at \$1.00 per pound.

	Amount of Candy	Price per Pound	Total Price
First Candy	x	1.00	$1.00x$
Second Candy	10	1.96	$1.96(10)$
Mixture	$x + 10$	1.60	$1.60(x + 10)$

Solve the equation:

$$1.00x + 1.96(10) = 1.60(x + 10)$$

Multiply by 10 to clear the decimals.
$$10x + 196 = 16(x + 10)$$
$$10x + 196 = 16x + 160$$
$$36 = 6x$$
$$6 = x$$

Use 6 pounds of the \$1.00 candy.

36. $\sqrt{288} = \sqrt{144 \cdot 2} = \sqrt{144}\sqrt{2} = 12\sqrt{2}$

37. $2\sqrt{32} - 5\sqrt{98} = 2\sqrt{16 \cdot 2} - 5\sqrt{49 \cdot 2}$

$\qquad\qquad = 2 \cdot 4\sqrt{2} - 5 \cdot 7\sqrt{2}$

$\qquad\qquad = 8\sqrt{2} - 35\sqrt{2}$

$\qquad\qquad = -27\sqrt{2}$

38. $\sqrt{2x+1} - \sqrt{x} = 1$

$\qquad \sqrt{2x+1} = 1 + \sqrt{x}$

$\qquad \left(\sqrt{2x+1}\right)^2 = \left(1 + \sqrt{x}\right)^2$

$\qquad 2x + 1 = 1 + 2\sqrt{x} + x$

$\qquad\qquad x = 2\sqrt{x}$

$\qquad\qquad (x)^2 = \left(2\sqrt{x}\right)^2$

$\qquad\qquad x^2 = 4x$

$\qquad x^2 - 4x = 0$

$\qquad x(x - 4) = 0$

$\qquad x = 0 \quad \text{or} \quad x = 4$

Check $x = 0:\quad \sqrt{1} - \sqrt{0} \overset{?}{=} 1 \quad$ *True*

Check $x = 4:\quad \sqrt{9} - \sqrt{4} \overset{?}{=} 1 \quad$ *True*

Solution set: $\{0, 4\}$

39. $(5 + 4i)(5 - 4i) = 5^2 - (4i)^2$

$\qquad\qquad = 25 - 16i^2$

$\qquad\qquad = 25 - 16(-1)$

$\qquad\qquad = 25 + 16 = 41$

40. $3x^2 - x - 1 = 0$

Here, $a = 3, b = -1, c = -1$.

Use the quadratic formula.

$$x = \frac{-b \pm \sqrt{b^2 - 4ac}}{2a}$$

$$x = \frac{-(-1) \pm \sqrt{(-1)^2 - 4(3)(-1)}}{2(3)}$$

$$= \frac{1 \pm \sqrt{1 + 12}}{6} = \frac{1 \pm \sqrt{13}}{6}$$

Solution set: $\left\{ \dfrac{1 + \sqrt{13}}{6}, \dfrac{1 - \sqrt{13}}{6} \right\}$

41. $k^2 + 2k - 8 > 0$

Solve the equation

$$k^2 + 2k - 8 = 0.$$
$$(k + 4)(k - 2) = 0$$

$$k + 4 = 0 \quad \text{or} \quad k - 2 = 0$$
$$k = -4 \quad \text{or} \quad\quad k = 2$$

The numbers -4 and 2 divide the number line into three regions.

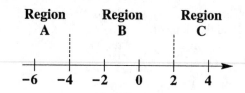

Test a number from each region in the inequality

$$k^2 + 2k - 8 > 0.$$

Region A: Let $k = -5$.

$\qquad 25 - 10 - 8 > 0 \qquad$?

$\qquad\qquad\qquad 7 > 0 \qquad\qquad$ *True*

Region B: Let $k = 0$.

$\qquad\qquad\qquad -8 > 0 \qquad\qquad$ *False*

Region C: Let $k = 3$.

$\qquad 9 + 6 - 8 > 0 \qquad$?

$\qquad\qquad\qquad 7 > 0 \qquad\qquad$ *True*

The numbers in Regions A and C, not including -4 or 2 because of $>$, are solutions.

Solution set: $(-\infty, -4) \cup (2, \infty)$

42. $x^4 - 5x^2 + 4 = 0$

Let $u = x^2$, so $u^2 = \left(x^2\right)^2 = x^4$.

$\qquad u^2 - 5u + 4 = 0$

$\qquad (u - 1)(u - 4) = 0$

$\qquad u - 1 = 0 \quad \text{or} \quad u - 4 = 0$

$\qquad\qquad u = 1 \quad \text{or} \quad\qquad u = 4$

To find x, substitute x^2 for u.

$\qquad x^2 = 1 \qquad \text{or} \quad x^2 = 4$

$\qquad x = \pm 1 \quad \text{or} \quad\quad x = \pm 2$

Solution set: $\{\pm 1, \pm 2\}$

43.   Let $x = $ one of the numbers;

$300 - x = $ the other number.

The product of the two numbers is given by

$$P = x(300 - x).$$

Writing this equation in standard form gives us

$$P = -x^2 + 300x.$$

Finding the maximum of the product is the same as finding the vertex of the graph of P. The x-value of the vertex is

$$x = -\frac{b}{2a} = -\frac{300}{2(-1)} = 150.$$

If x is 150, then $300 - x$ must also be 150. The two numbers are 150 and 150 and the product is $150 \cdot 150 = 22,500$.

44. $f(x) = \dfrac{1}{3}(x-1)^2 + 2$ is in

$f(x) = a(x-h)^2 + k$ form. The graph is a vertical parabola with vertex (h, k) at $(1, 2)$. Since $a = \dfrac{1}{3} > 0$, the graph opens upward.

Also, $|a| = \left|\dfrac{1}{3}\right| = \dfrac{1}{3} < 1$, so the graph is wider than the graph of $f(x) = x^2$. The points $\left(0, 2\dfrac{1}{3}\right)$, $(-2, 5)$, and $(4, 5)$ are also on the graph.

$$f(x) = \dfrac{1}{3}(x-1)^2 + 2$$

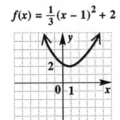

45. $f(x) = 2^x$

Make a table of values.

x	-2	-1	0	1	2
$f(x)$	$\dfrac{1}{4}$	$\dfrac{1}{2}$	1	2	4

Plot the ordered pairs from the table, and draw a smooth exponential curve through the points.

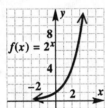

46. $5^{x+3} = \left(\dfrac{1}{25}\right)^{3x+2}$

$5^{x+3} = \left[\left(\dfrac{1}{5}\right)^2\right]^{3x+2}$

Write each side to the power of 5.

$5^{x+3} = \left(5^{-2}\right)^{(3x+2)}$

$5^{x+3} = 5^{-2(3x+2)}$

$x + 3 = -2(3x + 2)$

$x + 3 = -6x - 4$

$7x = -7$

$x = -1$

Solution set: $\{-1\}$

47. $f(x) = \log_3 x$

Make a table of values.

Powers of 3	3^{-2}	3^{-1}	3^0	3^1	3^2
x	$\dfrac{1}{9}$	$\dfrac{1}{3}$	1	3	9
y	-2	-1	0	1	2

Plot the ordered pairs and draw a smooth logarithmic curve through the points.

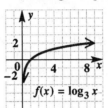

48. $\log_2 81 = \log_2 9^2 = 2\log_2 9$

$\approx 2(3.1699) = 6.3398$

49. $\log \dfrac{x^3 \sqrt{y}}{z}$

$= \log \dfrac{x^3 y^{1/2}}{z}$

$= \log\left(x^3 y^{1/2}\right) - \log z$ *Quotient rule*

$= \log x^3 + \log y^{1/2} - \log z$ *Product rule*

$= 3\log x + \dfrac{1}{2}\log y - \log z$ *Power rule*

50. $B(t) = 25{,}000e^{.2t}$

(a) At noon, $t = 0$.

$B(0) = 25{,}000e^{.2(0)}$

$= 25{,}000e^0 = 25{,}000(1) = 25{,}000$

$25{,}000$ bacteria are present at noon.

(b) At 1 P.M., $t = 1$.

$B(1) = 25{,}000e^{.2(1)} \approx 30{,}535$

About $30{,}500$ bacteria are present at 1 P.M.

(c) At 2 P.M., $t = 2$.

$B(2) = 25{,}000e^{.2(2)} \approx 37{,}296$

About $37{,}300$ bacteria are present at 2 P.M.

(d) At 5 P.M., $t = 5$.

$B(5) = 25{,}000e^{.2(5)} \approx 67{,}957$

About $68{,}000$ bacteria are present at 5 P.M.

CHAPTER 13 CONIC SECTIONS

Section 13.1

1. $x^2 + y^2 = 100$ can be written as

$$(x-0)^2 + (y-0)^2 = 10^2.$$

The center of the circle is $(0,0)$ and the radius of the circle is 10.

3. $$\frac{(y+2)^2}{25} + \frac{(x+3)^2}{49} = 1$$

can be written as

$$\frac{[x-(-3)]^2}{49} + \frac{[y-(-2)]^2}{25} = 1.$$

The center is $(-3, -2)$.

5. Center $(0,0)$; $r = 6$

Use the equation of a circle with radius r and center (h, k). Here, $h = 0$, $k = 0$, and $r = 6$.

$$(x-h)^2 + (y-k)^2 = r^2$$
$$(x-0)^2 + (y-0)^2 = 6^2$$
$$x^2 + y^2 = 36$$

7. Center $(-1, 3)$; $r = 4$

Substitute $h = -1$, $k = 3$, and $r = 4$ in the center-radius form of the equation of a circle.

$$(x-h)^2 + (y-k)^2 = r^2$$
$$[x-(-1)]^2 + (y-3)^2 = 4^2$$
$$(x+1)^2 + (y-3)^2 = 16$$

9. Center $(0, 4)$; $r = \sqrt{3}$

Substitute $h = 0$, $k = 4$, and $r = \sqrt{3}$ in the center-radius form of the equation of the circle.

$$(x-h)^2 + (y-k)^2 = r^2$$
$$(x-0)^2 + (y-4)^2 = \left(\sqrt{3}\right)^2$$
$$x^2 + (y-4)^2 = 3$$

11. By the vertical line test the set is not a function, because a vertical line may intersect the graph of an ellipse in two points.

13. $$x^2 + y^2 + 4x + 6y + 9 = 0$$

Rewrite the equation keeping only the variable terms on the left and grouping the x-terms and y-terms.

$$x^2 + 4x + y^2 + 6y = -9$$

Complete both squares on the left, and add the same constants to the right.

$$\left(x^2 + 4x + \underline{4}\right) + \left(y^2 + 6y + \underline{9}\right) = -9 + \underline{4} + \underline{9}$$
$$(x+2)^2 + (y+3)^2 = 4$$

From the form $(x-h)^2 + (y-k)^2 = r^2$, we have $h = -2$, $k = -3$, and $r = 2$. The center is $(-2, -3)$, and the radius r is 2.

15. $x^2 + y^2 + 10x - 14y - 7 = 0$

$$\left(x^2 + 10x \quad\right) + \left(y^2 - 14y \quad\right) = 7$$
$$\left(x^2 + 10x + \underline{25}\right) + \left(y^2 - 14y + \underline{49}\right)$$
$$= 7 + \underline{25} + \underline{49}$$
$$(x+5)^2 + (y-7)^2 = 81$$

The center is $(-5, 7)$, and the radius is $\sqrt{81} = 9$.

17. $3x^2 + 3y^2 - 12x - 24y + 12 = 0$

$$3\left(x^2 - 4x \quad\right) + 3\left(y^2 - 8y \quad\right) = -12$$

Divide by 3.

$$\left(x^2 - 4x \quad\right) + \left(y^2 - 8y \quad\right) = -4$$
$$\left(x^2 - 4x + \underline{4}\right) + \left(y^2 - 8y + \underline{16}\right)$$
$$= -4 + \underline{4} + \underline{16}$$
$$(x-2)^2 + (y-4)^2 = 16$$

The center is $(2, 4)$, and the radius is $\sqrt{16} = 4$.

19. $$x^2 + y^2 = 9$$
$$(x-0)^2 + (y-0)^2 = 3^2$$

Here, $h = 0$, $k = 0$, and $r = 3$, so the graph is a circle with center $(0,0)$ and radius 3.

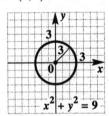

21. $$2y^2 = 10 - 2x^2$$
$$2x^2 + 2y^2 = 10$$
$$x^2 + y^2 = 5 \qquad \textit{Divide by 2.}$$
$$(x-0)^2 + (y-0)^2 = \left(\sqrt{5}\right)^2$$

Here, $h = 0$, $k = 0$, and $r = \sqrt{5} \approx 2.2$, so the graph is a circle with center $(0,0)$ and radius $\sqrt{5}$.

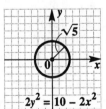

23. $(x+3)^2 + (y-2)^2 = 9$

Here, $h = -3$, $k = 2$, and $r = \sqrt{9} = 3$. The graph is a circle with center $(-3, 2)$ and radius 3.

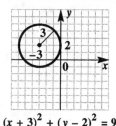

$$(x+3)^2 + (y-2)^2 = 9$$

25.
$$x^2 + y^2 - 4x - 6y + 9 = 0$$
$$(x^2 - 4x \quad) + (y^2 - 6y \quad) = -9$$

Complete the square for each variable.
$$\left(x^2 - 4x + \underline{4}\right) + \left(y^2 - 6y + \underline{9}\right)$$
$$= -9 + \underline{4} + \underline{9}$$
$$(x-2)^2 + (y-3)^2 = 4$$

Here, $h = 2$, $k = 3$, and $r = \sqrt{4} = 2$. The graph is a circle with center $(2, 3)$ and radius 2.

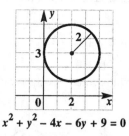

$$x^2 + y^2 - 4x - 6y + 9 = 0$$

27. This method works because the pencil is always the same distance from the fastened end. The fastened end works as the center, and the length of the string from the fastened end to the pencil is the radius.

29. A circular racetrack is most appropriate because the crawfish can move in any direction. Distance from the center determines the winner.

31. The equation $\dfrac{x^2}{9} + \dfrac{y^2}{25} = 1$ is of the form $\dfrac{x^2}{a^2} + \dfrac{y^2}{b^2} = 1$. The graph is an ellipse with $a^2 = 9$ and $b^2 = 25$, so $a = 3$ and $b = 5$. The x-intercepts are $(3, 0)$ and $(-3, 0)$. The y-intercepts are $(0, 5)$ and $(0, -5)$. Plot the intercepts, and draw the ellipse through them.

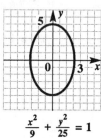

$$\frac{x^2}{9} + \frac{y^2}{25} = 1$$

33. $\dfrac{x^2}{36} = 1 - \dfrac{y^2}{16}$

$\dfrac{x^2}{36} + \dfrac{y^2}{16} = 1$ is in the form $\dfrac{x^2}{a^2} + \dfrac{y^2}{b^2} = 1$.

The graph is an ellipse with $a^2 = 36$ and $b^2 = 16$, so $a = 6$ and $b = 4$. The x-intercepts are $(6, 0)$ and $(-6, 0)$. The y-intercepts are $(0, 4)$ and $(0, -4)$. Plot the intercepts, and draw the ellipse through them.

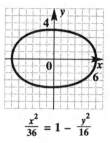

$$\frac{x^2}{36} = 1 - \frac{y^2}{16}$$

35. $\dfrac{y^2}{25} = 1 - \dfrac{x^2}{49}$

$\dfrac{x^2}{49} + \dfrac{y^2}{25} = 1$ is in the form $\dfrac{x^2}{a^2} + \dfrac{y^2}{b^2} = 1$.

The graph is an ellipse with $a^2 = 49$ and $b^2 = 25$, so $a = 7$ and $b = 5$. The x-intercepts are $(7, 0)$ and $(-7, 0)$. The y-intercepts are $(0, 5)$ and $(0, -5)$. Plot the intercepts, and draw the ellipse through them.

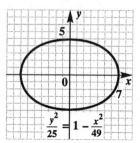

$$\frac{y^2}{25} = 1 - \frac{x^2}{49}$$

37. $\dfrac{(x+1)^2}{64} + \dfrac{(y-2)^2}{49} = 1$

This equation is of the form
$$\frac{(x-h)^2}{a^2} + \frac{(y-k)^2}{b^2} = 1,$$

so the center of the ellipse is at $(-1, 2)$. Since $a^2 = 64$, $a = 8$. Since $b^2 = 49$, $b = 7$. Add ± 8 to -1, and add ± 7 to 2 to find the points $(7, 2)$, $(-9, 2)$, $(-1, 9)$, and $(-1, -5)$. Plot the points, and draw the ellipse through them.

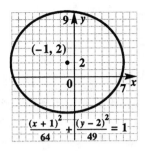

$$\frac{(x+1)^2}{64} + \frac{(y-2)^2}{49} = 1$$

39. $\dfrac{(x-2)^2}{16} + \dfrac{(y-1)^2}{9} = 1$

The center of the ellipse is at $(2, 1)$. Since $a^2 = 16$, $a = 4$. Since $b^2 = 9$, $b = 3$. Add ± 4 to 2, and add ± 3 to 1 to find the points $(6, 1)$, $(-2, 1)$, $(2, 4)$, and $(2, -2)$. Plot the points, and draw the ellipse through them.

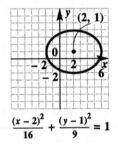

$$\frac{(x-2)^2}{16} + \frac{(y-1)^2}{9} = 1$$

41. $(x+2)^2 + (y-4)^2 = 16$

$$(y-4)^2 = 16 - (x+2)^2$$

Take the square root of each side.

$$y - 4 = \pm\sqrt{16 - (x+2)^2}$$
$$y = 4 \pm \sqrt{16 - (x+2)^2}$$

Therefore, the two functions used to obtain the graph were

$$y_1 = 4 + \sqrt{16 - (x+2)^2} \quad \text{and}$$
$$y_2 = 4 - \sqrt{16 - (x+2)^2}.$$

43. $x^2 + y^2 = 36$

$$y^2 = 36 - x^2$$

Take the square root of both sides.

$$y = \pm\sqrt{36 - x^2}$$

Therefore,

$$y_1 = \sqrt{36 - x^2} \quad \text{and} \quad y_2 = -\sqrt{36 - x^2}.$$

Use these two functions to obtain the graph.

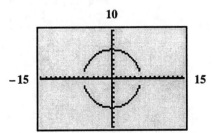

45. $\dfrac{x^2}{16} + \dfrac{y^2}{4} = 1$

$$\frac{y^2}{4} = 1 - \frac{x^2}{16}$$
$$y^2 = 4\left(1 - \frac{x^2}{16}\right)$$

Take the square root of both sides.

$$y = \pm 2\sqrt{1 - \frac{x^2}{16}}$$

Therefore,

$$y_1 = 2\sqrt{1 - \frac{x^2}{16}} \quad \text{and} \quad y_2 = -2\sqrt{1 - \frac{x^2}{16}}.$$

Use these two functions to obtain the graph.

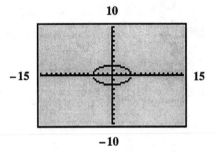

47. $\dfrac{x^2}{141.7^2} + \dfrac{y^2}{141.1^2} = 1$

(a) $c^2 = a^2 - b^2$, so

$$c = \sqrt{a^2 - b^2} = \sqrt{141.7^2 - 141.1^2}$$
$$= \sqrt{169.68} \approx 13.0$$

From the figure, the apogee is $a + c = 141.7 + 13.0 = 154.7$ million miles.

(b) The perigee is $a - c = 141.7 - 13.0 = 128.7$ million miles.

49. $\dfrac{x^2}{36} + \dfrac{y^2}{9} = 1$

$$c^2 = a^2 - b^2 = 36 - 9 = 27, \quad \text{so}$$
$$c = \sqrt{27} = 3\sqrt{3}.$$

The kidney stone and the source of the beam must be placed $3\sqrt{3}$ units from the center of the ellipse.

51. (a) The foci are $(c, 0)$ and $(-c, 0)$. The sum of the distances is $2a$. By the definition of an ellipse,

$$\sqrt{(x-c)^2 + (y-0)^2}$$
$$+ \sqrt{(x+c)^2 + (y-0)^2} = 2a$$
$$\sqrt{(x-c)^2 + y^2} + \sqrt{(x+c)^2 + y^2} = 2a$$
$$\sqrt{(x-c)^2 + y^2} = 2a - \sqrt{(x+c)^2 + y^2}.$$

Square both sides.

continued

$$(x-c)^2 + y^2 = 4a^2 - 4a\sqrt{(x+c)^2 + y^2}$$
$$+ (x+c)^2 + y^2$$
$$x^2 - 2xc + c^2 + y^2$$
$$= 4a^2 - 4a\sqrt{(x+c)^2 + y^2}$$
$$+ x^2 + 2xc + c^2 + y^2$$
$$-4xc = 4a^2 - 4a\sqrt{(x+c)^2 + y^2}$$
$$-4xc - 4a^2 = -4a\sqrt{(x+c)^2 + y^2}$$

Divide by -4.

$$xc + a^2 = a\sqrt{(x+c)^2 + y^2}$$

Square both sides.

$$x^2c^2 + 2xca^2 + a^4 = a^2\left[(x+c)^2 + y^2\right]$$
$$x^2c^2 + 2xca^2 + a^4 = a^2\left(x^2 + 2xc + c^2 + y^2\right)$$
$$x^2c^2 + 2xca^2 + a^4 = a^2x^2 + 2a^2xc + a^2c^2 + a^2y^2$$
$$x^2c^2 + a^4 = a^2x^2 + a^2c^2 + a^2y^2$$

Isolate the x^2- and y^2-terms on one side of the equals sign.

$$a^4 - a^2c^2 = a^2x^2 - x^2c^2 + a^2y^2$$
$$a^2\left(a^2 - c^2\right) = x^2\left(a^2 - c^2\right) + a^2y^2$$
$$1 = \frac{x^2(a^2 - c^2)}{a^2(a^2 - c^2)}$$
$$+ \frac{a^2y^2}{a^2(a^2 - c^2)}$$
$$1 = \frac{x^2}{a^2} + \frac{y^2}{a^2 - c^2}$$

(b) $\dfrac{x^2}{a^2} + \dfrac{y^2}{a^2 - c^2} = 1$

To find the x-intercept, let $y = 0$.

$$\frac{x^2}{a^2} + \frac{0^2}{a^2 - c^2} = 1$$
$$\frac{x^2}{a^2} = 1$$
$$x^2 = a^2$$
$$x = \pm a$$

The x-intercepts are $(a, 0)$ and $(-a, 0)$.

(c) Let $b^2 = a^2 - c^2$.
Therefore, the equation is
$$\frac{x^2}{a^2} + \frac{y^2}{b^2} = 1.$$

To find the y-intercepts, let $x = 0$.

$$\frac{0^2}{a^2} + \frac{y^2}{b^2} = 1$$
$$\frac{y^2}{b^2} = 1$$
$$y^2 = b^2$$
$$y = \pm b$$

The y-intercepts are $(0, b)$ and $(0, -b)$.

Section 13.2

1. $\dfrac{x^2}{25} + \dfrac{y^2}{9} = 1$

This is the standard form for the equation of an ellipse with x-intercepts $(5, 0)$ and $(-5, 0)$ and y-intercepts $(0, 3)$ and $(0, -3)$. This is graph C.

3. $\dfrac{x^2}{9} - \dfrac{y^2}{25} = 1$

This is the standard form for the equation of a hyperbola that opens left and right. Its x-intercepts are $(3, 0)$ and $(-3, 0)$. This is graph D.

5. If the equation of a hyperbola is in standard form (that is, equal to one), the hyperbola would open to the left and right if the x^2-term was positive. It would open up and down if the y^2-term was positive.

7. The equation $\dfrac{x^2}{16} - \dfrac{y^2}{9} = 1$ is in the form $\dfrac{x^2}{a^2} - \dfrac{y^2}{b^2} = 1$. The graph is a hyperbola with $a = 4$ and $b = 3$. The x-intercepts are $(4, 0)$ and $(-4, 0)$. There are no y-intercepts. The corners of the fundamental rectangle are $(4, 3)$, $(4, -3)$, $(-4, -3)$, and $(-4, 3)$. Extend the diagonals of the rectangle through these points to get the asymptotes. Graph a branch of the hyperbola through each intercept and approaching the asymptotes.

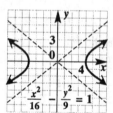

9. $\dfrac{y^2}{9} - \dfrac{x^2}{9} = 1$ is a hyperbola with $a = 3$ and $b = 3$. The y-intercepts are $(0, 3)$ and $(0, -3)$. There are no x-intercepts. One asymptote passes through $(3, 3)$ and $(-3, -3)$. The other asymptote passes through $(-3, 3)$ and $(3, -3)$. Draw the asymptotes and sketch the hyperbola through the intercepts and approaching the asymptotes.

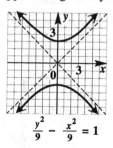

11. $\dfrac{x^2}{25} - \dfrac{y^2}{36} = 1$ is a hyperbola with $a = 5$ and $b = 6$. The x-intercepts are $(5, 0)$ and $(-5, 0)$. There are no y-intercepts. To sketch the graph, draw the diagonals of the fundamental rectangle with corners $(5, 6)$, $(5, -6)$, $(-5, -6)$ and $(-5, 6)$. Graph a branch of the hyperbola through each intercept and approaching the asymptotes.

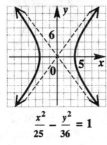

$$\dfrac{x^2}{25} - \dfrac{y^2}{36} = 1$$

13.
$$x^2 - y^2 = 16$$
$$\dfrac{x^2}{16} - \dfrac{y^2}{16} = 1 \qquad \textit{Divide by 16.}$$

This equation is in the form $\dfrac{x^2}{a^2} - \dfrac{y^2}{b^2} = 1$ with $a = 4$ and $b = 4$. The graph is a hyperbola with x-intercepts $(4, 0)$ and $(-4, 0)$ and no y-intercepts. One asymptote passes through $(4, 4)$ and $(-4, -4)$. The other asymptote passes through $(-4, 4)$ and $(4, -4)$. Sketch the graph through the intercepts and approaching the asymptotes.

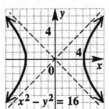

15.
$$4x^2 + y^2 = 16$$
$$\dfrac{x^2}{4} + \dfrac{y^2}{16} = 1 \qquad \textit{Divide by 16.}$$

This equation is in the form $\dfrac{x^2}{a^2} + \dfrac{y^2}{b^2} = 1$ with $a = 2$ and $b = 4$. The graph is an ellipse. The x-intercepts $(2, 0)$ and $(-2, 0)$. The y-intercepts are $(0, 4)$ and $(0, -4)$. Plot the intercepts and draw the ellipse through them.

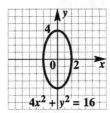

17.
$$y^2 = 36 - x^2$$
$$x^2 + y^2 = 36$$
$$(x - 0)^2 + (y - 0)^2 = 36$$

The graph is a circle with center at $(0, 0)$ and radius $\sqrt{36} = 6$.

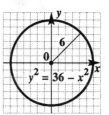

19.
$$9x^2 = 144 + 16y^2$$
$$9x^2 - 16y^2 = 144$$
$$\dfrac{x^2}{16} - \dfrac{y^2}{9} = 1 \qquad \textit{Divide by 144.}$$

The equation is a hyperbola in the form $\dfrac{x^2}{a^2} - \dfrac{y^2}{b^2} = 1$ with $a = 4$ and $b = 3$. The x-intercepts are $(4, 0)$ and $(-4, 0)$. There are no y-intercepts. To sketch the graph, draw the diagonals of the fundamental rectangle with corners $(4, 3)$, $(4, -3)$, $(-4, -3)$ and $(-4, 3)$. These are the asymptotes. Graph a branch of the hyperbola through each intercept approaching the asymptotes.

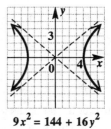

$$9x^2 = 144 + 16y^2$$

21.
$$y^2 = 4 + x^2$$
$$y^2 - x^2 = 4$$
$$\dfrac{y^2}{4} - \dfrac{x^2}{4} = 1 \qquad \textit{Divide by 4.}$$

The graph is a hyperbola with y-intercepts $(0, 2)$ and $(0, -2)$. One asymptote passes through $(2, 2)$ and $(-2, -2)$. The other asymptote passes through $(-2, 2)$ and $(2, -2)$.

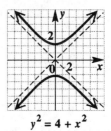

$$y^2 = 4 + x^2$$

23.
$$f(x) = -\sqrt{36 - x^2}$$

Replace $f(x)$ with y, and square both sides of the equation.

$$y = -\sqrt{36 - x^2}$$
$$y^2 = 36 - x^2$$
$$x^2 + y^2 = 36$$

This is a circle centered at the origin with radius $\sqrt{36} = 6$. Since $f(x)$, or y, represents a nonpositive square root in the original equation, $f(x)$ must be nonpositive. This restricts the graph to the bottom half of the circle.

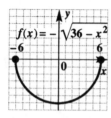

25.
$$f(x) = \sqrt{\frac{x + 4}{2}}$$

Replace $f(x)$ with y.

$$y = \sqrt{\frac{x + 4}{2}}$$

Square both sides.

$$y^2 = \frac{x + 4}{2}$$
$$2y^2 = x + 4$$
$$2y^2 - 4 = x$$
$$2(y - 0)^2 - 4 = x$$

This is a parabola that opens to the right with vertex $(-4, 0)$. However, $f(x)$, or y, is nonnegative in the original equation, so only the top half of the parabola is included in the graph.

x	-2	0	4
y	1	$\sqrt{2}$	2

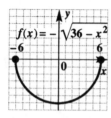

27. $\dfrac{(x - 2)^2}{4} - \dfrac{(y + 1)^2}{9} = 1$

is a hyperbola centered at $(2, -1)$, with $a = 2$ and $b = 3$. The x-intercepts are $(2 \pm 2, -1)$ or $(4, -1)$ and $(0, -1)$. The asymptotes are the extended diagonals of the rectangle with corners

$(2, 3)$, $(2, -3)$, $(-2, -3)$ and $(-2, 3)$ shifted 2 units right and 1 unit down, or $(4, 2)$, $(4, -4)$, $(0, -4)$ and $(0, 2)$. Draw the hyperbola.

29. $\dfrac{y^2}{36} - \dfrac{(x - 2)^2}{49} = 1$

is a hyperbola centered at $(2, 0)$ with $a = 7$, and $b = 6$. The y-intercepts are $(0, 0 \pm 6)$ or $(0, 6)$ and $(0, -6)$. The asymptotes are the extended diagonals of the rectangle with corners $(7, 6)$, $(7, -6)$, $(-7, -6)$, and $(-7, 6)$ shifted right 2 units, or $(9, 6)$, $(9, -6)$, $(-5, -6)$, and $(-5, 6)$. Draw the hyperbola.

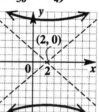

31. **(a)** $100x^2 + 324y^2 = 32{,}400$

Divide by $32{,}400$.

$$\frac{x^2}{324} + \frac{y^2}{100} = 1$$
$$\frac{x^2}{18^2} + \frac{y^2}{10^2} = 1$$

The height in the center is the y-coordinate of the positive y-intercept. The height is 10 meters.

(b) The width of the ellipse is the distance between the x-intercepts, $(-18, 0)$ and $(18, 0)$. The width across the bottom of the arch is $18 + 18 = 36$ meters.

33. $\dfrac{x^2}{g^2} - \dfrac{y^2}{g^2} = 1$

Think of the crossbar as the x-axis and the goal posts at $(g, 0)$ and $(-g, 0)$ as the x-intercepts. The asymptotes go through the center of the goal posts, $(0, 0)$, and on the line from $(-g, -g)$ to (g, g) and on the line from $(g, -g)$ to $(-g, g)$. Since $a = b = g$, the asymptotes are

$$y = \pm \frac{b}{a}x = \pm \frac{g}{g}x = \pm x.$$ These form a 45°

angle with the line through the goal posts. Most people can estimate a 45° angle fairly easily.

35. $y = .607\sqrt{383.9 + x^2}$

    (a) According to the graph, about 55% of women worked in 1985.

    (b) 1985 corresponds to $x = 85$.

$$y = .607\sqrt{383.9 + 85^2} \approx 52.95$$

According to the equation, about 53% of women worked in 1985.

37. $\dfrac{x^2}{9} - y^2 = 1$

$$-y^2 = 1 - \dfrac{x^2}{9}$$

$$y^2 = \dfrac{x^2}{9} - 1 \qquad \textit{Multiply by } -1.$$

Take the square root of both sides.

$$y = \pm\sqrt{\dfrac{x^2}{9} - 1}$$

The two functions used to obtain the graph were

$$y_1 = \sqrt{\dfrac{x^2}{9} - 1} \quad \text{and} \quad y_2 = -\sqrt{\dfrac{x^2}{9} - 1}.$$

39. $\dfrac{x^2}{25} - \dfrac{y^2}{49} = 1$

$$-\dfrac{y^2}{49} = 1 - \dfrac{x^2}{25}$$

$$\dfrac{y^2}{49} = \dfrac{x^2}{25} - 1 \qquad \textit{Multiply by } -1.$$

$$y^2 = 49\left(\dfrac{x^2}{25} - 1\right)$$

Take the square root of both sides.

$$y = \pm 7\sqrt{\dfrac{x^2}{25} - 1}$$

To obtain the graph, use the two functions

$$y_1 = 7\sqrt{\dfrac{x^2}{25} - 1} \quad \text{and} \quad y_2 = -7\sqrt{\dfrac{x^2}{25} - 1}.$$

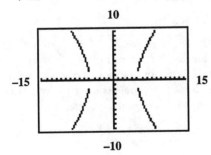

41. $\dfrac{y^2}{9} - x^2 = 1$

$$y^2 - 9x^2 = 9 \qquad \textit{Multiply by } 9.$$

$$y^2 = 9 + 9x^2$$

$$y^2 = 9(1 + x^2)$$

Take the square root of both sides.

$$y = \pm 3\sqrt{1 + x^2}$$

To obtain the graphs, use the two functions

$$y_1 = 3\sqrt{1 + x^2} \quad \text{and} \quad y_2 = -3\sqrt{1 + x^2}.$$

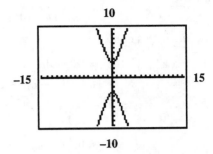

43. $\dfrac{x^2}{4} - y^2 = 1$

$$-y^2 = 1 - \dfrac{x^2}{4}$$

$$y^2 = \dfrac{x^2}{4} - 1 \qquad \textit{Multiply by } -1.$$

Take the square root of both sides.

$$y = \pm\sqrt{\dfrac{x^2}{4} - 1}$$

The positive square root is

$$y = \sqrt{\dfrac{x^2}{4} - 1}.$$

44. $\dfrac{x^2}{4} - y^2 = 1$

Here, $a^2 = 4$ and $b^2 = 1$, so $a = 2$ and $b = 1$. The equation of the asymptote with positive slope is $y = \dfrac{b}{a}x$, or $y = \dfrac{1}{2}x$.

45. $y = \sqrt{\dfrac{x^2}{4} - 1}$

$$y = \sqrt{\dfrac{50^2}{4} - 1} \qquad \textit{Let } x = 50.$$

$$= \sqrt{625 - 1} = \sqrt{624} \approx 24.98$$

46. $y = \dfrac{1}{2}x$

$$y = \dfrac{1}{2}(50) = 25 \quad \textit{Let } x = 50.$$

47. Because $24.98 < 25$, the graph of $y = \sqrt{\dfrac{x^2}{4} - 1}$ lies below the graph of $y = \dfrac{1}{2}x$ when $x = 50$.

48. If x-values larger than 50 are chosen, the y-values of the hyperbola will get closer to the y-values of the asymptote. The y-values on the hyperbola will always be less than the y-values on the line.

49. $\sqrt{(x-c)^2 + (y-0)^2}$

$-\sqrt{[x-(-c)]^2 + (y-0)^2} = 2a$

$\sqrt{(x-c)^2 + y^2} = 2a + \sqrt{(x+c)^2 + y^2}$

Square both sides.

$$(x-c)^2 + y^2 = 4a^2 + 4a\sqrt{(x+c)^2 + y^2}$$
$$+ (x+c)^2 + y^2$$
$$x^2 - 2xc + c^2 + y^2 = 4a^2 + 4a\sqrt{(x+c)^2 + y^2}$$
$$+ x^2 + 2xc + c^2 + y^2$$
$$-4xc = 4a^2 + 4a\sqrt{(x+c)^2 + y^2}$$

Divide by 4.

$$-xc = a^2 + a\sqrt{(x+c)^2 + y^2}$$
$$-xc - a^2 = a\sqrt{(x+c)^2 + y^2}$$

Square both sides again.

$$x^2c^2 + 2xca^2 + a^4 = a^2\left[(x+c)^2 + y^2\right]$$
$$x^2c^2 + 2xca^2 + a^4 = a^2\left(x^2 + 2xc + c^2 + y^2\right)$$
$$x^2c^2 + 2xca^2 + a^4 = a^2x^2 + 2a^2xc + a^2c^2 + a^2y^2$$
$$x^2c^2 + a^4 = a^2x^2 + a^2c^2 + a^2y^2$$

Isolate the x^2- and y^2-terms on one side of the equals sign.

$$x^2c^2 - a^2x^2 - a^2y^2 = a^2c^2 - a^4$$
$$x^2(c^2 - a^2) - a^2y^2 = a^2(c^2 - a^2)$$

Since $b^2 = c^2 - a^2$,

$$x^2b^2 - a^2y^2 = a^2b^2$$
$$\frac{x^2b^2}{a^2b^2} - \frac{a^2y^2}{a^2b^2} = \frac{a^2b^2}{a^2b^2} \qquad \textit{Divide by } a^2b^2.$$
$$\frac{x^2}{a^2} - \frac{y^2}{b^2} = 1.$$

Section 13.3

1. Substitute $x-1$ for y in the first equation. Then solve for x. Find the corresponding y-values by substituting back into $y = x - 1$. In the first equation, both variables are squared and in the second, both variables are to the first power, so the elimination method is not appropriate.

3. The line intersects the ellipse in exactly one point, so there is *one* point in the solution set of the system.

5. The line does not intersect the hyperbola, so the solution set is the empty set.

7. A line and a circle; no points

Draw any circle, and then draw a line that does not cross the circle.

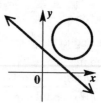

9. A line and an ellipse; no points

The line does not intersect the ellipse.

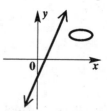

11. A line and a hyperbola; no points

Draw any hyperbola, and then draw a line between the two branches of the hyperbola that does not cross either branch.

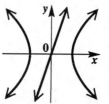

13. A line and a hyperbola; two points

Draw any hyperbola, and then draw a line that crosses both branches.

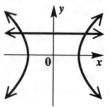

15. A circle and an ellipse; four points

Draw any ellipse, and then draw a circle with the same center whose radius is just large enough so that there are four points of intersection. (If the radius of the circle is too large or too small, there may be fewer points of intersection.)

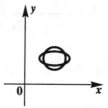

17. A parabola and an ellipse; four points

Draw any parabola, and then draw an ellipse large enough so that there are four points of intersection. (If the ellipse is too large or too small, there may be fewer points of intersection.)

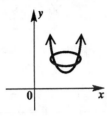

19. $y = 4x^2 - x$ (1)

$y = x$ (2)

Substitute x for y in equation (1).

$$y = 4x^2 - x \quad (1)$$
$$x = 4x^2 - x$$
$$0 = 4x^2 - 2x$$
$$0 = 2x(2x - 1)$$

$$2x = 0 \quad \text{or} \quad 2x - 1 = 0$$
$$x = 0 \quad \text{or} \quad x = \frac{1}{2}$$

Use equation (2) to find y for each x-value.

If $x = 0$, then $y = 0$.

If $x = \frac{1}{2}$, then $y = \frac{1}{2}$.

Solution set: $\left\{(0,0), \left(\frac{1}{2}, \frac{1}{2}\right)\right\}$

21. $y = x^2 + 6x + 9$ (1)

$x + y = 3$ (2)

Substitute $x^2 + 6x + 9$ for y in equation (2).

$$x + y = 3 \quad (2)$$
$$x + \left(x^2 + 6x + 9\right) = 3$$
$$x^2 + 7x + 9 = 3$$
$$x^2 + 7x + 6 = 0$$
$$(x + 6)(x + 1) = 0$$

$$x + 6 = 0 \quad \text{or} \quad x + 1 = 0$$
$$x = -6 \quad \text{or} \quad x = -1$$

Substitute these values for x in equation (2) and solve for y.

If $x = -6$, then

$$x + y = 3 \quad (2)$$
$$-6 + y = 3$$
$$y = 9.$$

If $x = -1$, then

$$x + y = 3 \quad (2)$$
$$-1 + y = 3$$
$$y = 4.$$

Solution set: $\{(-6, 9), (-1, 4)\}$

23. $x^2 + y^2 = 2$ (1)

$2x + y = 1$ (2)

Solve equation (2) for y.

$$y = 1 - 2x \quad (3)$$

Substitute $1 - 2x$ for y in equation (1).

$$x^2 + y^2 = 2 \quad (1)$$
$$x^2 + (1 - 2x)^2 = 2$$
$$x^2 + 1 - 4x + 4x^2 = 2$$
$$5x^2 - 4x - 1 = 0$$
$$(5x + 1)(x - 1) = 0$$

$$5x + 1 = 0 \quad \text{or} \quad x - 1 = 0$$
$$x = -\frac{1}{5} \quad \text{or} \quad x = 1$$

Use equation (3) to find y for each x-value.

If $x = \frac{1}{5}$, then

$$y = 1 - 2\left(-\frac{1}{5}\right) = 1 + \frac{2}{5} = \frac{7}{5}.$$

If $x = 1$, then

$$y = 1 - 2(1) = -1.$$

Solution set: $\left\{\left(-\frac{1}{5}, \frac{7}{5}\right), (1, -1)\right\}$

25. $xy = 4$ (1)

$3x + 2y = -10$ (2)

Solve equation (1) for y to get $y = \frac{4}{x}$.

Substitute $\frac{4}{x}$ for y in equation (2) to find x.

$$3x + 2y = -10 \quad (2)$$
$$3x + 2\left(\frac{4}{x}\right) = -10$$

Multiply by the LCD, x.

$$3x^2 + 8 = -10x$$
$$3x^2 + 10x + 8 = 0$$
$$(3x + 4)(x + 2) = 0$$

$$3x + 4 = 0 \quad \text{or} \quad x + 2 = 0$$
$$x = -\frac{4}{3} \quad \text{or} \quad x = -2$$

continued

Since $y = \dfrac{4}{x}$, if $x = -\dfrac{4}{3}$, then $y = \dfrac{4}{-\dfrac{4}{3}} = -3.$

If $x = -2$, then

$$y = \frac{4}{-2} = -2.$$

Solution set: $\left\{ (-2,-2), \left(-\dfrac{4}{3}, -3\right) \right\}$

27. $xy = -3$ (1)
$x + y = -2$ (2)

Solve equation (2) for y.

$$y = -x - 2. \quad (3)$$

Substitute $-x - 2$ for y in equation (1).

$$xy = -3 \quad (1)$$
$$x(-x - 2) = -3$$
$$-x^2 - 2x = -3$$
$$-x^2 - 2x + 3 = 0$$
$$x^2 + 2x - 3 = 0 \qquad \textit{Multiply by } -1.$$
$$(x + 3)(x - 1) = 0$$

$$x + 3 = 0 \quad \text{or} \quad x - 1 = 0$$
$$x = -3 \quad \text{or} \quad x = 1$$

Use equation (3) to find y for each x-value.

If $x = -3$, then

$$y = -(-3) - 2 = 1.$$

If $x = 1$, then

$$y = -(1) - 2 = -3.$$

Solution set: $\{(-3, 1), (1, -3)\}$

29. $y = 3x^2 + 6x$ (1)
$y = x^2 - x - 6$ (2)

Substitute $x^2 - x - 6$ for y in equation (1) to find x.

$$y = 3x^2 + 6x \qquad (1)$$
$$x^2 - x - 6 = 3x^2 + 6x$$
$$0 = 2x^2 + 7x + 6$$
$$0 = (2x + 3)(x + 2)$$

$$2x + 3 = 0 \quad \text{or} \quad x + 2 = 0$$
$$x = -\frac{3}{2} \quad \text{or} \quad x = -2$$

Substitute $-\dfrac{3}{2}$ for x in equation (1) to find y.

$$y = 3x^2 + 6x \qquad (1)$$
$$y = 3\left(-\frac{3}{2}\right)^2 + 6\left(-\frac{3}{2}\right)$$
$$= 3\left(\frac{9}{4}\right) + 6\left(-\frac{6}{4}\right)$$
$$= \frac{27}{4} - \frac{36}{4} = -\frac{9}{4}$$

Substitute -2 for x in equation (1) to find y.

$$y = 3x^2 + 6x \qquad (1)$$
$$y = 3(-2)^2 + 6(-2)$$
$$= 12 - 12 = 0$$

Solution set: $\left\{ \left(-\dfrac{3}{2}, -\dfrac{9}{4}\right), (-2, 0) \right\}$

31. $2x^2 - y^2 = 6$ (1)
$y = x^2 - 3$ (2)

Substitute $x^2 - 3$ for y in equation (1).

$$2x^2 - y^2 = 6 \quad (1)$$
$$2x^2 - (x^2 - 3)^2 = 6$$
$$2x^2 - (x^4 - 6x^2 + 9) = 6$$
$$-x^4 + 8x^2 - 9 = 6$$
$$-x^4 + 8x^2 - 15 = 0$$
$$x^4 - 8x^2 + 15 = 0 \qquad \textit{Multiply by } -1.$$

Let $z = x^2$, so $z^2 = x^4$.

$$z^2 - 8z + 15 = 0$$
$$(z - 3)(z - 5) = 0$$

$$z - 3 = 0 \quad \text{or} \quad z - 5 = 0$$
$$x = 3 \quad \text{or} \quad z = 5$$

Since $z = x^2$,

$$x^2 = 3 \quad \text{or} \quad x^2 = 5.$$

Use the square root property.

$$x = \pm\sqrt{3} \quad \text{or} \quad x = \pm\sqrt{5}$$

Use equation (2) to find y for each x-value.

If $x = \sqrt{3}$ or $-\sqrt{3}$, then

$$y = \left(\pm\sqrt{3}\right)^2 - 3 = 0.$$

(Note: We could substitute 3 for x^2 in (2) to obtain the same result.)

If $x = \sqrt{5}$ or $-\sqrt{5}$, then

$$y = \left(\pm\sqrt{5}\right)^2 - 3 = 2.$$

Solution set:

$$\left\{ \left(-\sqrt{3}, 0\right), \left(\sqrt{3}, 0\right), \left(-\sqrt{5}, 2\right), \left(\sqrt{5}, 2\right) \right\}$$

33. $3x^2 + 2y^2 = 12$ (1)

 $x^2 + 2y^2 = 4$ (2)

Multiply equation (2) by -1 and add the result to equation (1).

$$\begin{array}{rcll} 3x^2 & + & 2y^2 & = & 12 & (1) \\ -x^2 & - & 2y^2 & = & -4 & -1 \times (2) \\ \hline 2x^2 & & & = & 8 \\ & & x^2 & = & 4 \\ & & x & = & \pm 2 \end{array}$$

Substitute ± 2 for x in equation (2) to find y.

$$x^2 + 2y^2 = 4 \quad (2)$$
$$(\pm 2)^2 + 2y^2 = 4$$
$$4 + 2y^2 = 4$$
$$2y^2 = 0$$
$$y^2 = 0$$
$$y = 0$$

Solution set: $\{(-2, 0), (2, 0)\}$

35. $2x^2 + 3y^2 = 6$ (1)

 $x^2 + 3y^2 = 3$ (2)

Multiply equation (2) by -1 and add the result to equation (1).

$$\begin{array}{rcll} 2x^2 & + & 3y^2 & = & 6 & (1) \\ -x^2 & - & 3y^2 & = & -3 & -1 \times (2) \\ \hline x^2 & & & = & 3 \\ & & x & = & \pm \sqrt{3} \end{array}$$

Substitute $\pm \sqrt{3}$ for x in equation (2).

$$x^2 + 3y^2 = 3 \quad (2)$$
$$\left(\pm \sqrt{3}\right)^2 + 3y^2 = 3$$
$$3 + 3y^2 = 3$$
$$y^2 = 0$$
$$y = 0$$

Solution set: $\left\{\left(\sqrt{3}, 0\right), \left(-\sqrt{3}, 0\right)\right\}$

37. $2x^2 = 8 - 2y^2$ (1)

 $3x^2 = 24 - 4y^2$ (2)

Multiply equation (1) by -2 and add the result to equation (2).

$$\begin{array}{rcll} -4x^2 & = & -16 + 4y^2 & -2 \times (1) \\ 3x^2 & = & 24 - 4y^2 & (2) \\ \hline -x^2 & = & 8 \\ x^2 & = & -8 \\ x & = & \pm \sqrt{-8} = \pm 2i\sqrt{2} \end{array}$$

Substitute $\pm 2i\sqrt{2}$ for x in equation (1).

$$2x^2 = 8 - 2y^2$$
$$2\left(\pm 2i\sqrt{2}\right)^2 = 8 - 2y^2$$
$$2(-8) = 8 - 2y^2$$
$$-16 = 8 - 2y^2$$
$$2y^2 = 24$$
$$y^2 = 12$$
$$y = \pm \sqrt{12} = \pm 2\sqrt{3}$$

Since $2i\sqrt{2}$ can be paired with either $2\sqrt{3}$ or $-2\sqrt{3}$ and $-2i\sqrt{2}$ can be paired with either $2\sqrt{3}$ or $-2\sqrt{3}$, there are four possible solutions.

Solution set:

$$\left\{\left(-2i\sqrt{2}, -2\sqrt{3}\right), \left(-2i\sqrt{2}, 2\sqrt{3}\right),\right.$$
$$\left.\left(2i\sqrt{2}, -2\sqrt{3}\right), \left(2i\sqrt{2}, 2\sqrt{3}\right)\right\}$$

39. $x^2 + xy + y^2 = 15$ (1)

 $x^2 + y^2 = 10$ (2)

Multiply equation (2) by -1 and add the result to equation (1).

$$\begin{array}{rcll} x^2 + xy + y^2 & = & 15 & (1) \\ -x^2 - y^2 & = & -10 & -1 \times (2) \\ \hline xy & = & 5 \\ y & = & \dfrac{5}{x} \end{array}$$

Substitute $\dfrac{5}{x}$ for y in equation (2).

$$x^2 + y^2 = 10 \quad (2)$$
$$x^2 + \left(\frac{5}{x}\right)^2 = 10$$
$$x^2 + \frac{25}{x^2} = 10$$
$$x^4 + 25 = 10x^2 \quad \text{\textit{Multiply by} } x^2.$$
$$x^4 - 10x^2 + 25 = 0$$

Let $z = x^2$, so $z^2 = x^4$.

$$z^2 - 10z + 25 = 0$$
$$(z - 5)^2 = 0$$
$$z - 5 = 0$$
$$z = 5$$

Since $z = x^2$,

$$x^2 = 5, \text{ and } x = \pm \sqrt{5}.$$

Using the equation $y = \dfrac{5}{x}$, we get the following.

If $x = -\sqrt{5}$, then

continued

$$y = \frac{5}{-\sqrt{5}} = \frac{5 \cdot \sqrt{5}}{-\sqrt{5} \cdot \sqrt{5}} = \frac{5\sqrt{5}}{-5} = -\sqrt{5}.$$

Similarly, if $x = \sqrt{5}$, then $y = \sqrt{5}$.

Solution set: $\left\{ \left(-\sqrt{5}, -\sqrt{5} \right), \left(\sqrt{5}, \sqrt{5} \right) \right\}$

41. $3x^2 + 2xy - 3y^2 = 5 \quad (1)$
$-x^2 - 3xy + y^2 = 3 \quad (2)$

Multiply equation (2) by 3 and add the result to equation (1).

$$\begin{array}{rll} 3x^2 + 2xy - 3y^2 = & 5 & (1) \\ -3x^2 - 9xy + 3y^2 = & 9 & 3 \times (2) \\ \hline -7xy = & 14 & \end{array}$$

$$x = \frac{14}{-7y} = -\frac{2}{y}$$

Substitute $-\dfrac{2}{y}$ for x in equation (2).

$$-x^2 - 3xy + y^2 = 3 \qquad (2)$$

$$-\left(-\frac{2}{y} \right)^2 - 3\left(-\frac{2}{y} \right)y + y^2 = 3$$

$$-\left(\frac{4}{y^2} \right) + 6 + y^2 = 3$$

$$y^2 + 3 - \frac{4}{y^2} = 0$$

$$y^4 + 3y^2 - 4 = 0 \quad \textit{Multiply by } y^2.$$

$$(y^2 + 4)(y^2 - 1) = 0$$

$$\begin{array}{lcl} y^2 + 4 = 0 & \text{or} & y^2 - 1 = 0 \\ y^2 = -4 & & y^2 = 1 \\ y = \pm 2i & \text{or} & y = \pm 1 \end{array}$$

Since $x = -\dfrac{2}{y}$, substitute these values for y to find the values of x.

If $y = 2i$, then

$$x = -\frac{2}{2i} = -\frac{1}{i} = -\frac{1}{i} \cdot \frac{i}{i} = \frac{-i}{-1} = i.$$

If $y = -2i$, then

$$x = -\frac{2}{-2i} = \frac{1}{i} = \frac{1}{i} \cdot \frac{i}{i} = \frac{i}{-1} = -i.$$

If $y = 1$, then

$$x = -\frac{2}{1} = -2.$$

If $y = -1$, then

$$x = -\frac{2}{-1} = 2.$$

Solution set: $\left\{ (i, 2i), (-i, -2i), (2, -1), (-2, 1) \right\}$

43. $y = x^2 + 1 \quad (1)$
$x + y = 1 \quad (2)$

Substitute $x^2 + 1$ for y in equation (2) to get

$$x + (x^2 + 1) = 1$$
$$x^2 + x = 0$$
$$x(x + 1) = 0$$

$$x = 0 \quad \text{or} \quad x + 1 = 0$$
$$x = -1$$

From the graphing calculator screen, we already know one solution is $(0, 1)$.
Use equation (1) to find y for each $x = -1$.

$$y = (-1)^2 + 1 = 1 + 1 = 2$$

The other solution is $(-1, 2)$.

45. $y = \dfrac{1}{2}x^2 \quad (1)$
$x + y = 4 \quad (2)$

Replace y by $\dfrac{1}{2}x^2$ in equation (2).

$$x + \frac{1}{2}x^2 = 4$$
$$2x + x^2 = 8 \quad \textit{Multiply by 2.}$$
$$x^2 + 2x - 8 = 0$$
$$(x + 4)(x - 2) = 0$$

$$\begin{array}{lcl} x + 4 = 0 & \text{or} & x - 2 = 0 \\ x = -4 & \text{or} & x = 2 \end{array}$$

The graphing calculator screen shows the solution $(2, 2)$. If $x = -4$ in equation (1), then

$$y = \frac{1}{2}(-4)^2 = \frac{1}{2}(16) = 8,$$

and the other solution is $(-4, 8)$.

47. $xy = -6 \quad (1)$
$x + y = -1 \quad (2)$

Solve both equations for y.

$$y = -\frac{6}{x}$$
$$y = -x - 1$$

Graph $Y_1 = -\dfrac{6}{x}$ and $Y_2 = -x - 1$

on a graphing calculator to obtain the solution set $\left\{ (2, -3), (-3, 2) \right\}$.

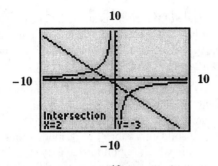

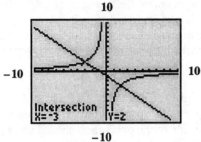

Now solve the system algebraically using substitution.

$$xy = -6 \quad (1)$$
$$x + y = -1 \quad (2)$$

Solve equation (2) for y.

$$y = -1 - x \quad (3)$$

Substitute $-1 - x$ for y in equation (1).

$$x(-1 - x) = -6$$
$$-x - x^2 = -6$$
$$0 = x^2 + x - 6$$
$$0 = (x + 3)(x - 2)$$

$$x + 3 = 0 \quad \text{or} \quad x - 2 = 0$$
$$x = -3 \quad \text{or} \quad x = 2$$

Substitute these values for x in (3) to find y.

If $x = -3$, then $y = -1 - (-3) = 2$.

If $x = 2$, then $y = -1 - 2 = -3$.

The solution set, $\{(2, -3), (-3, 2)\}$, is the same as that obtained using a graphing calculator.

49.
$$x^2 = 3x + 10$$
$$x^2 - 3x - 10 = 0$$
$$(x - 5)(x + 2) = 0$$

$$x - 5 = 0 \quad \text{or} \quad x + 2 = 0$$
$$x = 5 \quad \text{or} \quad x = -2$$

Solution set: $\{-2, 5\}$

50. The graph of $y = x^2$ is a parabola with vertex $(0, 0)$ that opens upward. The points $(1, 1)$, $(2, 4)$, and $(-2, 4)$ are also on the graph. The graph of

$y = 3x + 10$ is a line through intercepts $\left(-\dfrac{10}{3}, 0\right)$ and $(0, 10)$.

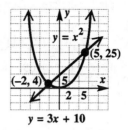

51. $y = x^2 \quad (1)$
$y = 3x + 10 \quad (2)$

Substitute x^2 for y in equation (2).

$$y = 3x + 10 \quad (2)$$
$$x^2 = 3x + 10$$
$$x^2 - 3x - 10 = 0$$
$$(x - 5)(x + 2) = 0$$

$$x - 5 = 0 \quad \text{or} \quad x + 2 = 0$$
$$x = 5 \quad \text{or} \quad x = -2$$

If $x = 5$, then $y = 5^2 = 25$.

If $x = -2$, then $y = (-2)^2 = 4$.

Solution set: $\{(-2, 4), (5, 25)\}$

52. They are exactly the same.

53. $y = x^2 - 3x - 10$

Complete the square.

$$\left[\frac{1}{2}(-3)\right]^2 = \left(-\frac{3}{2}\right)^2 = \frac{9}{4}$$
$$y = \left(x^2 - 3x + \frac{9}{4}\right) - \frac{9}{4} - 10$$
$$y = \left(x - \frac{3}{2}\right)^2 - \frac{49}{4}$$

This is the graph of a vertical parabola with vertex $\left(\dfrac{3}{2}, -\dfrac{49}{4}\right)$ that opens upward. The y-intercept is $(0, -10)$, and the x-intercepts are $(-2, 0)$ and $(5, 0)$.

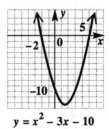

$y = x^2 - 3x - 10$

54. They are exactly the same.

55. Let $W =$ the width, and
$L =$ the length.

The formula for the area of a rectangle is $LW = A$, so

$$LW = 84. \quad (1)$$

The perimeter of a rectangle is given by $2L + 2W = P$, so

$$2L + 2W = 38. \quad (2)$$

Solve equation (2) for L to get

$$L = 19 - W. \quad (3)$$

Substitute $19 - W$ for L in equation (1).

$$
\begin{aligned}
LW &= 84 \quad (1) \\
(19 - W)W &= 84 \\
19W - W^2 &= 84 \\
-W^2 + 19W - 84 &= 0 \\
W^2 - 19W + 84 &= 0 \qquad \textit{Multiply by } -1. \\
(W - 7)(W - 12) &= 0
\end{aligned}
$$

$$
\begin{aligned}
W - 7 = 0 \quad &\text{or} \quad W - 12 = 0 \\
W = 7 \quad &\text{or} \qquad W = 12
\end{aligned}
$$

Using equation (3), with $W = 7$,

$$L = 19 - 7 = 12.$$

If $W = 12$, then $L = 7$, which are the same two numbers. Length must be greater than width, so the length is 12 feet and the width is 7 feet.

57. Men: $\quad y = .138x^2 + .064x + 451 \quad (1)$
Women: $y = 12.1x + 334 \qquad\qquad (2)$

Substitute $12.1x + 334$ for y in (1).
$12.1x + 334 = .138x^2 + .064x + 451$
$$0 = .138x^2 - 12.036x + 117$$
Use the quadratic formula.

$$x = \frac{-(-12.036) \pm \sqrt{(-12.036)^2 - 4(.138)(117)}}{2(.138)}$$

$$= \frac{12.036 \pm \sqrt{80.28196}}{.276}$$

These values are approximately 76 and 11. $x = 11$ corresponds to 1981 and $x = 76$ is out of the range for these models.

Substituting $x = 11$ into (2) gives us approximately 467 thousand bachelor's degrees awarded to women in 1981.

Section 13.4

1. $x^2 + y^2 < 25$
$y > -2$

The boundary, $x^2 + y^2 = 25$, is a dashed circle with center $(0, 0)$ and radius 5. When $(0, 0)$ is tested, a true statement, $0 < 25$, results, so the

inside of the circle is shaded. The graph of $y = -2$ is a dashed horizontal line through $(0, -2)$ with shading above the line, since $y > -2$. The correct answer is (c).

3. To graph the solution set of a nonlinear inequality, first graph the corresponding equality. This graph should be a dashed curve for $<$ or $>$ inequalities or a solid curve for $\leq$ or $\geq$ inequalities. Next, decide which region to shade by substituting any point not on the boundary (usually $(0, 0)$ is the easiest) into the inequality. If the statement is true, then shade that area. If the statement is false, then shade the other area.

5. $y \geq x^2 + 4$

This is an inequality whose boundary is a solid parabola, opening upward, with vertex $(0, 4)$. The inside of the parabola is shaded since $(0, 0)$ makes the inequality false. This is graph B.

7. $y < x^2 + 4$

This is an inequality whose boundary is a dashed parabola, opening upward, with vertex $(0, 4)$. The outside of the parabola is shaded since $(0, 0)$ makes the inequality true. This is graph A.

9.
$$
\begin{aligned}
y^2 &> 4 + x^2 \\
y^2 - x^2 &> 4 \\
\frac{y^2}{4} - \frac{x^2}{4} &> 1
\end{aligned}
$$

The boundary, $\dfrac{y^2}{4} - \dfrac{x^2}{4} = 1$, is a hyperbola with y-intercepts $(0, 2)$ and $(0, -2)$ and asymptotes formed by the diagonals of the fundamental rectangle with corners at $(2, 2)$, $(2, -2)$, $(-2, -2)$, and $(-2, 2)$. The hyperbola has dashed branches because of $>$. Test $(0, 0)$.

$$0^2 > 4 + 0^2 \quad ?$$
$$0 > 4 \qquad \textit{False}$$

Shade the sides of the hyperbola that do not contain $(0, 0)$. These are the regions inside the branches of the hyperbola.

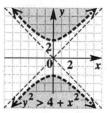

11. $y + 2 \geq x^2$
$y \geq x^2 - 2$
$y \geq (x - 0)^2 - 2$

Graph the solid vertical parabola $y = x^2 - 2$ with vertex $(0, -2)$. Two other points on the parabola are $(2, 2)$ and $(-2, 2)$. Test a point not on the parabola, say $(0,0)$, in $y \geq x^2 - 2$ to get $0 \geq -2$, a true statement. Shade that portion of the graph that contains the point $(0,0)$. This is the region inside the parabola.

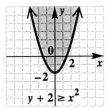

13. $$2y^2 \geq 8 - x^2$$
$$x^2 + 2y^2 \geq 8$$
$$\frac{x^2}{8} + \frac{y^2}{4} \geq 1$$

The boundary, $\frac{x^2}{8} + \frac{y^2}{4} = 1$, is the ellipse with intercepts $\left(2\sqrt{2}, 0\right)$, $\left(-2\sqrt{2}, 0\right)$, $(0, 2)$, and $(0, -2)$, drawn as a solid curve because of $\geq$. Test $(0,0)$.

$$2(0)^2 \geq 8 - 0^2 \quad ?$$
$$0 \geq 8 \qquad \textit{False}$$

Shade the region of the ellipse that does not contain $(0,0)$. This is the region outside the ellipse.

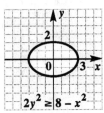

15. $y \leq x^2 + 4x + 2$

Graph the solid vertical parabola $y = x^2 + 4x + 2$. Use the vertex formula $x = \dfrac{-b}{2a}$ to obtain the vertex $(-2, -2)$. Two other points on the parabola are $(0, 2)$ and $(1, 7)$. Test a point not on the parabola, say $(0, 0)$, in $y \leq x^2 + 4x + 2$ to get $0 \leq 2$, a true statement. Shade outside the parabola, since this region contains $(0, 0)$.

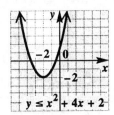

17. $$9x^2 > 16y^2 + 144$$
$$9x^2 - 16y^2 > 144$$
$$\frac{x^2}{16} - \frac{y^2}{9} > 1$$

The boundary, $\frac{x^2}{16} - \frac{y^2}{9} = 1$, is a hyperbola with x-intercepts $(4, 0)$ and $(-4, 0)$ and asymptotes formed by the diagonals of the fundamental rectangle with corners at $(4, 3)$, $(4, -3)$, $(-4, -3)$, and $(-4, 3)$. The hyperbola has dashed branches because of $>$. Test $(0, 0)$.

$$9(0)^2 > 16(0)^2 + 144 \quad ?$$
$$0 > 144 \qquad \textit{False}$$

Shade the sides of the hyperbola that do not contain $(0, 0)$. These are the regions inside the branches of the hyperbola.

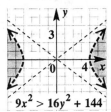

19. $$x^2 - 4 \geq -4y^2$$
$$x^2 + 4y^2 \geq 4$$
$$\frac{x^2}{4} + \frac{y^2}{1} \geq 1$$

Graph the solid ellipse $\frac{x^2}{4} + \frac{y^2}{1} = 1$ through the x-intercepts $(2, 0)$ and $(-2, 0)$ and y-intercepts $(0, 1)$ and $(0, -1)$. Test a point not on the ellipse, say $(0, 0)$, in $x^2 - 4 \geq -4y^2$ to get $-4 \geq 0$, a false statement. Shade outside the ellipse, since this region does *not* include $(0, 0)$.

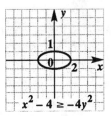

21. $x \leq -y^2 + 6y - 7$

Complete the square to find the vertex.

$$x = -y^2 + 6y - 7$$
$$\frac{x}{-1} = y^2 - 6y + 7$$
$$\frac{x}{-1} - 7 = y^2 - 6y$$
$$\frac{x}{-1} - 7 + 9 = y^2 - 6y + 9$$

continued

$$\frac{x}{-1} + 2 = (y - 3)^2$$
$$\frac{x}{-1} = (y - 3)^2 - 2$$
$$x = -(y - 3)^2 + 2$$

The boundary, $x = -(y - 3)^2 + 2$, is a solid horizontal parabola with vertex at $(2, 3)$ that opens to the left. Test $(0, 0)$.

$$0 \leq -0^2 + 6(0) - 7 \quad ?$$
$$0 \leq -7 \qquad \qquad \textit{False}$$

Shade the region of the parabola that does not contain $(0, 0)$. This is the region inside the parabola.

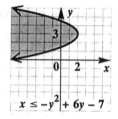

$$x \leq -y^2 + 6y - 7$$

23. $\quad 2x + 5y < 10$
$\qquad x - 2y < 4$

Graph $2x + 5y = 10$ as a dashed line through $(5, 0)$ and $(0, 2)$. Test $(0, 0)$.

$$2x + 5y < 10$$
$$2(0) + 5(0) < 10 \ ?$$
$$0 < 10 \ \textit{True}$$

Shade the region containing $(0, 0)$. Graph $x - 2y = 4$ as a dashed line through $(4, 0)$ and $(0, -2)$. Test $(0, 0)$.

$$x - 2y < 4$$
$$0 - 2(0) < 4 \ ?$$
$$0 < 4 \ \textit{True}$$

Shade the region containing $(0, 0)$. The graph of the system is the intersection of the two shaded regions.

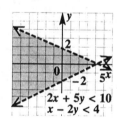

$$2x + 5y < 10$$
$$x - 2y < 4$$

25. $\quad 5x - 3y \leq 15$
$\qquad 4x + y \geq 4$

The boundary, $5x - 3y = 15$, is a solid line with intercepts $(3, 0)$ and $(0, -5)$. Test $(0, 0)$.

$$5(0) - 3(0) \leq 15 \ ?$$
$$0 \leq 15 \ \textit{True}$$

Shade the side of the line that contains $(0, 0)$. The boundary, $4x + y = 4$, is a solid line with intercepts $(1, 0)$ and $(0, 4)$. Test $(0, 0)$.

$$4(0) + 0 \geq 4 \ ?$$
$$0 \geq 4 \ \textit{False}$$

Shade the side of the line that does not contain $(0, 0)$.
The graph of the system is the intersection of the two shaded regions.

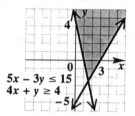

$$5x - 3y \leq 15$$
$$4x + y \geq 4$$

27. $\quad x \leq 5$
$\qquad y \leq 4$

Graph $x = 5$ as a solid vertical line through $(5, 0)$. Since $x \leq 5$, shade the left side of the line. Graph $y = 4$ as a solid horizontal line through $(0, 4)$. Since $y \leq 4$, shade below the line. The graph of the system is the intersection of the two shaded regions.

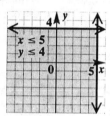

$$x \leq 5$$
$$y \leq 4$$

29. $\quad y > x^2 - 4$
$\qquad y < -x^2 + 3$

The boundary, $y = x^2 - 4$, is a dashed parabola with vertex $(0, -4)$ that opens upward. Test $(0, 0)$.

$$0 > 0^2 - 4 \qquad ?$$
$$0 > -4 \qquad \qquad \textit{True}$$

Shade the side of the parabola that contains $(0, 0)$. This is the region inside the parabola.
The boundary, $y = -x^2 + 3$, is a dashed parabola with vertex $(0, 3)$ that opens downward. Test $(0, 0)$.

$$0 < -0^2 + 3 \qquad ?$$
$$0 < 3 \qquad \qquad \textit{True}$$

Shade the side of the parabola that contains $(0, 0)$. This is the region inside the parabola.
The graph of the system is the intersection of the two shaded regions.

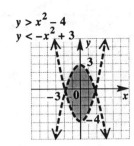

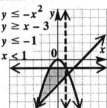

graph of the system.

31. $y^2 - x^2 \geq 4$
$-5 \leq y \leq 5$

Rewrite $y^2 - x^2 \geq 4$ as $\dfrac{y^2}{4} - \dfrac{x^2}{4} \geq 1$. Graph
the solid hyperbola through the y-intercepts $(0, 2)$
and $(0, -2)$. The asymptotes go through $(2, 2)$
and $(-2, -2)$ and through $(-2, 2)$ and $(2, -2)$.
Test $(0, 0)$ in $y^2 - x^2 \geq 4$ to get $0 \geq 4$, a false
statement. Shade above and below the two
branches of the hyperbola.
Graph solid horizontal lines through $(0, 5)$ and
$(0, -5)$. Since $-5 \leq y \leq 5$, shade the region
between the two horizontal lines.
The graph of the system is the intersection of the
shaded regions.

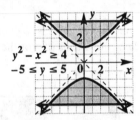

33. $y \leq -x^2$
$y \geq x - 3$
$y \leq -1$
$x < 1$

The boundary, $y = -x^2$, is a solid parabola with
vertex at $(0, 0)$ that opens downward. Test
$(0, -1)$.

$$-1 \leq -(0)^2 \quad ?$$
$$-1 \leq 0 \qquad \textit{True}$$

Shade the side of the parabola that contains
$(0, -1)$. This is the region inside the parabola.
The boundary, $y = x - 3$, is a solid line with
intercepts $(3, 0)$ and $(0, -3)$. Test $(0, 0)$.

$$0 \geq 0 - 3 \quad ?$$
$$0 \geq -3 \qquad \textit{True}$$

Shade the side of the line that contains $(0, 0)$.
For $y \leq -1$, shade below the solid horizontal line
$y = -1$.
For $x < 1$, shade to the left of the dashed vertical
line $x = 1$.
The intersection of the four shaded regions is the

35. $x^2 + y^2 > 36, x \geq 0$

This is a circle of radius 6 centered at the origin.
The graph is a dashed curve. Since $(0, 0)$ does not
satisfy the inequality, the shading will be outside
the circle. By including the restriction $x \geq 0$, we
consider only the shading to the right of, and
including, the y-axis.

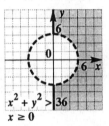

37. $x < y^2 - 3, x < 0$

Consider the equation.

$$x = y^2 - 3, \text{ or}$$
$$x = (y - 0)^2 - 3.$$

This is a parabola with vertex $(-3, 0)$, opening to
the right having the same shape as $x = y^2$. The
graph is a dashed curve. The shading will be
outside the parabola, since $(0, 0)$ does not satisfy
the inequality. By including the restriction $x < 0$,
we consider only the shading to the left of, but not
including, the y-axis. The y-axis is a dashed line.

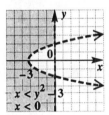

39. $4x^2 - y^2 > 16, x < 0$

Consider the equation

$$4x^2 - y^2 = 16, \text{ or}$$
$$\dfrac{x^2}{4} - \dfrac{y^2}{16} = 1.$$

This is a hyperbola with x-intercepts $(2, 0)$ and
$(-2, 0)$. The graph is a dashed curve. The
shading will be to the left and to the right of the
two branches of the hyperbola, since $(0, 0)$ does
not satisfy the inequality. By including the
restriction $x < 0$, we consider only the shading to
the left of the y-axis.   *continued*

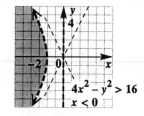

$4x^2 - y^2 > 16$
$x < 0$

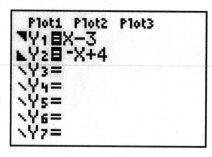

41. $x^2 + 4y^2 \geq 1$, $x \geq 0$, $y \geq 0$

Consider the equation

$$x^2 + 4y^2 = 1, \text{ or}$$

$$\frac{x^2}{1} + \frac{y^2}{\frac{1}{4}} = 1.$$

The graph is a solid ellipse with x-intercepts $(1, 0)$ and $(-1, 0)$ and y-intercepts $\left(0, \frac{1}{2}\right)$ and $\left(0, -\frac{1}{2}\right)$. The shading will be outside the ellipse, since $(0, 0)$ does not satisfy the inequality. By including the restrictions $x \geq 0$ and $y \geq 0$, we consider only the shading in quadrant I and portions of the x- and y-axis.

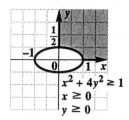

$x^2 + 4y^2 \geq 1$
$x \geq 0$
$y \geq 0$

43. $y > x^2 + 2$

$y = x^2 + 2$ is the graph of a parabola that opens upward and has been shifted up 2 units. The graph of $y > x^2 + 2$ consists of all points that are above the graph of $y = x^2 + 2$. The graph is shown in A.

45. $y > x^2 + 2$
$y < 5$

The graph of $y > x^2 + 2$ will be the region inside a parabola that opens upward and has been shifted two units up. The graph of $y < 5$ will be the region below the line $y = 5$. The graph of the system is choice B.

47. $y \geq x - 3$
$y \leq -x + 4$

The graphs of both inequalities include the points on the lines as part of the solution because of $\geq$ and $\leq$.
To produce the graph of the system on the TI-83, make the following Y-assignments:

To get the upper and lower darkened triangles to the left of Y_1 and Y_2, simply place the cursor in that spot and press $\boxed{\text{ENTER}}$ to cycle through the graphing choices.

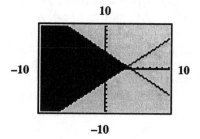

49. $y < x^2 + 4x + 4$
$y > -3$

The graphs do *not* include the points on the parabola or the line as part of the solution because of $<$ and $>$.

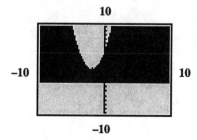

Chapter 13 Review Exercises

1.   Center $(-2, 4)$, radius 3
Here $h = -2$, $k = 4$, and $r = 3$, so the equation of the circle is

$$(x - h)^2 + (y - k)^2 = r^2$$
$$[x - (-2)]^2 + (y - 4)^2 = 3^2$$
$$(x + 2)^2 + (y - 4)^2 = 9.$$

2.   The graphed circle has center $(2, 1)$ and radius 1, so $h = 2$, $k = 1$, and $r = 1$. Then

$$(x - h)^2 + (y - k)^2 = r^2$$
$$(x - 2)^2 + (y - 1)^2 = 1^2$$
$$(x - 2)^2 + (y - 1)^2 = 1.$$

3. $x^2 + y^2 + 6x - 4y - 3 = 0$

Write the equation in
$$(x - h)^2 + (y - k)^2 = r^2$$
form by completing the squares on x and y.
$$\left(x^2 + 6x \quad\right) + \left(y^2 - 4y \quad\right) = 3$$
$$\left(x^2 + 6x + \underline{9}\right) + \left(y^2 - 4y + \underline{4}\right)$$
$$= 3 + \underline{9} + \underline{4}$$
$$(x + 3)^2 + (y - 2)^2 = 16$$
$$[x - (-3)]^2 + (y - 2)^2 = 16$$

The circle has center (h, k) at $(-3, 2)$ and radius
$\sqrt{16} = 4$.

4. $x^2 + y^2 - 8x - 2y + 13 = 0$

Write the equation in
$$(x - h)^2 + (y - k)^2 = r^2$$
form by completing the squares on x and y.
$$\left(x^2 - 8x \quad\right) + \left(y^2 - 2y \quad\right) = -13$$
$$\left(x^2 - 8x + \underline{16}\right) + \left(y^2 - 2y + \underline{1}\right)$$
$$= -13 + \underline{16} + \underline{1}$$
$$(x - 4)^2 + (y - 1)^2 = 4$$

The circle has center (h, k) at $(4, 1)$ and radius
$\sqrt{4} = 2$.

5. $2x^2 + 2y^2 + 4x + 20y = -34$
$$x^2 + y^2 + 2x + 10y = -17$$

Write the equation in
$$(x - h)^2 + (y - k)^2 = r^2$$
form by completing the squares on x and y.
$$\left(x^2 + 2x \quad\right) + \left(y^2 + 10y \quad\right) = -17$$
$$\left(x^2 + 2x + \underline{1}\right) + \left(y^2 + 10y + \underline{25}\right)$$
$$= -17 + \underline{1} + \underline{25}$$
$$(x + 1)^2 + (y + 5)^2 = 9$$
$$[x - (-1)]^2 + [y - (-5)]^2 = 9$$

The circle has center (h, k) at $(-1, -5)$ and radius
$\sqrt{9} = 3$.

6. $4x^2 + 4y^2 - 24x + 16y = 48$
$$x^2 + y^2 - 6x + 4y = 12$$

Write the equation in
$$(x - h)^2 + (y - k)^2 = r^2$$
form by completing the squares on x and y.
$$\left(x^2 - 6x \quad\right) + \left(y^2 + 4y \quad\right) = 12$$
$$\left(x^2 - 6x + \underline{9}\right) + \left(y^2 + 4y + \underline{4}\right)$$
$$= 12 + \underline{9} + \underline{4}$$
$$(x - 3)^2 + (y + 2)^2 = 25$$
$$(x - 3)^2 + [y - (-2)]^2 = 25$$

The circle has center (h, k) at $(3, -2)$ and radius
$\sqrt{25} = 5$.

7. $\dfrac{x^2}{16} + \dfrac{y^2}{9} = 1$ is in $\dfrac{x^2}{a^2} + \dfrac{y^2}{b^2} = 1$ form with
$a = 4$ and $b = 3$. The graph is an ellipse with
x-intercepts $(4, 0)$ and $(-4, 0)$ and y-intercepts
$(0, 3)$ and $(0, -3)$. Plot the intercepts, and draw
the ellipse through them.

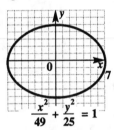

8. $\dfrac{x^2}{49} + \dfrac{y^2}{25} = 1$ is in $\dfrac{x^2}{a^2} + \dfrac{y^2}{b^2} = 1$ form with
$a = 7$ and $b = 5$. The graph is an ellipse with
x-intercepts $(7, 0)$ and $(-7, 0)$ and y-intercepts
$(0, 5)$ and $(0, -5)$. Plot the intercepts, and draw
the ellipse through them.

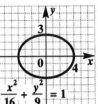

9. $\dfrac{x^2}{16} - \dfrac{y^2}{25} = 1$ is in $\dfrac{x^2}{a^2} - \dfrac{y^2}{b^2} = 1$ form with
$a = 4$ and $b = 5$. The graph is a hyperbola with
x-intercepts $(4, 0)$ and $(-4, 0)$ and asymptotes that
are the extended diagonals of the rectangle with
corners $(4, 5)$, $(4, -5)$, $(-4, -5)$, and $(-4, 5)$.
Graph a branch of the hyperbola through each
intercept approaching the asymptotes.

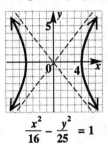

10. $\dfrac{y^2}{25} - \dfrac{x^2}{4} = 1$ is in $\dfrac{y^2}{b^2} - \dfrac{x^2}{a^2} = 1$ form with
$a = 2$ and $b = 5$. The graph is a hyperbola with
y-intercepts $(0, 5)$ and $(0, -5)$ and asymptotes that
are the extended diagonals of the rectangle with
corners $(2, 5)$, $(2, -5)$, $(-2, -5)$, and $(-2, 5)$.
Graph a branch of the hyperbola through each
intercept approaching the asymptotes.   *continued*

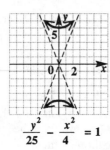

$$\frac{y^2}{25} - \frac{x^2}{4} = 1$$

11. $x^2 + 9y^2 = 9$ in $\frac{x^2}{a^2} + \frac{y^2}{b^2} = 1$ form is

$\frac{x^2}{9} + \frac{y^2}{1} = 1$ with $a = 3$ and $b = 1$. The graph

is an ellipse with x-intercepts $(3, 0)$ and $(-3, 0)$
and y-intercepts $(0, 1)$ and $(0, -1)$. Plot the
intercepts, and draw the ellipse through them.

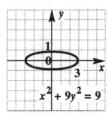

12. $f(x) = \sqrt{4 + x^2}$

Replace $f(x)$ with y.

$$y = \sqrt{4 + x^2}$$

Square both sides.

$$y^2 = 4 + x^2$$
$$y^2 - x^2 = 4$$
$$\frac{y^2}{4} - \frac{x^2}{4} = 1$$

The graph is a hyperbola that opens vertically with
vertices $(0, 2)$ and $(0, -2)$. The original equation
$f(x) = y = \sqrt{4 + x^2}$ indicates that y must be
positive. Therefore, only the upper half of the
hyperbola is part of the graph.

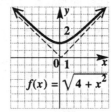

13. $(x - 2)^2 + (y + 3)^2 = 16$
$(x - 2)^2 + [y - (-3)]^2 = 4^2$

The graph is a circle with center $(2, -3)$ and
radius 4.

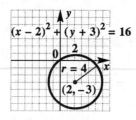

14. $y = 2x^2 - 3$
$y = 2(x - 0)^2 - 3$

The last equation is in $y = a(x - h)^2 + k$ form.
The graph is a parabola.

15. $y^2 = 2x^2 - 8$
$2x^2 - y^2 = 8$
$\frac{x^2}{4} - \frac{y^2}{8} = 1$

The last equation is in $\frac{x^2}{a^2} - \frac{y^2}{b^2} = 1$ form, so the

graph is a hyperbola.

16. $y^2 = 8 - 2x^2$
$2x^2 + y^2 = 8$
$\frac{x^2}{4} + \frac{y^2}{8} = 1$

The last equation is in $\frac{x^2}{a^2} + \frac{y^2}{b^2} = 1$ form, so the

graph is an ellipse.

17. $x = y^2 + 4$
$x = (y - 0)^2 + 4$

The last equation is in $x = a(y - k)^2 + h$ form,
so the graph is a parabola.

18. $x^2 + y^2 = 64$
$(x - 0)^2 + (y - 0)^2 = 8^2$

The last equation is in $(x - h)^2 + (y - k)^2 = r^2$
form. The graph is a circle.

19. The total distance on the horizontal axis is
$160 + 16,000 = 16,160$ km. This represents $2a$,

so $a = \frac{1}{2}(16,160) = 8080$. The distance from

Earth to the center of the ellipse is

$$8080 - 160 = 7920,$$

which is the value of c. From Exercise 47 in
Section 13.1, we know that $c^2 = a^2 - b^2$, so

$$b^2 = a^2 - c^2.$$
$$b^2 = 8080^2 - 7920^2$$
$$= 2,560,000$$

Thus, $b = \sqrt{2,560,000} = 1600$ and the equation
is

$$\frac{x^2}{8080^2} + \frac{y^2}{1600^2} = 1$$

$$\text{or } \frac{x^2}{65,286,400} + \frac{y^2}{2,560,000} = 1.$$

20. This hyperbola has center at the origin with foci $(-50, 0)$ and $(50, 0)$. The difference between the distances from any point on the hyperbola to the two foci is equal to $80 - 30 = 50$. From Exercise 49 in Section 13.2, the foci are $(-c, 0)$ and $(c, 0)$. Hence, $c = 50$. Since the difference between the distances from any points to the two foci is $2a$, $2a = 50$, or $a = 25$. Since

$$b^2 = c^2 - a^2,$$
$$b^2 = 50^2 - 25^2$$
$$= 2500 - 625 = 1875.$$

Since the equation of a hyperbola is

$$\frac{x^2}{a^2} - \frac{y^2}{b^2} = 1,$$

the equation for this hyperbola is

$$\frac{x^2}{25^2} - \frac{y^2}{1875} = 1 \text{ or } \frac{x^2}{625} - \frac{y^2}{1875} = 1.$$

21.
$$2y = 3x - x^2 \quad (1)$$
$$x + 2y = -12 \quad \quad (2)$$

Substitute $3x - x^2$ for $2y$ in equation (2).

$$x + 2y = -12 \quad (2)$$
$$x + (3x - x^2) = -12$$
$$-x^2 + 4x + 12 = 0$$
$$x^2 - 4x - 12 = 0$$
$$(x - 6)(x + 2) = 0$$

$$x - 6 = 0 \quad \text{or} \quad x + 2 = 0$$
$$x = 6 \quad \text{or} \quad x = -2$$

Substitute these values for x in equation (2) to find y.

If $x = 6$, then
$$x + 2y = -12 \quad (2)$$
$$6 + 2y = -12$$
$$2y = -18$$
$$y = -9.$$

If $x = -2$ then
$$x + 2y = -12 \quad (2)$$
$$-2 + 2y = -12$$
$$2y = -10$$
$$y = -5$$

Solution set: $\{(6, -9), (-2, -5)\}$

22.
$$y + 1 = x^2 + 2x \quad (1)$$
$$y + 2x = 4 \quad \quad (2)$$

Solve equation (2) for y.

$$y = 4 - 2x \quad (3)$$

Substitute $4 - 2x$ for y in (1).

$$(4 - 2x) + 1 = x^2 + 2x$$
$$0 = x^2 + 4x - 5$$
$$0 = (x + 5)(x - 1)$$

$$x + 5 = 0 \quad \text{or} \quad x - 1 = 0$$
$$x = -5 \quad \text{or} \quad x = 1$$

Substitute these values for x in equation (3) to find y.

If $x = -5$, then $y = 4 - 2(-5) = 14$.
If $x = 1$, then $y = 4 - 2(1) = 2$.

Solution set: $\{(1, 2), (-5, 14)\}$

23.
$$x^2 + 3y^2 = 28 \quad (1)$$
$$y - x = -2 \quad \quad (2)$$

Solve equation (2) for y.

$$y = x - 2$$

Substitute $x - 2$ for y in equation (1).

$$x^2 + 3y^2 = 28 \quad (1)$$
$$x^2 + 3(x - 2)^2 = 28$$
$$x^2 + 3(x^2 - 4x + 4) - 28 = 0$$
$$x^2 + 3x^2 - 12x + 12 - 28 = 0$$
$$4x^2 - 12x - 16 = 0$$
$$4(x^2 - 3x - 4) = 0$$
$$4(x - 4)(x + 1) = 0$$

$$x - 4 = 0 \quad \text{or} \quad x + 1 = 0$$
$$x = 4 \quad \text{or} \quad x = -1$$

Since $y = x - 2$, if $x = 4$, then $y = 4 - 2 = 2$.

If $x = -1$, then $y = -1 - 2 = -3$.

Solution set: $\{(4, 2), (-1, -3)\}$

24.
$$xy = 8 \quad (1)$$
$$x - 2y = 6 \quad (2)$$

Solve equation (2) for x.

$$x = 2y + 6 \quad (3)$$

Substitute $2y + 6$ for x in equation (1) to find y.

$$xy = 8 \quad (1)$$
$$(2y + 6)y = 8$$
$$2y^2 + 6y - 8 = 0$$
$$2(y^2 + 3y - 4) = 0$$
$$2(y + 4)(y - 1) = 0$$

$$y + 4 = 0 \quad \text{or} \quad y - 1 = 0$$
$$y = -4 \quad \text{or} \quad y = 1$$

continued

Substitute these values for y in equation (3) to find x.

If $y = -4$, then $x = 2(-4) + 6 = -2$.

If $y = 1$, then $x = 2(1) + 6 = 8$.

Solution set: $\{(-2, -4), (8, 1)\}$

25. $x^2 + y^2 = 6$ (1)

$x^2 - 2y^2 = -6$ (2)

Multiply equation (2) by -1 and add the result to equation (1).

$$
\begin{array}{rcll}
x^2 + y^2 &=& 6 & (1) \\
-x^2 + 2y^2 &=& 6 & -1 \times (2) \\
\hline
3y^2 &=& 12 & \\
y^2 &=& 4 &
\end{array}
$$

$$y = 2 \quad \text{or} \quad y = -2$$

Substitute these values for y in equation (1) to find x.

If $y = \pm 2$, then

$$
\begin{aligned}
x^2 + y^2 &= 6 \quad (1) \\
x^2 + (\pm 2)^2 &= 6 \\
x^2 + 4 &= 6 \\
x^2 &= 2.
\end{aligned}
$$

$$x = \sqrt{2} \quad \text{or} \quad x = -\sqrt{2}$$

Since each value of x can be paired with each value of y, there are four points and the solution set is

$$\left\{ \left(\sqrt{2}, 2\right), \left(-\sqrt{2}, 2\right), \left(\sqrt{2}, -2\right), \left(-\sqrt{2}, -2\right) \right\}.$$

26. $3x^2 - 2y^2 = 12$ (1)

$x^2 + 4y^2 = 18$ (2)

Multiply equation (1) by 2 and add the result to equation (2).

$$
\begin{array}{rcll}
6x^2 - 4y^2 &=& 24 & 2 \times (1) \\
x^2 + 4y^2 &=& 18 & (2) \\
\hline
7x^2 &=& 42 & \\
x^2 &=& 6 &
\end{array}
$$

$$x = \sqrt{6} \quad \text{or} \quad x = -\sqrt{6}$$

Substitute these values for x in equation (2) to find y.

If $x = \pm\sqrt{6}$, then

$$
\begin{aligned}
x^2 + 4y^2 &= 18. \quad (2) \\
\left(\pm\sqrt{6}\right)^2 + 4y^2 &= 18 \\
6 + 4y^2 &= 18 \\
4y^2 &= 12 \\
y^2 &= 3
\end{aligned}
$$

$$y = \sqrt{3} \quad \text{or} \quad y = -\sqrt{3}$$

The solution set is

$$\left\{ \left(\sqrt{6}, \sqrt{3}\right), \left(\sqrt{6}, -\sqrt{3}\right), \right.$$
$$\left. \left(-\sqrt{6}, \sqrt{3}\right), \left(-\sqrt{6}, -\sqrt{3}\right) \right\}.$$

27. A circle and a line can intersect in zero, one, or two points, so zero, one, or two solutions are possible.

28. A parabola and a hyperbola can intersect in zero, one, two, three, or four points, so zero, one, two, three, or four solutions are possible.

29. The graph shows that in about 1950 the altitude was reduced to approximately 4005 feet above sea level, causing the salinity to increase to about 6000 milligrams per liter.

30. $$9x^2 \geq 16y^2 + 144$$
$$9x^2 - 16y^2 \geq 144$$
$$\frac{x^2}{16} - \frac{y^2}{9} \geq 1$$

The boundary, $\frac{x^2}{16} - \frac{y^2}{9} = 1$, is a solid hyperbola with x-intercepts $(4, 0)$ and $(-4, 0)$. The asymptotes are the extended diagonals of the rectangle with corners $(4, 3)$, $(4, -3)$, $(-4, -3)$, and $(-4, 3)$. Test $(0, 0)$.

$$9(0)^2 \geq 16(0)^2 + 144 \quad ?$$
$$0 \geq 144 \qquad \text{False}$$

Shade the sides of the hyperbola that do not contain $(0, 0)$. These are the regions inside the branches of the hyperbola.

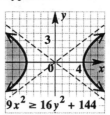

31. $$4x^2 + y^2 \geq 16$$
$$\frac{x^2}{4} + \frac{y^2}{16} \geq 1$$

The boundary, $\frac{x^2}{4} + \frac{y^2}{16} = 1$, is a solid ellipse with intercepts $(2, 0)$, $(-2, 0)$, $(0, 4)$, and $(0, -4)$. Test $(0, 0)$.

$$4(0)^2 + 0^2 \geq 16 \quad ?$$
$$0 \geq 16 \qquad \text{False}$$

Shade the side of ellipse that does not contain $(0, 0)$. This is the region outside the ellipse.

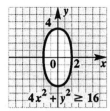

32. $y < -(x+2)^2 + 1$

The boundary, $y = -(x+2)^2 + 1$, is a dashed vertical parabola with vertex $(-2, 1)$. Since $a = -1 < 0$, the parabola opens downward. Also, $|a| = |-1| = 1$, so the graph has the same shape as the graph of $y = x^2$. Test $(0, 0)$.

$$0 < -(0+2)^2 + 1 \quad ?$$
$$0 < -(4) + 1 \quad ?$$
$$0 < -3 \qquad \textit{False}$$

Shade the side of the parabola that does not contain $(0, 0)$. This is the region inside the parabola.

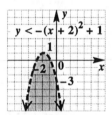

33. $\begin{aligned} 2x + 5y &\le 10 \\ 3x - y &\le 6 \end{aligned}$

The boundary, $2x + 5y = 10$ is a solid line with intercepts $(5, 0)$ and $(0, 2)$. Test $(0, 0)$.

$$2(0) + 5(0) \le 10 \quad ?$$
$$0 \le 10 \qquad \textit{True}$$

Shade the side of the line that contains $(0, 0)$. The boundary, $3x - y = 6$, is a solid line with intercepts $(2, 0)$ and $(0, -6)$. Test $(0, 0)$.

$$3(0) - 0 \le 6 \quad ?$$
$$0 \le 6 \qquad \textit{True}$$

Shade the side of the line that contains $(0, 0)$. The graph of the system is the intersection of the two shaded regions.

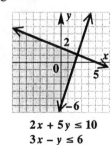

34. $\begin{aligned} |x| &\le 2 \\ |y| &> 1 \\ 4x^2 + 9y^2 &\le 36 \end{aligned}$

The equation of the boundary, $|x| = 2$, can be written as

$$x = -2 \quad \text{or} \quad x = 2.$$

The graph is these two solid vertical lines. Since $0 \le 2$ is true, the region between the lines, containing $(0, 0)$, is shaded.

The boundary, $|y| = 1$, consists of the two dashed horizontal lines $y = 1$ and $y = -1$. Since $0 > 1$ is false, the regions above and below the lines, not containing $(0, 0)$, are shaded.

The boundary given by

$$4x^2 + 9y^2 = 36$$
$$\text{or} \quad \frac{x^2}{9} + \frac{y^2}{4} = 1$$

is graphed as a solid ellipse with intercepts $(3, 0)$, $(-3, 0)$, $(0, 2)$, and $(0, -2)$. Test $(0, 0)$.

$$4(0)^2 + 9(0)^2 \le 36 \quad ?$$
$$0 \le 36 \qquad \textit{True}$$

The region inside the ellipse, containing $(0, 0)$, is shaded.

The graph of the system consists of the regions that include the common points of the three shaded regions.

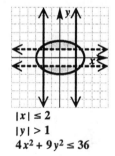

35. $\begin{aligned} 9x^2 &\le 4y^2 + 36 \\ x^2 + y^2 &\le 16 \end{aligned}$

The equation of the first boundary is

$$9x^2 = 4y^2 + 36$$
$$9x^2 - 4y^2 = 36$$
$$\frac{x^2}{4} - \frac{y^2}{9} = 1.$$

The graph is a solid hyperbola with x-intercepts $(2, 0)$ and $(-2, 0)$. The asymptotes are the extended diagonals of the rectangle with corners $(2, 3)$, $(2, -3)$, $(-2, -3)$, and $(-2, 3)$. Test $(0, 0)$.

continued

$$9(0)^2 \le 4(0)^2 + 36 \quad ?$$
$$0 \le 36 \qquad \text{True}$$

Shade the region between the branches of the hyperbola that contains $(0, 0)$.

The equation of the second boundary is $x^2 + y^2 = 16$. This is a solid circle with center $(0, 0)$ and radius 4. Test $(0, 0)$.

$$0^2 + 0^2 \le 16 \quad ?$$
$$0 \le 16 \qquad \text{True}$$

Shade the region inside the circle.

The graph of the system is the intersection of the shaded regions which is between the two branches of the hyperbola and inside the circle.

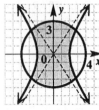

$$9x^2 \le 4y^2 + 36$$
$$x^2 + y^2 \le 16$$

36. $\dfrac{x^2}{64} + \dfrac{y^2}{25} = 1$ is in $\dfrac{x^2}{a^2} + \dfrac{y^2}{b^2} = 1$ form with $a = 8$ and $b = 5$. The graph is an ellipse with intercepts $(8, 0)$, $(-8, 0)$, $(0, 5)$, and $(0, -5)$. Plot the intercepts, and draw the ellipse through them.

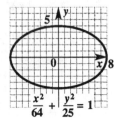

$$\frac{x^2}{64} + \frac{y^2}{25} = 1$$

37. $\dfrac{y^2}{4} - 1 = \dfrac{x^2}{9}$

$$\frac{y^2}{4} - \frac{x^2}{9} = 1$$

The equation is in $\dfrac{y^2}{b^2} - \dfrac{x^2}{a^2} = 1$ form with $a = 3$ and $b = 2$. The graph is a hyperbola with y-intercepts $(0, 2)$ and $(0, -2)$ and asymptotes that are the extended diagonals of the rectangle with corners $(3, 2)$, $(3, -2)$, $(-3, -2)$, and $(-3, 2)$. Draw a branch of the hyperbola through each intercept and approaching the asymptotes.

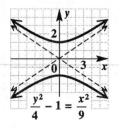

$$\frac{y^2}{4} - 1 = \frac{x^2}{9}$$

38. $x^2 + y^2 = 25$ is in $(x - h)^2 + (y - k)^2 = r^2$ form. The graph is a circle with center at $(0, 0)$ and radius 5.

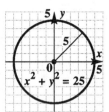

$$x^2 + y^2 = 25$$

39. $y = 2(x - 2)^2 - 3$ is in $y = a(x - h)^2 + k$ form. The graph is a parabola with vertex $(2, -3)$. Since $a = 2 > 0$, the parabola opens upward. Also, $|a| = |2| = 2 > 1$, so the graph is narrower than the graph of $y = x^2$. The points $(0, 5)$ and $(4, 5)$ are on the graph.

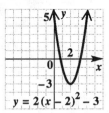

$$y = 2(x - 2)^2 - 3$$

40. $$f(x) = -\sqrt{16 - x^2}$$

Replace $f(x)$ with y.

$$y = -\sqrt{16 - x^2}$$

Square both sides.

$$y^2 = 16 - x^2$$
$$x^2 + y^2 = 16$$

This equation is the graph of a circle with center $(0, 0)$ and radius 4. Since $f(x)$ represents a nonpositive square root, $f(x)$ is nonpositive and its graph is the lower half of the circle.

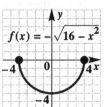

$$f(x) = -\sqrt{16 - x^2}$$

41.
$$f(x) = \sqrt{4 - x}$$

Replace $f(x)$ with y.
$$y = \sqrt{4 - x}$$

Square both sides.
$$y^2 = 4 - x$$
$$x = -y^2 + 4$$
$$x = -1(y - 0)^2 + 4$$

This equation is the graph of a horizontal parabola with vertex $(4, 0)$. Since $a = -1 < 0$, the graph opens to the left. Also, $|a| = |-1| = 1$, so the graph has the same shape as the graph of $y = x^2$. The points $(0, 2)$ and $(3, 1)$ are on the graph. Since $f(x)$ represents a square root, $f(x)$ is nonnegative and its graph is the upper half of the parabola.

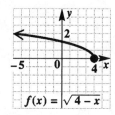

42. $3x + 2y \geq 0$
$$y \leq 4$$
$$x \leq 4$$

The boundary $3x + 2y = 0$ is a solid line through $(0, 0)$ and $(2, -3)$. Test $(0, 1)$.
$$3(0) + 2(1) \geq 0 \qquad ?$$
$$2 \geq 0 \qquad True$$

Shade the side of the line that contains $(0, 1)$. The boundary $y = 4$ is a solid horizontal line through $(0, 4)$. Since $y \leq 4$, shade below the line. The boundary $x = 4$ is a solid vertical line through $(4, 0)$. Since $x \leq 4$, shade the region to the left of the line.
The graph of the system is the intersection of the three shaded regions.

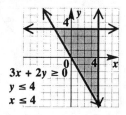

43. $4y > 3x - 12$
$$x^2 < 16 - y^2$$

The boundary $4y = 3x - 12$ is a dashed line with intercepts $(0, -3)$ and $(4, 0)$. Test $(0, 0)$.
$$4(0) > 3(0) - 12 \qquad ?$$
$$0 > -12 \qquad True$$

Shade the side of the line that contains $(0, 0)$. The boundary $x^2 = 16 - y^2$, or $x^2 + y^2 = 16$, is a dashed circle with center at $(0, 0)$ and radius 4. Test $(0, 0)$.
$$0^2 < 16 - 0^2 \qquad ?$$
$$0 < 16 \qquad True$$

Shade the region inside the circle.
The graph of the system is the intersection of the two shaded regions.

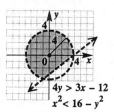

44. $x^2 + y^2 + ax + by + c = 0$

Let $x = 2$ and $y = 4$.
$$2^2 + 4^2 + a(2) + b(4) + c = 0$$
$$4 + 16 + 2a + 4b + c = 0$$
$$2a + 4b + c = -20$$

45. $x^2 + y^2 + ax + by + c = 0$

Let $x = 5$ and $y = 1$.
$$5^2 + 1^2 + a(5) + b(1) + c = 0$$
$$25 + 1 + 5a + b + c = 0$$
$$5a + b + c = -26$$

46. $x^2 + y^2 + ax + by + c = 0$

Let $x = -1$ and $y = 1$.
$$(-1)^2 + 1^2 + a(-1) + b(1) + c = 0$$
$$1 + 1 - a + b + c = 0$$
$$-a + b + c = -2$$

47.
$$2a + 4b + c = -20 \quad (1)$$
$$5a + b + c = -26 \quad (2)$$
$$-a + b + c = -2 \quad (3)$$

Multiply equation (3) by 2 and add the result to equation (1).
$$2a + 4b + c = -20 \quad (1)$$
$$\underline{-2a + 2b + 2c = -4} \qquad 2 \times (3)$$
$$6b + 3c = -24 \quad (4)$$

Multiply equation (3) by 5 and add the result to equation (2).
$$5a + b + c = -26 \quad (2)$$
$$\underline{-5a + 5b + 5c = -10} \qquad 5 \times (3)$$
$$6b + 6c = -36 \quad (5)$$

Multiply equation (4) by -1 and add the result to equation (5).

continued

$$-6b - 3c = 24 \qquad -1 \times (4)$$
$$\underline{6b + 6c = -36 \quad (5)}$$
$$3c = -12$$
$$c = -4$$

Substitute $c = -4$ into equation (4), and solve for b.

$$6b + 3c = -24 \quad (4)$$
$$6b + 3(-4) = -24$$
$$6b - 12 = -24$$
$$6b = -12$$
$$b = -2$$

Substitute $c = -4$ and $b = -2$ into equation (3), and solve for a.

$$-a + b + c = -2 \quad (3)$$
$$-a + (-2) + (-4) = -2$$
$$-a - 6 = -2$$
$$-a = 4$$
$$a = -4$$

The solution set is $\{(-4, -2, -4)\}$.

Therefore, the equation of the circle is

$$x^2 + y^2 - 4x - 2y - 4 = 0.$$

48. $x^2 + y^2 - 4x - 2y - 4 = 0$
$$\left(x^2 - 4x \quad\right) + \left(y^2 - 2y \quad\right) = 4$$

Complete the square for x and y.

$$\left(x^2 - 4x + \underline{4}\right) + \left(y^2 - 2y + \underline{1}\right) = 4 + \underline{4} + \underline{1}$$
$$(x - 2)^2 + (y - 1)^2 = 9$$

The center is $(2, 1)$ and the radius is $\sqrt{9} = 3$.

49. From Exercise 48, the equation

$$x^2 + y^2 - 4x - 2y - 4 = 0$$

is a circle with center $(2, 1)$ and radius 3.

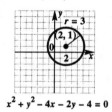

$$x^2 + y^2 - 4x - 2y - 4 = 0$$

Chapter 13 Test

1. The circle has center (h, k) at $(-4, 4)$ and radius 4. Here, $h = -4$, $k = 4$, and $r = 4$, so an equation is

$$(x - h)^2 + (y - k)^2 = r^2$$
$$[x - (-4)]^2 + (y - 4)^2 = 4^2$$
$$(x + 4)^2 + (y - 4)^2 = 16.$$

2. $x^2 + y^2 + 8x - 2y = 8$

To find the center and radius, complete the squares on x and y.

$$\left(x^2 + 8x \quad\right) + \left(y^2 - 2y \quad\right) = 8$$
$$\left(x^2 + 8x + \underline{16}\right) + \left(y^2 - 2y + \underline{1}\right) = 8 + \underline{16} + \underline{1}$$
$$(x + 4)^2 + (y - 1)^2 = 25$$

The graph is a circle with center $(-4, 1)$ and radius $\sqrt{25} = 5$.

3. $3x^2 + 3y^2 = 27$
$$x^2 + y^2 = 9 \quad \textit{Divide by 3.}$$

This is a *circle* centered at the origin with radius $\sqrt{9} = 3$.

4. $9x^2 + 4y^2 = 36$
$$\frac{x^2}{4} + \frac{y^2}{9} = 1 \quad \textit{Divide by 36.}$$

This is an *ellipse* with intercepts $(2, 0)$, $(-2, 0)$, $(0, 3)$, and $(0, -3)$.

5. $$9x^2 = 36 + 4y^2$$
$$9x^2 - 4y^2 = 36$$
$$\frac{x^2}{4} - \frac{y^2}{9} = 1 \qquad \textit{Divide by 36.}$$

This is a *hyperbola* that opens right and left with vertices $(2, 0)$ and $(-2, 0)$.

6. $x = 36 - 4y^2$ or $x = -4y^2 + 36$

This is a *parabola* that opens to the left. Its vertex is $(36, 0)$.

7. $x^2 + y^2 = 64$

This is a circle centered at the origin with radius $\sqrt{64} = 8$.

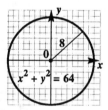

8. $4x^2 + 9y^2 = 36$
$$\frac{x^2}{9} + \frac{y^2}{4} = 1$$

The equation is in $\dfrac{x^2}{a^2} + \dfrac{y^2}{b^2} = 1$ form with $a = 3$ and $b = 2$. The graph is an ellipse with intercepts $(3, 0)$, $(-3, 0)$, $(0, 2)$, and $(0, -2)$. Plot these intercepts, and draw the ellipse through them.

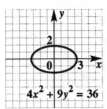

9. $16y^2 - 4x^2 = 64$

$$\frac{y^2}{4} - \frac{x^2}{16} = 1$$

The equation is in $\dfrac{y^2}{b^2} - \dfrac{x^2}{a^2} = 1$ form with $a = 4$ and $b = 2$. The graph is a hyperbola with y-intercepts $(0, 2)$ and $(0, -2)$ and asymptotes that are the extended diagonals of the rectangle with corners $(4, 2)$, $(4, -2)$, $(-4, -2)$, and $(-4, 2)$. Draw a branch of the hyperbola through each intercept and approaching the asymptotes.

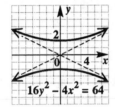

10. $f(x) = \sqrt{16 - x^2}$

Replace $f(x)$ with y.

$$y = \sqrt{16 - x^2}$$

Square both sides.

$$y^2 = 16 - x^2$$
$$x^2 + y^2 = 16$$

The graph of $x^2 + y^2 = 16$ is a circle of radius $\sqrt{16} = 4$ centered at the origin. Since $f(x)$ is nonnegative, only the top half of the circle is graphed.

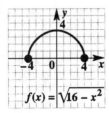

11. If a parabola and ellipse are graphed in the same plane, they can intersect at zero, one, two, three, or four points.

12. Draw any parabola, and then draw an ellipse that intersects the parabola in two points.

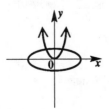

13. $2x - y = 9$ (1)
$xy = 5$ (2)

Solve equation (1) for y.

$$y = 2x - 9 \quad (3)$$

Substitute $2x - 9$ for y in equation (2).

$$\begin{aligned} xy &= 5 \quad (2) \\ x(2x - 9) &= 5 \\ 2x^2 - 9x &= 5 \\ 2x^2 - 9x - 5 &= 0 \\ (2x + 1)(x - 5) &= 0 \end{aligned}$$

$$2x + 1 = 0 \quad \text{or} \quad x - 5 = 0$$
$$x = -\frac{1}{2} \quad \text{or} \quad x = 5$$

Substitute these values for x in equation (3) to find y.

If $x = -\dfrac{1}{2}$, then $y = 2\left(-\dfrac{1}{2}\right) - 9 = -10$.

If $x = 5$, then $y = 2(5) - 9 = 1$.

Solution set: $\left\{\left(-\dfrac{1}{2}, -10\right), (5, 1)\right\}$

14. $x - 4 = 3y$ (1)
$x^2 + y^2 = 8$ (2)

Solve equation (1) for x.

$$x = 3y + 4$$

Substitute $3y + 4$ for x in equation (2).

$$\begin{aligned} x^2 + y^2 &= 8 \quad (2) \\ (3y + 4)^2 + y^2 &= 8 \\ 9y^2 + 24y + 16 + y^2 &= 8 \\ 10y^2 + 24y + 8 &= 0 \\ 2(5y^2 + 12y + 4) &= 0 \\ 2(5y + 2)(y + 2) &= 0 \end{aligned}$$

$$5y + 2 = 0 \quad \text{or} \quad y + 2 = 0$$
$$y = -\frac{2}{5} \quad \text{or} \quad y = -2$$

Since $x = 3y + 4$, substitute these values for y to find x.

If $y = -\dfrac{2}{5}$, then

$$x = 3\left(-\frac{2}{5}\right) + 4 = -\frac{6}{5} + 4 = \frac{14}{5}.$$

If $y = -2$, then $x = 3(-2) + 4 = -2$.

Solution set: $\left\{(-2, -2), \left(\dfrac{14}{5}, -\dfrac{2}{5}\right)\right\}$

15. $x^2 + y^2 = 25$ (1)
$x^2 - 2y^2 = 16$ (2)

Multiply equation (1) by 2 and add the result to equation (2).

$$
\begin{array}{rll}
2x^2 + 2y^2 &= 50 & 2 \times (1) \\
x^2 - 2y^2 &= 16 & (2) \\
\hline
3x^2 \qquad &= 66 & \\
x^2 &= 22 &
\end{array}
$$

$$x = \sqrt{22} \quad \text{or} \quad x = -\sqrt{22}$$

Substitute these values for x in equation (1).
If $x = \pm\sqrt{22}$, then

$$x^2 + y^2 = 25 \quad (1)$$
$$\left(\pm\sqrt{22}\right)^2 + y^2 = 25$$
$$22 + y^2 = 25$$
$$y^2 = 3$$

$$y = \sqrt{3} \quad \text{or} \quad y = -\sqrt{3},$$

Solution set: $\left\{\left(\sqrt{22}, \sqrt{3}\right), \left(\sqrt{22}, -\sqrt{3}\right),\right.$
$$\left.\left(-\sqrt{22}, \sqrt{3}\right), \left(-\sqrt{22}, -\sqrt{3}\right)\right\}$$

16. $y \le x^2 - 2$

The boundary, $y = x^2 - 2$, is a solid parabola in $y = a(x-h)^2 + k$ form with vertex (h, k) at $(0, -2)$. Since $a = 1 > 0$, the parabola opens upward. It also has the same shape as $y = x^2$. The points $(2, 2)$ and $(-2, 2)$ are on the graph. Test $(0, 0)$.

$$0 \le (0)^2 - 2 \quad ?$$
$$0 \le -2 \qquad \textit{False}$$

Shade the side of the parabola that does not contain $(0, 0)$. This is the region outside the parabola.

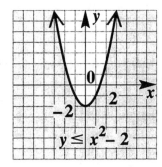

17. $2x - 5y \ge 12$
$3x + 4y \le 12$

To graph $2x - 5y \ge 12$, draw $2x - 5y = 12$ as a solid line through $(6, 0)$ and $\left(0, -\dfrac{12}{5}\right)$. Test $(0, 0)$ to get $0 \ge 12$, a false statement. Shade the side of the line not containing the origin. To graph $3x + 4y \le 12$, draw $3x + 4y = 12$ as a solid line through $(4, 0)$ and $(0, 3)$. Test $(0, 0)$ to get $0 \le 12$, a true statement. Shade the side of the line containing the origin.
The intersection of the shaded regions is the graph of the system.

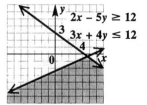

18. $x^2 + 25y^2 \le 25$
$x^2 + y^2 \le 9$

The first boundary, $\dfrac{x^2}{25} + \dfrac{y^2}{1} = 1$, is a solid ellipse with intercepts $(5, 0)$, $(-5, 0)$, $(0, 1)$, and $(0, -1)$. Test $(0, 0)$.

$$0^2 + 25 \cdot 0^2 \le 25 \quad ?$$
$$0 \le 25 \qquad \textit{True}$$

Shade the region inside the ellipse.
The second boundary, $x^2 + y^2 = 9$, is a solid circle with center $(0, 0)$ and radius 3. Test $(0, 0)$.

$$0^2 + 0^2 \le 9 \quad ?$$
$$0 \le 9 \qquad \textit{True}$$

Shade the region inside the circle. The solution of the system is the intersection of the two shaded regions.

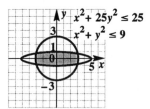

19. $\dfrac{x^2}{3352} + \dfrac{y^2}{3211} = 1$

$c^2 = a^2 - b^2 = 3352 - 3211,$
so $c^2 = 141$ and $c = \sqrt{141}.$

From Exercise 47 in Section 13.1, the apogee is $a + c = \sqrt{3352} + \sqrt{141} \approx 69.8$ million kilometers.

20. The perigee is $a - c = \sqrt{3352} - \sqrt{141} \approx 46.0$ million kilometers.

Cumulative Review Exercises Chapters 1–13

1. $-10 + |-5| - |3| + 4$
$= -10 + 5 - 3 + 4$
$= -5 + 1 = -4$

2. $4 - (2x + 3) + x = 5x - 3$
$4 - 2x - 3 + x = 5x - 3$
$-x + 1 = 5x - 3$
$-6x = -4$
$x = \dfrac{2}{3}$

Solution set: $\left\{\dfrac{2}{3}\right\}$

3. $-4k + 7 \geq 6k + 1$
$-10k \geq -6$

Divide by -10; reverse the direction of the inequality.

$k \leq \dfrac{-6}{-10}$
$k \leq \dfrac{3}{5}$

Solution set: $\left(-\infty, \dfrac{3}{5}\right]$

4. $|5m| - 6 = 14$
$|5m| = 20$

$5m = 20$ or $5m = -20$
$m = 4$ or $m = -4$

Solution set: $\{-4, 4\}$

5. $|2p - 5| > 15$

$2p - 5 > 15$ or $2p - 5 < -15$
$2p > 20$ $2p < -10$
$p > 10$ or $p < -5$

Solution set: $(-\infty, -5) \cup (10, \infty)$

6. Let $(x_1, y_1) = (2, 5)$ and $(x_2, y_2) = (-4, 1)$.
$m = \dfrac{y_2 - y_1}{x_2 - x_1} = \dfrac{1 - 5}{-4 - 2} = \dfrac{-4}{-6} = \dfrac{2}{3}$

7. $(5y - 3)^2 = (5y)^2 - 2(5y)3 + 3^2$
$= 25y^2 - 30y + 9$

8. $(2r + 7)(6r - 1)$
$= 12r^2 - 2r + 42r - 7$
$= 12r^2 + 40r - 7$

9. $\left(8x^4 - 4x^3 + 2x^2 + 13x + 8\right) \div (2x + 1)$

$$
\begin{array}{r}
4x^3 \;-\; 4x^2 \;+\; 3x \;+\; 5 \\
2x + 1 \overline{)\,8x^4 \;-\; 4x^3 \;+\; 2x^2 \;+\; 13x \;+\; 8} \\
\underline{8x^4 \;+\; 4x^3} \quad\quad\quad\quad\quad\quad\quad \\
-8x^3 \;+\; 2x^2 \quad\quad\quad\quad\quad \\
\underline{-8x^3 \;-\; 4x^2} \quad\quad\quad\quad\quad \\
6x^2 \;+\; 13x \quad\quad \\
\underline{6x^2 \;+\; 3x} \quad\quad \\
10x \;+\; 8 \\
\underline{10x \;+\; 5} \\
3
\end{array}
$$

The answer is

$$4x^3 - 4x^2 + 3x + 5 + \frac{3}{2x + 1}.$$

10. $12x^2 - 7x - 10 = (4x - 5)(3x + 2)$

11. $2y^4 + 5y^2 - 3$

Let $p = y^2$, so $p^2 = y^4$.

$2y^4 + 5y^2 - 3 = 2p^2 + 5p - 3$
$\qquad\qquad\qquad = (2p - 1)(p + 3)$

Now substitute y^2 for p.

$\qquad\qquad\qquad = \left(2y^2 - 1\right)\left(y^2 + 3\right)$

12. $z^4 - 1 = \left(z^2 + 1\right)\left(z^2 - 1\right)$
$\qquad\quad = \left(z^2 + 1\right)(z + 1)(z - 1)$

13. $a^3 - 27b^3 = a^3 - (3b)^3$
$\qquad\qquad = (a - 3b)\left(a^2 + 3ab + 9b^2\right)$

14. $\dfrac{5x - 15}{24} \cdot \dfrac{64}{3x - 9} = \dfrac{5(x - 3)}{3 \cdot 8} \cdot \dfrac{8 \cdot 8}{3(x - 3)}$

$\qquad\qquad = \dfrac{5 \cdot 8}{3 \cdot 3} = \dfrac{40}{9}$

15. $\dfrac{y^2 - 4}{y^2 - y - 6} \div \dfrac{y^2 - 2y}{y - 1}$

Multiply by the reciprocal.

$= \dfrac{y^2 - 4}{y^2 - y - 6} \cdot \dfrac{y - 1}{y^2 - 2y}$

Factor and simplify.

$= \dfrac{(y + 2)(y - 2)}{(y - 3)(y + 2)} \cdot \dfrac{(y - 1)}{y(y - 2)}$

$= \dfrac{y - 1}{y(y - 3)}$

16. $\dfrac{5}{c+5} - \dfrac{2}{c+3}$

The LCD is $(c+5)(c+3)$.

$$= \dfrac{5(c+3)}{(c+5)(c+3)} - \dfrac{2(c+5)}{(c+3)(c+5)}$$

$$= \dfrac{5c+15-2c-10}{(c+5)(c+3)}$$

$$= \dfrac{3c+5}{(c+5)(c+3)}$$

17. $\dfrac{p}{p^2+p} + \dfrac{1}{p^2+p} = \dfrac{p+1}{p^2+p}$

$$= \dfrac{p+1}{p(p+1)} = \dfrac{1}{p}$$

18. Let $x =$ the time to do the job working together.

Make a chart.

Worker	Rate	Time Together	Part of Job Done
Kareem	$\dfrac{1}{3}$	x	$\dfrac{x}{3}$
Jamal	$\dfrac{1}{2}$	x	$\dfrac{x}{2}$

Part done by Kareem plus part done by Jamal equals 1 whole job.

$$\dfrac{x}{3} \quad + \quad \dfrac{x}{2} \quad = \quad 1$$

Multiply by the LCD, 6.

$$6\left(\dfrac{x}{3} + \dfrac{x}{2}\right) = 6 \cdot 1$$

$$2x + 3x = 6$$

$$5x = 6$$

$$x = \dfrac{6}{5} = 1\dfrac{1}{5}$$

It takes $\dfrac{6}{5}$ or $1\dfrac{1}{5}$ hours to do the job together.

19. Through $(-3, -2)$; perpendicular to $2x - 3y = 7$

Write $2x - 3y = 7$ in slope-intercept form.

$$-3y = -2x + 7$$

$$y = \dfrac{2}{3}x - \dfrac{7}{3}$$

The slope is $\dfrac{2}{3}$. Perpendicular lines have slopes that are negative reciprocals of each other, so a line perpendicular to the given line will have slope $-\dfrac{3}{2}$. Let $m = -\dfrac{3}{2}$ and $(x_1, y_1) = (-3, -2)$ in the point-slope form.

$$y - y_1 = m(x - x_1)$$

$$y - (-2) = -\dfrac{3}{2}[x - (-3)]$$

$$y + 2 = -\dfrac{3}{2}(x + 3)$$

Multiply by 2 to clear the fraction.

$$2y + 4 = -3(x + 3)$$

$$2y + 4 = -3x - 9$$

$$3x + 2y = -13$$

20. $3x - y = 12 \quad (1)$
$2x + 3y = -3 \quad (2)$

Multiply equation (1) by 3 and add the result to equation (2).

$$\begin{array}{rcll} 9x - 3y &=& 36 & 3 \times (1) \\ 2x + 3y &=& -3 & (2) \\ \hline 11x &=& 33 & \\ x &=& 3 & \end{array}$$

Substitute 3 for x in equation (1) to find y.

$$3x - y = 12 \quad (1)$$

$$3(3) - y = 12$$

$$9 - y = 12$$

$$-y = 3$$

$$y = -3$$

Solution set: $\{(3, -3)\}$

21. $\begin{array}{rcl} x + y - 2z &=& 9 \quad (1) \\ 2x + y + z &=& 7 \quad (2) \\ 3x - y - z &=& 13 \quad (3) \end{array}$

Add equation (2) and equation (3).

$$\begin{array}{rcl} 2x + y + z &=& 7 \quad (2) \\ 3x - y - z &=& 13 \quad (3) \\ \hline 5x &=& 20 \\ x &=& 4 \end{array}$$

Multiply equation (1) by -1 and add the result to equation (2).

$$\begin{array}{rcl} -x - y + 2z &=& -9 \\ 2x + y + z &=& 7 \quad (2) \\ \hline x + 3z &=& -2 \quad (4) \end{array}$$

Substitute 4 for x in equation (4) to find z.

$$x + 3z = -2 \quad (4)$$

$$4 + 3z = -2$$

$$3z = -6$$

$$z = -2$$

Substitute 4 for x and -2 for z in equation (2) to find y.

$$2x + y + z = 7 \quad (2)$$

$$2(4) + y - 2 = 7$$

$$y + 6 = 7$$

$$y = 1$$

Solution set: $\{(4, 1, -2)\}$

22. $xy = -5$ (1)

$2x + y = 3$ (2)

Solve equation (2) for y.

$$y = -2x + 3 \quad (3)$$

Substitute $-2x + 3$ for y in equation (1).

$$\begin{aligned} xy &= -5 \quad (1) \\ x(-2x + 3) &= -5 \\ -2x^2 + 3x &= -5 \\ -2x^2 + 3x + 5 &= 0 \\ 2x^2 - 3x - 5 &= 0 \\ (2x - 5)(x + 1) &= 0 \end{aligned}$$

$$2x - 5 = 0 \quad \text{or} \quad x + 1 = 0$$

$$x = \frac{5}{2} \quad \text{or} \quad x = -1$$

Substitute these values for x in equation (3) to find y.

If $x = \dfrac{5}{2}$, then $y = -2\left(\dfrac{5}{2}\right) + 3 = -2$.

If $x = -1$, then $y = -2(-1) + 3 = 5$.

Solution set: $\left\{ (-1, 5), \left(\dfrac{5}{2}, -2\right) \right\}$

23. $\left(\dfrac{4}{3}\right)^{-1} = \left(\dfrac{3}{4}\right)^{1} = \dfrac{3}{4}$

24. $\dfrac{(2a)^{-2}a^4}{a^{-3}} = \dfrac{2^{-2}a^{-2}a^4}{a^{-3}} = \dfrac{2^{-2}a^2}{a^{-3}}$

$$= \dfrac{a^2 a^3}{2^2} = \dfrac{a^5}{4}$$

25. $4\sqrt[3]{16} - 2\sqrt[3]{54} = 4\sqrt[3]{8 \cdot 2} - 2\sqrt[3]{27 \cdot 2}$

$$= 4 \cdot 2\sqrt[3]{2} - 2 \cdot 3\sqrt[3]{2}$$

$$= 8\sqrt[3]{2} - 6\sqrt[3]{2} = 2\sqrt[3]{2}$$

26. $\dfrac{3\sqrt{5x}}{\sqrt{2x}} = \dfrac{3\sqrt{5x} \cdot \sqrt{2x}}{\sqrt{2x} \cdot \sqrt{2x}} = \dfrac{3\sqrt{10x^2}}{2x}$

$$= \dfrac{3x\sqrt{10}}{2x} = \dfrac{3\sqrt{10}}{2}$$

27. $\dfrac{5 + 3i}{2 - i}$

Multiply by the conjugate of the denominator.

$$= \dfrac{(5 + 3i)(2 + i)}{(2 - i)(2 + i)}$$

$$= \dfrac{10 + 5i + 6i + 3i^2}{4 - i^2}$$

$$= \dfrac{10 + 11i + 3(-1)}{4 - (-1)}$$

$$= \dfrac{7 + 11i}{5} = \dfrac{7}{5} + \dfrac{11}{5}i$$

28. $2\sqrt{k} = \sqrt{5k + 3}$

Square both sides.

$$\begin{aligned} 4k &= 5k + 3 \\ -k &= 3 \\ k &= -3 \end{aligned}$$

Since k must be nonnegative so that $\sqrt{k}$ is a real number, -3 cannot be a solution. The solution set is $\emptyset$.

29. $\begin{aligned} 10q^2 + 13q &= 3 \\ 10q^2 + 13q - 3 &= 0 \\ (5q - 1)(2q + 3) &= 0 \end{aligned}$

$$5q - 1 = 0 \quad \text{or} \quad 2q + 3 = 0$$

$$q = \dfrac{1}{5} \quad \text{or} \quad q = -\dfrac{3}{2}$$

Solution set: $\left\{ \dfrac{1}{5}, -\dfrac{3}{2} \right\}$

30. $(4x - 1)^2 = 8$

$$4x - 1 = \pm\sqrt{8}$$

$$4x = 1 \pm 2\sqrt{2}$$

$$x = \dfrac{1 \pm 2\sqrt{2}}{4}$$

Solution set: $\left\{ \dfrac{1 + 2\sqrt{2}}{4}, \dfrac{1 - 2\sqrt{2}}{4} \right\}$

31. $\log(2x) - \log(x - 1) = \log 3$

$$\log \dfrac{2x}{x - 1} = \log 3$$

$$\dfrac{2x}{x - 1} = 3$$

$$2x = 3(x - 1)$$

$$2x = 3x - 3$$

$$3 = x$$

Check $x = 3$: $\log 6 - \log 2 \overset{?}{=} \log 3$ *True*

Solution set: $\{3\}$

32. $2(x^2 - 3)^2 - 5(x^2 - 3) = 12$

Let $u = (x^2 - 3)$.

$$2u^2 - 5u = 12$$
$$2u^2 - 5u - 12 = 0$$
$$(2u + 3)(u - 4) = 0$$

$$2u + 3 = 0 \quad \text{or} \quad u - 4 = 0$$
$$u = -\frac{3}{2} \quad \text{or} \quad u = 4$$

Substitute $x^2 - 3$ for u to find x.

If $u = -\frac{3}{2}$, then

$$x^2 - 3 = -\frac{3}{2}$$
$$x^2 = \frac{3}{2}$$
$$x = \pm\sqrt{\frac{3}{2}} = \pm\frac{\sqrt{3}}{\sqrt{2}} \cdot \frac{\sqrt{2}}{\sqrt{2}} = \pm\frac{\sqrt{6}}{2}.$$

If $u = 4$, then

$$x^2 - 3 = 4$$
$$x^2 = 7$$
$$x = \pm\sqrt{7}.$$

Solution set: $\left\{ -\dfrac{\sqrt{6}}{2}, \dfrac{\sqrt{6}}{2}, -\sqrt{7}, \sqrt{7} \right\}$

33. Solve $F = \dfrac{kwv^2}{r}$ for v.
$$Fr = kwv^2$$
$$v^2 = \frac{Fr}{kw}$$

Take the square root of each side.

$$v = \pm\sqrt{\frac{Fr}{kw}} = \frac{\pm\sqrt{Fr}}{\sqrt{kw}} \cdot \frac{\sqrt{kw}}{\sqrt{kw}} = \frac{\pm\sqrt{Frkw}}{kw}$$

34. $3x + y = 5$

Write the equation in slope-intercept form, $y = mx + b$.

$$y = -3x + 5$$

The y-intercept is $(0, 5)$ and $m = -3$ or $\dfrac{-3}{1}$.

Plot $(0, 5)$. From $(0, 5)$, move down 3 units and right 1 unit. Draw the line through these two points.

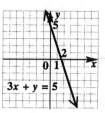

35. $f(x) = x^2 - 4x + 5$

The graph of f is a parabola. The x-value of the vertex is

$$x = \frac{-b}{2a} = \frac{-(-4)}{2(1)} = 2.$$

The y-value of the vertex is

$$f(2) = 2^2 - 4(2) + 5 = 1.$$

The vertex is at $(2, 1)$ and the parabola opens upward since $a = 1 > 0$. The y-intercept is $(0, 5)$ and by symmetry, the point $(4, 5)$ is also on the graph.

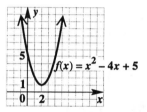

36. $f(x) = 3^{x-1}$

The graph of f is an increasing exponential.

x	-1	0	1	2	3
$f(x)$	$\frac{1}{9}$	$\frac{1}{3}$	1	3	9

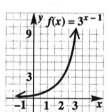

37. $\dfrac{x^2}{4} - \dfrac{y^2}{16} = 1$ is in $\dfrac{x^2}{a^2} - \dfrac{y^2}{b^2} = 1$ form.

The graph is a hyperbola with x-intercepts $(2, 0)$ and $(-2, 0)$ and asymptotes that are the extended diagonals of the rectangle with corners $(2, 4)$, $(2, -4)$, $(-2, -4)$, and $(-2, 4)$. Draw a branch of the hyperbola through each intercept approaching the asymptotes.

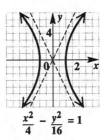

$$\frac{x^2}{4} - \frac{y^2}{16} = 1$$

38. $\frac{x^2}{25} + \frac{y^2}{16} \le 1$

The boundary, $\frac{x^2}{25} + \frac{y^2}{16} = 1$, is a solid ellipse in

$\frac{x^2}{a^2} + \frac{y^2}{b^2} = 1$ form with intercepts $(5,0)$, $(-5,0)$,

$(0,4)$, and $(0,-4)$. Test $(0,0)$.

$$\frac{0^2}{25} + \frac{0^2}{16} \le 1 \quad ?$$
$$0 \le 1 \quad True$$

Shade the region inside the ellipse.

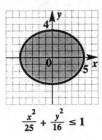

$$\frac{x^2}{25} + \frac{y^2}{16} \le 1$$

39. (a)
$$y = ax^2 + bx$$
$$61.7 = a(25)^2 + b(25)$$
$$61.7 = 625a + 25b \qquad (1)$$
$$106 = a(35)^2 + b(35)$$
$$106 = 1225a + 35b \qquad (2)$$

(b) Multiply (1) by -7 and (2) by 5 and add the results.

$$
\begin{array}{rll}
-431.9 = & -4375a - 175b & -7 \times (1) \\
530 = & 6125a + 175b & 5 \times (2) \\
\hline
98.1 = & 1750a &
\end{array}
$$

$$a = \frac{98.1}{1750} \approx .056$$

To find b, substitute $\frac{98.1}{1750}$ for a in (1).

$$61.7 = 625\left(\frac{98.1}{1750}\right) + 25b$$

$$25b = 61.7 - 625\left(\frac{98.1}{1750}\right)$$

$$b = \frac{61.7 - 625\left(\frac{98.1}{1750}\right)}{25}$$

$$\approx 1.067$$

(c)
$$y = ax^2 + bx$$
$$y = .056x^2 + 1.067x$$

(d) Let $x = 55$.

$$y = .056(55)^2 + 1.067(55)$$
$$= 228.085 \approx 228.1 \text{ feet}$$

40. (a) $y = 1.38(1.65)^x$
$$y = 1.38(1.65)^5 \quad Let \ x=5.$$
$$\approx 16.9 \text{ billion dollars}$$

(b) $2003 - 1995 = 8$

$$y = 1.38(1.65)^8 \quad Let \ x=8.$$
$$\approx 75.8 \text{ billion dollars}$$

CHAPTER 14 SEQUENCES AND SERIES

Section 14.1

1. The similarities are that both are defined by the same linear expression, and that points satisfying both lie in a straight line. The difference is that the domain of f consists of all real numbers, but the domain of the sequence is $\{1, 2, 3, \dots \}$. Some examples are $f(1) = 6$ and $a_1 = 6$ is a similarity, while $f\left(\dfrac{3}{2}\right) = 7$, but $a_{3/2}$ is not allowed.

3. $a_n = \dfrac{n+3}{n}$

To get a_1, the first term, replace n with 1.

$a_1 = \dfrac{1+3}{1} = \dfrac{4}{1} = 4$

To get a_2, the second term, replace n with 2.

$a_2 = \dfrac{2+3}{2} = \dfrac{5}{2}$

To get a_3, the third term, replace n with 3.

$a_3 = \dfrac{3+3}{3} = \dfrac{6}{3} = 2$

To get a_4, the fourth term, replace n with 4.

$a_4 = \dfrac{4+3}{4} = \dfrac{7}{4}$

To get a_5, the fifth term, replace n with 5.

$a_5 = \dfrac{5+3}{5} = \dfrac{8}{5}$

Answer: $4, \dfrac{5}{2}, 2, \dfrac{7}{4}, \dfrac{8}{5}$

5. $a_n = 3^n$

$a_1 = 3^1 = 3$

$a_2 = 3^2 = 9$

$a_3 = 3^3 = 27$

$a_4 = 3^4 = 81$

$a_5 = 3^5 = 243$

Answer: $3, 9, 27, 81, 243$

7. $a_n = \dfrac{1}{n^2}$

$a_1 = \dfrac{1}{1^2} = 1$

$a_2 = \dfrac{1}{2^2} = \dfrac{1}{4}$

$a_3 = \dfrac{1}{3^2} = \dfrac{1}{9}$

$a_4 = \dfrac{1}{4^2} = \dfrac{1}{16}$

$a_5 = \dfrac{1}{5^2} = \dfrac{1}{25}$

Answer: $1, \dfrac{1}{4}, \dfrac{1}{9}, \dfrac{1}{16}, \dfrac{1}{25}$

9. $a_n = (-1)^n$

$a_1 = (-1)^1 = -1$

$a_2 = (-1)^2 = 1$

$a_3 = (-1)^3 = -1$

$a_4 = (-1)^4 = 1$

$a_5 = (-1)^5 = -1$

Answer: $-1, 1, -1, 1, -1$

11. $a_n = -9n + 2$

$a_8 = -9(8) + 2 = -70$

13. $a_n = \dfrac{3n+7}{2n-5}$

$a_{14} = \dfrac{3(14)+7}{2(14)-5} = \dfrac{49}{23}$

15. $a_n = (n+1)(2n+3)$

$a_8 = (8+1)[2(8)+3]$
$= 9(19) = 171$

17. $4, 8, 12, 16, \dots$ can be written as
$4 \cdot 1, 4 \cdot 2, 4 \cdot 3, 4 \cdot 4, \dots$, so $a_n = 4n$.

19. $\dfrac{1}{3}, \dfrac{1}{9}, \dfrac{1}{27}, \dfrac{1}{81}, \dots$ can be written as
$\dfrac{1}{3^1}, \dfrac{1}{3^2}, \dfrac{1}{3^3}, \dfrac{1}{3^4}, \dots$, so $a_n = \dfrac{1}{3^n}$.

21. Make a table as follows:

Month	Interest	Payment	Unpaid balance
0			1000
1	$1000(.01) = 10$	$100 + 10 = 110$	$1000 - 100 = 900$
2	$900(.01) = 9$	$100 + 9 = 109$	$900 - 100 = 800$
3	$800(.01) = 8$	$100 + 8 = 108$	$800 - 100 = 700$
4	$700(.01) = 7$	$100 + 7 = 107$	$700 - 100 = 600$
5	$600(.01) = 6$	$100 + 6 = 106$	$600 - 100 = 500$
6	$500(.01) = 5$	$100 + 5 = 105$	$500 - 100 = 400$

The payments are $110, $109, $108, $107, $106, and $105; the unpaid balance is $400.

23. When new, the car is worth $20,000. Let $a_n =$ the value of the car after the nth year. The car retains $\dfrac{4}{5}$ of its value each year.

$a_1 = \dfrac{4}{5}(20,000) = \$16,000$ (value after the first year)

$a_2 = \dfrac{4}{5}(16,000) = \$12,800$

$a_3 = \dfrac{4}{5}(12,800) = \$10,240$

$a_4 = \dfrac{4}{5}(10,240) = \$8192$

$a_5 = \dfrac{4}{5}(8192) = \$6553.60$

The value of the car after 5 years is about $6554.

25. $\displaystyle\sum_{i=1}^{3} (i^2 + 2)$

$$= (1^2 + 2) + (2^2 + 2) + (3^2 + 2)$$
$$= 3 + 6 + 11 = 20$$

27. $\displaystyle\sum_{i=2}^{5} \frac{1}{i} = \frac{1}{2} + \frac{1}{3} + \frac{1}{4} + \frac{1}{5}$

$$= \frac{30}{60} + \frac{20}{60} + \frac{15}{60} + \frac{12}{60} = \frac{77}{60}$$

29. $\displaystyle\sum_{i=1}^{6} (-1)^i$

$$= (-1)^1 + (-1)^2 + (-1)^3 + (-1)^4$$
$$+ (-1)^5 + (-1)^6$$
$$= -1 + 1 - 1 + 1 - 1 + 1 = 0$$

31. $\displaystyle\sum_{i=3}^{7} (i - 3)(i + 2)$

$$= (3 - 3)(3 + 2) + (4 - 3)(4 + 2)$$
$$+ (5 - 3)(5 + 2) + (6 - 3)(6 + 2)$$
$$+ (7 - 3)(7 + 2)$$
$$= 0(5) + (1)(6) + (2)(7) + (3)(8)$$
$$+ (4)(9)$$
$$= 0 + 6 + 14 + 24 + 36 = 80$$

33. $\displaystyle\sum_{i=1}^{5} 2x \cdot i$

$$= 2x \cdot 1 + 2x \cdot 2 + 2x \cdot 3 + 2x \cdot 4$$
$$+ 2x \cdot 5$$
$$= 2x + 4x + 6x + 8x + 10x$$

35. $\displaystyle\sum_{i=1}^{5} i \cdot x^i = 1(x^1) + 2(x^2) + 3(x^3) + 4(x^4)$

$$+ 5(x^5)$$
$$= x + 2x^2 + 3x^3 + 4x^4 + 5x^5$$

37. $3 + 4 + 5 + 6 + 7$

$$= (1 + 2) + (2 + 2) + (3 + 2)$$
$$+ (4 + 2) + (5 + 2)$$
$$= \sum_{i=1}^{5} (i + 2)$$

39. $\dfrac{1}{2} + \dfrac{1}{3} + \dfrac{1}{4} + \dfrac{1}{5} + \dfrac{1}{6}$

$$= \frac{1}{1 + 1} + \frac{1}{2 + 1} + \frac{1}{3 + 1} + \frac{1}{4 + 1}$$
$$+ \frac{1}{5 + 1}$$
$$= \sum_{i=1}^{5} \frac{1}{i + 1}$$

41. A sequence is a list of terms in a specific order, while a series is the indicated sum of the terms of a sequence.

43. $\bar{x} = \dfrac{\displaystyle\sum_{i=1}^{n} x_i}{n} = \dfrac{\displaystyle\sum_{i=1}^{7} x_i}{7}$

$$= \frac{8 + 11 + 14 + 9 + 3 + 6 + 8}{7} = \frac{59}{7}$$

45. $\bar{x} = \dfrac{5 + 9 + 8 + 2 + 4 + 7 + 3 + 2}{8} = \dfrac{40}{8} = 5$

47. $\bar{x} = \dfrac{3427 + 3850 + 4558 + 5357 + 5761}{5}$

$$= \frac{22{,}953}{5} = 4590.6$$

The average number of funds available for this five-year period was about 4591.

49. $\displaystyle\sum_{i=1}^{6} (i^2 + 3i + 5)$

$$= \sum_{i=1}^{6} i^2 + \sum_{i=1}^{6} 3i + \sum_{i=1}^{6} 5$$

50. From Exercise 49, the second summation is

$$\sum_{i=1}^{6} 3i = 3 \sum_{i=1}^{6} i.$$

51. From Exercise 49, the third summation is

$$\sum_{i=1}^{6} 5 = 6 \cdot 5 = 30.$$

52. $1 + 2 + 3 + 4 + \cdots + n = \dfrac{n(n + 1)}{2}$

written in summation notation is

$$\sum_{i=1}^{n} i = \frac{n(n + 1)}{2}.$$

53. $1^2 + 2^2 + 3^2 + 4^2 + \cdots + n^2$

$$= \frac{n(n + 1)(2n + 1)}{6}$$

written in summation notation is

$$\sum_{i=1}^{n} i^2 = \frac{n(n + 1)(2n + 1)}{6}.$$

54. From Exercises 49–51,

$$\sum_{i=1}^{6} (i^2 + 3i + 5)$$

$$= \sum_{i=1}^{6} i^2 + 3 \sum_{i=1}^{6} i + 30.$$

Now apply the results of Exercises 52 & 53.

$$= \frac{6(6 + 1)(2 \cdot 6 + 1)}{6} + 3 \cdot \frac{6(6 + 1)}{2} + 30$$

$$= \frac{6(7)(13)}{6} + 3 \cdot \frac{6(7)}{2} + 30$$

$$= 91 + 63 + 30 = 184$$

55. $\displaystyle\sum_{i=1}^{12} \left(i^2 - i\right)$

$$= \sum_{i=1}^{12} i^2 - \sum_{i=1}^{12} i$$

$$= \frac{12(12+1)(2\cdot 12+1)}{6} - \frac{12(12+1)}{2}$$

$$= \frac{12(13)(25)}{6} - \frac{12(13)}{2}$$

$$= 650 - 78 = 572$$

56. $\displaystyle\sum_{i=1}^{20} \left(2 + i - i^2\right)$

$$\sum_{i=1}^{20} 2 + \sum_{i=1}^{20} i - \sum_{i=1}^{20} i^2$$

$$= 20\cdot 2 + \frac{20(20+1)}{2} - \frac{20(20+1)(2\cdot 20+1)}{6}$$

$$= 40 + 210 - 2870 = -2620$$

Section 14.2

1. An arithmetic sequence is a sequence (list) of numbers in a specific order such that there is a common difference between any two successive terms. For example, the sequence $1, 5, 9, 13, \ldots$ is arithmetic with difference $d = 5 - 1 = 9 - 5 = 13 - 9 = 4$. As another example, $2, -1, -4, -7, \ldots$ is an arithmetic sequence with $d = -3$.

3. $1, 2, 3, 4, 5, \ldots$
d is the difference between any two adjacent terms. Choose the terms 3 and 2.

$$d = 3 - 2 = 1$$

The terms 2 and 1 would give

$$d = 2 - 1 = 1,$$

the same result. Therefore, the common difference is $d = 1$.

Note: You should find the difference for all pairs of adjacent terms to determine if the sequence is arithmetic.

5. $2, -4, 6, -8, 10, -12, \ldots$
The difference between the first two terms is $-4 - 2 = -6$, but the difference between the second and third terms is $6 - (-4) = 10$. The differences are not the same so the sequence is *not arithmetic*.

7. $-10, -5, 0, 5, 10, \ldots$
Choose the terms 10 and 5, and find the difference.

$$d = 10 - 5 = 5$$

The terms -5 and -10 would give

$$d = -5 - (-10) = -5 + 10 = 5,$$

the same result. Therefore, the common difference is $d = 5$.

9. $3.42, 5.57, 7.72, 9.87, \ldots$
Choose the terms 5.57 and 3.42.

$$d = 5.57 - 3.42 = 2.15$$

The terms 7.72 and 5.57 give

$$d = 7.72 - 5.57 = 2.15,$$

the same result. Therefore, the common difference is $d = 2.15$.

11. $-\dfrac{5}{3}, -1, -\dfrac{1}{3}, \dfrac{1}{3}, \ldots$
Choose the terms $\dfrac{1}{3}$ and $-\dfrac{1}{3}$.

$$d = \frac{1}{3} - \left(-\frac{1}{3}\right) = \frac{2}{3}$$

The terms $-\dfrac{1}{3}$ and -1 give

$$d = -\frac{1}{3} - (-1) = -\frac{1}{3} + 1 = \frac{2}{3},$$

the same result. Therefore, the common difference is $d = \dfrac{2}{3}$.

13. $a_1 = 2, d = 5$
$$\begin{aligned} a_n &= a_1 + (n-1)d \\ &= 2 + (n-1)5 \\ &= 2 + 5n - 5 \\ &= 5n - 3 \end{aligned}$$

15. $3, \dfrac{15}{4}, \dfrac{9}{2}, \dfrac{21}{4}, \ldots$
To find d, subtract any two adjacent terms.

$$d = \frac{15}{4} - 3 = \frac{15}{4} - \frac{12}{4} = \frac{3}{4}$$

The first term is $a_1 = 3$. Now find a_n.
$$\begin{aligned} a_n &= a_1 + (n-1)d \\ &= 3 + (n-1)\left(\frac{3}{4}\right) \\ &= 3 + \frac{3}{4}n - \frac{3}{4} \\ &= \frac{3}{4}n + \frac{9}{4} \end{aligned}$$

17. $-3, 0, 3, \ldots$
To find d, subtract any two adjacent terms.

$$d = 0 - (-3) = 3$$

The first term is $a_1 = -3$. Now find a_n.
$$\begin{aligned} a_n &= a_1 + (n-1)d \\ &= -3 + (n-1)3 \\ &= -3 + 3n - 3 \\ &= 3n - 6 \end{aligned}$$

19. Given $a_1 = 4$ and $d = 3$; find a_{25}.
$$a_n = a_1 + (n - 1)d$$
$$a_{25} = 4 + (25 - 1)3$$
$$= 4 + 72 = 76$$

21. Given $2, 4, 6, \ldots$; find a_{24}.
Here, $a_1 = 2$ and $d = 4 - 2 = 2$.
Now find a_{24}.
$$a_n = a_1 + (n - 1)d$$
$$a_{24} = 2 + (24 - 1)2$$
$$= 2 + 46 = 48$$

23. Given $a_{12} = -45$ and $a_{10} = -37$; find a_1. Use $a_n = a_1 + (n - 1)d$ to write a system of equations.
$$a_{12} = a_1 + (12 - 1)d$$
$$-45 = a_1 + 11d \qquad (1)$$
$$a_{10} = a_1 + (10 - 1)d$$
$$-37 = a_1 + 9d \qquad (2)$$
To eliminate d, multiply equation (1) by -9 and equation (2) by 11. Then add the results.

$$
\begin{array}{rll}
405 &= -9a_1 - 99d & -9 \times (1) \\
-407 &= 11a_1 + 99d & 11 \times (2) \\
\hline
-2 &= 2a_1 & \\
-1 &= a_1 &
\end{array}
$$

25. $3, 5, 7, \ldots, 33$
Let n represent the number of terms in the sequence. So, $a_n = 33$, $a_1 = 3$, and $d = 5 - 3 = 2$.
$$a_n = a_1 + (n - 1)d$$
$$33 = 3 + (n - 1)2$$
$$33 = 3 + 2n - 2$$
$$33 = 2n + 1$$
$$32 = 2n$$
$$n = 16$$
The sequence has 16 terms.

27. $\dfrac{3}{4}, 3, \dfrac{21}{4}, \ldots, 12$
Let n represent the number of terms in the sequence. So, $a_n = 12$, $a_1 = \dfrac{3}{4}$, and $d = 3 - \dfrac{3}{4} = \dfrac{9}{4}$.
$$a_n = a_1 + (n - 1)d$$
$$12 = \frac{3}{4} + (n - 1)\left(\frac{9}{4}\right)$$
$$\frac{45}{4} = (n - 1)\left(\frac{9}{4}\right)$$
$$5 = n - 1 \qquad \textit{Multiply by } \frac{4}{9}.$$
$$6 = n$$
The sequence has 6 terms.

29. n represents the number of terms.

31. The sum of the left sides is $S + S = 2S$. For the right sides, $1 + 100 = 101$, $2 + 99 = 101$, $3 + 98 = 101$, and so on. The sum 101 appears 100 times. By multiplying, the sum of the right sides is $101 \cdot 100 = 10,100$.

32. $2S = 10,100$

33. Divide by 2 to obtain $S = 5050$.

34. The right sides now have a sum of 201 appearing 200 times. Thus, the sum is $201 \cdot 200 = 40,200$.
$$2S = 40,200$$
$$S = 20,100$$
Note that the pattern is $S = \dfrac{(n + 1)(n)}{2}$.

35. Find S_6 given $a_1 = 6$, $d = 3$, and $n = 6$.
$$S_n = \frac{n}{2}\big[2a_1 + (n - 1)d\big]$$
$$S_6 = \frac{6}{2}\big[2 \cdot 6 + (6 - 1)3\big]$$
$$= 3(12 + 5 \cdot 3)$$
$$= 3(27) = 81$$

37. Find S_6 given $a_1 = 7$, $d = -3$, and $n = 6$.
$$S_n = \frac{n}{2}\big[2a_1 + (n - 1)d\big]$$
$$S_6 = \frac{6}{2}\big[2 \cdot 7 + (6 - 1)(-3)\big]$$
$$= 3\big[14 + 5(-3)\big]$$
$$= 3(14 - 15) = 3(-1) = -3$$

39. Find S_6 given $a_n = 4 + 3n$.
$$a_1 = 4 + 3(1) = 7$$
$$a_6 = 4 + 3(6) = 22$$
$$S_n = \frac{n}{2}(a_1 + a_n)$$
$$S_6 = \frac{6}{2}(7 + 22)$$
$$= 3(29) = 87$$

41. $\displaystyle\sum_{i=1}^{10} (8i - 5)$
$$a_n = 8n - 5$$
$$a_1 = 8(1) - 5 = 3$$
$$a_{10} = 8(10) - 5 = 75$$
Use $S_n = \dfrac{n}{2}(a_1 + a_n)$ with $n = 10$, $a_1 = 3$, and $a_{10} = 75$.
$$S_{10} = \frac{10}{2}(3 + 75)$$
$$= 5(78) = 390$$

43. $\displaystyle\sum_{i=1}^{20} (2i - 5)$
$$a_n = 2n - 5$$
$$a_1 = 2(1) - 5 = -3$$
$$a_{20} = 2(20) - 5 = 35$$
Use $S_n = \dfrac{n}{2}(a_1 + a_n)$ with $n = 20$,

$a_1 = -3$, and $a_{20} = 35$.

$$S_{20} = \frac{20}{2}(-3 + 35)$$
$$= 10(32) = 320$$

45. $\sum_{i=1}^{250} i$

$a_n = n$, $a_1 = 1$, $a_{250} = 250$

Use $S_n = \frac{n}{2}(a_1 + a_n)$ with $n = 250$,

$a_1 = 1$, and $a_{250} = 250$.

$$S_{250} = \frac{250}{2}(1 + 250)$$
$$= 125(251) = 31,375$$

47. The sequence is $1, 2, 3, \ldots, 30$.

$$S_n = \frac{n}{2}(a_1 + a_n)$$
$$= \frac{30}{2}(1 + 30)$$
$$= 15(31) = 465$$

The account will have $465 deposited in it over the entire month.

49. Your salaries at six-month intervals form an arithmetic sequence with $a_1 = 1600$ and $d = 50$. Since your salary is increased every 6 months, or $2(5) = 10$ times, after 5 years your salary will equal the term a_{11}.

$$a_n = a_1 + (n - 1)d$$
$$a_{11} = 1600 + (11 - 1)50$$
$$= 1600 + 500 = 2100$$

Your salary will be $2100/month.

51. Given the sequence $20, 22, 24, \ldots$, for 25 terms, find a_{25}. Here,

$a_1 = 20$, $d = 22 - 20 = 2$, and $n = 25$.

$$a_n = a_1 + (n - 1)d$$
$$a_{25} = 20 + (25 - 1)2$$
$$= 20 + 48 = 68$$

There are 68 seats in the last row.

Now find S_{25}.

$$S_n = \frac{n}{2}(a_1 + a_n)$$
$$S_{25} = \frac{25}{2}(20 + 68)$$
$$= \frac{25}{2}(88) = 25(44) = 1100$$

There are 1100 seats in the section.

53. Given the sequence $35, 31, 27, \ldots$, can the sequence end in 1? If not, find the last positive value. If the sequence ends in 1, we can find n, a whole number.

$$d = 31 - 35 = -4$$
$$a_n = a_1 + (n - 1)d$$
$$1 = 35 + (n - 1)(-4)$$
$$1 = 35 - 4n + 4$$
$$-38 = -4n$$
$$9.5 = n$$

Since n is not a whole number, the sequence

cannot end in 1. The largest n possible is $n = 9$.

$$a_n = a_1 + (n - 1)d$$
$$a_9 = 35 + (9 - 1)(-4)$$
$$= 35 - 32 = 3$$

She can build 9 rows. There are 3 blocks in the last row.

55. $f(x) = mx + b$

$$f(1) = m(1) + b = m + b$$
$$f(2) = m(2) + b = 2m + b$$
$$f(3) = m(3) + b = 3m + b$$

56. The sequence $f(1)$, $f(2)$, $f(3)$ is an arithmetic sequence since the difference between any two adjacent terms is m.

57. The common difference is the difference between any two adjacent terms.

$$d = (3m + b) - (2m + b) = m$$

58. From Exercise 55, we know that $a_1 = m + b$. From Exercise 57, we know that $d = m$. Therefore,

$$a_n = a_1 + (n - 1)d$$
$$= (m + b) + (n - 1)m$$
$$= m + b + nm - m$$
$$a_n = mn + b.$$

Section 14.3

1. A geometric sequence is an ordered list of numbers such that each term after the first is obtained by multiplying the previous term by a constant, r, called the common ratio. For example, if the first term is 3 and $r = 4$, the sequence is $3, 12, 48, 192, \ldots$. If the first term is 2 and $r = -1$, the sequence is $2, -2, 2, -2, \ldots$.

3. $4, 8, 16, 32, \ldots$

To find r, choose any two adjacent terms and divide the second one by the first one.

$$r = \frac{8}{4} = 2$$

Notice that any two other adjacent terms could have been used with the same result. The common ratio is $r = 2$.

5. $\frac{1}{3}, \frac{2}{3}, \frac{3}{3}, \frac{4}{3}, \frac{5}{3}, \ldots$

Choose any two adjacent terms and divide the second by the first.

$$r = \frac{\frac{2}{3}}{\frac{1}{3}} = \frac{2}{3} \cdot 3 = 2$$

Confirm this result with any two other adjacent terms.

continued

$$r = \frac{\frac{3}{3}}{\frac{2}{3}} = 1 \cdot \frac{3}{2} = \frac{3}{2}$$

Since $2 \neq \frac{3}{2}$, the ratios are not the same. The sequence is *not geometric*.

7. $1, -3, 9, -27, 81, \ldots$

$$r = \frac{-3}{1} = -3$$
$$r = \frac{9}{-3} = -3$$

The common ratio is $r = -3$.

9. $1, -\frac{1}{2}, \frac{1}{4}, -\frac{1}{8}, \frac{1}{16}, \ldots$

$$r = \frac{-\frac{1}{2}}{1} = -\frac{1}{2}$$
$$r = \frac{\frac{1}{4}}{-\frac{1}{2}} = \frac{1}{4} \cdot (-2) = -\frac{1}{2}$$

The common ratio is $r = -\frac{1}{2}$.

11. Find a general term for $5, 10, \ldots$.
First, find r.

$$r = \frac{10}{5} = 2$$

Use $a_1 = 5$ and $r = 2$ to find a_n.

$$a_n = a_1 r^{n-1}$$
$$a_n = 5(2)^{n-1}$$

13. Find a general term for $\frac{1}{9}, \frac{1}{3}, \ldots$.
Here, $a_1 = \frac{1}{9}$. Find r.

$$r = \frac{\frac{1}{3}}{\frac{1}{9}} = \frac{1}{3} \cdot \frac{9}{1} = 3$$

Now find a_n.

$$a_n = a_1 r^{n-1}$$
$$a_n = \frac{1}{9}(3)^{n-1}$$
$$\text{or} \quad a_n = \frac{3^{n-1}}{9}$$

15. Find a general term for $10, -2, \ldots$.
Here, $a_1 = 10$. Find r.

$$r = \frac{-2}{10} = -\frac{1}{5}$$

Now find a_n.

$$a_n = a_1 r^{n-1}$$
$$a_n = 10\left(-\frac{1}{5}\right)^{n-1}$$

17. Given $2, 10, 50, \ldots$; find a_{10}.
First find the common ratio.

$$r = \frac{10}{2} = 5$$

Substitute $a_1 = 2$, $r = 5$, and $n = 10$ in the nth-term formula.

$$a_n = a_1 r^{n-1}$$
$$a_{10} = a_1(r)^{10-1}$$
$$= 2(5)^9$$

19. Given $\frac{1}{2}, \frac{1}{6}, \frac{1}{18}, \ldots$; find a_{12}.
First find the common ratio.

$$r = \frac{\frac{1}{6}}{\frac{1}{2}} = \frac{1}{6} \cdot 2 = \frac{1}{3}$$

Substitute $a_1 = \frac{1}{2}$, $r = \frac{1}{3}$, and $n = 12$ in the formula.

$$a_n = a_1 r^{n-1}$$
$$a_{12} = a_1(r)^{12-1}$$
$$= \left(\frac{1}{2}\right)\left(\frac{1}{3}\right)^{11}$$

21. Given $a_3 = \frac{1}{2}, a_7 = \frac{1}{32}$; find a_{25}.
Find a_1 and r using the general term $a_n = a_1 r^{n-1}$.

$$a_3 = a_1 r^{3-1}$$
$$\frac{1}{2} = a_1 r^2 \quad (1)$$
$$a_7 = a_1 r^{7-1}$$
$$\frac{1}{32} = a_1 r^6 \quad (2)$$

Solve (1) for a_1.

$$a_1 = \frac{1}{2r^2}$$

Substitute for a_1 in (2).

$$\frac{1}{32} = \frac{1}{2r^2} r^6$$
$$\frac{1}{16} = r^4$$
$$r^2 = \pm\frac{1}{4}$$

Since r^2 is positive,

$$r^2 = \frac{1}{4}.$$

Substitute $\frac{1}{4}$ for r^2 in (1).

$$\frac{1}{2} = a_1\left(\frac{1}{4}\right)$$

$$2 = a_1$$

Use $a_1 = 2$ and $r = \frac{1}{2}$ $\left(\text{or } -\frac{1}{2}\right)$ to find a_{25}.

$$a_{25} = a_1(r)^{25-1}$$

$$= 2\left(\frac{1}{2}\right)^{24}$$

$$= \frac{1}{2^{23}}$$

23. $\frac{1}{3} = .33333\ldots$

24. $\frac{2}{3} = .66666\ldots$

25. $\quad .33333\ldots$
$\underline{+ .66666\ldots}$
$\quad .99999\ldots$

26. $S = \dfrac{a_1}{1-r}$

$$= \frac{.9}{1-.1}$$

$$= \frac{.9}{.9} = 1$$

Therefore, $.99999\ldots = 1$.

27. $\dfrac{1}{3}, \dfrac{1}{9}, \dfrac{1}{27}, \dfrac{1}{81}, \dfrac{1}{243}$

Here, $a_1 = \dfrac{1}{3}$, $n = 5$, and

$$r = \frac{\frac{1}{9}}{\frac{1}{3}} = \frac{1}{9} \cdot 3 = \frac{1}{3}.$$

$$S_n = \frac{a_1(r^n - 1)}{r - 1}$$

$$S_5 = \frac{\frac{1}{3}\left[\left(\frac{1}{3}\right)^5 - 1\right]}{\frac{1}{3} - 1}$$

$$= \frac{\frac{1}{3}\left(\frac{1}{243} - 1\right)}{-\frac{2}{3}}$$

$$= \frac{\frac{1}{3}\left(-\frac{242}{243}\right)}{-\frac{2}{3}} = \frac{121}{243}$$

29. $-\dfrac{4}{3}, -\dfrac{4}{9}, -\dfrac{4}{27}, -\dfrac{4}{81}, -\dfrac{4}{243}, -\dfrac{4}{729}$

Here, $a_1 = -\dfrac{4}{3}$, $n = 6$, and

$$r = \frac{-\frac{4}{9}}{-\frac{4}{3}} = -\frac{4}{9} \cdot \left(-\frac{3}{4}\right) = \frac{1}{3}.$$

$$S_n = \frac{a_1(r^n - 1)}{r - 1}$$

$$S_6 = \frac{-\frac{4}{3}\left[\left(\frac{1}{3}\right)^6 - 1\right]}{\frac{1}{3} - 1}$$

$$= \frac{-\frac{4}{3}\left(\frac{1}{729} - 1\right)}{-\frac{2}{3}}$$

$$= \frac{-\frac{4}{3}\left(-\frac{728}{729}\right)}{-\frac{2}{3}} = -\frac{1456}{729} \approx -1.997$$

31. $\displaystyle\sum_{i=1}^{7} 4\left(\frac{2}{5}\right)^i$

Use $a_1 = 4\left(\dfrac{2}{5}\right) = \dfrac{8}{5}$, $n = 7$, and $r = \dfrac{2}{5}$.

$$S_n = \frac{a_1(r^n - 1)}{r - 1}$$

$$S_7 = \frac{\frac{8}{5}\left[\left(\frac{2}{5}\right)^7 - 1\right]}{\frac{2}{5} - 1}$$

$$= \frac{\frac{8}{5}\left[\left(\frac{2}{5}\right)^7 - 1\right]}{-\frac{3}{5}}$$

$$= -\frac{8}{3}\left[\left(\frac{2}{5}\right)^7 - 1\right] \approx 2.662$$

33. $\displaystyle\sum_{i=1}^{10} (-2)\left(\frac{3}{5}\right)^i$

Use $a_1 = (-2)\left(\dfrac{3}{5}\right) = -\dfrac{6}{5}$, $n = 10$, and $r = \dfrac{3}{5}$.

continued

$$S_n = \frac{a_1(r^n - 1)}{r - 1}$$

$$S_{10} = \frac{-\frac{6}{5}\left[\left(\frac{3}{5}\right)^{10} - 1\right]}{\frac{3}{5} - 1}$$

$$= \frac{-\frac{6}{5}\left[\left(\frac{3}{5}\right)^{10} - 1\right]}{-\frac{2}{5}}$$

$$= 3\left[\left(\frac{3}{5}\right)^{10} - 1\right] \approx -2.982$$

35. There are 22 deposits, so $n = 22$.

$$S = R\left[\frac{(1 + i)^n - 1}{i}\right]$$

$$= 1000\left[\frac{(1 + .095)^{22} - 1}{.095}\right]$$

$$= 66,988.91$$

There will be $66,988.91 in the account.

37. Quarterly deposits for 10 years give us $n = 4 \cdot 10 = 40$. The interest rate per period is $i = \frac{.07}{4} = .0175$.

$$S = R\left[\frac{(1 + i)^n - 1}{i}\right]$$

$$= 1200\left[\frac{(1 + .0175)^{40} - 1}{.0175}\right]$$

$$= 68,680.96$$

We now use the compound interest formula to determine the value of this money after 5 more years.

$$A = P\left(1 + \frac{r}{n}\right)^{nt}$$

$$= 68,680.96\left(1 + \frac{.09}{12}\right)^{12(5)}$$

$$= 107,532.48$$

The woman is also saving $300 per month, so we use the annuity formula to determine that value.

$$S = R\left[\frac{(1 + i)^n - 1}{i}\right]$$

$$= 300\left[\frac{\left(1 + \frac{.09}{12}\right)^{12(5)} - 1}{\frac{.09}{12}}\right]$$

$$= 22,627.24$$

Adding $22,627.24 to $107,532.48 gives a total of $130,159.72 in the account.

39. Find the sum if $a_1 = 6$ and $r = \frac{1}{3}$. Since $|r| < 1$, the sum exists.

$$S = \frac{a_1}{1 - r} = \frac{6}{1 - \frac{1}{3}} = \frac{6}{\frac{2}{3}} = 6 \cdot \frac{3}{2} = 9$$

41. Find the sum if $a_1 = 1000$ and $r = -\frac{1}{10}$. Since $|r| < 1$, the sum exists.

$$S = \frac{a_1}{1 - r} = \frac{1000}{1 - \left(-\frac{1}{10}\right)} = \frac{1000}{\frac{11}{10}}$$

$$= 1000 \cdot \frac{10}{11} = \frac{10,000}{11}$$

43. $\sum_{i=1}^{\infty} \frac{9}{8}\left(-\frac{2}{3}\right)^i$

$a_1 = \frac{9}{8}\left(-\frac{2}{3}\right)^1 = -\frac{3}{4}$ and $r = -\frac{2}{3}$.

Since $|r| < 1$, the sum exists.

$$S = \frac{a_1}{1 - r} = \frac{-\frac{3}{4}}{1 - \left(-\frac{2}{3}\right)} = \frac{-\frac{3}{4}}{\frac{5}{3}}$$

$$= -\frac{3}{4} \cdot \frac{3}{5} = -\frac{9}{20}$$

45. $\sum_{i=1}^{\infty} \frac{12}{5}\left(\frac{5}{4}\right)^i$

Since $|r| = \frac{5}{4} > 1$, the sum *does not exist.*

47. The ball is dropped from a height of 10 feet and will rebound $\frac{3}{5}$ of its original height.

Let a_n = the ball's height on the nth rebound.

$a_1 = 10$ and $r = \frac{3}{5}$.

Since we must find the height after the fourth bounce, $n = 5$ (since a_1 is the starting point).

Use $a_n = a_1 r^{n-1}$.

$$a_5 = 10\left(\frac{3}{5}\right)^{5-1} = 10\left(\frac{3}{5}\right)^4 \approx 1.3$$

The ball will rebound 1.3 feet after the fourth bounce.

49. This exercise can be modeled by a geometric sequence with $a_1 = 256$ and $r = \frac{1}{2}$. First we need to find n so that $a_n = 32$.

$$32 = a_1 r^{n-1}$$

$$32 = 256 \left(\frac{1}{2}\right)^{n-1}$$

$$\frac{1}{8} = \left(\frac{1}{2}\right)^{n-1}$$

$$\left(\frac{1}{2}\right)^3 = \left(\frac{1}{2}\right)^{n-1}$$

$$3 = n - 1$$

$$4 = n$$

Since n is 4, this means that 32 grams will be present on the day which corresponds to the 4th term of the sequence. That would be on day 3. To find what is left after the tenth day, we need to find a_{11} since we started with a_1.

$$a_{11} = a_1 r^{11-1} = 256 \left(\frac{1}{2}\right)^{10} = \frac{256}{1024} = \frac{1}{4}$$

There will be $\frac{1}{4}$ grams of the substance after 10 days.

51. **(a)** Here, $a_1 = 1.1$ billion and $r = 106\% = 1.06$. Since we must find the consumption after 5 years, $n = 6$ (since a_1 is the starting point).

$$a_6 = a_1 r^{n-1}$$
$$a_6 = 1.1(1.06)^{6-1}$$
$$= 1.1(1.06)^5 \approx 1.5$$

The community will use about 1.5 billion units 5 years from now.

(b) If consumption doubles, then the consumption would be $2a_1$.

$$2a_1 = a_1(1.06)^{n-1}$$
$$2 = (1.06)^{n-1}$$
$$\ln 2 = \ln(1.06)^{n-1}$$
$$\ln 2 = (n-1)\ln(1.06)$$
$$n - 1 = \frac{\ln 2}{\ln 1.06}$$
$$n - 1 \approx 12$$
$$n \approx 13$$

Since n is about 13, that would represent the 13th term of the sequence, which represents about 12 years after the start.

53. Since the machine depreciates by $\frac{1}{4}$ of its value, it retains $1 - \frac{1}{4} = \frac{3}{4}$ of its value. Since the cost of the machine new is $\$50,000$, $a_1 = 50,000$. We want the value after 8 years so since the original cost is a_1, we need to find a_9.

$$a_9 = a_1 r^{9-1}$$
$$= 50,000 \left(\frac{3}{4}\right)^8 \approx 5006$$

The machine's value after 8 years is about $\$5000$.

55. $g(x) = ab^x$
$$g(1) = ab^1 = ab, \ g(2) = ab^2, \ g(3) = ab^3$$

56. The sequence $g(1), g(2), g(3)$ is a geometric sequence because each term after the first is a constant multiple of the preceding term.

57. The common ratio is $r = \dfrac{ab^2}{ab} = b$.

58. From Exercise 55, $a_1 = ab$. From Exercise 57, $r = b$. Therefore,

$$a_n = a_1 r^{n-1}$$
$$= ab(b)^{n-1}$$
$$= ab^{1+n-1}$$
$$= ab^n.$$

Section 14.4

1. $2! = 2 \cdot 1 = 2$

3. $\dfrac{6!}{4! \, 2!} = \dfrac{6 \cdot 5 \cdot 4 \cdot 3 \cdot 2 \cdot 1}{(4 \cdot 3 \cdot 2 \cdot 1)(2 \cdot 1)} = \dfrac{6 \cdot 5}{2 \cdot 1} = 15$

5. $_6C_2 = \dfrac{6!}{2! \, (6-2)!} = \dfrac{6!}{2! \, 4!} = 15$, by Exercise 3.

7. $\dfrac{4!}{0! \, 4!} = \dfrac{4!}{(1)(4!)} = \dfrac{1}{1} = 1$

9. $5! + 2! = 5 \cdot 4 \cdot 3 \cdot 2 \cdot 1 + 2 \cdot 1$
$$= 120 + 2 = 122$$

11. 10 nCr 3 $= \dfrac{10!}{3! \, (10-3)!}$
$$= \dfrac{10!}{3! \, 7!}$$
$$= \dfrac{10 \cdot 9 \cdot 8 \cdot 7 \cdot 6 \cdot 5 \cdot 4 \cdot 3 \cdot 2 \cdot 1}{(3 \cdot 2 \cdot 1)(7 \cdot 6 \cdot 5 \cdot 4 \cdot 3 \cdot 2 \cdot 1)}$$
$$= \dfrac{10 \cdot 9 \cdot 8}{3 \cdot 2 \cdot 1} = 120$$

13. $(m + n)^4$
$$= m^4 + \frac{4!}{3! \, 1!} m^3 n^1 + \frac{4!}{2! \, 2!} m^2 n^2$$
$$+ \frac{4!}{1! \, 3!} m^1 n^3 + n^4$$
$$= m^4 + 4m^3 n + 6m^2 n^2 + 4mn^3 + n^4$$

15. $(a - b)^5$
$$= \left[a + (-b)\right]^5$$
$$= a^5 + \frac{5!}{4! \, 1!} a^4(-b)^1 + \frac{5!}{3! \, 2!} a^3(-b)^2$$
$$+ \frac{5!}{2! \, 3!} a^2(-b)^3 + \frac{5!}{1! \, 4!} a^1(-b)^4 + (-b)^5$$
$$= a^5 - 5a^4 b + 10a^3 b^2 - 10a^2 b^3 + 5ab^4 - b^5$$

17. $(2x + 3)^3$

$$= (2x)^3 + \frac{3!}{2!\,1!}(2x)^2(3)^1 + \frac{3!}{1!\,2!}(2x)(3)^2$$

$$+ (3)^3$$

$$= 8x^3 + 36x^2 + 54x + 27$$

19. $\left(\dfrac{x}{3} + 2y\right)^5$

$$= \left(\frac{x}{3}\right)^5 + \frac{5!}{4!\,1!}\left(\frac{x}{3}\right)^4(2y)^1$$

$$+ \frac{5!}{3!\,2!}\left(\frac{x}{3}\right)^3(2y)^2 + \frac{5!}{2!\,3!}\left(\frac{x}{3}\right)^2(2y)^3$$

$$+ \frac{5!}{1!\,4!}\left(\frac{x}{3}\right)^1(2y)^4 + (2y)^5$$

$$= \frac{x^5}{243} + \frac{10x^4y}{81} + \frac{40x^3y^2}{27} + \frac{80x^2y^3}{9}$$

$$+ \frac{80xy^4}{3} + 32y^5$$

21. $(mx - n^2)^3$

$$= \left[mx + (-n^2)\right]^3$$

$$= (mx)^3 + \frac{3!}{2!\,1!}(mx)^2(-n^2)^1$$

$$+ \frac{3!}{1!\,2!}(mx)^1(-n^2)^2 + (-n^2)^3$$

$$= m^3x^3 - 3m^2n^2x^2 + 3mn^4x - n^6$$

23. $(r + 2s)^{12}$

$$= r^{12} + \frac{12!}{11!\,1!}r^{11}(2s)^1 + \frac{12!}{10!\,2!}r^{10}(2s)^2$$

$$+ \frac{12!}{9!\,3!}r^9(2s)^3 + \cdots$$

The first four terms are
$r^{12} + 24r^{11}s + 264r^{10}s^2 + 1760r^9s^3$.

25. $(3x - y)^{14}$

$$= \left[3x + (-y)\right]^{14}$$

$$= (3x)^{14} + \frac{14!}{13!\,1!}(3x)^{13}(-y)^1$$

$$+ \frac{14!}{12!\,2!}(3x)^{12}(-y)^2$$

$$+ \frac{14!}{11!\,3!}(3x)^{11}(-y)^3 + \cdots$$

The first four terms are

$$3^{14}x^{14} - 14(3^{13})x^{13}y + 91(3^{12})x^{12}y^2$$

$$- 364(3^{11})x^{11}y^3.$$

27. $(t^2 + u^2)^{10}$

$$= (t^2)^{10} + \frac{10!}{9!\,1!}(t^2)^9(u^2)^1$$

$$+ \frac{10!}{8!\,2!}(t^2)^8(u^2)^2$$

$$+ \frac{10!}{7!\,3!}(t^2)^7(u^2)^3 + \cdots$$

The first four terms are

$$t^{20} + 10t^{18}u^2 + 45t^{16}u^4 + 120t^{14}u^6.$$

29. The rth term of the expansion of $(x + y)^n$ is

$$\frac{n!}{[n - (r-1)]!\,(r-1)!}(x)^{n-(r-1)}(y)^{r-1}.$$

Start with the exponent on y, which is 1 less than the term number r. In this case, we are looking for the fourth term, so $r = 4$ and $r - 1 = 3$. Thus, the fourth term of $(2m + n)^{10}$ is

$$\frac{10!}{(10-3)!\,3!}(2m)^{10-3}(n)^3$$

$$= \frac{10!}{7!\,3!}2^7m^7n^3$$

$$= 120(2^7)m^7n^3.$$

31. The seventh term of $\left(x + \dfrac{y}{2}\right)^8$ is

$$\frac{8!}{(8-6)!\,6!}(x)^{8-6}\left(\frac{y}{2}\right)^6$$

$$= \frac{8!}{6!\,2!}x^2\,\frac{y^6}{2^6}$$

$$= \frac{7x^2y^6}{16}.$$

33. The third term of $(k - 1)^9$ is

$$\frac{9!}{(9-2)!\,2!}k^{9-2}(-1)^2 = \frac{9!}{7!\,2!}k^7 = 36k^7.$$

35. The expansion of $(x^2 - 2y)^6$ has seven terms, so the middle term is the fourth. The fourth term of $(x^2 - 2y)^6$ is

$$\frac{6!}{(6-3)!\,3!}(x^2)^{6-3}(-2y)^3$$

$$= \frac{6!}{3!\,3!}(x^2)^3(-8y^3)$$

$$= 20x^6(-8y^3) = -160x^6y^3.$$

37. The term of the expansion of $(3x^3 - 4y^2)^5$ with x^9y^4 in it is the term with $(3x^3)^3(-4y^2)^2$, since $(x^3)^3(y^2)^2 = x^9y^4$. The term is

$$\frac{5!}{3!\,2!}(3x^3)^3(-4y^2)^2$$

$$= 10(27x^9)(16y^4) = 4320x^9y^4.$$

Chapter 14 Review Exercises

1. $a_n = 2n - 3$

$a_1 = 2(1) - 3 = -1$

$a_2 = 2(2) - 3 = 1$

$a_3 = 2(3) - 3 = 3$

$a_4 = 2(4) - 3 = 5$

Answer: $-1, 1, 3, 5$

2. $a_n = \dfrac{n-1}{n}$

$a_1 = \dfrac{1-1}{1} = 0$

$a_2 = \dfrac{2-1}{2} = \dfrac{1}{2}$

$a_3 = \dfrac{3-1}{3} = \dfrac{2}{3}$

$a_4 = \dfrac{4-1}{4} = \dfrac{3}{4}$

Answer: $0, \dfrac{1}{2}, \dfrac{2}{3}, \dfrac{3}{4}$

3. $a_n = n^2$

$a_1 = (1)^2 = 1$

$a_2 = (2)^2 = 4$

$a_3 = (3)^2 = 9$

$a_4 = (4)^2 = 16$

Answer: $1, 4, 9, 16$

4. $a_n = \left(\dfrac{1}{2}\right)^n$

$a_1 = \left(\dfrac{1}{2}\right)^1 = \dfrac{1}{2}$

$a_2 = \left(\dfrac{1}{2}\right)^2 = \dfrac{1}{4}$

$a_3 = \left(\dfrac{1}{2}\right)^3 = \dfrac{1}{8}$

$a_4 = \left(\dfrac{1}{2}\right)^4 = \dfrac{1}{16}$

Answer: $\dfrac{1}{2}, \dfrac{1}{4}, \dfrac{1}{8}, \dfrac{1}{16}$

5. $a_n = (n+1)(n-1)$

$a_1 = (1+1)(1-1) = 2(0) = 0$

$a_2 = (2+1)(2-1) = 3(1) = 3$

$a_3 = (3+1)(3-1) = 4(2) = 8$

$a_4 = (4+1)(4-1) = 5(3) = 15$

Answer: $0, 3, 8, 15$

6. $\displaystyle\sum_{i=1}^{5} i^2 x$

$= 1^2 x + 2^2 x + 3^2 x + 4^2 x + 5^2 x$

$= x + 4x + 9x + 16x + 25x$

7. $\displaystyle\sum_{i=1}^{6} (i+1)x^i$

$= (1+1)x^1 + (2+1)x^2 + (3+1)x^3$

$\quad + (4+1)x^4 + (5+1)x^5 + (6+1)x^6$

$= 2x + 3x^2 + 4x^3 + 5x^4 + 6x^5 + 7x^6$

8. $\displaystyle\sum_{i=1}^{4} (i+2)$

$= (1+2) + (2+2) + (3+2) + (4+2)$

$= 3 + 4 + 5 + 6 = 18$

9. $\displaystyle\sum_{i=1}^{6} 2^i$

$= 2^1 + 2^2 + 2^3 + 2^4 + 2^5 + 2^6$

$= 2 + 4 + 8 + 16 + 32 + 64 = 126$

10. $\displaystyle\sum_{i=4}^{7} \dfrac{i}{i+1}$

$= \dfrac{4}{4+1} + \dfrac{5}{5+1} + \dfrac{6}{6+1} + \dfrac{7}{7+1}$

$= \dfrac{4}{5} + \dfrac{5}{6} + \dfrac{6}{7} + \dfrac{7}{8} \qquad LCD = 2^3 \cdot 3 \cdot 5 \cdot 7 = 840$

$= \dfrac{672}{840} + \dfrac{700}{840} + \dfrac{720}{840} + \dfrac{735}{840} = \dfrac{2827}{840}$

11. $\bar{x} = \dfrac{\text{total}}{5}$, where $\text{total} = 684,588 + 680,913$

$\quad + 654,110 + 652,945 + 652,848.$

$\bar{x} = \dfrac{3,325,404}{5} = 665,080.8$

The average share volume is about $665,081$ (thousand) for the five given days.

12. $2, 5, 8, 11, \ldots$ is an *arithmetic* sequence with $d = 5 - 2 = 3$.

13. $-6, -2, 2, 6, 10, \ldots$ is an *arithmetic* sequence with $d = -2 - (-6) = 4$.

14. $\dfrac{2}{3}, -\dfrac{1}{3}, \dfrac{1}{6}, -\dfrac{1}{12}, \ldots$ is a *geometric* sequence with

$r = \dfrac{-\dfrac{1}{3}}{\dfrac{2}{3}} = -\dfrac{1}{3} \cdot \dfrac{3}{2} = -\dfrac{1}{2}$.

15. $-1, 1, -1, 1, -1, \ldots$ is a *geometric* sequence

with $r = \dfrac{1}{-1} = -1$.

16. $64, 32, 8, \dfrac{1}{2}, \ldots$

$32 - 64 = -32$ and $8 - 32 = -24$, so the sequence is not arithmetic.

$\dfrac{32}{64} \neq \dfrac{8}{32}$, so the sequence is not geometric.

Therefore, the sequence is *neither*.

17. $64, 32, 16, 8, \ldots$ is a *geometric* sequence with

$r = \dfrac{32}{64} = \dfrac{1}{2}$.

18. $10, 8, 6, 4, \ldots$ is an *arithmetic* sequence with $d = 8 - 10 = -2$.

19. Given $a_1 = -2$, $d = 5$; find a_{16}.
$$a_n = a_1 + (n-1)d$$
$$a_{16} = -2 + (16-1)5$$
$$= -2 + 15(5)$$
$$= -2 + 75 = 73$$

20. Given $a_6 = 12$, $a_8 = 18$; find a_{25}.
$$a_n = a_1 + (n-1)d$$
$$a_6 = a_1 + 5d, \quad \text{so} \quad 12 = a_1 + 5d \quad (1)$$
$$a_8 = a_1 + 7d, \quad \text{so} \quad 18 = a_1 + 7d \quad (2)$$
Multiply equation (1) by -1 and add the result to equation (2).
$$-12 = -a_1 - 5d \qquad -1 \times (1)$$
$$\underline{18 = \quad a_1 + 7d \quad (2)}$$
$$6 = \qquad 2d$$
$$3 = d$$
To find a_1, substitute $d = 3$ in equation (1).
$$12 = a_1 + 5d \qquad (1)$$
$$12 = a_1 + 5(3)$$
$$12 = a_1 + 15$$
$$-3 = a_1$$
Use $a_1 = -3$ and $d = 3$ to find a_{25}.
$$a_n = a_1 + (n-1)d$$
$$a_{25} = -3 + (25-1)3$$
$$= -3 + 24(3)$$
$$= -3 + 72 = 69$$

21. $a_1 = -4$, $d = -5$
$$a_n = a_1 + (n-1)d$$
$$a_n = -4 + (n-1)(-5)$$
$$= -4 - 5n + 5$$
$$= -5n + 1$$

22. $6, 3, 0, -3, \ldots$
To get the general term, a_n, find d.
$$d = 3 - 6 = -3$$
$$a_n = a_1 + (n-1)d$$
$$a_n = 6 + (n-1)(-3)$$
$$a_n = 6 - 3n + 3$$
$$a_n = -3n + 9$$

23. $7, 0, 13, \ldots, 49$
$a_1 = 7$, $d = 10 - 7 = 3$; find n, the number of terms.
$$a_n = a_1 + (n-1)d$$
$$49 = 7 + (n-1)(3)$$
$$42 = 3(n-1)$$
Divide by 3.
$$14 = n - 1$$
$$15 = n$$
There are 15 terms in this sequence.

24. $5, 1, -3, \ldots, -79$
$a_1 = 5$, $d = 1 - 5 = -4$; find n, the number of terms.
$$a_n = a_1 + (n-1)d$$
$$-79 = 5 + (n-1)(-4)$$
$$-79 = 9 - 4n$$
$$-88 = -4n$$
$$n = 22$$
There are 22 terms in this sequence.

25. Find S_8 if $a_1 = -2$ and $d = 6$.
$$S_n = \frac{n}{2}(a_1 + a_n); \text{ find } a_8 \text{ first.}$$
$$a_8 = a_1 + (8-1)d$$
$$= -2 + 7(6)$$
$$= -2 + 42 = 40$$
$$S_8 = \frac{8}{2}(-2 + 40)$$
$$= 4(38) = 152$$

26. Find S_8 if $a_n = -2 + 5n$.
$$a_1 = -2 + 5(1) = 3$$
$$a_8 = -2 + 5(8) = 38$$
$$S_n = \frac{n}{2}(a_1 + a_n)$$
$$S_8 = \frac{8}{2}(3 + 38) = 4(41) = 164$$

27. Find the general term for the geometric sequence $-1, -4, \ldots$.
$$a_1 = -1 \text{ and } r = \frac{-4}{-1} = 4.$$
$$a_n = a_1 r^{n-1}$$
$$a_n = -1(4)^{n-1}$$

28. $\dfrac{2}{3}, \dfrac{2}{15}, \ldots$
$$a_1 = \frac{2}{3} \text{ and } r = \frac{\frac{2}{15}}{\frac{2}{3}} = \frac{2}{15} \cdot \frac{3}{2} = \frac{1}{5}.$$
$$a_n = a_1 r^{n-1}$$
$$a_n = \frac{2}{3}\left(\frac{1}{5}\right)^{n-1}$$

29. Find a_{11} for $2, -6, 18, \ldots$.
$$a_1 = 2 \text{ and } r = \frac{-6}{2} = -3$$
$$a_n = a_1 r^{n-1}$$
$$a_{11} = 2(-3)^{11-1}$$
$$= 2(-3)^{10} = 118,098$$

30. Given $a_3 = 20$, $a_5 = 80$; find a_{10}.
$$a_n = a_1 r^{n-1}$$
For a_3, $\quad a_3 = a_1 r^{3-1}$
$$20 = a_1 r^2.$$

For a_5, $\qquad a_5 = a_1 r^{5-1}$
$\qquad\qquad 80 = a_1 r^4.$
The ratio of a_5 to a_3 is
$$\frac{80}{20} = \frac{a_1 r^4}{a_1 r^2}$$
$$4 = r^2$$
$$r = \pm 2.$$
Since $20 = a_1 r^2$, and $r^2 = 4$,
$$20 = a_1(4)$$
$$5 = a_1.$$
Now find a_{10}.
$$a_n = a_1 r^{n-1}$$
$$a_{10} = 5(\pm 2)^{10-1}$$
$$= 5(\pm 2)^9$$

Two answers are possible for a_{10}:

$5(2)^9 = 2560$ or $5(-2)^9 = -2560.$

31. $\displaystyle\sum_{i=1}^{5}\left(\frac{1}{4}\right)^i$

$a_1 = \dfrac{1}{4}, r = \dfrac{1}{4}, n = 5.$

$S_n = \dfrac{a_1(1 - r^n)}{1 - r}$

$$S_5 = \frac{\dfrac{1}{4}\left[1 - \left(\dfrac{1}{4}\right)^5\right]}{1 - \dfrac{1}{4}}$$

$$= \frac{\dfrac{1}{4}\left(1 - \dfrac{1}{1024}\right)}{\dfrac{3}{4}}$$

$$= \frac{1}{3}\left(\frac{1023}{1024}\right) = \frac{341}{1024}$$

32. $\displaystyle\sum_{i=1}^{8}\frac{3}{4}(-1)^i$

$a_1 = -\dfrac{3}{4}$ and $r = -1$, and $n = 8.$

$S_n = \dfrac{a_1(1 - r^n)}{1 - r}$

$$S_8 = \frac{-\dfrac{3}{4}\left[1 - (-1)^8\right]}{1 - (-1)}$$

$$= \frac{-\dfrac{3}{4}(1 - 1)}{2}$$

$$= \frac{-\dfrac{3}{4}(0)}{2} = 0$$

33. $\displaystyle\sum_{i=1}^{\infty}4\left(\frac{1}{5}\right)^i$

The terms are the infinite geometric sequence with
$a_1 = \dfrac{4}{5}$ and $r = \dfrac{1}{5}.$

$$S = \frac{a_1}{1 - r} = \frac{\dfrac{4}{5}}{1 - \dfrac{1}{5}} = \frac{\dfrac{4}{5}}{\dfrac{4}{5}} = 1$$

34. $\displaystyle\sum_{i=1}^{\infty}2(3)^i$

The terms are the infinite geometric sequence with
$a_1 = 6$ and $r = 3.$
$S = \dfrac{a_1}{1 - r}$ if $|r| < 1,$
but $r = 3$ so S does not exist, and, thus, the sum
does not exist.

35. $(2p - q)^5$

$= \left[2p + (-q)\right]^5$

$= (2p)^5 + \dfrac{5!}{4!\,1!}(2p)^4(-q)^1$

$\quad + \dfrac{5!}{3!\,2!}(2p)^3(-q)^2 + \dfrac{5!}{2!\,3!}(2p)^2(-q)^3$

$\quad + \dfrac{5!}{1!\,4!}(2p)^1(-q)^4 + (-q)^5$

$= 32p^5 + 5(16p^4)(-q) + 10(8p^3)q^2$

$\quad + 10(4p^2)(-q^3) + 5(2p)q^4 - q^5$

$= 32p^5 - 80p^4q + 80p^3q^2 - 40p^2q^3$

$\quad + 10pq^4 - q^5$

36. $(x^2 + 3y)^4$

$= (x^2)^4 + \dfrac{4!}{3!\,1!}(x^2)^3(3y)^1$

$\quad + \dfrac{4!}{2!\,2!}(x^2)^2(3y)^2$

$\quad + \dfrac{4!}{1!\,3!}(x^2)^1(3y)^3 + (3y)^4$

$= x^8 + 4(x^6)(3y) + 6(x^4)(9y^2)$

$\quad + 4(x^2)(27y^3) + 81y^4$

$= x^8 + 12x^6y + 54x^4y^2 + 108x^2y^3$

$\quad + 81y^4$

37. $\left(\sqrt{m} + \sqrt{n}\right)^4$

$= (\sqrt{m})^4 + \dfrac{4!}{3!\,1!}(\sqrt{m})^3(\sqrt{n})^1$

$\quad + \dfrac{4!}{2!\,2!}(\sqrt{m})^2(\sqrt{n})^2$

$\quad + \dfrac{4!}{1!\,3!}(\sqrt{m})^1(\sqrt{n})^3 + (\sqrt{n})^4$

$= m^2 + 4(m\sqrt{m})(\sqrt{n}) + 6(m)(n)$

$\quad + 4(\sqrt{m})(n\sqrt{n}) + n^2$

$= m^2 + 4m\sqrt{mn} + 6mn + 4n\sqrt{mn} + n^2$

38. The fourth term ($r = 4$, so $r - 1 = 3$) of $(3a + 2b)^{19}$ is

$$\frac{19!}{16!\,3!}(3a)^{16}(2b)^3$$
$$= 969(3)^{16}(a^{16})(8)b^3$$
$$= 7752(3)^{16}a^{16}b^3.$$

39. The twenty-third term ($r = 23$, so $r - 1 = 22$) of $(-2k + 3)^{25}$ is

$$= \frac{25!}{3!\,22!}(-2k)^3(3)^{22}$$
$$= -18,400(3)^{22}k^3.$$

40. The arithmetic sequence $1, 7, 13, \ldots$ has $a_1 = 1$ and $d = 7 - 1 = 6$.

$$a_n = a_1 + (n - 1)d$$
$$a_{40} = 1 + (40 - 1)6$$
$$= 1 + (39)6$$
$$= 1 + 234 = 235$$
$$a_{10} = 1 + (10 - 1)6$$
$$= 1 + (9)6 = 55$$
$$S_n = \frac{n}{2}(a_1 + a_n)$$
$$S_{10} = \frac{10}{2}(1 + 55)$$
$$= 5(56) = 280$$

41. The geometric sequence $-3, 6, -12, \ldots$ has

$$a_1 = -3 \text{ and } r = \frac{6}{-3} = -2.$$
$$a_n = a_1 r^{n-1}$$
$$a_{10} = -3(-2)^9$$
$$= -3(-512) = 1536$$
$$S_n = \frac{a_1(1 - r^n)}{1 - r}$$
$$S_{10} = \frac{-3\left[1 - (-2)^{10}\right]}{1 - (-2)}$$
$$= \frac{-3(1 - 1024)}{3}$$
$$= -(-1023) = 1023$$

42. $a_1 = 1, r = -3$

$$a_n = a_1 r^{n-1}$$
$$a_9 = 1(-3)^{9-1}$$
$$= (-3)^8 = 6561$$
$$S_n = \frac{a_1(r^n - 1)}{r - 1}$$
$$S_{10} = \frac{1\left[(-3)^{10} - 1\right]}{-3 - 1}$$
$$= -\frac{1}{4}(3^{10} - 1)$$
$$= -\frac{1}{4}(59,049 - 1)$$
$$= -14,762$$

43. The arithmetic sequence with $a_1 = -4$ and $d = 3$ is $-4, -1, 2, 5, \ldots$.

$$a_n = a_1 + (n - 1)d$$
$$a_{15} = -4 + (15 - 1)3$$
$$= -4 + 42 = 38$$

$$S_n = \frac{n}{2}[2a_1 + (n - 1)d]$$
$$S_{10} = \frac{10}{2}[2(-4) + (10 - 1)3]$$
$$= 5(-8 + 27)$$
$$= 5(19) = 95$$

44. $2, 7, 12, \ldots$

This is an arithmetic sequence with $a_1 = 2$ and $d = 7 - 2 = 5$.

$$a_n = a_1 + (n - 1)d$$
$$a_n = 2 + (n - 1)5$$
$$= 2 + 5n - 5$$
$$= 5n - 3$$

45. $2, 8, 32, \ldots$

This is a geometric sequence with $a_1 = 2$ and $r = \frac{8}{2} = 4$.

$$a_n = a_1 r^{n-1}$$
$$a_n = 2(4)^{n-1}$$

46. $27, 9, 3, \ldots$

This is a geometric sequence with $a_1 = 27$ and $r = \frac{9}{27} = \frac{1}{3}$.

$$a_n = a_1 r^{n-1}$$
$$a_n = 27\left(\frac{1}{3}\right)^{n-1}$$

47. $12, 9, 6, \ldots$

This is an arithmetic sequence with $a_1 = 12$ and $d = -3$.

$$a_n = a_1 + (n - 1)d$$
$$a_n = 12 + (n - 1)(-3)$$
$$= 12 - 3n + 3$$
$$= -3n + 15$$

48. The distances traveled in successive seconds are $3, 7, 11, 15, 19, \ldots$.

This is an arithmetic sequence.

$$S_n = \frac{n}{2}[2a_1 + (n - 1)d]$$

$$S_n = 210, \quad a_1 = 3, \quad d = 4$$
$$210 = \frac{n}{2}[2(3) + (n - 1)4]$$
$$420 = n(6 + 4n - 4)$$
$$420 = 6n + 4n^2 - 4n$$
$$0 = 4n^2 + 2n - 420$$
$$0 = 2n^2 + n - 210$$
$$0 = (2n + 21)(n - 10)$$

$$2n + 21 = 0 \quad \text{or} \quad n - 10 = 0$$

$$n = -\frac{21}{2} \quad \text{or} \quad n = 10$$

Discard $-\frac{21}{2}$ since time cannot be negative. It takes her 10 seconds.

49. $S = R\left[\dfrac{(1+i)^n - 1}{i}\right]$

$$S = 672\left[\dfrac{\left(1 + \dfrac{.08}{4}\right)^{4(7)} - 1}{\dfrac{.08}{4}}\right]$$

$$= 672\left[\dfrac{(1.02)^{28} - 1}{.02}\right]$$

$$= 24,898.41$$

The future value of the annuity is $24,898.41.

50. Since $100\% - 3\% = 97\% = .97$, the population after 1 year is $.97(50,000)$, after 2 years is $.97[.97(50,000)]$ or $(.97)^2(50,000)$, and after n years is $(.97)^n(50,000)$. After 6 years, the population is

$$(.97)^6(50,000) \approx 41,649 \approx 42,000.$$

51. $\left(\dfrac{1}{2}\right)^n$ is left after n strokes. So

$$\left(\dfrac{1}{2}\right)^7 = \dfrac{1}{128} = .0078125 \text{ is left after 7 strokes.}$$

52. (a) We can write the repeating decimal number .55555... as an infinite geometric sequence as follows:

$$\frac{5}{10} + \frac{5}{10}\left(\frac{1}{10}\right) + \frac{5}{10}\left(\frac{1}{10}\right)^2 + \frac{5}{10}\left(\frac{1}{10}\right)^3 + \cdots$$

(b) The common ratio r is

$$\frac{\frac{5}{10}\left(\frac{1}{10}\right)}{\frac{5}{10}} = \frac{1}{10}.$$

(c) Since $|r| < 1$, the sum exists.

$$S = \frac{a_1}{1-r} = \frac{\frac{5}{10}}{1 - \frac{1}{10}} = \frac{\frac{5}{10}}{\frac{9}{10}} = \frac{5}{9}$$

53. No, the sum cannot be found, because $r = 2$, and this value of r does not satisfy $|r| < 1$.

54. No, the terms must be successive, such as the first and second, or the second and the third.

Chapter 14 Test

1. $a_n = (-1)^n + 1$
$a_1 = (-1)^1 + 1 = 0$
$a_2 = (-1)^2 + 1 = 1 + 1 = 2$
$a_3 = (-1)^3 + 1 = -1 + 1 = 0$
$a_4 = (-1)^4 + 1 = 1 + 1 = 2$
$a_5 = (-1)^5 + 1 = -1 + 1 = 0$

Answer: 0, 2, 0, 2, 0

2. $a_1 = 4, d = 2$
$a_1 = 4$
$a_2 = a_1 + d = 4 + 2 = 6$
$a_3 = 6 + 2 = 8$
$a_4 = 8 + 2 = 10$
$a_5 = 10 + 2 = 12$

Answer: 4, 6, 8, 10, 12

3. $a_4 = 6, r = \dfrac{1}{2}$
$a_n = a_1 r^{n-1}$
$a_4 = a_1\left(\dfrac{1}{2}\right)^{4-1}$
$6 = a_1\left(\dfrac{1}{8}\right)$
$a_1 = 48$
$a_2 = \dfrac{1}{2}(48) = 24$
$a_3 = \dfrac{1}{2}(24) = 12$
$a_4 = \dfrac{1}{2}(12) = 6$
$a_5 = \dfrac{1}{2}(6) = 3$

Answer: 48, 24, 12, 6, 3

4. Given $a_1 = 6$ and $d = -2$; find a_4.
$a_n = a_1 + (n-1)d$
$a_4 = a_1 + (4-1)d$
$= 6 + (3)(-2)$
$= 6 - 6 = 0$

5. Given $a_5 = 16$ and $a_7 = 9$; find a_4.
This is a geometric sequence, so $a_6 = a_5 r$ and $a_7 = a_6 r$. Thus, $a_6 = \dfrac{a_7}{r}$ and
$$a_5 r = \frac{a_7}{r}$$
$$r^2 = \frac{a_7}{a_5}$$
$$r^2 = \frac{9}{16}$$

continued

$$r = \frac{3}{4} \quad \text{or} \quad r = -\frac{3}{4}.$$

$$a_5 = a_4 r \quad \text{so} \quad a_4 = \frac{a_5}{r}.$$

$$a_4 = \frac{16}{\frac{3}{4}} \quad \text{or} \quad \frac{16}{-\frac{3}{4}}$$

$$a_4 = \frac{64}{3} \quad \text{or} \quad a_4 = -\frac{64}{3}$$

6. Given the arithmetic sequence with $a_2 = 12$ and $a_3 = 15$; find S_5.

$$d = a_3 - a_2 = 15 - 12 = 3$$
$$a_1 = a_2 - 3 = 12 - 3 = 9$$
$$a_5 = a_4 + 3 = a_3 + 6 = 15 + 6 = 21$$
$$S_n = \frac{n}{2}(a_1 + a_n)$$
$$S_5 = \frac{5}{2}(9 + 21)$$
$$= \frac{5}{2}(30) = 75$$

7. Given the geometric sequence with $a_5 = 4$ and $a_7 = 1$; find S_5.

$$r^2 = \frac{a_7}{a_5} = \frac{1}{4}, \text{ so } r = \frac{1}{2} \text{ or } r = -\frac{1}{2}.$$

Use $a_7 = a_1 r^6$ to get $1 = a_1 \left(\frac{1}{2}\right)^6$, and so

$$a_1 = 64.$$
$$S_n = \frac{a_1(r^n - 1)}{r - 1}$$

$$S_5 = \frac{64\left[1 - \left(\frac{1}{2}\right)^5\right]}{1 - \frac{1}{2}} \quad \text{or } S_5 = \frac{64\left[1 - \left(-\frac{1}{2}\right)^5\right]}{1 - \left(-\frac{1}{2}\right)}$$

$$= 128\left(1 - \frac{1}{32}\right) \qquad = \frac{128}{3}\left(1 + \frac{1}{32}\right)$$

$$= 128\left(\frac{31}{32}\right) \qquad = \frac{128}{3}\left(\frac{33}{32}\right)$$

$$= 124 \qquad = 44$$

8. $\bar{x} = \dfrac{271.9 + 199.8 + 198.5 + 183.5 + 179.1}{5}$

$$= \frac{1032.8}{5} = 206.56$$

The average share volume for the five given stocks was about 206.6 million.

9. $S = R\left[\dfrac{(1 + i)^n - 1}{i}\right]$

$$= 4000 \left[\frac{\left(1 + \frac{.06}{4}\right)^{4(7)} - 1}{\frac{.06}{4}}\right]$$

$$= 4000 \left[\frac{(1.015)^{28} - 1}{.015}\right]$$

$$\approx 137,925.91$$

The account will have $\$137,925.91$ at the end of this term.

10. An infinite geometric series has a sum if $|r| < 1$, where r is the common ratio.

11. $\sum\limits_{i=1}^{5} (2i + 8)$

$$= [2(1) + 8] + [2(2) + 8] + [2(3) + 8]$$
$$\quad + [2(4) + 8] + [2(5) + 8]$$
$$= 10 + 12 + 14 + 16 + 18$$
$$= 70$$

12. $\sum\limits_{i=1}^{6} (3i - 5)$

Use the formula $S_6 = \dfrac{n}{2}(a_1 + a_6)$.

$$a_1 = 3(1) - 5 = -2$$
$$a_6 = 3(6) - 5 = 13$$
$$S_6 = \frac{6}{2}(-2 + 13) = 3(11) = 33$$

13. $\sum\limits_{i=1}^{500} i$

Use the formula $S_{500} = \dfrac{n}{2}(a_1 + a_{500})$ with $a_1 = 1$ and $a_{500} = 500$.

$$S_{500} = \frac{500}{2}(1 + 500)$$
$$= 250(501) = 125,250$$

14. $\sum\limits_{i=1}^{3} \dfrac{1}{2}(4^i) = \dfrac{1}{2}\sum\limits_{i=1}^{3} (4^i)$

$$= \frac{1}{2}\left[4 + 16 + 64\right] = 42$$

15. $\sum\limits_{i=1}^{\infty} \left(\dfrac{1}{4}\right)^i$

This is an infinite geometric series with

$$a_1 = \frac{1}{4} \text{ and } r = \frac{1}{4}.$$

Since $|r| = \dfrac{1}{4} < 1$, the sum is

$$S = \frac{a_1}{1 - r} = \frac{\frac{1}{4}}{1 - \frac{1}{4}} = \frac{\frac{1}{4}}{\frac{3}{4}}$$

$$= \frac{1}{4} \cdot \frac{4}{3} = \frac{1}{3}.$$

16. $\displaystyle\sum_{i=1}^{\infty} 6\left(\frac{3}{2}\right)^i$

This is an infinite geometric series with

$a_1 = 6\left(\dfrac{3}{2}\right)^1 = 9$ and $r = \dfrac{3}{2}$.

Since $|r| = \dfrac{3}{2} > 1$, the sum does not exist.

17. The amounts of unpaid balance during 15 months form an arithmetic sequence
$300, 280, 260, \ldots, 40, 20$
which is the sequence with $n = 15$, $a_1 = 300$, and $a_{15} = 20$. Find the sum of these balances.

$$S_n = \frac{n}{2}(a_1 + a_n)$$

$$S_{15} = \frac{15}{2}(300 + 20)$$

$$= \frac{15}{2}(320) = 2400$$

Since 1% interest is paid on this total, the interest paid is 1% of $2400 or $24. The sewing machine cost $300 (paid monthly at $20), so the total cost is $300 + $24 = $324.

18. The weekly populations form a geometric sequence with $a_1 = 20$ and $r = 3$ since the colony begins with 20 insects and triples each week. Find the general term of this geometric sequence.

$$a_n = a_1 r^{n-1}$$

$$a_n = 20(3)^{n-1}$$

We're assuming that from the beginning of July to the end of September is 12 weeks, so find a_{12}.

$$a_n = 20(3)^{n-1}$$

$$a_{12} = 20(3)^{11}$$

At the end of September, $20(3)^{11} = 3,542,940$ insects will be present in the colony.

19. The fifth term $(r = 5$, so $r - 1 = 4)$ of

$\left(2x - \dfrac{y}{3}\right)^{12}$ is

$$\frac{12!}{(12-4)!\,4!}(2x)^{12-4}\left(-\frac{y}{3}\right)^4$$

$$= \frac{12!}{8!\,4!}(2x)^8\left(-\frac{y}{3}\right)^4$$

$$= \frac{14,080x^8y^4}{9}.$$

20. $(3k - 5)^4$

$$= (3k)^4 + \frac{4!}{3!\,1!}(3k)^3(-5)^1$$

$$+ \frac{4!}{2!\,2!}(3k)^2(-5)^2 + \frac{4!}{1!\,3!}(3k)^1(-5)^3$$

$$+ (-5)^4$$

$$= 81k^4 - 4(27k^3)(5) + 6(9k^2)(25)$$

$$- 4(3k)(125) + 625$$

$$= 81k^4 - 540k^3 + 1350k^2 - 1500k + 625$$

Cumulative Review Exercises Chapters 1–14

In Exercises 1–4, let

$$P = \left\{-\frac{8}{3}, 10, 0, \sqrt{13}, -\sqrt{3}, \frac{45}{15}, \sqrt{-7}, .82, -3\right\}$$

1. The integers are $10, 0, \dfrac{45}{15}$ (or 3), and -3.

2. The rational numbers are
$-\dfrac{8}{3}, 10, 0, \dfrac{45}{15}, .82$, and -3.

3. The irrational numbers are $\sqrt{13}$ and $-\sqrt{3}$.

4. All are real numbers except $\sqrt{-7}$.

5. $|-7| + 6 - |-10| - (-8 + 3)$
$= 7 + 6 - 10 - (-5)$
$= 13 - 10 + 5 = 8$

6. $-15 - |-4| - 10 - |-6|$
$= -15 - 4 - 10 - 6 = -35$

7. $4(-6) + (-8)(5) - (-9)$
$= -24 - 40 + 9 = -55$

8. $9 - (5 + 3a) + 5a = -4(a - 3) - 7$
$9 - 5 - 3a + 5a = -4a + 12 - 7$
$4 + 2a = -4a + 5$
$6a = 1$
$a = \dfrac{1}{6}$

Solution set: $\left\{\dfrac{1}{6}\right\}$

9. $7m + 18 \le 9m - 2$
$-2m \le -20$
Divide by -2; reverse the direction of the inequality symbol.
$m \ge 10$
Solution set: $[10, \infty)$

10. $|4x - 3| = 21$

$4x - 3 = 21$ or $4x - 3 = -21$
$4x = 24$ $4x = -18$

$x = 6$ or $x = -\dfrac{18}{4} = -\dfrac{9}{2}$

Solution set: $\left\{-\dfrac{9}{2}, 6\right\}$

11. $\dfrac{x+3}{12} - \dfrac{x-3}{6} = 0$

Multiply by the LCD, 12.

$$12\left(\dfrac{x+3}{12} - \dfrac{x-3}{6}\right) = 12(0)$$

$$x + 3 - 2(x - 3) = 0$$
$$x + 3 - 2x + 6 = 0$$
$$9 - x = 0$$
$$9 = x$$

Check $x = 9$: $1 - 1 \overset{?}{=} 0$ *True*

Solution set: $\left\{9\right\}$

12. $2x > 8$ or $-3x > 9$

$x > 4$ or $x < -3$

Solution set: $(-\infty, -3) \cup (4, \infty)$

13. $|2m - 5| \geq 11$

$2m - 5 \geq 11$ or $2m - 5 \leq -11$

$2m \geq 16$ $2m \leq -6$

$m \geq 8$ or $m \leq -3$

Solution set: $(-\infty, -3] \cup [8, \infty)$

14. $(4p + 2)(5p - 3)$
$$= 20p^2 - 12p + 10p - 6$$
$$= 20p^2 - 2p - 6$$

15. $(3k - 7)^2 = (3k)^2 - 2(3k)(7) + 7^2$
$$= 9k^2 - 42k + 49$$

16. $(2m^3 - 3m^2 + 8m) - (7m^3 + 5m - 8)$
$$= 2m^3 - 7m^3 - 3m^2 + 8m - 5m + 8$$
$$= -5m^3 - 3m^2 + 3m + 8$$

17.

$$\begin{array}{r}
2t^3 \ + \ 3t^2 \ \ - \ 4t \ \ + \ 2 \\
3t - 2\overline{\smash{)}\,6t^4 \ + \ 5t^3 \ - \ 18t^2 \ + \ 14t \ - \ 1} \\
\underline{6t^4 \ - \ 4t^3} \\
9t^3 \ - \ 18t^2 \\
\underline{9t^3 \ - \ 6t^2} \\
- \ 12t^2 \ + \ 14t \\
\underline{- \ 12t^2 \ + \ 8t} \\
6t \ - \ 1 \\
\underline{6t \ - \ 4} \\
3
\end{array}$$

Remainder

Answer:

$$2t^3 + 3t^2 - 4t + 2 + \dfrac{3}{3t - 2}$$

18. $7x + x^3 = x(7 + x^2)$

19. $14y^2 + 13y - 12$

Look for two integers whose product is $(14)(-12) = -168$ and whose sum is 13. The required numbers are 21 and -8.

$14y^2 + 13y - 12$
$$= 14y^2 + 21y - 8y - 12$$
$$= 7y(2y + 3) - 4(2y + 3)$$
$$= (2y + 3)(7y - 4)$$

20. $6z^3 + 5z^2 - 4z = z(6z^2 + 5z - 4)$
$$= z(3z + 4)(2z - 1)$$

21. $49a^4 - 9b^2 = (7a^2)^2 - (3b)^2$
$$= (7a^2 + 3b)(7a^2 - 3b)$$

22. $c^3 + 27d^3 = c^3 + (3d)^3$
$$= (c + 3d)(c^2 - 3cd + 9d^2)$$

23. $64r^2 + 48rq + 9q^2$
$$= (8r)^2 + 2(8r)(3q) + (3q)^2$$
$$= (8r + 3q)^2$$

24. $2x^2 + x = 10$

$2x^2 + x - 10 = 0$

$(2x + 5)(x - 2) = 0$

$2x + 5 = 0$ or $x - 2 = 0$

$x = -\dfrac{5}{2}$ or $x = 2$

Solution set: $\left\{-\dfrac{5}{2}, 2\right\}$

25. $k^2 - k - 6 \leq 0$

Solve the equation

$$k^2 - k - 6 = 0.$$
$$(k - 3)(k + 2) = 0$$

$k - 3 = 0$ or $k + 2 = 0$

$k = 3$ or $k = -2$

The numbers -2 and 3 divide the number line into three regions.

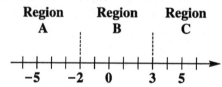

Test a number from each region in the original inequality.

$$k^2 - k - 6 \leq 0$$

Region A: Let $k = -3$.

$(-3)^2 - (-3) - 6 \leq 0$?

$6 \leq 0$ *False*

Region B: Let $k = 0$.

$0^2 - 0 - 6 \leq 0$?

$-6 \leq 0$ *True*

Region C: Let $k = 4$.

$4^2 - 4 - 6 \leq 0$?

$6 \leq 0$ *False*

The numbers in Region B, including the endpoints -2 and 3, are solutions.

Solution set: $\left[-2,\ 3\right]$

26. $\left(\dfrac{2}{3}\right)^{-2} = \left(\dfrac{3}{2}\right)^2 = \dfrac{3}{2} \cdot \dfrac{3}{2} = \dfrac{9}{4}$

27. $\dfrac{(3p^2)^3(-2p^6)}{4p^3(5p^7)} = \dfrac{3^3p^6(-2)p^6}{20p^{10}}$

$= \dfrac{-54p^{12}}{20p^{10}}$

$= -\dfrac{27}{10}p^{12-10}$

$= -\dfrac{27p^2}{10}$

28. $f(x) = \dfrac{2}{x^2 - 81} = \dfrac{2}{(x+9)(x-9)}$

The expression is undefined for $x = -9$ and $x = 9$ since division by 0 is not defined.

29. $\dfrac{x^2 - 16}{x^2 + 2x - 8} \div \dfrac{x - 4}{x + 7}$

$= \dfrac{x^2 - 16}{x^2 + 2x - 8} \cdot \dfrac{x + 7}{x - 4}$

$= \dfrac{(x+4)(x-4)(x+7)}{(x+4)(x-2)(x-4)}$

$= \dfrac{x + 7}{x - 2}$

30. $\dfrac{5}{p^2 + 3p} - \dfrac{2}{p^2 - 4p}$

$= \dfrac{5}{p(p+3)} - \dfrac{2}{p(p-4)}$

The LCD is $p(p+3)(p-4)$.

$= \dfrac{5(p-4)}{p(p+3)(p-4)} - \dfrac{2(p+3)}{p(p-4)(p+3)}$

$= \dfrac{5p - 20 - 2p - 6}{p(p+3)(p-4)}$

$= \dfrac{3p - 26}{p(p+3)(p-4)}$

31. $\dfrac{4}{x - 3} - \dfrac{6}{x + 3} = \dfrac{24}{x^2 - 9}$

$\dfrac{4}{x - 3} - \dfrac{6}{x + 3} = \dfrac{24}{(x+3)(x-3)}$

Multiply by the LCD, $(x + 3)(x - 3)$. $(x \neq \pm 3)$

$4(x + 3) - 6(x - 3) = 24$

$4x + 12 - 6x + 18 = 24$

$-2x + 30 = 24$

$-2x = -6$

$x = 3$

But $x \neq 3$.

Solution set: $\emptyset$

32. $6x^2 + 5x = 8$

$6x^2 + 5x - 8 = 0$

Use the quadratic formula with $a = 6$, $b = 5$, and $c = -8$.

$x = \dfrac{-b \pm \sqrt{b^2 - 4ac}}{2a}$

$x = \dfrac{-5 \pm \sqrt{5^2 - 4(6)(-8)}}{2(6)}$

$= \dfrac{-5 \pm \sqrt{25 + 192}}{12}$

$= \dfrac{-5 \pm \sqrt{217}}{12}$

Solution set: $\left\{ \dfrac{-5 + \sqrt{217}}{12},\ \dfrac{-5 - \sqrt{217}}{12} \right\}$

33. $\sqrt{3x - 2} = x$

$3x - 2 = x^2$ *Square*

$0 = x^2 - 3x + 2$

$0 = (x - 1)(x - 2)$

$x - 1 = 0$ or $x - 2 = 0$

$x = 1$ or $x = 2$

Check $x = 1$: $\sqrt{1} \overset{?}{=} 1$ *True*

Check $x = 2$: $\sqrt{4} \overset{?}{=} 2$ *True*

Solution set: $\left\{1,\ 2\right\}$

34. $f(x) = .07x + 135$

$f(2000) = .07(2000) + 135$

$= 140 + 135 = 275$

The weekly fee is \$275.

35. Let $(x_1, y_1) = (4, -5)$ and $(x_2, y_2) = (-12, -17)$. Then

$m = \dfrac{y_2 - y_1}{x_2 - x_1} = \dfrac{-17 - (-5)}{-12 - 4} = \dfrac{-12}{-16} = \dfrac{3}{4}.$

The slope is $\dfrac{3}{4}$.

36. To find the equation of the line through $(-2, 10)$ and parallel to $3x + y = 7$, find the slope of

$3x + y = 7$

$y = -3x + 7.$

The slope is -3, so a line parallel to it also has slope -3. Use $m = -3$ and $(x_1, y_1) = (-2, 10)$ in the point-slope form.

$y - y_1 = m(x - x_1)$

$y - 10 = -3\left[x - (-2)\right]$

$y - 10 = -3(x + 2)$

continued

Write in standard form.

$$y - 10 = -3x - 6$$
$$3x + y = 4$$

Alternative solution: The line must be of the form $3x + y = k$ since it is parallel to $3x + y = 7$. Substitute -2 for x and 10 for y to find k.

$$3(-2) + 10 = k$$
$$4 = k$$

The equation is $3x + y = 4$.

37. $x - 3y = 6$

Find the x- and y-intercepts. To find the x-intercept, let $y = 0$.

$$x - 3(0) = 6$$
$$x = 6$$

The x-intercept is $(6, 0)$.
To find the y-intercept, let $x = 0$.

$$0 - 3y = 6$$
$$y = -2$$

The y-intercept is $(0, -2)$.
Plot the intercepts and draw the line through them.

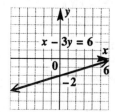

38. $4x - y < 4$

Graph the line $4x - y = 4$, which has intercepts $(0, -4)$ and $(1, 0)$, as a dashed line because the inequality involves $<$. Test $(0, 0)$, which yields $0 < 4$, a true statement. Shade the region on the side of the line that includes $(0, 0)$.

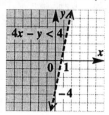

39. $2x + 5y = -19$ (1)
   $-3x + 2y = -19$ (2)

To eliminate x, multiply equation (1) by 3 and equation (2) by 2. Then add the results.

$$
\begin{array}{rl}
6x + 15y = -57 & \quad 3 \times (1) \\
-6x + \ 4y = -38 & \quad 2 \times (2) \\
\hline
19y = -95 & \\
y = -5 &
\end{array}
$$

Substitute -5 for y in equation (1) to find x.

$$
\begin{array}{rl}
2x + 5y = -19 & (1) \\
2x + 5(-5) = -19 & \\
2x - 25 = -19 & \\
2x = 6 & \\
x = 3 &
\end{array}
$$

Solution set : $\left\{ (3, -5) \right\}$

40. $\begin{array}{rcrcrcr} x & + & 2y & + & z & = & 8 \quad (1) \\ 2x & - & y & + & 3z & = & 15 \quad (2) \\ -x & + & 3y & - & 3z & = & -11 \quad (3) \end{array}$

To eliminate x, add equations (1) and (3).

$$
\begin{array}{rcrcrcr}
x & + & 2y & + & z & = & 8 \quad (1) \\
-x & + & 3y & - & 3z & = & -11 \quad (3) \\
\hline
 & & 5y & - & 2z & = & -3 \quad (4)
\end{array}
$$

To eliminate x again, multiply equation (3) by 2 and add the result to equation (2).

$$
\begin{array}{rcrcrcrl}
2x & - & y & + & 3z & = & 15 & (2) \\
-2x & + & 6y & - & 6z & = & -22 & \quad 2 \times (3) \\
\hline
 & & 5y & - & 3z & = & -7 & (5)
\end{array}
$$

Multiply equation (4) by -1 and add the result to equation (5).

$$
\begin{array}{rcrcrl}
-5y & + & 2z & = & 3 & \quad -1 \times (4) \\
5y & - & 3z & = & -7 & (5) \\
\hline
 & & -z & = & -4 & \\
 & & z & = & 4 &
\end{array}
$$

Substitute 4 for z in equation (5) to find y.

$$
\begin{array}{rl}
5y - 3z = -7 & (5) \\
5y - 3(4) = -7 & \\
5y - 12 = -7 & \\
5y = 5 & \\
y = 1 &
\end{array}
$$

Substitute 1 for y and 4 for z in equation (1) to find x.

$$
\begin{array}{rl}
x + 2y + z = 8 & (1) \\
x + 2(1) + 4 = 8 & \\
x + 6 = 8 & \\
x = 2 &
\end{array}
$$

Solution set : $\left\{ (2, 1, 4) \right\}$

41. $\begin{vmatrix} -3 & -2 \\ 6 & 9 \end{vmatrix} = -3(9) - (-2)(6)$

$$= -27 + 12 = -15$$

42. $\begin{vmatrix} 2 & 4 & 1 \\ 1 & 3 & 6 \\ 2 & 3 & -1 \end{vmatrix}$ Expand about row 1.

$$= 2\begin{vmatrix} 3 & 6 \\ 3 & -1 \end{vmatrix} - 4\begin{vmatrix} 1 & 6 \\ 2 & -1 \end{vmatrix} + 1\begin{vmatrix} 1 & 3 \\ 2 & 3 \end{vmatrix}$$
$$= 2(-3 - 18) - 4(-1 - 12) + 1(3 - 6)$$
$$= 2(-21) - 4(-13) + 1(-3)$$
$$= -42 + 52 - 3 = 7$$

43.   Let $x =$ the number of pounds of \$3 per pound nuts.

	Number of Pounds	Price per Pound	Value
\$3 nuts	x	3	$3x$
\$4.25 nuts	8	4.25	$4.25(8)$
Mixture	$x + 8$	4	$4(x + 8)$

The last column gives the equation.
$$3x + 4.25(8) = 4(x + 8)$$
$$3x + 34 = 4x + 32$$
$$2 = x$$
Use 2 pounds of the \$3 nuts.

44. $5\sqrt{72} - 4\sqrt{50} = 5\sqrt{36 \cdot 2} - 4\sqrt{25 \cdot 2}$
$$= 5 \cdot 6\sqrt{2} - 4 \cdot 5\sqrt{2}$$
$$= 30\sqrt{2} - 20\sqrt{2}$$
$$= 10\sqrt{2}$$

45. $(8 + 3i)(8 - 3i)$
$$= 8^2 - (3i)^2 = 64 - 9i^2$$
$$= 64 - 9(-1) = 64 + 9 = 73$$

46.   The graph of $f(x) = 9x + 5$ is a line. To find the inverse, replace $f(x)$ with y.
$$y = 9x + 5$$
Interchange x and y.
$$x = 9y + 5$$
Solve for y.
$$x - 5 = 9y$$
$$\frac{x - 5}{9} = y$$
Replace y with $f^{-1}(x)$.
$$f^{-1}(x) = \frac{x - 5}{9}$$

47.   Graph $g(x) = \left(\frac{1}{3}\right)^x$.

Make a table of values.

x	-2	-1	0	1	2
$g(x)$	9	3	1	$\frac{1}{3}$	$\frac{1}{9}$

Plot these points, and draw a smooth decreasing exponential curve through them.

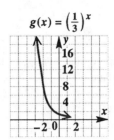

$$g(x) = \left(\frac{1}{3}\right)^x$$

48. $3^{2x-1} = 81$
$$3^{2x-1} = 3^4$$
$$2x - 1 = 4 \qquad \textit{Equate exponents}$$
$$2x = 5$$
$$x = \frac{5}{2}$$

Solution set: $\left\{\dfrac{5}{2}\right\}$

49.   Graph $y = \log_{1/3} x$.
Change to exponential form.
$$\left(\frac{1}{3}\right)^y = x$$
This is the inverse of the graph of
$$g(x) = y = \left(\frac{1}{3}\right)^x$$
in Exercise 47. To find points on the graph, interchange the x- and y-values in the table.

x	9	3	1	$\frac{1}{3}$	$\frac{1}{9}$
y	-2	-1	0	1	2

Plot these points, and draw a smooth decreasing logarithmic curve through them.

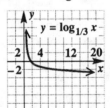

$$y = \log_{1/3} x$$

50. $\log_8 x + \log_8 (x + 2) = 1$
Use the product rule for logarithms.
$$\log_8 x(x + 2) = 1$$
Change to exponential form.
$$x(x + 2) = 8^1$$
$$x^2 + 2x - 8 = 0$$
$$(x + 4)(x - 2) = 0$$
$$x + 4 = 0 \qquad \text{or} \qquad x - 2 = 0$$
$$x = -4 \quad \text{or} \qquad x = 2$$
$x \neq -4$ because $\log_8(-4)$ does not exist, so the only answer is $x = 2$.

Solution set: $\left\{2\right\}$

51. $f(x) = 2(x - 2)^2 - 3$ is in
$f(x) = a(x - h)^2 + k$ form.

The graph is a vertical parabola with vertex (h, k) at $(2, -3)$. Since $a = 2 > 0$, the graph opens upward. Also, $|a| = |2| = 2 > 1$, so the graph is narrower than the graph of $f(x) = x^2$. The points $(0, 5)$ and $(4, 5)$ are on the graph.

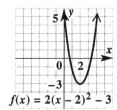

52. $\dfrac{x^2}{9} + \dfrac{y^2}{25} = 1$ is in $\dfrac{x^2}{a^2} + \dfrac{y^2}{b^2} = 1$ form with $a = 3$ and $b = 5$. The graph is an ellipse centered at $(0, 0)$ with x-intercepts $(3, 0)$ and $(-3, 0)$ and y-intercepts $(0, 5)$ and $(0, -5)$. Plot the intercepts and draw the ellipse through them.

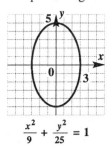

53. $x^2 - y^2 = 9$
$\dfrac{x^2}{9} - \dfrac{y^2}{9} = 1$

The graph is a hyperbola centered at $(0, 0)$ with x-intercepts $(3, 0)$ and $(-3, 0)$. The asymptotes are $y = \pm x$. Draw the right and left branches through the intercepts and approaching the asymptotes.

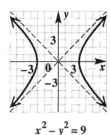

54.   Center at $(-5, 12)$; radius 9
Use the equation of a circle with $h = -5$, $k = 12$, and $r = 9$.
$$(x - h)^2 + (y - k)^2 = r^2$$
$$[x - (-5)]^2 + (y - 12)^2 = 9^2$$
$$(x + 5)^2 + (y - 12)^2 = 81$$

55. $a_n = 5n - 12$
$a_1 = 5(1) - 12 = 5 - 12 = -7$
$a_2 = 5(2) - 12 = 10 - 12 = -2$
$a_3 = 5(3) - 12 = 15 - 12 = 3$
$a_4 = 5(4) - 12 = 20 - 12 = 8$
$a_5 = 5(5) - 12 = 25 - 12 = 13$

Answer: $-7, -2, 3, 8, 13$

56. $a_1 = 8$, $d = 2$
$$S_n = \frac{n}{2}\big[2a_1 + (n - 1)d\big]$$
$$S_6 = \frac{6}{2}\big[2(8) + (6 - 1)2\big]$$
$$= 3(16 + 10) = 3(26) = 78$$

57. $15 - 6 + \dfrac{12}{5} - \dfrac{24}{25} + \cdots$
This is an infinite geometric series with
$a_1 = 15$ and $r = \dfrac{-6}{15} = -\dfrac{2}{5}$. The sum is
$$S = \frac{a_1}{1 - r} = \frac{15}{1 - \left(-\dfrac{2}{5}\right)}$$
$$= \frac{15}{\dfrac{7}{5}} = 15 \cdot \frac{5}{7} = \frac{75}{7}.$$

58. $\displaystyle\sum_{i=1}^{4} 3i = 3\sum_{i=1}^{4} i = 3\left[\frac{4(4 + 1)}{2}\right]$
$$= 3(10) = 30$$
We used the fact that $\displaystyle\sum_{i=1}^{n} i = \frac{n(n + 1)}{2}$.

59. $(2a - 1)^5$
$$= (2a)^5 + \frac{5!}{4!\,1!}(2a)^4(-1)^1$$
$$+ \frac{5!}{3!\,2!}(2a)^3(-1)^2 + \frac{5!}{2!\,3!}(2a)^2(-1)^3$$
$$+ \frac{5!}{1!\,4!}(2a)^1(-1)^4 + (-1)^5$$
$$= 32a^5 + 5(16a^4)(-1) + 10(8a^3)(1)$$
$$+ 10(4a^2)(-1) + 5(2a)(1) + (-1)$$
$$= 32a^5 - 80a^4 + 80a^3 - 40a^2 + 10a - 1$$

60.   The fourth term ($r = 4$, so $r - 1 = 3$) of
$\left(3x^4 - \dfrac{1}{2}y^2\right)^5$ is
$$\frac{5!}{(5 - 3)!\,3!}(3x^4)^{5-3}\left(-\frac{1}{2}y^2\right)^3$$
$$= \frac{5!}{2!\,3!}(3x^4)^2\left(-\frac{1}{2}\right)^3(y^2)^3$$
$$= 10(9x^8)\left(-\frac{1}{8}\right)y^6$$
$$= -\frac{45x^8y^6}{4}$$

APPENDIX A   REVIEW OF DECIMALS AND PERCENTS

1. $14.23 + 9.81 + 74.63 + 18.715$

Add in columns.

$$
\begin{array}{r}
14.230 \\
9.810 \\
74.630 \\
+\ 18.715 \\
\hline
117.385
\end{array}
\quad
\begin{array}{l}
\textit{Line up decimal points} \\
\textit{and attach zeros}
\end{array}
$$

3. $19.74 - 6.53$

Subtract in columns.

$$
\begin{array}{r}
19.74 \\
-\ 6.53 \\
\hline
13.21
\end{array}
\quad \textit{Line up decimal points}
$$

5. $219 - 68.51$

$$
\begin{array}{r}
219.00 \\
-\ 68.51 \\
\hline
150.49
\end{array}
\quad
\begin{array}{l}
\textit{Line up decimal points} \\
\textit{and attach zeros}
\end{array}
$$

7.
$$
\begin{array}{r}
48.960 \\
37.421 \\
+\ 9.720 \\
\hline
96.101
\end{array}
\quad \textit{Attach zeros}
$$

9.
$$
\begin{array}{r}
8.600 \\
-\ 3.751 \\
\hline
4.849
\end{array}
\quad \textit{Attach zeros}
$$

11. 39.6×4.2

Multiply as if the numbers were whole numbers.

$$
\begin{array}{r}
3\,9.6 \\
\times\quad 4.2 \\
\hline
7\,9\,2 \\
158\,4 \\
\hline
166.3\,2
\end{array}
\quad
\begin{array}{l}
\textit{1 decimal place} \\
\textit{1 decimal place} \\
\\
\\
\textit{2 decimal places}
\end{array}
$$

13. 42.1×3.9

$$
\begin{array}{r}
4\,2.1 \\
\times\quad 3.9 \\
\hline
37\,8\,9 \\
126\,3 \\
\hline
164.1\,9
\end{array}
\quad
\begin{array}{l}
\textit{1 decimal place} \\
\textit{1 decimal place} \\
\\
\\
\textit{2 decimal places}
\end{array}
$$

15.
$$
\begin{array}{r}
.042 \\
\times\quad 32 \\
\hline
84 \\
1\,26 \\
\hline
1.344
\end{array}
\quad
\begin{array}{l}
\textit{3 decimal places} \\
\textit{no decimal places} \\
\\
\\
\textit{3 decimal places}
\end{array}
$$

17. $24.84 \div 6$

The divisor is a whole number, so place the decimal point in the quotient above the decimal point in the dividend.

$$
\begin{array}{r}
4\,.\,1\,4 \\
6\,\overline{)\,24\,.\,8\,4} \\
\underline{24} \\
8 \\
\underline{6} \\
2\,4 \\
\underline{2\,4} \\
0
\end{array}
$$

19. $7.6266 \div 3.42$

Move the decimal point in 3.42 two places to the right to get a whole number. Then move the decimal point in 7.6266 two places to the right. Move the decimal point straight up and divide as with whole numbers.

$$
\begin{array}{r}
2.\,2\,3 \\
3.42.\,\overline{)\,7\,.62.\,66} \\
\underline{6\ 8\ 4} \\
78\ 6 \\
\underline{68\ 4} \\
10\ 26 \\
\underline{10\ 26} \\
0
\end{array}
$$

21. $2496 \div .52$

Move the decimal point in .52 and 2496 two places to the right. Bring the decimal point straight up and divide as with whole numbers.

$$
\begin{array}{r}
48\,00. \\
.52.\,\overline{)\,2496.00.} \\
\underline{208} \\
416 \\
\underline{416} \\
0
\end{array}
$$

23. $53\% = 53 \cdot 1\% = 53(.01) = .53$

25. $129\% = 129 \cdot 1\% = 129(.01) = 1.29$

27. $96\% = 96 \cdot 1\% = 96(.01) = .96$

29. $.9\% = .9 \cdot 1\% = .9(.01) = .009$

31. $.80 = 80(.01) = 80 \cdot 1\% = 80\%$

33. $.007 = .7(.01) = .7 \cdot 1\% = .7\%$

35. $.67 = 67(.01) = 67 \cdot 1\% = 67\%$

37. $.125 = 12.5(.01) = 12.5 \cdot 1\% = 12.5\%$

39.   What is 14% of 780?

14% of $780 = (.14)(780) = 109.2$

41. Find 22% of 1086.

$22\% \cdot 1086 = (.22)(1086) = 238.92$

43. 4 is what percent of 80?

As in Example 5(c), we can translate this sentence to symbols word by word.

$$4 = p \cdot 80$$

$$p = \frac{4}{80} = .05 = 5\%$$

4 is 5% of 80.

45. What percent of 5820 is 6402?

$$p \cdot 5820 = 6402$$

$$p = \frac{6402}{5820} = 1.1 \quad \textit{In decimal form}$$

Now convert 1.1 to a percent.

$1.1 = 110(.01) = 110 \cdot 1\% = 110\%$

6402 is 110% of 5820.

47. 121 is what percent of 484?

$$121 = p \cdot 484$$

$$p = \frac{121}{484} = .25 = 25\%$$

121 is 25% of 484.

49. Find 118% of 125.8.

118% of 125.8

$= (1.18)(125.8)$
$= 148.444$
$\approx 148.44 \quad \textit{Round to the nearest hundredth}$

51. What is 91.72% of 8546.95?

91.72% of 8546.95
$= (.9172)(8546.95)$
$= 7839.26254$
≈ 7839.26

53. What percent of 198.72 is 14.68?

To find the percent, divide 14.68 by 198.72.

$$\frac{14.68}{198.72} \approx .0739 \quad \begin{array}{l}\textit{Round to nearest} \\ \textit{ten-thousandth}\end{array}$$

Now convert .0739 to a percent.

$.0739 = 7.39(.01) = 7.39 \cdot 1\% = 7.39\%$

14.68 is 7.39% of 198.72.

55. $12\% \text{ of } 23,000 = (.12)(23,000)$
$= 2760$

She is earning $2760 per year.

57. $35\% \text{ of } 2300 = (.35)(2300)$
$= 805$

805 miles of the trip were by air.

59. $15\% \text{ of } 420 = (.15)(420)$
$= 63$

She could drive 63 extra miles.

61. Since the family spends 90% of the $2000 each month, they must save 10% of the $2000.

$$10\% \text{ of } \$2000 = (.10)(2000)$$
$$= \$200$$

Since there are 12 months in a year, the annual savings are

$$12 \cdot 200 = 2400,$$
$$\text{or } \$2400.$$

APPENDIX B   SETS

1. The set of all natural numbers less than 8 is

    $$\{1, 2, 3, 4, 5, 6, 7\}.$$

3. The set of seasons is

    $$\{\text{winter, spring, summer, fall}\}.$$

    The seasons may be written in any order within the braces.

5. To date, there have been no women presidents, so this set is the empty set, written $\emptyset$, or $\{\ \}$.

7. The set of letters of the alphabet between K and M is $\{L\}$.

9. The set of positive even integers is

    $$\{2, 4, 6, 8, 10, \dots\}.$$

11. The sets in Exercises 9 and 10 are infinite, since each contains an unlimited number of elements.

13. $5 \in \{1, 2, 5, 8\}$

    5 is an element of the set, so the statement is true.

15. $2 \in \{1, 3, 5, 7, 9\}$

    2 is not an element of the set, so the statement is false.

17. $7 \notin \{2, 4, 6, 8\}$

    7 is not an element of the set, so the statement is true.

19. $\{2, 4, 9, 12, 13\} = \{13, 12, 9, 4, 2\}$

    The two sets have exactly the same elements, so they are equal. The statement is true. (The order in which the elements are written does not matter.)

21. $A \subseteq U$

    $A = \{1, 3, 4, 5, 7, 8\}$

    $U = \{1, 2, 3, 4, 5, 6, 7, 8, 9, 10\}$

    Since all the elements of A are elements of U, the statement $A \subseteq U$ is true.

23. $\emptyset \subseteq A$

    Since the empty set contains no elements, the empty set is a subset of every set. The statement $\emptyset \subseteq A$ is true.

25. $C \subseteq A$

    $C = \{1, 3, 5, 7\}$

    $A = \{1, 3, 4, 5, 7, 8\}$

    Since all the elements of C are elements of A, the statement $C \subseteq A$ is true.

27. $D \subseteq B$

    $D = \{1, 2, 3\}$

    $B = \{2, 4, 6, 8\}$

    Since 1 and 3 are elements of D but are not elements of B, the statement $D \subseteq B$ is false.

29. $D \nsubseteq E$

    $D = \{1, 2, 3\}$

    $E = \{3, 7\}$

    Since 1 and 2 are elements of D and are not elements of E, D is not a subset of E, so the statement $D \nsubseteq E$ is true.

31. There are exactly 4 subsets of E.

    $E = \{3, 7\}$

    Since E has 2 elements, the number of subsets is $2^2 = 4$. The statement is true.

33. There are exactly 12 subsets of C.

    $C = \{1, 3, 5, 7\}$

    Since C has 4 elements, the number of subsets is $2^4 = 16$. The statement is false.

35. $\{4, 6, 8, 12\} \cap \{6, 8, 14, 17\} = \{6, 8\}$

    The symbol $\cap$ means the intersection of the two sets, which is the set of elements that belong to both sets. Since 6 and 8 are the only elements belonging to both sets, the statement is true.

37. $\{3, 1, 0\} \cap \{0, 2, 4\} = \{0\}$

    Only 0 belongs to both sets, so the statement is true.

39. $\{3, 9, 12\} \cap \emptyset = \{3, 9, 12\}$

    Since 3, 9, and 12 are not elements of the empty set, they are not in the intersection of the two sets. The intersection of any set with the empty set is the empty set. The statement is false.

41. $\{4, 9, 11, 7, 3\} \cup \{1, 2, 3, 4, 5\}$

    $= \{1, 2, 3, 4, 5, 7, 9, 11\}$

    The union of the two sets is the set of all elements that belong to either one of the sets or to both sets. The statement is true.

43. $\{3, 5, 7, 9\} \cup \{4, 6, 8\} = \emptyset$

    The union of the two sets is the set of all elements that belong to either one of the sets or to both sets.

    $\{3, 5, 7, 9\} \cup \{4, 6, 8\}$

    $= \{3, 4, 5, 6, 7, 8, 9\} \neq \emptyset$

    The statement is false.

45. A'

$U = \{a, b, c, d, e, f, g, h\}$

$A = \{a, b, c, d, e, f\}$

A' contains all elements in U that are not in A, so

$$A' = \{g, h\}.$$

47. C'

$U = \{a, b, c, d, e, f, g, h\}$

$C = \{a, f\}$

C' contains all elements in U that are not in C, so

$$C' = \{b, c, d, e, g, h\}.$$

49. $A \cap B$

$A = \{a, b, c, d, e, f\}$

$B = \{a, c, e\}$

The intersection of A and B is the set of all elements belonging to both A and B, so

$$A \cap B = \{a, c, e\} = B.$$

51. $A \cap D$

$A = \{a, b, c, d, e, f\}$

$D = \{d\}$

Since d is the only element in both A and D,

$$A \cap D = \{d\} = D.$$

53. $B \cap C$

$B = \{a, c, e\}$

$C = \{a, f\}$

Since a is the only element that belongs to both sets, the intersection is the set with a as its only element, so

$$B \cap C = \{a\}.$$

55. $B \cup D$

$B = \{a, c, e\}$

$D = \{d\}$

The union of B and D is the set of elements belonging to either B or D or both, so

$$B \cup D = \{a, c, d, e\}.$$

57. $C \cup B$

$C = \{a, f\}$

$B = \{a, c, e\}$

The union of C and B is the set of elements belonging to either C or B or both, so

$$C \cup B = \{a, c, e, f\}.$$

59. $A \cap \emptyset$

Since $\emptyset$ has no elements, there is no element that belongs to both A and $\emptyset$, so the intersection is the empty set. Thus,

$$A \cap \emptyset = \emptyset.$$

61. $A = \{a, b, c, d, e, f\}$ $C = \{a, f\}$

$B = \{a, c, e\}$ $D = \{d\}$

Disjoint sets are sets which have no elements in common.

B and D are disjoint since they have no elements in common. Also, C and D are disjoint since they have no elements in common.

APPENDIX C JOINT AND COMBINED VARIATION

1. The equation $y = \dfrac{3}{x}$ represents inverse variation. y varies inversely as x because x is in the denominator.

3. The equation $y = 10x^2$ represents direct variation. The number 10 is the constant of variation, and y varies directly as the square of x.

5. The equation $y = 3xz^4$ represents joint variation. y varies directly as x and z^4.

7. The equation $y = \dfrac{4x}{wz}$ represents combined variation. In the numerator, 4 is the constant of variation, and y varies directly as x. In the denominator, y varies inversely as w and z.

9. "x varies directly as y" means

$$x = ky$$

for some constant k.
Substitute $x = 9$ and $y = 3$ in the equation and solve for k.

$$x = ky$$
$$9 = k(3)$$
$$3 = k$$

So, $x = 3y$.
To find x when $y = 12$, substitute 12 for y in the equation.

$$x = 3y$$
$$x = 3(12)$$
$$x = 36$$

11. "z varies inversely as w" means

$$z = \dfrac{k}{w}$$

for some constant k. Since $z = 10$ when $w = .5$, substitute these values in the equation and solve for k.

$$z = \dfrac{k}{w}$$
$$10 = \dfrac{k}{.5}$$
$$5 = k$$

So, $z = \dfrac{5}{w}$.
To find z when $w = 8$, substitute 8 for w in the equation.

$$z = \dfrac{5}{w}$$
$$z = \dfrac{5}{8} \text{ or } .625$$

13. "p varies jointly as q and r^2" means

$$p = kqr^2$$

for some constant k. Given that $p = 200$ when $q = 2$ and $r = 3$, solve for k.

$$p = kqr^2$$
$$200 = k(2)(3)^2$$
$$200 = 18k$$
$$k = \dfrac{200}{18} = \dfrac{100}{9}$$

So, $p = \dfrac{100}{9}qr^2$.

Using $k = \dfrac{100}{9}$, $q = 5$, and $r = 2$, find p.

$$p = \dfrac{100}{9}qr^2$$
$$p = \dfrac{100}{9}(5)(2)^2$$
$$= \dfrac{100}{9}(20)$$
$$= \dfrac{2000}{9} \text{ or } 222\dfrac{2}{9}$$

15. Let $A =$ the amount of water emptied by a pipe in one hour
and $d =$ the diameter of the pipe.
A varies directly as d^2, so

$$A = kd^2$$

for some constant k. Since $A = 200$ when $d = 6$, substitute these values in the equation and solve for k.

$$A = kd^2$$
$$200 = k(6)^2$$
$$200 = 36k$$
$$k = \dfrac{200}{36} = \dfrac{50}{9}$$

So, $A = \dfrac{50}{9}d^2$.
When $d = 12$,

$$A = \dfrac{50}{9}d^2$$
$$A = \dfrac{50}{9}(12)^2$$
$$= \dfrac{50}{9}(144) = 800.$$

A 12-inch pipe would empty 800 gal of water in one hour.

17. Let $d =$ the distance and

$t =$ the time.

d varies directly as the square of t, so $d = kt^2$.

Let $d = -576$ and $t = 6$. (You could also use $d = 576$, but the negative sign indicates the direction of the body.)

$$-576 = k(6)^2$$
$$-576 = 36k$$
$$-16 = k$$

So, $d = -16t^2$.

Let $t = 4$.

$$d = -16(4)^2 = -256$$

The object fell 256 ft in the first 4 seconds.

19. Let $f =$ the frequency of a string in cycles per second

and $s =$ the length in feet.

f varies inversely as s, so

$$f = \frac{k}{s}$$

for some constant k. Since $f = 250$ when $k = 2$, substitute these values in the equation and solve for k.

$$f = \frac{k}{s}$$
$$250 = \frac{k}{2}$$
$$500 = k$$

So, $f = \dfrac{500}{s}$.

When $s = 5$,

$$f = \frac{500}{5} = 100.$$

The string would have a frequency of 100 $\dfrac{\text{cycles}}{\text{sec}}$.

21. Let $V =$ volume of gas,

$p =$ pressure, and

$t =$ temperature.

Then $V = \dfrac{kt}{p}$.

$V = 1.3$ when $t = 300$ and $p = 18$.

$$1.3 = \frac{k(300)}{18}$$
$$k = \frac{1.3(18)}{300} = .078$$

So, $V = \dfrac{.078t}{p}$.

Let $t = 340$ and $p = 24$.

$$V = \frac{.078(340)}{24} = 1.105$$

The volume is 1.105 L.

23. Let $L =$ the load,

$D =$ the diameter,

$H =$ the height.

Then $L = \dfrac{kD^4}{H^2}$.

Let $H = 9$, $D = 3$, and $L = 8$.

$$8 = \frac{k(3)^4}{9^2}$$
$$k = \frac{8(9)^2}{3^4} = 8$$

So, $L = \dfrac{8D^4}{H^2}$.

Let $H = 12$ and $D = 2$.

$$L = \frac{8(2)^4}{12^2} = \frac{8}{9}$$

The maximum load is $\dfrac{8}{9}$ ton.

25. Let $L =$ the load,

$w =$ the width,

$h =$ the height, and

$l =$ the length.

Then $L = \dfrac{kwh^2}{l}$.

Let $L = 360$, $l = 6$, $w = .1$, and $h = .06$.

$$360 = \frac{k(.1)(.06)^2}{6}$$
$$360 = .00006k$$
$$6,000,000 = k$$

So,

$$L = \frac{6,000,000wh^2}{l}.$$

Let $l = 16$, $w = .2$, and $h = .08$.

$$L = \frac{6,000,000(.2)(.08)^2}{16}$$
$$= \frac{7680}{16} = 480$$

The maximum load is 480 kg.

27. According to Example 3,

$$B = \frac{694w}{h^2}.$$

Let $w = 260$ and $h = 82$ (6 ft, 10 in = 82 in).

$$B = \frac{694(260)}{(82)^2} \approx 26.8 \approx 27$$

Chris Webber's BMI is about 27.

APPENDIX D SYNTHETIC DIVISION

1. Synthetic division provides a quick, easy way to divide a polynomial by a binomial of the form $x - k$.

3. $\dfrac{x^2 - 6x + 5}{x - 1}$

$$
\begin{array}{r|rrr}
1 & 1 & -6 & 5 \\
  &   & 1 & -5 \\
\hline
  & 1 & -5 & 0
\end{array}
$$
← *Coefficients of numerator*

$\downarrow \quad \downarrow$

$x \quad - \quad 5$

Write the answer from the bottom row.

Answer: $x - 5$

5. $\dfrac{4m^2 + 19m - 5}{m + 5}$

$m + 5 = m - (-5)$, so use -5.

$$
\begin{array}{r|rrr}
-5 & 4 & 19 & -5 \\
   &   & -20 & 5 \\
\hline
   & 4 & -1 & 0
\end{array}
$$

Answer: $4m - 1$

7. $\dfrac{2a^2 + 8a + 13}{a + 2}$

$a + 2 = a - (-2)$, so use -2.

$$
\begin{array}{r|rrr}
-2 & 2 & 8 & 13 \\
   &   & -4 & -8 \\
\hline
   & 2 & 4 & 5
\end{array}
$$
← *Remainder*

Answer: $2a + 4 + \dfrac{5}{a + 2}$

9. $(p^2 - 3p + 5) \div (p + 1)$

$$
\begin{array}{r|rrr}
-1 & 1 & -3 & 5 \\
   &   & -1 & 4 \\
\hline
   & 1 & -4 & 9
\end{array}
$$

Answer: $p - 4 + \dfrac{9}{p + 1}$

11. $\dfrac{4a^3 - 3a^2 + 2a - 3}{a - 1}$

$$
\begin{array}{r|rrrr}
1 & 4 & -3 & 2 & -3 \\
  &   & 4 & 1 & 3 \\
\hline
  & 4 & 1 & 3 & 0
\end{array}
$$

Answer: $4a^2 + a + 3$

13. $(x^5 - 2x^3 + 3x^2 - 4x - 2) \div (x - 2)$

Insert 0 for the missing x^4-term.

$$
\begin{array}{r|rrrrrr}
2 & 1 & 0 & -2 & 3 & -4 & -2 \\
  &   & 2 & 4 & 4 & 14 & 20 \\
\hline
  & 1 & 2 & 2 & 7 & 10 & 18
\end{array}
$$
← *Remainder*

Answer: $x^4 + 2x^3 + 2x^2 + 7x + 10 + \dfrac{18}{x - 2}$

15. $(-4r^6 - 3r^5 - 3r^4 + 5r^3 - 6r^2 + 3r + 3) \div (r - 1)$

$$
\begin{array}{r|rrrrrrr}
1 & -4 & -3 & -3 & 5 & -6 & 3 & 3 \\
  &   & -4 & -7 & -10 & -5 & -11 & -8 \\
\hline
  & -4 & -7 & -10 & -5 & -11 & -8 & -5
\end{array}
$$
← *Remainder*

Answer:

$-4r^5 - 7r^4 - 10r^3 - 5r^2 - 11r - 8 + \dfrac{-5}{r - 1}$

17. $(-3y^5 + 2y^4 - 5y^3 - 6y^2 - 1) \div (y + 2)$

Insert 0 for the missing y-term.

$$
\begin{array}{r|rrrrrr}
-2 & -3 & 2 & -5 & -6 & 0 & -1 \\
   &   & 6 & -16 & 42 & -72 & 144 \\
\hline
   & -3 & 8 & -21 & 36 & -72 & 143
\end{array}
$$
← *Remainder*

Answer:

$-3y^4 + 8y^3 - 21y^2 + 36y - 72 + \dfrac{143}{y + 2}$

19. $\dfrac{y^3 + 1}{y - 1} = \dfrac{y^3 + 0y^2 + 0y + 1}{y - 1}$

$$
\begin{array}{r|rrrr}
1 & 1 & 0 & 0 & 1 \\
  &   & 1 & 1 & 1 \\
\hline
  & 1 & 1 & 1 & 2
\end{array}
$$
← *Remainder*

Answer: $y^2 + y + 1 + \dfrac{2}{y - 1}$

21. $P(x) = 2x^3 - 4x^2 + 5x - 3; k = 2$

To find $P(2)$, divide the polynomial by $x - 2$. $P(2)$ will be the remainder.

$$
\begin{array}{r|rrrr}
2 & 2 & -4 & 5 & -3 \\
  &   & 4 & 0 & 10 \\
\hline
  & 2 & 0 & 5 & 7
\end{array}
$$
← *Remainder*

By the remainder theorem, $P(2) = 7$.

23. $P(r) = -r^3 - 5r^2 - 4r - 2; k = -4$

Divide by $r + 4$. The remainder is equal to $P(-4)$.

$$
\begin{array}{r|rrrr}
-4 & -1 & -5 & -4 & -2 \\
   &   & 4 & 4 & 0 \\
\hline
   & -1 & -1 & 0 & -2
\end{array}
$$
← *Remainder*

By the remainder theorem, $P(-4) = -2$.

25. $P(y) = 2y^3 - 4y^2 + 5y - 33; k = 3$

Divide by $y - 3$. The remainder is equal to $P(3)$.

$$
\begin{array}{r|rrrr}
3 & 2 & -4 & 5 & -33 \\
  &   & 6 & 6 & 33 \\
\hline
  & 2 & 2 & 11 & 0 \quad \leftarrow \quad Remainder
\end{array}
$$

By the remainder theorem, $P(3) = 0$.

27. By the remainder theorem, a zero remainder means that $P(k) = 0$; that is, k is a number that makes $P(x) = 0$.

29. Is $x = -2$ a solution of

$$x^3 - 2x^2 - 3x + 10 = 0?$$

To decide whether -2 is a solution to the given equation, divide the polynomial by $x + 2$.

$$
\begin{array}{r|rrrr}
-2 & 1 & -2 & -3 & 10 \\
   &   & -2 & 8 & -10 \\
\hline
   & 1 & -4 & 5 & 0 \quad \leftarrow \quad Remainder
\end{array}
$$

Since the remainder is 0, -2 is a solution of the equation.

31. Is $m = -2$ a solution of

$$m^4 + 2m^3 - 3m^2 + 8m - 8 = 0?$$

To decide whether -2 is a solution to the given equation, divide the polynomial by $m + 2$.

$$
\begin{array}{r|rrrrr}
-2 & 1 & 2 & -3 & 8 & -8 \\
   &   & -2 & 0 & 6 & -28 \\
\hline
   & 1 & 0 & -3 & 14 & -36 \quad \leftarrow \quad Remainder
\end{array}
$$

Since the remainder is not 0, -2 is not a solution of the equation.

33. Is $a = -2$ a solution of

$$3a^3 + 2a^2 - 2a + 11 = 0?$$

$$
\begin{array}{r|rrrr}
-2 & 3 & 2 & -2 & 11 \\
   &   & -6 & 8 & -12 \\
\hline
   & 3 & -4 & 6 & -1 \quad \leftarrow \quad Remainder
\end{array}
$$

Since the remainder is not 0, -2 is not a solution of the equation.

35. Is $x = -3$ a solution of

$$2x^3 - x^2 - 13x + 24 = 0?$$

$$
\begin{array}{r|rrrr}
-3 & 2 & -1 & -13 & 24 \\
   &   & -6 & 21 & -24 \\
\hline
   & 2 & -7 & 8 & 0 \quad \leftarrow \quad Remainder
\end{array}
$$

Since the remainder is 0, -3 is a solution of the equation.

APPENDIX E REVIEW OF EXPONENTS, POLYNOMIALS, AND FACTORING
(Transitions from Beginning to Intermediate Algebra)

1. $\left(a^4b^{-3}\right)\left(a^{-6}b^2\right) = \left(a^{4+(-6)}\right)\left(b^{-3+2}\right)$

$$= a^{-2}b^{-1}$$

$$= \frac{1}{a^2b}$$

3. $\left(5x^{-2}y\right)^2\left(2xy^4\right)^2 = \left(5^2x^{-4}y^2\right)\left(2^2x^2y^8\right)$

$$= 25 \cdot 4x^{-4+2}y^{2+8}$$

$$= 100x^{-2}y^{10}$$

$$= \frac{100y^{10}}{x^2}$$

5. $-6^0 + (-6)^0 = -1 \cdot 6^0 + 1$

$$= -1 \cdot 1 + 1$$

$$= -1 + 1 = 0$$

7. $\dfrac{\left(2w^{-1}x^2y^{-1}\right)^3}{\left(4w^5x^{-2}y\right)^2} = \dfrac{2^3w^{-3}x^6y^{-3}}{4^2w^{10}x^{-4}y^2}$

$$= \frac{8x^6x^4}{16w^{10}y^2w^3y^3}$$

$$= \frac{x^{10}}{2w^{13}y^5}$$

9. $\left(\dfrac{-4a^{-2}b^4}{a^3b^{-1}}\right)^{-3} = \left(\dfrac{a^3b^{-1}}{-4a^{-2}b^4}\right)^3$

$$= \frac{a^9b^{-3}}{(-4)^3a^{-6}b^{12}}$$

$$= \frac{a^9a^6}{-64b^{12}b^3}$$

$$= \frac{a^{15}}{-64b^{15}}$$

11. $\left(7x^{-4}y^2z^{-2}\right)^{-2}\left(7x^4y^{-1}z^3\right)^2$

$$= \left(7^{-2}x^8y^{-4}z^4\right)\left(7^2x^8y^{-2}z^6\right)$$

$$= \frac{7^2x^8z^4x^8z^6}{7^2y^4y^2}$$

$$= \frac{x^{16}z^{10}}{y^6}$$

13. $\left(2a^4 + 3a^3 - 6a^2 + 5a - 12\right)$

$$+ \left(-8a^4 + 8a^3 - 14a^2 + 21a - 3\right)$$

$$= (2 - 8)a^4 + (3 + 8)a^3 + (-6 - 14)a^2$$

$$+ (5 + 21)a + (-12 - 3)$$

$$= -6a^4 + 11a^3 - 20a^2 + 26a - 15$$

15. $\left(6x^3 - 12x^2 + 3x - 4\right)$

$$- \left(-2x^3 + 6x^2 - 3x + 12\right)$$

$$= 6x^3 - 12x^2 + 3x - 4 + 2x^3 - 6x^2 + 3x - 12$$

$$= 8x^3 - 18x^2 + 6x - 16$$

17. Add.

$$\begin{array}{r} 5x^2y + 2xy^2 + y^3 \\ -4x^2y - 3xy^2 + 5y^3 \\ \hline x^2y - xy^2 + 6y^3 \end{array}$$

19. $3\left(5x^2 - 12x + 4\right) - 2\left(9x^2 + 13x - 10\right)$

$$= 15x^2 - 36x + 12 - 18x^2 - 26x + 20$$

$$= -3x^2 - 62x + 32$$

21. Subtract.

$$\begin{array}{r} 6x^3 - 2x^2 + 3x - 1 \\ -4x^3 + 2x^2 - 6x + 3 \end{array}$$

Change the sign of each term in $-4x^3 + 2x^2 - 6x + 3$, and add.

$$\begin{array}{r} 6x^3 - 2x^2 + 3x - 1 \\ 4x^3 - 2x^2 + 6x - 3 \\ \hline 10x^3 - 4x^2 + 9x - 4 \end{array}$$

23. $(3x + 1)(2x - 7)$

$$= (3x)(2x) + (3x)(-7)$$

$$+ 1(2x) + 1(-7) \qquad FOIL$$

$$= 6x^2 - 21x + 2x - 7$$

$$= 6x^2 - 19x - 7$$

25. $(4x - 1)(x - 2)$

$$= 4x^2 - 8x - x + 2$$

$$= 4x^2 - 9x + 2$$

27. $(4t + 3)(4t - 3) = (4t)^2 - 3^2$

$$= 16t^2 - 9.$$

29. $(2y^2 + 4)(2y^2 - 4) = \left(2y^2\right)^2 - 4^2$

$$= 4y^4 - 16$$

31. $(4x - 3)^2 = (4x)^2 - 2(4x)(3) + 3^2$

$$= 16x^2 - 24x + 9$$

33. $(6r + 5y)^2 = (6r)^2 + 2(6r)(5y) + (5y)^2$

$$= 36r^2 + 60ry + 25y^2$$

35. $(c + 2d)\left(c^2 - 2cd + 4d^2\right)$

Multiply vertically.

$$\begin{array}{r} c^2 - 2cd + 4d^2 \\ c + 2d \\ \hline 2c^2d - 4cd^2 + 8d^3 \\ c^3 - 2c^2d + 4cd^2 \\ \hline c^3 \qquad\qquad\qquad + 8d^3 \end{array}$$

37.

$$\begin{array}{r} 16x^2 + 4x + 1 \\ 4x - 1 \\ \hline -16x^2 - 4x - 1 \\ 64x^3 + 16x^2 + 4x \\ \hline 64x^3 \qquad\qquad\qquad - 1 \end{array}$$

39.

$$
\begin{array}{r}
2t^2 \;+\; 5st \;-\; s^2 \\
7t \;+\; 5s \\
\hline
10st^2 \;+\; 25s^2t \;-\; 5s^3 \\
14t^3 \;+\; 35st^2 \;-\; 7s^2t \\
\hline
14t^3 \;+\; 45st^2 \;+\; 18s^2t \;-\; 5s^3
\end{array}
$$

41. $8x^3y^4 + 12x^2y^3 + 36xy^4$

The GCF is $4xy^3$.

$$= 4xy^3\left(2x^2y + 3x + 9y\right)$$

43. $x^2 - 2x - 15 = (x+3)(x-5)$

45. $2x^2 - 9x - 18 = (2x+3)(x-6)$

47. $36t^2 - 25 = (6t)^2 - 5^2$
$$= (6t+5)(6t-5)$$

49. $16t^2 + 24t + 9 = (4t)^2 + 2(4t)(3) + 3^2$
$$= (4t+3)^2$$

51. $4m^2p - 12mnp + 9n^2p$
$$= p\left(4m^2 - 12mn + 9n^2\right)$$
$$= p\left[(2m)^2 - 2(2m)(3n) + (3n)^2\right]$$
$$= p(2m - 3n)^2$$

53. $x^3 + 1$
$$= (x)^3 + (1)^3$$
$$= (x+1)\left[(x)^2 - (x)(1) + (1)^2\right] \quad \text{\textit{Sum of two cubes}}$$
$$= (x+1)\left(x^2 - x + 1\right)$$

55. $8t^3 + 125 = (2t)^3 + (5)^3$
$$= (2t+5)\left[(2t)^2 - (2t)(5) + (5)^2\right]$$
$$= (2t+5)\left(4t^2 - 10t + 25\right)$$

57. $t^6 - 125 = \left(t^2\right)^3 - (5)^3$
$$= \left(t^2 - 5\right)\left[\left(t^2\right)^2 + \left(t^2\right)(5) + (5)^2\right]$$
$$\textit{Difference of two cubes}$$
$$= \left(t^2 - 5\right)\left(t^4 + 5t^2 + 25\right)$$

59. $5xt + 15xr + 2yt + 6yr$
$$= (5xt + 15xr) + (2yt + 6yr)$$
$$= 5x(t + 3r) + 2y(t + 3r)$$
$$= (t + 3r)(5x + 2y)$$

61. $6ar + 12br - 5as - 10bs$
$$= (6ar + 12br) + (-5as - 10bs)$$
$$= 6r(a + 2b) - 5s(a + 2b)$$
$$= (a + 2b)(6r - 5s)$$

63. $t^4 - 1 = \left(t^2\right)^2 - 1^2$
$$= \left(t^2 + 1\right)\left(t^2 - 1\right)$$
$$= \left(t^2 + 1\right)(t + 1)(t - 1)$$

65. $4x^2 + 12xy + 9y^2 - 1$

The first three terms form a
perfect square trinomial.

$$= \left(4x^2 + 12xy + 9y^2\right) - 1$$
$$= \left[(2x)^2 + 2(2x)(3y) + (3y)^2\right] - 1$$
$$= (2x + 3y)^2 - 1^2$$

Now factor the difference of two squares.

$$= [(2x + 3y) + 1][(2x + 3y) - 1]$$
$$= (2x + 3y + 1)(2x + 3y - 1)$$